P9-CMS-629

# Introductory Algebra

## Eighth Edition

# Marvin L. Bittinger

Indiana University—Purdue University
at Indianapolis

 **ADDISON-WESLEY**

An imprint of Addison Wesley Longman, Inc.

Reading, Massachusetts • Menlo Park, California • New York • Harlow, England
Don Mills, Ontario • Sydney • Mexico City • Madrid • Amsterdam

| | |
|---|---|
| Publisher | Jason A. Jordan |
| Project Manager | Christine Poolos |
| Assistant Editor | Michelle Fowler |
| Managing Editor | Ron Hampton |
| Production Supervisor | Kathleen A. Manley |
| Production Assistant | Jane Estrella |
| Design Direction | Susan Carsten |
| Text Designer | Rebecca Lloyd Lemna |
| Editorial and Production Services | Martha Morong/Quadrata, Inc. |
| Art Editor | Janet Theurer |
| Marketing Managers | Craig Bleyer and Laura Rogers |
| Illustrators | Scientific Illustrators and Gayle Hayes |
| Compositor | The Beacon Group |
| Cover Designer | Jeannet Leendertse |
| Cover Photographs | © Photodisc, 1998; © Stephen Webster/PHOTONICA |
| Manufacturing Supervisor | Ralph Mattivello |

PHOTO CREDITS

**1, 38,** John Neubauer/PhotoEdit   **41,** John W. Banagan/The Image Bank   **67,** AP/Wide World Photos   **72,** John W. Banagan/The Image Bank   **111, 139, 141, 148,** AP/Wide World Photos   **151,** A. Lichtenstein/The Image Works   **172,** Tom Tracy/FPG International   **181,** Lee Snider/The Image Works   **183,** © 1996 Churchill & Klehr   **199,** Lee Snider/The Image Works   **215,** AP/Wide World Photos   **222,** Tom Stratman   **235,** AP/Wide World Photos   **257,** Bob Daemmrich/The Image Works   **268,** AP/Wide World Photos   **297,** Brian Spurlock   **317,** Judy Gelles/Stock Boston   **336,** Andrea Mohin/NYT Permissions   **370,** Judy Gelles/Stock Boston   **381, 429,** John Riley/Tony Stone Images   **431,** Ford Motor Company   **432,** AP/Wide World Photos   **433,** © 1990 Glenn Randall   **455, 464,** Bob Daemmrich/The Image Works   **493,** © 1998 Churchill & Klehr   **517,** M. Rangell/The Image Works   **523,** Churchill & Klehr   **525,** John Banagan/The Image Bank   **528,** © 1998 Churchill & Klehr   **541, 544,** Terje Rakke/The Image Bank   **547,** Eric Neurath/Stock Boston   **556,** David Young-Wolff/Tony Stone Images   **593,** Robert Severi/Gamma Liaison   **608,** © Scott Teven   **625,** Robert Severi/Gamma Liaison   **628,** AP/Wide World Photos   **640,** Frances M. Roberts

LIBRARY OF CONGRESS CATALOGING-IN-PUBLICATION DATA

Bittinger, Marvin L.
  Introductory algebra—8th ed./Marvin L. Bittinger.
      p.  cm.
  Includes index.
  ISBN 0-201-95959-3
  1. Algebra. I. Title.
QA152.2.B58   1998
512.9—dc21                          98-34318
                              CIP

Copyright © 1999 by Addison Wesley Longman, Inc. All rights reserved.
No part of this publication may be reproduced, stored in a retrieval system, or transmitted, in any form or by any means, electronic, mechanical, photocopying, recording, or otherwise, without the prior written permission of the publisher.
Printed in the United States of America.

   5 6 7 8 9 10—VH—01

# Contents

## Appendixes

# Preface

This text is in a series of texts that includes the following:

Bittinger: *Basic Mathematics,* Eighth Edition

Bittinger: *Fundamental Mathematics,* Second Edition

Bittinger: *Introductory Algebra,* Eighth Edition

Bittinger: *Intermediate Algebra,* Eighth Edition

Bittinger/Beecher: *Introductory and Intermediate Algebra: A Combined Approach*

*Introductory Algebra,* Eighth Edition, is a significant revision of the Seventh Edition, particularly with respect to design, art program, pedagogy, features, and supplements package. Its unique approach, which has been developed and refined over eight editions, continues to blend the following elements in order to bring students success:

- *Writing style.* The author writes in a clear easy-to-read style that helps students progress from concepts through examples and margin exercises to section exercises.

- *Problem-solving approach.* The basis for solving problems and real-data applications is a five-step process (*Familiarize, Translate, Solve, Check,* and *State*) introduced early in the text and used consistently throughout. This problem-solving approach provides students with a consistent framework for solving applications. (See pages 135, 369, and 432.)

- *Real data.* Real-data applications aid in motivating students by connecting the mathematics to their everyday lives. Extensive research was conducted to find new applications that relate mathematics to the real world.

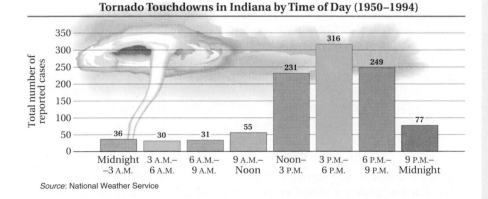

**Tornado Touchdowns in Indiana by Time of Day (1950–1994)**

*Source:* National Weather Service

Marianas
Trench

Puerto Rico
Trench

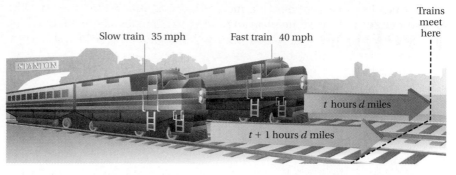

- **Art program.** The art program has been expanded to improve the visualization of mathematical concepts and to enhance the real-data applications.

Slow train 35 mph    Fast train 40 mph    Trains meet here

$t$ hours $d$ miles

$t + 1$ hours $d$ miles

- **Reviewer feedback.** The author solicits feedback from reviewers and students to help fulfill student and instructor needs.

- **Accuracy.** The manuscript is subjected to an extensive accuracy-checking process to eliminate errors.

- **Supplements package.** All ancillary materials are closely tied with the text and created by members of the author team to provide a complete and consistent package for both students and instructors.

# What's New in the Eighth Edition?

The style, format, and approach of the Seventh Edition have been strengthened in this new edition in a number of ways.

**Updated Applications**  Extensive research has been done to make the applications in the Eighth Edition even more up to date and realistic. A large number of the applications are new to this edition, and many are drawn from the fields of business and economics, life and physical sciences, social sciences, and areas of general interest such as sports and daily life. To encourage students to understand the relevance of mathematics, many applications are enhanced by graphs and drawings similar to those found in today's newspapers and magazines. Many applications are also titled for quick and easy reference, and most real-data applications are credited with a source line. (See pages 141, 182, 432, 608, and 640.)

**Improving Your Math Study Skills**  Occurring at least once in every chapter, and referenced in the table of contents, these mini-lessons provide students with concrete techniques to improve studying and test-taking. These features can be covered in their entirety at the beginning of the course, encouraging good study habits early on, or they can be used as they occur in the text, allowing students to learn them gradually. These features can also be used in conjunction with Marvin L. Bittinger's "Math Study Skills" Videotape, which is free to adopters. Please contact your Addison Wesley Longman sales consultant for details on how to obtain this videotape. (See pages 56, 244, and 422.)

**Calculator Spotlights**  Designed specifically for the beginning algebra student, these optional features include graphing-calculator instruction and practice exercises (see pages 503, 546, and 614). Answers to all Calculator Spotlight exercises appear at the back of the text.

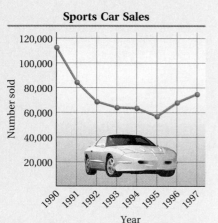

**Sports Car Sales**

*Source*: Autodata

**New Art and Design** To enhance the greater emphasis on real data and applications, we have extensively increased the number of pieces of technical and situational art (see pages 50, 137, 427, 432, and 530). The use of color has been carried out in a methodical and precise manner so that its use carries a consistent meaning, which enhances the readability of the text. For example, the use of both red and blue in mathematical art increases understanding of the concepts. When two lines are graphed using the same set of axes, one is usually red and the other blue. Note that equation labels are the same color as the corresponding line to aid in understanding. Answer lines have also been deleted from all section exercise sets to allow room for more exercises and additional art to better illustrate the exercises.

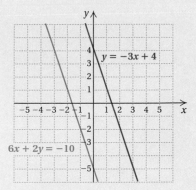

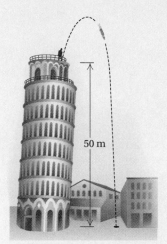

**World Wide Web Integration** The World Wide Web is a powerful resource available to more and more people every day. In an effort to get students more involved in using this source of information, we have added a World Wide Web address (www.mathmax.com) to every chapter opener (see pages 181, 455, and 593). Students can go to this page on the World Wide Web to further explore the subject matter of the chapter-opening application. Selected exercise sets, marked on the first page of the exercise set with an icon (see pages 77, 203, and 337), have additional practice-problem worksheets that can be downloaded from this site. Additional, more extensive, Summary and Review pages for each chapter, as well as other supplementary material, can also be downloaded for instructor and student use.

**Algebraic–Graphical Connections** To give students a better visual understanding of algebra, we have included algebraic–graphical connections in the Eighth Edition (see pages 199, 419, 498, and 595). This feature gives the algebra more meaning by connecting the algebra to a graphical interpretation.

**Collaborative Learning Features** An icon located at the end of an exercise set signals the existence of a Collaborative Learning Activity correlating to that section in Irene Doo's *Collaborative Learning Activities Manual* (see pages 258, 442, 478, and 586). Please contact your Addison Wesley Longman sales consultant for details on ordering this supplement.

**Exercises** The deletion of answer lines in the exercise sets has allowed us to include more exercises in the Eighth Edition. Exercises are paired, meaning that each even-numbered exercise is very much like the odd-numbered one that precedes it. This gives the instructor several options: If an instructor wants the student to have answers available, the odd-numbered exercises are assigned; if an instructor wants the student to practice (perhaps for a test), with no answers available, then the even-numbered exercises are assigned. In this way, each exercise set actually serves as two exercise sets. Answers to all odd-numbered exercises, with the exception of the Thinking and Writing exercises, and *all* Skill Maintenance exercises are provided at the back of the text. If an instructor wants the student to have access to all the answers, a complete answer book is available.

**Skill Maintenance Exercises** The Skill Maintenance exercises have been enhanced by the inclusion of 73% more exercises in this edition. These exercises focus on the four Objectives for Retesting listed at the beginning of each chapter, but they also review concepts from other sections of the text in order to prepare students for the Final Examination. Section and objective codes appear next to each Skill Maintenance exercise for easy reference. Answers to all Skill Maintenance exercises appear at the back of the book (see pages 279, 400, and 620).

**Synthesis Exercises**   These exercises now appear in every exercise set, Summary and Review, Chapter Test, and Cumulative Review. Synthesis exercises help build critical thinking skills by requiring students to synthesize or combine learning objectives from the section being studied as well as preceding sections in the book. In addition, the Critical Thinking feature from the Seventh Edition has been incorporated into the Synthesis exercises for the Eighth Edition (see pages 360, 486, and 564).

**Thinking and Writing Exercises** ◈ Two Thinking and Writing exercises (denoted by the maze icon) have been added to the Synthesis section of every exercise set and Summary and Review. Designed to develop comprehension of critical concepts, these exercises encourage students to both think and write about key mathematical ideas in the chapter (see pages 464, 506, and 628).

Content   We have made the following improvements to the content of *Introductory Algebra*.

- Graphing, formerly located in Chapter 7, has been divided into two chapters. Basic graphing of equations is now covered in Chapter 3 (*Graphs of Equations; Data Analysis*) and the concept of slope appears in Chapter 7 (*Graphs, Slope, and Applications*). This earlier introduction to graphing allows expanded coverage and integration of graphing in the rest of the text.

- Section 3.4 ("Applications and Data Analysis with Graphs") on data analysis and predictions is optional new content to the Eighth Edition. Students will learn to compare two sets of data using their means and to make predictions from a set of data using interpolation and extrapolation.

- Section 7.2 ("Equations of Lines") has been improved by emphasizing the use of $y = mx + b$ in determining equations of lines and by placing all of these topics in one section.

- To provide expanded preparation for later courses, the topic of functions is introduced in Chapter 10.

- Three new appendixes (*Factoring Sums or Differences of Cubes; Higher Roots;* and *Sets*) have been added to the Eighth Edition.

# Learning Aids

Interactive Worktext Approach   The pedagogy of this text is designed to provide an interactive learning experience between the student and the exposition, annotated examples, art, margin exercises, and exercise sets. This approach provides students with a clear set of learning objectives, involves them with the development of the material, and provides immediate and continual reinforcement and assessment.

> *Section objectives* are keyed by letter not only to section subheadings, but also to exercises in the exercise sets and Summary and Review, as well as answers to the Pretest, Chapter Test, and Cumulative Review questions. This enables students to easily find appropriate review material if they are unable to work a particular exercise.

- *Art program.* The art program has been expanded to improve the visualization of mathematical concepts and to enhance the real-data applications.

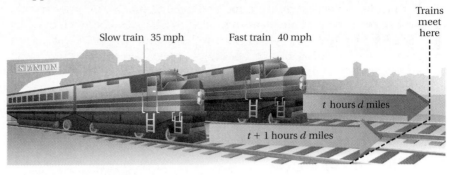

Slow train  35 mph    Fast train  40 mph    Trains meet here

$t$ hours $d$ miles

$t + 1$ hours $d$ miles

- *Reviewer feedback.* The author solicits feedback from reviewers and students to help fulfill student and instructor needs.

- *Accuracy.* The manuscript is subjected to an extensive accuracy-checking process to eliminate errors.

- *Supplements package.* All ancillary materials are closely tied with the text and created by members of the author team to provide a complete and consistent package for both students and instructors.

## What's New in the Eighth Edition?

The style, format, and approach of the Seventh Edition have been strengthened in this new edition in a number of ways.

**Updated Applications**  Extensive research has been done to make the applications in the Eighth Edition even more up to date and realistic. A large number of the applications are new to this edition, and many are drawn from the fields of business and economics, life and physical sciences, social sciences, and areas of general interest such as sports and daily life. To encourage students to understand the relevance of mathematics, many applications are enhanced by graphs and drawings similar to those found in today's newspapers and magazines. Many applications are also titled for quick and easy reference, and most real-data applications are credited with a source line. (See pages 141, 182, 432, 608, and 640.)

**Improving Your Math Study Skills**  Occurring at least once in every chapter, and referenced in the table of contents, these mini-lessons provide students with concrete techniques to improve studying and test-taking. These features can be covered in their entirety at the beginning of the course, encouraging good study habits early on, or they can be used as they occur in the text, allowing students to learn them gradually. These features can also be used in conjunction with Marvin L. Bittinger's "Math Study Skills" Videotape, which is free to adopters. Please contact your Addison Wesley Longman sales consultant for details on how to obtain this videotape. (See pages 56, 244, and 422.)

**Calculator Spotlights**  Designed specifically for the beginning algebra student, these optional features include graphing-calculator instruction and practice exercises (see pages 503, 546, and 614). Answers to all Calculator Spotlight exercises appear at the back of the text.

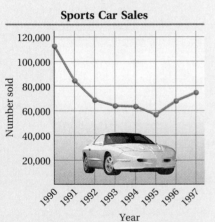

**Sports Car Sales**

Number sold / Year

120,000
100,000
80,000
60,000
40,000
20,000

1990 1991 1992 1993 1994 1995 1996 1997

*Source*: Autodata

# Preface

This text is in a series of texts that includes the following:

Bittinger: *Basic Mathematics*, Eighth Edition

Bittinger: *Fundamental Mathematics*, Second Edition

Bittinger: *Introductory Algebra*, Eighth Edition

Bittinger: *Intermediate Algebra*, Eighth Edition

Bittinger/Beecher: *Introductory and Intermediate Algebra: A Combined Approach*

*Introductory Algebra*, Eighth Edition, is a significant revision of the Seventh Edition, particularly with respect to design, art program, pedagogy, features, and supplements package. Its unique approach, which has been developed and refined over eight editions, continues to blend the following elements in order to bring students success:

- *Writing style.* The author writes in a clear easy-to-read style that helps students progress from concepts through examples and margin exercises to section exercises.

- *Problem-solving approach.* The basis for solving problems and real-data applications is a five-step process (*Familiarize, Translate, Solve, Check,* and *State*) introduced early in the text and used consistently throughout. This problem-solving approach provides students with a consistent framework for solving applications. (See pages 135, 369, and 432.)

- *Real data.* Real-data applications aid in motivating students by connecting the mathematics to their everyday lives. Extensive research was conducted to find new applications that relate mathematics to the real world.

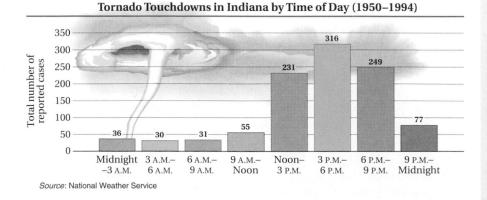

**Tornado Touchdowns in Indiana by Time of Day (1950–1994)**

*Source*: National Weather Service

Marianas Trench

−11,033 m

Puerto Rico Trench

−8648 m

Throughout the text, students are directed to numerous *margin exercises,* which provide immediate reinforcement of the concepts covered in each section.

**Review Material**   The Eighth Edition of *Introductory Algebra* continues to provide many opportunities for students to prepare for final assessment.

Now in a two-column format, the *Summary and Review* appears at the end of each chapter and provides an extensive set of review exercises. Reference codes beside each exercise or direction line preceding it allow the student to easily return to the objective being reviewed (see pages 375, 487, and 645).

Also included at the end of every chapter but Chapters R and 1 is a *Cumulative Review,* which reviews material from all preceding chapters. At the back of the text are answers to all Cumulative Review exercises, together with section and objective references, so that students know exactly what material to study if they miss a review exercise (see pages 315, 539, and 649).

*Objectives for Retesting* are covered in each Summary and Review and Chapter Test, and are also included in the Skill Maintenance exercises and in the Printed Test Bank (see pages 236, 382, and 456).

**For Extra Help**   Many valuable study aids accompany this text. Below the list of objectives found at the beginning of each section are references to appropriate videotape, audiotape, tutorial software, and CD-ROM programs to make it easy for the student to find the correct support materials.

**Testing**   The following assessment opportunities exist in the text.

The *Diagnostic Pretest,* provided at the beginning of the text, can place students in the appropriate chapter for their skill level by identifying familiar material and specific trouble areas (see page xxi).

*Chapter Pretests* can then be used to place students in a specific section of the chapter, allowing them to concentrate on topics with which they have particular difficulty (see pages 112, 182, and 542).

*Chapter Tests* allow students to review and test comprehension of chapter skills, as well as the four Objectives for Retesting from earlier chapters (see pages 229, 489, and 647).

Answers to all Diagnostic Pretest, Chapter Pretest, and Chapter Test questions are found at the back of the book, along with appropriate section and objective references.

# Supplements for the Instructor

### Annotated Instructor's Edition
0-201-33875-0

The *Annotated Instructor's Edition* is a specially bound version of the student text with answers to all margin exercises, exercise sets, and chapter tests printed in a special color near the corresponding exercises.

### Instructor's Solutions Manual
0-201-43411-3

The *Instructor's Solutions Manual* by Judith A. Penna contains brief worked-out solutions to all even-numbered exercises in the exercise sets and answers to all Thinking and Writing exercises.

---

**Objectives**

**a** Given the coordinates of two points on a line, find the slope of the line.

**b** Find the slope of a line from an equation.

**c** Find the slope or rate of change in an applied problem involving slope.

**For Extra Help**

TAPE 13    TAPE 13B    MAC WIN    CD-ROM

---

### Printed Test Bank/Instructor's Resource Guide
by Donna DeSpain
0-201-43412-1

The test-bank section of this supplement contains the following:

- Three alternate test forms for each chapter, with questions in the same topic order as the objectives presented in the chapter
- Five alternate test forms for each chapter, modeled after the Chapter Tests in the text
- Three alternate test forms for each chapter, designed for a 50-minute class period
- Two multiple-choice versions of each Chapter Test
- Two cumulative review tests for each chapter, with the exception of Chapters R and 1
- Eight final examinations: three with questions organized by chapter, three with questions scrambled as in the Cumulative Reviews, and two with multiple-choice questions
- Answers for the Chapter Tests and Final Examination

The resource-guide section contains the following:

- A conversion guide from the Seventh Edition to the Eighth Edition
- Extra practice exercises (with answers) for 40 of the most difficult topics in the text
- Critical Thinking exercises and answers
- Black-line masters of grids and number lines for transparency masters or test preparation
- Indexes to the videotapes and audiotapes that accompany the text
- Three-column chapter Summary and Review listing objectives, brief procedures, worked-out examples, multiple-choice problems similar to the example, and the answers to those problems
- Instructor support material for the CD-ROM

### Collaborative Learning Activities Manual
0-201-34575-7

The *Collaborative Learning Activities Manual*, written by Irene Doo of Austin Community College, features group activities that are tied to sections of the text via an icon . Instructions for classroom setup are also included in the manual.

### Answer Book
0-201-43413-x

The *Answer Book* contains answers to all exercises in the exercise sets in the text. Instructors can make quick reference to all answers or have quantities of these booklets made available for sale if they want students to have access to all the answers.

### TestGen-EQ
0-201-38147-8 (Windows), 0-201-38154-0 (Macintosh)

This test generation software is available in Windows and Macintosh versions. TestGen-EQ's friendly graphical interface enables instructors to easily view, edit, and add questions, transfer questions to tests, and print tests in a variety of fonts and forms. Search and sort features help the instructor quickly locate questions and arrange them in a preferred order.

Six question formats are available, including short-answer, true–false, multiple-choice, essay, matching, and bimodal formats. A built-in question editor gives the instructor the ability to create graphs, import graphics, insert mathematical symbols and templates, and insert variable numbers or text. Computerized testbanks include algorithmically defined problems organized according to each textbook. An "Export to HTML" feature lets instructors create practice tests for the World Wide Web. TestGen-EQ is free to qualifying adopters.

### QuizMaster-EQ
0-201-38147-8 (Windows), 0-201-38154-0 (Macintosh)

QuizMaster-EQ enables instructors to create and save tests and quizzes using TestGen-EQ so students can take them on a computer network. Instructors can set preferences for how and when tests are administered. QuizMaster-EQ automatically grades the exams and allows the instructor to view or print a variety of reports for individual students, classes, or courses. This software is available for both Windows and Macintosh and is fully networkable. QuizMaster-EQ is free to qualifying adopters.

# Supplements for the Student

### Student's Solutions Manual
0-201-34023-2

The *Student's Solutions Manual* by Judith A. Penna contains fully worked-out solutions with step-by-step annotations for all the odd-numbered exercises in the exercise sets in the text, with the exception of the Thinking and Writing exercises. It may be purchased by your students from Addison Wesley Longman.

### "Steps to Success" Videotapes
0-201-30360-4

*Steps to Success* is a complete revision of the existing series of videotapes, based on extensive input from both students and instructors. These videotapes feature an engaging team of mathematics teachers who present comprehensive coverage of each section of the text in a student-interactive format. The lecturers' presentations include examples and problems from the text and support an approach that emphasizes visualization and problem solving. A video icon ▭ at the beginning of each section references the appropriate videotape number. The videotapes are free to qualifying adopters.

### "Math Study Skills for Students" Videotape
0-201-84521-0

Designed to help students make better use of their math study time, this videotape help students improve retention of concepts and procedures taught in classes from basic mathematics through intermediate algebra. Through carefully-crafted graphics and comprehensive on-camera explanation, Marvin L. Bittinger helps viewers focus on study skills that are commonly overlooked.

### Audiotapes
0-201-43405-9

The audiotapes are designed to lead students through the material in each text section. Bill Saler explains solution steps to examples, cautions students about common errors, and instructs them at certain points to stop the tape and do exercises in the margin. He then reviews the margin-exercise solutions, pointing out potential errors. An audiotape icon ⌒ at the beginning of each section references the appropriate audiotape number. The audiotapes are free to qualifying adopters.

### InterAct Math Tutorial Software
0-201-38091-9 (Windows), 0-201-38098-6 (Macintosh)

*InterAct Math Tutorial Software* has been developed and designed by professional software engineers working closely with a team of experienced developmental-math teachers. This software includes exercises that are linked one-to-one with the odd-numbered exercises in the text and require the same computational and problem-solving skills as their companion exercises in the text. Each exercise has an example and an interactive guided solution that are designed to involve students in the solution process and to help them identify precisely where they are having trouble. In addition, the software recognizes common student errors and provides students with appropriate customized feedback. With its sophisticated answer recognition capabilities, *InterAct Math Tutorial Software* recognizes equivalent forms of the same answer for any kind of input. It also tracks for each section student activity and scores that can then be printed out. A disk icon ▦ at the beginning of each section identifies section coverage. Available for Windows and Macintosh computers, this software is free to qualifying adopters or can be bundled with books for sale to students.

### World Wide Web Supplement (www.mathmax.com)

This on-line supplement provides additional practice and learning resources for the student of introductory algebra. For each book chapter, students can find additional practice exercises, Web links for further exploration, and expanded Summary and Review pages that review and reinforce the concepts and skills learned throughout the chapter. In addition, students can download a plug-in for Addison Wesley Longman's *InterAct Math Tutorial Software* that allows them to access additional tutorial problems directly through their Web browser. Students and instructors can also learn about the other supplements available for the MathMax series via sample audio clips and complete descriptions of other services provided by Addison Wesley Longman.

### MathMax Multimedia CD-ROM
### for Introductory Algebra
0-201-39735-8 (Windows), 0-201-44457-7 (Macintosh)

The Introductory Algebra CD provides an active environment using graphics, animations, and audio narration to build on some of the unique and proven features of the MathMax series. Highlighting key concepts from the book, the content of the CD is tightly and consistently integrated with the *Introductory Algebra* text and retains references to the *Introductory Algebra* numbering scheme so that students can move smoothly between the CD and other *Introductory Algebra* supplements. The CD includes Addison Wesley Longman's *InterAct Math Tutorial Software* so that students can practice additional tutorial problems. An interactive

Summary and Review section allows students to review and practice what they have learned in each chapter; and multimedia presentations reiterate important study skills described throughout the book. A CD-ROM icon ❂ at the beginning of each section indicates section coverage. The Introductory Algebra CD is available for both Windows and Macintosh computers. Contact your Addison Wesley Longman sales consultant for a demonstration.

Your author and his team have committed themselves to publishing an accessible, clear, accomplishable, error-free book and supplements package that will provide the student with a successful learning experience and will foster appreciation and enjoyment of mathematics. As part of our continual effort to accomplish this goal, we welcome your comments and suggestions at the following email address:

Marv Bittinger
exponent@aol.com

# Acknowledgments

Many of you have helped to shape the Eighth Edition by reviewing, participating in telephone surveys and focus groups, filling out questionnaires, and spending time with us on your campuses. Our deepest appreciation to all of you and in particular to the following:

Dolores Anenson, *Merced College*
Martin Baker, *Parks College*
Sharon Balk, *Northeast Iowa Community College*
William Bordeaux, Jr., *Huntingdon College*
Richard Burns, *Springfield Technical Community College*
Mark Campbell, *Slippery Rock University*
Debra Caplinger Cross, *Alabama Southern Community College*
Irene Duranczyk, *Eastern Michigan University*
M. R. Eisfelder, *McHenry County College*
Linda Galloway, *Macon College*
Catherine Green, *Lawson State Community College*
Phil Green, *Skagit Valley College*
Don Griffin, *Greenville Technical College*
Mary Indelicado, *Normandale Community College*
Yvonne Jessee, *Mountain Empire Community College*
Juan Carlos Jiminez, *Springfield Technical Community College*
Joe Jordan, *John Tyler Community College*
Karen Knight, *Jefferson College*
Evelyn Kral, *Morgan Community College*
Lee M. Lacey, *Glendale Community College*
Ira Lansing, *College of Marin*
Christine Ledwith, *Florida Keys Community College—Upper Keys Center*
Linda Long, *Ricks College*
C. Vernon Marlin, *Southeastern Community College*
Ray Maruca, *Delaware County Community College*
Hubert McClure, *Tri-County Technical College*
Sharon McKindrick, *New Mexico State University—Grants Beach*
John Menzie, *Barstow College*
Rebecca Metzger
Sandra Miller, *Harrisburg Area Community College*
Mike Montano, *Riverside Community College—City Campus*
Charlie Montgomery, *Alabama Southern Community College*
Frank Mulvaney, *Delaware County Community College*
Linda Padilla, *Joliet Junior College*
Marilyn Platt, *Gatson College*
Carol Rardin, *Central Wyoming College*
Eugena Rohrberg, *Los Angeles Valley College*

Suzanne Rosenberger, *Harrisburg Area Community College*
Ned Schillow, *Lehigh Carbon Community College*
Minnie Shuler, *Gulf Coast Community College*
Lynn Siedenstrang, *Gray's Harbor College*
Lee Ann Spahr, *Durham Technical Community College*
Helen Stewart, *Brunswick Community College*
Tami Wellick, *McHenry County College*
Elaine Werner, *University of the Pacific*
Steve Wittel, *Augusta College*

We also wish to recognize the following people who wrote scripts, presented lessons on camera, and checked the accuracy of the videotapes:

Beth Burkenstock
Margaret Donlan, *University of Delaware*
David J. Ellenbogen, *Community College of Vermont*
Barbara Johnson, *Indiana University—Purdue University at Indianapolis*
Judith A. Penna, *Indiana University—Purdue University at Indianapolis*
Clen Vance, *Houston Community College*

---

I wish to thank Jason Jordan, my publisher and friend at Addison Wesley Longman, for his encouragement, for his marketing insight, and for providing me with the environment of creative freedom. The unwavering support of the Developmental Math group and the endless hours of hard work by Martha Morong and Janet Theurer have led to products of which I am immensely proud.

I also want to thank Judy Beecher, my co-author on many books and my developmental editor on this text. Her steadfast loyalty, vision, and encouragement have been invaluable. In addition to writing the Student's Solutions Manual, Judy Penna has continued to provide strong leadership in the preparation of the printed supplements, videotapes, and interactive CD-ROM. Other strong support has come from Donna DeSpain for the Printed Test Bank; Bill Saler for the audiotapes; Irene Doo for the Collaborative Learning Activities Manual; and Irene Doo, Barbara Johnson, Vera Preston, and Peggy Reijto for their accuracy checking.

M.L.B.

# Diagnostic Pretest

## Chapter R

Perform the indicated operations and simplify if possible.

**1.** $\dfrac{8}{9} \cdot \dfrac{3}{5}$ 
**2.** $\dfrac{1}{4} + \dfrac{2}{3}$ 
**3.** $4.94 \div 0.19$ 
**4.** $12.04 - 1.057$

## Chapter 1

Compute and simplify.

**5.** $3.8 + (-4.62) - (-2)$ 
**6.** $-9(1.3)$

**7.** A small business made a profit of \$135.97 on Tuesday. The next day, it had a loss of \$145.90. Find the total profit or loss.

**8.** Remove parentheses and simplify:
$$3[11(a - 2) - 2(3 - a)].$$

## Chapter 2

Solve.

**9.** $2(x - 1) = 4(x + 2)$ 
**10.** $4 - 13x \le 10x - 5$

**11.** A 36-in. string is cut into two pieces. One piece is three times as long as the other. How long are the pieces?

**12.** *Family Income.* A family spent \$270 one month on clothing. This was 18% of its income. What was the family's income?

## Chapter 3

Graph.

**13.** $y = -2x + 1$ 
**14.** $x = -2$

## Chapter 4

Simplify.

**15.** $\dfrac{x^2y^2}{x^{-2}y^3}$ 
**16.** $(-2x)^3(2x^4)^2$

**17.** Subtract: $(x^2 + 3x - 1) - (2x^2 - 5)$. 
**18.** Multiply: $(2x^2 + 3)(2x^2 - 3)$.

## Chapter 5

Factor completely.

**19.** $2x^2 - 162$ 
**20.** $5x^2 - 14x - 3$

Solve.

**21.** $x^2 + 3x = 10$ 
**22.** The width of a rectangle is 9 m less than the length. The area is 136 m$^2$. Find the width and the length.

**Chapter 6**

**23.** Divide and simplify:

$$\frac{2x^3 + 6x^2}{x^2 + 10x + 25} \div \frac{4x^3 - 36x}{x^2 + x - 20}.$$

**24.** Add and simplify:

$$\frac{1 - x}{x^2 + x} + \frac{x}{x^2 + 3x + 2}.$$

Solve.

**25.** $\dfrac{2}{x + 4} = \dfrac{1}{x}$

**26.** *Speed.* One car travels 15 mph faster than another. While one car travels 165 mi, the other travels 120 mi. How fast is each car traveling?

**Chapter 7**

**27.** Find the slope and the $y$-intercept of $2x + 3y = 8$.

**28.** Find an equation of the line containing the pair of points $(3, 2)$ and $(4, -1)$.

**29.** Graph: $x - 2y \le 6$.

**Chapter 8**

Solve.

**30.** $x + y = 5,$
$2x + 3y = 7$

**31.** $2x + 4y = 5,$
$3x - 2y = 9$

**32.** *Alcohol Solution.* Solution A is 20% alcohol and solution B is 50% alcohol. How much of each should be used in order to make 50 L of a solution that is 35% alcohol?

**33.** Two cars leave campus at the same time going in the same direction. One travels 56 mph and the other travels 62 mph. In how many hours will they be 60 mi apart?

**Chapter 9**

**34.** Multiply and simplify:
$$\sqrt{2x^2y} \cdot \sqrt{6xy^3}.$$

**35.** Divide and simplify:
$$\frac{\sqrt{5x^3}}{\sqrt{45xy^2}}.$$

**36.** Rationalize the denominator:

$$\frac{3}{2 - \sqrt{3}}.$$

**37.** Solve: $\sqrt{2x + 4} - 1 = 8$.

**Chapter 10**

Solve.

**38.** $3x^2 + 2x = 1$

**39.** $2x^2 + 10 = x$

**40.** The hypotenuse of a right triangle is 34 m. One leg is 14 m longer than the other. Find the lengths of the legs.

**41.** Graph: $y = x^2 - 4x + 1$.

# Introductory Algebra

## Eighth Edition

# R

# Prealgebra Review

## An Application

For health reasons, there must be 6 parts per billion (ppb) of chlorine in a swimming pool. What percent of total volume is this amount of chlorine?

This problem appears as Exercise 38 in the Summary and Review Exercises.

## The Mathematics

We find this percent as follows:

$$\frac{6}{1,000,000,000} = 0.000000006$$
$$= 0.0000006\%.$$

# Pretest: Chapter R

**1.** Find the prime factorization of 248.

**2.** Find the least common multiple of 12, 24, and 42.

**3.** Write an expression equivalent to $\frac{2}{3}$ by multiplying by 1 using $\frac{5}{5}$.

**4.** Write an expression equivalent to $\frac{11}{12}$ with a denominator of 48.

Simplify.

**5.** $\frac{46}{128}$

**6.** $\frac{28}{42}$

Compute and simplify.

**7.** $\frac{3}{5} \div \frac{6}{11}$

**8.** $\frac{3}{7} - \frac{1}{3}$

**9.** $\frac{3}{10} + \frac{1}{5}$

**10.** $\frac{4}{7} \cdot \frac{5}{12}$

**11.** Convert to fractional notation (do not simplify): 32.17.

**12.** Convert to decimal notation: $\frac{789}{10,000}$.

**13.** Add: $8.25 + 91 + 34.7862$.

**14.** Subtract: $230 - 17.95$.

**15.** Multiply: $34.78 \times 10.08$.

**16.** Divide: $78.12 \div 6.3$.

**17.** Convert to decimal notation: $\frac{13}{9}$.

**18.** Round to the nearest hundredth: 345.8395.

**19.** Round to the nearest tenth: 345.8395.

**20.** Convert to decimal notation: 11.6%.

**21.** Convert to fractional notation: 87%.

**22.** Convert to percent notation: $\frac{7}{8}$.

**23.** Write exponential notation: $5 \cdot 5 \cdot 5 \cdot 5$.

**24.** Evaluate: $2^3$.

**25.** Evaluate: $(1.1)^2$.

**26.** Calculate: $9 \cdot 3 + 24 \div 4 - 5^2 + 10$.

# R.1 Factoring and LCMs

## a | Factors and Prime Factorizations

We begin our review with *factoring*, which is a necessary skill for addition and subtraction with fractional notation. Factoring is also an important skill in algebra. You will eventually learn to factor algebraic expressions.

The numbers we will be factoring are from the set of **natural numbers:**

$$1, \quad 2, \quad 3, \quad 4, \quad 5, \quad \text{and so on.}$$

To **factor** a number means to express the number as a product. Consider the product $12 = 3 \cdot 4$. We say that 3 and 4 are **factors** of 12 and that $3 \cdot 4$ is a **factorization** of 12. Since $12 = 12 \cdot 1$, we also know that 12 and 1 are factors of 12 and that $12 \cdot 1$ is a factorization of 12.

**Example 1** Find all the factors of 12.

We first find some factorizations:

$$12 = 1 \cdot 12, \quad 12 = 2 \cdot 6, \quad 12 = 3 \cdot 4, \quad 12 = 2 \cdot 2 \cdot 3.$$

The factors of 12 are 1, 2, 3, 4, 6, and 12.

**Example 2** Find all the factors of 150.

We first find some factorizations:

$$150 = 1 \cdot 150, \quad 150 = 2 \cdot 75, \quad 150 = 3 \cdot 50, \quad 150 = 5 \cdot 30,$$
$$150 = 6 \cdot 25, \quad 150 = 10 \cdot 15, \quad 150 = 2 \cdot 5 \cdot 3 \cdot 5.$$

The factors of 150 are 1, 2, 3, 5, 6, 10, 15, 25, 30, 50, 75, and 150.

Note that the word "factor" is used both as a noun and as a verb. You **factor** when you express a number as a product. The numbers you multiply together to get the product are **factors**.

*Do Exercises 1–4 (in the margin at right).*

> A natural number that has *exactly two different factors,* itself and 1, is called a **prime number.**

**Example 3** Which of these numbers are prime? ⑦, 4, ⑪, 18, 1

7 is prime. It has exactly two different factors, 7 and 1.

4 is not prime. It has three different factors, 1, 2, and 4.

11 is prime. It has exactly two different factors, 11 and 1.

18 is not prime. It has factors 1, 2, 3, 6, 9, and 18.

1 is not prime. It does not have two *different* factors.

**Objectives**

a | Find all the factors of numbers and find prime factorizations of numbers.

b | Find the LCM of two or more numbers using prime factorizations.

**For Extra Help**

TAPE 1   TAPE 1A   MAC   CD-ROM
                    WIN

Find all the factors of the number.

**1.** 9   $9 \cdot 1, 3 \cdot 3$

**2.** 16   $8 \cdot 2, 16 \cdot 1$
$4 \cdot 4$

**3.** 24
$8 \cdot 3, 6 \cdot 4, 2 \cdot 12$
$24 \cdot 1$

**4.** 180
$90 \cdot 2, 10 \cdot 18,$
$4 \cdot 60, 6 \cdot 30 \cdot$
$12 \cdot 15, 20 \cdot 9$

*Answers on page A-1*

A TABLE OF PRIMES

2, 3, 5, 7, 11, 13, 17, 19, 23, 29, 31, 37, 41, 43, 47, 53, 59, 61, 67, 71, 73, 79, 83, 89, 97, 101, 103, 107, 109, 113, 127, 131, 137, 139, 149, 151, 157

**5.** Which of these numbers are prime?

8,  6,  13,  14,  1

Find the prime factorization.

**6.** 48

**7.** 50

**8.** 770

*Answers on page A-1*

In the margin at left is a table of the prime numbers from 2 to 157. There are more extensive tables, but these prime numbers will be the most helpful to you in this text.

***Do Exercise 5.***

If a natural number, other than 1, is not prime, we call it **composite.** Every composite number can be factored into a product of prime numbers. Such a factorization is called a **prime factorization.**

**Example 4**   Find the prime factorization of 36.

We begin by factoring 36 any way we can. One way is like this:

$$36 = 4 \cdot 9.$$

The factors 4 and 9 are not prime, so we factor them:

$$36 = 4 \cdot 9$$
$$= 2 \cdot 2 \cdot 3 \cdot 3$$

The factors in the last factorization are all prime, so we now have the *prime factorization* of 36. Note that 1 is *not* part of this factorization because it is not prime.

Another way to find the prime factorization of 36 is like this:

$$36 = 2 \cdot 18 = 2 \cdot 3 \cdot 6 = 2 \cdot 3 \cdot 2 \cdot 3.$$

In effect, we begin factoring any way we can think of and keep factoring until all factors are prime. Using a **factor tree** might also be helpful.

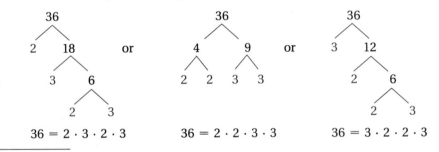

No matter which way we begin, the result is the same: The prime factorization of 36 contains two factors of 2 and two factors of 3. Every composite number has a *unique* prime factorization.

**Example 5**   Find the prime factorization of 60.

This time, we use the list of primes from the table. We go through the table until we find a prime that is a factor of 60. The first such prime is 2.

$$60 = 2 \cdot 30$$

We keep dividing by 2 until it is not possible to do so.

$$60 = 2 \cdot 2 \cdot 15$$

Now we go to the next prime in the table that is a factor of 60. It is 3.

$$60 = 2 \cdot 2 \cdot 3 \cdot 5$$

Each factor in $2 \cdot 2 \cdot 3 \cdot 5$ is a prime. Thus this is the prime factorization.

***Do Exercises 6–8.***

## b  Least Common Multiples

Least common multiples are used to add and subtract with fractional notation.

The **multiples** of a number all have that number as a factor. For example, the multiples of 2 are

$$2, \quad 4, \quad 6, \quad 8, \quad 10, \quad 12, \quad 14, \quad 16, \ldots.$$

We could name each of them in such a way as to show 2 as a factor. For example, $14 = 2 \cdot 7$.

The multiples of 3 all have 3 as a factor:

$$3, \quad 6, \quad 9, \quad 12, \quad 15, \quad 18, \ldots.$$

Two or more numbers always have many multiples in common. From lists of multiples, we can find common multiples.

**Example 6**  Find the common multiples of 2 and 3.

We make lists of their multiples and circle the multiples that appear in both lists.

2, 4, ⑥, 8, 10, ⑫, 14, 16, ⑱, 20, 22, ㉔, 26, 28, ㉚, 32, 34, ㊱, ...;
3, ⑥, 9, ⑫, 15, ⑱, 21, ㉔, 27, ㉚, 33, ㊱, ...

The common multiples of 2 and 3 are

$$6, \quad 12, \quad 18, \quad 24, \quad 30, \quad 36, \ldots.$$

*Do Exercises 9 and 10.*

In Example 6, we found common multiples of 2 and 3. The *least*, or smallest, of those common multiples is 6. We abbreviate **least common multiple** as **LCM**.

There are several methods that work well for finding the LCM of several numbers. Some of these do not work well in algebra, especially when we consider expressions with variables such as $4ab$ and $12abc$. We now review a method that will work in arithmetic *and in algebra as well*. To see how it works, let's look at the prime factorizations of 9 and 15 in order to find the LCM:

$$9 = 3 \cdot 3, \qquad 15 = 3 \cdot 5.$$

Any multiple of 9 must have *two* 3's as factors. Any multiple of 15 must have *one* 3 and *one* 5 as factors. The smallest number satisfying all of these conditions is

$$\underbrace{3 \cdot 3}_{} \cdot 5 = 45.$$

Two 3's; 9 is a factor

One 3, one 5; 15 is a factor

The LCM must have all the factors of 9 and all the factors of 15, but the factors cannot be repeated when they are common to both numbers.

---

To find the LCM of several numbers:

**a)** Write the prime factorization of each number.

**b)** Form the LCM by writing the product of the different factors from step (a), using each factor the greatest number of times that it occurs in any one factorization.

---

9. Find the common multiples of 3 and 5 by making lists of multiples.

3, 6, 9, 12, ⑮ 18, 21, 24

5, 10, ⑮, 20, 25

10. Find the common multiples of 9 and 15 by making lists of multiples.

9, 18, 24, 36, ㊺, 54

15, 30, ㊺, 60, 75

*Answers on page A-1*

Find the LCM by factoring.

**11.** 8 and 10

*8,16,24,32,40*
*10,20,30,40*
*(40)*

**12.** 18 and 27

*18,36,54,72*
*27,54*
*(54)*

**13.** Find the LCM of 18, 24, and 30.

*18,36,54,72,90,108,120*
*24,48,72,96,120*
*30,60,90,120*
*(120)*

Find the LCM.

**14.** 3,   18

*(54)*
*(54)*

**15.** 12,   24

Find the LCM.

**16.** 4,   9

**17.** 5,   6,   7

*Answers on page A-1*

**Example 7**   Find the LCM of 40 and 100.

**a)** We find the prime factorizations:

$$40 = 2 \cdot 2 \cdot 2 \cdot 5,$$
$$100 = 2 \cdot 2 \cdot 5 \cdot 5.$$

**b)** We write 2 as a factor three times (the greatest number of times that it occurs in any one factorization). We write 5 as a factor two times (the greatest number of times that it occurs in any one factorization).

The LCM is $2 \cdot 2 \cdot 2 \cdot 5 \cdot 5$, or 200.

*Do Exercises 11 and 12.*

**Example 8**   Find the LCM of 27, 90, and 84.

**a)** We factor:

$$27 = 3 \cdot 3 \cdot 3,$$
$$90 = 2 \cdot 3 \cdot 3 \cdot 5,$$
$$84 = 2 \cdot 2 \cdot 3 \cdot 7.$$

**b)** We write 2 as a factor two times, 3 three times, 5 one time, and 7 one time.

The LCM is $2 \cdot 2 \cdot 3 \cdot 3 \cdot 3 \cdot 5 \cdot 7$, or 3780.

*Do Exercise 13.*

**Example 9**   Find the LCM of 7 and 21.

Since 7 is prime, it has no prime factorization. It still, however, must be a factor of the LCM:

$$7 = 7,$$
$$21 = 3 \cdot 7.$$

The LCM is $7 \cdot 3$, or 21.

> If one number is a factor of another, then the LCM is the larger of the two numbers.

*Do Exercises 14 and 15.*

**Example 10**   Find the LCM of 8 and 9.

We have

$$8 = 2 \cdot 2 \cdot 2,$$
$$9 = 3 \cdot 3.$$

The LCM is $2 \cdot 2 \cdot 2 \cdot 3 \cdot 3$, or 72.

> If two or more numbers have no common prime factor, then the LCM is the product of the numbers.

*Do Exercises 16 and 17.*

# Exercise Set R.1

Always review the objectives before doing an exercise set. See page 3. Note how the objectives are keyed to the exercises.

**a** Find all the factors of the number.

**1.** 20

**2.** 36

**3.** 72

**4.** 81

Find the prime factorization of the number.

**5.** 15

**6.** 14

**7.** 22

**8.** 33

**9.** 9

**10.** 25

**11.** 49

**12.** 121

**13.** 18

**14.** 24

**15.** 40

**16.** 56

**17.** 90

**18.** 120

**19.** 210

**20.** 330

**21.** 91

**22.** 143

**23.** 119

**24.** 221

**b** Find the prime factorization of the numbers. Then find the LCM.

**25.** 4,  5

**26.** 18,  40

**27.** 24,  36

**28.** 24,  27

**29.** 3,  15

**30.** 20,  40

**31.** 30,  40

**32.** 50,  60

**33.** 13,  23

**34.** 12,  18

**35.** 18,  30

**36.** 45,  72

**37.** 30,  36

**38.** 30,  50

**39.** 24,  30

**40.** 60,  70

**41.** 17,  29

**42.** 18,  24

**43.** 12,  28

**44.** 35,  45

**45.** 2, 3, 5       **46.** 3, 5, 7       **47.** 24, 36, 12       **48.** 8, 16, 22

**49.** 5, 12, 15       **50.** 12, 18, 40       **51.** 6, 12, 18       **52.** 24, 35, 45

---

## Synthesis

Exercises designed as *synthesis exercises* differ from those found in the main body of the exercise set. The icon ◈ denotes synthesis exercises that are writing exercises, which are meant to be answered in one or more complete sentences. Because answers to writing exercises often vary, they are not listed at the back of the book. Exercises marked with a 🖩 are meant to be solved using a calculator. These and the other synthesis exercises will often challenge you to put together two or more objectives at once.

**53.** ◈ Describe the notation of a prime number as though you were talking to a classmate.

**54.** ◈ Explain a method for finding a composite number that contains exactly two factors other than itself and 1.

**55.** Consider the numbers 8 and 12. Determine whether each of the following is the LCM of 8 and 12. Tell why or why not.

     **a)** $2 \cdot 2 \cdot 3 \cdot 3$       **b)** $2 \cdot 2 \cdot 3$       **c)** $2 \cdot 3 \cdot 3$       **d)** $2 \cdot 2 \cdot 2 \cdot 3$

🖩 Use a calculator to find the LCM of the numbers.

**56.** 288, 324                           **57.** 2700, 7800

*Planetary Orbits and LCMs.* The earth, Jupiter, Saturn, and Uranus all revolve around the sun. The earth takes 1 yr, Jupiter 12 yr, Saturn 30 yr, and Uranus 84 yr. On a certain night, you look at all the planets and wonder how many years it will be before they all have the same position again. To find out, you find the LCM of 12, 30, and 84. It will be that number of years.

**58.** How often will Saturn and Uranus appear in the same position?

**59.** How often will Jupiter and Saturn appear in the same position?

**60.** How often will Jupiter, Saturn, and Uranus appear in the same position?

Copyright © 1999 Addison Wesley Longman

---

# R.2 Fractional Notation

We now review fractional notation and its use with addition, subtraction, multiplication, and division of *arithmetic numbers.*

## a Equivalent Expressions and Fractional Notation

An example of **fractional notation** for a number is

$$\frac{2}{3} \begin{matrix} \leftarrow \text{Numerator} \\ \leftarrow \text{Denominator} \end{matrix}$$

The top number is called the **numerator**, and the bottom number is called the **denominator**.

The **whole numbers** consist of the natural numbers and 0:

$$0, \quad 1, \quad 2, \quad 3, \quad 4, \quad 5, \ldots .$$

The **arithmetic numbers,** also called the **nonnegative rational numbers,** consist of the whole numbers and the fractions, such as $\frac{2}{3}$ and $\frac{9}{5}$. The arithmetic numbers can also be described as follows.

> The **arithmetic numbers** are the whole numbers and the fractions, such as $\frac{3}{4}$, $\frac{6}{5}$, or 8. All of these numbers can be named with fractional notation $\frac{a}{b}$, where $a$ and $b$ are whole numbers and $b \neq 0$.

Note that all whole numbers can be named with fractional notation. For example, we can name the whole number 8 as $\frac{8}{1}$. We call 8 and $\frac{8}{1}$ **equivalent expressions.**

Being able to find an equivalent expression is critical to a study of algebra. Some simple but powerful properties of numbers that allow us to find equivalent expressions are the identity properties of 0 and 1. ✳

> **THE IDENTITY PROPERTY OF 0**
>
> For any number $a$,
>
> $$a + 0 = a.$$
>
> (Adding 0 to any number gives that same number.)

> **THE IDENTITY PROPERTY OF 1**
>
> For any number $a$,
>
> $$a \cdot 1 = a.$$
>
> (Multiplying any number by 1 gives that same number.)

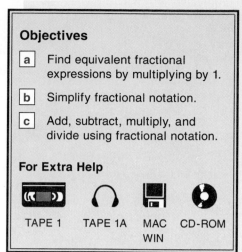

**Objectives**

a Find equivalent fractional expressions by multiplying by 1.

b Simplify fractional notation.

c Add, subtract, multiply, and divide using fractional notation.

**For Extra Help**

TAPE 1    TAPE 1A    MAC WIN    CD-ROM

1. Write a fractional expression equivalent to $\frac{2}{3}$ with a denominator of 12.

2. Write a fractional expression equivalent to $\frac{3}{4}$ with a denominator of 28.

3. Multiply by 1 to find three different fractional expressions for $\frac{7}{8}$.

Here are some ways to name the number 1:

$$\frac{5}{5}, \quad \frac{3}{3}, \quad \text{and} \quad \frac{26}{26}.$$

The following property allows us to find equivalent fractional expressions, that is, find other names for arithmetic numbers.

> **EQUIVALENT EXPRESSIONS FOR 1**
>
> For any number $a$, $a \neq 0$,
>
> $$\frac{a}{a} = 1.$$

We can use the identity property of 1 and the preceding result to find equivalent fractional expressions.

**Example 1**   Write a fractional expression equivalent to $\frac{2}{3}$ with a denominator of 15.

Note that $15 = 3 \cdot 5$. We want fractional notation for $\frac{2}{3}$ that has a denominator of 15, but the denominator 3 is missing a factor of 5. We multiply by 1, using $\frac{5}{5}$ as an equivalent expression for 1. Recall from arithmetic that to multiply with fractional notation, we multiply numerators and denominators:

$$\frac{2}{3} = \frac{2}{3} \cdot 1 \qquad \text{Using the identity property of 1}$$

$$= \frac{2}{3} \cdot \frac{5}{5} \qquad \text{Using } \tfrac{5}{5} \text{ for 1}$$

$$= \frac{10}{15}. \qquad \text{Multiplying numerators and denominators}$$

*Do Exercises 1–3.*

## b   Simplifying Expressions

We know that $\frac{1}{2}, \frac{2}{4}, \frac{4}{8}$, and so on, all name the same number. Any arithmetic number can be named in many ways. The **simplest fractional notation** is the notation that has the smallest numerator and denominator. We call the process of finding the simplest fractional notation **simplifying**. We reverse the process of Example 1 by first factoring the numerator and the denominator. Then we factor the fractional expression and remove a factor of 1 using the identity property of 1.

**Example 2**   Simplify: $\frac{10}{15}$.

$$\frac{10}{15} = \frac{2 \cdot 5}{3 \cdot 5} \qquad \text{Factoring the numerator and the denominator. In this case, each is the prime factorization.}$$

$$= \frac{2}{3} \cdot \frac{5}{5} \qquad \text{Factoring the fractional expression}$$

$$= \frac{2}{3} \cdot 1$$

$$= \frac{2}{3} \qquad \text{Using the identity property of 1 (removing a factor of 1)}$$

**Example 3** Simplify: $\dfrac{36}{24}$.

$$\dfrac{36}{24} = \dfrac{2 \cdot 3 \cdot 2 \cdot 3}{2 \cdot 2 \cdot 3 \cdot 2}$$     **Factoring the numerator and the denominator**

$$= \dfrac{2 \cdot 3 \cdot 2}{2 \cdot 3 \cdot 2} \cdot \dfrac{3}{2}$$     **Factoring the fractional expression**

$$= 1 \cdot \dfrac{3}{2}$$

$$= \dfrac{3}{2}$$     **Removing a factor of 1**

It is always a good idea to check at the end to see if you have indeed factored out all the common factors of the numerator and the denominator.

## Canceling

Canceling is a shortcut that you may have used to remove a factor of 1 when working with fractional notation. With *great* concern, we mention it as a possible way to speed up your work. You should use canceling only when removing common factors in numerators and denominators. Each common factor allows us to remove a factor of 1 in a product. Canceling *cannot* be done in sums or when adding expressions together. Our concern is that "canceling" be performed with care and understanding. Example 3 might have been done faster as follows:

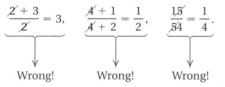

*CAUTION!* The difficulty with canceling is that it is often applied incorrectly in situations like these:

$$\dfrac{\cancel{2} + 3}{\cancel{2}} = 3, \qquad \dfrac{\cancel{4} + 1}{\cancel{4} + 2} = \dfrac{1}{2}, \qquad \dfrac{1\cancel{5}}{\cancel{5}4} = \dfrac{1}{4}.$$

Wrong!      Wrong!      Wrong!

In each of these situations, the expressions canceled out were *not* factors of 1. Factors are parts of products. For example, in $2 \cdot 3$, 2 and 3 are factors, but in $2 + 3$, 2 and 3 are *not* factors. **If you can't factor, you can't cancel! If in doubt, don't cancel!**

Examples of correct ways to cancel are as follows:

$$\dfrac{2 + 3}{2} = \dfrac{5}{2}, \qquad \dfrac{4 + 1}{4 + 2} = \dfrac{5}{6}, \qquad \dfrac{15}{54} = \dfrac{\cancel{3} \cdot 5}{\cancel{3} \cdot 18} = \dfrac{5}{18}.$$

Correct      Correct      Correct

*Do Exercises 4–6.*

---

Simplify.

**4.** $\dfrac{18}{45}$

**5.** $\dfrac{38}{18}$

**6.** $\dfrac{72}{27}$

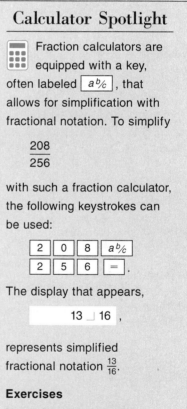

**Calculator Spotlight**

Fraction calculators are equipped with a key, often labeled $\boxed{a^b\!/_c}$, that allows for simplification with fractional notation. To simplify

$$\dfrac{208}{256}$$

with such a fraction calculator, the following keystrokes can be used:

$$\boxed{2}\ \boxed{0}\ \boxed{8}\ \boxed{a^b\!/_c}$$
$$\boxed{2}\ \boxed{5}\ \boxed{6}\ \boxed{=}.$$

The display that appears,

$$\boxed{13 \sqcup 16},$$

represents simplified fractional notation $\frac{13}{16}$.

**Exercises**

Use a fraction calculator to simplify each of the following.

**1.** $\dfrac{35}{40}$      **2.** $\dfrac{84}{90}$

**3.** $\dfrac{690}{835}$      **4.** $\dfrac{42}{150}$

*Answers on page A-2*

Simplify.

**7.** $\dfrac{27}{54}$

**8.** $\dfrac{48}{12}$

Multiply and simplify.

**9.** $\dfrac{6}{5} \cdot \dfrac{25}{12}$

**10.** $\dfrac{3}{8} \cdot \dfrac{5}{3} \cdot \dfrac{7}{2}$

*Answers on page A-2*

The number of factors in the numerator and the denominator may not always be the same. If not, we can always insert the number 1 as a factor. The identity property of 1 allows us to do that.

**Example 4**   Simplify: $\dfrac{18}{72}$.

$$\frac{18}{72} = \frac{2 \cdot 9}{8 \cdot 9} = \frac{2}{8} = \frac{2 \cdot 1}{2 \cdot 4} = \frac{1}{4},$$

or

$$\frac{18}{72} = \frac{1 \cdot 18}{4 \cdot 18} = \frac{1}{4}$$

**Example 5**   Simplify: $\dfrac{72}{9}$.

$$\frac{72}{9} = \frac{8 \cdot 9}{1 \cdot 9} \qquad \text{Factoring and inserting a factor of 1 in the denominator}$$

$$= \frac{8 \cdot 9}{1 \cdot 9} \qquad \text{Removing a factor of 1: } \frac{9}{9} = 1$$

$$= \frac{8}{1}$$

$$= 8 \qquad \text{Simplifying}$$

*Do Exercises 7 and 8.*

## c   Multiplication, Addition, Subtraction, and Division

After we have performed an operation of multiplication, addition, subtraction, or division, the answer may or may not be in simplified form. We simplify, if at all possible.

### Multiplication

To multiply using fractional notation, we multiply the numerators to get the new numerator, and we multiply the denominators to get the new denominator.

**Example 6**   Multiply and simplify: $\dfrac{5}{6} \cdot \dfrac{9}{25}$.

$$\frac{5}{6} \cdot \frac{9}{25} = \frac{5 \cdot 9}{6 \cdot 25} \qquad \text{Multiplying numerators and denominators}$$

$$= \frac{1 \cdot 5 \cdot 3 \cdot 3}{2 \cdot 3 \cdot 5 \cdot 5} \qquad \text{Factoring the numerator and the denominator}$$

$$= \frac{3 \cdot 5 \cdot 1 \cdot 3}{3 \cdot 5 \cdot 2 \cdot 5} \qquad \text{Removing a factor of 1: } \frac{3 \cdot 5}{3 \cdot 5} = 1$$

$$= \frac{3}{10} \qquad \text{Simplifying}$$

*Do Exercises 9 and 10.*

## Addition

When denominators are the same, we can add by adding the numerators and keeping the same denominator.

**Example 7** Add and simplify: $\dfrac{4}{8} + \dfrac{5}{8}$.

The common denominator is 8. We add the numerators and keep the common denominator:

$$\frac{4}{8} + \frac{5}{8} = \frac{4+5}{8} = \frac{9}{8}.$$

In arithmetic, we generally write $\frac{9}{8}$ as $1\frac{1}{8}$. In algebra, you will find that *improper* symbols such as $\frac{9}{8}$ are more useful and are quite *proper* for our purposes.

When denominators are different, we use the identity property of 1 and multiply to find a common denominator. The smallest such denominator is called the lowest or **least common denominator.** That number is the least common multiple of the original denominators. The least common denominator is often abbreviated **LCD.**

**Example 8** Add and simplify: $\dfrac{3}{8} + \dfrac{5}{12}$.

The LCM of the denominators, 8 and 12, is 24. Thus the LCD is 24. We multiply each fraction by 1 to obtain the LCD:

$$\frac{3}{8} + \frac{5}{12} = \frac{3}{8} \cdot \frac{3}{3} + \frac{5}{12} \cdot \frac{2}{2}$$

Multiplying by 1. Since $3 \cdot 8 = 24$, we multiply the first number by $\frac{3}{3}$. Since $2 \cdot 12 = 24$, we multiply the second number by $\frac{2}{2}$.

$$= \frac{9}{24} + \frac{10}{24}$$

$$= \frac{9+10}{24}$$

Adding the numerators and keeping the same denominator

$$= \frac{19}{24}.$$

*Do Exercises 11–14.*

## Subtraction

When subtracting, we also multiply by 1 to obtain the LCD. After we have made the denominators the same, we can subtract by subtracting the numerators and keeping the same denominator.

**Example 9** Subtract and simplify: $\dfrac{9}{8} - \dfrac{4}{5}$.

$$\frac{9}{8} - \frac{4}{5} = \frac{9}{8} \cdot \frac{5}{5} - \frac{4}{5} \cdot \frac{8}{8} \qquad \text{The LCD is 40.}$$

$$= \frac{45}{40} - \frac{32}{40} = \frac{45-32}{40}$$

$$= \frac{13}{40}$$

Add and simplify.

**11.** $\dfrac{4}{5} + \dfrac{3}{5}$

**12.** $\dfrac{5}{6} + \dfrac{7}{6}$

**13.** $\dfrac{5}{6} + \dfrac{7}{10}$

**14.** $\dfrac{1}{4} + \dfrac{1}{2}$

*Answers on page A-2*

Subtract and simplify.

**15.** $\dfrac{7}{8} - \dfrac{2}{5}$

**16.** $\dfrac{5}{12} - \dfrac{2}{9}$

Find the reciprocal.

**17.** $\dfrac{4}{11}$

**18.** $\dfrac{15}{7}$

**19.** 5

**20.** $\dfrac{1}{3}$

**21.** Divide by multiplying by 1:
$$\dfrac{\frac{3}{5}}{\frac{4}{7}}.$$

**Example 10** Subtract and simplify: $\dfrac{7}{10} - \dfrac{1}{5}$.

$$\dfrac{7}{10} - \dfrac{1}{5} = \dfrac{7}{10} - \dfrac{1}{5} \cdot \dfrac{2}{2} \qquad \text{The LCD is 10.}$$

$$= \dfrac{7}{10} - \dfrac{2}{10} = \dfrac{7-2}{10}$$

$$= \dfrac{5}{10} = \dfrac{1 \cdot 5}{2 \cdot 5} = \dfrac{1}{2} \qquad \text{Removing a factor of 1: } \dfrac{5}{5} = 1$$

*Do Exercises 15 and 16.*

## Reciprocals

Two numbers whose product is 1 are called **reciprocals**, or **multiplicative inverses,** of each other. All the arithmetic numbers, except zero, have reciprocals.

## Examples

**11.** The reciprocal of $\frac{2}{3}$ is $\frac{3}{2}$ because $\frac{2}{3} \cdot \frac{3}{2} = \frac{6}{6} = 1$.

**12.** The reciprocal of 9 is $\frac{1}{9}$ because $9 \cdot \frac{1}{9} = \frac{9}{9} = 1$.

**13.** The reciprocal of $\frac{1}{4}$ is 4 because $\frac{1}{4} \cdot 4 = \frac{4}{4} = 1$.

*Do Exercises 17–20.*

## Reciprocals and Division

Reciprocals and the number 1 can be used to justify a fast way to divide arithmetic numbers. We multiply by 1, carefully choosing the expression for 1.

**Example 14** Divide $\dfrac{2}{3}$ by $\dfrac{7}{5}$.

This is a symbol for 1.

$$\dfrac{2}{3} \div \dfrac{7}{5} = \dfrac{\frac{2}{3}}{\frac{7}{5}} = \dfrac{\frac{2}{3}}{\frac{7}{5}} \cdot \dfrac{\frac{5}{7}}{\frac{5}{7}} \qquad \text{Multiplying by } \dfrac{\frac{5}{7}}{\frac{5}{7}}. \text{ We use } \tfrac{5}{7} \text{ because it is the reciprocal of } \tfrac{7}{5}.$$

$$= \dfrac{\frac{2}{3} \cdot \frac{5}{7}}{\frac{7}{5} \cdot \frac{5}{7}} \qquad \text{Multiplying numerators and denominators}$$

$$= \dfrac{\frac{10}{21}}{\frac{35}{35}} = \dfrac{\frac{10}{21}}{1} \qquad \tfrac{35}{35} = 1$$

$$= \dfrac{10}{21} \qquad \text{Simplifying}$$

After multiplying, we had a denominator of $\frac{35}{35}$, or 1. That was because we used $\frac{5}{7}$, the reciprocal of the divisor, for both the numerator and the denominator of the symbol for 1.

*Do Exercise 21.*

When multiplying by 1 to divide, we get a denominator of 1. What do we get in the numerator? In Example 14, we got $\frac{2}{3} \cdot \frac{5}{7}$. This is the product of $\frac{2}{3}$, the dividend, and $\frac{5}{7}$, the reciprocal of the divisor.

> To divide, multiply by the reciprocal of the divisor:
> $$\frac{a}{b} \div \frac{c}{d} = \frac{a}{b} \cdot \frac{d}{c}.$$

**Example 15**   Divide by multiplying by the reciprocal of the divisor: $\frac{1}{2} \div \frac{3}{5}$.

$$\frac{1}{2} \div \frac{3}{5} = \frac{1}{2} \cdot \frac{5}{3} \qquad \text{$\frac{5}{3}$ is the reciprocal of $\frac{3}{5}$}$$

$$= \frac{5}{6} \qquad \text{\textbf{Multiplying}}$$

After dividing, simplification is often possible and should be done.

**Example 16**   Divide and simplify: $\frac{2}{3} \div \frac{4}{9}$.

$$\frac{2}{3} \div \frac{4}{9} = \frac{2}{3} \cdot \frac{9}{4} \qquad \text{$\frac{9}{4}$ is the reciprocal of $\frac{4}{9}$}$$

$$= \frac{2 \cdot 3 \cdot 3}{3 \cdot 2 \cdot 2} \qquad \text{\textbf{Removing a factor of 1: } } \frac{2 \cdot 3}{2 \cdot 3} = 1$$

$$= \frac{3}{2}$$

*Do Exercises 22–24.*

**Example 17**   Divide and simplify: $\frac{5}{6} \div 30$.

$$\frac{5}{6} \div 30 = \frac{5}{6} \div \frac{30}{1} = \frac{5}{6} \cdot \frac{1}{30} = \frac{5 \cdot 1}{6 \cdot 30} = \frac{5 \cdot 1}{\underbrace{6 \cdot 5 \cdot 6}} = \frac{1}{6 \cdot 6} = \frac{1}{36}$$

$$\text{\textbf{Removing a factor of 1: } } \frac{5}{5} = 1$$

**Example 18**   Divide and simplify: $24 \div \frac{3}{8}$.

$$24 \div \frac{3}{8} = \frac{24}{1} \div \frac{3}{8} = \frac{24}{1} \cdot \frac{8}{3} = \frac{24 \cdot 8}{1 \cdot 3} = \frac{\underbrace{3 \cdot 8 \cdot 8}}{1 \cdot 3} = \frac{8 \cdot 8}{1} = 64$$

$$\text{\textbf{Removing a factor of 1: } } \frac{3}{3} = 1$$

*Do Exercises 25 and 26.*

Divide by multiplying by the reciprocal of the divisor. Then simplify.

**22.** $\frac{4}{3} \div \frac{7}{2}$

**23.** $\frac{5}{4} \div \frac{3}{2}$

**24.** $\dfrac{\frac{2}{9}}{\frac{5}{12}}$

Divide and simplify.

**25.** $\frac{7}{8} \div 56$

**26.** $36 \div \frac{4}{9}$

*Answers on page A-2*

# Improving Your Math Study Skills

## Homework

Throughout this textbook, you will find a feature called "Improving Your Math Study Skills." At least one such topic is included in each chapter. Each topic title is listed in the table of contents beginning on p. iii. Here we consider your homework study skills.

### Before Doing Your Homework

- **Setting.** Consider doing your homework as soon as possible after class, before you forget what you learned in the lecture. Research has shown that after 24 hours, most people forget about half of what is in their short-term memory. To avoid this "automatic" forgetting, you need to transfer the knowledge into long-term memory.

  Try to set a specific time for your homework. Then choose a location that is quiet and uninterrupted. Some students find it helpful to listen to music when doing homework. Research has shown that classical music creates the best atmosphere for studying: Give it a try!

- **Reading.** Before you begin doing the homework exercises, you should reread the assigned material in the textbook. You may also want to look over your class notes again and rework some of the examples given in class.

  You should not read a math textbook as you would a novel or history textbook. Math texts are not meant to be read passively. Work the examples on your own paper as you read them. Mark what you do not understand. Reread any paragraph as you see the need, and look up any sections referenced.

- **Doing the margin exercises.** Be sure to stop and do the margin exercises when directed. We cannot overemphasize the importance of this.

### While Doing Your Homework

- **Write legibly.** Write legibly so that you can easily check over your work. Clearly label each section and each exercise. Your legible writing will also be appreciated should your homework be collected and graded. Tutors and instructors are more helpful if they can see and understand all the steps in your work.

- **Use notebook paper if extra workspace is needed.** The text you are using is a workbook. You might consider tearing out the pages and placing them in a three-ring notebook. Then you can organize your homework and class notes in appropriate places between the text material. You want to be able to go over your homework when studying for a test. Therefore, you need to be able to easily access any problem in your homework notebook.

- **Show all the steps.** Be sure to show all the steps in your work. This avoids the common difficulty of trying to do too much "in your head" and also provides you and your instructor with an effective means of checking your work.

- **Check answers.** When you are finished with your homework, check the answers to the odd-numbered exercises at the back of the book or in the *Student's Solutions Manual* and make corrections. If you do not understand why an answer is wrong, mark it so you can ask questions in class or during the instructor's office hours.

- **Form a study group.** For some students, forming a study group can be helpful. Many times, two heads are better than one. Also, it is true that "to teach is to learn." Thus, when you explain a concept to your classmate, you often gain a better understanding of the concept yourself. If you do study in a group, resist the temptation to waste time by socializing.

  If you work regularly with someone, be careful not to become dependent on that person. Always allow some time to work on your own so that you are able to learn even when your study partner is not available.

### After Doing Your Homework

- **Daily review.** If you complete your homework several days before the next class, review your work every day. This will keep the material fresh in your mind. You should also review the work immediately before the next class so that you can ask questions as needed.

- **Extra practice.** The best way to learn math concepts is to perform practice exercises repeatedly. This is the "drill-and-practice" part of learning math that comes when you do your homework. It cannot be overlooked if you want to succeed in your study of math.

# Exercise Set R.2

**a** Write an equivalent expression for each of the following. Use the indicated name for 1.

**1.** $\dfrac{3}{4}$ $\left(\text{Use } \dfrac{3}{3} \text{ for } 1.\right)$

**2.** $\dfrac{5}{6}$ $\left(\text{Use } \dfrac{10}{10} \text{ for } 1.\right)$

**3.** $\dfrac{3}{5}$ $\left(\text{Use } \dfrac{20}{20} \text{ for } 1.\right)$

**4.** $\dfrac{8}{9}$ $\left(\text{Use } \dfrac{4}{4} \text{ for } 1.\right)$

**5.** $\dfrac{13}{20}$ $\left(\text{Use } \dfrac{8}{8} \text{ for } 1.\right)$

**6.** $\dfrac{13}{32}$ $\left(\text{Use } \dfrac{40}{40} \text{ for } 1.\right)$

Write an equivalent expression with the given denominator.

**7.** $\dfrac{7}{8}$ (Denominator: 24)

**8.** $\dfrac{5}{6}$ (Denominator: 48)

**9.** $\dfrac{5}{4}$ (Denominator: 16)

**10.** $\dfrac{2}{9}$ (Denominator: 54)

**b** Simplify.

**11.** $\dfrac{18}{27}$

**12.** $\dfrac{49}{56}$

**13.** $\dfrac{56}{14}$

**14.** $\dfrac{48}{27}$

**15.** $\dfrac{6}{42}$

**16.** $\dfrac{13}{104}$

**17.** $\dfrac{56}{7}$

**18.** $\dfrac{132}{11}$

**19.** $\dfrac{19}{76}$

**20.** $\dfrac{17}{51}$

**21.** $\dfrac{100}{20}$

**22.** $\dfrac{150}{25}$

**23.** $\dfrac{425}{525}$

**24.** $\dfrac{625}{325}$

**25.** $\dfrac{2600}{1400}$

**26.** $\dfrac{4800}{1600}$

**27.** $\dfrac{8 \cdot x}{6 \cdot x}$

**28.** $\dfrac{13 \cdot v}{39 \cdot v}$

**c** Compute and simplify.

**29.** $\dfrac{1}{3} \cdot \dfrac{1}{4}$

**30.** $\dfrac{15}{16} \cdot \dfrac{8}{5}$

**31.** $\dfrac{15}{4} \cdot \dfrac{3}{4}$

**32.** $\dfrac{10}{11} \cdot \dfrac{11}{10}$

**33.** $\dfrac{1}{3} + \dfrac{1}{3}$

**34.** $\dfrac{1}{4} + \dfrac{1}{3}$

**35.** $\dfrac{4}{9} + \dfrac{13}{18}$

**36.** $\dfrac{4}{5} + \dfrac{8}{15}$

**37.** $\dfrac{3}{10} + \dfrac{8}{15}$  **38.** $\dfrac{9}{8} + \dfrac{7}{12}$  **39.** $\dfrac{5}{4} - \dfrac{3}{4}$  **40.** $\dfrac{12}{5} - \dfrac{2}{5}$  **41.** $\dfrac{11}{12} - \dfrac{3}{8}$

**42.** $\dfrac{15}{16} - \dfrac{5}{12}$  **43.** $\dfrac{11}{12} - \dfrac{2}{5}$  **44.** $\dfrac{15}{16} - \dfrac{2}{3}$  **45.** $\dfrac{7}{6} \div \dfrac{3}{5}$  **46.** $\dfrac{7}{5} \div \dfrac{3}{4}$

**47.** $\dfrac{8}{9} \div \dfrac{4}{15}$  **48.** $\dfrac{3}{4} \div \dfrac{3}{7}$  **49.** $\dfrac{1}{8} \div \dfrac{1}{4}$  **50.** $\dfrac{1}{20} \div \dfrac{1}{5}$  **51.** $\dfrac{\frac{13}{12}}{\frac{39}{5}}$

**52.** $\dfrac{\frac{17}{6}}{\frac{3}{8}}$  **53.** $100 \div \dfrac{1}{5}$  **54.** $78 \div \dfrac{1}{6}$  **55.** $\dfrac{3}{4} \div 10$  **56.** $\dfrac{5}{6} \div 15$

---

## Skill Maintenance

This heading indicates that the exercises that follow are *skill maintenance exercises,* which review any skill previously studied in the text. You can expect such exercises in every exercise set. Answers to *all* skill maintenance exercises are found at the back of the book. If you miss an exercise, restudy the objective shown in blue.

Find the prime factorization.   [R.1a]

**57.** 28  **58.** 56  **59.** 1000  **60.** 192

Find the LCM.   [R.1b]

**61.** 16,  24  **62.** 28,  49,  56  **63.** 48,  64,  96  **64.** 25,  75,  150

---

## Synthesis

**65.** ◈ A student incorrectly insists that $\frac{12}{35} \div \frac{4}{5}$ is $\frac{7}{3}$. What mistake is the student likely making?

**66.** ◈ Explain in your own words when it *is* possible to *cancel* and when it is *not* possible to cancel.

Simplify.

**67.** $\dfrac{192}{256}$  **68.** $\dfrac{p \cdot q}{r \cdot q}$  **69.** $\dfrac{64 \cdot a \cdot b}{16 \cdot a \cdot b}$

**70.** $\dfrac{4 \cdot 9 \cdot 24}{2 \cdot 8 \cdot 15}$  **71.** $\dfrac{36 \cdot (2 \cdot h)}{8 \cdot (9 \cdot h)}$

Copyright © 1999 Addison Wesley Longman

# R.3 Decimal Notation

Let's say that the cost of a sound system is

$1768.95.

This amount is given in **decimal notation.** The following place-value chart shows the place value of each digit in 1768.95.

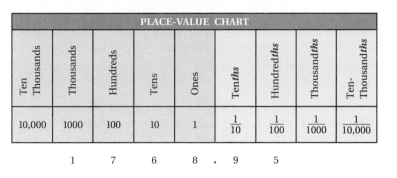

| PLACE-VALUE CHART | | | | | | | | |
|---|---|---|---|---|---|---|---|---|
| Ten Thousands | Thousands | Hundreds | Tens | Ones | Ten*ths* | Hundred*ths* | Thousand*ths* | Ten-Thousand*ths* |
| 10,000 | 1000 | 100 | 10 | 1 | $\frac{1}{10}$ | $\frac{1}{100}$ | $\frac{1}{1000}$ | $\frac{1}{10,000}$ |
| | 1 | 7 | 6 | 8 . | 9 | 5 | | |

**Objectives**

a  Convert from decimal notation to fractional notation.

b  Add, subtract, multiply, and divide using decimal notation.

c  Round numbers to a specified decimal place.

**For Extra Help**

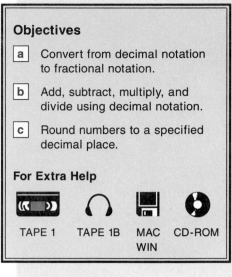

TAPE 1    TAPE 1B    MAC WIN    CD-ROM

## a  Converting from Decimal Notation to Fractional Notation

Decimals are defined in terms of fractions—for example,

$$0.1 = \frac{1}{10}, \qquad 0.6875 = \frac{6875}{10,000}, \qquad 53.47 = \frac{5347}{100}.$$

We see that the number of zeros in the denominator is the same as the number of decimal places in the number. From these examples, we obtain the following procedure for converting from decimal notation to fractional notation.

| | | |
|---|---|---|
| To convert from decimal notation to fractional notation: | | |
| **a)** Count the number of decimal places. | 4.98 | 2 places |
| **b)** Move the decimal point that many places to the right. | 4.98. | Move 2 places. |
| **c)** Write the result over a denominator with that number of zeros. | $\frac{498}{100}$ | 2 zeros |

**Example 1**  Convert 0.876 to fractional notation. Do not simplify.

$$0.876 \qquad 0.876. \qquad 0.876 = \frac{876}{1000}$$

3 places       3 places          3 zeros

Convert to fractional notation. Do not simplify.

**1.** 0.568

**2.** 2.3

**3.** 89.04

*Answers on page A-2*

Convert to decimal notation.

**4.** $\dfrac{4131}{1000}$

**5.** $\dfrac{4131}{10,000}$

**6.** $\dfrac{573}{100}$

Add.

**7.** $69 + 1.785 + 213.67$

**8.** $17.95 + 14.68 + 236$

*Answers on page A-2*

---

**Example 2**   Convert 1.5018 to fractional notation. Do not simplify.

$$1.5018 \qquad 1.5018. \qquad 1.5018 = \dfrac{15,018}{10,000}$$

4 places         4 zeros

***Do Exercises 1–3 on the preceding page.***

---

To convert from fractional notation to decimal notation when the denominator is a number like 10, 100, or 1000:

**a)** Count the number of zeros.          $\dfrac{8679}{1000}$

3 zeros

**b)** Move the decimal point that number of places to the left. Leave off the denominator.          8.679.

Move 3 places.

---

**Example 3**   Convert to decimal notation: $\dfrac{123,067}{10,000}$.

$$\dfrac{123,067}{10,000} \qquad 12.3067. \qquad \dfrac{123,067}{10,000} = 12.3067$$

4 zeros         4 places

***Do Exercises 4–6.***

## **b** | Addition, Subtraction, Multiplication, and Division

### Addition

Adding with decimal notation is similar to adding whole numbers. First we line up the decimal points. Then we add the thousandths, then the hundredths, and so on, carrying if necessary.

**Example 4**   Add: $74 + 26.46 + 0.998$.

$$\begin{array}{r} \scriptstyle 1\ \ 1\ \ 1 \\ 7\,4. \\ 2\,6.4\,6 \\ +\ \ \ \ 0.9\,9\,8 \\ \hline 1\,0\,1.4\,5\,8 \end{array}$$

You can place extra zeros to the right of any decimal point so that there are the same number of decimal places, but this is not necessary. If you did, the preceding problem would look like this:

$$\begin{array}{r} \scriptstyle 1\ \ 1\ \ 1 \\ 7\,4.0\,0\,0 \\ 2\,6.4\,6\,0 \\ +\ \ \ \ 0.9\,9\,8 \\ \hline 1\,0\,1.4\,5\,8 \end{array}$$

***Do Exercises 7 and 8.***

## Subtraction

> Subtracting with decimal notation is similar to subtracting whole numbers. First we line up the decimal points. Then we subtract the thousandths, then the hundredths, the tenths, and so on, borrowing if necessary. Extra zeros can be added if needed.

### Examples

**5.** Subtract: $76.14 - 18.953$.

$$
\begin{array}{r}
{\scriptstyle 15\ 10\ 13} \\
{\scriptstyle 6\ \ 5\ \ 0\ \ 3\ \ 10} \\
7\,6.1\,4\,0 \\
-\ 1\,8.9\,5\,3 \\
\hline
5\,7.1\,8\,7
\end{array}
$$

**6.** Subtract: $200 - 0.68$.

$$
\begin{array}{r}
{\scriptstyle 1\ 9\ 9\ 9\ 10} \\
2\,0\,0.0\,0 \\
-\ \ \ \ \ 0.6\,8 \\
\hline
1\,9\,9.3\,2
\end{array}
$$

*Do Exercises 9–12.*

## Multiplication

Look at this product.

$$5.14 \times 0.8 = \frac{514}{100} \times \frac{8}{10} = \frac{514 \times 8}{100 \times 10} = \frac{4112}{1000} = 4.112$$

2 places    1 place    3 places

We can also do this more quickly by multiplying the whole numbers 8 and 514 and then determining the position of the decimal point.

> To multiply with decimal notation:
>
> a) Ignore the decimal points and multiply as whole numbers.
> b) Place the decimal point in the result of step (a) by adding the number of decimal places in the original factors.

**Example 7**  Multiply: $5.14 \times 0.8$.

a) Ignore the decimal points and multiply as whole numbers.

$$
\begin{array}{r}
{\scriptstyle 1\ \ \ 3} \\
5.1\,4 \\
\times\ \ \ \ \ 0.8 \\
\hline
4\,1\,1\,2
\end{array}
$$

b) Place the decimal point in the result of step (a) by adding the number of decimal places in the original factors.

$$
\begin{array}{r}
5.1\,4 \longleftarrow \text{2 decimal places} \\
\times\ \ \ \ \ 0.8 \longleftarrow \text{1 decimal place} \\
\hline
4.1\,1\,2
\end{array}
$$

3 decimal places

*Do Exercises 13–15.*

---

Subtract.

**9.** $29.35 - 1.674$

**10.** $92.375 - 27.692$

**11.** $100 - 0.41$

**12.** $240 - 0.117$

Multiply.

**13.**
$$
\begin{array}{r}
6.5\,2 \\
\times\ \ \ 0.9 \\
\hline
\end{array}
$$

**14.**
$$
\begin{array}{r}
6.5\,2 \\
\times\,0.0\,9 \\
\hline
\end{array}
$$

**15.**
$$
\begin{array}{r}
5\,6.7\,6 \\
\times\,0.9\,0\,8 \\
\hline
\end{array}
$$

*Answers on page A-2*

Divide.

**16.** 7 ) 3 4 2.3

**17.** 1 6 ) 2 5 3.1 2

Divide.

**18.** 2 5 ) 3 2

**19.** 3 8 ) 6 8 2.1

*Answers on page A-2*

## Division

Note that $37.6 \div 8 = 4.7$ because $8 \times 4.7 = 37.6$. If we write this as

```
        4.7
8 ) 3 7.6
    3 2
      5 6
      5 6
        0
```

we see how the following method can be used to divide by a whole number.

> **To divide by a whole number:**
> **a)** Place the decimal point in the quotient directly above the decimal point in the dividend.
> **b)** Divide as whole numbers.

**Example 8**   Divide: $216.75 \div 25$.

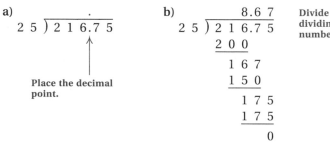

a)
```
              .
2 5 ) 2 1 6.7 5
```
Place the decimal point.

b)
```
              8.6 7
2 5 ) 2 1 6.7 5
      2 0 0
        1 6 7
        1 5 0
          1 7 5
          1 7 5
              0
```
Divide as though dividing whole numbers.

*Do Exercises 16 and 17.*

Sometimes it is helpful to write extra zeros to the right of the decimal point. Doing so does not change the answer. Remember that the decimal point for a whole number, though not normally written, is to the right of the number.

**Example 9**   Divide: $54 \div 8$.

a)
```
          .
8 ) 5 4.
```

b)
```
      6.7 5
8 ) 5 4.0 0
    4 8 ↓
      6 0
      5 6 ↓
        4 0
        4 0
          0
```
Extra zeros are written to the right of the decimal point as needed.

*Do Exercises 18 and 19.*

> **To divide when the divisor is not a whole number:**
> **a)** Move the decimal point in the divisor as many places to the right as it takes to make it a whole number. Move the decimal point in the dividend the same number of places to the right and place the decimal point in the quotient.
> **b)** Divide as whole numbers, inserting zeros if necessary.

**Example 10** Divide: $83.79 \div 0.098$.

a)

$$0.0\,9\,8.\,\overline{)\,8\,3.7\,9\,0.}$$

b)

$$
\begin{array}{r}
8\,5\,5.\phantom{0} \\
0.0\,9\,8_\wedge\,\overline{)\,8\,3.7\,9\,0_\wedge} \\
7\,8\,4\phantom{00} \\
\hline
5\,3\,9\phantom{0} \\
4\,9\,0\phantom{0} \\
\hline
4\,9\,0 \\
4\,9\,0 \\
\hline
0
\end{array}
$$

*Do Exercises 20 and 21.*

## Converting from Fractional Notation to Decimal Notation

To convert from fractional notation to decimal notation when the denominator is not a number like 10, 100, or 1000, we divide the numerator by the denominator.

**Example 11** Convert to decimal notation: $\dfrac{5}{16}$.

$$
\begin{array}{r}
0.3\,1\,2\,5 \\
1\,6\,\overline{)\,5.0\,0\,0\,0} \\
4\,8\phantom{.0000} \\
\hline
2\,0\phantom{000} \\
1\,6\phantom{000} \\
\hline
4\,0\phantom{00} \\
3\,2\phantom{00} \\
\hline
8\,0\phantom{0} \\
8\,0\phantom{0} \\
\hline
0
\end{array}
$$

If we get a remainder of 0, the decimal terminates. Decimal notation for $\frac{5}{16}$ is 0.3125.

**Example 12** Convert to decimal notation: $\dfrac{7}{12}$.

$$
\begin{array}{r}
0.5\,8\,3\,3 \\
1\,2\,\overline{)\,7.0\,0\,0\,0} \\
6\,0\phantom{.0000} \\
\hline
1\,0\,0\phantom{000} \\
9\,6\phantom{000} \\
\hline
4\,0\phantom{00} \\
3\,6\phantom{00} \\
\hline
4\,0\phantom{0} \\
3\,6\phantom{0} \\
\hline
4
\end{array}
$$

The number 4 repeats as a remainder, so the digits will repeat in the quotient. Therefore,

$$\frac{7}{12} = 0.583333\ldots.$$

Divide.

**20.** $0.0\,2\,4\,\overline{)\,2\,0.5\,4\,4}$

**21.** $4.6\,\overline{)\,3.9\,1}$

Convert to decimal notation.

**22.** $\dfrac{5}{8}$

**23.** $\dfrac{2}{3}$

**24.** $\dfrac{84}{11}$

*Answers on page A-2*

Round to the nearest tenth.

**25.** 2.76

**26.** 13.85

**27.** 7.009

Round to the nearest hundredth.

**28.** 7.834

**29.** 34.675

**30.** 0.025

Round to the nearest thousandth.

**31.** 0.9434

**32.** 8.0038

**33.** 43.1119

**34.** 37.4005

Round 7459.3549 to the nearest:

**35.** thousandth.

**36.** hundredth.

**37.** tenth.

**38.** one.

**39.** ten.

*Answers on page A-2*

Instead of dots, we often put a bar over the repeating part—in this case, only the 3. Thus,

$$\frac{7}{12} = 0.58\overline{3}.$$

***Do Exercises 22–24 on the preceding page.***

### c | Rounding

When working with decimal notation, we often shorten notation by **rounding**. Although there are many rules for rounding, we will use the following.

> To round to a certain place:
>
> **a)** Locate the digit in that place.
> **b)** Consider the digit to its right.
> **c)** If the digit to the right is 5 or higher, round up; if the digit to the right is less than 5, round down.

**Example 13**  Round 3872.2459 to the nearest tenth.

**a)** We locate the digit in the tenths place.

3 8 7 2.2 4 5 9
           ↑

**b)** Then we consider the next digit to the right.

3 8 7 2.2 4 5 9
             ↑

**c)** Since that digit is less than 5, we round down.

3 8 7 2.2 ← This is the answer.

Note that 3872.3 is *not* a correct answer to Example 13. It is incorrect to round from the ten-thousandths place over, as follows:

3872.246,    3872.25,    3872.3

**Example 14**  Round 3872.2459 to the nearest thousandth, hundredth, tenth, one, ten, hundred, and thousand.

| | |
|---|---|
| thousandth: | 3872.246 |
| hundredth: | 3872.25 |
| tenth: | 3872.2 |
| one: | 3872 |
| ten: | 3870 |
| hundred: | 3900 |
| thousand: | 4000 |

***Do Exercises 25–39.***

In rounding, we sometimes use the symbol ≈, which means "is approximately equal to." Thus,

46.124 ≈ 46.1.

# Exercise Set R.3

**a** Convert to fractional notation. Do not simplify.

**1.** 5.3

**2.** 2.7

**3.** 0.67

**4.** 0.93

**5.** 2.0007

**6.** 4.0008

**7.** 7889.8

**8.** 1122.3

Convert to decimal notation.

**9.** $\dfrac{1}{10}$

**10.** $\dfrac{1}{100}$

**11.** $\dfrac{1}{10,000}$

**12.** $\dfrac{1}{1000}$

**13.** $\dfrac{9999}{1000}$

**14.** $\dfrac{39}{10,000}$

**15.** $\dfrac{4578}{10,000}$

**16.** $\dfrac{94}{100,000}$

**b** Add.

**17.**
$$\begin{array}{r} 4\,1\,5.7\,8 \\ +\;\;\;\;2\,9.1\,6 \\ \hline \end{array}$$

**18.**
$$\begin{array}{r} 7\,0\,8.9\,9 \\ +\;\;\;\;7\,5.4\,8 \\ \hline \end{array}$$

**19.**
$$\begin{array}{r} 2\,3\,4.0\,0\,0 \\ +1\,5\,6.6\,1\,7 \\ \hline \end{array}$$

**20.**
$$\begin{array}{r} 1\,3\,4\,5.1\,2 \\ +\;\;\;\;5\,6\,6.9\,8 \\ \hline \end{array}$$

**21.** 85 + 67.95 + 2.774

**22.** 119 + 43.74 + 18.876

**23.** 17.95 + 16.99 + 28.85

**24.** 14.59 + 16.79 + 19.95

Subtract.

**25.**
$$\begin{array}{r} 7\,8.1\,1\,0 \\ -4\,5.8\,7\,6 \\ \hline \end{array}$$

**26.**
$$\begin{array}{r} 1\,4.0\,8\,0 \\ -\;\;\;9.1\,9\,9 \\ \hline \end{array}$$

**27.**
$$\begin{array}{r} 3\,8.7 \\ -1\,1.8\,6\,5 \\ \hline \end{array}$$

**28.**
$$\begin{array}{r} 3\,0\,0. \\ -\;\;2\,4.6\,7\,7 \\ \hline \end{array}$$

**29.** 57.86 − 9.95

**30.** 2.6 − 1.08

**31.** 3 − 1.0807

**32.** 5 − 3.4051

Multiply.

**33.**
$$\begin{array}{r} 7.3\,4 \\ \times\;\;\;1.8 \\ \hline \end{array}$$

**34.**
$$\begin{array}{r} 6.5\,5 \\ \times\;\;\;3.2 \\ \hline \end{array}$$

**35.**
$$\begin{array}{r} 0.8\,6 \\ \times0.9\,3 \\ \hline \end{array}$$

**36.**
$$\begin{array}{r} 0.0\,2\,8 \\ \times7.4\,0\,9 \\ \hline \end{array}$$

**37.**
$$\begin{array}{r} 1\,7.9\,5 \\ \times\;\;\;\;\;1\,0 \\ \hline \end{array}$$

**38.**
$$\begin{array}{r} 1\,8.9\,4 \\ \times\;\;\;\;0.1 \\ \hline \end{array}$$

**39.**
$$\begin{array}{r} 0.4\,5\,7 \\ \times\;\;\;3.0\,8 \\ \hline \end{array}$$

**40.**
$$\begin{array}{r} 0.0\,0\,2\,4 \\ \times\;\;\;0.0\,1\,5 \\ \hline \end{array}$$

**41.**
$$\begin{array}{r} 3.6\,4\,2 \\ \times\;\;\;0.9\,9 \\ \hline \end{array}$$

**42.**
$$\begin{array}{r} 2\,8\,7.4 \\ \times\;\;\;1.0\,8 \\ \hline \end{array}$$

Divide.

**43.** $7\,2\,\overline{)\,1\,6\,5.6}$      **44.** $5.2\,\overline{)\,4\,4.2}$      **45.** $8.5\,\overline{)\,4\,4.2}$      **46.** $7.8\,\overline{)\,7\,2.5\,4}$

**47.** $9.9\,\overline{)\,0.2\,2\,7\,7}$      **48.** $1\,0\,0\,\overline{)\,9\,5}$      **49.** $0.6\,4\,\overline{)\,1\,2}$      **50.** $1.6\,\overline{)\,7\,5}$

**51.** $1.0\,5\,\overline{)\,6\,9\,3}$      **52.** $2\,5\,\overline{)\,4}$      **53.** $8.6\,\overline{)\,5.8\,4\,8}$      **54.** $0.4\,7\,\overline{)\,0.1\,2\,2\,2}$

Convert to decimal notation.

**55.** $\dfrac{11}{32}$      **56.** $\dfrac{17}{32}$      **57.** $\dfrac{13}{11}$      **58.** $\dfrac{17}{12}$

**59.** $\dfrac{5}{9}$      **60.** $\dfrac{5}{6}$      **61.** $\dfrac{19}{9}$      **62.** $\dfrac{9}{11}$

c   Round to the nearest hundredth, tenth, one, ten, and hundred.

**63.** 745.06534      **64.** 317.18565      **65.** 6780.50568      **66.** 840.15493

Round to the nearest cent and to the nearest dollar (nearest one).

**67.** $17.988      **68.** $20.492      **69.** $346.075      **70.** $4.718

Round to the nearest dollar.

**71.** $16.95      **72.** $17.50      **73.** $189.50      **74.** $567.24

Divide and round to the nearest ten-thousandth, thousandth, hundredth, tenth, and one.

**75.** $\dfrac{1000}{81}$      **76.** $\dfrac{23}{17}$      **77.** $\dfrac{23}{39}$      **78.** $\dfrac{8467}{5603}$

---

**Skill Maintenance**

Calculate.  [R.2c]

**79.** $\dfrac{7}{8} + \dfrac{5}{32}$      **80.** $\dfrac{15}{16} - \dfrac{11}{12}$      **81.** $\dfrac{15}{16} \cdot \dfrac{11}{12}$      **82.** $\dfrac{15}{32} \div \dfrac{3}{8}$

---

**Synthesis**

**83.** ◈ A student rounds 536.448 to the nearest one and gets 537. Explain the possible error.

**84.** ◈ A student insists that $5.367 \div 0.1$ is 0.5367. How could you convince this student that a mistake has been made?

# R.4 Percent Notation

## a | Converting to Decimal Notation

The average family spends 28% of its income for food. What does this mean? It means that of every $100 earned, $28 is spent for food. Thus, 28% is a ratio of 28 to 100.

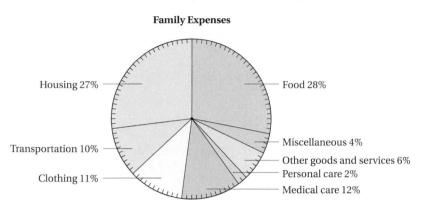

**Family Expenses**

Housing 27%
Food 28%
Transportation 10%
Miscellaneous 4%
Other goods and services 6%
Personal care 2%
Clothing 11%
Medical care 12%

*Source*: Bureau of Labor Statistics

The percent symbol % means "per hundred." We can regard the percent symbol as part of a name for a number. For example,

$$28\% \quad \text{is defined to mean} \quad 28 \times 0.01, \quad \text{or} \quad 28 \times \frac{1}{100}, \quad \text{or} \quad \frac{28}{100}.$$

> $n\%$ means $n \times 0.01$, or $n \times \frac{1}{100}$, or $\frac{n}{100}$.

**Example 1** Convert 78.5% to decimal notation.

$78.5\% = 78.5 \times 0.01$    **Replacing % with × 0.01**

$\phantom{78.5\%} = 0.785$

> To convert from percent notation to decimal notation, move the decimal point *two* places to the left and drop the percent symbol.

**Example 2** Convert 43.67% to decimal notation.

43.67%    0.43.67    43.67% = 0.4367

**Move the decimal point two places to the left.**

*Do Exercises 1 and 2.*

## b | Converting to Fractional Notation

**Example 3** Convert 88% to fractional notation.

$88\% = 88 \times \frac{1}{100}$    **Replacing % with $\times \frac{1}{100}$**

$\phantom{88\%} = \frac{88}{100}$    **Multiplying. You need not simplify.**

### Objectives

a | Convert from percent notation to decimal notation.

b | Convert from percent notation to fractional notation.

c | Convert from decimal notation to percent notation.

d | Convert from fractional notation to percent notation.

### For Extra Help

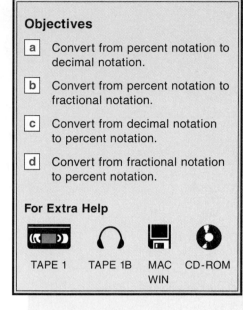

TAPE 1    TAPE 1B    MAC WIN    CD-ROM

Convert to decimal notation.

1. 87.3%

2. 100%

*Answers on page A-2*

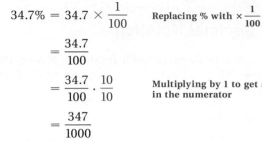

Convert to fractional notation.

**3.** 53%

**4.** 45.6%

**5.** 0.23%

Convert to percent notation.

**6.** 6.77

**7.** 0.9944

Convert to percent notation.

**8.** $\dfrac{1}{4}$

**9.** $\dfrac{7}{8}$

**10.** $\dfrac{2}{3}$

*Answers on page A-2*

**Example 4**  Convert 34.7% to fractional notation.

$$34.7\% = 34.7 \times \frac{1}{100} \qquad \text{Replacing \% with } \times \frac{1}{100}$$

$$= \frac{34.7}{100}$$

$$= \frac{34.7}{100} \cdot \frac{10}{10} \qquad \text{Multiplying by 1 to get a whole number in the numerator}$$

$$= \frac{347}{1000}$$

*Do Exercises 3–5.*

## c | Converting from Decimal Notation

By applying the definition of percent in reverse, we can convert from decimal notation to percent notation. We multiply by 1, expressing it as 100 × 0.01 and replacing × 0.01 with %.

**Example 5**  Convert 0.93 to percent notation.

$$0.93 = 0.93 \times 1 \qquad \text{Identity property of 1}$$

$$= 0.93 \times 100\% \qquad \text{Expressing 1 as 100\%}$$

$$= 0.93 \times (100 \times 0.01) \qquad \text{Expressing 100\% as 100 × 0.01}$$

$$= (0.93 \times 100) \times 0.01$$

$$= 93 \times 0.01$$

$$= 93\% \qquad \text{Replacing × 0.01 with \%}$$

> To convert from decimal notation to percent notation, move the decimal point *two* places to the right and write a percent symbol.

**Example 6**  Convert 0.032 to percent notation.

$$0.032 \qquad 0.03.2 \qquad 0.032 = 3.2\%$$

Move the decimal point two places to the right.

*Do Exercises 6 and 7.*

## d | Converting from Fractional Notation

We can also convert from fractional notation to percent notation by converting first to decimal notation. Then we move the decimal point two places to the *right* and write a percent symbol.

**Example 7**  Convert $\dfrac{5}{8}$ to percent notation.

$$\frac{5}{8} = 0.625 = 62.5\%$$

*Do Exercises 8–10.*

# Exercise Set R.4

**a** Convert to decimal notation.

**1.** 63%  **2.** 64%  **3.** 94.1%  **4.** 34.6%

**5.** 1%  **6.** 100%  **7.** 0.61%  **8.** 125%

**9.** 240%  **10.** 0.73%  **11.** 3.25%  **12.** 2.3%

**b** Convert to fractional notation.

**13.** 60%  **14.** 40%  **15.** 28.9%  **16.** 37.5%

**17.** 110%  **18.** 120%  **19.** 0.042%  **20.** 0.68%

**21.** 250%  **22.** 3.2%  **23.** 3.47%  **24.** 12.557%

**c** Convert to percent notation.

**25.** 1  **26.** 8.56  **27.** 0.996  **28.** 0.83

**29.** 0.0047  **30.** 2  **31.** 0.072  **32.** 1.34

**33.** 9.2  **34.** 0.013  **35.** 0.0068  **36.** 0.675

**37.** $\dfrac{1}{6}$

**38.** $\dfrac{1}{5}$

**39.** $\dfrac{13}{20}$

**40.** $\dfrac{14}{25}$

**41.** $\dfrac{29}{100}$

**42.** $\dfrac{123}{100}$

**43.** $\dfrac{8}{10}$

**44.** $\dfrac{7}{10}$

**45.** $\dfrac{3}{5}$

**46.** $\dfrac{17}{50}$

**47.** $\dfrac{2}{3}$

**48.** $\dfrac{7}{8}$

**49.** $\dfrac{7}{4}$

**50.** $\dfrac{3}{8}$

**51.** $\dfrac{3}{4}$

**52.** $\dfrac{99.4}{100}$

---

**Skill Maintenance**

Convert to decimal notation.   [R.3b]

**53.** $\dfrac{9}{4}$

**54.** $\dfrac{11}{8}$

**55.** $\dfrac{17}{12}$

**56.** $\dfrac{8}{9}$

**57.** $\dfrac{10}{11}$

**58.** $\dfrac{17}{11}$

Calculate.   [R.3b]

**59.** $23.458 \times 7.03$

**60.** $7.8\overline{)440.154}$

**61.** $809.569 + 86.99$

**62.** $809.569 - 86.99$

---

**Synthesis**

**63.** ◈ ▦ What would you do to an entry on a calculator in order to get percent notation?

**64.** ◈ Is it always best to convert from fractional notation to percent notation by first finding decimal notation? Why or why not?

Simplify. Express the answer in percent notation.

**65.** $18\% + 14\%$

**66.** $84\% - 12\%$

**67.** $1 - 30\%$

**68.** $92\% - 10\%$

**69.** $27 \times 100\%$

**70.** $42\% - (1 - 58\%)$

**71.** $3(1 + 15\%)$

**72.** $7(1\% + 13\%)$

**73.** $\dfrac{100\%}{40}$

**74.** $\dfrac{3}{4} + 20\%$

# R.5 Exponential Notation and Order of Operations

## a | Exponential Notation

Exponents provide a shorter way of writing products. An abbreviation for a product in which the factors are the same is called a **power**. For

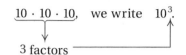

$10 \cdot 10 \cdot 10$, we write $10^3$.

3 factors

This is read "ten to the third power." We call the number 3 an **exponent** and we say that 10 is the **base**. An exponent of 2 or greater tells how many times the base is used as a factor. For example,

$$a \cdot a \cdot a \cdot a = a^4.$$

In this case, the exponent is 4 and the base is $a$. An expression for a power is called **exponential notation.**

This is the exponent.

$$a^n$$

This is the base.

**Example 1** Write exponential notation for $10 \cdot 10 \cdot 10 \cdot 10 \cdot 10$.

$$10 \cdot 10 \cdot 10 \cdot 10 \cdot 10 = 10^5$$

*Do Exercises 1–3.*

## b | Evaluating Exponential Expressions

**Example 2** Evaluate: $5^2$.

$$5^2 = 5 \cdot 5 = 25$$

**Example 3** Evaluate: $3^4$.

We have

$$3^4 = 3 \cdot 3 \cdot 3 \cdot 3 = 9 \cdot 9 = 81.$$

We could also carry out the calculation as follows:

$$3^4 = 3 \cdot 3 \cdot 3 \cdot 3 = 9 \cdot 3 \cdot 3 = 27 \cdot 3 = 81.$$

> **EXPONENTIAL NOTATION**
>
> For any natural number $n$ greater than or equal to 2,
>
> $n$ factors
>
> $$b^n = \overbrace{b \cdot b \cdot b \cdot b \cdots b}.$$

*Do Exercises 4–6.*

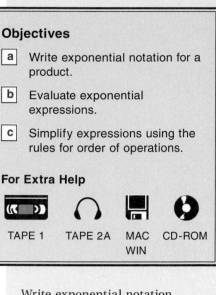

**Objectives**

**a** Write exponential notation for a product.

**b** Evaluate exponential expressions.

**c** Simplify expressions using the rules for order of operations.

**For Extra Help**

TAPE 1   TAPE 2A   MAC WIN   CD-ROM

Write exponential notation.

**1.** $4 \cdot 4 \cdot 4$

**2.** $6 \cdot 6 \cdot 6 \cdot 6 \cdot 6$

**3.** $1.08 \times 1.08$

Evaluate.

**4.** $10^4$

**5.** $8^3$

**6.** $(1.1)^3$

*Answers on page A-2*

## c | Order of Operations

What does $5 \times 2 + 4$ mean? If we multiply 5 by 2 and add 4, we get 14. If we add 2 and 4 and multiply by 5, we get 30. Since our results are different, we see that the order in which we carry out operations is important. To tell which operation to do first, we can use grouping symbols such as parentheses ( ), or brackets [ ], or braces { }. For example,

$$(3 \times 5) + 6 = 15 + 6 = 21,$$

but

$$3 \times (5 + 6) = 3 \times 11 = 33.$$

Grouping symbols tell us what to do first. If there are no grouping symbols, we have agreements about the order in which operations should be done.

---

**RULES FOR ORDER OF OPERATIONS**

**1.** Do all calculations within grouping symbols before operations outside.

**2.** Evaluate all exponential expressions.

**3.** Do all multiplications and divisions in order from left to right.

**4.** Do all additions and subtractions in order from left to right.

---

**Example 4**    Calculate: $15 - 2 \times 5 + 3$.

$$
\begin{aligned}
15 - 2 \times 5 + 3 &= 15 - 10 + 3 && \text{Multiplying} \\
&= 5 + 3 && \text{Subtracting} \\
&= 8 && \text{Adding}
\end{aligned}
$$

---

## Calculator Spotlight

 Exponents and Powers.   Eventually in this text, we will be using only graphing calculators in working the Calculator Spotlights. Because of this and because there is extensive variance in keystroke procedures among calculators, we will henceforth focus on the keystrokes for a graphing calculator such as a TI-82 or a TI-83. For the use of any other calculator, consult its particular manual.

To find $3^5$, we press the following keystrokes:

[ 3 ] [ ^ ] [ 5 ] [ ENTER ] .

The answer 243 is displayed as follows:

|            243 . |

To find $\left(\frac{5}{8}\right)^3$, we press

[ ( ] [ 5 ] [ ÷ ] [ 8 ] [ ) ] [ ^ ] [ 3 ] [ ENTER ] .

The result is 0.244140625. Since raising a number to the second power is so common, the calculator has a special "$x$-squared" key, [ $x^2$ ]. To find $1.5^2$, we press

[ 1 ] [ . ] [ 5 ] [ $x^2$ ] [ ENTER ] .

The display will show 2.25:

|            2.25 . |

**Exercises**

Evaluate.

**1.** $4^5$        **2.** $7^9$

**3.** $19^2$       **4.** $5.718^2$

**5.** $1.8^4$       **6.** $23.04^3$

**7.** $\left(\frac{17}{32}\right)^5$       **8.** $\left(\frac{17}{32}\right)^9$

Always calculate within parentheses first. When there are exponents and no parentheses, simplify powers before multiplying or dividing.

**Example 5**  Calculate: $(3 \times 4)^2$.

$$(3 \times 4)^2 = (12)^2 \qquad \text{Working within parentheses first}$$
$$= 144 \qquad \text{Evaluating the exponential expression}$$

**Example 6**  Calculate: $3 \times 4^2$.

$$3 \times 4^2 = 3 \times 16 \qquad \text{Evaluating the exponential expression}$$
$$= 48 \qquad \text{Multiplying}$$

Note that $(3 \times 4)^2 \neq 3 \times 4^2$.

**Example 7**  Calculate: $7 + 3 \times 4^2 - 29$.

$$7 + 3 \times 4^2 - 29 = 7 + 3 \times 16 - 29 \qquad \text{There are no parentheses, so we find } 4^2 \text{ first.}$$
$$= 7 + 48 - 29 \qquad \text{Multiplying second}$$
$$= 55 - 29 \qquad \text{Adding}$$
$$= 26 \qquad \text{Subtracting}$$

*Do Exercises 7–10.*

**Example 8**  Calculate: $2.56 \div 1.6 \div 0.4$.

$$2.56 \div 1.6 \div 0.4 = 1.6 \div 0.4 \qquad \text{Doing the divisions in order from left to right}$$
$$= 4 \qquad \text{Doing the second division}$$

**Example 9**  Calculate: $1000 \cdot \dfrac{1}{10} \div \dfrac{4}{5}$.

$$1000 \cdot \frac{1}{10} \div \frac{4}{5} = 100 \div \frac{4}{5} \qquad \text{Doing the multiplication}$$
$$= 125 \qquad \text{Dividing}$$

*Do Exercises 11 and 12.*

Sometimes combinations of grouping symbols are used, as in

$$5[4 + (8 - 2)].$$

The rules still apply. We begin with the innermost grouping symbols—in this case, the parentheses—and work to the outside.

**Example 10**  Calculate: $5[4 + (8 - 2)]$.

$$5[4 + (8 - 2)] = 5[4 + 6] \qquad \text{Subtracting within the parentheses first}$$
$$= 5[10] \qquad \text{Adding inside the brackets}$$
$$= 50 \qquad \text{Multiplying}$$

Calculate.

**7.** $18 - 4 \times 3 + 7$

**8.** $(2 \times 5)^3$

**9.** $2 \times 5^3$

**10.** $8 + 2 \times 5^3 - 4 \cdot 20$

Calculate.

**11.** $51.2 \div 0.64 \div 40$

**12.** $1000 \div \dfrac{1}{10} \cdot \dfrac{4}{5}$

*Answers on page A-2*

Calculate.

**13.** $4[(8 - 3) + 7]$

**14.** $\dfrac{13(10 - 6) + 4 \cdot 9}{5^2 - 3^2}$

A fraction bar can play the role of a grouping symbol.

**Example 11**  Calculate: $\dfrac{12(9 - 7) + 4 \cdot 5}{3^4 + 2^3}$.

An equivalent expression with brackets as grouping symbols is

$$[12(9 - 7) + 4 \cdot 5] \div [3^4 + 2^3].$$

What this shows, in effect, is that we do the calculations first in the numerator and then in the denominator, and divide the results:

$$\frac{12(9 - 7) + 4 \cdot 5}{3^4 + 2^3} = \frac{12(2) + 4 \cdot 5}{81 + 8} = \frac{24 + 20}{89} = \frac{44}{89}.$$

*Do Exercises 13 and 14.*

---

## Calculator Spotlight

 Order of Operations and Grouping Symbols.

Let's consider simplifying the expression

$$3 + 4 \cdot 2.$$

We know that by following the rules for order of operations, we have

$$3 + 4 \cdot 2 = 3 + 8 = 11.$$

That is, we multiply first and then add. Most graphing calculators have built-in procedures to follow the rules for order of operations. Using the scientific keys on a graphing calculator, we can evaluate this expression by pressing the following keystrokes:

$\boxed{3}\ \boxed{+}\ \boxed{4}\ \boxed{\times}\ \boxed{2}\ \boxed{\text{ENTER}}$.

We obtain the answer 11.

Most graphing calculators provide keys for grouping symbols—for example, $\boxed{(}$ and $\boxed{)}$. To evaluate an expression like $7(13 - 2) - 40$, we press the following keys:

$\boxed{7}\ \boxed{(}\ \boxed{1}\ \boxed{3}\ \boxed{-}\ \boxed{2}\ \boxed{)}\ \boxed{-}$
$\boxed{4}\ \boxed{0}\ \boxed{\text{ENTER}}$.

The answer is 37.

To simplify an expression like

$$\frac{38 + 142}{47 - 2},$$

we think of rewriting it with grouping symbols as

$$(38 + 142) \div (47 - 2).$$

Thus we press

$\boxed{(}\ \boxed{3}\ \boxed{8}\ \boxed{+}\ \boxed{1}\ \boxed{4}\ \boxed{2}\ \boxed{)}\ \boxed{\div}$
$\boxed{(}\ \boxed{4}\ \boxed{7}\ \boxed{-}\ \boxed{2}\ \boxed{)}\ \boxed{\text{ENTER}}$.

The answer is 4.

**Exercises**

Calculate.

**1.** $36 \div 2 \cdot 3 - 4 \cdot 4$

**2.** $68 - 8 \div 4 + 3 \cdot 5$

**3.** $36 \div (2 \cdot 3 - 4) \cdot 4$

**4.** $(15 + 3)^3 + 4(12 - 7)^2$

**5.** $50.6 - 8.9 \times 3.01 + 4(5^2 - 24.7)$

**6.** $3.2 + 4.7[159.3 - 2.1(60.3 - 59.4)]$

**7.** $\{(150 \cdot 5) \div [(3 \cdot 16) \div (8 \cdot 3)]\} + 25(12 \div 4)$

**8.** $\left(\dfrac{28}{89} + 42.8 \times 17.01\right)^3 \div \left(\dfrac{678}{119} - \dfrac{23.2}{46.08}\right)^2$

**9.** $\dfrac{178 - 38}{5 + 30}$

**10.** $\dfrac{311 - 17^2}{13 - 2}$

**11.** $785 - \dfrac{5^4 - 285}{17 + 3 \cdot 51}$

**12.** $12^5 - 12^4 + 11^5 \div 11^3 - 10.2^2$

**13.** What result do you get if you ignore the parentheses when evaluating $(39 + 141) \div (47 - 2)$? How did the calculator do the calculation?

---

*Answers on page A-2*

# Exercise Set R.5

**a** Write exponential notation.

**1.** $5 \times 5 \times 5 \times 5$

**2.** $3 \times 3 \times 3 \times 3 \times 3$

**3.** $10 \cdot 10 \cdot 10$

**4.** $1 \cdot 1 \cdot 1$

**5.** $10 \times 10 \times 10 \times 10 \times 10$

**6.** $18 \cdot 18$

**b** Evaluate.

**7.** $7^2$

**8.** $4^3$

**9.** $9^5$

**10.** $12^4$

**11.** $10^2$

**12.** $1^5$

**13.** $1^4$

**14.** $(1.8)^2$

**15.** $(2.3)^2$

**16.** $(0.1)^3$

**17.** $(0.2)^3$

**18.** $(14.8)^2$

**19.** $(20.4)^2$

**20.** $\left(\dfrac{4}{5}\right)^2$

**21.** $\left(\dfrac{3}{8}\right)^2$

**22.** $2^4$

**23.** $5^3$

**24.** $(1.4)^3$

**25.** $1000 \times (1.02)^3$

**26.** $2000 \times (1.06)^2$

**c** Calculate.

**27.** $9 + 2 \times 8$

**28.** $14 + 6 \times 6$

**29.** $9 \times 8 + 7 \times 6$

**30.** $30 \times 5 + 2 \times 2$

**31.** $39 - 4 \times 2 + 2$

**32.** $14 - 2 \times 6 + 7$

**33.** $9 \div 3 + 16 \div 8$

**34.** $32 - 8 \div 4 - 2$

**35.** $7 + 10 - 10 \div 2$

**36.** $(5 \cdot 4)^2$

**37.** $(6 \cdot 3)^2$

**38.** $3 \cdot 2^3$

**39.** $4 \cdot 5^2$

**40.** $(7 + 3)^2$

**41.** $(8 + 2)^3$

**42.** $7 + 2^2$

**43.** $6 + 4^2$

**44.** $(5 - 2)^2$

**45.** $(3 - 2)^2$

**46.** $10 - 3^2$

**Exercise Set R.5**

**35**

**47.** $4^3 \div 8 - 4$      **48.** $20 + 4^3 \div 8 - 4$      **49.** $120 - 3^3 \cdot 4 \div 6$      **50.** $7 \times 3^4 + 18$

**51.** $6[9 + (3 + 4)]$      **52.** $8[(13 + 6) - 11]$      **53.** $8 + (7 + 9)$      **54.** $(8 + 7) + 9$

**55.** $15(4 + 2)$      **56.** $15 \cdot 4 + 15 \cdot 2$      **57.** $12 - (8 - 4)$      **58.** $(12 - 8) - 4$

**59.** $1000 \div 100 \div 10$      **60.** $256 \div 32 \div 4$      **61.** $2000 \div \dfrac{3}{50} \cdot \dfrac{3}{2}$      **62.** $400 \times 0.64 \div 3.2$

**63.** $\dfrac{80 - 6^2}{9^2 + 3^2}$      **64.** $\dfrac{5^2 + 4^3 - 3}{9^2 - 2^2 + 1^5}$      **65.** $\dfrac{3(6 + 7) - 5 \cdot 4}{6 \cdot 7 + 8(4 - 1)}$

**66.** $\dfrac{20(8 - 3) - 4(10 - 3)}{10(6 + 2) + 2(5 + 2)}$      **67.** $8 \cdot 2 - (12 - 0) \div 3 - (5 - 2)$      **68.** $95 - 2^3 \cdot 5 \div (24 - 4)$

---

**Skill Maintenance**

Find percent notation.   [R.4d]                  Simplify.   [R.2b]

**69.** $\dfrac{5}{16}$      **70.** $\dfrac{17}{32}$          **71.** $\dfrac{125}{325}$      **72.** $\dfrac{64}{96}$

**73.** Find the prime factorization of 48.   [R.1a]          **74.** Find the LCM of 12, 24, and 56.   [R.1b]

---

**Synthesis**

**75.** ◈ The expression $(3 \cdot 4)^2$ contains parentheses. Are they necessary? Why or why not?

**76.** ◈ The expression $9 - (4 \cdot 2)$ contains parentheses. Are they necessary? Why or why not?

Write each of the following with a single exponent.

**77.** $\dfrac{10^5}{10^3}$      **78.** $\dfrac{10^7}{10^2}$      **79.** $5^4 \cdot 5^2$      **80.** $\dfrac{2^8}{8^2}$

**81.** *Five 5's.* We can use five 5's and any combination of grouping symbols to represent the numbers 0 through 10. For example,

$$0 = 5 \cdot 5 \cdot 5(5 - 5), \qquad 1 = \frac{5 + 5}{5} - \frac{5}{5}, \qquad 2 = \frac{5 \cdot 5 - 5}{5 + 5}.$$

Often more than one way to make a representation is possible. Use five 5's to represent the numbers 3 through 10.

---

Collaborative
Learning Manual

Use the order of operations to modify an expression.

Copyright © 1999 Addison Wesley Longman

# Summary and Review Exercises: Chapter R

## Important Properties and Formulas

*Identity Property of 0:*  $a + 0 = a$

*Identity Property of 1:*  $a \cdot 1 = a$

*Equivalent Expressions for 1:*  $\dfrac{a}{a} = 1, \quad a \neq 0$

$n\% = n \times 0.01 = n \times \dfrac{1}{100} = \dfrac{n}{100}$

*Exponential Notation:*  $a^n = \underbrace{a \cdot a \cdot a \cdots a}_{n \text{ factors}}$

The review exercises that follow are for practice. Answers are at the back of the book. If you miss an exercise, restudy the objective indicated in blue next to the exercise or direction line that precedes it.

Find the prime factorization.  [R.1a]

**1.** 92

**2.** 1400

Find the LCM.  [R.1b]

**3.** 13,  32

**4.** 5,  18,  45

Write an equivalent expression using the indicated number for 1.  [R.2a]

**5.** $\dfrac{2}{5}$  $\left( \text{Use } \dfrac{6}{6} \text{ for 1.} \right)$

**6.** $\dfrac{12}{23}$  $\left( \text{Use } \dfrac{8}{8} \text{ for 1.} \right)$

Write an equivalent expression with the given denominator.  [R.2a]

**7.** $\dfrac{5}{8}$  (Denominator: 64)

**8.** $\dfrac{13}{12}$  (Denominator: 84)

Simplify.  [R.2b]

**9.** $\dfrac{20}{48}$

**10.** $\dfrac{1020}{1820}$

Compute and simplify.  [R.2c]

**11.** $\dfrac{4}{9} + \dfrac{5}{12}$

**12.** $\dfrac{3}{4} \div 3$

**13.** $\dfrac{2}{3} - \dfrac{1}{15}$

**14.** $\dfrac{9}{10} \cdot \dfrac{16}{5}$

**15.** Convert to fractional notation: 17.97.  [R.3a]

**16.** Convert to decimal notation: $\dfrac{2337}{10,000}$.  [R.3a]

Add.  [R.3b]

**17.**
$$\begin{array}{r} 2\ 3\ 4\ 4.5\ 6 \\ +\ \ \ \ \ 9\ 8.3\ 4\ 5 \\ \hline \end{array}$$

**18.** $6.04 + 78 + 1.9898$

Subtract.  [R.3b]

**19.** $20.4 - 11.058$

**20.**
$$\begin{array}{r} 7\ 8\ 9.0\ 3\ 2 \\ -\ 6\ 5\ 5.7\ 6\ 8 \\ \hline \end{array}$$

Multiply.  [R.3b]

**21.**
$$\begin{array}{r} 1\ 7.9\ 5 \\ \times \ \ \ \ \ 2\ 4 \\ \hline \end{array}$$

**22.**
$$\begin{array}{r} 5\ 6.9\ 5 \\ \times \ \ \ 1.9\ 4 \\ \hline \end{array}$$

Divide.  [R.3b]

**23.** $2.8\overline{)1\ 5\ 5.6\ 8}$

**24.** $5\ 2\overline{)2\ 3.4}$

**25.** Convert to decimal notation: $\dfrac{19}{12}$.  [R.3b]

**26.** Round to the nearest tenth: 34.067.  [R.3c]

**27.** Convert to decimal notation: 4.7%.  [R.4a]

**28.** Convert to fractional notation: 60%.  [R.4b]

Convert to percent notation.

**29.** 0.886  [R.4c]

**30.** $\dfrac{5}{8}$  [R.4d]

**31.** $\dfrac{29}{25}$  [R.4d]

**32.** Write exponential notation: $6 \cdot 6 \cdot 6$.  [R.5a]

**33.** Evaluate: $(1.06)^2$.  [R.5b]

Calculate and compare answers to Exercises 34–36.
[R.5c]

**34.** $120 - 6^2 \div 4 + 8$

**35.** $(120 - 6^2) \div 4 + 8$

**36.** $(120 - 6^2) \div (4 + 8)$

**37.** Calculate: $\dfrac{4(18 - 8) + 7 \cdot 9}{9^2 - 8^2}$.  [R.5c]

**Synthesis**

**38.** *Swimming Pool.* For health reasons, there must be 6 parts per billion (ppb) of chlorine in a swimming pool. What percent of total volume is this amount of chlorine?  [R.4d]

# Test: Chapter R

**1.** Find the prime factorization of 300.

**2.** Find the LCM of 15, 24, and 60.

**3.** Write an expression equivalent to $\frac{3}{7}$ using $\frac{7}{7}$ as a name for 1.

**4.** Write an equivalent expression with the given denominator:

$$\frac{11}{16}. \quad \text{(Denominator: 48)}$$

Simplify.

**5.** $\frac{16}{24}$

**6.** $\frac{925}{1525}$

Compute and simplify.

**7.** $\frac{10}{27} \div \frac{8}{3}$

**8.** $\frac{9}{10} - \frac{5}{8}$

**9.** Convert to fractional notation (do not simplify): 6.78.

**10.** Convert to decimal notation: $\frac{1895}{1000}$.

**11.** Add: $7.14 + 89 + 2.8787$.

**12.** Subtract: $1800 - 3.42$.

## Answers

**13.** _____

**14.** _____

**15.** _____

**16.** _____

**17.** _____

**18.** _____

**19.** _____

**20.** _____

**21.** _____

**22.** _____

**23.** _____

**24.** _____

**25.** _____

**13.** Multiply:   $\begin{array}{r} 1\ 2\ 3.6 \\ \times\quad 3.5\ 2 \end{array}$

**14.** Divide: $7.2\,\overline{)\,1\ 1.5\ 2\,}$.

**15.** Convert to decimal notation: $\dfrac{23}{11}$.

**16.** Round 234.7284 to the nearest tenth.

**17.** Round 234.7284 to the nearest thousandth.

**18.** Convert to decimal notation: 0.7%.

**19.** Convert to fractional notation: 91%.

**20.** Convert to percent notation: $\dfrac{11}{25}$.

**21.** Evaluate: $5^4$.

**22.** Evaluate: $(1.2)^2$.

**23.** Calculate: $200 - 2^3 + 5 + 10$.

**24.** Calculate: $8000 \div 0.16 \div 2.5$.

### Synthesis

**25.** Simplify: $\dfrac{13,860}{42,000}$.

# 1

# Introduction to Real Numbers and Algebraic Expressions

## Introduction

In this chapter, we consider the number system used most often in algebra. It is called the real-number system. We will learn to add, subtract, multiply, and divide real numbers and to manipulate certain expressions. Such manipulation will be important when we solve equations and applied problems in Chapter 2.

## An Application

The casino game of blackjack makes use of many card-counting systems to give players a winning edge if the count becomes negative. One such system is called *High–Low*, first developed by Harvey Dubner in 1963. Each card counts as −1, 0, or 1 as follows:

   2, 3, 4, 5, 6     count as +1;
   7, 8, 9          count as 0;
   10, J, Q, K, A  count as −1.

Find the final count on the sequence of cards

   K, A, 2, 4, 5, 10, J, 8, Q, K, 5.

**Source:** Patterson, Jerry L., *Casino Gambling.* New York: Perigee, 1982

## The Mathematics

We add the following numbers:

$$(-1) + (-1) + 1 + 1 + 1 + (-1) + (-1) + 0 + (-1) + (-1) + 1$$
$$= -2.$$

The numbers in red are negative numbers.

This problem appears as Exercise 120 in Exercise Set 1.4.

**World Wide Web** For more information, visit us at www.mathmax.com

# Pretest: Chapter 1

**1.** Evaluate $x/2y$ for $x = 5$ and $y = 8$.

**2.** Write an algebraic expression: Seventy-eight percent of some number.

**3.** Find the area of a rectangle when the length is 22.5 ft and the width is 16 ft.

**4.** Find $-x$ when $x = -12$.

Use either $<$ or $>$ for ▨ to write a true sentence.

**5.** $0$ ▨ $-5$

**6.** $10$ ▨ $-5$

**7.** $-35$ ▨ $-45$

**8.** $-\dfrac{2}{3}$ ▨ $\dfrac{4}{5}$

Find the absolute value.

**9.** $|-12|$

**10.** $|2.3|$

**11.** $|0|$

Find the opposite, or additive inverse.

**12.** $5.4$

**13.** $-\dfrac{2}{3}$

Find the reciprocal.

**14.** $10$

**15.** $-\dfrac{2}{3}$

Compute and simplify.

**16.** $-9 + (-8)$

**17.** $20.2 - (-18.4)$

**18.** $-\dfrac{5}{6} - \dfrac{3}{10}$

**19.** $-11.5 + 6.5$

**20.** $-9(-7)$

**21.** $\dfrac{5}{8}\left(-\dfrac{2}{3}\right)$

**22.** $-19.6 \div 0.2$

**23.** $-56 \div (-7)$

**24.** $12 - (-6) + 14 - 8$

**25.** $20 - 10 \div 5 + 2^3$

Multiply.

**26.** $9(z - 2)$

**27.** $-2(2a + b - 5c)$

Factor.

**28.** $4x - 12$

**29.** $6y - 9z - 18$

Simplify.

**30.** $3y - 7 - 2(2y + 3)$

**31.** $\{2[3(y + 1) - 4] - [5(y - 3) - 5]\}$

**32.** Write an inequality with the same meaning as $x > 12$.

## Objectives for Retesting

The objectives to be tested in addition to the material in this chapter are as follows.

[R.1a, b] Find all the factors of numbers, find prime factorizations of numbers, and find the LCM of two or more numbers using prime factorizations.

[R.2b, c] Simplify fractional notation, and add, subtract, multiply, and divide using fractional notation.

[R.4a, d] Convert from percent notation to decimal notation, and convert from fractional notation to percent notation.

[R.5b, c] Evaluate exponential expressions, and simplify expressions using the rules for order of operations.

# 1.1 Introduction to Algebra

Many types of problems require the use of equations in order to be solved effectively. The study of algebra involves the use of equations to solve problems. Equations are constructed from algebraic expressions. The purpose of this section is to introduce you to the types of expressions encountered in algebra.

## a | Evaluating Algebraic Expressions

In arithmetic, you have worked with expressions such as

$$49 + 75, \quad 8 \times 6.07, \quad 29 - 14, \quad \text{and} \quad \frac{5}{6}.$$

In algebra, we use certain letters for numbers and work with *algebraic expressions* such as

$$x + 75, \quad 8 \times y, \quad 29 - t, \quad \text{and} \quad \frac{a}{b}.$$

Sometimes a letter can represent various numbers. In that case, we call the letter a **variable**. Let $a$ = your age. Then $a$ is a variable since $a$ changes from year to year. Sometimes a letter can stand for just one number. In that case, we call the letter a **constant**. Let $b$ = your date of birth. Then $b$ is a constant.

Where do algebraic expressions occur? Most often we encounter them when we are solving applied problems. For example, consider the bar graph shown at right, one that we might find in a book or magazine. Suppose we want to know how much longer the diameter of Earth is than the diameter of Mars.

In algebra, we translate the problem into an equation. It might be done as follows.

$$\underbrace{\text{Diameter of Mars}} \quad \underbrace{\text{plus}} \quad \underbrace{\text{How much}} \quad \underbrace{\text{is}} \quad \underbrace{\text{Diameter of Earth}}$$
$$4217 \qquad\qquad + \qquad\qquad x \qquad = \qquad\qquad 7927$$

Note that we have an algebraic expression on the left of the equals sign. To find the number $x$, we can subtract 4217 on both sides of the equation:

$$4217 + x = 7927$$
$$4217 + x - 4217 = 7927 - 4217$$
$$x = 3710.$$

The value of $x$ gives us the answer, 3710 miles.

In arithmetic, you probably would do this subtraction right away without considering an equation. In algebra, more complex problems are difficult to solve without first solving an equation.

***Do Exercise 1.***

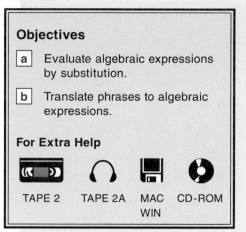

**Objectives**

**a** | Evaluate algebraic expressions by substitution.

**b** | Translate phrases to algebraic expressions.

**For Extra Help**

TAPE 2      TAPE 2A      MAC WIN      CD-ROM

**1.** Translate this problem to an equation. Use the graph below.

How much longer is the diameter of Venus than the diameter of Pluto?

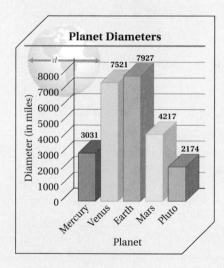

*Answer on page A-3*

**2.** Evaluate $a + b$ for $a = 38$ and $b = 26$.

**3.** Evaluate $x - y$ for $x = 57$ and $y = 29$.

**4.** Evaluate $4t$ for $t = 15$.

**5.** Find the area of a rectangle when $l$ is 24 ft and $w$ is 8 ft.

---

### Calculator Spotlight

 Evaluating Algebraic Expressions. To evaluate the expression $12m/n$ for $m = 8$ and $n = 16$, we first write (or think)

$$\frac{12 \cdot 8}{16}.$$

Then we press

$( \boxed{1}\,\boxed{2}\,\boxed{\times}\,\boxed{8}\,)$
$\boxed{\div}\,\boxed{1}\,\boxed{6}\,\boxed{ENTER}.$

The answer is 6.

**Exercises**

Evaluate.

**1.** $\dfrac{12m}{n}$, for $m = 82.3$ and $n = 16.56$

**2.** $a + b$, for $a = 8.2$ and $b = 3.7$

**3.** $b + a$, for $a = 8.2$ and $b = 3.7$

**4.** $27xy$, for $x = 12.7$ and $y = 100.4$

**5.** $3a + 2b$, for $a = 8.2$ and $b = 3.7$

**6.** $2a + 3b$, for $a = 8.2$ and $b = 3.7$

---

*Answers on page A-3*

---

An **algebraic expression** consists of variables, constants, numerals, and operation signs. When we replace a variable with a number, we say that we are **substituting** for the variable. This process is called **evaluating the expression.**

**Example 1**   Evaluate $x + y$ for $x = 37$ and $y = 29$.

We substitute 37 for $x$ and 29 for $y$ and carry out the addition:

$$x + y = 37 + 29 = 66.$$

The number 66 is called the **value** of the expression.

Algebraic expressions involving multiplication can be written in several ways. For example, "8 times $a$" can be written as $8 \times a$, $8 \cdot a$, $8(a)$, or simply $8a$. Two letters written together without an operation symbol, such as $ab$, also indicates a multiplication.

**Example 2**   Evaluate $3y$ for $y = 14$.

$$3y = 3(14) = 42$$

*Do Exercises 2–4.*

**Example 3**   The area $A$ of a rectangle of length $l$ and width $w$ is given by the formula $A = lw$. Find the area when $l$ is 24.5 in. and $w$ is 16 in.

We substitute 24.5 in. for $l$ and 16 in. for $w$ and carry out the multiplication:

$$\begin{aligned}
A = lw &= (24.5 \text{ in.})(16 \text{ in.}) \\
&= (24.5)(16)(\text{in.})(\text{in.}) \\
&= 392 \text{ in}^2, \text{ or } 392 \text{ square inches.}
\end{aligned}$$

*Do Exercise 5.*

Algebraic expressions involving division can also be written in several ways. For example, "8 divided by $t$" can be written as $8 \div t$, $\dfrac{8}{t}$, or $8/t$, where the fraction bar is a division symbol.

**Example 4**   Evaluate $\dfrac{a}{b}$ for $a = 63$ and $b = 9$.

We substitute 63 for $a$ and 9 for $b$ and carry out the division:

$$\frac{a}{b} = \frac{63}{9} = 7.$$

**Example 5**   Evaluate $\dfrac{12m}{n}$ for $m = 8$ and $n = 16$.

$$\frac{12m}{n} = \frac{12 \cdot 8}{16} = \frac{96}{16} = 6$$

*Do Exercises 6 and 7 on the following page.*

**Example 6** *Motorcycle Travel.* Ed takes a trip on his motorcycle. He wants to travel 660 mi on a particular day. The time $t$, in hours, that it takes to travel 660 mi is given by

$$t = \frac{660}{r},$$

where $r$ is the speed of Ed's motorcycle. Find the time of travel if the speed $r$ is 60 mph.

We substitute 60 for $r$ and carry out the division:

$$t = \frac{660}{r} = \frac{660}{60} = 11 \text{ hr.}$$

*Do Exercise 8.*

**6.** Evaluate $a/b$ for $a = 200$ and $b = 8$.

## b  Translating to Algebraic Expressions

In algebra, we translate problems to equations. The different parts of an equation are translations of word phrases to algebraic expressions. It is easier to translate if we know that certain words often translate to certain operation symbols.

**7.** Evaluate $10p/q$ for $p = 40$ and $q = 25$.

| KEY WORDS | | | |
|---|---|---|---|
| **Addition (+)** | **Subtraction (−)** | **Multiplication (·)** | **Division (÷)** |
| add<br>sum<br>plus<br>more than<br>increased by | subtract<br>difference<br>minus<br>less than<br>decreased by<br>take from | multiply<br>product<br>times<br>twice<br>of | divide<br>quotient<br>divided by |

**Example 7**  Translate to an algebraic expression:

Twice (or two times) some number.

Think of some number, say, 8. What number is twice 8? It is 16. How did you get 16? You multiplied by 2. Do the same thing using a variable. We can use any variable we wish, such as $x$, $y$, $m$, or $n$. Let's use $y$ to stand for some number. If we multiply by 2, we get an expression

$$y \times 2, \quad 2 \times y, \quad 2 \cdot y, \quad \text{or} \quad 2y.$$

**Example 8**  Translate to an algebraic expression:

Thirty-eight percent of some number.

The word "of" translates to a multiplication symbol, so we get the following expressions as a translation:

$$38\% \cdot n, \quad 0.38 \times n, \quad \text{or} \quad 0.38n.$$

**8.** *Motorcycle Travel.* Find the time it takes to travel 660 mi if the speed is 55 mph.

*Answers on page A-3*

Translate to an algebraic expression.

**9.** Eight less than some number

**10.** Eight more than some number

**11.** Four less than some number

**12.** Half of a number

**13.** Six more than eight times some number

**14.** The difference of two numbers

**15.** Fifty-nine percent of some number

**16.** Two hundred less than the product of two numbers

**17.** The sum of two numbers

*Answers on page A-3*

**Example 9** Translate to an algebraic expression:

Seven less than some number.

We let

$x$ represent the number.

Now if the number were 23, then the translation would be $23 - 7$. If we knew the number to be 345, then the translation would be $345 - 7$. If the number is $x$, then the translation is

$x - 7$.

---

*CAUTION!* Note that $7 - x$ is *not* a correct translation of the expression in Example 9. The expression $7 - x$ is a translation of "seven minus some number" or "some number less than seven."

---

**Example 10** Translate to an algebraic expression:

Eighteen more than a number.

We let

$t$ = the number.

Now if the number were 26, then the translation would be $26 + 18$. If we knew the number to be 174, then the translation would be $174 + 18$. If the number is $t$, then the translation is

$t + 18$,   or   $18 + t$.

**Example 11** Translate to an algebraic expression:

A number divided by 5.

We let

$m$ = the number.

Now if the number were 76, then the translation would be $76 \div 5$, or 76/5, or $\frac{76}{5}$. If the number were 213, then the translation would be $213 \div 5$, or 213/5, or $\frac{213}{5}$. If the number is $m$, then the translation is

$m \div 5$,     $m/5$,   or   $\dfrac{m}{5}$.

**Example 12** Translate each of the following phrases to an algebraic expression.

| Phrase | Algebraic Expression |
|---|---|
| Five more than some number | $n + 5$, or $5 + n$ |
| Half of a number | $\frac{1}{2}t$, $\frac{t}{2}$, or $t/2$ |
| Five more than three times some number | $3p + 5$, or $5 + 3p$ |
| The difference of two numbers | $x - y$ |
| Six less than the product of two numbers | $mn - 6$ |
| Seventy-six percent of some number | $76\%z$, or $0.76z$ |

*Do Exercises 9–17.*

# Exercise Set 1.1

 **a** Substitute to find values of the expressions in each of the following applied problems.

**1.** *Enrollment Costs.* At Emmett Community College, it costs $600 to enroll in the 8 A.M. section of Elementary Algebra. Suppose that the variable $n$ stands for the number of students who enroll. Then $600n$ stands for the total amount of money collected for this course. How much is collected if 34 students enroll? 78 students? 250 students?

**2.** *Commuting Time.* It takes Erin 24 min less time to commute to work than it does George. Suppose that the variable $x$ stands for the time it takes George to get to work. Then $x - 24$ stands for the time it takes Erin to get to work. How long does it take Erin to get to work if it takes George 56 min? 93 min? 105 min?

**3.** The area $A$ of a triangle with base $b$ and height $h$ is given by $A = \frac{1}{2}bh$. Find the area when $b = 45$ m (meters) and $h = 86$ m.

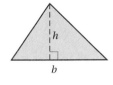

**4.** The area $A$ of a parallelogram with base $b$ and height $h$ is given by $A = bh$. Find the area of the parallelogram when the height is 15.4 cm (centimeters) and the base is 6.5 cm.

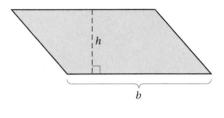

**5.** *Distance Traveled.* A driver who drives at a speed of $r$ mph for $t$ hr will travel a distance $d$ mi given by $d = rt$ mi. How far will a driver travel at a speed of 65 mph for 4 hr?

**6.** *Simple Interest.* The simple interest $I$ on a principal of $P$ dollars at interest rate $r$ for time $t$, in years, is given by $I = Prt$. Find the simple interest on a principal of $4800 at 9% for 2 yr. (*Hint:* 9% = 0.09.)

Evaluate.

**7.** $8x$, for $x = 7$

**8.** $6y$, for $y = 7$

**9.** $\dfrac{a}{b}$, for $a = 24$ and $b = 3$

**10.** $\dfrac{p}{q}$, for $p = 16$ and $q = 2$

**11.** $\dfrac{3p}{q}$, for $p = 2$ and $q = 6$

**12.** $\dfrac{5y}{z}$, for $y = 15$ and $z = 25$

**13.** $\dfrac{x + y}{5}$, for $x = 10$ and $y = 20$

**14.** $\dfrac{p + q}{2}$, for $p = 2$ and $q = 16$

**15.** $\dfrac{x - y}{8}$, for $x = 20$ and $y = 4$

**16.** $\dfrac{m - n}{5}$, for $m = 16$ and $n = 6$

**b** Translate to an algebraic expression.

**17.** 7 more than $b$

**18.** 9 more than $t$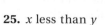

**19.** 12 less than $c$

**20.** 14 less than $d$

**21.** 4 increased by $q$

**22.** 13 increased by $z$

**23.** $b$ more than $a$

**24.** $c$ more than $d$

**25.** $x$ less than $y$

**26.** $c$ less than $h$

**27.** $x$ added to $w$

**28.** $s$ added to $t$

**29.** $m$ subtracted from $n$

**30.** $p$ subtracted from $q$

**31.** The sum of $r$ and $s$

**32.** The sum of $a$ and $b$

**33.** Twice $z$

**34.** Three times $q$

**35.** 3 multiplied by $m$

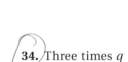

**36.** The product of 8 and $t$

**37.** The product of 89% and some number

**38.** 67% of some number

**39.** A driver drove at a speed of 55 mph for $t$ hours. How far did the driver travel?

**40.** An executive assistant has $d$ dollars before going to an office supply store. He bought some fax paper for $18.95. How much did he have after the purchase?

---

**Skill Maintenance**

Find the prime factorization.    [R.1a]

**41.** 54

**42.** 32

**43.** 108

**44.** 192

Find the LCM.    [R.1b]

**45.** 6,  18

**46.** 6,  24,  32

**47.** 10,  20,  30

**48.** 16,  24

---

**Synthesis**

**49.** ◈ If the length of a rectangle is doubled, does the area double? Why or why not?

**50.** ◈ If the height and the base of a triangle are doubled, what happens to the area? Explain.

Translate to an algebraic expression.

**51.** Some number $x$ plus three times $y$

**52.** Some number $a$ plus 2 plus $b$

**53.** A number that is 3 less than twice $x$

**54.** Your age in 5 years, if you are $a$ years old now

Copyright © 1999 Addison Wesley Longman

# 1.2 The Real Numbers

A **set** is a collection of objects. (See Appendix A for more on sets.) For our purposes, we will most often be considering sets of numbers. One way to name a set uses what is called **roster notation.** For example, roster notation for the set containing the numbers 0, 2, and 5 is {0, 2, 5}.

Sets that are parts of other sets are called **subsets**. In this section, we become acquainted with the set of *real numbers* and its various subsets.

Two important subsets of the real numbers are listed below using roster notation.

> **Natural numbers** = {1, 2, 3, ...}. These are the numbers used for counting. *or "Counting Numbers"*

> **Whole numbers** = {0, 1, 2, 3, ...}. This is the set of natural numbers with 0 included.

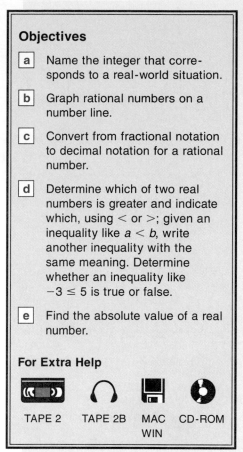

**Objectives**

**a** Name the integer that corresponds to a real-world situation.

**b** Graph rational numbers on a number line.

**c** Convert from fractional notation to decimal notation for a rational number.

**d** Determine which of two real numbers is greater and indicate which, using < or >; given an inequality like $a < b$, write another inequality with the same meaning. Determine whether an inequality like $-3 \leq 5$ is true or false.

**e** Find the absolute value of a real number.

**For Extra Help**

TAPE 2    TAPE 2B    MAC WIN    CD-ROM

We can represent these sets on a number line. The natural numbers are those to the right of zero. The whole numbers are the natural numbers and zero.

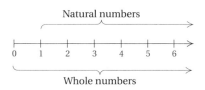

We create a new set, called the *integers,* by starting with the whole numbers, 0, 1, 2, 3, and so on. For each natural number 1, 2, 3, and so on, we obtain a new number to the left of zero on the number line:

For the number 1, there will be an *opposite* number −1 (negative 1).

For the number 2, there will be an *opposite* number −2 (negative 2).

For the number 3, there will be an *opposite* number −3 (negative 3), and so on.

The **integers** consist of the whole numbers and these new numbers. We picture them on a number line as follows.

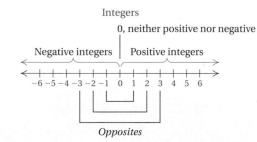

*Opposites*

We call these new numbers to the left of 0 **negative integers.** The natural numbers are also called **positive integers.** Zero is neither positive nor negative. We call −1 and 1 **opposites** of each other. Similarly, −2 and 2 are opposites, −3 and 3 are opposites, −100 and 100 are opposites, and 0 is its own opposite. Pairs of opposite numbers like −3 and 3 are equidistant from 0. The integers extend infinitely on the number line to the left and right of zero.

> The set of **integers** = {..., −5, −4, −3, −2, −1, 0, 1, 2, 3, 4, 5, ...}.

## a | Integers and the Real World

Integers correspond to many real-world problems and situations. The following examples will help you get ready to translate problem situations that involve integers to mathematical language.

**Example 1** Tell which integer corresponds to this situation: The temperature is 3 degrees below zero.

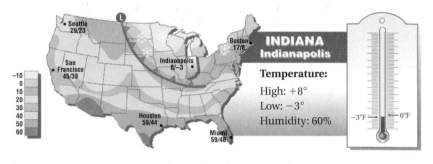

The integer −3 corresponds to the situation. The temperature is −3°.

**Example 2** *Jeopardy.* Tell which integer corresponds to this situation: A contestant missed a $600 question on the television game show "Jeopardy."

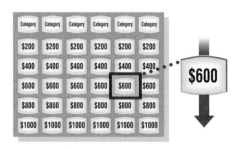

Missing a $600 question means −600.

Missing a $600 question causes a $600 loss on the score—that is, the contestant earns −600 dollars.

**Example 3** *Elevation.* Tell which integer corresponds to this situation: The lowest point in New Orleans is 8 ft below sea level.

The integer −8 corresponds to the situation. The elevation is −8 ft.

**Example 4** *Dow Jones Industrial Average.* Tell which integers correspond to this situation: The largest daily decrease in the Dow Jones Industrial Average was 508 points; the largest increase was 187 (**Source:** *The Guinness Book of Records*).

The integer $-508$ corresponds to a decrease in the average. The integer 187 corresponds to an increase in the average.

*Do Exercises 1–4.*

## b | The Rational Numbers

We created the set of integers by obtaining a negative number for each natural number. To create a larger number system, called the set of **rational numbers,** we consider quotients of integers with nonzero divisors. The following are rational numbers:

$$\frac{2}{3}, \quad -\frac{2}{3}, \quad \frac{7}{1}, \quad 4, \quad -3, \quad 0, \quad \frac{23}{-8}, \quad 2.4, \quad -0.17, \quad 10\frac{1}{2}.$$

The number $-\frac{2}{3}$ (read "negative two-thirds") can also be named $\frac{2}{-3}$ or $\frac{-2}{3}$. The number 2.4 can be named $\frac{24}{10}$ or $\frac{12}{5}$, and $-0.17$ can be named $-\frac{17}{100}$.

Note that this new set of numbers, the rational numbers, contains the whole numbers, the integers, and the arithmetic numbers (also called the nonnegative rational numbers). We can describe the set of rational numbers using **set-builder notation,** as follows.

> The set of **rational numbers** $= \left\{\dfrac{a}{b} \,\middle|\, a \text{ and } b \text{ are integers and } b \neq 0\right\}$.
>
> $\left(\text{This is read "the set of numbers } \dfrac{a}{b}, \text{ where } a \text{ and } b \text{ are integers and } b \neq 0."\right)$

We picture the rational numbers on a number line as follows. There is a point on the line for every rational number.

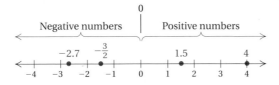

To **graph** a number means to find and mark its point on the number line. Some rational numbers are graphed in the preceding figure.

**Example 5** Graph: $\frac{5}{2}$.

The number $\frac{5}{2}$ can be named $2\frac{1}{2}$, or 2.5. Its graph is halfway between 2 and 3.

Tell which integers correspond to the given situation.

**1.** The halfback gained 8 yd on the first down. The quarterback was sacked for a 5-yd loss on the second down.

**2.** The highest temperature ever recorded in the United States was 134° in Death Valley on July 10, 1913. The coldest temperature ever recorded in the United States was 80° below zero in Prospect Creek, Alaska, in January 1971.

**3.** At 10 sec before liftoff, ignition occurs. At 156 sec after liftoff, the first stage is detached from the rocket.

**4.** A submarine dove 120 ft, rose 50 ft, and then dove 80 ft.

*Answers on page A-3*

Graph on a number line.

**5.** $-\dfrac{7}{2}$

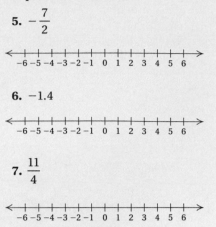

**6.** $-1.4$

**7.** $\dfrac{11}{4}$

## Calculator Spotlight

 Negative Numbers on a Calculator. To enter a negative number on a graphing calculator, we use an opposite key $\boxed{(-)}$. Note that this is different from the subtraction key. To enter $-5$, we press

$$\boxed{(-)}\;\boxed{5}\;\boxed{\text{ENTER}}.$$

The display then reads

$$\boxed{\qquad\qquad -5}\;.$$

To enter $-\dfrac{7}{8}$, we press

$$\boxed{(-)}\;\boxed{7}\;\boxed{/}\;\boxed{8}$$
$$\boxed{\text{ENTER}}.$$

The number is displayed in decimal notation:

$$\boxed{\qquad\qquad -.875}\;.$$

**Exercises**

Press the appropriate keys so that your calculator displays the given number.

**1.** $-3$     **2.** $-508$

**3.** $-0.17$     **4.** $-\dfrac{5}{8}$

*Answers on page A-3*

**Example 6**   Graph: $-3.2$.

The graph of $-3.2$ is $\frac{2}{10}$ of the way from $-3$ to $-4$.

**Example 7**   Graph: $\frac{13}{8}$.

The number $\frac{13}{8}$ can be named $1\frac{5}{8}$, or $1.625$. The graph is about $\frac{6}{10}$ of the way from 1 to 2.

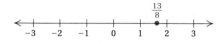

*Do Exercises 5–7.*

## c   Notation for Rational Numbers

Each rational number can be named using fractional or decimal notation.

**Example 8**   Convert to decimal notation: $-\frac{5}{8}$.

We first find decimal notation for $\frac{5}{8}$. Since $\frac{5}{8}$ means $5 \div 8$, we divide.

$$
\begin{array}{r}
0.6\,2\,5 \\
8\,)\overline{\,5.0\,0\,0} \\
\underline{4\,8}\phantom{00} \\
2\,0\phantom{0} \\
\underline{1\,6}\phantom{0} \\
4\,0 \\
\underline{4\,0} \\
0
\end{array}
$$

Thus, $\frac{5}{8} = 0.625$, so $-\frac{5}{8} = -0.625$.

Decimal notation for $-\frac{5}{8}$ is $-0.625$. We consider $-0.625$ to be a **terminating decimal.** Decimal notation for some numbers repeats.

**Example 9**   Convert to decimal notation: $\frac{7}{11}$.

We divide.

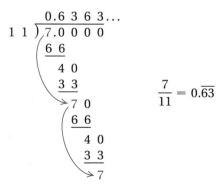

$$\frac{7}{11} = 0.\overline{63}$$

We can abbreviate repeating decimal notation by writing a bar over the repeating part—in this case, $0.\overline{63}$.

The following are other examples to show how each rational number can be named using fractional or decimal notation:

$$0 = \frac{0}{8}, \qquad \frac{27}{100} = 0.27, \qquad -8\frac{3}{4} = -8.75, \qquad \frac{-13}{6} = -2.1\overline{6}.$$

*Do Exercises 8–10.*

## d | The Real Numbers and Order

Every rational number has a point on the number line. However, there are some points on the line for which there is no rational number. These points correspond to what are called **irrational numbers.**

What kinds of numbers are irrational? One example is the number $\pi$, which is used in finding the area and the circumference of a circle: $A = \pi r^2$ and $C = 2\pi r$.

Another example of an irrational number is the square root of 2, named $\sqrt{2}$.

It is the length of the diagonal of a square with sides of length 1. It is also the number that when multiplied by itself gives 2. There is no rational number that can be multiplied by itself to get 2. But the following are rational *approximations*:

1.4 is an approximation of $\sqrt{2}$ because $(1.4)^2 = 1.96$;

1.41 is a better approximation because $(1.41)^2 = 1.9881$;

1.4142 is an even better approximation because $(1.4142)^2 = 1.99996164.$

We can find rational approximations for square roots using a calculator.

Decimal notation for rational numbers *either* terminates *or* repeats. Decimal notation for irrational numbers *neither* terminates *nor* repeats. Some other examples of irrational numbers are $\sqrt{3}$, $-\sqrt{8}$, $\sqrt{11}$, and 0.121221222122221.... Whenever we take the square root of a number that is not a perfect square, we will get an irrational number.

The rational numbers and the irrational numbers together correspond to all the points on a number line and make up what is called the **real-number system.**

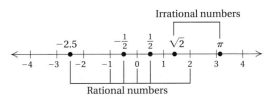

> The set of **real numbers** = The set of all numbers corresponding to points on the number line.

Convert to decimal notation.

**8.** $-\frac{3}{8}$

**9.** $-\frac{6}{11}$

**10.** $\frac{4}{3}$

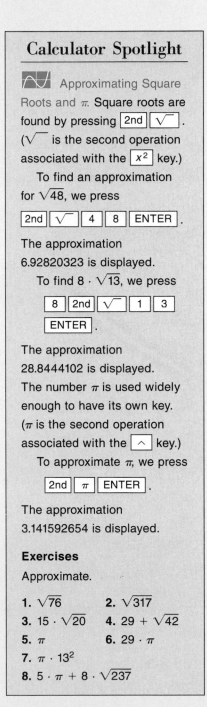

## Calculator Spotlight

Approximating Square Roots and $\pi$. Square roots are found by pressing [2nd] [$\sqrt{\ }$]. ($\sqrt{\ }$ is the second operation associated with the [$x^2$] key.)

To find an approximation for $\sqrt{48}$, we press

[2nd] [$\sqrt{\ }$] [4] [8] [ENTER].

The approximation 6.92820323 is displayed.

To find $8 \cdot \sqrt{13}$, we press

[8] [2nd] [$\sqrt{\ }$] [1] [3] [ENTER].

The approximation 28.8444102 is displayed.

The number $\pi$ is used widely enough to have its own key. ($\pi$ is the second operation associated with the [$\wedge$] key.)

To approximate $\pi$, we press

[2nd] [$\pi$] [ENTER].

The approximation 3.141592654 is displayed.

**Exercises**

Approximate.

**1.** $\sqrt{76}$      **2.** $\sqrt{317}$

**3.** $15 \cdot \sqrt{20}$      **4.** $29 + \sqrt{42}$

**5.** $\pi$      **6.** $29 \cdot \pi$

**7.** $\pi \cdot 13^2$

**8.** $5 \cdot \pi + 8 \cdot \sqrt{237}$

*Answers on page A-3*

Use either $<$ or $>$ for ▨ to write a true sentence.

**11.** $-3$ ▨ $7$

**12.** $-8$ ▨ $-5$

**13.** $7$ ▨ $-10$

**14.** $3.1$ ▨ $-9.5$

**15.** $-\dfrac{2}{3}$ ▨ $-1$

**16.** $-\dfrac{11}{8}$ ▨ $\dfrac{23}{15}$

**17.** $-\dfrac{2}{3}$ ▨ $-\dfrac{5}{9}$

**18.** $-4.78$ ▨ $-5.01$

The real numbers consist of the rational numbers and the irrational numbers. The following figure shows the relationships among various kinds of numbers.

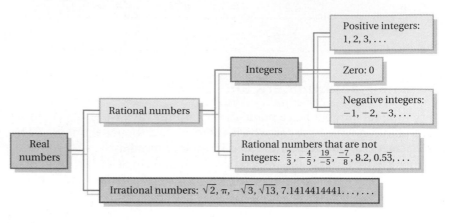

## Order

Real numbers are named in order on the number line, with larger numbers named farther to the right. For any two numbers on the line, the one to the left is less than the one to the right.

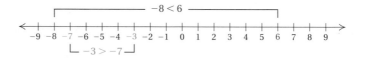

We use the symbol $<$ to mean "**is less than.**" The sentence $-8 < 6$ means "$-8$ is less than 6." The symbol $>$ means "**is greater than.**" The sentence $-3 > -7$ means "$-3$ is greater than $-7$." The sentences $-8 < 6$ and $-3 > -7$ are **inequalities**.

**Examples**   Use either $<$ or $>$ for ▨ to write a true sentence.

**10.** $2$ ◁ $9$        Since 2 is to the left of 9, 2 is less than 9, so $2 < 9$.

**11.** $-7$ ◁ $3$       Since $-7$ is to the left of 3, we have $-7 < 3$.

**12.** $6$ ▷ $-12$     Since 6 is to the right of $-12$, then $6 > -12$.

**13.** $-18$ ◁ $-5$     Since $-18$ is to the left of $-5$, we have $-18 < -5$.

**14.** $-2.7$ ▨ $-\dfrac{3}{2}$     The answer is $-2.7 < -\dfrac{3}{2}$.

**15.** $1.5$ ▨ $-2.7$     The answer is $1.5 > -2.7$.

**16.** $1.38$ ▨ $1.83$     The answer is $1.38 < 1.83$.

**17.** $-3.45$ ▨ $1.32$     The answer is $-3.45 < 1.32$.

**18.** $-4$ ▨ $0$     The answer is $-4 < 0$.

**19.** $5.8$ ▨ $0$     The answer is $5.8 > 0$.

**20.** $\dfrac{5}{8}$ ▨ $\dfrac{7}{11}$     We convert to decimal notation: $\dfrac{5}{8} = 0.625$ and $\dfrac{7}{11} = 0.6363\ldots$. Thus, $\dfrac{5}{8} < \dfrac{7}{11}$.

*Do Exercises 11–18.*

Note that both $-8 < 6$ and $6 > 8$ arc truc. Every true inequality yields another true inequality when we interchange the numbers or variables and reverse the direction of the inequality sign.

> $a < b$ also has the meaning $b > a$.

**Examples** Write another inequality with the same meaning.

**21.** $a < -5$   The inequality $-5 > a$ has the same meaning.
**22.** $-3 > -8$   The inequality $-8 < -3$ has the same meaning.

A helpful mental device is to think of an inequality sign as an "arrow" with the arrow pointing to the smaller number.

*Do Exercises 19 and 20.*

Note that all positive real numbers are greater than zero and all negative real numbers are less than zero.

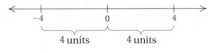

> If $b$ is a positive real number, then $b > 0$.
> If $a$ is a negative real number, then $a < 0$.

Expressions like $a \le b$ and $b \ge a$ are also inequalities. We read $a \le b$ as "**$a$ is less than or equal to $b$.**" We read $a \ge b$ as "**$a$ is greater than or equal to $b$.**"

**Examples** Write true or false for each statement.

**23.** $-3 \le 5.4$   True since $-3 < 5.4$ is true
**24.** $-3 \le -3$   True since $-3 = -3$ is true
**25.** $-5 \ge 1\frac{2}{3}$   False since neither $-5 > 1\frac{2}{3}$ nor $-5 = 1\frac{2}{3}$ is true

*Do Exercises 21–23.*

## e Absolute Value

From the number line, we see that numbers like 4 and $-4$ are the same distance from zero. Distance is always a nonnegative number. We call the distance from zero on a number line the **absolute value** of the number.

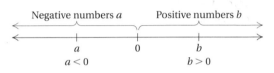

> The **absolute value** of a number is its distance from zero on a number line. We use the symbol $|x|$ to represent the absolute value of a number $x$.

---

Write another inequality with the same meaning.

**19.** $-5 < 7$

**20.** $x > 4$

Write true or false.

**21.** $-4 \le -6$

**22.** $7.8 \ge 7.8$

**23.** $-2 \le \dfrac{3}{8}$

*Answers on page A-3*

Find the absolute value.

**24.** $|8|$    8

**25.** $|0|$    0

**26.** $|-9|$    9

**27.** $\left|-\dfrac{2}{3}\right|$    $\frac{2}{3}$

**28.** $|5.6|$    5.6

> To find absolute value:
>
> **a)** If a number is negative, make it positive.
> **b)** If a number is positive or zero, leave it alone.

**Examples**   Find the absolute value.

**26.** $|-7|$    The distance of $-7$ from 0 is 7, so $|-7| = 7$.
**27.** $|12|$    The distance of 12 from 0 is 12, so $|12| = 12$.
**28.** $|0|$    The distance of 0 from 0 is 0, so $|0| = 0$.
**29.** $\left|\dfrac{3}{2}\right| = \dfrac{3}{2}$
**30.** $|-2.73| = 2.73$

*Do Exercises 24–28.*

# Improving Your Math Study Skills

## Tips for Using This Textbook

Throughout this textbook, you will find a feature called "Improving Your Math Study Skills." At least one such topic is included in each chapter. Each topic title is listed in the table of contents beginning on p. iii. You saw the first of these study skill features in Section R.2 on p. 16.

One of the most important ways to improve your math study skills is to learn the proper use of the textbook. Here we highlight a few points that we consider most helpful.

- **Be sure to note the special symbols**  **and so on, that correspond to the objectives you are to be able to perform.** They appear in many places throughout the text. The first time you see them is in the margin at the beginning of each section. The second time is in the subheadings of each section, and the third time is in the exercise set. You will also find them next to the skill maintenance exercises in each exercise set and in the review exercises at the end of the chapter, as well as in the answers to the chapter tests and the cumulative reviews. These objective symbols allow you to refer back whenever you need to review a topic.

- **Note the symbols in the margin under the list of objectives at the beginning of each section.** These refer to the many distinctive study aids that accompany the book.

- **Read and study each step of each example.** The examples include important side comments that explain each step. These carefully chosen

examples and notes prepare you for success in the exercise set.

- **Stop and do the margin exercises as you study a section.** When our students come to us troubled about how they are doing in the course, the first question we ask is "Are you doing the margin exercises when directed to do so?" This is one of the most effective ways to enhance your ability to learn mathematics from this text. Don't deprive yourself of its benefits!

- **When you study the book, don't mark the points that you think are important, but mark the points you do not understand!** This book includes many design features that highlight important points. Use your efforts to mark where you are having trouble. Then when you go to class, a math lab, or a tutoring session, you will be prepared to ask questions that home in on your difficulties rather than spending time going over what you already understand.

- **If you are having trouble, consider using the *Student's Solutions Manual,*** which contains worked-out solutions to the odd-numbered exercises in the exercise sets.

- **Try to keep one section ahead of your syllabus.** If you study ahead of your lectures, you can concentrate on what is being explained in them, rather than trying to write everything down. You can then take notes only of special points or of questions related to what is happening in class.

*Answers on page A-3*

# Exercise Set 1.2

**a** Tell which integers correspond to the situation.

1. *Elevation.* The Dead Sea, between Jordan and Israel, is 1286 ft below sea level; Mt. Rainier in Washington State is 13,804 ft above sea level.

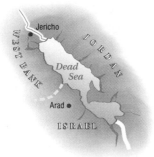

2. Amy's golf score was 3 under par; Juan's was 7 over par.

3. On Wednesday, the temperature was 24° above zero. On Thursday, it was 2° below zero.

4. A student deposited her tax refund of $750 in a savings account. Two weeks later, she withdrew $125 to pay sorority fees.

5. *U.S. Public Debt.* Recently, the total public debt of the United States was about $5.2 trillion (**Source:** U.S. Department of the Treasury).

6. *Birth and Death Rates.* Recently, the world birth rate was 27 per thousand. The death rate was 9.7 per thousand. (**Source:** United Nations Population Fund)

**b** Graph the number on the number line.

7. $\dfrac{10}{3}$

$$\overset{\longleftarrow \;|\;|\;|\;|\;|\;|\;|\;|\;|\;|\;|\;|\;|\;\longrightarrow}{\phantom{x}-6\;-5\;-4\;-3\;-2\;-1\;\;\;0\;\;\;1\;\;\;2\;\;\;3\;\;\;4\;\;\;5\;\;\;6}$$

8. $-\dfrac{17}{4}$

$$\overset{\longleftarrow \;|\;|\;|\;|\;|\;|\;|\;|\;|\;|\;|\;|\;|\;\longrightarrow}{\phantom{x}-6\;-5\;-4\;-3\;-2\;-1\;\;\;0\;\;\;1\;\;\;2\;\;\;3\;\;\;4\;\;\;5\;\;\;6}$$

9. $-5.2$

$$\overset{\longleftarrow \;|\;|\;|\;|\;|\;|\;|\;|\;|\;|\;|\;|\;|\;\longrightarrow}{\phantom{x}-6\;-5\;-4\;-3\;-2\;-1\;\;\;0\;\;\;1\;\;\;2\;\;\;3\;\;\;4\;\;\;5\;\;\;6}$$

10. $4.78$

$$\overset{\longleftarrow \;|\;|\;|\;|\;|\;|\;|\;|\;|\;|\;|\;|\;|\;\longrightarrow}{\phantom{x}-6\;-5\;-4\;-3\;-2\;-1\;\;\;0\;\;\;1\;\;\;2\;\;\;3\;\;\;4\;\;\;5\;\;\;6}$$

**c** Convert to decimal notation.

11. $-\dfrac{7}{8}$      12. $-\dfrac{1}{8}$      13. $\dfrac{5}{6}$      14. $\dfrac{5}{3}$      15. $\dfrac{7}{6}$      16. $\dfrac{5}{12}$

17. $\dfrac{2}{3}$      18. $\dfrac{1}{4}$      19. $-\dfrac{1}{2}$      20. $\dfrac{5}{8}$      21. $\dfrac{1}{10}$      22. $-\dfrac{7}{20}$

**d** Use either $<$ or $>$ for ▨ to write a true sentence.

23. $8 \;▨\; 0$      24. $3 \;▨\; 0$      25. $-8 \;▨\; 3$      26. $6 \;▨\; -6$

27. $-8 \;▨\; 8$      28. $0 \;▨\; -9$      29. $-8 \;▨\; -5$      30. $-4 \;▨\; -3$

31. $-5 \;▨\; -11$      32. $-3 \;▨\; -4$      33. $-6 \;▨\; -5$      34. $-10 \;▨\; -14$

35. $2.14 \;▨\; 1.24$      36. $-3.3 \;▨\; -2.2$      37. $-14.5 \;▨\; 0.011$      38. $17.2 \;▨\; -1.67$

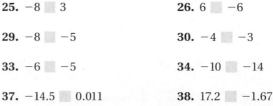

**39.** $-12.88$ ▢ $-6.45$    **40.** $-14.34$ ▢ $-17.88$    **41.** $\dfrac{5}{12}$ ▢ $\dfrac{11}{25}$    **42.** $-\dfrac{13}{16}$ ▢ $-\dfrac{5}{9}$

Write true or false.

**43.** $-3 \geq -11$    **44.** $5 \leq -5$    **45.** $0 \geq 8$    **46.** $-5 \leq 7$

Write an inequality with the same meaning.

**47.** $-6 > x$    **48.** $x < 8$    **49.** $-10 \leq y$    **50.** $12 \geq t$

**e** Find the absolute value.

**51.** $|-3|$    **52.** $|-7|$    **53.** $|10|$    **54.** $|11|$    **55.** $|0|$    **56.** $|-4|$

**57.** $|-24|$    **58.** $|325|$    **59.** $\left|-\dfrac{2}{3}\right|$    **60.** $\left|-\dfrac{10}{7}\right|$    **61.** $\left|\dfrac{0}{4}\right|$    **62.** $|14.8|$

---

**Skill Maintenance**

Convert to decimal notation.    [R.4a]

**63.** $63\%$    **64.** $8.3\%$    **65.** $110\%$    **66.** $22.76\%$

Convert to percent notation.    [R.4d]

**67.** $\dfrac{3}{4}$    **68.** $\dfrac{5}{8}$    **69.** $\dfrac{5}{6}$    **70.** $\dfrac{19}{32}$

---

**Synthesis**

**71.** ◈ ▦ When Jennifer's calculator gives a decimal approximation for $\sqrt{2}$ and that approximation is promptly squared, the result is 2. Yet, when that same approximation is entered by hand and then squared, the result is not exactly 2. Why do you suppose this happens?

**72.** ◈ How many rational numbers are there between 0 and 1? Why?

List in order from the least to the greatest.

**73.** $-\dfrac{2}{3}, \ \dfrac{1}{2}, \ -\dfrac{3}{4}, \ -\dfrac{5}{6}, \ \dfrac{3}{8}, \ \dfrac{1}{6}$

**74.** $-8\dfrac{7}{8}, \ 7^1, \ -5, \ |-6|, \ 4, \ |3|, \ -8\dfrac{5}{8}, \ -100, \ 0, \ 1^7, \ \dfrac{14}{4}, \ \dfrac{-67}{8}$

Given that $0.3\overline{3} = \frac{1}{3}$ and $0.6\overline{6} = \frac{2}{3}$, express each of the following as a quotient or ratio of two integers.

**75.** $0.1\overline{1}$    **76.** $0.9\overline{9}$    **77.** $5.5\overline{5}$

---

# 1.3 Addition of Real Numbers

In this section, we consider addition of real numbers. First, to gain an understanding, we add using a number line. Then we consider rules for addition.

Addition of numbers can be illustrated on a number line. To do the addition $a + b$, we start at $a$, and then move according to $b$.

a) If $b$ is positive, we move to the right.

b) If $b$ is negative, we move to the left.

c) If $b$ is 0, we stay at $a$.

**Example 1** Add: $3 + (-5)$.

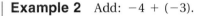
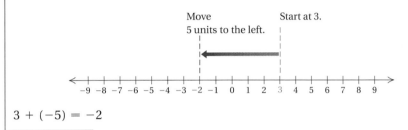

$3 + (-5) = -2$

**Example 2** Add: $-4 + (-3)$.

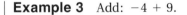
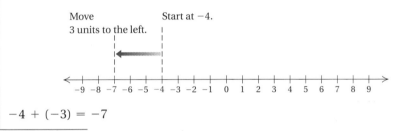

$-4 + (-3) = -7$

**Example 3** Add: $-4 + 9$.

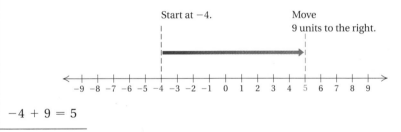

$-4 + 9 = 5$

**Example 4** Add: $-5.2 + 0$.

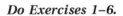
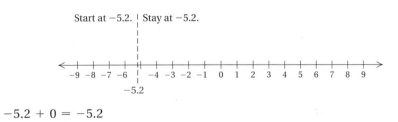

$-5.2 + 0 = -5.2$

*Do Exercises 1–6.*

## Objectives

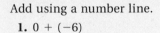

**a** Add real numbers without using a number line.

**b** Find the opposite, or additive inverse, of a real number.

## For Extra Help

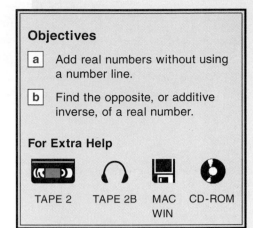

TAPE 2    TAPE 2B    MAC WIN    CD-ROM

Add using a number line.

**1.** $0 + (-6)$

**2.** $1 + (-4)$

**3.** $-3 + (-5)$

**4.** $-3 + 7$

**5.** $-5.4 + 5.4$

**6.** $-\dfrac{5}{2} + \dfrac{1}{2}$

*Answers on page A-4*

Add without using a number line.

**7.** $-5 + (-6)$

**8.** $-9 + (-3)$

**9.** $-4 + 6$

**10.** $-7 + 3$

**11.** $5 + (-7)$

**12.** $-20 + 20$

**13.** $-11 + (-11)$

**14.** $10 + (-7)$

**15.** $-0.17 + 0.7$

**16.** $-6.4 + 8.7$

**17.** $-4.5 + (-3.2)$

**18.** $-8.6 + 2.4$

**19.** $\dfrac{5}{9} + \left(-\dfrac{7}{9}\right)$

**20.** $-\dfrac{1}{5} + \left(-\dfrac{3}{4}\right)$

*Answers on page A-4*

---

**a** **Adding Without a Number Line**

You may have noticed some patterns in the preceding examples. These lead us to rules for adding without using a number line that are more efficient for adding larger numbers.

> **RULES FOR ADDITION OF REAL NUMBERS**
>
> 1. *Positive numbers*: Add the same as arithmetic numbers. The answer is positive.
> 2. *Negative numbers*: Add absolute values. The answer is negative.
> 3. *A positive and a negative number*: Subtract the smaller absolute value from the larger. Then:
>    a) If the positive number has the greater absolute value, the answer is positive.
>    b) If the negative number has the greater absolute value, the answer is negative.
>    c) If the numbers have the same absolute value, the answer is 0.
> 4. *One number is zero*: The sum is the other number.

Rule 4 is known as the **identity property of 0.** It says that for any real number $a$, $a + 0 = a$.

**Examples**   Add without using a number line.

**5.** $-12 + (-7) = -19$ — Two negatives. *Think*: Add the absolute values, 12 and 7, getting 19. Make the answer *negative*, $-19$.

**6.** $-1.4 + 8.5 = 7.1$ — The absolute values are 1.4 and 8.5. The difference is 7.1. The positive number has the larger absolute value, so the answer is *positive*, 7.1.

**7.** $-36 + 21 = -15$ — The absolute values are 36 and 21. The difference is 15. The negative number has the larger absolute value, so the answer is *negative*, $-15$.

**8.** $1.5 + (-1.5) = 0$ — The numbers have the same absolute value. The sum is 0.

**9.** $-\dfrac{7}{8} + 0 = -\dfrac{7}{8}$ — One number is zero. The sum is $-\frac{7}{8}$.

**10.** $-9.2 + 3.1 = -6.1$

**11.** $-\dfrac{3}{2} + \dfrac{9}{2} = \dfrac{6}{2} = 3$

**12.** $-\dfrac{2}{3} + \dfrac{5}{8} = -\dfrac{16}{24} + \dfrac{15}{24} = -\dfrac{1}{24}$

*Do Exercises 7–20.*

Suppose we want to add several numbers, some positive and some negative, as follows. How can we proceed?

$$15 + (-2) + 7 + 14 + (-5) + (-12)$$

We can change grouping and order as we please when adding. For instance, we can group the positive numbers together and the negative numbers together and add them separately. Then we add the two results.

**Example 13** Add: $15 + (-2) + 7 + 14 + (-5) + (-12)$.

**a)** $15 + 7 + 14 = 36$        Adding the positive numbers

**b)** $-2 + (-5) + (-12) = -19$        Adding the negative numbers

**c)** $36 + (-19) = 17$        Adding the results

We can also add the numbers in any other order we wish, say, from left to right as follows:

$$
\begin{aligned}
15 + (-2) + 7 + 14 + (-5) + (-12) &= 13 + 7 + 14 + (-5) + (-12) \\
&= 20 + 14 + (-5) + (-12) \\
&= 34 + (-5) + (-12) \\
&= 29 + (-12) \\
&= 17
\end{aligned}
$$

*Do Exercises 21–24.*

## b   Opposites, or Additive Inverses

Suppose we add two numbers that are **opposites**, such as 6 and $-6$. The result is 0. When opposites are added, the result is always 0. Such numbers are also called **additive inverses.** Every real number has an opposite, or additive inverse.

> Two numbers whose sum is 0 are called **opposites**, or **additive inverses,** of each other.

**Examples** Find the opposite of each number.

**14.** 34      The opposite of 34 is $-34$ because $34 + (-34) = 0$.

**15.** $-8$      The opposite of $-8$ is 8 because $-8 + 8 = 0$.

**16.** 0      The opposite of 0 is 0 because $0 + 0 = 0$.

**17.** $-\dfrac{7}{8}$      The opposite of $-\dfrac{7}{8}$ is $\dfrac{7}{8}$ because $-\dfrac{7}{8} + \dfrac{7}{8} = 0$.

*Do Exercises 25–30.*

To name the opposite, we use the symbol $-$, as follows.

> The opposite, or additive inverse, of a number $a$ can be named $-a$ (read "the opposite of $a$," or "the additive inverse of $a$").

Note that if we take a number, say, 8, and find its opposite, $-8$, and then find the opposite of the result, we will have the original number, 8, again.

Add.

**21.** $(-15) + (-37) + 25 + 42 + (-59) + (-14)$

**22.** $42 + (-81) + (-28) + 24 + 18 + (-31)$

**23.** $-2.5 + (-10) + 6 + (-7.5)$

**24.** $-35 + 17 + 14 + (-27) + 31 + (-12)$

Find the opposite.

**25.** $-4$

**26.** 8.7

**27.** $-7.74$

**28.** $-\dfrac{8}{9}$

**29.** 0

**30.** 12

*Answers on page A-4*

Find $-x$ and $-(-x)$ when $x$ is each of the following.

**31.** 14

**32.** 1

**33.** $-19$

**34.** $-1.6$

**35.** $\dfrac{2}{3}$

**36.** $-\dfrac{9}{8}$

Find the opposite. (Change the sign.)

**37.** $-4$

**38.** $-13.4$

**39.** 0

**40.** $\dfrac{1}{4}$

Answers on page A-4

> The opposite of the opposite of a number is the number itself. (The additive inverse of the additive inverse of a number is the number itself.) That is, for any number $a$,
>
> $$-(-a) = a.$$

**Example 18**   Find $-x$ and $-(-x)$ when $x = 16$.

If $x = 16$, then $-x = -16$.   **The opposite of 16 is $-16$.**

If $x = 16$, then $-(-x) = -(-16) = 16$.   **The opposite of the opposite of 16 is 16.**

**Example 19**   Find $-x$ and $-(-x)$ when $x = -3$.

If $x = -3$, then $-x = -(-3) = 3$.

If $x = -3$, then $-(-x) = -(-(-3)) = -3$.

Note that in Example 19 we used a second set of parentheses to show that we are substituting the negative number $-3$ for $x$. Symbolism like $--x$ is not considered meaningful.

*Do Exercises 31–36.*

A symbol such as $-8$ is usually read "negative 8." It could be read "the additive inverse of 8," because the additive inverse of 8 is negative 8. It could also be read "the opposite of 8," because the opposite of 8 is $-8$. Thus a symbol like $-8$ can be read in more than one way. A symbol like $-x$, which has a variable, should be read "the opposite of $x$" or "the additive inverse of $x$" and *not* "negative $x$," because we do not know whether $x$ represents a positive number, a negative number, or 0. You can check this in Examples 18 and 19.

We can use the symbolism $-a$ to restate the definition of opposite, or additive inverse.

> For any real number $a$, the **opposite**, or **additive inverse**, of $a$, which is $-a$, is such that
>
> $$a + (-a) = (-a) + a = 0.$$

## Signs of Numbers

A negative number is sometimes said to have a "negative sign." A positive number is said to have a "positive sign." When we replace a number with its opposite, we can say that we have "changed its sign."

**Examples**   Find the opposite. (Change the sign.)

**20.** $-3$ $\quad -(-3) = 3$ $\quad$ **The opposite of $-3$ is 3.**

**21.** $-10$ $\quad -(-10) = 10$

**22.** $0$ $\quad -(0) = 0$

**23.** $14$ $\quad -(14) = -14$

*Do Exercises 37–40.*

# Exercise Set 1.3

**a** Add. Do not use a number line except as a check.

**1.** $2 + (-9)$  **2.** $-5 + 2$  **3.** $-11 + 5$  **4.** $4 + (-3)$  **5.** $-6 + 6$

**6.** $8 + (-8)$  **7.** $-3 + (-5)$  **8.** $-4 + (-6)$  **9.** $-7 + 0$  **10.** $-13 + 0$

**11.** $0 + (-27)$  **12.** $0 + (-35)$  **13.** $17 + (-17)$  **14.** $-15 + 15$  **15.** $-17 + (-25)$

**16.** $-24 + (-17)$  **17.** $18 + (-18)$  **18.** $-13 + 13$  **19.** $-28 + 28$  **20.** $11 + (-11)$

**21.** $8 + (-5)$  **22.** $-7 + 8$  **23.** $-4 + (-5)$  **24.** $10 + (-12)$  **25.** $13 + (-6)$

**26.** $-3 + 14$  **27.** $-25 + 25$  **28.** $50 + (-50)$  **29.** $53 + (-18)$  **30.** $75 + (-45)$

**31.** $-8.5 + 4.7$  **32.** $-4.6 + 1.9$  **33.** $-2.8 + (-5.3)$  **34.** $-7.9 + (-6.5)$  **35.** $-\dfrac{3}{5} + \dfrac{2}{5}$

**36.** $-\dfrac{4}{3} + \dfrac{2}{3}$  **37.** $-\dfrac{2}{9} + \left(-\dfrac{5}{9}\right)$  **38.** $-\dfrac{4}{7} + \left(-\dfrac{6}{7}\right)$  **39.** $-\dfrac{5}{8} + \dfrac{1}{4}$  **40.** $-\dfrac{5}{6} + \dfrac{2}{3}$

**41.** $-\dfrac{5}{8} + \left(-\dfrac{1}{6}\right)$  **42.** $-\dfrac{5}{6} + \left(-\dfrac{2}{9}\right)$  **43.** $-\dfrac{3}{8} + \dfrac{5}{12}$  **44.** $-\dfrac{7}{16} + \dfrac{7}{8}$

**45.** $76 + (-15) + (-18) + (-6)$

**46.** $29 + (-45) + 18 + 32 + (-96)$

**47.** $-44 + \left(-\dfrac{3}{8}\right) + 95 + \left(-\dfrac{5}{8}\right)$

**48.** $24 + 3.1 + (-44) + (-8.2) + 63$

**49.** $98 + (-54) + 113 + (-998) + 44 + (-612)$

**50.** $-458 + (-124) + 1025 + (-917) + 218$

b  Find the opposite, or additive inverse.

**51.** $24$

**52.** $-64$

**53.** $-26.9$

**54.** $48.2$

Find $-x$ when $x$ is each of the following.

**55.** $8$

**56.** $-27$

**57.** $-\dfrac{13}{8}$

**58.** $\dfrac{1}{236}$

Find $-(-x)$ when $x$ is each of the following.

**59.** $-43$

**60.** $39$

**61.** $\dfrac{4}{3}$

**62.** $-7.1$

Find the opposite. (Change the sign.)

**63.** $-24$

**64.** $-12.3$

**65.** $-\dfrac{3}{8}$

**66.** $10$

---

**Skill Maintenance**

Convert to decimal notation.  [R.4a]

**67.** $57\%$

**68.** $49\%$

**69.** $52.9\%$

**70.** $71.3\%$

Convert to percent notation.  [R.4d]

**71.** $\dfrac{5}{4}$

**72.** $\dfrac{1}{8}$

**73.** $\dfrac{13}{25}$

**74.** $\dfrac{13}{32}$

---

**Synthesis**

**75.** ◆ Without actually performing the addition, explain why the sum of all integers from $-50$ to $50$ is $0$.

**76.** ◆ Explain in your own words why the sum of two negative numbers is always negative.

**77.** For what numbers $x$ is $-x$ negative?

**78.** For what numbers $x$ is $-x$ positive?

Add.

**79.** ▦ $-3496 + (-2987)$

**80.** ▦ $2708 + (-3749)$

Tell whether the sum is positive, negative, or zero.

**81.** If $a$ is positive and $b$ is negative, $-a + b$ is

_____ .

**82.** If $a = b$ and $a$ and $b$ are negative, $-a + (-b)$ is

_____ .

Add integers using a variety of methods.

# 1.4 Subtraction of Real Numbers

## a | Subtraction

We now consider subtraction of real numbers. Subtraction is defined as follows.

> The difference $a - b$ is the number that when added to $b$ gives $a$.

For example, $45 - 17 = 28$ because $28 + 17 = 45$. Let's consider an example whose answer is a negative number.

**Example 1** Subtract: $5 - 8$.

*Think*: $5 - 8$ is the number that when added to 8 gives 5. What number can we add to 8 to get 5? The number must be negative. The number is $-3$:

$$5 - 8 = -3.$$

That is, $5 - 8 = -3$ because $5 = -3 + 8$.

*Do Exercises 1–3.*

The definition above does *not* provide the most efficient way to do subtraction. From that definition, however, we can develop a faster way to subtract. Look for a pattern in the following examples.

| Subtractions | Adding an Opposite |
|---|---|
| $5 - 8 = -3$ | $5 + (-8) = -3$ |
| $-6 - 4 = -10$ | $-6 + (-4) = -10$ |
| $-7 - (-10) = 3$ | $-7 + 10 = 3$ |
| $-7 - (-2) = -5$ | $-7 + 2 = -5$ |

*Do Exercises 4–7.*

Perhaps you have noticed that we can subtract by adding the opposite of the number being subtracted. This can always be done.

> For any real numbers $a$ and $b$,
> $$a - b = a + (-b).$$
> (To subtract, add the opposite, or additive inverse, of the number being subtracted.)

This is the method generally used for quick subtraction of real numbers.

**Objectives**

a | Subtract real numbers and simplify combinations of additions and subtractions.

b | Solve applied problems involving addition and subtraction of real numbers.

**For Extra Help**

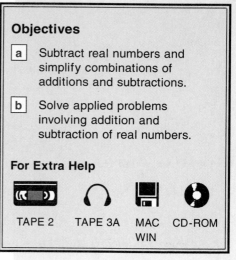

TAPE 2    TAPE 3A    MAC WIN    CD-ROM

Subtract.

1. $-6 - 4$
   *Think*: What number can be added to 4 to get $-6$?

2. $-7 - (-10)$
   *Think*: What number can be added to $-10$ to get $-7$?

3. $-7 - (-2)$
   *Think*: What number can be added to $-2$ to get $-7$?

Complete the addition and compare with the subtraction.

4. $4 - 6 = -2$;
   $4 + (-6) =$ _____

5. $-3 - 8 = -11$;
   $-3 + (-8) =$ _____

6. $-5 - (-9) = 4$;
   $-5 + 9 =$ _____

7. $-5 - (-3) = -2$;
   $-5 + 3 =$ _____

*Answers on page A-4*

Subtract.

**8.** $2 - 8$

**9.** $-6 - 10$

**10.** $12.4 - 5.3$

**11.** $-8 - (-11)$

**12.** $-8 - (-8)$

**13.** $\dfrac{2}{3} - \left(-\dfrac{5}{6}\right)$

Read each of the following. Then subtract by adding the opposite of the number being subtracted.

**14.** $3 - 11$

**15.** $12 - 5$

**16.** $-12 - (-9)$

**17.** $-12.4 - 10.9$

**18.** $-\dfrac{4}{5} - \left(-\dfrac{4}{5}\right)$

Simplify.

**19.** $-6 - (-2) - (-4) - 12 + 3$

**20.** $9 - (-6) + 7 - 11 - 14 - (-20)$

**21.** $-9.6 + 7.4 - (-3.9) - (-11)$

*Answers on page A-4*

**Examples**   Subtract.

**2.** $2 - 6 = 2 + (-6) = -4$   The opposite of 6 is −6. We change the subtraction to addition and add the opposite.

**3.** $4 - (-9) = 4 + 9 = 13$   The opposite of −9 is 9. We change the subtraction to addition and add the opposite.

**4.** $-4.2 - (-3.6) = -4.2 + 3.6 = -0.6$   Adding the opposite. *Check:* $-0.6 + (-3.6) = -4.2$.

**5.** $-\dfrac{1}{2} - \left(-\dfrac{3}{4}\right) = -\dfrac{1}{2} + \dfrac{3}{4} = \dfrac{1}{4}$   Adding the opposite. *Check:* $\frac{1}{4} + \left(-\frac{3}{4}\right) = -\frac{1}{2}$.

*Do Exercises 8–13.*

**Examples**   Read each of the following. Then subtract by adding the opposite of the number being subtracted.

**6.** $3 - 5;$
$3 - 5 = 3 + (-5) = -2$   Read "three minus five is three plus the opposite of five"

**7.** $\dfrac{1}{8} - \dfrac{7}{8};$
$\dfrac{1}{8} - \dfrac{7}{8} = \dfrac{1}{8} + \left(-\dfrac{7}{8}\right) = -\dfrac{6}{8},$ or $-\dfrac{3}{4}$   Read "one-eighth minus seven-eighths is one-eighth plus the opposite of seven-eighths"

**8.** $-4.6 - (-9.8);$
$-4.6 - (-9.8) = -4.6 + 9.8 = 5.2$   Read "negative four point six minus negative nine point eight is negative four point six plus the opposite of negative nine point eight"

**9.** $-\dfrac{3}{4} - \dfrac{7}{5};$
$-\dfrac{3}{4} - \dfrac{7}{5} = -\dfrac{3}{4} + \left(-\dfrac{7}{5}\right) = -\dfrac{15}{20} + \left(-\dfrac{28}{20}\right) = -\dfrac{43}{20}$   Read "negative three-fourths minus seven-fifths is negative three-fourths plus the opposite of seven-fifths"

*Do Exercises 14–18.*

When several additions and subtractions occur together, we can make them all additions.

**Examples**   Simplify.

**10.** $8 - (-4) - 2 - (-4) + 2 = 8 + 4 + (-2) + 4 + 2$   Adding the opposites where subtraction is indicated
$= 16$

**11.** $8.2 - (-6.1) + 2.3 - (-4) = 8.2 + 6.1 + 2.3 + 4$
$= 20.6$

*Do Exercises 19–21.*

## b | Applications and Problem Solving

Let's now see how we can use addition and subtraction of real numbers to solve applied problems.

**Example 12** *Home-Run Differential.* In baseball the difference between the number of home runs hit by a team's players and the number given up by its pitchers is called the *home-run differential*, that is,

$$\text{Home run differential} = \frac{\text{Number of}}{\text{home runs}} - \frac{\text{Number of home}}{\text{runs allowed}}.$$

Teams strive for a positive home-run differential.

**a)** In a recent year, Atlanta hit 197 home runs and gave up 120. Find its home-run differential.

**b)** In a recent year, San Francisco hit 153 home runs and gave up 194. Find its home-run differential.

We solve as follows.

**a)** We subtract 120 from 197 to find the home-run differential for Atlanta:

Home-run differential = 197 − 120 = 77.

**b)** We subtract 194 from 153 to find the home-run differential for San Francisco:

Home-run differential = 153 − 194 = −41.

*Do Exercises 22 and 23.*

---

**22.** *Home-Run Differential.* Complete the following table to find the home-run differentials for all the major-league baseball teams.

| National League | | | |
|---|---|---|---|
| | HRs | HRs allowed | Diff. |
| Atlanta | 197 | 120 | +77 |
| Florida | 150 | 113 | |
| Los Angeles | 150 | 125 | |
| Cincinnati | 191 | 167 | |
| Colorado | 221 | 198 | |
| San Diego | 147 | 138 | |
| Montreal | 148 | 152 | |
| Cubs | 175 | 184 | |
| Mets | 147 | 159 | |
| Houston | 129 | 154 | |
| Philadelphia | 132 | 160 | |
| St. Louis | 142 | 173 | |
| San Francisco | 153 | 194 | −41 |
| Pittsburgh | 138 | 183 | |

| American League | | | |
|---|---|---|---|
| | HRs | HRs allowed | Diff. |
| Texas | 221 | 168 | |
| Baltimore | 257 | 209 | |
| Cleveland | 218 | 173 | |
| Oakland | 243 | 205 | |
| Seattle | 245 | 216 | |
| Boston | 209 | 185 | |
| White Sox | 195 | 174 | |
| Yankees | 162 | 143 | |
| Toronto | 177 | 187 | |
| California | 192 | 219 | |
| Milwaukee | 178 | 213 | |
| Detroit | 204 | 241 | |
| Kansas City | 123 | 176 | |
| Minnesota | 118 | 233 | |

**23.** *Temperature Extremes.* In Churchill, Manitoba, Canada, the average daily low temperature in January is −31°C. The average daily low temperature in Key West, Florida, is 19°C. How much higher is the average daily low temperature in Key West, Florida?

*Answers on page A-4*

# Improving Your Math Study Skills

## Classwork: Before and During Class

### Before Class

### Textbook

- Check your syllabus (or ask your instructor) to find out which sections will be covered during the next class. Then be sure to read these sections *before* class. Although you may not understand all the concepts, you will at least be familiar with the material, which will help you follow the discussion during class.

- This book makes use of color, shading, and design elements to highlight important concepts, so you do not need to highlight these. Instead, it is more productive for you to note trouble spots with either a highlighter or Post-It notes. Then use these marked points as possible questions for clarification by your instructor at the appropriate time.

### Homework

- Review the previous day's homework just before class. This will refresh your memory on the concepts covered in the last class, and again provide you with possible questions to ask your instructor.

### During Class

### Class Seating

- If possible, choose a seat at the front of the class. In most classes, the more serious students tend to sit there so you will probably be able to concentrate better if you do the same. You should also avoid sitting next to noisy or distracting students.

- If your instructor uses an overhead projector, choose a seat that will give you an unobstructed view of the screen.

### Taking Notes

- This textbook has been written and laid out so that it represents a quality set of notes at the same time that it teaches. Thus you might not need to take notes in class. Just watch, listen, and ask yourself questions as the class moves along, rather than racing to keep up your note-taking.

  However, if you still feel more comfortable taking your own notes, consider using the following two-column method. Divide your page in half vertically so that you have two columns side by side. Write down what is on the board in the left column; then, in the right column, write clarifying comments or questions.

- If you have any difficulty keeping up with the instructor, use abbreviations to speed up your note-taking. Consider standard abbreviations like "Ex" for "Example," "$\approx$" for "approximately equal to," or "$\therefore$" for "therefore." Create your own abbreviations as well.

- Another shortcut for note-taking is to write only the beginning of a word, leaving space for the rest. Be sure you write enough of the word to know what it means later on!

# Exercise Set 1.4

**a** Subtract.

**1.** $2 - 9$

**2.** $3 - 8$

**3.** $0 - 4$

**4.** $0 - 9$

**5.** $-8 - (-2)$

**6.** $-6 - (-8)$

**7.** $-11 - (-11)$

**8.** $-6 - (-6)$

**9.** $12 - 16$

**10.** $14 - 19$

**11.** $20 - 27$

**12.** $30 - 4$

**13.** $-9 - (-3)$

**14.** $-7 - (-9)$

**15.** $-40 - (-40)$

**16.** $-9 - (-9)$

**17.** $7 - 7$

**18.** $9 - 9$

**19.** $7 - (-7)$

**20.** $4 - (-4)$

**21.** $8 - (-3)$

**22.** $-7 - 4$

**23.** $-6 - 8$

**24.** $6 - (-10)$

**25.** $-4 - (-9)$

**26.** $-14 - 2$

**27.** $1 - 8$

**28.** $2 - 8$

**29.** $-6 - (-5)$

**30.** $-4 - (-3)$

**31.** $8 - (-10)$

**32.** $5 - (-6)$

**33.** $0 - 10$

**34.** $0 - 18$

**35.** $-5 - (-2)$

**36.** $-3 - (-1)$

**37.** $-7 - 14$

**38.** $-9 - 16$

**39.** $0 - (-5)$

**40.** $0 - (-1)$

**41.** $-8 - 0$

**42.** $-9 - 0$

**43.** $7 - (-5)$

**44.** $7 - (-4)$

**45.** $2 - 25$

**46.** $18 - 63$

**47.** $-42 - 26$

**48.** $-18 - 63$

**49.** $-71 - 2$

**50.** $-49 - 3$

**51.** $24 - (-92)$

**52.** $48 - (-73)$

**53.** $-50 - (-50)$

**54.** $-70 - (-70)$

**55.** $-\dfrac{3}{8} - \dfrac{5}{8}$

**56.** $\dfrac{3}{9} - \dfrac{9}{9}$

**57.** $\dfrac{3}{4} - \dfrac{2}{3}$

**58.** $\dfrac{5}{8} - \dfrac{3}{4}$

**59.** $-\dfrac{3}{4} - \dfrac{2}{3}$

**60.** $-\dfrac{5}{8} - \dfrac{3}{4}$

**61.** $-\dfrac{5}{8} - \left(-\dfrac{3}{4}\right)$

**62.** $-\dfrac{3}{4} - \left(-\dfrac{2}{3}\right)$

**63.** $6.1 - (-13.8)$

**64.** $1.5 - (-3.5)$

**65.** $-2.7 - 5.9$

**66.** $-3.2 - 5.8$

**67.** $0.99 - 1$

**68.** $0.87 - 1$

**69.** $-79 - 114$

**70.** $-197 - 216$

**71.** $0 - (-500)$

**72.** $500 - (-1000)$

**73.** $-2.8 - 0$

**74.** $6.04 - 1.1$

**75.** $7 - 10.53$

**76.** $8 - (-9.3)$

**77.** $\dfrac{1}{6} - \dfrac{2}{3}$

**78.** $-\dfrac{3}{8} - \left(-\dfrac{1}{2}\right)$

**79.** $-\dfrac{4}{7} - \left(-\dfrac{10}{7}\right)$

**80.** $\dfrac{12}{5} - \dfrac{12}{5}$

**81.** $-\dfrac{7}{10} - \dfrac{10}{15}$

**82.** $-\dfrac{4}{18} - \left(-\dfrac{2}{9}\right)$

**83.** $\dfrac{1}{5} - \dfrac{1}{3}$

**84.** $-\dfrac{1}{7} - \left(-\dfrac{1}{6}\right)$

Simplify.

**85.** $18 - (-15) - 3 - (-5) + 2$

**86.** $22 - (-18) + 7 + (-42) - 27$

**87.** $-31 + (-28) - (-14) - 17$

**88.** $-43 - (-19) - (-21) + 25$

**89.** $-34 - 28 + (-33) - 44$

**90.** $39 + (-88) - 29 - (-83)$

**91.** $-93 - (-84) - 41 - (-56)$

**92.** $84 + (-99) + 44 - (-18) - 43$

**93.** $-5 - (-30) + 30 + 40 - (-12)$

**94.** $14 - (-50) + 20 - (-32)$

**95.** $132 - (-21) + 45 - (-21)$

**96.** $81 - (-20) - 14 - (-50) + 53$

**b** Solve.

**97.** *Ocean Depth.* The deepest point in the Pacific Ocean is the Marianas Trench, with a depth of 11,033 m. The deepest point in the Atlantic Ocean is the Puerto Rico Trench, with a depth of 8648 m. What is the difference in the elevation of the two trenches?

**98.** *Depth of Offshore Oil Wells.* In 1993, the elevation of the world's deepest offshore oil well was −2860 ft. By 1998, the deepest well is expected to be 360 ft deeper. (***Source:*** *New York Times,* 12/7/94, p. D1.) What will be the elevation of the deepest well in 1998?

Marianas Trench

Puerto Rico Trench

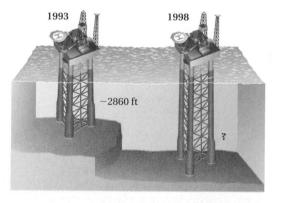

**99.** Laura has a charge of $476.89 on her credit card, but she then returns a sweater that cost $128.95. How much does she now owe on her credit card?

**100.** Chris has $720 in a checking account. He writes a check for $970 to pay for a sound system. What is the balance in his checking account?

**101.** *Temperature Records.* The greatest recorded temperature change in one day occurred in Browning, Montana, where the temperature fell from 44°F to −56°F (**Source:** *The Guinness Book of Records*). How much did the temperature drop?

**102.** *Low Points on Continents.* The lowest point in Africa is Lake Assal, which is 515 ft below sea level. The lowest point in South America is the Valdes Peninsula, which is 132 ft below sea level. How much lower is Lake Assal than the Valdes Peninsula?

$$= -132 - ^-515$$
$$-132 + 515 = \boxed{383}$$ lower

$$\begin{array}{r} 515 \\ -132 \\ \hline 383 \end{array}$$

---

**Skill Maintenance**

**103.** Evaluate: $5^3$.  [R.5b]

**104.** Find the prime factorization of 864.  [R.1a]

**105.** Simplify: $256 \div 64 \div 2^3 + 100$.  [R.5c]

**106.** Simplify: $5 \cdot 6 + (7 \cdot 2)^2$.  [R.5c]

**107.** Convert to decimal notation: 58.3%.  [R.4a]

**108.** Simplify: $\dfrac{164}{256}$.  [R.2b]

---

**SYNTHESIS**

**109.** ◆ If a negative number is subtracted from a positive number, will the result always be positive? Why or why not?

**110.** ◆ Write a problem for a classmate to solve. Design the problem so that the solution is "The temperature dropped to −9°."

Subtract.

**111.** ▦ 123,907 − 433,789

**112.** ▦ 23,011 − (−60,432)

Tell whether the statement is true or false for all integers $a$ and $b$. If false, show why.

**113.** $a - 0 = 0 - a$

**114.** $0 - a = a$

**115.** If $a \neq b$, then $a - b \neq 0$.

**116.** If $a = -b$, then $a + b = 0$.

**117.** If $a + b = 0$, then $a$ and $b$ are opposites.

**118.** If $a - b = 0$, then $a = -b$.

**119.** Maureen is a stockbroker. She kept track of the changes in the stock market over a period of 5 weeks. By how many points had the market risen or fallen over this time?

| Week 1 | Week 2 | Week 3 | Week 4 | Week 5 |
|--------|--------|--------|--------|--------|
| Down 13 pts | Down 16 pts | Up 36 pts | Down 11 pts | Up 19 pts |

**120.** *Blackjack Counting System.* The casino game of blackjack makes use of many card-counting systems to give players a winning edge if the count becomes negative. One such system is called *High–Low*, first developed by Harvey Dubner in 1963. Each card counts as −1, 0, or 1 as follows:

2, 3, 4, 5, 6      count as +1;
7, 8, 9      count as 0;
10, J, Q, K, A      count as −1.

(**Source:** Patterson, Jerry L., *Casino Gambling.* New York: Perigee, 1982)

**a)** Find the final count on the sequence of cards

     K, A, 2, 4, 5, 10, J, 8, Q, K, 5.

**b)** Does the player have a winning edge?

Subtract integers using tiles.

# 1.5 Multiplication of Real Numbers

## a  Multiplication

Multiplication of real numbers is very much like multiplication of arithmetic numbers. The only difference is that we must determine whether the answer is positive or negative.

### Multiplication of a Positive Number and a Negative Number

To see how to multiply a positive number and a negative number, consider the pattern of the following.

This number decreases by 1 each time.

$$4 \cdot 5 = 20$$
$$3 \cdot 5 = 15$$
$$2 \cdot 5 = 10$$
$$1 \cdot 5 = 5$$
$$0 \cdot 5 = 0$$
$$-1 \cdot 5 = -5$$
$$-2 \cdot 5 = -10$$
$$-3 \cdot 5 = -15$$

This number decreases by 5 each time.

*Do Exercise 1.*

According to this pattern, it looks as though the product of a negative number and a positive number is negative. That is the case, and we have the first part of the rule for multiplying numbers.

> To multiply a positive number and a negative number, multiply their absolute values. The answer is negative.

**Examples**  Multiply.

**1.** $8(-5) = -40$   **2.** $-\dfrac{1}{3} \cdot \dfrac{5}{7} = -\dfrac{5}{21}$   **3.** $(-7.2)5 = -36$

*Do Exercises 2–7.*

### Multiplication of Two Negative Numbers

How do we multiply two negative numbers? Again, we look for a pattern.

This number decreases by 1 each time.

$$4 \cdot (-5) = -20$$
$$3 \cdot (-5) = -15$$
$$2 \cdot (-5) = -10$$
$$1 \cdot (-5) = -5$$
$$0 \cdot (-5) = 0$$
$$-1 \cdot (-5) = 5$$
$$-2 \cdot (-5) = 10$$
$$-3 \cdot (-5) = 15$$

This number increases by 5 each time.

*Do Exercise 8.*

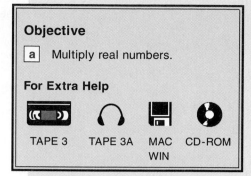

**Objective**

a  Multiply real numbers.

**For Extra Help**

TAPE 3    TAPE 3A    MAC WIN    CD-ROM

**1.** Complete, as in the example.

$$4 \cdot 10 = 40$$
$$3 \cdot 10 = 30$$
$$2 \cdot 10 =$$
$$1 \cdot 10 =$$
$$0 \cdot 10 =$$
$$-1 \cdot 10 =$$
$$-2 \cdot 10 =$$
$$-3 \cdot 10 =$$

Multiply.

**2.** $-3 \cdot 6$

**3.** $20 \cdot (-5)$

**4.** $4 \cdot (-20)$

**5.** $-\dfrac{2}{3} \cdot \dfrac{5}{6}$

**6.** $-4.23(7.1)$

**7.** $\dfrac{7}{8}\left(-\dfrac{4}{5}\right)$

**8.** Complete, as in the example.

$$3 \cdot (-10) = -30$$
$$2 \cdot (-10) = -20$$
$$1 \cdot (-10) =$$
$$0 \cdot (-10) =$$
$$-1 \cdot (-10) =$$
$$-2 \cdot (-10) =$$
$$-3 \cdot (-10) =$$

*Answers on page A-4*

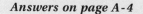

**Multiply.**

**9.** $-9 \cdot (-3)$

**10.** $-16 \cdot (-2)$

**11.** $-7 \cdot (-5)$

**12.** $-\dfrac{4}{7}\left(-\dfrac{5}{9}\right)$

**13.** $-\dfrac{3}{2}\left(-\dfrac{4}{9}\right)$

**14.** $-3.25(-4.14)$

**Multiply.**

**15.** $5(-6)$

**16.** $(-5)(-6)$

**17.** $(-3.2) \cdot 0$

**18.** $\left(-\dfrac{4}{5}\right)\left(\dfrac{10}{3}\right)$

*Answers on page A-4*

According to the pattern, it appears that the product of two negative numbers is positive. That is actually so, and we have the second part of the rule for multiplying real numbers.

> To multiply two negative numbers, multiply their absolute values. The answer is positive.

### Do Exercises 9–14.

The following is another way to consider the rules we have for multiplication.

> To multiply two real numbers:
>
> **a)** Multiply the absolute values.
> **b)** If the signs are the same, the answer is positive.
> **c)** If the signs are different, the answer is negative.

## Multiplication by Zero

The only case that we have not considered is multiplying by zero. As with other numbers, the product of any real number and 0 is 0.

> **THE MULTIPLICATION PROPERTY OF ZERO**
> For any real number $a$,
> $$a \cdot 0 = 0.$$
> (The product of 0 and any real number is 0.)

**Examples** Multiply.

**4.** $(-3)(-4) = 12$

**5.** $-1.6(2) = -3.2$

**6.** $-19 \cdot 0 = 0$

**7.** $\left(-\dfrac{5}{6}\right)\left(-\dfrac{1}{9}\right) = \dfrac{5}{54}$

### Do Exercises 15–18.

## Multiplying More Than Two Numbers

When multiplying more than two real numbers, we can choose order and grouping as we please.

**Examples** Multiply.

**8.** $-8 \cdot 2(-3) = -16(-3)$     **Multiplying the first two numbers**
$$= 48$$

**9.** $-8 \cdot 2(-3) = 24 \cdot 2$     **Multiplying the negatives. Every pair of negative numbers gives a positive product.**
$$= 48$$

**10.** $-3(-2)(-5)(4) = 6(-5)(4)$     **Multiplying the first two numbers**
$$= (-30)4$$
$$= -120$$

**11.** $\left(-\dfrac{1}{2}\right)(8)\left(-\dfrac{2}{3}\right)(-6) = (-4)4$     **Multiplying the first two numbers and the last two numbers**
$$= -16$$

**12.** $-5 \cdot (-2) \cdot (-3) \cdot (-6) = 10 \cdot 18$
$$= 180$$

**13.** $(-3)(-5)(-2)(-3)(-6) = (-30)(18)$
$$= -540$$

We can see the following pattern in the results of Examples 12 and 13.

> The product of an even number of negative numbers is positive.
> The product of an odd number of negative numbers is negative.

*Do Exercises 19–24.*

Let's compare the expressions $(-x)^2$ and $-x^2$.

**Example 14** Evaluate $(-x)^2$ and $-x^2$ for $x = 5$.

$(-x)^2 = (-5)^2 = (-5)(-5) = 25$;     **Substitute 5 for $x$. Then evaluate the power.**

$-x^2 = -(5)^2 = -25$     **Substitute 5 for $x$. Evaluate the power. Then find the opposite.**

The expressions $(-x)^2$ and $-x^2$ are *not* equivalent. That is, they do not have the same value for every allowable replacement of the variable by a real number. To find $(-x)^2$, we take the opposite and then square. To find $-x^2$, we find the square and then take the opposite.

**Example 15** Evaluate $2x^2$ for $x = 3$ and $x = -3$.

$2x^2 = 2(3)^2 = 2(9) = 18$;

$2x^2 = 2(-3)^2 = 2(9) = 18$

*Do Exercises 25–27.*

---

Multiply.

**19.** $5 \cdot (-3) \cdot 2$

**20.** $-3 \times (-4.1) \times (-2.5)$

**21.** $-\dfrac{1}{2} \cdot \left(-\dfrac{4}{3}\right) \cdot \left(-\dfrac{5}{2}\right)$

**22.** $-2 \cdot (-5) \cdot (-4) \cdot (-3)$

**23.** $(-4)(-5)(-2)(-3)(-1)$

**24.** $(-1)(-1)(-2)(-3)(-1)(-1)$

**25.** Evaluate $(-x)^2$ and $-x^2$ for $x = 2$.

**26.** Evaluate $(-x)^2$ and $-x^2$ for $x = 3$.

**27.** Evaluate $3x^2$ for $x = 4$ and for $x = -4$.

*Answers on page A-4*

# Improving Your Math Study Skills

## Studying for Tests and Making the Most of Tutoring Sessions

This math study skill feature focuses on the very important task of test preparation.

### *Test-Taking Tips*

- **Make up your own test questions as you study.** You have probably become accustomed by now to the section and objective codes that appear throughout the book. After you have done your homework over a particular objective, write one or two questions on your own that you think might be on a test. You will be amazed at the insight this will provide. You are actually carrying out a task similar to what a teacher does in preparing an exam.

- **Do an overall review of the chapter focusing on the objectives and the examples.** This should be accompanied by a study of any class notes you may have taken.

- **Do the review exercises at the end of the chapter.** Check your answers at the back of the book. If you have trouble with an exercise, use the objective symbol as a guide to go back and do further study of that objective. These review exercises are very much like a sample test.

- **Do the chapter test at the end of the chapter.** This is like taking a second sample test. Check the answers and objective symbols at the back of the book.

- **Ask former students for old exams.** Working such exams can be very helpful and allows you to see what various professors think is important.

- **When taking a test, read each question carefully and try to do all the questions the first time through, but pace yourself.** Answer all the questions, and mark those to recheck if you have time at the end. Very often, your first hunch will be correct.

- **Try to write your test in a neat and orderly manner.** Very often, your instructor tries to give you partial credit when grading an exam. If your test paper is sloppy and disorderly, it is difficult to verify the partial credit. Doing your work neatly can ease such a task for the instructor. Try using an erasable pen to make your writing darker and therefore more readable.

- **What about the student who says, "I could do the work at home, but on the test I made silly mistakes"?** Yes, all of us, including instructors, make silly computational mistakes in class, on homework, and on tests. But your instructor, if he or she has taught for some time, is probably aware that 90% of students who make such comments in truth do not have the required depth of knowledge of the subject matter, and such silly mistakes often are a sign that the student has not mastered the material. There is no way we can make that analysis for you. It will have to be unraveled by some careful soul searching on your part or by a conference with your instructor.

### *Making the Most of Tutoring and Help Sessions*

Often you will determine that a tutoring session would be helpful. The following comments may help you to make the most of such sessions.

- **Work on the topics before you go to the help or tutoring session. Do not go to such sessions viewing yourself as an empty cup and the tutor as a magician who will pour in the learning.** The primary source of your ability to learn is within you. We have seen so many students over the years go to help or tutoring sessions with no advanced preparation. You are often wasting your time and perhaps your money if you are paying for such sessions. Go to class, study the textbook, and mark trouble spots. Then use the help and tutoring sessions to deal with these difficulties most efficiently.

- **Do not be afraid to ask questions in these sessions!** The more you talk to your tutor, the more the tutor can help you with your difficulties.

- **Try being a "tutor" yourself.** Explaining a topic to someone else—a classmate, your instructor—is often the best way to learn it.

# Exercise Set 1.5

**a** Multiply.

**1.** $-4 \cdot 2$

**2.** $-3 \cdot 5$

**3.** $-8 \cdot 6$

**4.** $-5 \cdot 2$

**5.** $8 \cdot (-3)$

**6.** $9 \cdot (-5)$

**7.** $-9 \cdot 8$

**8.** $-10 \cdot 3$

**9.** $-8 \cdot (-2)$

**10.** $-2 \cdot (-5)$

**11.** $-7 \cdot (-6)$

**12.** $-9 \cdot (-2)$

**13.** $15 \cdot (-8)$

**14.** $-12 \cdot (-10)$

**15.** $-14 \cdot 17$

**16.** $-13 \cdot (-15)$

**17.** $-25 \cdot (-48)$

**18.** $39 \cdot (-43)$

**19.** $-3.5 \cdot (-28)$

**20.** $97 \cdot (-2.1)$

**21.** $9 \cdot (-8)$

**22.** $7 \cdot (-9)$

**23.** $4 \cdot (-3.1)$

**24.** $3 \cdot (-2.2)$

**25.** $-5 \cdot (-6)$

**26.** $-6 \cdot (-4)$

**27.** $-7 \cdot (-3.1)$

**28.** $-4 \cdot (-3.2)$

**29.** $\dfrac{2}{3} \cdot \left(-\dfrac{3}{5}\right)$

**30.** $\dfrac{5}{7} \cdot \left(-\dfrac{2}{3}\right)$

**31.** $-\dfrac{3}{8} \cdot \left(-\dfrac{2}{9}\right)$

**32.** $-\dfrac{5}{8} \cdot \left(-\dfrac{2}{5}\right)$

**33.** $-6.3 \times 2.7$

**34.** $-4.1 \times 9.5$

**35.** $-\dfrac{5}{9} \cdot \dfrac{3}{4}$

**36.** $-\dfrac{8}{3} \cdot \dfrac{9}{4}$

**37.** $7 \cdot (-4) \cdot (-3) \cdot 5$

**38.** $9 \cdot (-2) \cdot (-6) \cdot 7$

**39.** $-\dfrac{2}{3} \cdot \dfrac{1}{2} \cdot \left(-\dfrac{6}{7}\right)$

**40.** $-\dfrac{1}{8} \cdot \left(-\dfrac{1}{4}\right) \cdot \left(-\dfrac{3}{5}\right)$

**41.** $-3 \cdot (-4) \cdot (-5)$

**42.** $-2 \cdot (-5) \cdot (-7)$

**43.** $-2 \cdot (-5) \cdot (-3) \cdot (-5)$

**44.** $-3 \cdot (-5) \cdot (-2) \cdot (-1)$

**45.** $\dfrac{1}{5}\left(-\dfrac{2}{9}\right)$

**46.** $-\dfrac{3}{5}\left(-\dfrac{2}{7}\right)$

**47.** $-7 \cdot (-21) \cdot 13$

**48.** $-14 \cdot (34) \cdot 12$

**49.** $-4 \cdot (-1.8) \cdot 7$

**50.** $-8 \cdot (-1.3) \cdot (-5)$

**51.** $-\dfrac{1}{9}\left(-\dfrac{2}{3}\right)\left(\dfrac{5}{7}\right)$

**52.** $-\dfrac{7}{2}\left(-\dfrac{5}{7}\right)\left(-\dfrac{2}{5}\right)$

**53.** $4 \cdot (-4) \cdot (-5) \cdot (-12)$

**54.** $-2 \cdot (-3) \cdot (-4) \cdot (-5)$

**55.** $0.07 \cdot (-7) \cdot 6 \cdot (-6)$

**56.** $80 \cdot (-0.8) \cdot (-90) \cdot (-0.09)$

**57.** $\left(-\dfrac{5}{6}\right)\left(\dfrac{1}{8}\right)\left(-\dfrac{3}{7}\right)\left(-\dfrac{1}{7}\right)$

**58.** $\left(\dfrac{4}{5}\right)\left(-\dfrac{2}{3}\right)\left(-\dfrac{15}{7}\right)\left(\dfrac{1}{2}\right)$

**59.** $(-14) \cdot (-27) \cdot 0$

**60.** $7 \cdot (-6) \cdot 5 \cdot (-4) \cdot 3 \cdot (-2) \cdot 1 \cdot 0$

**61.** $(-8)(-9)(-10)$

**62.** $(-7)(-8)(-9)(-10)$

**63.** $(-6)(-7)(-8)(-9)(-10)$

**64.** $(-5)(-6)(-7)(-8)(-9)(-10)$

**65.** Evaluate $(-3x)^2$ and $-3x^2$ for $x = 7$.

**66.** Evaluate $(-2x)^2$ and $-2x^2$ for $x = 3$.

**67.** Evaluate $5x^2$ for $x = 2$ and for $x = -2$.

**68.** Evaluate $2x^2$ for $x = 5$ and for $x = -5$.

---

**Skill Maintenance**

**69.** Find the LCM of 36 and 60.   [R.1b]

**70.** Find the prime factorization of 4608.   [R.1a]

Simplify.   [R.2b]

**71.** $\dfrac{26}{39}$

**72.** $\dfrac{48}{54}$

**73.** $\dfrac{264}{484}$

**74.** $\dfrac{1025}{6625}$

---

**Synthesis**

**75.** ◈ Multiplication can be thought of as repeated addition. Using this concept and a number line, explain why $3 \cdot (-5) = -15$.

**76.** ◈ What rule have we developed that would tell you the sign of $(-7)^8$ and $(-7)^{11}$ without doing the computations? Explain.

**77.** After diving 95 m below the surface, a diver rises at a rate of 7 meters per minute for 9 min. What is the diver's new elevation?

**78.** Jo wrote seven checks for $13 each. If she began with a balance of $68 in her account, what was her balance after having written the checks?

**79.** What must be true of $a$ and $b$ if $-ab$ is to be (a) positive? (b) zero? (c) negative?

**80.** Evaluate $-6(3x - 5y) + z$ for $x = -2$, $y = -4$, and $z = 5$.

Copyright © 1999 Addison Wesley Longman

# 1.6 Division of Real Numbers

We now consider division of real numbers. The definition of division results in rules for division that are the same as those for multiplication.

## a | Division of Integers

> The quotient $\dfrac{a}{b}$ (or $a \div b$) is the number, if there is one, that when multiplied by $b$ gives $a$.

Let's use the definition to divide integers.

**Examples** Divide, if possible. Check your answer.

**1.** $14 \div (-7) = -2$     *Think*: What number multiplied by $-7$ gives 14? That number is $-2$. *Check*: $(-2)(-7) = 14$.

**2.** $\dfrac{-32}{-4} = 8$     *Think*: What number multiplied by $-4$ gives $-32$? That number is 8. *Check*: $8(-4) = -32$.

**3.** $\dfrac{-10}{7} = -\dfrac{10}{7}$     *Think*: What number multiplied by 7 gives $-10$? That number is $-\frac{10}{7}$. *Check*: $-\frac{10}{7} \cdot 7 = -10$.

**4.** $\dfrac{-17}{0}$ is **undefined**.     *Think*: What number multiplied by 0 gives $-17$? There is no such number because the product of 0 and *any* number is 0.

The rules for division are the same as those for multiplication.

> To multiply or divide two real numbers:
>
> **a)** Multiply or divide the absolute values.
>
> **b)** If the signs are the same, the answer is positive.
>
> **c)** If the signs are different, the answer is negative.

*Do Exercises 1–6.*

### Division by Zero

Example 4 shows why we cannot divide $-17$ by 0. We can use the same argument to show why we cannot divide any nonzero number $b$ by 0. Consider $b \div 0$. We look for a number that when multiplied by 0 gives $b$. There is no such number because the product of 0 and any number is 0. Thus we cannot divide a nonzero number $b$ by 0.

On the other hand, if we divide 0 by 0, we look for a number $r$ such that $0 \cdot r = 0$. But $0 \cdot r = 0$ for any number $r$. Thus it appears that $0 \div 0$ could be any number we choose. Getting any answer we want when we divide 0 by 0 would be very confusing. Thus we agree that division by zero is undefined.

> Division by 0 is undefined.
>
> $a \div 0$ is undefined for all real numbers $a$.
>
> 0 divided by a nonzero number $a$ is 0.
>
> $0 \div a = 0, \quad a \neq 0$.

## Objectives

**a** Divide integers.

**b** Find the reciprocal of a real number.

**c** Divide real numbers.

### For Extra Help

TAPE 3    TAPE 3B    MAC WIN    CD-ROM

Divide.

**1.** $6 \div (-3)$

*Think*: What number multiplied by $-3$ gives 6?

**2.** $\dfrac{-15}{-3}$

*Think*: What number multiplied by $-3$ gives $-15$?

**3.** $-24 \div 8$

*Think*: What number multiplied by 8 gives $-24$?

**4.** $\dfrac{-48}{-6}$

**5.** $\dfrac{30}{-5}$

**6.** $\dfrac{30}{-7}$

*Answers on page A-4*

Divide, if possible.

**7.** $\dfrac{-5}{0}$

**8.** $\dfrac{0}{-3}$

Find the reciprocal.

**9.** $\dfrac{2}{3}$

**10.** $-\dfrac{5}{4}$

**11.** $-3$

**12.** $-\dfrac{1}{5}$

**13.** $1.6$

**14.** $\dfrac{1}{2/3}$

*Answers on page A-4*

---

For example, $\frac{0}{4} = 0$, $\frac{4}{0}$ is undefined, and $\frac{0}{0}$ is undefined.

*Do Exercises 7 and 8.*

##  Reciprocals

When two numbers like $\frac{1}{2}$ and 2 are multiplied, the result is 1. Such numbers are called **reciprocals** of each other. Every nonzero real number has a reciprocal, also called a **multiplicative inverse.**

> Two numbers whose product is 1 are called **reciprocals,** or **multiplicative inverses,** of each other.

**Examples**   Find the reciprocal.

**5.** $\dfrac{7}{8}$   The reciprocal of $\dfrac{7}{8}$ is $\dfrac{8}{7}$ because $\dfrac{7}{8} \cdot \dfrac{8}{7} = 1.$

**6.** $-5$   The reciprocal of $-5$ is $-\dfrac{1}{5}$ because $-5\left(-\dfrac{1}{5}\right) = 1.$

**7.** $3.9$   The reciprocal of $3.9$ is $\dfrac{1}{3.9}$ because $3.9\left(\dfrac{1}{3.9}\right) = 1.$

**8.** $-\dfrac{1}{2}$   The reciprocal of $-\dfrac{1}{2}$ is $-2$ because $\left(-\dfrac{1}{2}\right)(-2) = 1.$

**9.** $-\dfrac{2}{3}$   The reciprocal of $-\dfrac{2}{3}$ is $-\dfrac{3}{2}$ because $\left(-\dfrac{2}{3}\right)\left(-\dfrac{3}{2}\right) = 1.$

**10.** $\dfrac{1}{3/4}$   The reciprocal of $\dfrac{1}{3/4}$ is $\dfrac{3}{4}$ because $\left(\dfrac{1}{3/4}\right)\left(\dfrac{3}{4}\right) = 1.$

> For $a \neq 0$, the reciprocal of $a$ can be named $\dfrac{1}{a}$ and the reciprocal of $\dfrac{1}{a}$ is $a$.
>
> The reciprocal of a nonzero number $\dfrac{a}{b}$ can be named $\dfrac{b}{a}$.
>
> The number 0 has no reciprocal.

*Do Exercises 9–14.*

The reciprocal of a positive number is also a positive number, because their product must be the positive number 1. The reciprocal of a negative number is also a negative number, because their product must be the positive number 1.

> The reciprocal of a number has the same sign as the number itself.

---

***CAUTION!***   It is important *not* to confuse *opposite* with *reciprocal*. Keep in mind that the opposite, or additive inverse, of a number is what we add to the number to get 0. The reciprocal, or multiplicative inverse, is what we multiply the number by to get 1.

Compare the following.

| Number | Opposite (Change the Sign.) | Reciprocal (Invert But Do Not Change the Sign.) |
|---|---|---|
| $-\dfrac{3}{8}$ | $\dfrac{3}{8}$ | $-\dfrac{8}{3}$ |
| $19$ | $-19$ | $\dfrac{1}{19}$ |
| $\dfrac{18}{7}$ | $-\dfrac{18}{7}$ | $\dfrac{7}{18}$ |
| $-7.9$ | $7.9$ | $-\dfrac{1}{7.9}$, or $-\dfrac{10}{79}$ |
| $0$ | $0$ | Undefined |

$\left(-\dfrac{3}{8}\right)\left(-\dfrac{8}{3}\right) = 1$

$-\dfrac{3}{8} + \dfrac{3}{8} = 0$

*Do Exercise 15.*

## c │ Division of Real Numbers

We know that we can subtract by adding an opposite. Similarly, we can divide by multiplying by a reciprocal.

 For any real numbers $a$ and $b$, $b \neq 0$,

$$a \div b = \frac{a}{b} = a \cdot \frac{1}{b}.$$

(To divide, we can multiply by the reciprocal of the divisor.)

**Examples** Rewrite the division as a multiplication.

**11.** $-4 \div 3$     $-4 \div 3$ is the same as $-4 \cdot \dfrac{1}{3}$

**12.** $\dfrac{6}{-7}$     $\dfrac{6}{-7} = 6\left(-\dfrac{1}{7}\right)$

**13.** $\dfrac{x+2}{5}$     $\dfrac{x+2}{5} = (x+2)\dfrac{1}{5}$    **Parentheses are necessary here.**

**14.** $\dfrac{-17}{1/b}$     $\dfrac{-17}{1/b} = -17 \cdot b$

**15.** $\dfrac{3}{5} \div \left(-\dfrac{9}{7}\right)$     $\dfrac{3}{5} \div \left(-\dfrac{9}{7}\right) = \dfrac{3}{5}\left(-\dfrac{7}{9}\right)$

*Do Exercises 16–20.*

When actually doing division calculations, we sometimes multiply by a reciprocal and we sometimes divide directly. With fractional notation, it is usually better to multiply by a reciprocal. With decimal notation, it is usually better to divide directly.

**15.** Complete the following table.

| Number | Opposite | Reciprocal |
|---|---|---|
| $\dfrac{2}{3}$ | | |
| $-\dfrac{5}{4}$ | | |
| $0$ | | |
| $1$ | | |
| $-8$ | | |
| $-4.5$ | | |

Rewrite the division as a multiplication.

**16.** $\dfrac{4}{7} \div \left(-\dfrac{3}{5}\right)$

**17.** $\dfrac{5}{-8}$

**18.** $\dfrac{a-b}{7}$

**19.** $\dfrac{-23}{1/a}$

**20.** $-5 \div 7$

*Answers on page A-4*

Divide by multiplying by the reciprocal of the divisor.

**21.** $\dfrac{4}{7} \div \left(-\dfrac{3}{5}\right)$

**22.** $-\dfrac{8}{5} \div \dfrac{2}{3}$

**23.** $-\dfrac{12}{7} \div \left(-\dfrac{3}{4}\right)$

**24.** Divide: $21.7 \div (-3.1)$.

Find two equal expressions for the number with negative signs in different places.

**25.** $\dfrac{-5}{6}$

**26.** $-\dfrac{8}{7}$

**27.** $\dfrac{10}{-3}$

*Answers on page A-4*

**Examples**   Divide by multiplying by the reciprocal of the divisor.

**16.** $\dfrac{2}{3} \div \left(-\dfrac{5}{4}\right) = \dfrac{2}{3} \cdot \left(-\dfrac{4}{5}\right) = -\dfrac{8}{15}$

**17.** $-\dfrac{5}{6} \div \left(-\dfrac{3}{4}\right) = -\dfrac{5}{6} \cdot \left(-\dfrac{4}{3}\right) = \dfrac{20}{18} = \dfrac{10 \cdot 2}{9 \cdot 2} = \dfrac{10}{9} \cdot \dfrac{2}{2} = \dfrac{10}{9}$

> *Caution!*   Be careful not to change the sign when taking a reciprocal!

**18.** $-\dfrac{3}{4} \div \dfrac{3}{10} = -\dfrac{3}{4} \cdot \left(\dfrac{10}{3}\right) = -\dfrac{30}{12} = -\dfrac{5}{2} \cdot \dfrac{6}{6} = -\dfrac{5}{2}$

With decimal notation, it is easier to carry out long division than to multiply by the reciprocal.

**Examples**   Divide.

**19.** $-27.9 \div (-3) = \dfrac{-27.9}{-3} = 9.3$   Do the long division $3\overline{)27.9}$.
The answer is positive.

**20.** $-6.3 \div 2.1 = -3$   Do the long division $2.1\overline{)6.3}$.
The answer is negative.

*Do Exercises 21–24.*

Consider the following:

**1.** $\dfrac{2}{3} = \dfrac{2}{3} \cdot 1 = \dfrac{2}{3} \cdot \dfrac{-1}{-1} = \dfrac{2(-1)}{3(-1)} = \dfrac{-2}{-3}$.   Thus, $\dfrac{2}{3} = \dfrac{-2}{-3}$.

**2.** $-\dfrac{2}{3} = -1 \cdot \dfrac{2}{3} = \dfrac{-1}{1} \cdot \dfrac{2}{3} = \dfrac{-1 \cdot 2}{1 \cdot 3} = \dfrac{-2}{3}$.   Thus, $-\dfrac{2}{3} = \dfrac{-2}{3}$.

$\dfrac{-2}{3} = \dfrac{-2}{3} \cdot 1 = \dfrac{-2}{3} \cdot \dfrac{-1}{-1} = \dfrac{-2(-1)}{3(-1)} = \dfrac{2}{-3}$.   Thus, $\dfrac{-2}{3} = \dfrac{2}{-3}$.

We can use the following properties to make sign changes in fractional notation.

> For any numbers $a$ and $b$, $b \neq 0$:
>
> **1.** $\dfrac{-a}{-b} = \dfrac{a}{b}$
>
> (The opposite of a number $a$ divided by the opposite of another number $b$ is the same as the quotient of the two numbers $a$ and $b$.)
>
> **2.** $\dfrac{-a}{b} = \dfrac{a}{-b} = -\dfrac{a}{b}$
>
> (The opposite of a number $a$ divided by another number $b$ is the same as the number $a$ divided by the opposite of the number $b$, and both are the same as the opposite of $a$ *divided by* $b$.)

*Do Exercises 25–27.*

# Exercise Set 1.6

**a** Divide, if possible. Check each answer.

**1.** $48 \div (-6)$

**2.** $\dfrac{42}{-7}$

**3.** $\dfrac{28}{-2}$

**4.** $24 \div (-12)$

**5.** $\dfrac{-24}{8}$

**6.** $-18 \div (-2)$

**7.** $\dfrac{-36}{-12}$

**8.** $-72 \div (-9)$

**9.** $\dfrac{-72}{9}$

**10.** $\dfrac{-50}{25}$

**11.** $-100 \div (-50)$

**12.** $\dfrac{-200}{8}$

**13.** $-108 \div 9$

**14.** $\dfrac{-63}{-7}$

**15.** $\dfrac{200}{-25}$

**16.** $-300 \div (-16)$

**17.** $\dfrac{75}{0}$

**18.** $\dfrac{0}{-5}$

**19.** $\dfrac{-23}{-2}$

**20.** $\dfrac{-23}{0}$

**b** Find the reciprocal.

**21.** $\dfrac{15}{7}$

**22.** $\dfrac{3}{8}$

**23.** $-\dfrac{47}{13}$

**24.** $-\dfrac{31}{12}$

**25.** $13$

**26.** $-10$

**27.** $4.3$

**28.** $-8.5$

**29.** $\dfrac{1}{-7.1}$

**30.** $\dfrac{1}{-4.9}$

**31.** $\dfrac{p}{q}$

**32.** $\dfrac{s}{t}$

**33.** $\dfrac{1}{4y}$

**34.** $\dfrac{-1}{8a}$

**35.** $\dfrac{2a}{3b}$

**36.** $\dfrac{-4y}{3x}$

**c** Rewrite the division as a multiplication.

**37.** $4 \div 17$

**38.** $5 \div (-8)$

**39.** $\dfrac{8}{-13}$

**40.** $-\dfrac{13}{47}$

**41.** $\dfrac{13.9}{-1.5}$

**42.** $-\dfrac{47.3}{21.4}$

**43.** $\dfrac{x}{\frac{1}{y}}$

**44.** $\dfrac{13}{x}$

**45.** $\dfrac{3x + 4}{5}$

**46.** $\dfrac{4y - 8}{-7}$

**47.** $\dfrac{5a - b}{5a + b}$

**48.** $\dfrac{2x + x^2}{x - 5}$

Divide.

**49.** $\dfrac{3}{4} \div \left(-\dfrac{2}{3}\right)$

**50.** $\dfrac{7}{8} \div \left(-\dfrac{1}{2}\right)$

**51.** $-\dfrac{5}{4} \div \left(-\dfrac{3}{4}\right)$

**52.** $-\dfrac{5}{9} \div \left(-\dfrac{5}{6}\right)$

**53.** $-\dfrac{2}{7} \div \left(-\dfrac{4}{9}\right)$

**54.** $-\dfrac{3}{5} \div \left(-\dfrac{5}{8}\right)$

**55.** $-\dfrac{3}{8} \div \left(-\dfrac{8}{3}\right)$

**56.** $-\dfrac{5}{8} \div \left(-\dfrac{6}{5}\right)$

**57.** $-6.6 \div 3.3$

**58.** $-44.1 \div (-6.3)$

**59.** $\dfrac{-11}{-13}$

**60.** $\dfrac{-1.9}{20}$

**61.** $\dfrac{48.6}{-3}$

**62.** $\dfrac{-17.8}{3.2}$

**63.** $\dfrac{-9}{17 - 17}$

**64.** $\dfrac{-8}{-5 + 5}$

---

**Skill Maintenance**

**65.** Simplify: $\dfrac{264}{468}$.  [R.2b]

**66.** Convert to decimal notation: 47.7%.  [R.4a]

**67.** Simplify: $2^3 - 5 \cdot 3 + 8 \cdot 10 \div 2$.  [R.5c]

**68.** Add and simplify: $\dfrac{2}{3} + \dfrac{5}{6}$.  [R.2c]

**69.** Convert to percent notation: $\dfrac{7}{8}$.  [R.4d]

**70.** Simplify: $\dfrac{40}{60}$.  [R.2b]

**71.** Divide and simplify: $\dfrac{12}{25} \div \dfrac{32}{75}$.  [R.2c]

**72.** Multiply and simplify: $\dfrac{12}{25} \cdot \dfrac{32}{75}$.  [R.2c]

---

**Synthesis**

**73.** ◆ Explain how multiplication can be used to justify why a negative number divided by a positive number is negative.

**74.** ◆ Explain how multiplication can be used to justify why a negative number divided by a negative number is positive.

**75.** ◆ 🖩 Find the reciprocal of $-10.5$. What happens if you take the reciprocal of the result?

**76.** Determine those real numbers $a$ for which the opposite of $a$ is the same as the reciprocal of $a$.

Tell whether the expression represents a positive number or a negative number when $a$ and $b$ are negative.

**77.** $\dfrac{-a}{b}$

**78.** $\dfrac{-a}{-b}$

**79.** $-\left(\dfrac{a}{-b}\right)$

**80.** $-\left(\dfrac{-a}{b}\right)$

**81.** $-\left(\dfrac{-a}{-b}\right)$

Copyright © 1999 Addison Wesley Longman

# 1.7 Properties of Real Numbers

## a | Equivalent Expressions

In solving equations and doing other kinds of work in algebra, we manipulate expressions in various ways. For example, instead of

$$x + x,$$

we might write

$$2x,$$

knowing that the two expressions represent the same number for any allowable replacement of $x$. In that sense, the expressions $x + x$ and $2x$ are **equivalent**, as are $3/x$ and $3x/x^2$, even though 0 is not an allowable replacement because division by 0 is undefined.

> Two expressions that have the same value for all allowable replacements are called **equivalent**.

The expressions $x + 3x$ and $5x$ are *not* equivalent.

### Do Exercises 1 and 2.

In this section, we will consider several laws of real numbers that will allow us to find equivalent expressions. The first two laws are the *identity properties of 0 and 1.*

> **THE IDENTITY PROPERTY OF 0**
>
> For any real number $a$,
> $$a + 0 = 0 + a = a.$$
> (The number 0 is the *additive identity*.)

> **THE IDENTITY PROPERTY OF 1**
>
> For any real number $a$,
> $$a \cdot 1 = 1 \cdot a = a.$$
> (The number 1 is the *multiplicative identity*.)

We often refer to the use of the identity property of 1 as "multiplying by 1." We can use this method to find equivalent fractional expressions. Recall from arithmetic that to multiply with fractional notation, we multiply numerators and denominators. (See also Section R.2.)

### Objectives

**a** Find equivalent fractional expressions and simplify fractional expressions.

**b** Use the commutative and associative laws to find equivalent expressions.

**c** Use the distributive laws to multiply expressions like 8 and $x - y$.

**d** Use the distributive laws to factor expressions like $4x - 12 + 24y$.

**e** Collect like terms.

### For Extra Help

| TAPE 3 | TAPE 3B | MAC WIN | CD-ROM |

Complete the table by evaluating each expression for the given values.

1.

| | $x + x$ | $2x$ |
|---|---|---|
| $x = 3$ | $3+3=6$ | $6$ |
| $x = -6$ | $-6+6=-12$ | $-6$ |
| $x = 4.8$ | $4.8+4.8$ | $9.6$ |

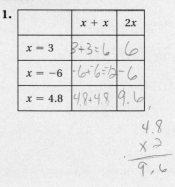

2.

| | $x + 3x$ | $5x$ |
|---|---|---|
| $x = 2$ | $2+3(2)$ | $10$ |
| $x = -6$ | $-6+3(-6)$ | $-30$ |
| $x = 4.8$ | $4.8+3(4.8)$ | |

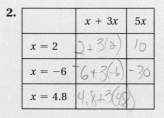

*Answers on page A-5*

**3.** Write a fractional expression equivalent to $\frac{3}{4}$ with a denominator of 8.

**4.** Write a fractional expression equivalent to $\frac{3}{4}$ with a denominator of $4t$.

Simplify.

**5.** $\dfrac{3y}{4y}$

**6.** $-\dfrac{16m}{12m}$

**7.** Evaluate $x + y$ and $y + x$ for $x = -2$ and $y = 3$.

**8.** Evaluate $xy$ and $yx$ for $x = -2$ and $y = 5$.

*Answers on page A-5*

**Example 1**   Write a fractional expression equivalent to $\frac{2}{3}$ with a denominator of $3x$.

Note that $3x = 3 \cdot x$. We want fractional notation for $\frac{2}{3}$ that has a denominator of $3x$, but the denominator 3 is missing a factor of $x$. Thus we multiply by 1, using $x/x$ as an equivalent expression for 1:

$$\frac{2}{3} = \frac{2}{3} \cdot 1 = \frac{2}{3} \cdot \frac{x}{x} = \frac{2x}{3x}.$$

The expressions $2/3$ and $2x/3x$ are equivalent. They have the same value for any allowable replacement. Note that $2x/3x$ is undefined for a replacement of 0, but for all nonzero real numbers, the expressions $2/3$ and $2x/3x$ have the same value.

*Do Exercises 3 and 4.*

In algebra, we consider an expression like $2/3$ to be "simplified" from $2x/3x$. To find such simplified expressions, we use the identity property of 1 to remove a factor of 1. (See also Section R.2.)

**Example 2**   Simplify: $-\dfrac{20x}{12x}$.

$$-\frac{20x}{12x} = -\frac{5 \cdot 4x}{3 \cdot 4x} \qquad \text{We look for the largest factor common to both the numerator and the denominator and factor each.}$$

$$= -\frac{5}{3} \cdot \frac{4x}{4x} \qquad \text{Factoring the fractional expression}$$

$$= -\frac{5}{3} \cdot 1 \xleftarrow{\phantom{xx}} \quad \frac{4x}{4x} = 1$$

$$= -\frac{5}{3} \qquad \text{Removing a factor of 1 using the identity property of 1}$$

*Do Exercises 5 and 6.*

---

## b   The Commutative and Associative Laws

### The Commutative Laws

Let's examine the expressions $x + y$ and $y + x$, as well as $xy$ and $yx$.

**Example 3**   Evaluate $x + y$ and $y + x$ for $x = 4$ and $y = 3$.

We substitute 4 for $x$ and 3 for $y$ in both expressions:

$$x + y = 4 + 3 = 7; \qquad y + x = 3 + 4 = 7.$$

**Example 4**   Evaluate $xy$ and $yx$ for $x = 23$ and $y = 12$.

We substitute 23 for $x$ and 12 for $y$ in both expressions:

$$xy = 23 \cdot 12 = 276; \qquad yx = 12 \cdot 23 = 276.$$

*Do Exercises 7 and 8.*

Note that the expressions

$$x + y \quad \text{and} \quad y + x$$

have the same values no matter what the variables stand for. Thus they are equivalent. Therefore, when we add two numbers, the order in which we add does not matter. Similarly, the expressions $xy$ and $yx$ are equivalent. They also have the same values, no matter what the variables stand for. Therefore, when we multiply two numbers, the order in which we multiply does not matter.

The following are examples of general patterns or laws.

> **THE COMMUTATIVE LAWS**
>
> *Addition.* For any numbers $a$ and $b$,
>
> $$a + b = b + a.$$
>
> (We can change the order when adding without affecting the answer.)
>
> *Multiplication.* For any numbers $a$ and $b$,
>
> $$ab = ba.$$
>
> (We can change the order when multiplying without affecting the answer.)

Using a commutative law, we know that $x + 2$ and $2 + x$ are equivalent. Similarly, $3x$ and $x(3)$ are equivalent. Thus, in an algebraic expression, we can replace one with the other and the result will be equivalent to the original expression.

**Example 5** Use the commutative laws to write an expression equivalent to $y + 5$, $ab$, and $7 + xy$.

An expression equivalent to $y + 5$ is $5 + y$ by the commutative law of addition.

An expression equivalent to $ab$ is $ba$ by the commutative law of multiplication.

An expression equivalent to $7 + xy$ is $xy + 7$ by the commutative law of addition. Another expression equivalent to $7 + xy$ is $7 + yx$ by the commutative law of multiplication.

*Do Exercises 9–11.*

## The Associative Laws

Now let's examine the expressions $a + (b + c)$ and $(a + b) + c$. Note that these expressions involve the use of parentheses as *grouping* symbols, and they also involve three numbers. Calculations within parentheses are to be done first.

**Example 6** Calculate and compare: $3 + (8 + 5)$ and $(3 + 8) + 5$.

$$
\begin{aligned}
3 + (8 + 5) &= 3 + 13 \\
&= 16;
\end{aligned}
$$
Calculating within parentheses first; adding the 8 and 5

$$
\begin{aligned}
(3 + 8) + 5 &= 11 + 5 \\
&= 16
\end{aligned}
$$
Calculating within parentheses first; adding the 3 and 8

Use a commutative law to write an equivalent expression.

**9.** $x + 9$

**10.** $pq$

**11.** $xy + t$

*Answers on page A-5*

**1.7   Properties of Real Numbers**

**87**

**12.** Calculate and compare:

$8 + (9 + 2)$ and $(8 + 9) + 2$.

**13.** Calculate and compare:

$10 \cdot (5 \cdot 3)$ and $(10 \cdot 5) \cdot 3$.

Use an associative law to write an equivalent expression.

**14.** $r + (s + 7)$

**15.** $9(ab)$

*Answers on page A-5*

---

The two expressions in Example 6 name the same number. Moving the parentheses to group the additions differently does not affect the value of the expression.

**Example 7**   Calculate and compare: $3 \cdot (4 \cdot 2)$ and $(3 \cdot 4) \cdot 2$.

$$3 \cdot (4 \cdot 2) = 3 \cdot 8 = 24; \qquad (3 \cdot 4) \cdot 2 = 12 \cdot 2 = 24$$

*Do Exercises 12 and 13.*

You may have noted that when only addition is involved, parentheses can be placed any way we please without affecting the answer. When only multiplication is involved, parentheses also can be placed any way we please without affecting the answer.

> **THE ASSOCIATIVE LAWS**
>
> *Addition.*   For any numbers $a$, $b$, and $c$,
> $$a + (b + c) = (a + b) + c.$$
> (Numbers can be grouped in any manner for addition.)
>
> *Multiplication.*   For any numbers $a$, $b$, and $c$,
> $$a \cdot (b \cdot c) = (a \cdot b) \cdot c.$$
> (Numbers can be grouped in any manner for multiplication.)

**Example 8**   Use an associative law to write an expression equivalent to $(y + z) + 3$ and $8(xy)$.

An equivalent expression is $y + (z + 3)$ by the associative law of addition.

An equivalent expression is $(8x)y$ by the associative law of multiplication.

*Do Exercises 14 and 15.*

The associative laws say parentheses can be placed any way we please when only additions or only multiplications are involved. Thus we often omit them. For example,

$$x + (y + 2) \quad \text{means} \quad x + y + 2, \quad \text{and} \quad (lw)h \quad \text{means} \quad lwh.$$

## Using the Commutative and Associative Laws Together

**Example 9**   Use the commutative and associative laws to write at least three expressions equivalent to $(x + 5) + y$.

a) $(x + 5) + y = x + (5 + y)$     Using the associative law first and then using the commutative law
$\phantom{(x + 5) + y} = x + (y + 5)$

b) $(x + 5) + y = y + (x + 5)$     Using the commutative law first and then the commutative law again
$\phantom{(x + 5) + y} = y + (5 + x)$

c) $(x + 5) + y = (5 + x) + y$     Using the commutative law first and then the associative law
$\phantom{(x + 5) + y} = 5 + (x + y)$

**Example 10** Use the commutative and associative laws to write at least three expressions equivalent to $(3x)y$.

**a)** $(3x)y = 3(xy)$     Using the associative law first and then
         $= 3(yx)$     using the commutative law

**b)** $(3x)y = y(3x)$     Using the commutative law twice
         $= y(x3)$

**c)** $(3x)y = (x3)y$     Using the commutative law, and then
         $= x(3y)$     the associative law, and then
         $= x(y3)$     the commutative law again

*Do Exercises 16 and 17.*

## c | The Distributive Laws

The *distributive laws* are the basis of many procedures in both arithmetic and algebra. They are probably the most important laws that we use to manipulate algebraic expressions. The distributive law of multiplication over addition involves two operations: addition and multiplication.

Let's begin by considering a multiplication problem from arithmetic:

$$
\begin{array}{r}
4\ 5 \\
\times\ \ \ 7 \\
\hline
3\ 5 \leftarrow \text{This is } 7 \cdot 5. \\
2\ 8\ 0 \leftarrow \text{This is } 7 \cdot 40. \\
\hline
3\ 1\ 5 \leftarrow \text{This is the sum } 7 \cdot 40 + 7 \cdot 5.
\end{array}
$$

To carry out the multiplication, we actually added two products. That is,

$$7 \cdot 45 = 7(40 + 5) = 7 \cdot 40 + 7 \cdot 5.$$

Let's examine this further. If we wish to multiply a sum of several numbers by a factor, we can either add and then multiply, or multiply and then add.

**Example 11** Compute in two ways: $5 \cdot (4 + 8)$.

**a)**    $5 \cdot (4 + 8)$     Adding within parentheses first, and then multiplying

    $= 5 \cdot\ \ \ 12$
    $= 60$

**b)**    $(5 \cdot 4) + (5 \cdot 8)$     Distributing the multiplication to terms within parentheses first and then adding

   $=\ \ \ 20\ \ +\ \ 40$
   $=\ \ \ 60$

*Do Exercises 18–20.*

> ▶ **THE DISTRIBUTIVE LAW OF MULTIPLICATION OVER ADDITION**
>
> For any numbers $a$, $b$, and $c$,
> $$a(b + c) = ab + ac.$$

Use the commutative and associative laws to write at least three equivalent expressions.

**16.** $4(tu)$

**17.** $r + (2 + s)$

Compute.

**18. a)** $7 \cdot (3 + 6)$

   **b)** $(7 \cdot 3) + (7 \cdot 6)$

**19. a)** $2 \cdot (10 + 30)$

   **b)** $(2 \cdot 10) + (2 \cdot 30)$

**20. a)** $(2 + 5) \cdot 4$

   **b)** $(2 \cdot 4) + (5 \cdot 4)$

*Answers on page A-5*

Calculate.

**21. a)** $4(5 - 3)$

**b)** $4 \cdot 5 - 4 \cdot 3$

**22. a)** $-2 \cdot (5 - 3)$

**b)** $-2 \cdot 5 - (-2) \cdot 3$

**23. a)** $5 \cdot (2 - 7)$

**b)** $5 \cdot 2 - 5 \cdot 7$

What are the terms of the expression?

**24.** $5x - 8y + 3$

**25.** $-4y - 2x + 3z$

Multiply.

**26.** $3(x - 5)$

**27.** $5(x + 1)$

**28.** $\dfrac{3}{5}(p + q - t)$

*Answers on page A-5*

In the statement of the distributive law, we know that in an expression such as $ab + ac$, the multiplications are to be done first according to the rules for order of operations. (See Section R.5.) So, instead of writing $(4 \cdot 5) + (4 \cdot 7)$, we can write $4 \cdot 5 + 4 \cdot 7$. However, in $a(b + c)$, we cannot omit the parentheses. If we did, we would have $ab + c$, which means $(ab) + c$. For example, $3(4 + 2) = 18$, but $3 \cdot 4 + 2 = 14$.

There is another distributive law that relates multiplication and subtraction. This law says that to multiply by a difference, we can either subtract and then multiply, or multiply and then subtract.

---

> **THE DISTRIBUTIVE LAW OF MULTIPLICATION OVER SUBTRACTION**
>
> For any numbers $a$, $b$, and $c$,
>
> $$a(b - c) = ab - ac.$$

---

We often refer to "*the* distributive law" when we mean *either* or *both* of these laws.

*Do Exercises 21–23.*

What do we mean by the *terms* of an expression? **Terms** are separated by addition signs. If there are subtraction signs, we can find an equivalent expression that uses addition signs.

**Example 12**   What are the terms of $3x - 4y + 2z$?

We have

$$3x - 4y + 2z = 3x + (-4y) + 2z. \qquad \text{Separating parts with + signs}$$

The terms are $3x$, $-4y$, and $2z$.

*Do Exercises 24 and 25.*

The distributive laws are a basis for a procedure in algebra called **multiplying**. In an expression like $8(a + 2b - 7)$, we multiply each term inside the parentheses by 8:

$$8(a + 2b - 7) = 8 \cdot a + 8 \cdot 2b - 8 \cdot 7 = 8a + 16b - 56.$$

**Examples**   Multiply.

**13.** $9(x - 5) = 9x - 9(5)$   Using the distributive law of multiplication over subtraction

$\qquad\qquad = 9x - 45$

**14.** $\frac{2}{3}(w + 1) = \frac{2}{3} \cdot w + \frac{2}{3} \cdot 1$   Using the distributive law of multiplication over addition

$\qquad\qquad = \frac{2}{3}w + \frac{2}{3}$

**15.** $\frac{4}{3}(s - t + w) = \frac{4}{3}s - \frac{4}{3}t + \frac{4}{3}w$   Using both distributive laws

*Do Exercises 26–28.*

**Example 16** Multiply: $-4(x - 2y + 3z)$.

$$-4(x - 2y + 3z) = -4 \cdot x - (-4)(2y) + (-4)(3z) \quad \text{Using both distributive laws}$$

$$= -4x - (-8y) + (-12z) \quad \text{Multiplying}$$

$$= -4x + 8y - 12z$$

We can also do this problem by first finding an equivalent expression with all plus signs and then multiplying:

$$-4(x - 2y + 3z) = -4[x + (-2y) + 3z]$$

$$= -4 \cdot x + (-4)(-2y) + (-4)(3z)$$

$$= -4x + 8y - 12z.$$

*Do Exercises 29–31.*

## d | Factoring

**Factoring** is the reverse of multiplying. To factor, we can use the distributive laws in reverse:

$$ab + ac = a(b + c) \quad \text{and} \quad ab - ac = a(b - c).$$

> To **factor** an expression is to find an equivalent expression that is a product.

Look at Example 13. To *factor* $9x - 45$, we find an equivalent expression that is a product, $9(x - 5)$. When all the terms of an expression have a factor in common, we can "factor it out" using the distributive laws. Note the following.

$9x$ has the factors $9, -9, 3, -3, 1, -1, x, -x, 3x, -3x, 9x, -9x$;

$-45$ has the factors $1, -1, 3, -3, 5, -5, 9, -9, 15, -15, 45, -45$

We generally remove the largest common factor. In this case, that factor is 9. Thus,

$$9x - 45 = 9 \cdot x - 9 \cdot 5$$

$$= 9(x - 5).$$

Remember that an expression has been factored when we have found an equivalent expression that is a product.

**Examples** Factor.

**17.** $5x - 10 = 5 \cdot x - 5 \cdot 2 \quad \text{Try to do this step mentally.}$

$\qquad = 5(x - 2) \quad \text{You can check by multiplying.}$

**18.** $ax - ay + az = a(x - y + z)$

**19.** $9x + 27y - 9 = 9 \cdot x + 9 \cdot 3y - 9 \cdot 1 = 9(x + 3y - 1)$

Multiply.

**29.** $-2(x - 3)$

**30.** $5(x - 2y + 4z)$

**31.** $-5(x - 2y + 4z)$

*Answers on page A-5*

Factor.

**32.** $6x - 12$

**33.** $3x - 6y + 9$

**34.** $bx + by - bz$

**35.** $16a - 36b + 42$

**36.** $\dfrac{3}{8}x - \dfrac{5}{8}y + \dfrac{7}{8}$

**37.** $-12x + 32y - 16z$

Collect like terms.

**38.** $6x - 3x$

**39.** $7x - x$

**40.** $x - 9x$

**41.** $x - 0.41x$

**42.** $5x + 4y - 2x - y$

**43.** $3x - 7x - 11 + 8y + 4 - 13y$

**44.** $-\dfrac{2}{3} - \dfrac{3}{5}x + y + \dfrac{7}{10}x - \dfrac{2}{9}y$

*Answers on page A-5*

**Examples**   Factor. Try to write just the answer, if you can.

**20.** $5x - 5y = 5(x - y)$

**21.** $-3x + 6y - 9z = -3(x - 2y + 3z)$

We usually factor out a negative when the first term is negative. The way we factor can depend on the situation in which we are working. We might also factor the expression in Example 21 as follows:

$$-3x + 6y - 9z = 3(-x + 2y - 3z).$$

**22.** $18z - 12x - 24 = 6(3z - 2x - 4)$

**23.** $\frac{1}{2}x + \frac{3}{2}y - \frac{1}{2} = \frac{1}{2}(x + 3y - 1)$

Remember that you can always check factoring by multiplying. Keep in mind that an expression is factored when it is written as a product.

*Do Exercises 32–37.*

## e   Collecting Like Terms

Terms such as $5x$ and $-4x$, whose variable factors are exactly the same, are called **like terms.** Similarly, numbers, such as $-7$ and $13$, are like terms. Also, $3y^2$ and $9y^2$ are like terms because the variables are raised to the same power. Terms such as $4y$ and $5y^2$ are not like terms, and $7x$ and $2y$ are not like terms.

The process of **collecting like terms** is also based on the distributive laws. We can apply the distributive law when a factor is on the right because of the commutative law of multiplication.

**Examples**   Collect like terms. Try to write just the answer, if you can.

**24.** $4x + 2x = (4 + 2)x = 6x$    **Factoring out the $x$ using a distributive law**

**25.** $2x + 3y - 5x - 2y = 2x - 5x + 3y - 2y$

$$= (2 - 5)x + (3 - 2)y = -3x + y$$

**26.** $3x - x = (3 - 1)x = 2x$

**27.** $x - 0.24x = 1 \cdot x - 0.24x = (1 - 0.24)x = 0.76x$

**28.** $x - 6x = 1 \cdot x - 6 \cdot x = (1 - 6)x = -5x$

**29.** $4x - 7y + 9x - 5 + 3y - 8 = 13x - 4y - 13$

**30.** $\frac{2}{3}a - b + \frac{4}{5}a + \frac{1}{4}b - 10 = \frac{2}{3}a - 1 \cdot b + \frac{4}{5}a + \frac{1}{4}b - 10$

$$= \left(\frac{2}{3} + \frac{4}{5}\right)a + \left(-1 + \frac{1}{4}\right)b - 10$$

$$= \left(\frac{10}{15} + \frac{12}{15}\right)a + \left(-\frac{4}{4} + \frac{1}{4}\right)b - 10$$

$$= \frac{22}{15}a - \frac{3}{4}b - 10$$

*Do Exercises 38–44.*

# Exercise Set 1.7

**a** Find an equivalent expression with the given denominator.

**1.** $\dfrac{3}{5}$;   $5y$

**2.** $\dfrac{5}{8}$;   $8t$

**3.** $\dfrac{2}{3}$;   $15x$

**4.** $\dfrac{6}{7}$;   $14y$

Simplify.

**5.** $-\dfrac{24a}{16a}$

**6.** $-\dfrac{42t}{18t}$

**7.** $-\dfrac{42ab}{36ab}$

**8.** $-\dfrac{64pq}{48pq}$

**b** Write an equivalent expression. Use a commutative law.

**9.** $y + 8$

**10.** $x + 3$

**11.** $mn$

**12.** $ab$

**13.** $9 + xy$

**14.** $11 + ab$

**15.** $ab + c$

**16.** $rs + t$

Write an equivalent expression. Use an associative law.

**17.** $a + (b + 2)$

**18.** $3(vw)$

**19.** $(8x)y$

**20.** $(y + z) + 7$

**21.** $(a + b) + 3$

**22.** $(5 + x) + y$

**23.** $3(ab)$

**24.** $(6x)y$

Use the commutative and associative laws to write three equivalent expressions.

**25.** $(a + b) + 2$

**26.** $(3 + x) + y$

**27.** $5 + (v + w)$

**28.** $6 + (x + y)$

**29.** $(xy)3$

**30.** $(ab)5$

**31.** $7(ab)$

**32.** $5(xy)$

**c** Multiply.

**33.** $2(b + 5)$

**34.** $4(x + 3)$

**35.** $7(1 + t)$

**36.** $4(1 + y)$

**37.** $6(5x + 2)$

**38.** $9(6m + 7)$

**39.** $7(x + 4 + 6y)$

**40.** $4(5x + 8 + 3p)$

**41.** $7(x - 3)$

**42.** $15(y - 6)$

**43.** $-3(x - 7)$

**44.** $1.2(x - 2.1)$

**45.** $\dfrac{2}{3}(b - 6)$

**46.** $\dfrac{5}{8}(y + 16)$

**47.** $7.3(x - 2)$

**48.** $5.6(x - 8)$

**49.** $-\dfrac{3}{5}(x - y + 10)$

**50.** $-\dfrac{2}{3}(a + b - 12)$

**51.** $-9(-5x - 6y + 8)$

**52.** $-7(-2x - 5y + 9)$

**53.** $-4(x - 3y - 2z)$

**54.** $8(2x - 5y - 8z)$

**55.** $3.1(-1.2x + 3.2y - 1.1)$

**56.** $-2.1(-4.2x - 4.3y - 2.2)$

List the terms of the expression.

**57.** $4x + 3z$

**58.** $8x - 1.4y$

**59.** $7x + 8y - 9z$

**60.** $8a + 10b - 18c$

$\boxed{\text{d}}$  Factor. Check by multiplying.

**61.** $2x + 4$

**62.** $5y + 20$

**63.** $30 + 5y$

**64.** $7x + 28$

**65.** $14x + 21y$

**66.** $18a + 24b$

**67.** $5x + 10 + 15y$

**68.** $9a + 27b + 81$

**69.** $8x - 24$

**70.** $10x - 50$

**71.** $32 - 4y$

**72.** $24 - 6m$

**73.** $8x + 10y - 22$

**74.** $9a + 6b + 15$

**75.** $ax - a$

**76.** $by - 9b$

**77.** $ax - ay - az$

**78.** $cx + cy - cz$

**79.** $18x - 12y + 6$

**80.** $-14x + 21y + 7$

**81.** $\dfrac{2}{3}x - \dfrac{5}{3}y + \dfrac{1}{3}$

**82.** $\dfrac{3}{5}a + \dfrac{4}{5}b - \dfrac{1}{5}$

**e** Collect like terms.

**83.** $9a + 10a$

**84.** $12x + 2x$

**85.** $10a - a$

**86.** $-16x + x$

**87.** $2x + 9z + 6x$

**88.** $3a - 5b + 7a$

**89.** $7x + 6y^2 + 9y^2$

**90.** $12m^2 + 6q + 9m^2$

**91.** $41a + 90 - 60a - 2$

**92.** $42x - 6 - 4x + 2$

**93.** $23 + 5t + 7y - t - y - 27$

**94.** $45 - 90d - 87 - 9d + 3 + 7d$

**95.** $\dfrac{1}{2}b + \dfrac{1}{2}b$

**96.** $\dfrac{2}{3}x + \dfrac{1}{3}x$

**97.** $2y + \dfrac{1}{4}y + y$

**98.** $\dfrac{1}{2}a + a + 5a$

**99.** $11x - 3x$

**100.** $9t - 17t$

**101.** $6n - n$

**102.** $10t - t$

**103.** $y - 17y$

**104.** $3m - 9m + 4$

**105.** $-8 + 11a - 5b + 6a - 7b + 7$

**106.** $8x - 5x + 6 + 3y - 2y - 4$

**107.** $9x + 2y - 5x$

**108.** $8y - 3z + 4y$

**109.** $11x + 2y - 4x - y$

**110.** $13a + 9b - 2a - 4b$

**111.** $2.7x + 2.3y - 1.9x - 1.8y$

**112.** $6.7a + 4.3b - 4.1a - 2.9b$

**113.** $\dfrac{13}{2}a + \dfrac{9}{5}b - \dfrac{2}{3}a - \dfrac{3}{10}b - 42$

**114.** $\dfrac{11}{4}x + \dfrac{2}{3}y - \dfrac{4}{5}x - \dfrac{1}{6}y + 12$

---

## Skill Maintenance

**115.** Add and simplify: $\dfrac{11}{12} + \dfrac{15}{16}$.   [R.2c]

**116.** Subtract and simplify: $\dfrac{7}{8} - \dfrac{2}{3}$.   [R.2c]

**117.** Find the LCM of 16, 18, and 24.   [R.1b]

**118.** Convert to percent notation: $\dfrac{3}{10}$.   [R.4d]

**119.** Subtract and simplify: $\dfrac{1}{8} - \dfrac{1}{3}$.   [R.2c]

**120.** Find the LCM of 12, 15, and 20.   [R.1b]

---

## Synthesis

**121.** ◈ The distributive law was introduced before the discussion on collecting like terms. Why do you think this was done?

**122.** ◈ Find two different expressions for the total area of the two rectangles shown below. Explain the equivalence of the expressions in terms of the distributive law.

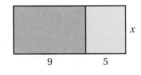

Tell whether the expressions are equivalent. Explain.

**123.** $3t + 5$ and $3 \cdot 5 + t$

**124.** $4x$ and $x + 4$

**125.** $5m + 6$ and $6 + 5m$

**126.** $(x + y) + z$ and $z + (x + y)$

Collect like terms, if possible, and factor the result.

**127.** $q + qr + qrs + qrst$

**128.** $21x + 44xy + 15y - 16x - 8y - 38xy + 2y + xy$

---

Use the commutative and associative laws to
add a series of numbers.

# 1.8 Simplifying Expressions; Order of Operations

We now expand our ability to manipulate expressions by first considering opposites of sums and differences. Then we simplify expressions involving parentheses.

 ## a Opposites of Sums

What happens when we multiply a real number by $-1$? Consider the following products:

$$-1(7) = -7, \qquad -1(-5) = 5, \qquad -1(0) = 0.$$

From these examples, it appears that when we multiply a number by $-1$, we get the opposite, or additive inverse, of that number.

> ### The Property of $-1$
> For any real number $a$,
> $$-1 \cdot a = -a.$$
> (Negative one times $a$ is the opposite, or additive inverse, of $a$.)

The property of $-1$ enables us to find certain expressions equivalent to opposites of sums.

**Examples** Find an equivalent expression without parentheses.

**1.** $-(3 + x) = -1(3 + x)$    **Using the property of $-1$**
$= -1 \cdot 3 + (-1)x$    **Using a distributive law, multiplying each term by $-1$**
$= -3 + (-x)$    **Using the property of $-1$**
$= -3 - x$

**2.** $-(3x + 2y + 4) = -1(3x + 2y + 4)$    **Using the property of $-1$**
$= -1(3x) + (-1)(2y) + (-1)4$    **Using a distributive law**
$= -3x - 2y - 4$    **Using the property of $-1$**

*Do Exercises 1 and 2.*

Suppose we want to remove parentheses in an expression like

$$-(x - 2y + 5).$$

We can first rewrite any subtractions inside the parentheses as additions. Then we take the opposite of each term:

$$-(x - 2y + 5) = -[x + (-2y) + 5]$$
$$= -x + 2y - 5.$$

The most efficient method for removing parentheses is to replace each term in the parentheses with its opposite ("change the sign of every term"). Doing so for $-(x - 2y + 5)$, we obtain $-x + 2y - 5$ as an equivalent expression.

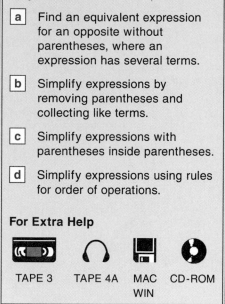

### Objectives

**a** Find an equivalent expression for an opposite without parentheses, where an expression has several terms.

**b** Simplify expressions by removing parentheses and collecting like terms.

**c** Simplify expressions with parentheses inside parentheses.

**d** Simplify expressions using rules for order of operations.

### For Extra Help

TAPE 3    TAPE 4A    MAC WIN    CD-ROM

Find an equivalent expression without parentheses.

**1.** $-(x + 2)$

**2.** $-(5x + 2y + 8)$

*Answers on page A-5*

Find an equivalent expression without parentheses. Try to do this in one step.

**3.** $-(6 - t)$

**4.** $-(x - y)$

**5.** $-(-4a + 3t - 10)$

**6.** $-(18 - m - 2n + 4z)$

Remove parentheses and simplify.

**7.** $5x - (3x + 9)$

**8.** $5y - 2 - (2y - 4)$

Remove parentheses and simplify.

**9.** $6x - (4x + 7)$

**10.** $8y - 3 - (5y - 6)$

**11.** $(2a + 3b - c) - (4a - 5b + 2c)$

*Answers on page A-5*

**Examples** Find an equivalent expression without parentheses.

**3.** $-(5 - y) = -5 + y$    **Changing the sign of each term**

**4.** $-(2a - 7b - 6) = -2a + 7b + 6$

**5.** $-(-3x + 4y + z - 7w - 23) = 3x - 4y - z + 7w + 23$

*Do Exercises 3–6.*

## b | Removing Parentheses and Simplifying

When a sum is added, as in $5x + (2x + 3)$, we can simply remove, or drop, the parentheses and collect like terms because of the associative law of addition:

$$5x + (2x + 3) = 5x + 2x + 3 = 7x + 3.$$

On the other hand, when a sum is subtracted, as in $3x - (4x + 2)$, no "associative" law applies. However, we can subtract by adding an opposite. We then remove parentheses by changing the sign of each term inside the parentheses and collecting like terms.

**Example 6** Remove parentheses and simplify.

$$\begin{aligned} 3x - (4x + 2) &= 3x + [-(4x + 2)] && \text{Adding the opposite of } (4x + 2) \\ &= 3x + (-4x - 2) && \text{Changing the sign of each term inside the parentheses} \\ &= 3x - 4x - 2 \\ &= -x - 2 && \text{Collecting like terms} \end{aligned}$$

*Do Exercises 7 and 8.*

In practice, the first three steps of Example 6 are usually combined by changing the sign of each term in parentheses and then collecting like terms.

**Examples** Remove parentheses and simplify.

**7.** $5y - (3y + 4) = 5y - 3y - 4$    **Removing parentheses by changing the sign of every term inside the parentheses**

$\phantom{5y - (3y + 4)} = 2y - 4$    **Collecting like terms**

**8.** $3y - 2 - (2y - 4) = 3y - 2 - 2y + 4 = y + 2$

**9.** $(3a + 4b - 5) - (2a - 7b + 4c - 8)$
$\phantom{xx} = 3a + 4b - 5 - 2a + 7b - 4c + 8$
$\phantom{xx} = a + 11b - 4c + 3$

*Do Exercises 9–11.*

Next, consider subtracting an expression consisting of several terms multiplied by a number other than 1 or −1.

**Example 10** Remove parentheses and simplify.

$$\begin{aligned} x - 3(x + y) &= x + [-3(x + y)] && \text{Adding the opposite of } 3(x + y) \\ &= x + [-3x - 3y] && \text{Multiplying } x + y \text{ by } -3 \\ &= x - 3x - 3y \\ &= -2x - 3y && \text{Collecting like terms} \end{aligned}$$

**Examples** Remove parentheses and simplify.

**11.** $3y - 2(4y - 5) = 3y - 8y + 10$   Multiplying each term in parentheses by −2
$$= -5y + 10$$

**12.** $(2a + 3b - 7) - 4(-5a - 6b + 12)$
$$= 2a + 3b - 7 + 20a + 24b - 48$$
$$= 22a + 27b - 55$$

**13.** $2y - \frac{1}{3}(9y - 12) = 2y - 3y + 4$
$$= -y + 4$$

*Do Exercises 12–15.*

## c   Parentheses Within Parentheses

In addition to parentheses, some expressions contain other grouping symbols such as brackets [ ] and braces { }.

> When more than one kind of grouping symbol occurs, do the computations in the innermost ones first. Then work from the inside out.

**Examples** Simplify.

**14.** $[3 - (7 + 3)] = [3 - 10]$   Computing 7 + 3
$$= -7$$

**15.** $\{8 - [9 - (12 + 5)]\} = \{8 - [9 - 17]\}$   Computing 12 + 5
$$= \{8 - [-8]\}$$   Computing 9 − 17
$$= 8 + 8$$
$$= 16$$

**16.** $\left[(-4) \div \left(-\frac{1}{4}\right)\right] \div \frac{1}{4} = [(-4) \cdot (-4)] \div \frac{1}{4}$   Working within the brackets computing $(-4) \div \left(-\frac{1}{4}\right)$
$$= 16 \div \frac{1}{4}$$
$$= 16 \cdot 4$$
$$= 64$$

**17.** $4(2 + 3) - \{7 - [4 - (8 + 5)]\}$
$$= 4 \cdot 5 - \{7 - [4 - 13]\}$$   Working with the innermost parentheses first
$$= 20 - \{7 + [+9]\}$$   Computing 4 · 5 and 4 − 13
$$= 20 - 16$$   Computing 7 − [−9]
$$= 4$$

*Do Exercises 16–19.*

**Example 18** Simplify.

$$[5(x + 2) - 3x] - [3(y + 2) - 7(y - 3)]$$
$$= [5x + 10 - 3x] - [3y + 6 - 7y + 21]$$   Working with the innermost parentheses first
$$= [2x + 10] - [-4y + 27]$$   Collecting like terms within brackets
$$= 2x + 10 + 4y - 27$$   Removing brackets
$$= 2x + 4y - 17$$   Collecting like terms

*Do Exercise 20.*

---

Remove parentheses and simplify.

**12.** $y - 9(x + y)$

**13.** $5a - 3(7a - 6)$

**14.** $4a - b - 6(5a - 7b + 8c)$

**15.** $5x - \frac{1}{4}(8x + 28)$

Simplify.

**16.** $12 - (8 + 2)$

**17.** $\{9 - [10 - (13 + 6)]\}$

**18.** $[24 \div (-2)] \div (-2)$

**19.** $5(3 + 4) - \{8 - [5 - (9 + 6)]\}$

**20.** Simplify:

$$[3(x + 2) + 2x] - [4(y + 2) - 3(y - 2)].$$

*Answers on page A-5*

Simplify.

**21.** $23 - 42 \cdot 30$

**22.** $32 \div 8 \cdot 2$

**23.** $52 \cdot 5 + 5^3 - (4^2 - 48 \div 4)$

**24.** $\dfrac{5 - 10 - 5 \cdot 23}{2^3 + 3^2 - 7}$

**d** | **Order of Operations**

When several operations are to be done in a calculation or a problem, we apply the same rules that we did in Section R.5. We repeat them here for review. (If you did not study that section earlier, you should do so now.)

---

**RULES FOR ORDER OF OPERATIONS**

**1.** Do all calculations within parentheses before operations outside.

**2.** Evaluate all exponential expressions.

**3.** Do all multiplications and divisions in order from left to right.

**4.** Do all additions and subtractions in order from left to right.

---

These rules are consistent with the way in which most computers and scientific calculators perform calculations.

**Example 19**   Simplify: $-34 \cdot 56 - 17$.

There are no parentheses or powers, so we start with the third step.

$$-34 \cdot 56 - 17 = -1904 - 17 \qquad \text{Carrying out all multiplications and divisions in order from left to right}$$

$$= -1921 \qquad \text{Carrying out all additions and subtractions in order from left to right}$$

**Example 20**   Simplify: $2^4 + 51 \cdot 4 - (37 + 23 \cdot 2)$.

$$2^4 + 51 \cdot 4 - (37 + 23 \cdot 2)$$

$$= 2^4 + 51 \cdot 4 - (37 + 46) \qquad \text{Following the rules for order of operations within the parentheses first}$$

$$= 2^4 + 51 \cdot 4 - 83 \qquad \text{Completing the addition inside parentheses}$$

$$= 16 + 51 \cdot 4 - 83 \qquad \text{Evaluating exponential expressions}$$

$$= 16 + 204 - 83 \qquad \text{Doing all multiplications}$$

$$= 220 - 83 \qquad \text{Doing all additions and subtractions in order from left to right}$$

$$= 137$$

A fraction bar can play the role of a grouping symbol, although such a symbol is not as evident as the others.

**Example 21**   Simplify: $\dfrac{-64 \div (-16) \div (-2)}{2^3 - 3^2}$.

An equivalent expression with brackets as grouping symbols is

$$[-64 \div (-16) \div (-2)] \div [2^3 - 3^2].$$

This shows, in effect, that we can do the calculations in the numerator and then in the denominator, and divide the results:

$$\frac{-64 \div (-16) \div (-2)}{2^3 - 3^2} = \frac{4 \div (-2)}{8 - 9} = \frac{-2}{-1} = 2.$$

*Do Exercises 21–24.*

# Calculator Spotlight

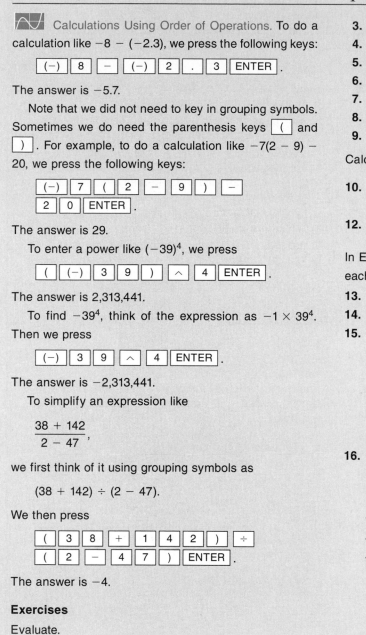

Calculations Using Order of Operations. To do a calculation like $-8 - (-2.3)$, we press the following keys:

[(−)] [8] [−] [(−)] [2] [.] [3] [ENTER] .

The answer is $-5.7$.

Note that we did not need to key in grouping symbols. Sometimes we do need the parenthesis keys [ ( ] and [ ) ] . For example, to do a calculation like $-7(2 - 9) - 20$, we press the following keys:

[(−)] [7] [ ( ] [2] [−] [9] [ ) ] [−] [2] [0] [ENTER] .

The answer is 29.

To enter a power like $(-39)^4$, we press

[ ( ] [(−)] [3] [9] [ ) ] [^] [4] [ENTER] .

The answer is 2,313,441.

To find $-39^4$, think of the expression as $-1 \times 39^4$. Then we press

[(−)] [3] [9] [^] [4] [ENTER] .

The answer is $-2,313,441$.

To simplify an expression like

$$\frac{38 + 142}{2 - 47},$$

we first think of it using grouping symbols as

$$(38 + 142) \div (2 - 47).$$

We then press

[ ( ] [3] [8] [+] [1] [4] [2] [ ) ] [÷]
[ ( ] [2] [−] [4] [7] [ ) ] [ENTER] .

The answer is $-4$.

## Exercises

Evaluate.

1. $-8 + 4(7 - 9) + 5$
2. $-3[2 + (-5)]$
3. $7[4 - (-3)] + 5[3^2 - (-4)]$
4. $(-7)^6$
5. $(-17)^5$
6. $(-104)^3$
7. $-7^6$
8. $-17^5$
9. $-104^3$

Calculate.

10. $\dfrac{38 - 178}{5 + 30}$

11. $\dfrac{311 - 17^2}{2 - 13}$

12. $785 - \dfrac{285 - 5^4}{17 + 3 \cdot 51}$

In Exercises 13 and 14, place one of $+$, $-$, $\times$, and $\div$ in each blank to make a true sentence.

13. $-32 \;\blacksquare\; (88 \;\blacksquare\; 29) = -1888$

14. $3^5 \;\blacksquare\; 10^2 \;\blacksquare\; 5^2 = -2257$

15. Consider the numbers 2, 4, 6, and 8. Assume that each can be placed in a blank in the following.

$$\blacksquare + \blacksquare \cdot \blacksquare - \blacksquare = ?$$

What placement of the numbers in the blanks yields the largest number? Explain why there are two answers.

16. Consider the numbers 3, 5, 7, and 9. Assume that each can be placed in a blank in the following.

$$\blacksquare + \blacksquare^2 \cdot \blacksquare - \blacksquare = ?$$

What placement of the numbers in the blanks yields the largest number? Explain why, unlike Exercise 15, there is just one answer.

# Improving Your Math Study Skills

## Learning Resources and Time Management

Two other topics to consider in enhancing your math study skills are learning resources and time management.

### Learning Resources

- **Textbook supplements.** Are you aware of all the supplements that exist for this textbook? Many details are given in the preface. Now that you are more familiar with the book, let's discuss them.

  1. The *Student's Solutions Manual* contains worked-out solutions to the odd-numbered exercises in the exercise sets. Consider obtaining a copy if you are having trouble. It should be your first choice if you can make an additional purchase.

  2. An extensive set of *videotapes* supplement this text. These may be available to you on your campus at a learning center or math lab. Check with your instructor.

  3. *Tutorial software* also accompanies the text. If not available in the campus learning center, you might order it by calling the number 1-800-322-1377.

- **The Internet.** Our on-line World Wide Web supplement provides additional practice resources. If you have internet access, you can reach this site through the address:

    http://www.mathmax.com

  It contains many helpful ideas as well as many links to other resources for learning mathematics.

- **Your college or university.** Your own college or university probably has resources to enhance your math learning.

  1. For example, is there a learning lab or tutoring center for drop-in tutoring?

  2. Are there special lab classes or formal tutoring sessions tailored for the specific course you are taking?

  3. Perhaps there is a bulletin board or network where you can locate the names of experienced private tutors.

- **Your instructor.** Although it might seem obvious to ask your instructor for help, many students fail to use this valuable resource. Learn what your instructor's office hours are and meet with your instructor at those times when you need extra help.

### Time Management

- **Juggling time.** Have reasonable expectations about the time you need to study math. Unreasonable expectations may lead to lower grades and frustrations. Working 40 hours per week and taking 12 hours of credit is equivalent to working two full-time jobs. Can you handle such a load? As a rule of thumb, your ratio of work hours to credit load should be about 40/3, 30/6, 20/9, 10/12, and 5/14. Budget about 2–3 hours of study time outside of class for each hour that you spend in class.

- **Daily schedule.** Make an hour-by-hour schedule of your typical week. Include work, college, home, personal, sleep, study, and leisure times. Be realistic about the amount of time needed for sleep and home duties. If possible, try to schedule time for study when you are most alert.

# Exercise Set 1.8

**a** Find an equivalent expression without parentheses.

**1.** $-(2x + 7)$     **2.** $-(8x + 4)$     **3.** $-(5x - 8)$     **4.** $-(4x - 3)$

**5.** $-(4a - 3b + 7c)$     **6.** $-(x - 4y - 3z)$     **7.** $-(6x - 8y + 5)$     **8.** $-(4x + 9y + 7)$

**9.** $-(3x - 5y - 6)$     **10.** $-(6a - 4b - 7)$     **11.** $-(-8x - 6y - 43)$     **12.** $-(-2a + 9b - 5c)$

**b** Remove parentheses and simplify.

**13.** $9x - (4x + 3)$     **14.** $4y - (2y + 5)$     **15.** $2a - (5a - 9)$

**16.** $12m - (4m - 6)$     **17.** $2x + 7x - (4x + 6)$     **18.** $3a + 2a - (4a + 7)$

**19.** $2x - 4y - 3(7x - 2y)$     **20.** $3a - 9b - 1(4a - 8b)$     **21.** $15x - y - 5(3x - 2y + 5z)$

**22.** $4a - b - 4(5a - 7b + 8c)$     **23.** $(3x + 2y) - 2(5x - 4y)$     **24.** $(-6a - b) - 5(2b + a)$

**25.** $(12a - 3b + 5c) - 5(-5a + 4b - 6c)$     **26.** $(-8x + 5y - 12) - 6(2x - 4y - 10)$

**c** Simplify.

**27.** $[9 - 2(5 - 4)]$

**28.** $[6 - 5(8 - 4)]$

**29.** $8[7 - 6(4 - 2)]$

**30.** $10[7 - 4(7 - 5)]$

**31.** $[4(9 - 6) + 11] - [14 - (6 + 4)]$

**32.** $[7(8 - 4) + 16] - [15 - (7 + 8)]$

**33.** $[10(x + 3) - 4] + [2(x - 1) + 6]$

**34.** $[9(x + 5) - 7] + [4(x - 12) + 9]$

**35.** $[7(x + 5) - 19] - [4(x - 6) + 10]$

**36.** $[6(x + 4) - 12] - [5(x - 8) + 14]$

**37.** $3\{[7(x - 2) + 4] - [2(2x - 5) + 6]\}$

**38.** $4\{[8(x - 3) + 9] - [4(3x - 2) + 6]\}$

**39.** $4\{[5(x - 3) + 2] - 3[2(x + 5) - 9]\}$

**40.** $3\{[6(x - 4) + 5] - 2[5(x + 8) - 3]\}$

**d** Simplify.

**41.** $8 - 2 \cdot 3 - 9$

**42.** $8 - (2 \cdot 3 - 9)$

**43.** $(8 - 2 \cdot 3) - 9$

**44.** $(8 - 2)(3 - 9)$

**45.** $[(-24) \div (-3)] \div \left(-\frac{1}{2}\right)$

**46.** $[32 \div (-2)] \div (-2)$

**47.** $16 \cdot (-24) + 50$

**48.** $10 \cdot 20 - 15 \cdot 24$

**49.** $2^4 + 2^3 - 10$

**50.** $40 - 3^2 - 2^3$

**51.** $5^3 + 26 \cdot 71 - (16 + 25 \cdot 3)$

**52.** $4^3 + 10 \cdot 20 + 8^2 - 23$

**53.** $4 \cdot 5 - 2 \cdot 6 + 4$

**54.** $4 \cdot (6 + 8)/(4 + 3)$

**55.** $4^3/8$

**56.** $5^3 - 7^2$

**57.** $8(-7) + 6(-5)$

**58.** $10(-5) + 1(-1)$

**59.** $19 - 5(-3) + 3$

**60.** $14 - 2(-6) + 7$

**61.** $9 \div (-3) + 16 \div 8$

**62.** $-32 - 8 \div 4 - (-2)$

**63.** $6 - 4^2$

**64.** $(2 - 5)^2$

**65.** $(3 - 8)^2$

**66.** $3 - 3^2$

**67.** $12 - 20^3$

**68.** $20 + 4^3 \div (-8)$

**69.** $2 \cdot 10^3 - 5000$

**70.** $-7(3^4) + 18$

**71.** $6[9 - (3 - 4)]$

**72.** $8[(6 - 13) - 11]$

**73.** $-1000 \div (-100) \div 10$

**74.** $256 \div (-32) \div (-4)$

**75.** $8 - (7 - 9)$

**76.** $(8 - 7) - 9$

**77.** $\dfrac{10 - 6^2}{9^2 + 3^2}$

**78.** $\dfrac{5^2 - 4^3 - 3}{9^2 - 2^2 - 1^5}$

**79.** $\dfrac{3(6-7)-5\cdot 4}{6\cdot 7-8(4-1)}$

**80.** $\dfrac{20(8-3)-4(10-3)}{10(2-6)-2(5+2)}$

**81.** $\dfrac{2^3-3^2+12\cdot 5}{-32\div(-16)\div(-4)}$

**82.** $\dfrac{|3-5|^2-|7-13|}{|12-9|+|11-14|}$

---

## Skill Maintenance

**83.** Find the prime factorization of 236.   [R.1a]

**84.** Find the LCM of 28 and 36.   [R.1b]

**85.** Divide and simplify: $\dfrac{2}{3}\div\dfrac{5}{12}$.   [R.2c]

**86.** Multiply and simplify: $\dfrac{2}{3}\cdot\dfrac{5}{12}$.   [R.2c]

Evaluate.   [R.5b]

**87.** $3^4$

**88.** $10^3$

**89.** $10^2$

**90.** $15^2$

---

## Synthesis

**91.** ◈ Some students use the memory device PEMDAS ("Please Excuse My Dear Aunt Sally") to remember the rules for the order of operations. Explain how this can be done.

**92.** ◈ Determine whether $|-x|$ and $|x|$ are equivalent. Explain.

Find an equivalent expression by enclosing the last three terms in parentheses preceded by a minus sign.

**93.** $6y+2x-3a+c$

**94.** $x-y-a-b$

**95.** $6m+3n-5m+4b$

Simplify.

**96.** $z-\{2z-[3z-(4z-5z)-6z]-7z\}-8z$

**97.** $\{x-[f-(f-x)]+[x-f]\}-3x$

**98.** $x-\{x-1-[x-2-(x-3-\{x-4-[x-5-(x-6)]\})]\}$

**99.** 🖩 Use your calculator to do the following.

a) Evaluate $x^2+3$ for $x=7$, for $x=-7$, and for $x=-5.013$.

b) Evaluate $1-x^2$ for $x=5$, for $x=-5$, and for $x=-10.455$.

**100.** Express $3^3+3^3+3^3$ as a power of 3.

Use the order of operations as a group to simplify expressions.

Collaborative
Learning Manual

Copyright © 1999 Addison Wesley Longman

# Summary and Review Exercises: Chapter 1

## Important Properties and Formulas

### Properties of the Real-Number System

| | |
|---|---|
| *The Commutative Laws:* | $a + b = b + a, \quad ab = ba$ |
| *The Associative Laws:* | $a + (b + c) = (a + b) + c, \quad a(bc) = (ab)c$ |
| *The Identity Properties:* | For every real number $a$, $a + 0 = a$ and $a \cdot 1 = a$. |
| *The Inverse Properties:* | For each real number $a$, there is an opposite $-a$, such that $a + (-a) = 0$. |
| | For each nonzero real number $a$, there is a reciprocal $\dfrac{1}{a}$, such that $a\left(\dfrac{1}{a}\right) = 1$. |
| *The Distributive Laws:* | $a(b + c) = ab + ac, \quad a(b - c) = ab - ac$ |

The review exercises that follow are for practice. Answers are at the back of the book. If you miss an exercise, restudy the objective indicated in blue after the exercise or the direction line that precedes it. Beginning with this chapter, certain objectives, from four particular sections of preceding chapters, will be retested on the chapter test. The objectives to be tested in addition to the material in this chapter are [R.1a, b], [R.2b, c], [R.4a, d], and [R.5b, c].

**1.** Evaluate $\dfrac{x - y}{3}$ for $x = 17$ and $y = 5$.   [1.1a]

**2.** Translate to an algebraic expression:   [1.1b]
Nineteen percent of some number.

**3.** Tell which integers correspond to this situation: [1.2a]
David has a debt of $45 and Joe has $72 in his savings account.

**4.** Find: $|-38|$.   [1.2e]

Graph the number on a number line.   [1.2b]

**5.** $-2.5$

**6.** $\dfrac{8}{9}$

Use either $<$ or $>$ for ▨ to write a true sentence. [1.2d]

**7.** $-3$ ▨ $10$

**8.** $-1$ ▨ $-6$

**9.** $0.126$ ▨ $-12.6$

**10.** $-\dfrac{2}{3}$ ▨ $-\dfrac{1}{10}$

Find the opposite.   [1.3b]

**11.** $3.8$

**12.** $-\dfrac{3}{4}$

Find the reciprocal.   [1.6b]

**13.** $\dfrac{3}{8}$

**14.** $-7$

**15.** Find $-x$ when $x = -34$.   [1.3b]

**16.** Find $-(-x)$ when $x = 5$.   [1.3b]

Compute and simplify.

**17.** $4 + (-7)$   [1.3a]

**18.** $6 + (-9) + (-8) + 7$   [1.3a]

**19.** $-3.8 + 5.1 + (-12) + (-4.3) + 10$   [1.3a]

**20.** $-3 - (-7)$   [1.4a]

**21.** $-\dfrac{9}{10} - \dfrac{1}{2}$   [1.4a]

**22.** $-3.8 - 4.1$   [1.4a]

**23.** $-9 \cdot (-6)$   [1.5a]

**24.** $-2.7(3.4)$   [1.5a]

**25.** $\dfrac{2}{3} \cdot \left(-\dfrac{3}{7}\right)$   [1.5a]

**26.** $3 \cdot (-7) \cdot (-2) \cdot (-5)$   [1.5a]

**27.** $35 \div (-5)$   [1.6a]

**28.** $-5.1 \div 1.7$   [1.6c]

**29.** $-\dfrac{3}{11} \div \left(-\dfrac{4}{11}\right)$   [1.6c]

**30.** $(-3.4 - 12.2) - 8(-7)$   [1.8d]

**31.** $\dfrac{-12(-3) - 2^3 - (-9)(-10)}{3 \cdot 10 + 1}$   [1.8d]

Solve.  [1.4b]

**32.** On the first, second, and third downs, a football team had these gains and losses: 5-yd gain, 12-yd loss, and 15-yd gain, respectively. Find the total gain (or loss).

**33.** Kaleb's total assets are $170. He borrows $300. What are his total assets now?

Multiply.  [1.7c]

**34.** $5(3x - 7)$

**35.** $-2(4x - 5)$

**36.** $10(0.4x + 1.5)$

**37.** $-8(3 - 6x)$

Factor.  [1.7d]

**38.** $2x - 14$

**39.** $6x - 6$

**40.** $5x + 10$

**41.** $12 - 3x$

Collect like terms.  [1.7e]

**42.** $11a + 2b - 4a - 5b$

**43.** $7x - 3y - 9x + 8y$

**44.** $6x + 3y - x - 4y$

**45.** $-3a + 9b + 2a - b$

Remove parentheses and simplify.

**46.** $2a - (5a - 9)$  [1.8b]

**47.** $3(b + 7) - 5b$  [1.8b]

**48.** $3[11 - 3(4 - 1)]$  [1.8c]

**49.** $2[6(y - 4) + 7]$  [1.8c]

**50.** $[8(x + 4) - 10] - [3(x - 2) + 4]$  [1.8c]

**51.** $5\{[6(x - 1) + 7] - [3(3x - 4) + 8]\}$  [1.8c]

Write true or false.  [1.2d]

**52.** $-9 \le 11$

**53.** $-11 \ge -3$

**54.** Write another inequality with the same meaning as $-3 < x$.  [1.2d]

___

**Skill Maintenance**

**55.** Divide and simplify: $\dfrac{11}{12} \div \dfrac{7}{10}$.  [R.2c]

**56.** Compute and simplify: $\dfrac{5^3 - 2^4}{5 \cdot 2 + 2^3}$.  [R.5c]

**57.** Find the prime factorization of 648.  [R.1a]

**58.** Convert to percent notation: $\dfrac{5}{8}$.  [R.4d]

**59.** Convert to decimal notation: 5.67%.  [R.4a]

**60.** Find the LCM of 15, 27, and 30.  [R.1b]

___

**Synthesis**

Simplify.  [1.2e], [1.4a], [1.6a], [1.8d]

**61.** $-\left| \dfrac{7}{8} - \left( -\dfrac{1}{2} \right) - \dfrac{3}{4} \right|$

**62.** $(|2.7 - 3| + 3^2 - |-3|) \div (-3)$

**63.** $2000 - 1990 + 1980 - 1970 + \cdots + 20 - 10$

**64.** Find a formula for the perimeter of the following figure.  [1.7e]

# Test: Chapter 1

**1.** Evaluate $\dfrac{3x}{y}$ for $x = 10$ and $y = 5$.

**2.** Write an algebraic expression: Nine less than some number.

**3.** Find the area of a triangle when the height $h$ is 30 ft and the base $b$ is 16 ft.

Use either $<$ or $>$ for ▓ to write a true sentence.

**4.** $-4$ ▓ $0$

**5.** $-3$ ▓ $-8$

**6.** $-0.78$ ▓ $-0.87$

**7.** $-\dfrac{1}{8}$ ▓ $\dfrac{1}{2}$

Find the absolute value.

**8.** $|-7|$

**9.** $\left|\dfrac{9}{4}\right|$

**10.** $|-2.7|$

Find the opposite.

**11.** $\dfrac{2}{3}$

**12.** $-1.4$

**13.** Find $-x$ when $x = -8$.

Find the reciprocal.

**14.** $-2$

**15.** $\dfrac{4}{7}$

Compute and simplify.

**16.** $3.1 - (-4.7)$

**17.** $-8 + 4 + (-7) + 3$

**18.** $-\dfrac{1}{5} + \dfrac{3}{8}$

**19.** $2 - (-8)$

**20.** $3.2 - 5.7$

**21.** $\dfrac{1}{8} - \left(-\dfrac{3}{4}\right)$

Answers

1. _____

2. _____

3. _____

4. _____

5. _____

6. _____

7. _____

8. _____

9. _____

10. _____

11. _____

12. _____

13. _____

14. _____

15. _____

16. _____

17. _____

18. _____

19. _____

20. _____

21. _____

22.

23.

24.

25.

26.

27.

28.

29.

30.

31.

32.

33.

34.

35.

36.

37.

38.

39.

40.

41.

42.

43.

44.

45.

46.

**22.** $4 \cdot (-12)$

**23.** $-\dfrac{1}{2} \cdot \left(-\dfrac{3}{8}\right)$

**24.** $-45 \div 5$

**25.** $-\dfrac{3}{5} \div \left(-\dfrac{4}{5}\right)$

**26.** $4.864 \div (-0.5)$

**27.** $-2(16) - |2(-8) - 5^3|$

**28.** *Antarctica Highs and Lows.* The continent of Antarctica, which lies in the southern hemisphere, experiences winter in July. The average high temperature is $-67°F$ and the average low temperature is $-81°F$. How much higher is the average high than the average low?

Multiply.

**29.** $3(6 - x)$

**30.** $-5(y - 1)$

Factor.

**31.** $12 - 22x$

**32.** $7x + 21 + 14y$

Simplify.

**33.** $6 + 7 - 4 - (-3)$

**34.** $5x - (3x - 7)$

**35.** $4(2a - 3b) + a - 7$

**36.** $4\{3[5(y - 3) + 9] + 2(y + 8)\}$

**37.** $256 \div (-16) \div 4$

**38.** $2^3 - 10[4 - (-2 + 18)3]$

**39.** Write an inequality with the same meaning as $x \le -2$.

---

**Skill Maintenance**

**40.** Evaluate: $(1.2)^3$.

**41.** Convert to percent notation: $\dfrac{1}{8}$.

**42.** Find the prime factorization of 280.

**43.** Find the LCM of 16, 20, and 30.

---

**Synthesis**

Simplify.

**44.** $|-27 - 3(4)| - |-36| + |-12|$

**45.** $a - \{3a - [4a - (2a - 4a)]\}$

**46.** Find a formula for the perimeter of the following figure.

Copyright © 1999 Addison Wesley Longman

# 2

# Solving Equations and Inequalities

## An Application

The perimeter of an NBA basketball court is 288 ft. The length is 44 ft longer than the width. (**Source**: National Basketball Association) Find the dimensions of the court.

This problem appears as Example 5 in Section 2.4.

## The Mathematics

The perimeter $P$ of a rectangle is given by the formula $2l + 2w = P$, where $l =$ the length and $w =$ the width. To translate the problem, we substitute $w + 44$ for $l$ and 288 for $P$:

$$2l + 2w = P$$
$$\underbrace{2(w + 44) + 2w = 288.}$$

To find the dimensions, we first solve this equation.

**World Wide Web**   For more information, visit us at www.mathmax.com

# Pretest: Chapter 2

Solve.

**1.** $-7x = 49$

**2.** $4y + 9 = 2y + 7$

**3.** $6a - 2 = 10$

**4.** $4 + x = 12$

**5.** $7 - 3(2x - 1) = 40$

**6.** $\dfrac{4}{9}x - 1 = \dfrac{7}{8}$

**7.** $1 + 2(a + 3) = 3(2a - 1) + 6$

**8.** $-3x \leq 18$

**9.** $y + 5 > 1$

**10.** $5 - 2a < 7$

**11.** $3x + 4 \geq 2x + 7$

**12.** $8y < -18$

**13.** Solve for $G$: $P = 3KG$.

**14.** Solve for $a$: $A = \dfrac{3a - b}{b}$.

Solve.

**15.** The perimeter of the ornate frame of an oil painting is 146 in. The width is 5 in. less than the length. Find the dimensions.

**16.** Money is invested in a savings account at 4.25% simple interest. After 1 year, there is $479.55 in the account. How much was originally invested?

**17.** The sum of three consecutive integers is 246. Find the integers.

**18.** When 18 is added to six times a number, the result is less than 120. For what numbers is this possible?

Graph on a number line.

**19.** $x > -3$

**20.** $x \leq 4$

## Objectives for Retesting

The objectives to be tested in addition to the material in this chapter are as follows.

[1.1a]  Evaluate algebraic expressions by substitution.
[1.1b]  Translate phrases to algebraic expressions.
[1.3a]  Add real numbers.
[1.8b]  Simplify expressions by removing parentheses and collecting like terms.

# 2.1 Solving Equations: The Addition Principle

## a | Equations and Solutions

In order to solve problems, we must learn to solve equations.

> An **equation** is a number sentence that says that the expressions on either side of the equals sign, =, represent the same number.

Here are some examples:

$$3 + 2 = 5, \quad 14 - 10 = 1 + 3, \quad x + 6 = 13, \quad 3x - 2 = 7 - x.$$

Equations have expressions on each side of the equals sign. The sentence "$14 - 10 = 1 + 3$" asserts that the expressions $14 - 10$ and $1 + 3$ name the same number.

Some equations are true. Some are false. Some are neither true nor false.

**Examples** Determine whether the equation is true, false, or neither.

**1.** $3 + 2 = 5$      The equation is *true*.

**2.** $7 - 2 = 4$      The equation is *false*.

**3.** $x + 6 = 13$      The equation is *neither* true nor false, because we do not know what number $x$ represents.

*Do Exercises 1–3.*

> Any replacement for the variable that makes an equation true is called a **solution** of the equation. To solve an equation means to find *all* of its solutions.

One way to determine whether a number is a solution of an equation is to evaluate the expression on each side of the equals sign by substitution. If the values are the same, then the number is a solution.

**Example 4** Determine whether 7 is a solution of $x + 6 = 13$.

We have

$$
\begin{array}{ll}
x + 6 = 13 & \text{Writing the equation} \\
\overline{7 + 6 \; ? \; 13} & \text{Substituting 7 for } x \\
\quad 13 \; | & \text{TRUE}
\end{array}
$$

Since the left-hand and the right-hand sides are the same, we have a solution. No other number makes the equation true, so the only solution is the number 7.

## Objectives

**a**   Determine whether a given number is a solution of a given equation.

**b**   Solve equations using the addition principle.

### For Extra Help

TAPE 4    TAPE 4A    MAC WIN    CD-ROM

Determine whether the equation is true, false, or neither.

**1.** $5 - 8 = -4$

**2.** $12 + 6 = 18$

**3.** $x + 6 = 7 - x$

*Answers on page A-6*

Determine whether the given number is a solution of the given equation.

**4.** 8;   $x + 4 = 12$

**5.** 0;   $x + 4 = 12$

**6.** −3;   $7 + x = -4$

**7.** Solve using the addition principle:

$$x + 2 = 11.$$

*Answers on page A-6*

**Example 5**   Determine whether 19 is a solution of $7x = 141$.

We have

$$\begin{array}{c|c} 7x = 141 & \text{Writing the equation} \\ \hline 7(19) \ ? \ 141 & \text{Substituting 19 for } x \\ 133 \ | & \text{FALSE} \end{array}$$

Since the left-hand and the right-hand sides are not the same, we do not have a solution.

*Do Exercises 4–6.*

## $\boxed{\text{b}}$   Using the Addition Principle

Consider the equation

$$x = 7.$$

We can easily see that the solution of this equation is 7. If we replace $x$ with 7, we get

$$7 = 7, \quad \text{which is true.}$$

Now consider the equation of Example 4:

$$x + 6 = 13.$$

In Example 4, we discovered that the solution of this equation is also 7, but the fact that 7 is the solution is not as obvious. We now begin to consider principles that allow us to start with an equation and end up with an *equivalent equation*, like $x = 7$, in which the variable is alone on one side and for which the solution is easy to find.

> Equations with the same solutions are called **equivalent equations.**

One of the principles that we use in solving equations involves adding. An equation $a = b$ says that $a$ and $b$ stand for the same number. Suppose this is true, and we add a number $c$ to the number $a$. We get the same answer if we add $c$ to $b$, because $a$ and $b$ are the same number.

> **THE ADDITION PRINCIPLE**
> For any real numbers $a$, $b$, and $c$,
> $$a = b \quad \text{is equivalent to} \quad a + c = b + c.$$

Let's again solve the equation $x + 6 = 13$ using the addition principle. We want to get $x$ alone on one side. To do so, we use the addition principle, choosing to add −6 because $6 + (-6) = 0$:

$$\begin{array}{ll} x + 6 = 13 & \\ x + 6 + (-6) = 13 + (-6) & \text{Using the addition principle:} \\ & \text{adding } -6 \text{ on both sides} \\ x + 0 = 7 & \text{Simplifying} \\ x = 7. & \text{Identity property of 0: } x + 0 = x \end{array}$$

*Do Exercise 7.*

When we use the addition principle, we sometimes say that we "add the same number on both sides of the equation." This is also true for subtraction, since we can express every subtraction as an addition. That is, since

$$a - c = b - c \quad \text{is equivalent to} \quad a + (-c) = b + (-c),$$

the addition principle tells us that we can "subtract the same number on both sides of the equation."

**Example 6** Solve: $x + 5 = -7$.

We have

$$x + 5 = -7$$
$$x + 5 - 5 = -7 - 5 \qquad \text{Using the addition principle: adding } -5 \text{ on both sides or subtracting 5 on both sides}$$
$$x + 0 = -12 \qquad \text{Simplifying}$$
$$x = -12. \qquad \text{Identity property of 0}$$

We can see that the solution of $x = -12$ is the number $-12$. To check the answer, we substitute $-12$ in the original equation.

CHECK:
$$\frac{x + 5 = -7}{-12 + 5 \; ? \; -7}$$
$$\qquad -7 \; | \qquad \text{TRUE}$$

The solution of the original equation is $-12$.

In Example 6, to get $x$ alone, we used the addition principle and subtracted 5 on both sides. This eliminated the 5 on the left. We started with $x + 5 = -7$, and, using the addition principle, we found a simpler equation $x = -12$ for which it was easy to "*see*" the solution. The equations $x + 5 = -7$ and $x = -12$ are *equivalent*.

*Do Exercise 8.*

Now we solve an equation with a subtraction using the addition principle.

**Example 7** Solve: $a - 4 = 10$.

We have

$$a - 4 = 10$$
$$a - 4 + 4 = 10 + 4 \qquad \text{Using the addition principle: adding 4 on both sides}$$
$$a + 0 = 14 \qquad \text{Simplifying}$$
$$a = 14. \qquad \text{Identity property of 0}$$

CHECK:
$$\frac{a - 4 = 10}{14 - 4 \; ? \; 10}$$
$$\qquad 10 \; | \qquad \text{TRUE}$$

The solution is 14.

*Do Exercise 9.*

8. Solve using the addition principle, subtracting 7 on both sides:

$$x + 7 = 2.$$

9. Solve: $t - 3 = 19$.

*Answers on page A-6*

Solve.

**10.** $8.7 = n - 4.5$

**11.** $y + 17.4 = 10.9$

Solve.

**12.** $x + \dfrac{1}{2} = -\dfrac{3}{2}$

**13.** $t - \dfrac{13}{4} = \dfrac{5}{8}$

*Answers on page A-6*

**Example 8**   Solve: $-6.5 = y - 8.4$.

We have

$$-6.5 = y - 8.4$$
$$-6.5 + 8.4 = y - 8.4 + 8.4 \qquad \text{Using the addition principle:}$$
$$\text{adding 8.4 to eliminate } -8.4 \text{ on the right}$$
$$1.9 = y.$$

CHECK:
$$\frac{-6.5 = y - 8.4}{-6.5 \ ? \ 1.9 - 8.4}$$
$$| \ -6.5 \qquad \text{TRUE}$$

The solution is 1.9.

Note that equations are reversible. That is, if $a = b$ is true, then $b = a$ is true. Thus when we solve $-6.5 = y - 8.4$, we can reverse it and solve $y - 8.4 = -6.5$ if we wish.

*Do Exercises 10 and 11.*

**Example 9**   Solve: $-\dfrac{2}{3} + x = \dfrac{5}{2}$.

We have

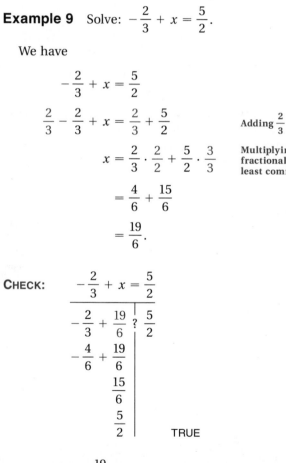

$$-\frac{2}{3} + x = \frac{5}{2}$$
$$\frac{2}{3} - \frac{2}{3} + x = \frac{2}{3} + \frac{5}{2} \qquad \text{Adding } \frac{2}{3}$$
$$x = \frac{2}{3} \cdot \frac{2}{2} + \frac{5}{2} \cdot \frac{3}{3} \qquad \text{Multiplying by 1 to obtain equivalent fractional expressions with the least common denominator 6}$$
$$= \frac{4}{6} + \frac{15}{6}$$
$$= \frac{19}{6}.$$

CHECK:
$$\frac{-\dfrac{2}{3} + x = \dfrac{5}{2}}{-\dfrac{2}{3} + \dfrac{19}{6} \ ? \ \dfrac{5}{2}}$$
$$-\dfrac{4}{6} + \dfrac{19}{6}$$
$$\dfrac{15}{6}$$
$$\dfrac{5}{2} \qquad \text{TRUE}$$

The solution is $\dfrac{19}{6}$.

*Do Exercises 12 and 13.*

# Exercise Set 2.1

**a**  Determine whether the given number is a solution of the given equation.

**1.** 15;  $x + 17 = 32$     **2.** 35;  $t + 17 = 53$     **3.** 21;  $x - 7 = 12$     **4.** 36;  $a - 19 = 17$

**5.** −7;  $6x = 54$     **6.** −9;  $8y = -72$     **7.** 30;  $\dfrac{x}{6} = 5$     **8.** 49;  $\dfrac{y}{8} = 6$

**9.** 19;  $5x + 7 = 107$     **10.** 9;  $9x + 5 = 86$     **11.** −11;  $7(y - 1) = 63$     **12.** −18;  $x + 3 = 3 + x$

**b**  Solve using the addition principle. Don't forget to check!

**13.** $x + 2 = 6$     **14.** $y + 4 = 11$     **15.** $x + 15 = -5$     **16.** $t + 10 = 44$     **17.** $x + 6 = -8$

**CHECK:** $\dfrac{x + 2 = 6}{\phantom{?}}$     **CHECK:** $\dfrac{y + 4 = 11}{\phantom{?}}$     **CHECK:** $\dfrac{x + 15 = -5}{\phantom{?}}$     **CHECK:** $\dfrac{t + 10 = 44}{\phantom{?}}$     **CHECK:** $\dfrac{x + 6 = -8}{\phantom{?}}$

**18.** $z + 9 = -14$     **19.** $x + 16 = -2$     **20.** $m + 18 = -13$     **21.** $x - 9 = 6$     **22.** $x - 11 = 12$

**23.** $x - 7 = -21$     **24.** $x - 3 = -14$     **25.** $5 + t = 7$     **26.** $8 + y = 12$     **27.** $-7 + y = 13$

**28.** $-8 + y = 17$     **29.** $-3 + t = -9$     **30.** $-8 + t = -24$     **31.** $x + \dfrac{1}{2} = 7$     **32.** $24 = -\dfrac{7}{10} + r$

**33.** $12 = a - 7.9$

**34.** $2.8 + y = 11$

**35.** $r + \dfrac{1}{3} = \dfrac{8}{3}$

**36.** $t + \dfrac{3}{8} = \dfrac{5}{8}$

**37.** $m + \dfrac{5}{6} = -\dfrac{11}{12}$

**38.** $x + \dfrac{2}{3} = -\dfrac{5}{6}$

**39.** $x - \dfrac{5}{6} = \dfrac{7}{8}$

**40.** $y - \dfrac{3}{4} = \dfrac{5}{6}$

**41.** $-\dfrac{1}{5} + z = -\dfrac{1}{4}$

**42.** $-\dfrac{1}{8} + y = -\dfrac{3}{4}$

**43.** $7.4 = x + 2.3$

**44.** $8.4 = 5.7 + y$

**45.** $7.6 = x - 4.8$

**46.** $8.6 = x - 7.4$

**47.** $-9.7 = -4.7 + y$

**48.** $-7.8 = 2.8 + x$

**49.** $5\dfrac{1}{6} + x = 7$

**50.** $5\dfrac{1}{4} = 4\dfrac{2}{3} + x$

**51.** $q + \dfrac{1}{3} = -\dfrac{1}{7}$

**52.** $52\dfrac{3}{8} = -84 + x$

---

**Skill Maintenance**

**53.** Add: $-3 + (-8)$.  [1.3a]

**54.** Subtract: $-3 - (-8)$.  [1.4a]

**55.** Multiply: $-\dfrac{2}{3} \cdot \dfrac{5}{8}$.  [1.5a]

**56.** Divide: $-\dfrac{3}{7} \div \left(-\dfrac{9}{7}\right)$.  [1.6c]

**57.** Divide: $\dfrac{2}{3} \div \left(-\dfrac{4}{9}\right)$.  [1.6c]

**58.** Add: $-8.6 + 3.4$.  [1.3a]

Translate to an algebraic expression. [1.1b]

**59.** Liza had \$50 before paying $x$ dollars for a pizza. How much does she have left?

**60.** Donnie drove his S-10 pickup truck 65 mph for $t$ hours. How far did he drive?

---

**Synthesis**

**61.** ◈ Explain the difference between equivalent expressions and equivalent equations.

**62.** ◈ When solving an equation using the addition principle, how do you determine which number to add or subtract on both sides of the equation?

Solve.

**63.** ▦ $-356.788 = -699.034 + t$

**64.** $-\dfrac{4}{5} + \dfrac{7}{10} = x - \dfrac{3}{4}$

**65.** $x + \dfrac{4}{5} = -\dfrac{2}{3} - \dfrac{4}{15}$

**66.** $8 - 25 = 8 + x - 21$

**67.** $16 + x - 22 = -16$

**68.** $x + x = x$

**69.** $x + 3 = 3 + x$

**70.** $x + 4 = 5 + x$

**71.** $-\dfrac{3}{2} + x = -\dfrac{5}{17} - \dfrac{3}{2}$

**72.** $|x| = 5$

**73.** $|x| + 6 = 19$

Copyright © 1999 Addison Wesley Longman

# 2.2 Solving Equations: The Multiplication Principle

## a | Using the Multiplication Principle

Suppose that $a = b$ is true, and we multiply $a$ by some number $c$. We get the same answer if we multiply $b$ by $c$, because $a$ and $b$ are the same number.

> **THE MULTIPLICATION PRINCIPLE**
>
> For any real numbers $a$, $b$, and $c$, $c \neq 0$,
>
> $\quad a = b \quad$ is equivalent to $\quad a \cdot c = b \cdot c$.

When using the multiplication principle, we sometimes say that we "multiply on both sides of the equation by the same number."

**Example 1**   Solve: $5x = 70$.

To get $x$ alone, we multiply by the *multiplicative inverse*, or *reciprocal*, of 5. Then we get the *multiplicative identity* 1 times $x$, or $1 \cdot x$, which simplifies to $x$. This allows us to eliminate 5 on the left.

$$5x = 70 \qquad \text{The reciprocal of 5 is } \tfrac{1}{5}.$$

$$\frac{1}{5} \cdot 5x = \frac{1}{5} \cdot 70 \qquad \begin{array}{l}\text{Multiplying by } \tfrac{1}{5} \text{ to get } 1 \cdot x \text{ and} \\ \text{eliminate 5 on the left}\end{array}$$

$$1 \cdot x = 14 \qquad \text{Simplifying}$$

$$x = 14 \qquad \text{Identity property of 1: } 1 \cdot x = x$$

CHECK: 
$$\begin{array}{c} 5x = 70 \\ \hline 5 \cdot 14 \; ? \; 70 \\ 70 \; | \qquad \text{TRUE} \end{array}$$

The solution is 14.

The multiplication principle also tells us that we can "divide on both sides of the equation by a nonzero number." This is because division is the same as multiplying by a reciprocal. That is,

$$\frac{a}{c} = \frac{b}{c} \quad \text{is equivalent to} \quad a \cdot \frac{1}{c} = b \cdot \frac{1}{c}, \quad \text{when } c \neq 0.$$

In an expression like $5x$ in Example 1, the number 5 is called the **coefficient**. Example 1 could be done as follows, dividing by 5, the coefficient of $x$, on both sides.

**1.** Solve. Multiply on both sides.

$$6x = 90$$

**2.** Solve. Divide on both sides.

$$4x = -7$$

**3.** Solve: $-6x = 108$.

**4.** Solve: $-x = -10$.

**Example 2** Solve: $5x = 70$.

We have

$$5x = 70$$

$$\frac{5x}{5} = \frac{70}{5} \qquad \text{Dividing by 5 on both sides}$$

$$1 \cdot x = 14 \qquad \text{Simplifying}$$

$$x = 14. \qquad \text{Identity property of 1}$$

*Do Exercises 1 and 2.*

**Example 3** Solve: $-4x = 92$.

We have

$$-4x = 92$$

$$\frac{-4x}{-4} = \frac{92}{-4} \qquad \text{Using the multiplication principle. Dividing by } -4 \text{ on both sides is the same as multiplying by } -\frac{1}{4}.$$

$$1 \cdot x = -23 \qquad \text{Simplifying}$$

$$x = -23. \qquad \text{Identity property of 1}$$

**CHECK:**

$$\frac{-4x = 92}{-4(-23) \; ? \; 92}$$
$$92 \; | \qquad \text{TRUE}$$

The solution is $-23$.

*Do Exercise 3.*

**Example 4** Solve: $-x = 9$.

We have

$$-x = 9$$

$$-1 \cdot x = 9 \qquad \text{Using the property of } -1: \; -x = -1 \cdot x$$

$$\frac{-1 \cdot x}{-1} = \frac{9}{-1} \qquad \text{Dividing by } -1$$

$$1 \cdot x = -9$$

$$x = -9.$$

**CHECK:**

$$\frac{-x = 9}{-(-9) \; ? \; 9}$$
$$9 \; | \qquad \text{TRUE}$$

The solution is $-9$.

*Do Exercise 4.*

*Answers on page A-6*

In practice, it is generally more convenient to "divide" on both sides of the equation if the coefficient of the variable is in decimal notation or is an integer. If the coefficient is in fractional notation, it is more convenient to "multiply" by a reciprocal.

**Example 5**   Solve: $\dfrac{3}{8} = -\dfrac{5}{4}x$.

$$\frac{3}{8} = -\frac{5}{4}x$$

The reciprocal of $-\frac{5}{4}$ is $-\frac{4}{5}$. There is no sign change.

$$-\frac{4}{5} \cdot \frac{3}{8} = -\frac{4}{5} \cdot \left(-\frac{5}{4}x\right)$$

Multiplying by $-\frac{4}{5}$ to get $1 \cdot x$ and eliminate $-\frac{5}{4}$ on the right

$$-\frac{12}{40} = 1 \cdot x$$

$$-\frac{3}{10} = 1 \cdot x \qquad \text{Simplifying}$$

$$-\frac{3}{10} = x \qquad \text{Identity property of 1}$$

CHECK:   
$$\frac{3}{8} = -\frac{5}{4}x$$

$$\frac{3}{8} \;?\; -\frac{5}{4}\left(-\frac{3}{10}\right)$$

$$\frac{3}{8} \qquad \text{TRUE}$$

The solution is $-\dfrac{3}{10}$.

Note that equations are reversible. That is, if $a = b$ is true, then $b = a$ is true. Thus when we solve $\frac{3}{8} = -\frac{5}{4}x$, we can reverse it and solve $-\frac{5}{4}x = \frac{3}{8}$ if we wish.

*Do Exercise 5.*

**Example 6**   Solve: $1.16y = 9744$.

$$1.16y = 9744$$

$$\frac{1.16y}{1.16} = \frac{9744}{1.16} \qquad \text{Dividing by 1.16}$$

$$y = \frac{9744}{1.16}$$

$$= 8400$$

CHECK:   
$$1.16y = 9744$$

$$1.16(8400) \;?\; 9744$$

$$9744 \qquad \text{TRUE}$$

The solution is 8400.

*Do Exercises 6 and 7.*

---

**5.** Solve: $\dfrac{2}{3} = -\dfrac{5}{6}y$.

Solve.

**6.** $1.12x = 8736$

**7.** $6.3 = -2.1y$

*Answers on page A-6*

The Multiplication Principle

**121**

**8.** Solve: $-14 = \dfrac{-y}{2}$.

Now we solve an equation with a division using the multiplication principle. Consider an equation like $-y/9 = 14$. In Chapter 1, we learned that a division can be expressed as multiplication by the reciprocal of the divisor. Thus,

$$\frac{-y}{9} \quad \text{is equivalent to} \quad \frac{1}{9}(-y).$$

The reciprocal of $\frac{1}{9}$ is 9. Then, using the multiplication principle, we multiply by 9 on both sides. This is shown in the following example.

**Example 7**  Solve: $\dfrac{-y}{9} = 14$.

$$\frac{-y}{9} = 14$$

$$\frac{1}{9}(-y) = 14$$

$$9 \cdot \frac{1}{9}(-y) = 9 \cdot 14 \qquad \text{\textbf{Multiplying by 9 on both sides}}$$

$$-y = 126$$

$$-1 \cdot (-y) = -1 \cdot 126 \qquad \text{\textbf{Multiplying by }}-1\text{\textbf{, or dividing}} \atop \text{\textbf{by }}-1\text{\textbf{, on both sides}}$$

$$y = -126$$

CHECK:
$$\frac{-y}{9} = 14$$

$$\frac{-(-126)}{9} \; ? \; 14$$

$$\frac{126}{9}$$

$$14 \qquad \text{TRUE}$$

The solution is $-126$.

***Do Exercise 8.***

# Exercise Set 2.2

**a** Solve using the multiplication principle. Don't forget to check!

**1.** $6x = 36$

CHECK: $\dfrac{6x = 36}{\quad \big|\; ?}$

**2.** $3x = 51$

CHECK: $\dfrac{3x = 51}{\quad \big|\; ?}$

**3.** $5x = 45$

CHECK: $\dfrac{5x = 45}{\quad \big|\; ?}$

**4.** $8x = 72$

CHECK: $\dfrac{8x = 72}{\quad \big|\; ?}$

$8(9) \overset{?}{\,\big|\,} 72$

$72 \div 8 \;\big|\; 9$

$9 \div 72 \quad 8$

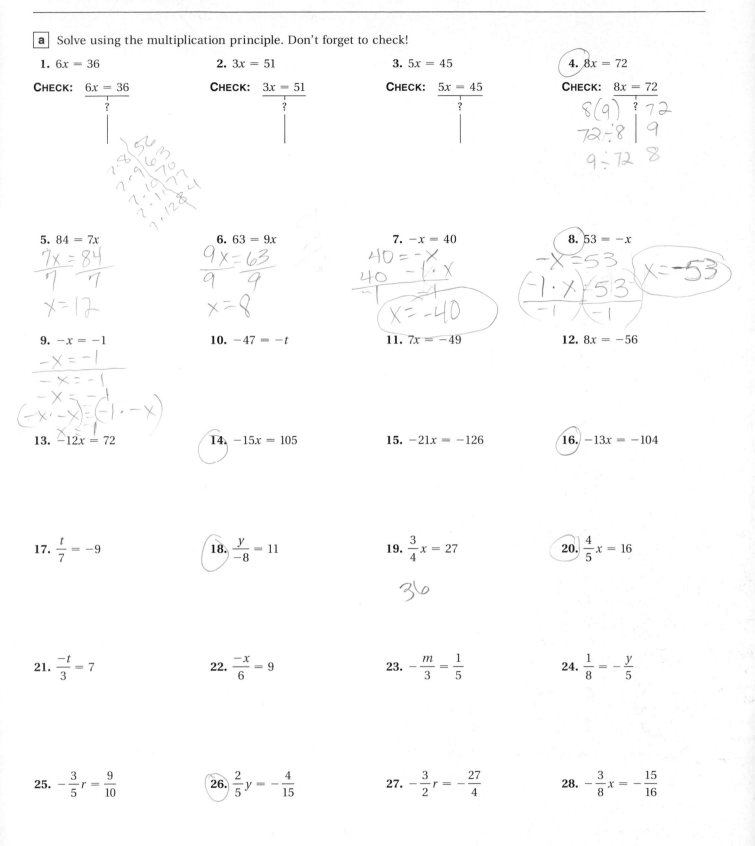

**5.** $84 = 7x$

$\dfrac{7x}{7} = \dfrac{84}{7}$

$x = 12$

**6.** $63 = 9x$

$\dfrac{9x}{9} = \dfrac{63}{9}$

$x = 8$

**7.** $-x = 40$

$40 = -x$

$40 \quad -1 \cdot x$

$x = -40$

**8.** $53 = -x$

$-x = 53$

$\dfrac{-1 \cdot x}{-1} = \dfrac{53}{-1}$   $x = -53$

**9.** $-x = -1$

$-x = -1$

$-x = -1$

$-x = -1$

$(-x \cdot -x) = (-1 \cdot -x)$

$x = 1$

**10.** $-47 = -t$

**11.** $7x = -49$

**12.** $8x = -56$

**13.** $-12x = 72$

**14.** $-15x = 105$

**15.** $-21x = -126$

**16.** $-13x = -104$

**17.** $\dfrac{t}{7} = -9$

**18.** $\dfrac{y}{-8} = 11$

$36$

**19.** $\dfrac{3}{4}x = 27$

**20.** $\dfrac{4}{5}x = 16$

**21.** $\dfrac{-t}{3} = 7$

**22.** $\dfrac{-x}{6} = 9$

**23.** $-\dfrac{m}{3} = \dfrac{1}{5}$

**24.** $\dfrac{1}{8} = -\dfrac{y}{5}$

**25.** $-\dfrac{3}{5}r = \dfrac{9}{10}$

**26.** $\dfrac{2}{5}y = -\dfrac{4}{15}$

**27.** $-\dfrac{3}{2}r = -\dfrac{27}{4}$

**28.** $-\dfrac{3}{8}x = -\dfrac{15}{16}$

**29.** $6.3x = 44.1$

**30.** $2.7y = 54$

**31.** $-3.1y = 21.7$

**32.** $-3.3y = 6.6$

**33.** $38.7m = 309.6$

**34.** $29.4m = 235.2$

**35.** $-\dfrac{2}{3}y = -10.6$

**36.** $-\dfrac{9}{7}y = 12.06$

---

**Skill Maintenance**

Collect like terms.  [1.7e]

**37.** $3x + 4x$

**38.** $6x + 5 - 7x$

**39.** $-4x + 11 - 6x + 18x$

**40.** $8y - 16y - 24y$

Remove parentheses and simplify.  [1.8b]

**41.** $3x - (4 + 2x)$

**42.** $2 - 5(x + 5)$

**43.** $8y - 6(3y + 7)$

**44.** $-2a - 4(5a - 1)$

Translate to an algebraic expression.  [1.1b]

**45.** Patty drives her van for 8 hr at a speed of $r$ mph. How far does she drive?

**46.** A triangle has a height of 10 meters and a base of $b$ meters. What is the area of the triangle?

---

**Synthesis**

**47.** ◈ When solving an equation using the multiplication principle, how do you determine by what number to multiply or divide on both sides of the equation?

**48.** ◈ Are the equations $x = 5$ and $x^2 = 25$ equivalent? Why or why not?

Solve.

**49.** ▦ $-0.2344m = 2028.732$

**50.** $0 \cdot x = 0$

**51.** $0 \cdot x = 9$

**52.** $4|x| = 48$

**53.** $2|x| = -12$

Solve for $x$.

**54.** $ax = 5a$

**55.** $3x = \dfrac{b}{a}$

**56.** $cx = a^2 + 1$

**57.** $\dfrac{a}{b}x = 4$

**58.** A student makes a calculation and gets an answer of 22.5. On the last step, the student multiplies by 0.3 when a division by 0.3 should have been done. What is the correct answer?

---

## 2.3 Using the Principles Together

### a | Applying Both Principles

Consider the equation $3x + 4 = 13$. It is more complicated than those we discussed in the preceding two sections. In order to solve such an equation, we first isolate the $x$-term, $3x$, using the addition principle. Then we apply the multiplication principle to get $x$ by itself.

**Example 1** Solve: $3x + 4 = 13$.

$$3x + 4 = 13$$
$$3x + 4 - 4 = 13 - 4$$ Using the addition principle: subtracting 4 on both sides

First, isolate the $x$-term. → $3x = 9$   Simplifying

$$\frac{3x}{3} = \frac{9}{3}$$ Using the multiplication principle: dividing by 3 on both sides

Then isolate $x$. → $x = 3$   Simplifying

CHECK:
$$\frac{3x + 4 = 13}{3 \cdot 3 + 4 \; ? \; 13}$$
$$9 + 4 \;\Big|$$
$$13 \;\Big|\quad\text{TRUE}$$

We use the rules for order of operations to carry out the check. We find the product $3 \cdot 3$. Then we add 4.

The solution is 3.

*Do Exercise 1.*

**Example 2** Solve: $-5x - 6 = 16$.

$$-5x - 6 = 16$$
$$-5x - 6 + 6 = 16 + 6$$ Adding 6 on both sides
$$-5x = 22$$
$$\frac{-5x}{-5} = \frac{22}{-5}$$ Dividing by $-5$ on both sides
$$x = -\frac{22}{5}, \text{ or } -4\frac{2}{5}$$ Simplifying

CHECK:
$$\frac{-5x - 6 = 16}{-5\left(-\frac{22}{5}\right) - 6 \; ? \; 16}$$
$$22 - 6 \;\Big|$$
$$16 \;\Big|\quad\text{TRUE}$$

The solution is $-\frac{22}{5}$.

*Do Exercises 2 and 3.*

**Objectives**

a | Solve equations using both the addition and the multiplication principles.

b | Solve equations in which like terms may need to be collected.

c | Solve equations by first removing parentheses and collecting like terms.

**For Extra Help**

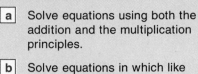

TAPE 4   TAPE 4B   MAC WIN   CD-ROM

1. Solve: $9x + 6 = 51$.

Solve.
2. $8x - 4 = 28$

3. $-\frac{1}{2}x + 3 = 1$

*Answers on page A-6*

**4.** Solve: $-18 - m = -57$.

Solve.

**5.** $-4 - 8x = 8$

**6.** $41.68 = 4.7 - 8.6y$

Solve.

**7.** $4x + 3x = -21$

**8.** $x - 0.09x = 728$

*Answers on page A-6*

**Example 3**   Solve: $45 - t = 13$.

$$45 - t = 13$$
$$-45 + 45 - t = -45 + 13 \qquad \text{Adding } -45 \text{ on both sides}$$
$$-t = -32$$
$$-1 \cdot t = -32 \qquad \text{Using the property of } -1: -t = -1 \cdot t$$
$$\frac{-1 \cdot t}{-1} = \frac{-32}{-1} \qquad \begin{array}{l}\text{Dividing by } -1 \text{ on both sides (You could have}\\\text{multiplied by } -1 \text{ on both sides instead. That}\\\text{would also change the sign on both sides.)}\end{array}$$
$$t = 32$$

The number 32 checks and is the solution.

*Do Exercise 4.*

**Example 4**   Solve: $16.3 - 7.2y = -8.18$.

$$16.3 - 7.2y = -8.18$$
$$-16.3 + 16.3 - 7.2y = -16.3 + (-8.18) \qquad \text{Adding } -16.3 \text{ on both sides}$$
$$-7.2y = -24.48$$
$$\frac{-7.2y}{-7.2} = \frac{-24.48}{-7.2} \qquad \text{Dividing by } -7.2 \text{ on both sides}$$
$$y = 3.4$$

CHECK:
$$\begin{array}{c|c} 16.3 - 7.2y = -8.18 \\ \hline 16.3 - 7.2(3.4) \;?\; -8.18 \\ 16.3 - 24.48 \\ -8.18 \mid \qquad \text{TRUE} \end{array}$$

The solution is 3.4.

*Do Exercises 5 and 6.*

## b  Collecting Like Terms

If there are like terms on one side of the equation, we collect them before using the addition or the multiplication principle.

**Example 5**   Solve: $3x + 4x = -14$.

$$3x + 4x = -14$$
$$7x = -14 \qquad \text{Collecting like terms}$$
$$\frac{7x}{7} = \frac{-14}{7} \qquad \text{Dividing by 7 on both sides}$$
$$x = -2$$

The number $-2$ checks, so the solution is $-2$.

*Do Exercises 7 and 8.*

If there are like terms on opposite sides of the equation, we get them on the same side by using the addition principle. Then we collect them. In other words, we get all terms with a variable on one side and all numbers on the other.

**Example 6**  Solve: $2x - 2 = -3x + 3$.

$$2x - 2 = -3x + 3$$

$2x - 2 + 2 = -3x + 3 + 2$  **Adding 2**

$2x = -3x + 5$  **Collecting like terms**

$2x + 3x = -3x + 3x + 5$  **Adding** $3x$

$5x = 5$  **Simplifying**

$\dfrac{5x}{5} = \dfrac{5}{5}$  **Dividing by 5**

$x = 1$  **Simplifying**

CHECK:

$$\begin{array}{c|c} 2x - 2 = -3x + 3 \\ \hline 2 \cdot 1 - 2 \ ? \ -3 \cdot 1 + 3 \\ 2 - 2 \ \big| \ -3 + 3 \\ 0 \ \big| \ 0 \end{array} \quad \text{TRUE}$$

The solution is 1.

*Do Exercise 9.*

In Example 6, we used the addition principle to get all terms with a variable on one side and all numbers on the other side. Then we collected like terms and proceeded as before. If there are like terms on one side at the outset, they should be collected before proceeding.

**Example 7**  Solve: $6x + 5 - 7x = 10 - 4x + 3$.

$$6x + 5 - 7x = 10 - 4x + 3$$

$-x + 5 = 13 - 4x$  **Collecting like terms**

$4x - x + 5 = 13 - 4x + 4x$  **Adding** $4x$ **to get all terms with a variable on one side**

$3x + 5 = 13$  **Simplifying; that is, collecting like terms**

$3x + 5 - 5 = 13 - 5$  **Subtracting 5**

$3x = 8$  **Simplifying**

$\dfrac{3x}{3} = \dfrac{8}{3}$  **Dividing by 3**

$x = \dfrac{8}{3}$  **Simplifying**

The number $\frac{8}{3}$ checks, so it is the solution.

*Do Exercises 10–12.*

**9.** Solve: $7y + 5 = 2y + 10$.

Solve.

**10.** $5 - 2y = 3y - 5$

**11.** $7x - 17 + 2x = 2 - 8x + 15$

**12.** $3x - 15 = 5x + 2 - 4x$

*Answers on page A-6*

**13.** Solve: $\dfrac{7}{8}x - \dfrac{1}{4} + \dfrac{1}{2}x = \dfrac{3}{4} + x.$

## Clearing Fractions and Decimals

In general, equations are easier to solve if they do not contain fractions or decimals. Consider, for example,

$$\frac{1}{2}x + 5 = \frac{3}{4} \quad \text{and} \quad 2.3x + 7 = 5.4.$$

If we multiply by 4 on both sides of the first equation and by 10 on both sides of the second equation, we have

$$4\left(\frac{1}{2}x + 5\right) = 4 \cdot \frac{3}{4} \quad \text{and} \quad 10(2.3x + 7) = 10 \cdot 5.4$$

or

$$4 \cdot \frac{1}{2}x + 4 \cdot 5 = 4 \cdot \frac{3}{4} \quad \text{and} \quad 10 \cdot 2.3x + 10 \cdot 7 = 10 \cdot 5.4$$

or

$$2x + 20 = 3 \quad \text{and} \quad 23x + 70 = 54.$$

The first equation has been "cleared of fractions" and the second equation has been "cleared of decimals." Both resulting equations are equivalent to the original equations and are easier to solve. *It is your choice* whether to clear fractions or decimals, but doing so often eases computations.

The easiest way to clear an equation of fractions is to multiply *every term on both sides* by the **least common multiple of all the denominators.**

**Example 8**   Solve: $\dfrac{2}{3}x - \dfrac{1}{6} + \dfrac{1}{2}x = \dfrac{7}{6} + 2x.$

The number 6 is the least common multiple of all the denominators. We multiply by 6 on both sides.

$$6\left(\frac{2}{3}x - \frac{1}{6} + \frac{1}{2}x\right) = 6\left(\frac{7}{6} + 2x\right) \qquad \text{Multiplying by 6 on both sides}$$

$$6 \cdot \frac{2}{3}x - 6 \cdot \frac{1}{6} + 6 \cdot \frac{1}{2}x = 6 \cdot \frac{7}{6} + 6 \cdot 2x \qquad \begin{array}{l}\text{Using the distributive law}\\ (\textit{Caution!} \text{ Be sure to multiply } \textit{all}\\ \text{the terms by 6.)}\end{array}$$

$$4x - 1 + 3x = 7 + 12x \qquad \begin{array}{l}\text{Simplifying. Note that the}\\ \text{fractions are cleared.}\end{array}$$

$$7x - 1 = 7 + 12x \qquad \text{Collecting like terms}$$

$$7x - 1 - 12x = 7 + 12x - 12x \qquad \text{Subtracting } 12x$$

$$-5x - 1 = 7 \qquad \text{Collecting like terms}$$

$$-5x - 1 + 1 = 7 + 1 \qquad \text{Adding 1}$$

$$-5x = 8 \qquad \text{Collecting like terms}$$

$$\frac{-5x}{-5} = \frac{8}{-5} \qquad \text{Dividing by } -5$$

$$x = -\frac{8}{5}$$

The number $-\dfrac{8}{5}$ checks, so it is the solution.

*Do Exercise 13.*

*Answer on page A-6*

To illustrate clearing decimals, we repeat Example 4, but this time we clear the equation of decimals first. Compare both methods.

To clear an equation of decimals, we count the greatest number of decimal places in any one number. If the greatest number of decimal places is 1, we multiply by 10; if it is 2, we multiply by 100; and so on.

**Example 9** Solve: $16.3 - 7.2y = -8.18$.

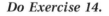

The greatest number of decimal places in any one number is *two*. Multiplying by 100, which has *two* 0's, will clear all decimals.

$$100(16.3 - 7.2y) = 100(-8.18)$$ Multiplying by 100 on both sides

$$100(16.3) - 100(7.2y) = 100(-8.18)$$ Using the distributive law

$$1630 - 720y = -818$$ Simplifying

$$1630 - 720y - 1630 = -818 - 1630$$ Subtracting 1630 on both sides

$$-720y = -2448$$ Collecting like terms

$$\frac{-720y}{-720} = \frac{-2448}{-720}$$ Dividing by $-720$ on both sides

$$y = 3.4$$

The number 3.4 checks, so it is the solution.

*Do Exercise 14.*

### c | Equations Containing Parentheses

To solve certain kinds of equations that contain parentheses, we first use the distributive laws to remove the parentheses. Then we proceed as before.

**Example 10** Solve: $4x = 2(12 - 2x)$.

$$4x = 2(12 - 2x)$$

$$4x = 24 - 4x$$ Using the distributive law to multiply and remove parentheses

$$4x + 4x = 24 - 4x + 4x$$ Adding $4x$ to get all the $x$-terms on one side

$$8x = 24$$ Collecting like terms

$$\frac{8x}{8} = \frac{24}{8}$$ Dividing by 8

$$x = 3$$

CHECK:

$$\begin{array}{c|c} 4x = 2(12 - 2x) \\ \hline 4 \cdot 3 \ ? \ 2(12 - 2 \cdot 3) \\ 12 \ \bigm| \ 2(12 - 6) \\ \bigm| \ 2 \cdot 6 \\ \bigm| \ 12 \end{array}$$  TRUE

We use the rules for order of operations to carry out the calculations on each side of the equation.

The solution is 3.

*Do Exercises 15 and 16.*

**14.** Solve: $41.68 = 4.7 - 8.6y$.

Solve.

**15.** $2(2y + 3) = 14$

$$4y + 6 = 14$$
$$\quad -6 \quad -6$$
$$\frac{4y}{4} = \frac{8}{4}$$
$$y = 2$$

**16.** $5(3x - 2) = 35$

$$15x - 10 = 35$$
$$\quad +10 \quad +10$$
$$\frac{15x}{15} = \frac{45}{15}$$
$$x = 3$$

*Answers on page A-6*

Solve.

**17.** $3(7 + 2x) = 30 + 7(x - 1)$

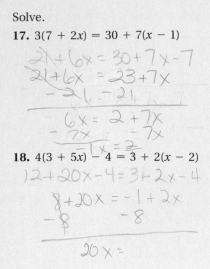

**18.** $4(3 + 5x) - 4 = 3 + 2(x - 2)$

$12 + 20x - 4 = 3 + 2x - 4$

$8 + 20x = -1 + 2x$

$\phantom{8}-8 \phantom{+ 20x = -1 +} -8$

$\phantom{8 +} 20x =$

## Calculator Spotlight

Checking Possible Solutions. To check possible solutions on a calculator, we substitute and carry out the calculations on each side. Let's check the solution of Example 11. We first substitute −2 on the left:

$2 - 5(-2 + 5).$

We carry out this computation as shown in the Calculator Spotlight in Section 1.8. We get −13. Then we substitute −2 on the right:

$3(-2 - 2) - 1.$

Carrying out the calculations, we again get −13.

### Exercises

1. Check the solution of Margin Exercise 17.
2. Check the solution of Margin Exercise 18.
3. Check the solution of Example 9.

Here is a procedure for solving the types of equation discussed in this section.

> **AN EQUATION-SOLVING PROCEDURE**
>
> 1. Multiply on both sides to clear the equation of fractions or decimals. (This is optional, but it can ease computations.)
> 2. If parentheses occur, multiply to remove them using the *distributive laws.*
> 3. Collect like terms on each side, if necessary.
> 4. Get all terms with variables on one side and all numbers (constant terms) on the other side, using the *addition principle.*
> 5. Collect like terms again, if necessary.
> 6. Multiply or divide to solve for the variable, using the *multiplication principle.*
> 7. Check all possible solutions in the original equation.

**Example 11**   Solve: $2 - 5(x + 5) = 3(x - 2) - 1.$

$$2 - 5(x + 5) = 3(x - 2) - 1$$

$$2 - 5x - 25 = 3x - 6 - 1 \qquad \text{Using the distributive laws to multiply and remove parentheses}$$

$$-5x - 23 = 3x - 7 \qquad \text{Collecting like terms}$$

$$-5x - 23 + 5x = 3x - 7 + 5x \qquad \text{Adding } 5x$$

$$-23 = 8x - 7 \qquad \text{Collecting like terms}$$

$$-23 + 7 = 8x - 7 + 7 \qquad \text{Adding } 7$$

$$-16 = 8x \qquad \text{Collecting like terms}$$

$$\frac{-16}{8} = \frac{8x}{8} \qquad \text{Dividing by 8}$$

$$-2 = x$$

**CHECK:**

$$\begin{array}{c|c} \multicolumn{2}{c}{2 - 5(x + 5) = 3(x - 2) - 1} \\ \hline 2 - 5(-2 + 5) \ ? \ 3(-2 - 2) - 1 \\ 2 - 5(3) \ \bigm| \ 3(-4) - 1 \\ 2 - 15 \ \bigm| \ -12 - 1 \\ -13 \ \bigm| \ -13 \qquad \text{TRUE} \end{array}$$

The solution is −2.

*Do Exercises 17 and 18.*

*Answers on page A-6*

# Exercise Set 2.3

*show check xtra point.*

**a** Solve. Don't forget to check!

**1.** $5x + 6 = 31$

CHECK: $\dfrac{5x + 6 = 31}{\overset{?}{\vert}}$

**2.** $7x + 6 = 13$

CHECK: $\dfrac{7x + 6 = 13}{\overset{?}{\vert}}$

**3.** $8x + 4 = 68$

CHECK: $\dfrac{8x + 4 = 68}{\overset{?}{\vert}}$

**4.** $4y + 10 = 46$

CHECK: $\dfrac{4y + 10 = 46}{\overset{?}{\vert}}$

**5.** $4x - 6 = 34$

**6.** $5y - 2 = 53$

**7.** $3x - 9 = 33$

**8.** $4x - 19 = 5$

**9.** $7x + 2 = -54$

**10.** $5x + 4 = -41$

**11.** $-45 = 3 + 6y$

**12.** $-91 = 9t + 8$

**13.** $-4x + 7 = 35$

**14.** $-5x - 7 = 108$

**15.** $-7x - 24 = -129$

**16.** $-6z - 18 = -132$

**b** Solve.

**17.** $5x + 7x = 72$

CHECK: $\dfrac{5x + 7x = 72}{\overset{?}{\vert}}$

**18.** $8x + 3x = 55$

CHECK: $\dfrac{8x + 3x = 55}{\overset{?}{\vert}}$

**19.** $8x + 7x = 60$

CHECK: $\dfrac{8x + 7x = 60}{\overset{?}{\vert}}$

**20.** $8x + 5x = 104$

CHECK: $\dfrac{8x + 5x = 104}{\overset{?}{\vert}}$

**21.** $4x + 3x = 42$

**22.** $7x + 18x = 125$

**23.** $-6y - 3y = 27$

**24.** $-5y - 7y = 144$

**f 25.** $-7y - 8y = -15$

**26.** $-10y - 3y = -39$

**27.** $x + \dfrac{1}{3}x = 8$

**28.** $x + \dfrac{1}{4}x = 10$

**29.** $10.2y - 7.3y = -58$    **30.** $6.8y - 2.4y = -88$    **31.** $8y - 35 = 3y$    **32.** $4x - 6 = 6x$

**33.** $8x - 1 = 23 - 4x$    **34.** $5y - 2 = 28 - y$    **35.** $2x - 1 = 4 + x$    **36.** $4 - 3x = 6 - 7x$

**37.** $6x + 3 = 2x + 11$    **38.** $14 - 6a = -2a + 3$

**39.** $5 - 2x = 3x - 7x + 25$    **40.** $-7z + 2z - 3z - 7 = 17$

**41.** $4 + 3x - 6 = 3x + 2 - x$    **42.** $5 + 4x - 7 = 4x - 2 - x$

**43.** $4y - 4 + y + 24 = 6y + 20 - 4y$    **44.** $5y - 7 + y = 7y + 21 - 5y$

Solve. Clear fractions or decimals first.

**45.** $\dfrac{7}{2}x + \dfrac{1}{2}x = 3x + \dfrac{3}{2} + \dfrac{5}{2}x$    **46.** $\dfrac{7}{8}x - \dfrac{1}{4} + \dfrac{3}{4}x = \dfrac{1}{16} + x$

**47.** $\dfrac{2}{3} + \dfrac{1}{4}t = \dfrac{1}{3}$    **48.** $-\dfrac{3}{2} + x = -\dfrac{5}{6} - \dfrac{4}{3}$

**49.** $\dfrac{2}{3} + 3y = 5y - \dfrac{2}{15}$    **50.** $\dfrac{1}{2} + 4m = 3m - \dfrac{5}{2}$

**51.** $\dfrac{5}{3} + \dfrac{2}{3}x = \dfrac{25}{12} + \dfrac{5}{4}x + \dfrac{3}{4}$    **52.** $1 - \dfrac{2}{3}y = \dfrac{9}{5} - \dfrac{y}{5} + \dfrac{3}{5}$

**53.** $2.1x + 45.2 = 3.2 - 8.4x$    **54.** $0.96y - 0.79 = 0.21y + 0.46$

Copyright © 1999 Addison Wesley Longman

**55.** $1.03 - 0.62x = 0.71 - 0.22x$

**56.** $1.7t + 8 \quad 1.62t = 0.4t - 0.32 + 8$

**57.** $\dfrac{2}{7}x - \dfrac{1}{2}x = \dfrac{3}{4}x + 1$

**58.** $\dfrac{5}{16}y + \dfrac{3}{8}y = 2 + \dfrac{1}{4}y$

| c | Solve.

**59.** $3(2y - 3) = 27$

**60.** $8(3x + 2) = 30$

**61.** $40 = 5(3x + 2)$

**62.** $9 = 3(5x - 2)$

**63.** $2(3 + 4m) - 9 = 45$

**64.** $5x + 5(4x - 1) = 20$

**65.** $5r - (2r + 8) = 16$

**66.** $6b - (3b + 8) = 16$

**67.** $6 - 2(3x - 1) = 2$

**68.** $10 - 3(2x - 1) = 1$

**69.** $5(d + 4) = 7(d - 2)$

**70.** $3(t - 2) = 9(t + 2)$

$$3t - 6 = 9t + 18$$
$$+6 \qquad +6$$
$$3t = 9t + 24$$
$$-9t \quad -9t$$

**71.** $8(2t + 1) = 4(7t + 7)$

**72.** $7(5x - 2) = 6(6x - 1)$

**73.** $3(r - 6) + 2 = 4(r + 2) - 21$

**74.** $5(t + 3) + 9 = 3(t - 2) + 6$

**75.** $19 - (2x + 3) = 2(x + 3) + x$

$$13 - 2c - 2 = 2c + 4 + 3c$$

**76.** $13 - (2c + 2) = 2(c + 2) + 3c$

$$13 - 2c - 2 = 2c + 4 + 3c$$
$$11 - 2c = 5c + 4$$
$$+2c = +2c$$
$$11 = 7c + 4$$

**77.** $2[4 - 2(3 - x)] - 1 = 4[2(4x - 3) + 7] - 25$

**78.** $5[3(7 - t) - 4(8 + 2t)] - 20 = -6[2(6 + 3t) - 4]$

$$11 = 7c + 4$$
$$-4 = -4$$
$$\dfrac{7}{7} = \dfrac{7c}{7} \qquad \boxed{c = 1}$$

**79.** $0.7(3x + 6) = 1.1 - (x + 2)$

**80.** $0.9(2x + 8) = 20 - (x + 5)$

**81.** $a + (a - 3) = (a + 2) - (a + 1)$

**82.** $0.8 - 4(b - 1) = 0.2 + 3(4 - b)$

---

**Skill Maintenance**

**83.** Divide: $-22.1 \div 3.4$.  [1.6c]

**84.** Factor: $7x - 21 - 14y$.  [1.7d]

**85.** Use $<$ or $>$ for ▮ to write a true sentence:  [1.2d]
$-15$ ▮ $-13$.

**86.** Find $-(-x)$ when $x = -14$.  [1.3b]

**87.** Add: $-22.1 + 3.4$.  [1.3a]

**88.** Subtract: $-22.1 - 3.4$.  [1.4a]

Translate to an algebraic expression.  [1.1b]

**89.** A number $c$ is divided by 8.

**90.** A parallelogram with height $h$ has a base length of 13.4. What is the area of the parallelogram?

---

**Synthesis**

**91.** ◈ What procedure would you follow to solve an equation like $0.23x + \frac{17}{3} = -0.8 + \frac{3}{4}x$? Could your procedure be streamlined? If so, how?

**92.** ◈ Consider any equation of the form $ax + b = c$. Describe a procedure that can be used to solve for $x$.

Solve.

**93.** ▦ $0.008 + 9.62x - 42.8 = 0.944x + 0.0083 - x$

**94.** $\dfrac{y - 2}{3} = \dfrac{2 - y}{5}$

**95.** $0 = y - (-14) - (-3y)$

**96.** $3x = 4x$

**97.** $\dfrac{5 + 2y}{3} = \dfrac{25}{12} + \dfrac{5y + 3}{4}$

**98.** ▦ $0.05y - 1.82 = 0.708y - 0.504$

**99.** $-2y + 5y = 6y$

**100.** $\dfrac{1}{4}(8y + 4) - 17 = -\dfrac{1}{2}(4y - 8)$

**101.** $\dfrac{1}{3}(6x + 24) - 20 = -\dfrac{1}{4}(12x - 72)$

**102.** $\dfrac{2}{3}\left(\dfrac{7}{8} - 4x\right) - \dfrac{5}{8} = \dfrac{3}{8}$

**103.** $\dfrac{3}{4}\left(3x - \dfrac{1}{2}\right) - \dfrac{2}{3} = \dfrac{1}{3}$

**104.** $\dfrac{4 - 3x}{7} = \dfrac{2 + 5x}{49} - \dfrac{x}{14}$

**105.** Solve the equation $4x - 8 = 32$ by first using the addition principle. Then solve it by first using the multiplication principle.

Solve linear equations as a group.

Chapter 2  Solving Equations
and Inequalities

Collaborative
Learning Manual

# 2.4 Applications and Problem Solving

## a | Five Steps for Solving Problems

We have studied many new equation-solving tools in this chapter. We now use them for applications and problem solving. The following five-step strategy can be very helpful in solving problems.

> **FIVE STEPS FOR PROBLEM SOLVING IN ALGEBRA**
>
> 1. *Familiarize* yourself with the problem situation.
> 2. *Translate* the problem to an equation.
> 3. *Solve* the equation.
> 4. *Check* the answer in the original problem.
> 5. *State* the answer to the problem clearly.

**Objective**

a | Solve applied problems by translating to equations.

**For Extra Help**

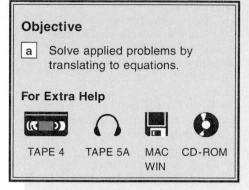

TAPE 4    TAPE 5A    MAC WIN    CD-ROM

Of the five steps, the most important is probably the first one: becoming familiar with the problem situation. The table in the margin lists some hints for familiarization.

**Example 1** *Subway Sandwich.* Subway is a national restaurant firm that serves sandwiches prepared in buns of length 18 in. (**Source**: Subway Restaurants). Suppose Jenny, Demi, and Sarah buy one of these sandwiches and take it back to their room. Since they have different appetites, Jenny cuts the sandwich in such a way that Demi gets half of what Jenny gets and Sarah gets three-fourths of what Jenny gets. Find the length of each person's sandwich.

1. **Familiarize.** We first make a drawing. Because the sandwich lengths are expressed in terms of Jenny's sandwich, we let

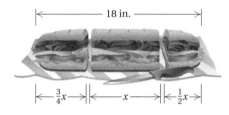

$x$ = the length of Jenny's sandwich.

Then $\frac{1}{2}x$ = the length of Demi's sandwich

and $\frac{3}{4}x$ = the length of Sarah's sandwich.

2. **Translate.** From the statement of the problem and the drawing, we see that the lengths add up to 18 in. That gives us our translation:

| Length of Jenny's sandwich | plus | Length of Demi's sandwich | plus | Length of Sarah's sandwich | is | Total length |
|:--:|:--:|:--:|:--:|:--:|:--:|:--:|
| ↓ | ↓ | ↓ | ↓ | ↓ | ↓ | ↓ |
| $x$ | $+$ | $\frac{1}{2}x$ | $+$ | $\frac{3}{4}x$ | $=$ | $18$ |

> To familiarize yourself with a problem situation:
>
> - If a problem is given in words, read it carefully. Reread the problem, perhaps aloud. Try to verbalize the problem as if you were explaining it to someone else.
> - Make a drawing and label it with known information, using specific units if given. Also, indicate unknown information.
> - Choose a variable (or variables) to represent the unknown and clearly state what the variable represents. Be descriptive! For example, let $L$ = length, $d$ = distance, and so on.
> - Find further information. Look up formulas or definitions with which you are not familiar. (Geometric formulas appear on the inside front cover of this text.) Consult a reference librarian or an expert in the field.
> - Create a table that lists all the information you have available. Look for patterns that may help in the translation to an equation.
> - Guess or estimate the answer.

**1.** *Rocket Sections.* A rocket is divided into three sections: the payload and navigation section in the top, the fuel section in the middle, and the rocket engine section in the bottom. The top section is one-sixth the length of the bottom section. The middle section is one-half the length of the bottom section. The total length is 240 ft. Find the length of each section.

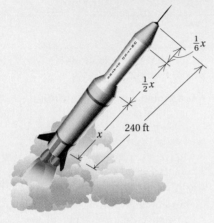

**2.** If 5 is subtracted from three times a certain number, the result is 10. What is the number?

*Answers on page A-7*

**3. Solve.**   We solve the equation by clearing fractions as follows:

$$x + \frac{1}{2}x + \frac{3}{4}x = 18 \qquad \text{The LCM of all the denominators is 4.}$$

$$4\left(x + \frac{1}{2}x + \frac{3}{4}x\right) = 4 \cdot 18 \qquad \text{Multiplying by the LCM, 4}$$

$$4 \cdot x + 4 \cdot \frac{1}{2}x + 4 \cdot \frac{3}{4}x = 4 \cdot 18 \qquad \text{Using the distributive law}$$

$$4x + 2x + 3x = 72 \qquad \text{Simplifying}$$

$$9x = 72 \qquad \text{Collecting like terms}$$

$$\frac{9x}{9} = \frac{72}{9} \qquad \text{Dividing by 9}$$

$$x = 8.$$

**4. Check.**   Do we have an answer to the *problem*? If the length of Jenny's sandwich is 8 in., then the length of Demi's sandwich is $\frac{1}{2} \cdot 8$ in., or 4 in., and the length of Sarah's sandwich is $\frac{3}{4} \cdot 8$ in., or 6 in. These lengths add up to 18 in. Our answer checks.

**5. State.**   The length of Jenny's sandwich is 8 in., the length of Demi's sandwich is 4 in., and the length of Sarah's sandwich is 6 in.

*Do Exercise 1.*

**Example 2**   Five plus three more than a number is nineteen. What is the number?

**1. Familiarize.**   Let $x =$ the number. Then "three more than a number" translates to $x + 3$, and "five plus three more than a number" translates to $5 + (x + 3)$.

**2. Translate.**   The familiarization leads us to the following translation:

| Five | plus | Three more than a number | is | Nineteen. |
|:---:|:---:|:---:|:---:|:---:|
| ↓ | ↓ | ↓ | ↓ | ↓ |
| 5 | + | $(x + 3)$ | = | 19. |

**3. Solve.**   We solve the equation:

$$5 + (x + 3) = 19$$

$$x + 8 = 19 \qquad \text{Collecting like terms}$$

$$x + 8 - 8 = 19 - 8 \qquad \text{Subtracting 8}$$

$$x = 11.$$

**4. Check.**   Be sure to check your answer in the original wording of the problem, not in the equation that you solved. This will enable you to check for errors in the translation as well. Three more than 11 is 14. Adding 5 to 14, we get 19. This checks.

**5. State.**   The number is 11.

*Do Exercise 2.*

Recall that the

Set of integers $= \{\ldots, -5, -4, -3, -2, -1, 0, 1, 2, 3, 4, 5, \ldots\}$.

Before we solve the next problem, we need to learn some additional terminology regarding integers.

The following are examples of **consecutive integers:** 16, 17, 18, 19, 20; and −31, −30, −29, −28. Note that consecutive integers can be represented in the form $x$, $x + 1$, $x + 2$, and so on.

The following are examples of **consecutive even integers:** 16, 18, 20, 22, 24; and −52, −50, −48, −46. Note that consecutive even integers can be represented in the form $x$, $x + 2$, $x + 4$, and so on.

The following are examples of **consecutive odd integers:** 21, 23, 25, 27, 29; and −71, −69, −67, −65. Note that consecutive odd integers can be represented in the form $x$, $x + 2$, $x + 4$, and so on.

**Example 3**  *Interstate Mile Markers.* If you are traveling on a U.S. interstate highway, you will notice numbered markers every mile to tell your location in case of an accident or other emergency. In many states, the numbers on the markers increase from west to east. (***Source:*** Federal Highway Administration, Ed Rotalewski) The sum of two consecutive mile markers on I-70 in Kansas is 559. Find the numbers on the markers.

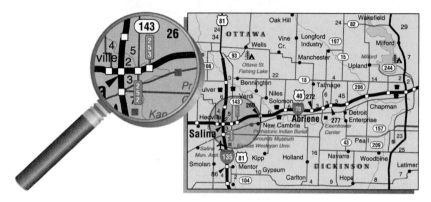

1. **Familiarize.**  The numbers on the mile markers are consecutive positive integers. Thus if we let $x =$ the smaller number, then $x + 1 =$ the larger number.

   To become familiar with the problem, we can make a table. First, we guess a value for $x$; then we find $x + 1$. Finally, we add the two numbers and check the sum. From the table, we see that the first marker should be between 252 and 302. You might actually solve the problem this way, but let's work on developing our algebra skills.

2. **Translate.**  We reword the problem and translate as follows.

   | First integer | plus | Second integer | is | 559 | Rewording |
   |:---:|:---:|:---:|:---:|:---:|---|
   | ↓ | ↓ | ↓ | ↓ | ↓ | |
   | $x$ | $+$ | $(x + 1)$ | $=$ | 559 | Translating |

3. **Solve.**  We solve the equation:

$$x + (x + 1) = 559$$
$$2x + 1 = 559 \quad \text{Collecting like terms}$$
$$2x + 1 - 1 = 559 - 1 \quad \text{Subtracting 1}$$
$$2x = 558$$
$$\frac{2x}{2} = \frac{558}{2} \quad \text{Dividing by 2}$$
$$x = 279.$$

If $x$ is 279, then $x + 1$ is 280.

**3.** *Interstate Mile Markers.* The sum of two consecutive mile markers on I-90 in upstate New York is 627 (***Source:*** New York State Department of Transportation). (On I-90 in New York, the marker numbers *increase* from east to west.) Find the numbers on the markers.

| $x$ | $x + 1$ | Sum of $x$ and $x + 1$ |
|:---:|:---:|:---:|
| 114 | 115 | 229 |
| 252 | 253 | 505 |
| 302 | 303 | 605 |

*Answer on page A-7*

**4.** *IKON Copiers.* The law firm in Example 4 decides to raise its budget to $2400 for the 3-month period. How many copies can they make for $2400?

**4. Check.** Our possible answers are 279 and 280. These are consecutive positive integers and 279 + 280 = 559, so the answers check.

**5. State.** The mile markers are 279 and 280.

*Do Exercise 3 on the preceding page.*

**Example 4** *IKON Copiers.* IKON Office Solutions rents a Canon GP30F copier for $240 per month plus 1.8¢ per copy. A law firm needs to lease a copy machine for use during a special case that they anticipate will take 3 months. If they allot a budget of $1500, how many copies can they make?

*Source*: IKON Office Solutions, Keith Palmer

**1. Familiarize.** Suppose that the law firm makes 20,000 copies. Then the cost is

Monthly charges  plus  Copy charges

or

3($240)  plus  Cost per copy  times  Number of copies

$720  +  $0.018  ·  20,000,

which is $1080. This process familiarizes us with the way in which a calculation is made. Note that we convert 1.8¢ to $0.018 so that all information is in the same unit, dollars. Otherwise, we will not get the correct answer.

We let $c$ = the number of copies that can be made for $1500.

**2. Translate.** We reword the problem and translate as follows.

Monthly costs  plus  Cost per copy  times  Number of copies  is  Cost

3($240)  +  $0.018  ·  $c$  =  $1500

**3. Solve.** We solve the equation:

$$3(240) + 0.018c = 1500$$
$$720 + 0.018c = 1500$$
$$1000(720 + 0.018c) = 1000 \cdot 1500 \quad \text{Multiplying by 1000 on both sides to clear decimals}$$
$$1000(720) + 1000(0.018c) = 1,500,000 \quad \text{Using the distributive law}$$
$$720,000 + 18c = 1,500,000 \quad \text{Simplifying}$$
$$720,000 + 18c - 720,000 = 1,500,000 - 720,000 \quad \text{Subtracting 720,000}$$
$$18c = 780,000$$
$$\frac{18c}{18} = \frac{780,000}{18} \quad \text{Dividing by 18}$$
$$c \approx 43,333. \quad \text{Rounding to the nearest one. "≈" means "is approximately equal to."}$$

*Answer on page A-7*

**4. Check.** We check in the original problem. The cost for 43,333 pages is 43,333($0.018) = $779.994. The rental for 3 months is 3($240) = $720. The total cost is then $779.994 + $720 ≈ $1499.99, which is just about the $1500 allotted.

**5. State.** The law firm can make 43,333 copies on the copy rental allotment of $1500.

*Do Exercise 4 on the preceding page.*

**Example 5** *Perimeter of NBA Court.* The perimeter of an NBA basketball court is 288 ft. The length is 44 ft longer than the width. (*Source:* National Basketball Association) Find the dimensions of the court.

**1. Familiarize.** We first make a drawing.

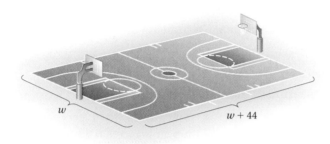

We let $w$ = the width of the rectangle. Then $w + 44$ = the length. The perimeter $P$ of a rectangle is the distance around the rectangle and is given by the formula $2l + 2w = P$, where

$l$ = the length and $w$ = the width.

**2. Translate.** To translate the problem, we substitute $w + 44$ for $l$ and 288 for $P$:

$$2l + 2w = P$$
$$2(w + 44) + 2w = 288.$$

**3. Solve.** We solve the equation:

$$2(w + 44) + 2w = 288$$
$$2 \cdot w + 2 \cdot 44 + 2w = 288 \qquad \text{Using the distributive law}$$
$$4w + 88 = 288 \qquad \text{Collecting like terms}$$
$$4w + 88 - 88 = 288 - 88 \qquad \text{Subtracting 88}$$
$$4w = 200$$
$$\frac{4w}{4} = \frac{200}{4} \qquad \text{Dividing by 4}$$
$$w = 50.$$

Thus possible dimensions are

$w = 50$ ft and $l = w + 44 = 50 + 44$, or 94 ft.

**4. Check.** If the width is 50 ft and the length is 94 ft, then the perimeter is 2(50 ft) + 2(94 ft), or 288 ft. This checks.

**5. State.** The width is 50 ft and the length is 94 ft.

*Do Exercise 5.*

**5.** *Perimeter of High School Basketball Court.* The perimeter of a standard high school basketball court is 268 ft. The length is 34 ft longer than the width. Find the dimensions of the court.

*Answer on page A-7*

**6.** The second angle of a triangle is three times as large as the first. The third angle measures 30° more than the first angle. Find the measures of the angles.

**Example 6** *Cross Section of a Roof.* In a triangular cross section of a roof, the second angle is twice as large as the first angle. The measure of the third angle is 20° greater than that of the first angle. How large are the angles?

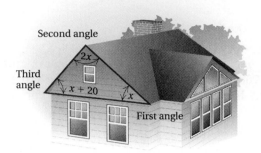

1. **Familiarize.** We first make a drawing as shown above. We let

$$\text{measure of first angle} = x.$$

Then   measure of second angle $= 2x$

and       measure of third angle $= x + 20$.

2. **Translate.** To translate, we need to recall a geometric fact. (You might, as part of step 1, look it up in a geometry book or in the list of formulas on the inside front cover.) Remember, the measures of the angles of a triangle total 180°.

| Measure of first angle | plus | Measure of second angle | plus | Measure of third angle | is | 180° |
|:---:|:---:|:---:|:---:|:---:|:---:|:---:|
| ↓ | ↓ | ↓ | ↓ | ↓ | ↓ | ↓ |
| $x$ | $+$ | $2x$ | $+$ | $(x + 20)$ | $=$ | 180° |

3. **Solve.** We solve the equation:

$$x + 2x + (x + 20) = 180$$
$$4x + 20 = 180$$
$$4x + 20 - 20 = 180 - 20$$
$$4x = 160$$
$$\frac{4x}{4} = \frac{160}{4}$$
$$x = 40.$$

Possible measures for the angles are as follows:

First angle:          $x = 40°$;

Second angle:       $2x = 2(40) = 80°$;

Third angle:     $x + 20 = 40 + 20 = 60°$.

4. **Check.** Consider our answers: 40°, 80°, and 60°. The second is twice the first and the third is 20° greater than the first. The sum is 180°. The angles check.

5. **State.** The measures of the angles are 40°, 80°, and 60°.

---

*CAUTION!* Units are important in answers. Remember to include them, where appropriate.

---

*Do Exercise 6.*

**Example 7** *Nike, Inc.* The equation

$$y = 0.69606x + 1.68722$$

can be used to approximate the total revenue $y$, in billions of dollars, of Nike, Inc., in year $x$, where

> $x = 0$ corresponds to 1990,
>
> $x = 1$ corresponds to 1991,
>
> $x = 2$ corresponds to 1992,
>
> $x = 10$ corresponds to 2000,
>
> and so on.

(**Source**: Nike, Inc.) (This equation was developed from a procedure called *regression*. Its discussion belongs to a later course.)

**a)** Find the total revenue in 1999 and 2008.

**b)** In what year will the total revenue be about $12.82418 billion?

  Since a formula has been given, we will not use the five-step problem-solving strategy.

**a)** To find the total revenue for 1999, note first that $1999 - 1990 = 9$. We substitute 9 for $x$:

$$y = 0.69606x + 1.68722$$
$$= 0.69606(9) + 1.68722$$
$$= \$7.95176.$$

To find the total revenue for 2008, note that $2008 - 1990 = 18$. We substitute 18 for $x$:

$$y = 0.69606x + 1.68722$$
$$= 0.69606(18) + 1.68722$$
$$= \$14.2163.$$

Thus the total revenue will be $7.95176 billion in 1999 and $14.2163 billion in 2008.

**b)** To determine the year in which the total revenue will be about $12.82418 billion, we first substitute 12.82418 for $y$. Then we solve for $x$. Note that we have not cleared decimals because the numbers have the same number of decimal places. (You may choose to do so.)

$$y = 0.69606x + 1.68722$$
$$12.82418 = 0.69606x + 1.68722$$
$$12.82418 - 1.68722 = 0.69606x + 1.68722 - 1.68722$$
$$11.13696 = 0.69606x$$
$$\frac{11.13696}{0.69606} = \frac{0.69606x}{0.69606}$$
$$16 = x$$

The number 16 is the number of years *after* 1990. To find that year, we add 16 to 1990: $1990 + 16 = 2006$. Thus, assuming the equation continues to be valid, the total revenue of Nike, Inc., will be $12.82418 billion in 2006.

*Do Exercise 7.*

**7.** *Nike, Inc.* Referring to Example 7:

**a)** Find the total revenue in 2002 and 2010.

**b)** In what year will the total revenue be about $9.34388 billion?

*Answer on page A-7*

# Improving Your Math Study Skills

## Extra Tips on Problem Solving

The following tips, which are focused on problem solving, summarize some points already considered and propose some new ones.

- Get in the habit of using all five steps for problem solving.

1. *Familiarize* **yourself with the problem situation.** Some suggestions for this are given on p. 135.

2. *Translate* **the problem to an equation.** As you study more mathematics, you will find that the translation may be to some other kind of mathematical language, such as an inequality.

3. *Solve* **the equation.** If the translation is to some other kind of mathematical language, you would carry out some kind of mathematical manipulation—in the case of an inequality, you would solve it.

4. *Check* **the answer in the original problem.** This does not mean to check in the translated equation. It means to go back to the original worded problem.

5. *State* **the answer to the problem clearly.**

For Step 4, some further comment on checking is appropriate. *You may be able to translate to an equation and to solve the equation, but you may find that none of the solutions of the equation is the solution of the original problem.* To see how this can happen, consider this example.

### Example

The sum of two consecutive even integers is 537. Find the integers.

1. *Familiarize.* Suppose we let $x =$ the first number. Then $x + 2 =$ the second number.

2. *Translate.* The problem can be translated to the following equation: $x + (x + 2) = 537$.

3. *Solve.* We solve the equation as follows:

$$2x + 2 = 537$$
$$2x = 535$$
$$x = \frac{535}{2}, \text{ or } 267.5.$$

4. *Check.* Then $x + 2 = 269.5$. However, the numbers are not only not even, but they are not integers.

5. *State.* The problem has no solution.

The following are some other tips.

- **To be good at problem solving, do lots of problems.** Learning to solve problems is similar to learning other skills such as golf. At first you may not be successful, but the more you practice and work at improving your skills, the more successful you will become. For problem solving, do more than just two or three odd-numbered assigned problems. Do them all, and if you have time, do the even-numbered problems as well. Then find another book on the same subject and do problems in that book.

- **Look for patterns when solving problems.** You will eventually see patterns in similar kinds of problems. For example, there is a pattern in the way that you solve problems involving consecutive integers.

- **When translating to an equation, or some other mathematical language, consider the dimensions of the variables and the constants in the equation.** The variables that represent length should all be in the same unit, those that represent money should all be in dollars or in cents, and so on.

# Exercise Set 2.4

**a** Solve.

**1.** Two times a number added to 85 is 117. Find the number.

$2x + 85 = 117$
$\quad -85 \quad -85 \qquad x = 16$
$\overline{\quad 2x = 32\quad}$

**2.** Eight times a number plus 7 is 2559. Find the number.

$8 \cdot x + 7 = 2559$

**3.** Three less than twice a number is −4. Find the number.

$2x - 3 = -4$
$\quad + 3 = +3$
$\overline{\quad 2x = -1\quad} \qquad \boxed{x = -\dfrac{1}{2}}$
$\quad \dfrac{2x}{2} = \dfrac{-1}{2}$

**4.** Seven less than four times a number is −27. Find the number.

$4x - 7 = -27$

**5.** When 17 is subtracted from four times a certain number, the result is 211. What is the number?

**6.** When 36 is subtracted from five times a certain number, the result is 374. What is the number?

**7.** A 240-in. pipe is cut into two pieces. One piece is three times the length of the other. Find the lengths of the pieces.

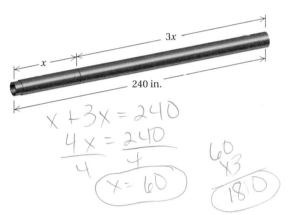

$x + 3x = 240$
$\dfrac{4x = 240}{4 \qquad 4}$
$\boxed{x = 60}$

$\begin{array}{r} 60 \\ \times 3 \\ \hline \boxed{180} \end{array}$

**8.** A 72-in. board is cut into two pieces. One piece is 2 in. longer than the other. Find the lengths of the pieces.

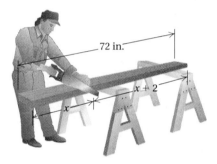

**9.** *Statue of Liberty.* The height of the Eiffel Tower is 974 ft, which is about 669 ft higher than the Statue of Liberty. What is the height of the Statue of Liberty?

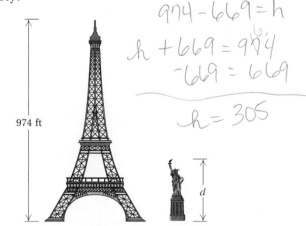

$974 - 669 = h$
$h + 669 = 974$
$\quad -669 = 669$
$\overline{\qquad h = 305}$

**10.** *Area of Lake Ontario.* The area of Lake Superior is about four times the area of Lake Ontario. The area of Lake Superior is 30,172 mi². What is the area of Lake Ontario?

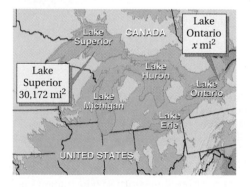

**11.** *Wheaties.* Recently, the cost of four 18-oz boxes of Wheaties cereal was $11.56. What was the cost of one box?

**12.** *Women's Dresses.* In a recent year, the total amount spent on women's blouses was $6.5 billion. This was $0.2 billion more than what was spent on women's dresses. How much was spent on women's dresses?

**13.** If you double a certain number and then add 16, you get $\frac{2}{3}$ of the original number. What is the original number?

$2x + 16 = \frac{2}{3}x$

$-12 = x$

**14.** If you double a certain number and then add 85, you get $\frac{3}{4}$ of the original number. What is the original number?

**15.** *Iditarod Race.* The Iditarod sled dog race extends for 1049 mi from Anchorage to Nome (**Source**: Iditarod Trail Commission). If a musher is twice as far from Anchorage as from Nome, how much of the race has the musher completed?

Chukchi Sea

Nome

Alaska
Anchorage

The 1049–mile
Iditarod race route

0 100 200
miles

Gulf of Alaska

**16.** *Home Remodeling.* In a recent year, Americans spent a total of $35 billion to remodel bathrooms and kitchens. Twice as much was spent on kitchens as bathrooms. How much was spent on each?

**17.** *Consecutive Page Numbers.* The sum of the page numbers on the facing pages of a book is 573. What are the page numbers?

$x$      $x + 1$

**18.** *Consecutive Post Office Box Numbers.* The sum of the numbers on two consecutive post office boxes is 547. What are the numbers?

$x$   $x + 1$

**19.** The numbers on Sam's three raffle tickets are consecutive integers. The sum of the numbers is 126. What are the numbers?

$x + (x+1) + (x+2) = 126$

$3x + 3 = 126$
$\quad\; -3 \quad\; -3$

$\dfrac{3x}{3} = \dfrac{123}{3}$

$x = 41$   42
    43

**20.** The ages of Whitney, Wesley, and Wanda are consecutive integers. The sum of their ages is 108. What are their ages?

**21.** The sum of three consecutive odd integers is 189. What are the integers?

**22.** Three consecutive integers are such that the first plus one-half the second plus seven less than twice the third is 2101. What are the integers?

Copyright © 1999 Addison Wesley Longman

**23.** *Standard Billboard Sign.* A standard rectangular highway billboard sign has a perimeter of 124 ft. The length is 6 ft more than three times the width. Find the dimensions.

$2w + 2(3w+6) = 124$    $w = 14$
$2w + 6w + 12 = 124$    $\times 3$
$8w + 12 = 124$    $\frac{2}{42+6}$
$\phantom{8w} \;\; -12 \;\; -12$

**24.** *Two by Four.* The perimeter of a cross section of a "two-by-four" piece of lumber is $10\frac{1}{2}$ in. The length is twice the width. Find the actual dimensions of the cross section of a two-by-four.

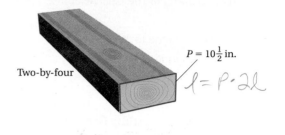

$P = 10\frac{1}{2}$ in.

$\ell = P \cdot 2\ell$

**25.** *Parking Costs.* A hospital parking lot charges $1.50 for the first hour or part thereof, and $1.00 for each additional hour or part thereof. A weekly pass costs $27.00 and allows unlimited parking for 7 days. Suppose that each visit Ed makes to the hospital lasts $1\frac{1}{2}$ hr. What is the minimum number of times that Ed would have to visit per week to make it worthwhile for him to buy the pass?

48'

**26.** *Van Rental.* Value Rent-A-Car rents vans at a daily rate of $84.95 plus 60 cents per mile. Molly rents a van to deliver electrical parts to her customers. She is allotted a daily budget of $320. How many miles can she drive for $320?

**27.** The second angle of a triangular field is three times as large as the first angle. The third angle is 40° greater than the first angle. How large are the angles?

**28.** *Triangular Parking Lot.* The second angle of a triangular parking lot is four times as large as the first angle. The third angle is 45° less than the sum of the other two angles. How large are the angles?

**29.** *Triangular Backyard.* A home has a triangular backyard. The second angle of the triangle is 5° more than the first angle. The third angle is 10° more than three times the first angle. Find the angles of the triangular yard.

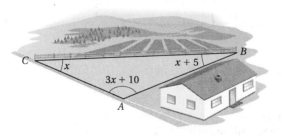

**30.** *Boarding Stable.* A rancher needs to form a triangular horse pen using ropes next to a stable. The second angle is three times the first angle. The third angle is 15° less than the first angle. Find the angles of the triangular pen.

**31.** *Coca-Cola Co.* The equation

$$y = 66.2x + 460.2$$

can be used to approximate the total revenue $y$, in millions of dollars, of the Coca-Cola Co., in year $x$, where

$x = 0$ corresponds to 1990,

$x = 2$ corresponds to 1992,

$x = 10$ corresponds to 2000,

and so on.

(**Source:** Coca-Cola Bottling Consolidated)

**a)** Find the total revenue in 1999, 2000, and 2010.
**b)** In what year will the total revenue be about $1254.6 million?

**32.** *Running Records in the 200-m Dash.* The equation

$$R = -0.028t + 20.8$$

can be used to predict the world record in the 200-m dash, where $R$ = the record in seconds, and $t$ = the number of years since 1920 (**Source:** International Amateur Athletic Federation).

**a)** Predict the record in 2000 and 2010.
**b)** In what year will the record be 18.0 sec?

---

**Skill Maintenance**

Calculate.

**33.** $-\dfrac{4}{5} - \dfrac{3}{8}$  [1.4a]

**34.** $-\dfrac{4}{5} + \dfrac{3}{8}$  [1.3a]

**35.** $-\dfrac{4}{5} \cdot \dfrac{3}{8}$  [1.5a]

**36.** $-\dfrac{4}{5} \div \dfrac{3}{8}$  [1.6c]

**37.** $-25.6 \div (-16)$  [1.6c]

**38.** $-25.6(-16)$  [1.5a]

**39.** $-25.6 - (-16)$  [1.4a]

**40.** $-25.6 + (-16)$  [1.3a]

---

**Synthesis**

**41.** ◆ A fellow student claims to be able to solve most of the problems in this section by guessing. Is there anything wrong with this approach? Why or why not?

**43.** Apples are collected in a basket for six people. One-third, one-fourth, one-eighth, and one-fifth are given to four people, respectively. The fifth person gets ten apples with one apple remaining for the sixth person. Find the original number of apples in the basket.

**45.** 🖩 The area of this triangle is 2.9047 in². Find $x$.

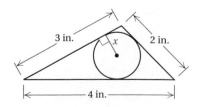

**42.** ◆ Write a problem for a classmate to solve so that it can be translated to the equation

$$\tfrac{2}{3}x + (x + 5) + x = 375.$$

**44.** A student scored 78 on a test that had 4 seven-point fill-ins and 24 three-point multiple-choice questions. The student had one fill-in wrong. How many multiple-choice questions did the student answer correctly?

**46.** A storekeeper goes to the bank to get $10 worth of change. She requests twice as many quarters as half dollars, twice as many dimes as quarters, three times as many nickels as dimes, and no pennies or dollars. How many of each coin did the storekeeper get?

Copyright © 1999 Addison Wesley Longman

# 2.5 Applications with Percents

**a** Many applied problems involve percents. We can use our knowledge of equations and the problem-solving process to solve such problems. For background on percent notation, see Section R.4.

**Example 1** What percent of 45 is 15?

1. **Familiarize.** This type of problem is stated so explicitly that we can proceed directly to the translation. We first let $x$ = the percent.
2. **Translate.** We translate as follows:

$$\underbrace{\text{What percent}}_{x} \quad \underset{\cdot}{\text{of}} \quad \underset{45}{\text{45}} \quad \underset{=}{\text{is}} \quad \underset{15.}{\text{15?}}$$

3. **Solve.** We solve the equation:

$$x \cdot 45 = 15$$

$$\frac{x \cdot 45}{45} = \frac{15}{45} \qquad \text{Dividing by 45}$$

$$x = \frac{1}{3} \qquad \text{Simplifying}$$

$$= 33\frac{1}{3}\%. \qquad \text{Changing fractional notation to percent notation}$$

4. **Check.** We check by finding $33\frac{1}{3}\%$ of 45:

$$33\frac{1}{3}\% \cdot 45 = \frac{1}{3} \cdot \overset{15}{\cancel{45}} = 15.$$

5. **State.** The answer is $33\frac{1}{3}\%$.

*Do Exercises 1 and 2.*

**Example 2** 3 is 16% of what number?

1. **Familiarize.** This problem is stated so explicitly that we can proceed directly to the translation. We let $y$ = the number that we are taking 16% of.
2. **Translate.** The translation is as follows:

$$\underset{3}{\text{3}} \quad \underset{=}{\text{is}} \quad \underset{16\%}{\text{16\%}} \quad \underset{\cdot}{\text{of}} \quad \underset{y}{\text{what?}}$$

3. **Solve.** We solve the equation:

$$3 = 16\% \cdot y$$

$$3 = 0.16y \qquad \text{Converting to decimal notation}$$

$$0.16y = 3 \qquad \text{Reversing the equation}$$

$$\frac{0.16y}{0.16} = \frac{3}{0.16} \qquad \text{Dividing by 0.16}$$

$$y = 18.75.$$

## Objective

**a** Solve applied problems involving percent.

### For Extra Help

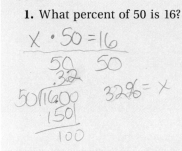

TAPE 5     TAPE 5A     MAC     CD-ROM
                       WIN

Solve.

1. What percent of 50 is 16?

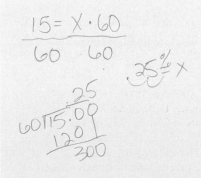

2. 15 is what percent of 60?

*Answers on page A-7*

Solve.

**3.** 45 is 20 percent of what number?

$45 = 20\% \cdot X$

$\dfrac{45}{20} = \dfrac{20\%}{20\%}$

$2.25 = X$

$225$

$2014800$

$480$

**4.** 120 percent of what number is 60?

$\dfrac{120\% \cdot X = 60}{120} \quad \dfrac{}{120}$

$X = .5$

$120$
$\times .5$
$60.0$ ✓

$\begin{array}{r} 225 \\ \times .20 \\ \hline 45.00 \end{array}$ ✓

$\begin{array}{r} 1.20\sqrt{6000} \\ 600 \\ \hline \end{array}$

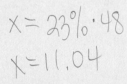

Solve.

**5.** What is 23% of 48?

$X = 23\% \cdot 48$

$X = 11.04$

**6.** Referring to Example 3, determine how many deaths by lightning occurred near telephone poles.

*Answers on page A-7*

**4. Check.** We check by finding 16% of 18.75:

$$16\% \times 18.75 = 0.16 \times 18.75 = 3.$$

**5. State.** The answer is 18.75.

*Do Exercises 3 and 4.*

Perhaps you have noticed that to handle percents in problems such as those in Examples 1 and 2, you can convert to decimal notation before continuing.

**Example 3** *Locations of Deaths by Lightning.* The circle graph below shows the various locations of people who are struck and killed by lightning. It is known that in the United States, 3327 people were killed by lightning from 1959 to 1996. How many were killed in fields or ballparks?

**Number of Deaths Due to Lightning**

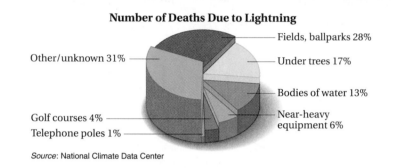

Fields, ballparks 28%
Other/unknown 31%
Under trees 17%
Bodies of water 13%
Near-heavy equipment 6%
Golf courses 4%
Telephone poles 1%

*Source*: National Climate Data Center

**1. Familiarize.** We first write down the information.

Total number of lightning strikes: 3327

Percent killed in fields or ballparks: 28%

We let $x$ = the number of people killed in fields or ballparks. It seems reasonable that we would take 28% of 3327. This leads us to rewording and translating the problem.

**2. Translate.** The translation is as follows.

| 28% | of | 3327 | is | what? | Rewording |
|-----|-----|------|-----|-------|-----------|
| ↓ | ↓ | ↓ | ↓ | ↓ | |
| 28% | · | 3327 | = | $x$ | Translating |

**3. Solve.** We solve the equation:

$$28\% \cdot 3327 = x$$

$0.28 \times 3327 = x$     Converting 28% to decimal notation

$932 \approx x.$     Multiplying and rounding to the nearest one

**4. Check.** The check is actually the computation we use to solve the equation:

$$28\% \cdot 3327 = 0.28 \times 3327 = 931.56 \approx 932.$$

**5. State.** About 932 lightning deaths occurred in fields or ballparks.

*Do Exercises 5–7. (Exercise 7 is on the following page.)*

**Example 4** *Simple Interest.* An investment is made at 6% simple interest for 1 year. It grows to $768.50. How much was originally invested (the principal)?

**1. Familiarize.** Suppose that $100 was invested. Recalling the formula for simple interest, $I = Prt$, we know that the interest for 1 year on $100 at 6% simple interest is given by $I = \$100 \cdot 6\% \cdot 1 = \$6$. Then, at the end of the year, the amount in the account is found by adding the principal and the interest:

$$\text{Principal} \quad + \quad \text{Interest} \quad = \quad \text{Amount}$$
$$\downarrow \qquad\qquad \downarrow \qquad\qquad \downarrow$$
$$\$100 \quad + \quad \$6 \quad = \quad \$106.$$

In this problem, we are working backward. We are trying to find the principal, which is the original investment. We let $x =$ the principal.

**2. Translate.** We reword the problem and then translate.

$$\text{Principal} \quad + \quad \text{Interest} \quad = \quad \text{Amount}$$
$$\downarrow \qquad\qquad \downarrow \qquad\qquad \downarrow$$
$$x \quad + \quad 6\%x \quad = \quad 768.50 \qquad \text{\textbf{Interest is 6\% of the principal.}}$$

**3. Solve.** We solve the equation:

$$x + 6\%x = 768.50$$
$$x + 0.06x = 768.50 \qquad \text{\textbf{Converting to decimal notation}}$$
$$1x + 0.06x = 768.50 \qquad \text{\textbf{Identity property of 1}}$$
$$1.06x = 768.50 \qquad \text{\textbf{Collecting like terms}}$$
$$\frac{1.06x}{1.06} = \frac{768.50}{1.06} \qquad \text{\textbf{Dividing by 1.06}}$$
$$x = 725.$$

**4. Check.** We check by taking 6% of $725 and adding it to $725:

$$6\% \times \$725 = 0.06 \times 725 = \$43.50.$$

Then $725 + $43.50 = $768.50, so $725 checks.

**5. State.** The original investment was $725.

*Do Exercise 8.*

---

**7.** The area of Arizona is 19% of the area of Alaska. The area of Alaska is 586,400 mi². What is the area of Arizona?

**8.** An investment is made at 7% simple interest for 1 year. It grows to $8988. How much was originally invested (the principal)?

*Answers on page A-7*

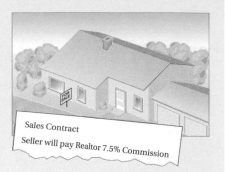

Sales Contract
Seller will pay Realtor 7.5% Commission

**Example 5** *Selling a Home.* The Fowlers are selling their home. They want to clear $115,625 after paying a $7\frac{1}{2}$% commission to a realtor. For how much must they sell the house?

1. **Familiarize.** Suppose the Fowlers sold the house for $120,000. We can determine the $7\frac{1}{2}$% commission by taking $7\frac{1}{2}$% of $120,000:

$$7\tfrac{1}{2}\% \text{ of } \$120,000 = 0.075(\$120,000) = \$9000.$$

Subtracting this commission from $120,000 would leave the Fowlers with

$$\$120,000 - \$9000 = \$111,000.$$

This shows us that in order for the Fowlers to clear $115,625, the house must be sold for more than $120,000. To determine exactly what the sale price would need to be, we could check more guesses. Instead, let's take advantage of our algebra skills. We let $x$ = the selling price of the house. Because the commission is $7\frac{1}{2}$%, the realtor receives $7\frac{1}{2}\%x$.

2. **Translate.** We reword and translate.

| Selling price | minus | Commission | is | Amount cleared | Rewording |
|:---:|:---:|:---:|:---:|:---:|:---|
| ↓ | ↓ | ↓ | ↓ | ↓ | |
| $x$ | $-$ | $7\frac{1}{2}\%x$ | $=$ | 115,625 | Translating |

3. **Solve.** We solve the equation:

$$x - 7\tfrac{1}{2}\%x = 115{,}625$$
$$1x - 0.075x = 115{,}625 \qquad \text{\small Converting to decimal notation}$$
$$(1 - 0.075)x = 115{,}625 \qquad \text{\small Collecting like terms}$$
$$0.925x = 115{,}625$$
$$\frac{0.925x}{0.925} = \frac{115{,}625}{0.925} \qquad \text{\small Dividing by 0.925}$$
$$x = 125{,}000.$$

4. **Check.** To check, we first find $7\frac{1}{2}$% of $125,000, calculating as we did in the *Familiarize* step:

$$7\tfrac{1}{2}\% \text{ of } \$125,000 = 0.075(\$125,000) = \$9375. \qquad \text{\small This is the commission.}$$

Then we subtract the commission to find the amount cleared:

$$\$125,000 - \$9375 = \$115,625.$$

Since, after the commission, the Fowlers are left with $115,625, our answer checks. Note that the sale price of $125,000 is greater than $120,000, as predicted in the *Familiarize* step.

5. **State.** The Fowlers need to sell their house for $125,000.

---

*CAUTION!* The problem in Example 5 is easy to solve with algebra. Without algebra, it is not. A common error in such a problem is to take $7\frac{1}{2}$% of the sale price and then subtract or add. Note that $7\frac{1}{2}$% of the selling price $\left(7\frac{1}{2}\% \cdot \$125,000 = \$9375\right)$ is not equal to $7\frac{1}{2}$% of the price that the Fowlers wanted to clear $\left(7\frac{1}{2}\% \cdot \$115,625 \approx \$8671.88\right)$.

---

*Do Exercise 9.*

9. The price of a suit was decreased to a sale price of $526.40. This was a 20% reduction. What was the former price?

*Answer on page A-7*

# Exercise Set 2.5

**a** Solve.

**1.** What percent of 180 is 36?

**2.** What percent of 76 is 19?

**3.** 45 is 30% of what number?

**4.** 20.4 is 24% of what number?

**5.** What number is 65% of 840?

**6.** What number is 1% of 1,000,000?

**7.** 30 is what percent of 125?

**8.** 57 is what percent of 300?

**9.** 12% of what number is 0.3?

**10.** 7 is 175% of what number?

**11.** 2 is what percent of 40?

**12.** 40 is 2% of what number?

*National Hamburger Sales.* The circle graph below shows hamburger sales by various restaurants in 1996. The total sales were $39 billion.

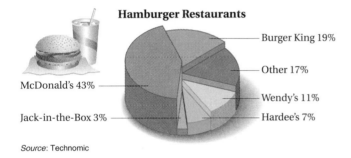

**Hamburger Restaurants**

Burger King 19%

Other 17%

McDonald's 43%

Wendy's 11%

Jack-in-the-Box 3%

Hardee's 7%

*Source*: Technomic

**13.** What was the total amount of hamburger sales, in dollars, by McDonald's?

**14.** What was the total amount of hamburger sales, in dollars, by Wendy's?

**15.** *Junk Mail.* The U.S. Postal Service reports that we open and read 78% of the junk mail that we receive (*Source*: U.S. Postal Service). A sports instructional videotape company sends out 10,500 advertising brochures.

   a) How many of the brochures can it expect to be opened and read?
   b) The company sells videos to 189 of the people who receive the brochure. What percent of the 10,500 people who receive the brochure buy the video?

**16.** *FBI Applications.* The FBI annually receives 16,000 applications for agents. It accepts 600 of these applicants. (*Source*: Federal Bureau of Investigation) What percent does it accept?

**17.** Leon left a $4 tip for a meal that cost $25.

    **a)** What percent of the cost of the meal was the tip?

    **b)** What was the total cost of the meal including the tip?

**18.** Selena left a $12.76 tip for a meal that cost $58.

    **a)** What percent of the cost of the meal was the tip?

    **b)** What was the total cost of the meal including the tip?

**19.** Leon left a 15% tip for a meal that cost $25.

    **a)** How much was the tip?

    **b)** What was the total cost of the meal including the tip?

**20.** Selena left a 15% tip for a meal that cost $58.

    **a)** How much was the tip?

    **b)** What was the total cost of the meal including the tip?

**21.** Leon left a 15% tip of $4.32 for a meal.

    **a)** What was the cost of the meal before the tip?

    **b)** What was the total cost of the meal including the tip?

**22.** Selena left a 15% tip of $8.40 for a meal.

    **a)** What was the cost of the meal before the tip?

    **b)** What was the total cost of the meal including the tip?

**23.** Leon left a 15% tip for a meal. The total cost of the meal, including the tip, was $41.40. What was the cost of the meal before the tip was added?

**24.** Selena left an 18% tip for a meal. The total cost of the meal, including the tip, was $40.71. What was the cost of the meal before the tip was added?

**25.** In a medical study of a group of pregnant women with "poor" diets, 16 of the women, or 8%, had babies who were in good or excellent health. How many women were in the original study?

**26.** In a medical study of a group of pregnant women with "good-to-excellent" diets, 285 of the women, or 95%, had babies who were in good or excellent health. How many women were in the original study?

Copyright © 1999 Addison Wesley Longman

**27.** *Life Insurance for Smokers vs. Nonsmokers.* The premium for a $100,000 life insurance policy for a female nonsmoker, age 22, is about $166 per year. The premium for a smoker is 170% of the premium for a nonsmoker. What is the premium for a smoker?

**28.** *Catching Colds from Kissing.* In a medical study, it was determined that if 800 people kiss someone else who has a cold, only 56 will actually catch the cold. What percent is this?

**29.** *Body Fat.* The author of this text exercises regularly at a local YMCA that recently offered a body-fat percentage test to its members. The device used measures the passage of a very low voltage of electricity through the body. The author's body-fat percentage was found to be 19.8% and he weighs 214 lb. What part, in pounds, of his body weight is fat?

**30.** *Calories Burned.* The author of this text exercises regularly at a local YMCA. The readout on a stairmaster machine tells him that if he exercises for 24 min, he will burn 356 calories. He decides to increase his time on the machine in order to lose more weight.

**a)** By what percent has he increased his time on the stairmaster if he exercises for 30 min?

**b)** How many calories does he burn if he exercises for 30 min? 40 min?

**31.** *Lightning Deaths Under Trees.* Referring to Example 3, determine how many deaths by lightning occurred under trees from 1959 to 1996.

**32.** *Lightning Deaths on Golf Courses.* Referring to Example 3, determine how many deaths by lightning occurred on golf courses from 1959 to 1996.

**33.** *Major League Baseball.* In 1997, the Boston Red Sox made the highest increase in the average price of a ticket to a baseball game. The price was $17.69 and represented an increase of 15% from the preceding season. (*Source:* Major League Baseball) What was the average price of a ticket the preceding season?

**34.** *Major League Baseball.* The Atlanta Braves opened a new ballpark, Turner Field, in 1997. Attending a game represented the costliest outing in major league baseball that year. To take a family of four to a game, buy four small soft drinks, two small beers, and four hot dogs, and add in the cost of parking, two game programs, and two twill caps, the average cost would be $129.16, an increase of 6% over the year before. (*Source:* Major League Baseball) What did this outing cost the year before?

**35.** An investment was made at 6% simple interest for 1 year. It grows to $8268. How much was originally invested?

**36.** Money is borrowed at 6.2% simple interest. After 1 year, $6945.48 pays off the loan. How much was originally borrowed?

**37.** After a 40% reduction, a shirt is on sale for $34.80. What was the original price (that is, the price before reduction)?

**38.** After a 34% reduction, a blouse is on sale for $42.24. What was the original price?

---

**Skill Maintenance**

Compute. [R.3b]

**39.** $9.076 \div 0.05$

**40.** $9.076 \times 0.05$

**41.** $1.089 + 10.89 + 0.1089$

**42.** $1000.23 - 156.0893$

Evaluate. [1.1a]

**43.** $x - y$, for $x = 58$ and $y = 42$

**44.** $8t$, for $t = 23.7$

**45.** $\dfrac{6a}{b}$, for $a = 25$ and $b = 15$

**46.** $\dfrac{a + b}{8}$, for $a = 45.6$ and $b = 102.3$

---

**Synthesis**

**47.** ◈ Comment on the following quote by Yogi Berra, a famous Major League Hall of Fame baseball player: "Ninety percent of hitting is mental. The other half is physical."

**48.** ◈ Erin returns a tent that she bought during a storewide 35% off sale that has ended. She is offered store credit for 125% of what she paid (not to be used on sale items). Is this fair to Erin? Why or why not?

**49.** It has been determined that at the age of 15, a boy has reached 96.1% of his final adult height. Jaraan is 6 ft, 4 in. at the age of 15. What will his final adult height be?

**50.** It has been determined that at the age of 10, a girl has reached 84.4% of her final adult height. Dana is 4 ft, 8 in. at the age of 10. What will her final adult height be?

**51.** In one city, a sales tax of 9% was added to the price of gasoline as registered on the pump. Suppose a driver asked for $10 worth of gas. The attendant filled the tank until the pump read $9.10 and charged the driver $10. Something was wrong. Use algebra to correct the error.

Collaborative Learning Manual

Calculate the sale price and the original price of discounted items.

# 2.6 Formulas

## a | Evaluating and Solving Formulas

A **formula** is a "recipe" for doing a certain type of calculation. Formulas are often given as equations. Here is an example of a formula that has to do with weather: $M = \frac{1}{5}n$. You see a flash of lightning. After a few seconds you hear the thunder associated with that flash. How far away was the lightning?

Your distance from the storm is $M$ miles. You can find that distance by counting the number of seconds $n$ that it takes the sound of the thunder to reach you and then multiplying by $\frac{1}{5}$.

**Example 1** *Storm Distance.* Consider the formula $M = \frac{1}{5}n$. It takes 10 sec for the sound of thunder to reach you after you have seen a flash of lightning. How far away is the storm?

We substitute 10 for $n$ and calculate $M$: $M = \frac{1}{5}n = \frac{1}{5}(10) = 2$. The storm is 2 mi away.

*Do Exercise 1.*

Suppose that we think we know how far we are from the storm and want to check by calculating the number of seconds it should take the sound of the thunder to reach us. We could substitute a number for $M$—say, 2—and solve for $n$:

$$2 = \frac{1}{5}n$$
$$10 = n. \qquad \text{Multiplying by 5}$$

However, if we wanted to do this repeatedly, it might be easier to solve for $n$ by getting it alone on one side. We "solve" the formula for $n$.

**Example 2** Solve for $n$: $M = \frac{1}{5}n$.

We have

$$M = \frac{1}{5}n \qquad \text{We want this letter alone.}$$
$$5 \cdot M = 5 \cdot \frac{1}{5}n \qquad \text{Multiplying by 5 on both sides}$$
$$5M = n.$$

In the above situation for $M = 2$, $n = 5(2)$, or 10.

*Do Exercise 2.*

To see how the addition and multiplication principles apply to formulas, compare the following.

**A.** Solve.

$$5x + 2 = 12$$
$$5x = 12 - 2$$
$$5x = 10$$
$$x = \frac{10}{5} = 2$$

**B.** Solve.

$$5x + 2 = 12$$
$$5x = 12 - 2$$
$$x = \frac{12 - 2}{5}$$

**C.** Solve for $x$.

$$ax + b = c$$
$$ax = c - b$$
$$x = \frac{c - b}{a}$$

In (A), we solved as we did before. In (B), we did not carry out the calculations. In (C), we could not carry out the calculations because we had unknown numbers.

### Objective

a | Evaluate formulas and solve a formula for a specified letter.

### For Extra Help

TAPE 5    TAPE 5B    MAC    CD-ROM
                     WIN

1. Suppose that it takes the sound of thunder 14 sec to reach you. How far away is the storm?

2. Solve for $I$: $E = IR$.
(This is a formula from electricity relating voltage $E$, current $I$, and resistance $R$.)

*Answers on page A-7*

**3.** Solve for $D$: $C = \pi D$.

(This is a formula for the circumference $C$ of a circle of diameter $D$.)

**4.** *Averages.* Solve for $c$:

$$A = \frac{a + b + c + d}{4}.$$

**5.** Use the formula of Example 5.

**a)** Estimate the weight of a yellow tuna that is 7 ft long and has a girth of about 54 in.

**b)** Solve the formula for $L$.

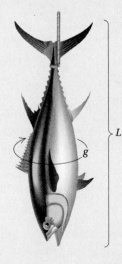

**Example 3**   *Circumference.* Solve for $r$: $C = 2\pi r$. This is a formula for the circumference $C$ of a circle of radius $r$.

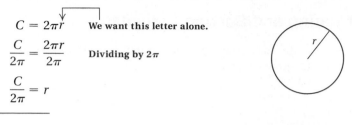

$$C = 2\pi r \qquad \text{We want this letter alone.}$$

$$\frac{C}{2\pi} = \frac{2\pi r}{2\pi} \qquad \text{Dividing by } 2\pi$$

$$\frac{C}{2\pi} = r$$

> To solve a formula for a given letter, identify the letter and:
>
> 1. Multiply on both sides to clear fractions or decimals, if that is needed.
> 2. Collect like terms on each side, if necessary.
> 3. Get all terms with the letter to be solved for on one side of the equation and all other terms on the other side.
> 4. Collect like terms again, if necessary.
> 5. Solve for the letter in question.

**Example 4**   Solve for $a$: $A = \dfrac{a + b + c}{3}$. This is a formula for the average $A$ of three numbers $a$, $b$, and $c$.

$$A = \frac{a + b + c}{3} \qquad \text{We want the letter } a \text{ alone.}$$

$$3A = a + b + c \qquad \text{Multiplying by 3 to clear the fraction}$$

$$3A - b - c = a \qquad \text{Subtracting } b \text{ and } c$$

**Do Exercises 3 and 4.**

**Example 5**   *Estimating the Weight of a Fish.* An ancient fisherman's formula for estimating the weight of a fish is

$$W = \frac{Lg^2}{800},$$

where $W$ is the weight in pounds, $L$ is the length in inches, and $g$ is the girth (distance around the midsection) in inches.

**a)** Estimate the weight of a great bluefin tuna that is 8 ft long and has a girth of about 76 in.

**b)** Solve the formula for $g^2$.

We solve as follows:

**a)** We substitute 96 for $L$ (8 ft = 96 in.) and 76 for $g$. Then we calculate $W$.

$$W = \frac{Lg^2}{800} = \frac{96 \cdot 76^2}{800} \approx 693 \text{ lb}$$

The tuna weighs about 693 lb.

**b)**
$$W = \frac{Lg^2}{800} \qquad \text{We want to get } g^2 \text{ alone.}$$

$$800W = Lg^2 \qquad \text{Multiplying by 800}$$

$$\frac{800W}{L} = g^2 \qquad \text{Dividing by } L$$

**Do Exercise 5.**

# Exercise Set 2.6

**a** Solve for the given letter.

**1.** *Area of a Parallelogram*:

$A = bh$, for $h$

(Area $A$, base $b$, height $h$)

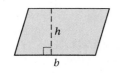

**2.** *Distance Formula*:

$d = rt$, for $r$

(Distance $d$, speed $r$, time $t$)

Speed, $r$     Time, $t$

Distance, $d$

**3.** *Perimeter of a Rectangle*:

$P = 2l + 2w$, for $w$

(Perimeter $P$, length $l$, width $w$)

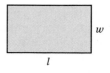

**4.** *Area of a Circle*:

$A = \pi r^2$, for $r^2$

(Area $A$, radius $r$)

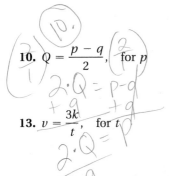

**5.** *Average of Two Numbers*:

$A = \dfrac{a + b}{2}$, for $a$

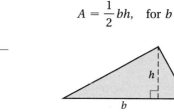

**6.** *Area of a Triangle*:

$A = \dfrac{1}{2}bh$, for $b$

**7.** *Force*:

$F = ma$, for $a$

(Force $F$, mass $m$, acceleration, $a$)

**8.** *Simple Interest*:

$I = Prt$, for $P$

(Interest $I$, principal $P$, interest rate $r$, time $t$)

**9.** *Relativity*:

$E = mc^2$, for $c^2$

(Energy $E$, mass $m$, speed of light $c$)

**10.** $Q = \dfrac{p - q}{2}$, for $p$

**11.** $Ax + By = c$, for $x$

**12.** $Ax + By = c$, for $y$

**13.** $v = \dfrac{3k}{t}$, for $t$

**14.** $P = \dfrac{ab}{c}$, for $c$

**15.** *Furnace Output.* The formula

$b = 30a$

is used in New England to estimate the minimum furnace output, $b$, in Btu's, for a modern house with $a$ square feet of flooring.

**a)** Determine the minimum furnace output for a 1900-ft$^2$ modern house.

**b)** Solve the formula for $a$.

**16.** *Surface Area of a Cube.* The surface area of a cube with side $s$ is given by

$A = 6s^2$.

**a)** Determine the surface area of a cube with 3-in. sides.

**b)** Solve the formula for $s^2$.

**17.** *Full-Time Equivalent Students.* Colleges accommodate students who need to take different total-credit-hour loads. They determine the number of "full-time-equivalent" students, $F$, using the formula

$$F = \frac{n}{15},$$

where $n$ is the total number of credits students enroll in for a given semester.

a) Determine the number of full-time equivalent students on a campus in which students register for 21,345 credits.
b) Solve the formula for $n$.

**18.** *Young's Rule in Medicine.* Young's rule for determining the amount of a medicine dosage for a child is given by the formula

$$c = \frac{ad}{a + 12},$$

where $a$ is the child's age and $d$ is the usual adult dosage. (*Warning!* Do not apply this formula without checking with a physician!)

a) The usual adult dosage of medication for an adult is 250 mg. Find the dosage for a child of age 2.
b) Solve the formula for $d$. (*Source*: Olsen, June L., et al., *Medical Dosage Calculations*, 6th ed. Reading, MA: Addison Wesley Longman, p. A-31.)

**19.** *Female Caloric Needs.* The number of calories $K$ needed each day by a moderately active woman who weighs $w$ pounds, is $h$ inches tall, and is $a$ years old can be estimated by the formula

$$K = 917 + 6(w + h - a).$$

(*Source*: Parker, M., *She Does Math.* Mathematical Association of America, p. 96.)

a) Elaine is moderately active, weighs 120 lb, is 67 in. tall, and is 23 yr old. What are her caloric needs?
b) Solve the formula for $a$, for $h$, and for $w$.

**20.** *Male Caloric Needs.* The number of calories $K$ needed each day by a moderately active man who weighs $w$ kilograms, is $h$ centimeters tall, and is $a$ years old can be estimated by the formula

$$K = 19.18w + 7h - 9.52a + 92.4.$$

(*Source*: Parker, M., *She Does Math.* Mathematical Association of America, p. 96.)

a) Marv is moderately active, weighs 97 kg, is 185 cm tall, and is 55 yr old. What are his caloric needs?
b) Solve the formula for $a$, for $h$, and for $w$.

---

**Skill Maintenance**

**21.** Convert to decimal notation: $\dfrac{23}{25}$.  [R.3a]

**22.** Add: $-23 + (-67)$.  [1.3a]

**23.** Subtract: $-45.8 - (-32.6)$.  [1.4a]

**24.** Remove parentheses and simplify:  [1.8b]
$$4a - 8b - 5(5a - 4b).$$

**25.** Add: $-\dfrac{2}{3} + \dfrac{5}{6}$.  [1.3a]

**26.** Subtract: $-\dfrac{2}{3} - \dfrac{5}{6}$.  [1.4a]

---

**Synthesis**

**27.** ◆ Devise an application in which it would be useful to solve the equation $d = rt$ for $r$. (See Exercise 2.)

**28.** ◆ The equations

$$P = 2l + 2w \quad \text{and} \quad w = \frac{P}{2} - l$$

are equivalent formulas involving the perimeter $P$, the length $l$, and the width $w$ of a rectangle. Devise a problem for which the second of the two formulas would be more useful.

Solve.

**29.** $A = \dfrac{1}{2}ah + \dfrac{1}{2}bh$, for $b$; for $h$

**30.** $P = 4m + 7mn$, for $m$

**31.** In $A = lw$, $l$ and $w$ both double. What is the effect on $A$?

**32.** In $P = 2a + 2b$, $P$ doubles. Do $a$ and $b$ necessarily both double?

**33.** In $A = \frac{1}{2}bh$, $b$ increases by 4 units and $h$ does not change. What happens to $A$?

**34.** Solve for $F$:

$$D = \frac{1}{E + F}.$$

Copyright © 1999 Addison Wesley Longman

# 2.7 Solving Inequalities

We now extend our equation-solving principles to the solving of inequalities.

## a Solutions of Inequalities

In Section 1.2, we defined the symbols > (greater than), < (less than), ≥ (greater than or equal to), and ≤ (less than or equal to). For example, $3 \leq 4$ and $3 \leq 3$ are both true, but $-3 \leq -4$ and $0 \geq 2$ are both false.

An **inequality** is a number sentence with >, <, ≥, or ≤ as its verb—for example,

$$-4 > t, \quad x < 3, \quad 2x + 5 \geq 0, \quad \text{and} \quad -3y + 7 \leq -8.$$

Some replacements for a variable in an inequality make it true and some make it false.

> A replacement that makes an inequality true is called a **solution**. The set of all solutions is called the **solution set.** When we have found the set of all solutions of an inequality, we say that we have **solved** the inequality.

**Examples**  Determine whether the number is a solution of $x < 2$.

**1.** $-2.7$  Since $-2.7 < 2$ is true, $-2.7$ is a solution.

**2.** 2  Since $2 < 2$ is false, 2 is not a solution.

**Examples**  Determine whether the number is a solution of $y \geq 6$.

**3.** 6  Since $6 \geq 6$ is true, 6 is a solution.

**4.** $-\frac{4}{3}$  Since $-\frac{4}{3} \geq 6$ is false, $-\frac{4}{3}$ is not a solution.

*Do Exercises 1 and 2.*

## b Graphs of Inequalities

Some solutions of $x < 2$ are 0.45, $-8.9$, $-\pi$, $\frac{5}{8}$, and so on. In fact, there are infinitely many real numbers that are solutions. Because we cannot list them all individually, it is helpful to make a drawing that represents all the solutions.

A **graph** of an inequality is a drawing that represents its solutions. An inequality in one variable can be graphed on a number line. An inequality in two variables can be graphed on a coordinate plane; we will study such graphs in Chapter 7.

We first graph inequalities in one variable on a number line.

**Example 5**  Graph: $x < 2$.

The solutions of $x < 2$ are all those numbers less than 2. They are shown on the graph by shading all points to the left of 2. The open circle at 2 indicates that 2 is not part of the graph.

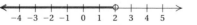

## Objectives

**a** Determine whether a given number is a solution of an inequality.

**b** Graph an inequality on a number line.

**c** Solve inequalities using the addition principle.

**d** Solve inequalities using the multiplication principle.

**e** Solve inequalities using the addition and multiplication principles together.

### For Extra Help

TAPE 5   TAPE 5B   MAC   CD-ROM
WIN

Determine whether each number is a solution of the inequality.

**1.** $x > 3$
   a) 2    b) 0

   c) $-5$    d) 15.4

   e) 3    f) $-\frac{2}{5}$

**2.** $x \leq 6$
   a) 6    b) 0

   c) $-4.3$    d) 25

   e) $-6$    f) $\frac{5}{8}$

*Answers on page A-7*

Graph.

**3.** $x \le 4$

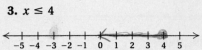

**4.** $x > -2$

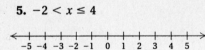

**5.** $-2 < x \le 4$

**Example 6** Graph: $x \ge -3$.

The solutions of $x \ge -3$ are shown on the number line by shading the point for $-3$ and all points to the right of $-3$. The closed circle at $-3$ indicates that $-3$ *is* part of the graph.

**Example 7** Graph: $-3 \le x < 2$.

The inequality $-3 \le x < 2$ is read "$-3$ is less than or equal to $x$ *and* $x$ is less than 2," or "$x$ is greater than or equal to $-3$ *and* $x$ is less than 2." In order to be a solution of this inequality, a number must be a solution of both $-3 \le x$ and $x < 2$. The number 1 is a solution, as are $-1.7$, 0, 1.5, and $\frac{3}{8}$. We can see from the graphs below that the solution set consists of the numbers that overlap in the two solution sets in Examples 5 and 6:

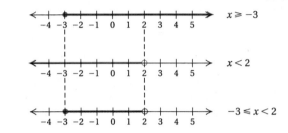

The open circle at 2 means that 2 is *not* part of the graph. The closed circle at $-3$ means that $-3$ *is* part of the graph. The other solutions are shaded.

*Do Exercises 3–5.*

## c | Solving Inequalities Using the Addition Principle

Consider the true inequality $3 < 7$. If we add 2 on both sides, we get another true inequality:

$$3 + 2 < 7 + 2, \quad \text{or} \quad 5 < 9.$$

Similarly, if we add $-4$ on both sides of $x + 4 < 10$, we get an *equivalent* inequality:

$$x + 4 + (-4) < 10 + (-4),$$

or

$$x < 6.$$

To say that $x + 4 < 10$ and $x < 6$ are **equivalent** is to say that they have the same solution set. For example, the number 3 is a solution of $x + 4 < 10$. It is also a solution of $x < 6$. The number $-2$ is a solution of $x < 6$. It is also a solution of $x + 4 < 10$. Any solution of one is a solution of the other—they are equivalent.

> **THE ADDITION PRINCIPLE FOR INEQUALITIES**
>
> For any real numbers $a$, $b$, and $c$:
>
> $a < b$    is equivalent to    $a + c < b + c$;
>
> $a > b$    is equivalent to    $a + c > b + c$;
>
> $a \leq b$    is equivalent to    $a + c \leq b + c$;
>
> $a \geq b$    is equivalent to    $a + c \geq b + c$.
>
> In other words, when we add or subtract the same number on both sides of an inequality, the direction of the inequality symbol is not changed.

As with equation solving, when solving inequalities, our goal is to isolate the variable on one side. Then it is easier to determine the solution set.

**Example 8**   Solve: $x + 2 > 8$. Then graph.

We use the addition principle, subtracting 2 on both sides:

$$x + 2 - 2 > 8 - 2$$
$$x > 6.$$

From the inequality $x > 6$, we can determine the solutions directly. Any number greater than 6 makes the last sentence true and is a solution of that sentence. Any such number is also a solution of the original sentence. Thus the inequality is solved. The graph is as follows:

We cannot check all the solutions of an inequality by substitution, as we can check solutions of equations, because there are too many of them. A partial check can be done by substituting a number greater than 6 — say, 7 — into the original inequality:

$$\frac{x + 2 > 8}{7 + 2 \mid 8}$$
$$9 \mid \quad \text{TRUE}$$

Since $9 > 8$ is true, 7 is a solution. Any number greater than 6 is a solution.

**Example 9**   Solve: $3x + 1 \leq 2x - 3$. Then graph.

We have

$$3x + 1 \leq 2x - 3$$
$$3x + 1 - 1 \leq 2x - 3 - 1 \qquad \text{Subtracting 1}$$
$$3x \leq 2x - 4 \qquad \text{Simplifying}$$
$$3x - 2x \leq 2x - 4 - 2x \qquad \text{Subtracting } 2x$$
$$x \leq -4. \qquad \text{Simplifying}$$

The graph is as follows:

Solve. Then graph.

**6.** $x + 3 > 5$

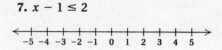

**7.** $x - 1 \leq 2$

**8.** $5x + 1 < 4x - 2$

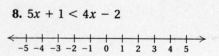

*Answers on page A-7*

Solve.

**9.** $x + \dfrac{2}{3} \geq \dfrac{4}{5}$

In Example 9, any number less than or equal to $-4$ is a solution. The following are some solutions:

$$-4, \quad -5, \quad -6, \quad -\dfrac{13}{3}, \quad -204.5, \quad \text{and} \quad -18\pi.$$

Besides drawing a graph, we can also describe all the solutions of an inequality using **set notation.** We could just begin to list them in a set using roster notation (see p. 49), as follows:

$$\{-4, -5, -6, -4.1, -204.5, -18\pi, \ldots\}.$$

We can never list them all this way, however. Seeing this set without knowing the inequality makes it difficult for us to know what real numbers we are considering. There is, however, another kind of notation that we can use. It is

$$\{x \mid x \leq -4\},$$

which is read

"The set of all $x$ such that $x$ is less than or equal to $-4$."

This shorter notation for sets is called **set-builder notation** (see Section 1.2). From now on, we will use this notation when solving inequalities.

*Do Exercises 6–8 on the preceding page.*

**Example 10**   Solve: $x + \frac{1}{3} > \frac{5}{4}$.

We have

$$x + \tfrac{1}{3} > \tfrac{5}{4}$$

$$x + \tfrac{1}{3} - \tfrac{1}{3} > \tfrac{5}{4} - \tfrac{1}{3} \qquad \text{Subtracting } \tfrac{1}{3}$$

**10.** $5y + 2 \leq -1 + 4y$

$$x > \tfrac{5}{4} \cdot \tfrac{3}{3} - \tfrac{1}{3} \cdot \tfrac{4}{4} \qquad \text{\textbf{Multiplying by 1 to obtain a common denominator}}$$

$$x > \tfrac{15}{12} - \tfrac{4}{12}$$

$$x > \tfrac{11}{12}.$$

Any number greater than $\frac{11}{12}$ is a solution. The solution set is

$$\left\{x \mid x > \tfrac{11}{12}\right\},$$

which is read

"The set of all $x$ such that $x$ is greater than $\frac{11}{12}$."

When solving inequalities, you may obtain an answer like $7 < x$. Recall from Chapter 1 that this has the same meaning as $x > 7$. Thus the solution set can be described as $\{x \mid 7 < x\}$ or as $\{x \mid x > 7\}$. The latter is used most often.

*Do Exercises 9 and 10.*

*Answers on page A-7*

## d | Solving Inequalities Using the Multiplication Principle

There is a multiplication principle for inequalities that is similar to that for equations, but it must be modified. When we are multiplying on both sides by a negative number, the direction of the inequality symbol must be changed. Let's see what happens. Consider the true inequality $3 < 7$. If we multiply on both sides by a *positive* number, like 2, we get another true inequality:

$$3 \cdot 2 < 7 \cdot 2, \quad \text{or} \quad 6 < 14. \qquad \text{True}$$

If we multiply on both sides by a *negative* number, like $-2$, and we do not change the direction of the inequality symbol, we get a *false* inequality:

$$3 \cdot (-2) < 7 \cdot (-2), \quad \text{or} \quad -6 < -14. \qquad \text{False}$$

The fact that $6 < 14$ is true but $-6 < -14$ is false stems from the fact that the negative numbers, in a sense, mirror the positive numbers. That is, whereas 14 is to the *right* of 6 on a number line, the number $-14$ is to the *left* of $-6$. Thus, if we reverse (change the direction of) the inequality symbol, we get a *true* inequality: $-6 > -14$.

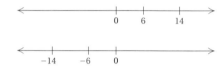

> **THE MULTIPLICATION PRINCIPLE FOR INEQUALITIES**
>
> For any real numbers $a$ and $b$, and any *positive* number $c$:
>
> $\qquad a < b$   is equivalent to   $ac < bc$;
>
> $\qquad a > b$   is equivalent to   $ac > bc$.
>
> For any real numbers $a$ and $b$, and any *negative* number $c$:
>
> $\qquad a < b$   is equivalent to   $ac > bc$;
>
> $\qquad a > b$   is equivalent to   $ac < bc$.
>
> Similar statements hold for $\leq$ and $\geq$.
>
> In other words, when we multiply or divide by a positive number on both sides of an inequality, the direction of the inequality symbol stays the same. When we multiply or divide by a negative number on both sides of an inequality, the direction of the inequality symbol is reversed.

**Example 11**   Solve: $4x < 28$. Then graph.

We have

$$4x < 28$$

$$\frac{4x}{4} < \frac{28}{4} \qquad \text{Dividing by 4}$$

$$\text{The symbol stays the same.}$$

$$x < 7. \qquad \text{Simplifying}$$

The solution set is $\{x \mid x < 7\}$. The graph is as follows:

**Do Exercises 11 and 12.**

Solve. Then graph.

**11.** $8x < 64$

**12.** $5y \geq 160$

Solve.

**13.** $-4x \le 24$

**14.** $-5y > 13$

**15.** Solve: $7 - 4x < 8$.

**Example 12**   Solve: $-2y < 18$. Then graph.

We have

$$-2y < 18$$

$$\frac{-2y}{-2} > \frac{18}{-2} \qquad \text{Dividing by } -2$$

The symbol must be reversed!

$$y > -9. \qquad \text{Simplifying}$$

The solution set is $\{y \mid y > -9\}$. The graph is as follows:

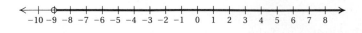

*Do Exercises 13 and 14.*

**e**   **Using the Principles Together**

All of the equation-solving techniques used in Sections 2.1–2.3 can be used with inequalities provided we remember to reverse the inequality symbol when multiplying or dividing on both sides by a negative number.

**Example 13**   Solve: $6 - 5y > 7$.

We have

$$6 - 5y > 7$$

$$-6 + 6 - 5y > -6 + 7 \qquad \text{Adding } -6. \text{ The symbol stays the same.}$$

$$-5y > 1 \qquad \text{Simplifying}$$

$$\frac{-5y}{-5} < \frac{1}{-5} \qquad \text{Dividing by } -5$$

The symbol must be reversed.

$$y < -\frac{1}{5}. \qquad \text{Simplifying}$$

The solution set is $\left\{y \mid y < -\frac{1}{5}\right\}$.

*Do Exercise 15.*

**Example 14**   Solve: $8y - 5 > 17 - 5y$.

$$-17 + 8y + 5 > -17 + 17 - 5y \qquad \text{Adding } -17. \text{ The symbol stays the same.}$$

$$8y - 22 > -5y \qquad \text{Simplifying}$$

$$-8y + 8y - 22 > -8y - 5y \qquad \text{Adding } -8y$$

$$-22 > -13y \qquad \text{Simplifying}$$

$$\frac{-22}{-13} < \frac{-13y}{-13} \qquad \text{Dividing by } -13$$

The symbol must be reversed.

$$\frac{22}{13} < y.$$

The solution set is $\left\{y \mid \frac{22}{13} < y\right\}$, or $\left\{y \mid y > \frac{22}{13}\right\}$.

We can often solve inequalities in such a way as to avoid having to reverse the inequality symbol. We add so that after like terms have been collected, the coefficient of the variable term is positive. We show this by solving the inequality in Example 14 a different way.

**Example 15**   Solve: $8y - 5 > 17 - 5y$.

Note that if we add $5y$ on both sides, the coefficient of the $y$-term will be positive after like terms have been collected.

$$8y - 5 + 5y > 17 - 5y + 5y \qquad \text{Adding } 5y$$
$$13y - 5 > 17 \qquad \text{Simplifying}$$
$$13y - 5 + 5 > 17 + 5 \qquad \text{Adding 5}$$
$$13y > 22 \qquad \text{Simplifying}$$
$$\frac{13y}{13} > \frac{22}{13} \qquad \text{Dividing by 13}$$
$$y > \frac{22}{13}$$

The solution set is $\{y \mid y > \frac{22}{13}\}$.

*Do Exercises 16 and 17.*

**Example 16**   Solve: $3(x - 2) - 1 < 2 - 5(x + 6)$.

$$3(x - 2) - 1 < 2 - 5(x + 6)$$
$$3x - 6 + 1 < 2 - 5x - 30 \qquad \begin{array}{l}\text{Using the distributive law to multiply}\\\text{and remove parentheses}\end{array}$$
$$3x - 7 < -5x - 28 \qquad \text{Simplifying}$$
$$3x + 5x < -28 + 7 \qquad \begin{array}{l}\text{Adding } 5x \text{ and 7 to get all } x\text{-terms on one}\\\text{side and all other terms on the other side}\end{array}$$
$$8x < -21 \qquad \text{Simplifying}$$
$$x < \frac{-21}{8}, \text{ or } -\frac{21}{8} \qquad \text{Dividing by 8}$$

The solution set is $\{x \mid x < -\frac{21}{8}\}$.

*Do Exercise 18.*

**Example 17**   Solve: $16.3 - 7.2p \le -8.18$.

The greatest number of decimal places in any one number is *two*. Multiplying by 100, which has two 0's, will clear decimals. Then we proceed as before.

$$16.3 - 7.2p \le -8.18$$
$$100(16.3 - 7.2p) \le 100(-8.18) \qquad \text{Multiplying by 100}$$
$$100(16.3) - 100(7.2p) \le 100(-8.18) \qquad \text{Using the distributive law}$$
$$1630 - 720p \le -818 \qquad \text{Simplifying}$$
$$1630 - 720p - 1630 \le -818 - 1630 \qquad \text{Subtracting 1630}$$
$$-720p \le -2448 \qquad \text{Simplifying}$$
$$\frac{-720p}{-720} \ge \frac{-2448}{-720} \qquad \text{Dividing by } -720$$

The symbol must be reversed.

$$p \ge 3.4$$

The solution set is $\{p \mid p \ge 3.4\}$.

**16.** Solve: $24 - 7y \le 11y - 14$.

$\{y \mid y \ge \frac{19}{9}\}$

**17.** Solve. Use a method like the one used in Example 15.

$$24 - 7y \le 11y - 14$$

$\{y \mid y \ge \frac{19}{9}\}$

**18.** Solve:

$$3(7 + 2x) \le 30 + 7(x - 1).$$

$\{x \mid x \ge -2\}$

*Answers on page A-7*

**19.** Solve:

$2.1x + 43.2 \geq 1.2 - 8.4x.$

$\{x \mid x \geq -4\}$

**20.** Solve:

$\dfrac{3}{4} + x < \dfrac{7}{8}x - \dfrac{1}{4} + \dfrac{1}{2}x.$

*Do Exercise 19.*

**Example 18** Solve: $\dfrac{2}{3}x - \dfrac{1}{6} + \dfrac{1}{2}x > \dfrac{7}{6} + 2x.$

The number 6 is the least common multiple of all the denominators. Thus we multiply by 6 on both sides.

$$\dfrac{2}{3}x - \dfrac{1}{6} + \dfrac{1}{2}x > \dfrac{7}{6} + 2x$$

$$6\left(\dfrac{2}{3}x - \dfrac{1}{6} + \dfrac{1}{2}x\right) > 6\left(\dfrac{7}{6} + 2x\right)$$     **Multiplying by 6 on both sides**

$$6 \cdot \dfrac{2}{3}x - 6 \cdot \dfrac{1}{6} + 6 \cdot \dfrac{1}{2}x > 6 \cdot \dfrac{7}{6} + 6 \cdot 2x$$     **Using the distributve law**

$$4x - 1 + 3x > 7 + 12x$$     **Simplifying**

$$7x - 1 > 7 + 12x$$     **Collecting like terms**

$$7x - 1 - 12x > 7 + 12x - 12x$$     **Subtracting 12x**

$$-5x - 1 > 7$$     **Collecting like terms**

$$-5x - 1 + 1 > 7 + 1$$     **Adding 1**

$$-5x > 8$$     **Simplifying**

$$\dfrac{-5x}{-5} < \dfrac{8}{-5}$$     **Dividing by −5**

    **The symbol must be reversed.**

$$x < -\dfrac{8}{5}$$

The solution set is $\left\{x \mid x < -\dfrac{8}{5}\right\}.$

*Do Exercise 20.*

# Exercise Set 2.7

**a** Determine whether each number is a solution of the given inequality.

**1.** $x > -4$
  a) 4
  b) 0
  c) $-4$
  d) 6
  e) 5.6

**2.** $x \leq 5$
  a) 0
  b) 5
  c) $-1$
  d) $-5$
  e) $7\frac{1}{4}$

**3.** $x \geq 6.8$
  a) $-6$
  b) 0
  c) 6
  d) 8
  e) $-3\frac{1}{2}$

**4.** $x < 8$
  a) 8
  b) $-10$
  c) 0
  d) 11
  e) $-4.7$

**b** Graph on a number line.

**5.** $x > 4$

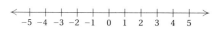

**6.** $x < 0$

**7.** $t < -3$

**8.** $y > 5$

**9.** $m \geq -1$

**10.** $x \leq -2$

**11.** $-3 < x \leq 4$

**12.** $-5 \leq x < 2$

**13.** $0 < x < 3$

**14.** $-5 \leq x \leq 0$

**c** Solve using the addition principle. Then graph.

**15.** $x + 7 > 2$

**16.** $x + 5 > 2$

**17.** $x + 8 \leq -10$

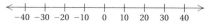

**18.** $x + 8 \leq -11$

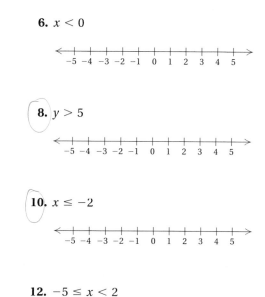

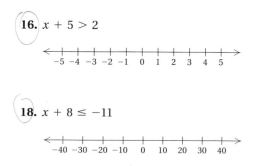

Solve using the addition principle.

**19.** $y - 7 > -12$

**20.** $y - 9 > -15$

**21.** $2x + 3 > x + 5$

**22.** $2x + 4 > x + 7$

**23.** $3x + 9 \leq 2x + 6$

**24.** $3x + 18 \leq 2x + 16$

**25.** $5x - 6 < 4x - 2$

**26.** $9x - 8 < 8x - 9$

**27.** $-9 + t > 5$

**28.** $-8 + p > 10$

**29.** $y + \dfrac{1}{4} \leq \dfrac{1}{2}$

**30.** $x - \dfrac{1}{3} \leq \dfrac{5}{6}$

**31.** $x - \dfrac{1}{3} > \dfrac{1}{4}$

**32.** $x + \dfrac{1}{8} > \dfrac{1}{2}$

**d** Solve using the multiplication principle. Then graph.

**33.** $5x < 35$

**34.** $8x \geq 32$

**35.** $-12x > -36$

**36.** $-16x > -64$

Solve using the multiplication principle.

**37.** $5y \geq -2$

**38.** $3x < -4$

**39.** $-2x \leq 12$

**40.** $-3x \leq 15$

**41.** $-4y \geq -16$

**42.** $-7x < -21$

**43.** $-3x < -17$

**44.** $-5y > -23$

**45.** $-2y > \dfrac{1}{7}$

**46.** $-4x \leq \dfrac{1}{9}$

**47.** $-\dfrac{6}{5} \leq -4x$

**48.** $-\dfrac{7}{9} > 63x$

Solve using the addition and multiplication principles.

**49.** $4 + 3x < 28$

**50.** $3 + 4y < 35$

**51.** $3x - 5 \leq 13$

**52.** $5y - 9 \leq 21$

**53.** $13x - 7 < -46$

**54.** $8y - 6 < -54$

**55.** $30 > 3 - 9x$

**56.** $48 > 13 - 7y$

**57.** $4x + 2 - 3x \leq 9$

**58.** $15x + 5 - 14x \leq 9$

**59.** $-3 < 8x + 7 - 7x$

**60.** $-8 < 9x + 8 - 8x - 3$

**61.** $6 - 4y > 4 - 3y$

**62.** $9 - 8y > 5 - 7y + 2$

**63.** $5 - 9y \leq 2 - 8y$

**64.** $6 - 18x \leq 4 - 12x - 5x$

**65.** $19 - 7y - 3y < 39$

**66.** $18 - 6y - 4y < 63 + 5y$

**67.** $2.1x + 45.2 > 3.2 - 8.4x$

**68.** $0.96y - 0.79 \leq 0.21y + 0.46$

**69.** $\dfrac{x}{3} - 2 \leq 1$

**70.** $\dfrac{2}{3} + \dfrac{x}{5} < \dfrac{4}{15}$

**71.** $\dfrac{y}{5} + 1 \leq \dfrac{2}{5}$

**72.** $\dfrac{3x}{4} - \dfrac{7}{8} \geq -15$

**73.** $3(2y - 3) < 27$

**74.** $4(2y - 3) > 28$

**75.** $2(3 + 4m) - 9 \geq 45$

**76.** $3(5 + 3m) - 8 \leq 88$

**77.** $8(2t + 1) > 4(7t + 7)$

**78.** $7(5y - 2) > 6(6y - 1)$

**79.** $3(r - 6) + 2 < 4(r + 2) - 21$

**80.** $5(x + 3) + 9 \leq 3(x - 2) + 6$

**81.** $0.8(3x + 6) \geq 1.1 - (x + 2)$

**82.** $0.4(2x + 8) \geq 20 - (x + 5)$

**83.** $\dfrac{5}{3} + \dfrac{2}{3}x < \dfrac{25}{12} + \dfrac{5}{4}x + \dfrac{3}{4}$

**84.** $1 - \dfrac{2}{3}y \geq \dfrac{9}{5} - \dfrac{y}{5} + \dfrac{3}{5}$

## Skill Maintenance

Add or subtract.  [1.3a], [1.4a]

**85.** $-56 + (-18)$

**86.** $-2.3 + 7.1$

**87.** $-\dfrac{3}{4} + \dfrac{1}{8}$

**88.** $8.12 - 9.23$

**89.** $-56 - (-18)$

**90.** $-\dfrac{3}{4} - \dfrac{1}{8}$

**91.** $-2.3 - 7.1$

**92.** $-8.12 + 9.23$

Simplify.

**93.** $5 - 3^2 + (8 - 2)^2 \cdot 4$  [1.8d]

**94.** $10 \div 2 \cdot 5 - 3^2 + (-5)^2$  [1.8d]

**95.** $5(2x - 4) - 3(4x + 1)$  [1.8b]

**96.** $9(3 + 5x) - 4(7 + 2x)$  [1.8b]

## Synthesis

**97.** ◈ Are the inequalities $3x - 4 < 10 - 4x$ and $2(x - 5) > 3(2x - 6)$ equivalent? Why or why not?

**98.** ◈ Explain in your own words why it is necessary to reverse the inequality symbol when multiplying on both sides of an inequality by a negative number.

**99.** Determine whether each number is a solution of the inequality $|x| < 3$.

    a) 0             b) $-2$

    c) $-3$        d) 4

    e) 3            f) 1.7

    g) $-2.8$

**100.** Graph $|x| < 3$ on a number line.

$$\xleftarrow{\quad\;\;\;\;\;\;\;\;\;\;\;\;\;\;\;\;\;\;\;\;\;\;\;\;\;\;\;\;}\xrightarrow{\quad}$$
$$-5 \;-4 \;-3 \;-2 \;-1 \;\;\; 0 \;\;\; 1 \;\;\; 2 \;\;\; 3 \;\;\; 4 \;\;\; 5$$

Solve.

**101.** $x + 3 \leq 3 + x$

**102.** $x + 4 < 3 + x$

Solve linear inequalities as a group.

Collaborative
Learning Manual

Copyright © 1999 Addison Wesley Longman

# 2.8 Applications and Problem Solving with Inequalities

We can use inequalities to solve certain types of problems.

### Objectives

**a** Translate number sentences to inequalities.

**b** Solve applied problems using inequalities.

### For Extra Help

TAPE 5    TAPE 6A    MAC WIN    CD-ROM

## a Translating to Inequalities

First let's practice translating sentences to inequalities.

**Examples**  Translate to an inequality.

**1.** A number is less than 5.

$$x < 5$$

**2.** A number is greater than or equal to $3\frac{1}{2}$.

$$y \geq 3\frac{1}{2}$$

**3.** He can earn, at most, $34,000.

$$E \leq \$34,000$$

**4.** The number of compact disc players sold in this city in a year is at least 2700.

$$C \geq 2700$$

**5.** 12 more than twice a number is less than 37.

$$2x + 12 < 37$$

*Do Exercises 1–5.*

## b Solving Problems

**Example 6**  *Test Scores.* A pre-med student is taking a chemistry course in which four tests are to be given. To get an A, she must average at least 90 on the four tests. The student got scores of 91, 86, and 89 on the first three tests. Determine (in terms of an inequality) what scores on the last test will allow her to get an A.

**1. Familiarize.**  Let's try some guessing. Suppose the student gets a 92 on the last test. The average of the four scores is their sum divided by the number of tests, 4, and is given by

$$\frac{91 + 86 + 89 + 92}{4} = 89.5.$$

In order for this average to be *at least* 90, it must be greater than or equal to 90. Since $89.5 \geq 90$ is false, a score of 92 will not give the student an A. But there are scores that will give an A. To find them, we translate to an inequality and solve. Let $x$ = the student's score on the last test.

**2. Translate.**  The average of the four scores must be *at least* 90. This means that it must be greater than or equal to 90. Thus we can translate the problem to the inequality

$$\frac{91 + 86 + 89 + x}{4} \geq 90.$$

Translate.

**1.** A number is less than or equal to 8.

**2.** A number is greater than $-2$.

**3.** That car can be driven at most 180 mph.

**4.** The price of that car is at least $5800.

**5.** Twice a number minus 32 is greater than 5.

*Answers on page A-8*

**6.** *Test Scores.* A student is taking a literature course in which four tests are to be given. To get a B, he must average at least 80 on the four tests. The student got scores of 82, 76, and 78 on the first three tests. Determine (in terms of an inequality) what scores on the last test will allow him to get at least a B.

**7.** *Gold Temperatures.* Gold stays solid at Fahrenheit temperatures below 1945.4°. Determine (in terms of an inequality) those Celsius temperatures for which gold stays solid. Use the formula given in Example 7.

*Answers on page A-8*

**3. Solve.** We solve the inequality. We first multiply by 4 to clear the fraction.

$$4\left(\frac{91 + 86 + 89 + x}{4}\right) \geq 4 \cdot 90 \qquad \text{Multiplying by 4}$$

$$91 + 86 + 89 + x \geq 360$$

$$266 + x \geq 360 \qquad \text{Collecting like terms}$$

$$x \geq 94 \qquad \text{Subtracting 266}$$

The solution set is $\{x \mid x \geq 94\}$.

**4. Check.** We can obtain a partial check by substituting a number greater than or equal to 94. We leave it to the student to try 95 in a manner similar to what was done in the *Familiarize* step.

**5. State.** Any score that is at least 94 will give the student an A.

*Do Exercise 6.*

**Example 7** *Butter Temperatures.* Butter stays solid at Fahrenheit temperatures below 88°. The formula

$$F = \tfrac{9}{5}C + 32$$

can be used to convert Celsius temperatures $C$ to Fahrenheit temperatures $F$. Determine (in terms of an inequality) those Celsius temperatures for which butter stays solid.

**1. Familiarize.** Let's make a guess. We try a Celsius temperature of 40°. We substitute and find $F$:

$$F = \tfrac{9}{5}C + 32 = \tfrac{9}{5}(40) + 32 = 72 + 32 = 104°.$$

This is higher than 88°, so 40° is *not* a solution. To find the solutions, we need to solve an inequality.

**2. Translate.** The Fahrenheit temperature $F$ is to be less than 88. We have the inequality

$$F < 88.$$

To find the Celsius temperatures $C$ that satisfy this condition, we substitute $\tfrac{9}{5}C + 32$ for $F$, which gives us the following inequality:

$$\tfrac{9}{5}C + 32 < 88.$$

**3. Solve.** We solve the inequality:

$$\tfrac{9}{5}C + 32 < 88$$

$$5\left(\tfrac{9}{5}C + 32\right) < 5(88) \qquad \text{Multiplying by 5 to clear the fraction}$$

$$5\left(\tfrac{9}{5}C\right) + 5(32) < 440 \qquad \text{Using a distributive law}$$

$$9C + 160 < 440 \qquad \text{Simplifying}$$

$$9C < 280 \qquad \text{Subtracting 160}$$

$$C < \frac{280}{9} \qquad \text{Dividing by 9}$$

$$C < 31.1. \qquad \text{Dividing and rounding to the nearest tenth}$$

The solution set of the inequality is $\{C \mid C < 31.1°\}$.

**4. Check.** The check is left to the student.

**5. State.** Butter stays solid at Celsius temperatures below 31.1°.

*Do Exercise 7.*

# Exercise Set 2.8

**a** Translate to an inequality.

**1.** A number is greater than 8.

**2.** A number is less than 5.

**3.** A number is less than or equal to −4.

**4.** A number is greater than or equal to 18.

**5.** The number of people is at least 1300.

**6.** The cost is at most $4857.95.

**7.** The amount of acid is not to exceed 500 liters.

**8.** The cost of gasoline is no less than 94 cents per gallon.

**9.** Two more than three times a number is less than 13.

**10.** Five less than one-half a number is greater than 17.

**b** Solve.

**11.** *Test Scores.* Your quiz grades are 73, 75, 89, and 91. Determine (in terms of an inequality) what scores on the last quiz will allow you to get an average quiz grade of at least 85.

**12.** *Body Temperatures.* The human body is considered to be fevered when its temperature is higher than 98.6°F. Using the formula given in Example 7, determine (in terms of an inequality) those Celsius temperatures for which the body is fevered.

**13.** *World Records in the 1500-m Run.* The formula

$$R = -0.075t + 3.85$$

can be used to predict the world record in the 1500-m run *t* years after 1930. Determine (in terms of an inequality) those years for which the world record will be less than 3.5 min.

**14.** *World Records in the 200-m Dash.* The formula

$$R = -0.028t + 20.8$$

can be used to predict the world record in the 200-m dash *t* years after 1920. Determine (in terms of an inequality) those years for which the world record will be less than 19.0 sec.

**15.** *Sizes of Envelopes.* Rhetoric Advertising is a direct-mail company. It determines that for a particular campaign, it can use any envelope with a fixed width of $3\frac{1}{2}$ in. and an area of at least $17\frac{1}{2}$ in². Determine (in terms of an inequality) those lengths that will satisfy the company constraints.

**16.** *Sizes of Packages.* An overnight delivery service accepts packages of up to 165 in. in length and girth combined. (Girth is the distance around the package.) A package has a fixed girth of 53 in. Determine (in terms of an inequality) those lengths for which a package is acceptable.

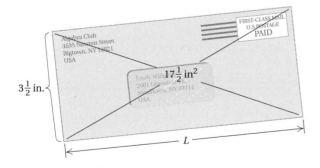

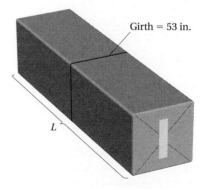

**17.** Find all numbers such that the sum of the number and 15 is less than four times the number.

**18.** Find all numbers such that three times the number minus ten times the number is greater than or equal to eight times the number.

**19.** *Black Angus Calves.* Black Angus calves weigh about 75 lb at birth and gain about 2 lb per day for the first few weeks. Determine (in terms of an inequality) those days for which the calf's weight is more than 125 lb.

**20.** *IKON Copiers.* IKON Office Solutions rents a Canon GP30F copier for $240 per month plus 1.8¢ per copy (*Source*: Ikon Office Solutions, Keith Palmer). A catalog publisher needs to lease a copy machine for use during a special project that they anticipate will take 3 months. They decide to rent the copier, but must stay within a budget of $5400 for copies. Determine (in terms of an inequality) the number of copies they can make per month and still remain within budget.

**21.** One side of a triangle is 2 cm shorter than the base. The other side is 3 cm longer than the base. What lengths of the base will allow the perimeter to be greater than 19 cm?

**22.** The perimeter of a rectangular swimming pool is not to exceed 70 ft. The length is to be twice the width. What widths will meet these conditions?

**23.** Dirk's Electric made 17 customer calls last week and 22 calls this week. How many calls must be made next week in order to maintain an average of at least 20 calls for the three-week period?

**24.** Ginny and Jill do volunteer work at a hospital. Jill worked 3 hr more than Ginny, and together they worked more than 27 hr. What possible number of hours did each work?

**25.** A family's air conditioner needs freon. The charge for a service call is a flat fee of $70 plus $60 an hour. The freon costs $35. The family has at most $150 to pay for the service call. Determine (in terms of an inequality) those lengths of time of the call that will allow the family to stay within its $150 budget.

**26.** A student is shopping for a new pair of jeans and two sweaters of the same kind. He is determined to spend no more than $120.00 for the outfit. He buys jeans for $21.95. What is the most that the student can spend for each sweater?

**27.** *Skippy Reduced-Fat Peanut Butter.* In order for a food to be advertised as "reduced fat," it must have at least 25% less fat than the regular food of that type. Reduced-fat Skippy Peanut Butter contains 12 g of fat per serving. What can you conclude about how much fat is in regular Skippy peanut butter?

**28.** A landscaping company is laying out a triangular flower bed. The height of the triangle is 16 ft. What lengths of the base will make the area at least 200 ft$^2$?

---

**Skill Maintenance**

Simplify.

**29.** $-3 + 2(-5)^2(-3) - 7$   [1.8d]

**30.** $3x + 2[4 - 5(2x - 1)]$   [1.8c]

**31.** $23(2x - 4) - 15(10 - 3x)$   [1.8b]

**32.** $256 \div 64 \div 4^2$   [1.8d]

---

**Synthesis**

**33.** ◆ Chassman and Bem booksellers offers a preferred customer card for $25. The card entitles a customer to a 10% discount on all purchases for a period of 1 year. Under what circumstances would an individual save money by buying a card?

**34.** ◆ After 9 quizzes, Brenda's average is 84. Is it possible for her to improve her average by two points with the next quiz? Why or why not?

Copyright © 1999 Addison Wesley Longman

# Summary and Review Exercises: Chapter 2

## Important Properties and Formulas

*The Addition Principle for Equations*: For any real numbers $a$, $b$, and $c$: $a = b$ is equivalent to $a + c = b + c$.

*The Multiplication Principle for Equations*: For any real numbers $a$, $b$, and $c$, $c \neq 0$: $a = b$ is equivalent to $a \cdot c = b \cdot c$.

*The Addition Principle for Inequalities*: For any real numbers $a$, $b$, and $c$:
$a < b$ is equivalent to $a + c < b + c$;
$a > b$ is equivalent to $a + c > b + c$;
$a \leq b$ is equivalent to $a + c \leq b + c$;
$a \geq b$ is equivalent to $a + c \geq b + c$.

*The Multiplication Principle for Inequalities*: For any real numbers $a$ and $b$, and any *positive* number $c$:
$a < b$ is equivalent to $ac < bc$;  $a > b$ is equivalent to $ac > bc$.

For any real numbers $a$ and $b$, and any *negative* number $c$:
$a < b$ is equivalent to $ac > bc$;  $a > b$ is equivalent to $ac < bc$.

The objectives to be tested in addition to the material in this chapter are [1.1a], [1.1b], [1.3a], and [1.8b].

Solve.  [2.1b]

**1.** $x + 5 = -17$

**2.** $n - 7 = -6$

**3.** $x - 11 = 14$

**4.** $y - 0.9 = 9.09$

Solve.  [2.2a]

**5.** $-\dfrac{2}{3}x = -\dfrac{1}{6}$

**6.** $-8x = -56$

**7.** $-\dfrac{x}{4} = 48$

**8.** $15x = -35$

**9.** $\dfrac{4}{5}y = -\dfrac{3}{16}$

Solve.  [2.3a]

**10.** $5 - x = 13$

**11.** $\dfrac{1}{4}x - \dfrac{5}{8} = \dfrac{3}{8}$

Solve.  [2.3b]

**12.** $5t + 9 = 3t - 1$

**13.** $7x - 6 = 25x$

**14.** $14y = 23y - 17 - 10$

**15.** $0.22y - 0.6 = 0.12y + 3 - 0.8y$

**16.** $\dfrac{1}{4}x - \dfrac{1}{8}x = 3 - \dfrac{1}{16}x$

Solve.  [2.3c]

**17.** $4(x + 3) = 36$

**18.** $3(5x - 7) = -66$

**19.** $8(x - 2) = 5(x + 4)$

**20.** $-5x + 3(x + 8) = 16$

Determine whether the given number is a solution of the inequality $x \leq 4$.  [2.7a]

**21.** $-3$

**22.** $7$

**23.** $4$

Solve. Write set notation for the answers.  [2.7c, d, e]

**24.** $y + \dfrac{2}{3} \geq \dfrac{1}{6}$

**25.** $9x \geq 63$

**26.** $2 + 6y > 14$

**27.** $7 - 3y \geq 27 + 2y$

**28.** $3x + 5 < 2x - 6$

**29.** $-4y < 28$

**30.** $3 - 4x < 27$

**31.** $4 - 8x < 13 + 3x$

**32.** $-3y \geq -21$

**33.** $-4x \leq \dfrac{1}{3}$

Graph on a number line.  [2.7b, e]

**34.** $4x - 6 < x + 3$

**35.** $-2 < x \leq 5$

**36.** $y > 0$

Solve.  [2.6a]

**37.** $C = \pi d$, for $d$

**38.** $V = \dfrac{1}{3}Bh$, for $B$

**39.** $A = \dfrac{a + b}{2}$, for $a$

Solve.  [2.4a]

**40.** *Dimensions of Wyoming.* The state of Wyoming is roughly in the shape of a rectangle whose perimeter is 1280 mi. The length is 90 mi more than the width. Find the dimensions.

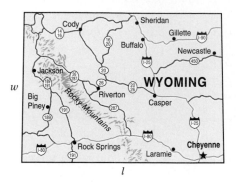

**41.** If 14 is added to a certain number, the result is 41. Find the number.

**42.** *Interstate Mile Markers.* The sum of two mile markers on I-5 in California is 691. Find the numbers on the markers.

**43.** An entertainment center sold for $2449 in June. This was $332 more than the cost in February. Find the cost in February.

**44.** Ty is paid a commission of $4 for each appliance he sells. One week, he received $108 in commissions. How many appliances did he sell?

**45.** The measure of the second angle of a triangle is 50° more than that of the first angle. The measure of the third angle is 10° less than twice the first angle. Find the measures of the angles.

Solve.  [2.5a]

**46.** After a 30% reduction, a bread maker is on sale for $154. What was the marked price (the price before the reduction)?

**47.** A hotel manager's salary is $30,000, which is a 15% increase over the previous year's salary. What was the previous salary (to the nearest dollar)?

**48.** A tax-exempt charity received a bill of $145.90 for a sump pump. The bill incorrectly included sales tax of 5%. How much does the charity actually owe?

Solve.

**49.** *Test Scores.* Your test grades are 71, 75, 82, and 86. What is the lowest grade that you can get on the next test and still have an average test score of at least 80?  [2.8b]

**50.** The length of a rectangle is 43 cm. What widths will make the perimeter greater than 120 cm? [2.8a]

**51.** *Estimating the Weight of a Fish.* An ancient fisherman's formula for estimating the weight of a fish is

$$W = \frac{Lg^2}{800},$$

where $W$ is the weight in pounds, $L$ is the length in inches, and $g$ is the girth (distance around the midsection) in inches.  [2.6a]

a) Estimate the weight of a salmon that is 3 ft long and has a girth of about 13.5 in.
b) Solve the formula for $L$.

**Skill Maintenance**

**52.** Evaluate $\dfrac{a + b}{4}$ for $a = 16$ and $b = 25$.  [1.1a]

**53.** Translate to an algebraic expression:  [1.1b]

Tricia drives her car at 58 mph for $t$ hours. How far has she driven?

**54.** Add: $-12 + 10 + (-19) + (-24)$.  [1.3a]

**55.** Remove parentheses and simplify: $5x - 8(6x - y)$. [1.8b]

**Synthesis**

**56.** ◈ Would it be better to receive a 5% raise and then an 8% raise or the other way around? Why?

**57.** ◈ Are the inequalities $x > -5$ and $-x < 5$ equivalent? Why or why not?

Solve.

**58.** $2|x| + 4 = 50$  [1.2e], [2.3a]

**59.** $|3x| = 60$  [1.2e], [2.2a]

**60.** $y = 2a - ab + 3$, for $a$  [2.6a]

# Test: Chapter 2

Answers

1. _____
2. _____
3. _____
4. _____
5. _____
6. _____
7. _____
8. _____
9. _____
10. _____
11. _____
12. _____
13. _____
14. _____
15. _____
16. _____
17. _____
18. _____
19. _____
20. _____
21. _____
22. _____

Solve.

1. $x + 7 = 15$

2. $t - 9 = 17$

3. $3x = -18$

4. $-\dfrac{4}{7}x = -28$

5. $3t + 7 = 2t - 5$

6. $\dfrac{1}{2}x - \dfrac{3}{5} = \dfrac{2}{5}$

7. $8 - y = 16$

8. $-\dfrac{2}{5} + x = -\dfrac{3}{4}$

9. $3(x + 2) = 27$

10. $-3x - 6(x - 4) = 9$

11. $0.4p + 0.2 = 4.2p - 7.8 - 0.6p$

Solve. Write set notation for the answers.

12. $x + 6 \leq 2$

13. $14x + 9 > 13x - 4$

14. $12x \leq 60$

15. $-2y \geq 26$

16. $-4y \leq -32$

17. $-5x \geq \dfrac{1}{4}$

18. $4 - 6x > 40$

19. $5 - 9x \geq 19 + 5x$

Graph on a number line.

20. $y \leq 9$

21. $6x - 3 < x + 2$

22. $-2 \leq x \leq 2$

23. _____

24. _____

25. _____

26. _____

27. _____

28. _____

29. a) _____

b) _____

30. _____

31. _____

32. _____

33. _____

34. _____

35. _____

36. _____

37. _____

38. _____

Solve.

23. The perimeter of a rectangular photograph is 36 cm. The length is 4 cm greater than the width. Find the width and the length.

24. If you triple a number and then subtract 14, you get two-thirds of the original number. What is the original number?

25. The numbers on three raffle tickets are consecutive integers whose sum is 7530. Find the integers.

26. Money is invested in a savings account at 5% simple interest. After 1 year, there is $924 in the account. How much was originally invested?

27. An 8-m board is cut into two pieces. One piece is 2 m longer than the other. How long are the pieces?

28. Solve $A = 2\pi rh$ for $r$.

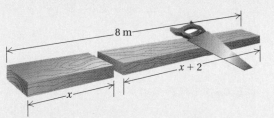

29. *Male Caloric Needs.* The number of calories $K$ needed each day by a moderately active man who weighs $w$ kilograms, is $h$ centimeters tall, and is $a$ years old can be estimated by the formula

$$K = 19.18w + 7h - 9.52a + 92.4.$$

a) David is moderately active, weighs 89 kg, is 180 cm tall, and is 43 yr old. What are his caloric needs?
b) Solve the formula for $w$.

***Source:*** Parker, M., *She Does Math.* Mathematical Association of America, p. 96.

30. Find all numbers such that six times the number is greater than the number plus 30.

31. The width of a rectangle is 96 yd. Find all possible lengths such that the perimeter of the rectangle will be at least 540 yd.

---

**Skill Maintenance**

32. Add: $\dfrac{2}{3} + \left(-\dfrac{8}{9}\right)$.

33. Evaluate $\dfrac{4x}{y}$ for $x = 2$ and $y = 3$.

34. Translate to an algebraic expression: Seventy-three percent of $p$.

35. Simplify: $2x - 3y - 5(4x - 8y)$.

---

**Synthesis**

36. Solve $c = \dfrac{1}{a - d}$ for $d$.

37. Solve: $3|w| - 8 = 37$.

38. A movie theater had a certain number of tickets to give away. Five people got the tickets. The first got one-third of the tickets, the second got one-fourth of the tickets, and the third got one-fifth of the tickets. The fourth person got eight tickets, and there were five tickets left for the fifth person. Find the total number of tickets given away.

Copyright © 1999 Addison Wesley Longman

# Cumulative Review: Chapters 1–2

Evaluate.

**1.** $\dfrac{y - x}{4}$, for $y = 12$ and $x = 6$

**2.** $\dfrac{3x}{y}$, for $x = 5$ and $y = 4$

**3.** $x - 3$, for $x = 3$

**4.** Translate to an algebraic expression:  Four less than twice $w$.

Use $<$ or $>$ for ▩ to write a true sentence.

**5.** $-4$ ▩ $-6$

**6.** $0$ ▩ $-5$

**7.** $-8$ ▩ $7$

**8.** Find the opposite and the reciprocal of $\dfrac{2}{5}$.

Find the absolute value.

**9.** $|3|$

**10.** $\left| -\dfrac{3}{4} \right|$

**11.** $|0|$

Compute and simplify.

**12.** $-6.7 + 2.3$

**13.** $-\dfrac{1}{6} - \dfrac{7}{3}$

**14.** $-\dfrac{5}{8}\left( -\dfrac{4}{3} \right)$

**15.** $(-7)(5)(-6)(-0.5)$

**16.** $81 \div (-9)$

**17.** $-10.8 \div 3.6$

**18.** $-\dfrac{4}{5} \div -\dfrac{25}{8}$

Multiply.

**19.** $5(3x + 5y + 2z)$

**20.** $4(-3x - 2)$

**21.** $-6(2y - 4x)$

Factor.

**22.** $64 + 18x + 24y$

**23.** $16y - 56$

**24.** $5a - 15b + 25$

Collect like terms.

**25.** $9b + 18y + 6b + 4y$

**26.** $3y + 4 + 6z + 6y$

**27.** $-4d - 6a + 3a - 5d + 1$

**28.** $3.2x + 2.9y - 5.8x - 8.1y$

Simplify.

**29.** $7 - 2x - (-5x) - 8$

**30.** $-3x - (-x + y)$

**31.** $-3(x - 2) - 4x$

**32.** $10 - 2(5 - 4x)$

**33.** $[3(x + 6) - 10] - [5 - 2(x - 8)]$

Solve.

**34.** $x + 1.75 = 6.25$

**35.** $\dfrac{5}{2}y = \dfrac{2}{5}$

**36.** $-2.6 + x = 8.3$

**37.** $4\dfrac{1}{2} + y = 8\dfrac{1}{3}$

**38.** $-\dfrac{3}{4}x = 36$

**39.** $-2.2y = -26.4$

**40.** $5.8x = -35.96$

**41.** $-4x + 3 = 15$

**42.** $-3x + 5 = -8x - 7$

**43.** $4y - 4 + y = 6y + 20 - 4y$

**44.** $-3(x - 2) = -15$

**45.** $\dfrac{1}{3}x - \dfrac{5}{6} = \dfrac{1}{2} + 2x$

**46.** $-3.7x + 6.2 = -7.3x - 5.8$

**47.** $3x - 1 < 2x + 1$

**48.** $5 - y \le 2y - 7$

**49.** $3y + 7 > 5y + 13$

**50.** $A = \dfrac{1}{2}h(b + c)$, for $h$

**51.** $Q = \dfrac{p - q}{2}$, for $q$

Solve.

**52.** If 25 is subtracted from a certain number, the result is 129. Find the number.

**53.** Lance and Rocky purchased rollerblades for a total of $107. Lance paid $17 more for his rollerblades than Rocky did. What did Rocky pay?

**54.** Money is invested in a savings account at 8% simple interest. After 1 year, there is $1134 in the account. How much was originally invested?

**55.** A 143-m wire is cut into three pieces. The second piece is 3 m longer than the first. The third is four-fifths as long as the first. How long is each piece?

**56.** Your test grades are 75, 82, 86, and 79. Determine (in terms of an inequality) what scores on the last test will allow you to get an average test score of at least 80.

**57.** After a 25% reduction, a tie is on sale for $18.45. What was the price before reduction?

---

### Synthesis

**58.** An engineer's salary at the end of a year is $38,563.20. This reflects a 4% salary increase and a later 3% cost-of-living adjustment during the year. What was the salary at the beginning of the year?

**59.** Nadia needs to use a copier to reduce a drawing to fit on a page. The original drawing is 9 in. long and it must fit into a space that is 6.3 in. long. By what percent should she reduce the drawing on the copier?

Solve.

**60.** $4|x| - 13 = 3$

**61.** $4(x + 2) = 4(x - 2) + 16$

**62.** $0(x + 3) + 4 = 0$

**63.** $\dfrac{2 + 5x}{4} = \dfrac{11}{28} + \dfrac{8x + 3}{7}$

**64.** $5(7 + x) = (x + 7)5$

**65.** $p = \dfrac{2}{m + Q}$, for $Q$

**3**

# Graphs of Equations; Data Analysis

| An Application | The Mathematics |
| --- | --- |

The cost *y*, in dollars, of mailing a FedEx Priority Overnight package weighing 1 lb or more is given by the equation

$$y = 2.085x + 15.08,$$

where *x* = the number of pounds (**Source**: Federal Express Corporation). Graph the equation and then use the graph to estimate the cost of mailing a $6\frac{1}{2}$-lb package.

This problem appears as Example 8 in Section 3.2.

The graph is shown below. It appears that the cost of mailing a $6\frac{1}{2}$-lb package is about $29.

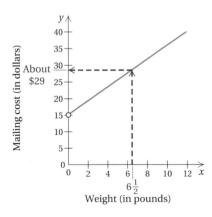

# Pretest: Chapter 3

Graph on a plane.

**1.** $y = -x$

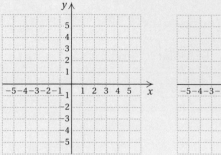

**2.** $x = -4$

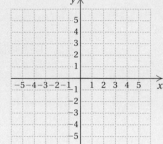

**3.** $4x - 5y = 20$

**4.** $y = \dfrac{2}{3}x - 1$

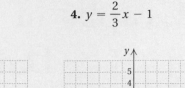
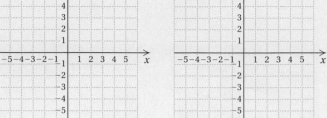

**5.** In which quadrant is the point $(-4, -1)$ located?

**6.** Determine whether the ordered pair $(-4, -1)$ is a solution of $4x - 5y = 20$.

**7.** Find the intercepts of the graph of $4x - 5y = 20$.

**8.** Find the $y$-intercept of $y = 3x - 8$.

**9.** *Price of Printing.* The price $P$, in cents, of a photocopied and bound lab manual is given by

$$P = \frac{7}{2}n + 20,$$

where $n =$ the number of pages in the manual. Graph the equation and then use the graph to estimate the price of an 85-page manual.

**10.** *Blood Alcohol Levels of Drivers in Fatal Accidents.* Find the mean, the median, and the mode of this set of data:

0.18, 0.17, 0.21, 0.16, 0.18.

**11.** *Height of Girls.* Use extrapolation to estimate the missing data.

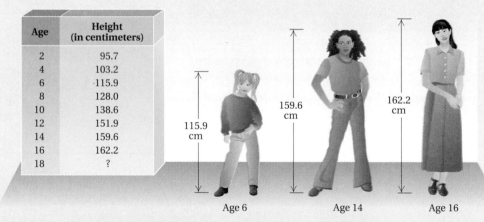

| Age | Height (in centimeters) |
|-----|-------------------------|
| 2 | 95.7 |
| 4 | 103.2 |
| 6 | 115.9 |
| 8 | 128.0 |
| 10 | 138.6 |
| 12 | 151.9 |
| 14 | 159.6 |
| 16 | 162.2 |
| 18 | ? |

115.9 cm — Age 6

159.6 cm — Age 14

162.2 cm — Age 16

*Source*: Kempe, C. Henry, et al (eds.), *Current Pediatric Diagnosis & Treatment 1987.* Norwalk, CT: Appleton & Lange, 1987

## Objectives for Retesting

The objectives to be retested in addition to the material in this chapter are as follows.

[R.3c]  Round numbers to a specified decimal place.
[1.2c]  Convert from fractional notation to decimal notation for a rational number.
[1.2e]  Find the absolute value of a real number.
[2.5a]  Solve applied problems involving percent.

# 3.1 Graphs and Applications

Often data are available regarding an application in mathematics that we are reviewing. We can use graphs to show the data and extract information about the data that can lead to making analyses and predictions.

Today's print and electronic media make extensive use of graphs. This is due in part to the ease with which some graphs can be prepared by computer and in part to the large quantity of information that a graph can display. We first consider applications with circle, bar, and line graphs.

 **a** **Applications with Graphs**

### Circle Graphs

*Circle graphs* and *pie graphs*, or *charts*, are often used to show what percent of a whole each particular item in a group represents.

**Example 1** *U.S. Soft-Drink Retail Sales.* The following circle graph shows the percentages of sales of various soft drinks in the United States in a recent year.

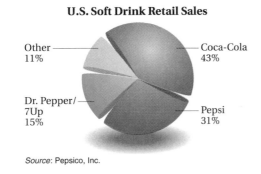

**U.S. Soft Drink Retail Sales**

Other 11%
Coca-Cola 43%
Dr. Pepper/ 7Up 15%
Pepsi 31%

*Source*: Pepsico, Inc.

Total soft-drink sales in the United States that year reached $54 billion. What were the sales of Pepsi?

1. **Familiarize.** The graph shows that 31% of the soft-drink sales were of Pepsi. We let

$y$ = the amount spent on Pepsi.

2. **Translate.** We reword and translate the problem as follows.

What is 31% of $54?    **Rewording**
$\downarrow$   $\downarrow$   $\downarrow$   $\downarrow$   $\downarrow$
$y$   =   31%   ·   54    **Translating**

3. **Solve.** We solve the equation by carrying out the computation on the right:

$y = 31\% \cdot 54 = 0.31 \cdot 54 = \$16.74.$

4. **Check.** We leave the check to the student.

5. **State.** Pepsi accounted for $16.74 billion of the soft-drink sales that particular year.

*Do Exercise 1.*

## Objectives

**a** Solve applied problems involving circle, bar, and line graphs.

**b** Plot points associated with ordered pairs of numbers.

**c** Determine the quadrant in which a point lies.

**d** Find the coordinates of a point on a graph.

### For Extra Help

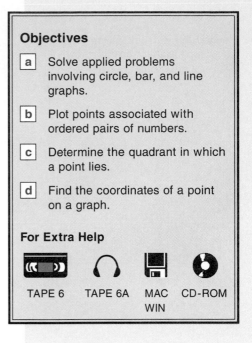

TAPE 6    TAPE 6A    MAC WIN    CD-ROM

1. Referring to Example 1, determine the soft-drink sales of Coca-Cola.

*Answer on page A-9*

**2.** *Tornado Touchdowns.*
Referring to Example 2,
determine the following.

**a)** During which interval did
the smallest number of
touchdowns occur?

## Bar Graphs

*Bar graphs* are convenient for showing comparisons. In every bar graph, certain categories are paired with certain numbers. Example 2 pairs intervals of time with the total number of reported cases of tornado touchdowns.

**Example 2** *Tornado Touchdowns.* The following bar graph shows the total number of tornado touchdowns by time of day in Indiana from 1950–1994.

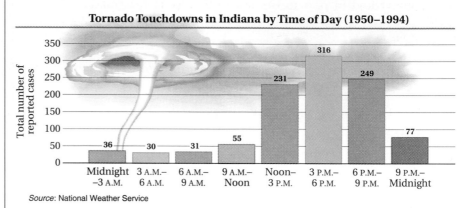

**Tornado Touchdowns in Indiana by Time of Day (1950–1994)**

*Source*: National Weather Service

**a)** During which interval of time did the greatest number of tornado touchdowns occur?

**b)** During which intervals was the number of tornado touchdowns greater than 200?

We solve as follows.

**a)** In this bar graph, the values are written at the top of the bars. We see that 316 is the greatest number. We look at the bottom of that bar on the horizontal scale and see that the time interval of greatest occurrence is 3 P.M.–6 P.M.

**b)** We locate 200 on the vertical scale and move across the graph or draw a horizontal line. We note that the value on three bars exceeds 200. Then we look down at the horizontal scale and see that the corresponding time intervals are noon–3 P.M., 3 P.M.–6 P.M., and 6 P.M.–9 P.M.

**b)** During which intervals was
the number of touchdowns
less than 60?

*Do Exercise 2.*

*Answers on page A-9*

## Line Graphs

*Line graphs* are often used to show change over time. Certain points are plotted to represent given information. When segments are drawn to connect the points, a line graph is formed.

Sometimes it is impractical to begin the listing of horizontal or vertical values with zero. When this happens, as in Example 3, the symbol ⌇ is used to indicate a break in the list of values.

**Example 3**   *Exercise and Pulse Rate.* The following line graph shows the relationship between a person's resting pulse rate and months of regular exercise.

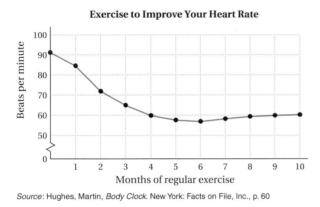

**Exercise to Improve Your Heart Rate**

*Source*: Hughes, Martin, *Body Clock*. New York: Facts on File, Inc., p. 60

a) How many months of regular exercise are required to lower the pulse rate to its lowest point?

b) How many months of regular exercise are needed to achieve a pulse rate of 65 beats per minute?

We solve as follows.

a) The lowest point on the graph occurs above the number 6. Thus after 6 months of regular exercise, the pulse rate has been lowered as much as possible.

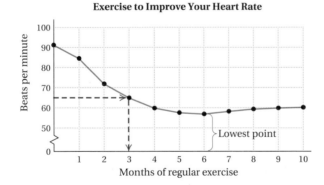

**Exercise to Improve Your Heart Rate**

b) We locate 65 on the vertical scale and then move right until we reach the line. At that point, we move down to the horizontal scale and read the information we are seeking. The pulse rate is 65 beats per minute after 3 months of regular exercise.

*Do Exercise 3.*

**3.** *Exercise and Pulse Rate.* Referring to Example 3, determine the following.

a) About how many months of regular exercise are needed to achieve a pulse rate of about 72 beats per minute?

b) What pulse rate has been achieved after 10 months of exercise?

*Answers on page A-9*

Plot these points on the graph below.

**4.** (4, 5)          **5.** (5, 4)

**6.** (−2, 5)          **7.** (−3, −4)

**8.** (5, −3)          **9.** (−2, −1)

**10.** (0, −3)          **11.** (2, 0)

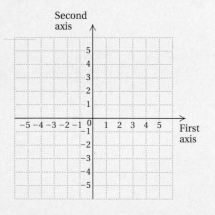

## b | Plotting Ordered Pairs

The line graph in Example 3 is formed from a collection of points. Each point pairs a number of months of exercise with a pulse rate.

In Chapter 2, we graphed numbers and inequalities in one variable on a line. To enable us to graph an equation that contains two variables, we now learn to graph number pairs on a plane.

On a number line, each point is the graph of a number. On a plane, each point is the graph of a number pair. We use two perpendicular number lines called **axes**. They cross at a point called the **origin**. The arrows show the positive directions.

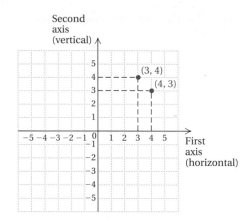

Consider the ordered pair (3, 4). The numbers in an ordered pair are called **coordinates**. In (3, 4), the **first coordinate** is 3 and the **second coordinate** is 4. To plot (3, 4), we start at the origin and move horizontally to the 3. Then we move up vertically 4 units and make a "dot."

The point (4, 3) is also plotted. Note that (3, 4) and (4, 3) give different points. The order of the numbers in the pair is indeed important. They are called **ordered pairs** because it makes a difference which number comes first. The coordinates of the origin are (0, 0) even though it is usually labeled either with the number 0 or not at all.

**Example 4**   Plot the point (−5, 2).

The first number, −5, is negative. Starting at the origin, we move −5 units in the horizontal direction (5 units to the left). The second number, 2, is positive. We move 2 units in the vertical direction (up).

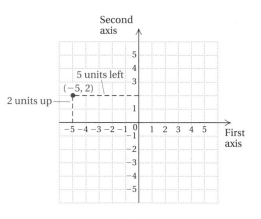

*Do Exercises 4–11.*

## c | Quadrants

This figure shows some points and their coordinates. In region I (the *first quadrant*), both coordinates of any point are positive. In region II (the *second quadrant*), the first coordinate is negative and the second positive. In region III (the *third quadrant*), both coordinates are negative. In region IV (the *fourth quadrant*), the first coordinate is positive and the second is negative.

**Example 5** In which quadrant, if any, are the points $(-4, 5)$, $(5, -5)$, $(2, 4)$, $(-2, -5)$, and $(-5, 0)$ located?

The point $(-4, 5)$ is in the second quadrant. The point $(5, -5)$ is in the fourth quadrant. The point $(2, 4)$ is in the first quadrant. The point $(-2, -5)$ is in the third quadrant. The point $(-5, 0)$ is on an axis and is *not* in any quadrant.

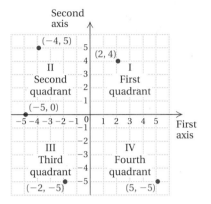

*Do Exercises 12–18.*

## d | Finding Coordinates

To find the coordinates of a point, we see how far to the right or left of zero it is located and how far up or down.

**Example 6** Find the coordinates of points $A$, $B$, $C$, $D$, $E$, $F$, and $G$.

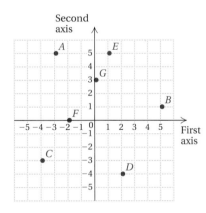

Point $A$ is 3 units to the left (horizontal direction) and 5 units up (vertical direction). Its coordinates are $(-3, 5)$. The coordinates of the other points are as follows:

B: $(5, 1)$;   C: $(-4, -3)$;   D: $(2, -4)$;

E: $(1, 5)$;   F: $(-2, 0)$;   G: $(0, 3)$.

*Do Exercise 19.*

---

**12.** What can you say about the coordinates of a point in the third quadrant?

**13.** What can you say about the coordinates of a point in the fourth quadrant?

In which quadrant, if any, is the point located?

**14.** $(5, 3)$

**15.** $(-6, -4)$

**16.** $(10, -14)$

**17.** $(-13, 9)$

**18.** $(0, -3)$

**19.** Find the coordinates of points $A$, $B$, $C$, $D$, $E$, $F$, and $G$ on the graph below.

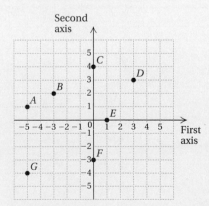

*Answers on page A-9*

# Improving Your Math Study Skills

## Better Test Taking

How often do you make the following statement after taking a test: "I was able to do the homework, but I froze during the test"? Instructors have heard this comment for years, and in most cases, it is merely a coverup for a lack of proper study habits. Here are two related tips, however, to help you with this difficulty. Both are intended to make test taking less stressful by getting you to practice good test-taking habits on a daily basis.

- **Treat *every* homework exercise as if it were a test question.** If you had to work a problem at your job with no backup answer provided, what would you do? You would probably work it very deliberately, checking and rechecking every step. You might work it more than one time, or you might try to work it another way to check the result. Try to use this approach when doing your homework. Treat every exercise as though it were a test question and no answers were provided at the back of the book.

- **Be sure that you do questions without answers as part of every homework assignment whether or not the instructor has assigned them!** One reason a test may seem such a different task is that questions on a test lack answers. That is the reason for taking a test: to see if you can do the questions without assistance. As part of your test preparation, be sure you do some exercises for which you do not have the answers. Thus when you take a test, you are doing a more familiar task.

The purpose of doing your homework using these approaches is to give you more test-taking practice beforehand. Let's make a sports analogy here. At a basketball game, the players take lots of practice shots before the game. They play the first half, go to the locker room, and come out for the second half. What do they do before the second half, even though they have just played 20 minutes of basketball? They shoot baskets again! We suggest the same approach here. Create more and more situations in which you practice taking test questions by treating each homework exercise like a test question and by doing exercises for which you have no answers. Good luck! Please send me an e-mail (exponent@aol.com) and let me know how it works for you.

# Exercise Set 3.1

**a** Solve.

*Driving While Intoxicated (DWI).* State laws have determined that a blood alcohol level of at least 0.10% or higher indicates that an individual has consumed too much alcohol to drive safely. The following bar graph shows the number of drinks that a person of a certain weight would need to consume in order to reach a blood alcohol level of 0.10%. A 12-oz beer, a 5-oz glass of wine, or a cocktail containing $1\frac{1}{2}$ oz of distilled liquor all count as one drink. Use the bar graph for Exercises 1–6.

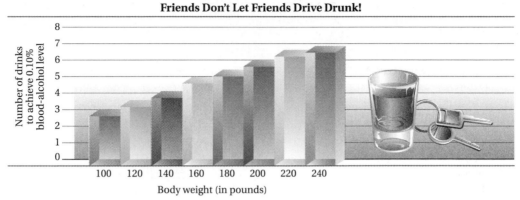

**Friends Don't Let Friends Drive Drunk!**

Source: *Neighborhood Digest*, 7, no. 12

1. Approximately how many drinks would a 200-lb person have consumed if he or she had a blood alcohol level of 0.10%?

2. What can be concluded about the weight of someone who can consume 4 drinks without reaching a blood alcohol level of 0.10%?

3. What can be concluded about the weight of someone who can consume 6 drinks without reaching a blood alcohol level of 0.10%?

4. Approximately how many drinks would a 160-lb person have consumed if he or she had a blood alcohol level of 0.10%?

5. What can be concluded about the weight of someone who has consumed $3\frac{1}{2}$ drinks without reaching a blood alcohol level of 0.10%?

6. What can be concluded about the weight of someone who has consumed $4\frac{1}{2}$ drinks without reaching a blood alcohol level of 0.10%?

*Cost of Raising a Child.* A family is in a $32,800–$55,500 income bracket. The following pie chart shows the various costs involved for a family in a $32,800–$55,500 income bracket in raising a child to the age of 18. Use the pie chart for Exercises 7–10.

**Cost of Raising a Child to the Age of 18**

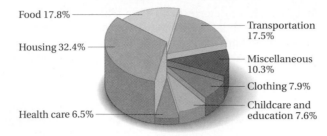

Food 17.8%
Housing 32.4%
Health care 6.5%
Transportation 17.5%
Miscellaneous 10.3%
Clothing 7.9%
Childcare and education 7.6%

*Source*: U.S. Department of Agriculture, Food, Nutrition, and Consumer Service

**7.** What percent of the total expense is for housing?

**8.** What percent of the total expense is for health care?

**9.** It costs a total of about $136,320 to raise a child to the age of 18. How much of this cost is for child care and education?

**10.** It costs a total of about $136,320 to raise a child to the age of 18. How much of this cost is for transportation?

*MADD (Mothers Against Drunk Driving).* Despite efforts by groups such as MADD, the number of alcohol-related deaths is rising after many years of decline. The data in the following graph show the number of deaths from 1989 to 1995. Use this graph for Exercises 11–16.

**Number of Alcohol-Related Traffic Deaths**

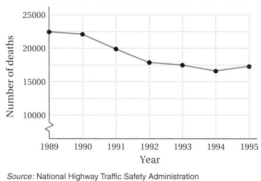

*Source*: National Highway Traffic Safety Administration

**11.** About how many alcohol-related deaths occurred in 1991?

**12.** About how many alcohol-related deaths occurred in 1995?

**13.** In what year did the lowest number of deaths occur?

**14.** In what years did fewer than 18,000 deaths occur?

**15.** By how much did the number of alcohol-related deaths increase from 1994 to 1995?

**16.** By how much did the number of alcohol-related deaths decrease from 1989 to 1994?

**b**

**17.** Plot these points.

(2, 5)   (−1, 3)   (3, −2)   (−2, −4)

(0, 4)   (0, −5)   (5, 0)      (−5, 0)

**18.** Plot these points.

(4, 4)   (−2, 4)   (5, −3)   (−5, −5)

(0, 4)   (0, −4)   (3, 0)      (−4, 0)

**c** In which quadrant is the point located?

**19.** (−5, 3)

**20.** (1, −12)

**21.** (100, −1)

**22.** (−2.5, 35.6)

**23.** (−6, −29)

**24.** (3.6, 105.9)

**25.** (3.8, 9.2)

**26.** (−895, −492)

**27.** $\left(-\dfrac{1}{3}, \dfrac{15}{7}\right)$

**28.** $\left(-\dfrac{2}{3}, -\dfrac{9}{8}\right)$

**29.** $\left(12\dfrac{7}{8}, -1\dfrac{1}{2}\right)$

**30.** $\left(23\dfrac{5}{8}, 81.74\right)$

**31.** In quadrant III, first coordinates are always _____ and second coordinates are always _____ .

**32.** In quadrant II, _2nd_ coordinates are always positive and _1st_ coordinates are always negative.

**33.** In quadrant IV, _____ coordinates are always negative and _____ coordinates are always positive.

**34.** In quadrant I, first coordinates are always _____ and second coordinates are always _____ .

In Exercises 35–38, tell in which quadrant(s) the point can be located.

**35.** The first coordinate is positive.

**36.** The second coordinate is negative.

**37.** The first and second coordinates are equal.

**38.** The first coordinate is the additive inverse of the second coordinate.

d

**39.** Find the coordinates of points *A*, *B*, *C*, *D*, and *E*.

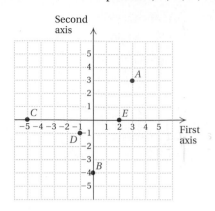

**40.** Find the coordinates of points *A*, *B*, *C*, *D*, and *E*.

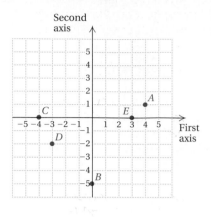

---

**Skill Maintenance**

Find the absolute value.   [1.2e]

**41.** $|-12|$

**42.** $|4.89|$

**43.** $|0|$

**44.** $\left|-\frac{4}{5}\right|$

Solve.   [2.5a]

**45.** *Baseball Salaries.* In 1997, the total amount spent on the salaries of major-league baseball players soared to $1.06 billion. This was a 17.7% increase over the total amount spent in 1996. (**Source:** Major League Baseball) How much was spent in 1996?

**46.** Erin left a 15% tip for a meal. The total cost of the meal, including the tip, was $21.16. What was the cost of the meal before the tip was added?

---

**Synthesis**

**47.** ◈ The sales of snow skis are highest in the winter months and lowest in the summer months. Sketch a line graph that might show the sales of a ski store and explain how an owner might use such a graph in decision making.

**48.** ◈ The graph in Example 3 tends to flatten out. Explain why the graph does not continue to decrease downward.

**49.** The points $(-1, 1)$, $(4, 1)$, and $(4, -5)$ are three vertices of a rectangle. Find the coordinates of the fourth vertex.

**50.** Three parallelograms share the vertices $(-2, -3)$, $(-1, 2)$, and $(4, -3)$. Find the fourth vertex of each parallelogram.

**51.** Graph eight points such that the sum of the coordinates in each pair is 6.

**52.** Graph eight points such that the first coordinate minus the second coordinate is 1.

**53.** Find the perimeter of a rectangle whose vertices have coordinates $(5, 3)$, $(5, -2)$, $(-3, -2)$, and $(-3, 3)$.

**54.** Find the area of a triangle whose vertices have coordinates $(0, 9)$, $(0, -4)$, and $(5, -4)$.

Collaborative
Learning Manual

Practice finding and plotting ordered pairs by playing a variation of the game Battleship.

# 3.2 Graphing Linear Equations

We have seen how circle, bar, and line graphs can be used to represent the data in an application. Now we begin to learn how graphs can be used to represent solutions of equations.

## a | Solutions of Equations

When an equation contains two variables, the solutions of the equation are *ordered pairs* in which each number in the pair corresponds to a letter in the equation. Unless stated otherwise, to determine whether a pair is a solution, we use the first number in each pair to replace the variable that occurs first alphabetically.

**Example 1** Determine whether each of the following pairs is a solution of $4q - 3p = 22$: $(2, 7)$ and $(-1, 6)$.

For $(2, 7)$, we substitute 2 for $p$ and 7 for $q$ (using alphabetical order of variables):

$$\frac{4q - 3p = 22}{4 \cdot 7 - 3 \cdot 2\ ?\ 22}$$
$$28 - 6$$
$$22 \qquad \text{TRUE}$$

Thus, $(2, 7)$ is a solution of the equation.

For $(-1, 6)$, we substitute $-1$ for $p$ and 6 for $q$:

$$\frac{4q - 3p = 22}{4 \cdot 6 - 3 \cdot -1\ ?\ 22}$$
$$24 + 3$$
$$27 \qquad \text{FALSE}$$

Thus, $(-1, 6)$ is *not* a solution of the equation.

*Do Exercises 1 and 2.*

**Example 2** Show that the pairs $(3, 7)$, $(0, 1)$, and $(-3, -5)$ are solutions of $y = 2x + 1$. Then graph the three points and use the graph to determine another pair that is a solution.

To show that a pair is a solution, we substitute, replacing $x$ with the first coordinate and $y$ with the second coordinate of each pair:

$$\frac{y = 2x + 1}{7\ ?\ 2 \cdot 3 + 1} \qquad \frac{y = 2x + 1}{1\ ?\ 2 \cdot 0 + 1}$$
$$6 + 1 \qquad\qquad\qquad 0 + 1$$
$$7 \quad \text{TRUE} \qquad\qquad 1 \quad \text{TRUE}$$

$$\frac{y = 2x + 1}{-5\ ?\ 2(-3) + 1}$$
$$-6 + 1$$
$$-5 \qquad \text{TRUE}$$

In each of the three cases, the substitution results in a true equation. Thus the pairs are all solutions.

## Objectives

a  Determine whether an ordered pair is a solution of an equation with two variables.

b  Graph linear equations of the type $y = mx + b$ and $Ax + By = C$, identifying the $y$-intercept.

c  Solve applied problems involving graphs of linear equations.

### For Extra Help

TAPE 6    TAPE 6B    MAC WIN    CD-ROM

1. Determine whether $(2, -4)$ is a solution of $4q - 3p = 22$.

2. Determine whether $(2, -4)$ is a solution of $7a + 5b = -6$.

*Answers on page A-9*

**3.** Use the graph in Example 2 to find at least two more points that are solutions of $y = 2x + 1$.

We plot the points as shown at right. The order of the points follows the alphabetical order of the variables. That is, $x$ comes before $y$, so $x$-values are first coordinates and $y$-values are second coordinates. Similarly, we also label the horizontal axis as the $x$-axis and the vertical axis as the $y$-axis.

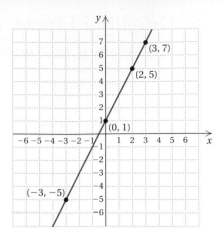

Note that the three points appear to "line up." That is, they appear to be on a straight line. Will other points that line up with these points also represent solutions of $y = 2x + 1$? To find out, we use a straightedge and lightly sketch a line passing through $(3, 7)$, $(0, 1)$, and $(-3, -5)$.

The line appears to also pass through $(2, 5)$. Let's see if this pair is a solution of $y = 2x + 1$:

$$\begin{array}{c|c} y = 2x + 1 \\ \hline 5 \;?\; 2 \cdot 2 + 1 \\ \phantom{5}|\; 4 + 1 \\ \phantom{5}|\; 5 \qquad\qquad \text{TRUE} \end{array}$$

Thus, $(2, 5)$ is a solution.

*Do Exercise 3.*

Example 2 leads us to suspect that any point on the line that passes through $(3, 7)$, $(0, 1)$, and $(-3, -5)$ represents a solution of $y = 2x + 1$. In fact, every solution of $y = 2x + 1$ is represented by a point on that line and every point on that line represents a solution. The line is said to be the *graph* of the equation.

> The **graph** of an equation is a drawing that represents all its solutions.

## b | Graphs of Linear Equations

Equations like $y = 2x + 1$ and $4q - 3p = 22$ are said to be **linear** because the graph of each equation is a straight line. In general, any equation equivalent to one of the form $y = mx + b$ or $Ax + By = C$, where $m$, $b$, $A$, $B$, and $C$ are constants (not variables) and $A$ and $B$ are not both 0, is linear.

To graph a linear equation:

**1.** Select a value for one variable and calculate the corresponding value of the other variable. Form an ordered pair using alphabetical order as indicated by the variables.

**2.** Repeat step (1) to obtain at least two other ordered pairs. Two points are essential to determine a straight line. A third point serves as a check.

**3.** Plot the ordered pairs and draw a straight line passing through the points.

*Answer on page A-9*

In general, calculating three (or more) ordered pairs is not difficult for equations of the form $y = mx + b$. We simply substitute values for $x$ and calculate the corresponding values for $y$.

**Example 3**   Graph: $y = 2x$.

First, we find some ordered pairs that are solutions. We choose *any* number for $x$ and then determine $y$ by substitution. Since $y = 2x$, we find $y$ by doubling $x$. Suppose that we choose 3 for $x$. Then

$$y = 2x = 2 \cdot 3 = 6.$$

We get a solution: the ordered pair (3, 6).
    Suppose that we choose 0 for $x$. Then

$$y = 2x = 2 \cdot 0 = 0.$$

We get another solution: the ordered pair (0, 0).
    For a third point, we make a negative choice for $x$. We now have enough points to plot the line, but if we wish, we can compute more. If a number takes us off the graph paper, we either do not use it or we use larger paper or rescale the axes. Continuing in this manner, we create a table like the one shown below.
    Now we plot these points. We draw the line, or graph, with a straight-edge and label it $y = 2x$.

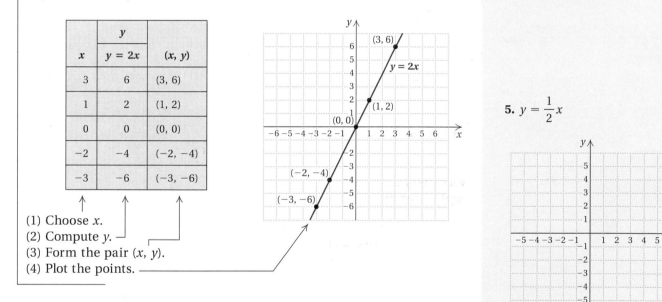

| $x$ | $y$ $y = 2x$ | $(x, y)$ |
|-----|-----|----------|
| 3 | 6 | (3, 6) |
| 1 | 2 | (1, 2) |
| 0 | 0 | (0, 0) |
| −2 | −4 | (−2, −4) |
| −3 | −6 | (−3, −6) |

(1) Choose $x$.
(2) Compute $y$.
(3) Form the pair $(x, y)$.
(4) Plot the points.

*Do Exercises 4 and 5.*

**4.** $y = -2x$

**5.** $y = \dfrac{1}{2}x$

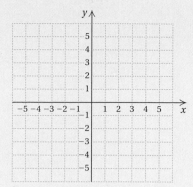

*Answers on page A-10*

Graph.

**6.** $y = 2x + 3$

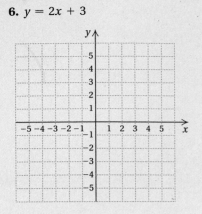

**7.** $y = -\dfrac{1}{2}x - 3$

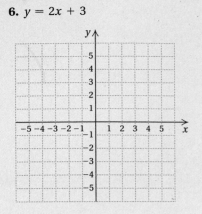

**Example 4** Graph: $y = -3x + 1$.

We select a value for $x$, compute $y$, and form an ordered pair. Then we repeat the process for other choices of $x$.

If $x = 2$,    then $y = -3 \cdot 2 + 1 = -5$,    and $(2, -5)$ is a solution.

If $x = 0$,    then $y = -3 \cdot 0 + 1 = 1$,    and $(0, 1)$ is a solution.

If $x = -1$,   then $y = -3 \cdot (-1) + 1 = 4$,   and $(-1, 4)$ is a solution.

Results are often listed in a table, as shown below. The points corresponding to each pair are then plotted.

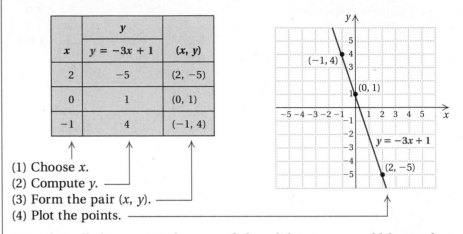

|   |   | $y$ |   |
|---|---|---|---|
| $x$ | $y = -3x + 1$ | $(x, y)$ |
| 2 | $-5$ | $(2, -5)$ |
| 0 | 1 | $(0, 1)$ |
| $-1$ | 4 | $(-1, 4)$ |

(1) Choose $x$.
(2) Compute $y$.
(3) Form the pair $(x, y)$.
(4) Plot the points.

Note that all three points line up. If they did not, we would know that we had made a mistake. When only two points are plotted, a mistake is harder to detect. We use a ruler or other straightedge to draw a line through the points. Every point on the line represents a solution of $y = -3x + 1$.

*Do Exercises 6 and 7.*

In Example 3, we saw that $(0, 0)$ is a solution of $y = 2x$. It is also the point at which the graph crosses the $y$-axis. Similarly, in Example 4, we saw that $(0, 1)$ is a solution of $y = -3x + 1$. It is also the point at which the graph crosses the $y$-axis. A generalization can be made: If $x$ is replaced with 0 in the equation $y = mx + b$, then the corresponding $y$-value is $m \cdot 0 + b$, or $b$. Thus any equation of the form $y = mx + b$ has a graph that passes through the point $(0, b)$. Since $(0, b)$ is the point at which the graph crosses the $y$-axis, it is called the **$y$-intercept**. Sometimes, for convenience, we simply refer to $b$ as the $y$-intercept.

The graph of the equation $y = mx + b$ passes through the **$y$-intercept** $(0, b)$.

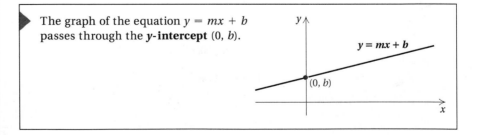

*Answers on page A-10*

**Example 5**  Graph $y = \frac{2}{5}x + 4$ and identify the $y$-intercept.

We select a value for $x$, compute $y$, and form an ordered pair. Then we repeat the process for other choices of $x$. In this case, using multiples of 5 avoids fractions.

If $x = 0$,    then $y = \frac{2}{5} \cdot 0 + 4 = 4$,    and $(0, 4)$ is a solution.

If $x = 5$,    then $y = \frac{2}{5} \cdot 5 + 4 = 6$,    and $(5, 6)$ is a solution.

If $x = -5$,   then $y = \frac{2}{5} \cdot (-5) + 4 = 2$,   and $(-5, 2)$ is a solution.

The following table lists these solutions. Next, we plot the points and see that they form a line. Finally, we draw and label the line.

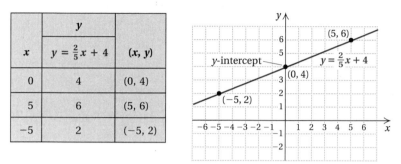

| $x$ | $y = \frac{2}{5}x + 4$ | $(x, y)$ |
|-----|------|----------|
| 0 | 4 | $(0, 4)$ |
| 5 | 6 | $(5, 6)$ |
| $-5$ | 2 | $(-5, 2)$ |

We see that $(0, 4)$ is a solution of $y = \frac{2}{5}x + 4$. It is the $y$-intercept. Because the equation is in the form $y = mx + b$, we can read the $y$-intercept directly from the equation as follows:

$$y = \frac{2}{5}x + 4 \qquad (0, 4) \text{ is the } y\text{-intercept.}$$

**Do Exercises 8 and 9.**

Calculating ordered pairs is generally easiest when $y$ is isolated on one side of the equation, as in $y = mx + b$. To graph an equation in which $y$ is not isolated, we can use the addition and multiplication principles to solve for $y$ (see Sections 2.3 and 2.6).

**Example 6**  Graph $3y + 5x = 0$ and identify the $y$-intercept.

To find an equivalent equation in the form $y = mx + b$, we solve for $y$:

$$3y + 5x = 0$$
$$3y + 5x - 5x = 0 - 5x \qquad \text{Subtracting } 5x$$
$$3y = -5x \qquad \text{Collecting like terms}$$
$$\frac{3y}{3} = \frac{-5x}{3} \qquad \text{Dividing by 3}$$
$$y = -\frac{5}{3}x.$$

Graph the equation and identify the $y$-intercept.

**8.** $y = \frac{3}{5}x + 2$

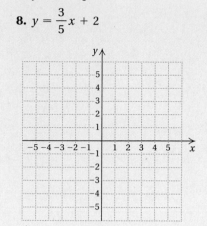

**9.** $y = -\frac{3}{5}x - 1$

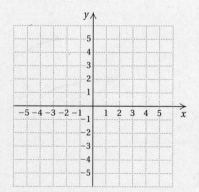

*Answers on page A-10*

Graph the equation and identify the *y*-intercept.

**10.** $5y + 4x = 0$

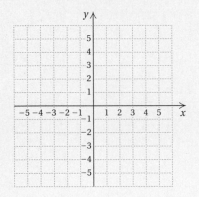

**11.** $4y = 3x$

Because all the equations above are equivalent, we can use $y = -\frac{5}{3}x$ to draw the graph of $3y + 5x = 0$. To graph $y = -\frac{5}{3}x$, we select *x*-values and compute *y*-values. In this case, if we select multiples of 3, we can avoid fractions.

$$\text{If } x = 0, \quad \text{then } y = -\frac{5}{3} \cdot 0 = 0.$$

$$\text{If } x = 3, \quad \text{then } y = -\frac{5}{3} \cdot 3 = -5.$$

$$\text{If } x = -3, \quad \text{then } y = -\frac{5}{3} \cdot (-3) = 5.$$

We list these solutions in a table. Next, we plot the points and see that they form a line. Finally, we draw and label the line. The *y*-intercept is (0, 0).

| *x* | *y* |
|-----|-----|
| 0 | 0 |
| 3 | −5 |
| −3 | 5 |

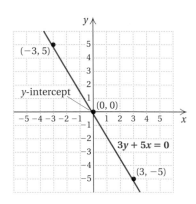

*Do Exercises 10 and 11.*

**Example 7**  Graph $4y + 3x = -8$ and identify the *y*-intercept.

To find an equivalent equation in the form $y = mx + b$, we solve for *y*:

$$4y + 3x = -8$$
$$4y + 3x - 3x = -8 - 3x \qquad \textbf{Subtracting } 3x$$
$$4y = -3x - 8 \qquad \textbf{Simplifying}$$
$$\frac{1}{4} \cdot 4y = \frac{1}{4} \cdot (-3x - 8) \qquad \textbf{Multiplying by } \tfrac{1}{4} \textbf{ or dividing by 4}$$
$$y = \frac{1}{4} \cdot (-3x) - \frac{1}{4} \cdot 8 \qquad \textbf{Using the distributive law}$$
$$= -\frac{3}{4}x - 2. \qquad \textbf{Simplifying}$$

*Answers on page A-10*

Thus, $4y + 3x = -8$ is equivalent to $y = -\frac{3}{4}x - 2$. The $y$-intercept is $(0, -2)$. We find two other pairs using multiples of 4 for $x$ to avoid fractions. We then complete and label the graph as shown.

| x | y |
|---|---|
| 0 | -2 |
| 4 | -5 |
| -4 | 1 |

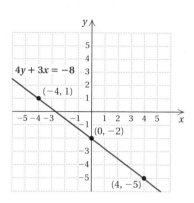

*Do Exercises 12 and 13.*

# c Applications of Linear Equations

Mathematical concepts become more understandable through visualization. Throughout this text, you will occasionally see the heading **AG** Algebraic–Graphical Connection, as in Example 8, which follows. In this feature, the algebraic approach is enhanced and expanded with a graphical connection. Relating a solution of an equation to a graph can often give added meaning to the algebraic solution.

**Example 8** *FedEx Mailing Costs.* The cost $y$, in dollars, of mailing a FedEx Priority Overnight package weighing 1 lb or more is given by the equation

$$y = 2.085x + 15.08,$$

where $x = $ the number of pounds (***Source:*** Federal Express Corporation).

a) Find the cost of mailing packages weighing 2 lb, 5 lb, and 7 lb.

b) Graph the equation and then use the graph to estimate the cost of mailing a $6\frac{1}{2}$-lb package.

c) If a package costs $177.71 to mail, how much does it weigh?

Graph the equation and identify the $y$-intercept.

**12.** $5y - 3x = -10$

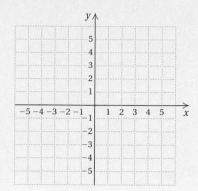

**13.** $5y + 3x = 20$

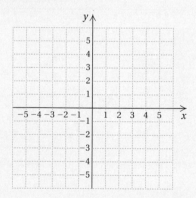

*Answers on page A-10*

**14.** *Value of a Color Copier.* The value of Dupliographic's color copier is given by

$$v = -0.68t + 3.4,$$

where $v$ = the value, in thousands of dollars, $t$ years from the date of purchase.

**a)** Find the value after 1 yr, 2 yr, 4 yr, and 5 yr.

**b)** Graph the equation and use the graph to estimate the value of the copier after $2\frac{1}{2}$ yr.

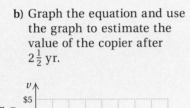

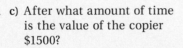

**c)** After what amount of time is the value of the copier $1500?

*Answers on page A-10*

We solve as follows.

**a)** We substitute 2, 5, and 7 for $x$ and then calculate $y$:

If $x = 2$,  then $y = 2.085(2) + 15.08 = \$19.25$.

If $x = 5$,  then $y = 2.085(5) + 15.08 \approx \$25.51$.

If $x = 7$,  then $y = 2.085(7) + 15.08 \approx \$29.68$.

## 🖥 Algebraic–Graphical Connection

**b)** We have three ordered pairs from (a). We plot these points and see that they line up. Thus our calculations are probably correct. Since zero and negative $x$-values have no meaning in this problem, we use an open circle at (0, 15.08) when drawing the graph.

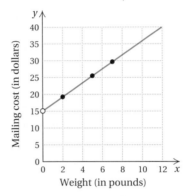

To estimate the cost of mailing a $6\frac{1}{2}$-lb package, we need to determine what $y$-value is paired with $x = 6\frac{1}{2}$. We locate the point on the line that is above $6\frac{1}{2}$ and then find the value on the $y$-axis that corresponds to that point. It appears that the cost of mailing a $6\frac{1}{2}$-lb package is about \$29.

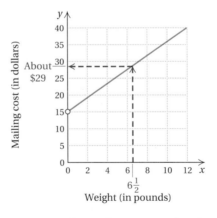

To obtain a more accurate cost, we can simply substitute into the equation:

$$y = 2.085(6.5) + 15.08 \approx \$28.63.$$

**c)** We substitute $177.71 for $y$ and then solve for $x$:

$$y = 2.085x + 15.08$$
$$177.71 = 2.085x + 15.08 \qquad \text{Substituting}$$
$$162.63 = 2.085x \qquad \text{Subtracting 15.08}$$
$$78 = x. \qquad \text{Dividing by 2.085}$$

*Do Exercise 14 on the preceding page.*

Many equations in two variables have graphs that are not straight lines. Three such graphs are shown below. As before, each graph represents the solutions of the given equation. We are not going to develop methods of doing such graphing at this time, although such *nonlinear graphs* can be created very easily using a graphing calculator. We will cover such graphs in the optional Calculator Spotlights.

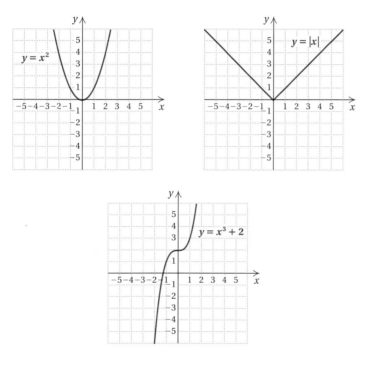

# Calculator Spotlight

Introduction to the Use of a Graphing
Calculator: Windows and Graphs

**Viewing Windows.** One feature common to all graphers is the **viewing window**. Windows are described by four numbers, [**L, R, B, T**], which represent the **L**eft and **R**ight endpoints of the $x$-axis and the **B**ottom and **T**op endpoints of the $y$-axis. A WINDOW feature is used to set these dimensions. Below is a window setting of $[-20, 20, -5, 5]$ with axis scaling denoted as $\text{Xscl} = 5$ and $\text{Yscl} = 1$. The notation $\text{Xres} = 1$ indicates the number of pixels (black rectangular dots). We will usually leave it at 1 and not refer to it unless needed. On some graphers, a setting of $[-10, 10, -10, 10]$, $\text{Xscl} = 1$, $\text{Yscl} = 1$ is considered **standard**.

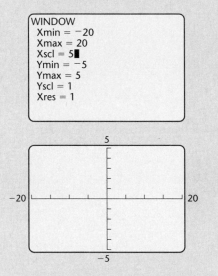

The primary use for a grapher is to graph equations. For example, let's graph the equation $y = x^2 - 3x - 5$. The equation can be entered using the $\boxed{y=}$ key.

$y = x^2 - 3x - 5$

To graph an equation like $4y + 3x = -8$, most graphers require that the equation be solved for $y$, that is, "$y = \ldots$." Thus we must rewrite the equation $4y + 3x = -8$ as

$$y = \frac{-3x - 8}{4}, \quad \text{or} \quad y = -\frac{3}{4}x - 2,$$

as we did in Example 7. We then enter this equation as $y = -(3/4)x - 2$.

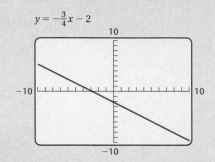

## Exercises

Use a grapher to graph each of the following equations. Select the standard window $[-10, 10, -10, 10]$ and axis scaling $\text{Xscl} = 1$, $\text{Yscl} = 1$.

**1.** $y = 2x + 1$      **2.** $y = -3x + 1$

**3.** $y = \frac{2}{5}x + 4$      **4.** $y = -\frac{3}{5}x - 1$

**5.** $y = 2.085x + 15.08$      **6.** $y = -\frac{4}{5}x + \frac{13}{7}$

**7.** $2x + 3y = 18$      **8.** $5y + 3x = 4$

**9.** $y = x^2$      **10.** $y = 0.5x^2$

**11.** $y = 8 - x^2$      **12.** $y = 4 - 3x - x^2$

**13.** $y = 5x^2 - 3x - 10$      **14.** $y = x^3 + 2$

**15.** $y = |x|$   (On most graphers, this is entered as $y = \text{abs}(x)$.)

**16.** $y = |x - 5|$      **17.** $y = |x| - 5$

**18.** $y = 8 - |x|$

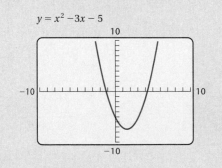

# Exercise Set 3.2

**a** Determine whether the given point is a solution of the equation.

**1.** (2, 9); $y = 3x - 1$

**2.** (1, 7); $y = 2x + 5$

**3.** (4, 2); $2x + 3y = 12$

**4.** (0, 5); $5x - 3y = 15$

**5.** (3, −1); $3a - 4b = 13$

**6.** (−5, 1); $2p - 3q = -13$

In Exercises 7–12, an equation and two ordered pairs are given. Show that each pair is a solution. Then use the graph of the two points to determine another solution. Answers may vary.

**7.** $y = x - 5$; (4, −1) and (1, −4)

**8.** $y = x + 3$; (−1, 2) and (3, 6)

**9.** $y = \dfrac{1}{2}x + 3$; (4, 5) and (−2, 2)

**10.** $3x + y = 7$; (2, 1) and (4, −5)

**11.** $4x - 2y = 10$; (0, −5) and (4, 3)

**12.** $6x - 3y = 3$; (1, 1) and (−1, −3)

**b** Graph the equation and identify the $y$-intercept.

**13.** $y = x + 1$

**14.** $y = x - 1$

**15.** $y = x$

**16.** $y = -x$

**17.** $y = \dfrac{1}{2}x$

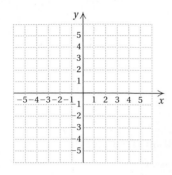

**18.** $y = \dfrac{1}{3}x$

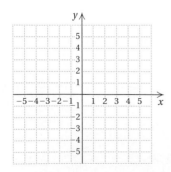

**19.** $y = x - 3$

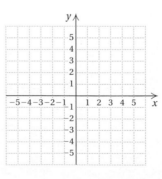

**20.** $y = x + 3$

$x + 3 = y$

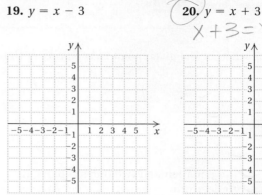

**21.** $y = 3x - 2$

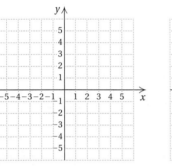

**22.** $y = 2x + 2$

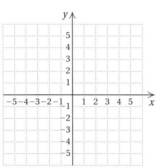

**23.** $y = \dfrac{1}{2}x + 1$

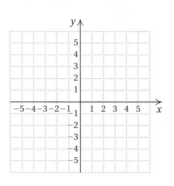

**24.** $y = \dfrac{1}{3}x - 4$

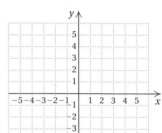

**25.** $x + y = -5$

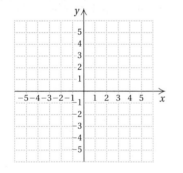

**26.** $x + y = 4$

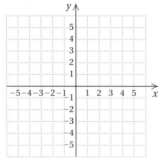

**27.** $y = \dfrac{5}{3}x - 2$

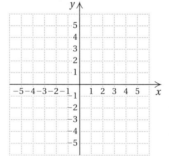

**28.** $y = \dfrac{5}{2}x + 3$

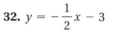

**29.** $x + 2y = 8$

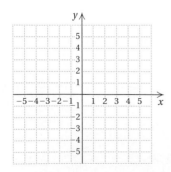

**30.** $x + 2y = -6$

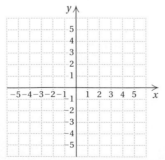

**31.** $y = \dfrac{3}{2}x + 1$

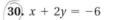

**32.** $y = -\dfrac{1}{2}x - 3$

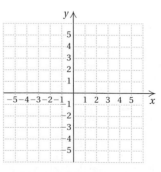

Copyright © 1999 Addison Wesley Longman

**33.** $8x - 2y = -10$

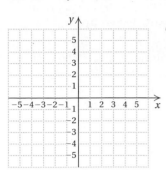

**34.** $6x - 3y = 9$

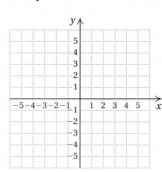

**35.** $8y + 2x = -4$

**36.** $6y + 2x = 8$

---

[c] Solve.

**37.** *Value of Computer Software.* The value $V$, in dollars, of a shopkeeper's inventory software program is given by

$$V = -50t + 300,$$

where $t$ = the number of years since the shopkeeper first bought the program.

**a)** Find the value of the software after 0 yr, 4 yr, and 6 yr.

**b)** Graph the equation and then use the graph to estimate the value of the software after 5 yr.

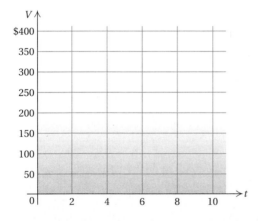

**c)** After how many years is the value of the software $150?

**38.** *College Costs.* The cost $T$, in dollars, of tuition and fees at many community colleges can be approximated by

$$T = 120c + 100,$$

where $c$ = the number of credits for which a student registers (***Source***: Community College of Vermont).

**a)** Find the cost of tuition for a student who takes 8 hr, 12 hr, and 15 hr.

**b)** Graph the equation and then use the graph to estimate the cost of tuition for a student who takes 9 hr.

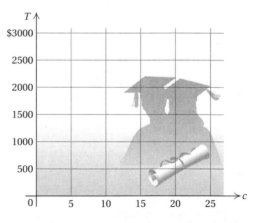

**c)** Estimate how many hours a student can take for $1420.

**39.** *Tea Consumption.* The number of gallons $N$ of tea consumed each year by the average U.S. consumer can be approximated by

$$N = 0.1d + 7,$$

where $d$ = the number of years since 1991 (**Source:** *Statistical Abstract of the United States*).

a) Find the number of gallons of tea consumed in 1992 ($n = 1$), 1995 ($n = 4$), 1999 ($n = 8$), and 2001 ($n = 10$).

b) Graph the equation and use the graph to estimate what the tea consumption was in 1997.

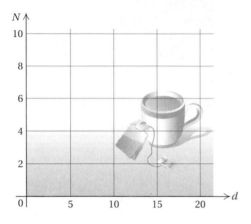

c) In what year will tea consumption be about 9 gal?

**40.** *Record Temperature Drop.* On 22 January 1943, the temperature $T$, in degrees Fahrenheit, in Spearfish, South Dakota, could be approximated by

$$T = -2.15m + 54,$$

where $m$ = the number of minutes since 9:00 that morning (**Source:** *Information Please Almanac,* 1996).

a) Find the temperature at 9:01 A.M., 9:08 A.M., and 9:20 A.M.

b) Graph the equation and use the graph to estimate the temperature at 9:15 A.M.

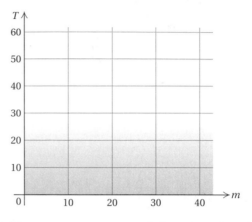

c) The temperature stopped dropping when it reached $-4°$. At what time did this occur? (*Note:* The linear equation could not be used after that time.)

---

**Skill Maintenance**

Round to the nearest thousand.   [R.3c]

**41.** 2567.03

**42.** 124,748

**43.** 293.4572

**44.** 6,078,124

Convert to decimal notation.   [1.2c]

**45.** $-\dfrac{7}{8}$

**46.** $\dfrac{23}{32}$

**47.** $\dfrac{117}{64}$

**48.** $-\dfrac{27}{12}$

---

**Synthesis**

**49.** ◈ The equations $3x + 4y = 8$ and $y = -\frac{3}{4}x + 2$ are equivalent. Which equation is easier to graph and why?

**50.** ◈ Referring to Exercise 40, discuss why the linear equation no longer described the temperature after the temperature reached $-4°$.

In Exercises 51–54, find an equation for the graph shown.

**51.**

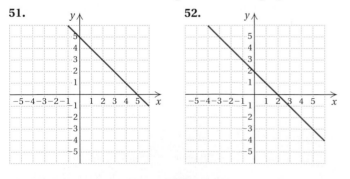

**52.**

**53.**

**54.**

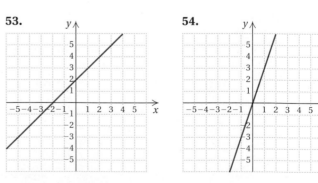

Copyright © 1999 Addison Wesley Longman

# 3.3 More with Graphing and Intercepts

## a Graphing Using Intercepts

In Section 3.2, we graphed linear equations of the form $Ax + By = C$ by first solving for $y$ to find an equivalent equation in the form $y = mx + b$. We did so because it is then easier to calculate the $y$-value that corresponds to a given $x$-value. Another convenient way to graph $Ax + By = C$ is to use **intercepts**. Look at the graph of $-2x + y = 4$ shown below.

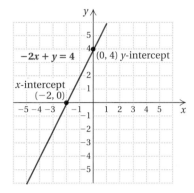

The $y$-intercept is (0, 4). It occurs where the line crosses the $y$-axis and thus will always have 0 as the first coordinate. The $x$-intercept is $(-2, 0)$. It occurs where the line crosses the $x$-axis and thus will always have 0 as the second coordinate.

*Do Exercise 1.*

We find intercepts as follows.

> The **$y$-intercept** is (0, $b$). To find $b$, let $x = 0$ and solve the original equation for $y$.
>
> The **$x$-intercept** is ($a$, 0). To find $a$, let $y = 0$ and solve the original equation for $x$.

Now let's draw a graph using intercepts.

**Example 1** Consider $4x + 3y = 12$. Find the intercepts. Then graph the equation using the intercepts.

To find the $y$-intercept, we let $x = 0$. Then we solve for $y$:

$$4 \cdot 0 + 3y = 12$$
$$3y = 12$$
$$y = 4.$$

Thus, (0, 4) is the $y$-intercept. Note that finding this intercept amounts to covering up the $x$-term and solving the rest of the equation.

To find the $x$-intercept, we let $y = 0$. Then we solve for $x$:

$$4x + 3 \cdot 0 = 12$$
$$4x = 12$$
$$x = 3.$$

## Objectives

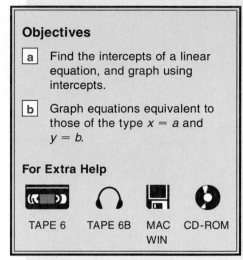

a Find the intercepts of a linear equation, and graph using intercepts.

b Graph equations equivalent to those of the type $x = a$ and $y = b$.

**For Extra Help**

TAPE 6    TAPE 6B    MAC WIN    CD-ROM

1. Look at the graph shown below.

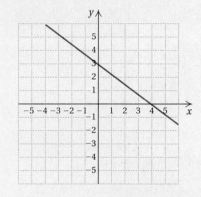

a) Find the coordinates of the $y$-intercept.

b) Find the coordinates of the $x$-intercept.

*Answers on page A-12*

For each equation, find the intercepts. Then graph the equation using the intercepts.

**2.** $2x + 3y = 6$

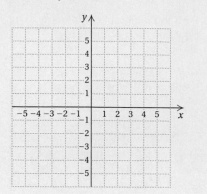

**3.** $3y - 4x = 12$

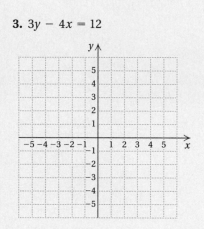

Graph.

**4.** $x = 5$

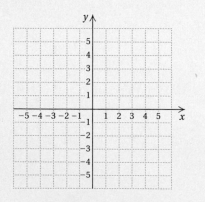

*Answers on page A-12*

Thus, (3, 0) is the $x$-intercept. Note that finding this intercept amounts to covering up the $y$-term and solving the rest of the equation.

We plot these points and draw the line, or graph.

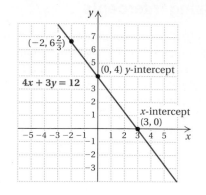

A third point should be used as a check. We substitute any convenient value for $x$ and solve for $y$. In this case, we choose $x = -2$. Then

$$4(-2) + 3y = 12 \qquad \text{Substituting } -2 \text{ for } x$$
$$-8 + 3y = 12$$
$$3y = 12 + 8 = 20$$
$$y = \frac{20}{3}, \text{ or } 6\frac{2}{3}. \qquad \text{Solving for } y$$

It appears that the point $\left(-2, 6\frac{2}{3}\right)$ is on the graph, though graphing fractional values can be inexact. The graph is probably correct.

Graphs of equations of the type $y = mx$ pass through the origin. Thus the $x$-intercept and the $y$-intercept are the same, (0, 0). In such cases, we must calculate another point in order to complete the graph. Another point would also have to be calculated if a check is desired.

***Do Exercises 2 and 3.***

**b  Equations Whose Graphs Are Horizontal or Vertical Lines**

**Example 2**  Graph: $y = 3$.

Consider $y = 3$. We can also think of this equation as $0 \cdot x + y = 3$. No matter what number we choose for $x$, we find that $y$ is 3. We make up a table with all 3's in the $y$-column.

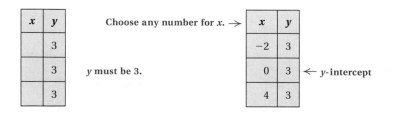

When we plot the ordered pairs $(-2, 3)$, $(0, 3)$, and $(4, 3)$ and connect the points, we will obtain a horizontal line. Any ordered pair $(x, 3)$ is a solution. So the line is parallel to the $x$-axis with $y$-intercept $(0, 3)$.

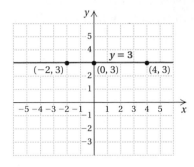

## Example 3 Graph: $x = -4$.

Consider $x = -4$. We can also think of this equation as $x + 0 \cdot y = -4$. We make up a table with all $-4$'s in the $x$-column.

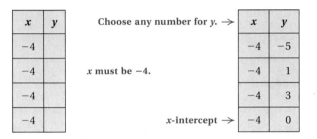

When we plot the ordered pairs $(-4, -5)$, $(-4, 1)$, $(-4, 3)$, and $(-4, 0)$ and connect the points, we will obtain a vertical line. Any ordered pair $(-4, y)$ is a solution. So the line is parallel to the $y$-axis with $x$-intercept $(-4, 0)$.

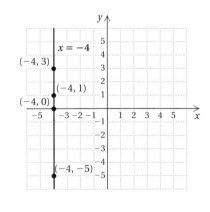

> The graph of $y = b$ is a **horizontal line.** The $y$-intercept is $(0, b)$.
>
> The graph of $x = a$ is a **vertical line.** The $x$-intercept is $(a, 0)$.

*Do Exercises 4–7. (Exercise 4 is on the preceding page.)*

Graph.

**5.** $y = -2$

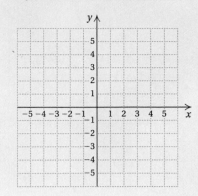

**6.** $x = 0$

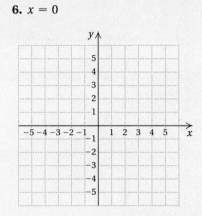

**7.** $x = -3$

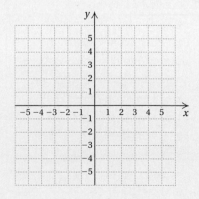

*Answers on page A-12*

## Calculator Spotlight

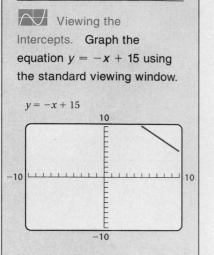

Viewing the Intercepts. **Graph the equation $y = -x + 15$ using the standard viewing window.**

$y = -x + 15$

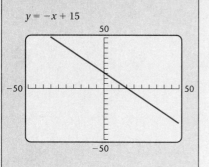

Note that the graph is barely visible in the upper right-hand corner, and neither intercept can be seen. To better view the intercepts, we can try different window settings.

$y = -x + 15$

### Exercises

Find the intercepts of the equation using algebra (algebraically). Then graph the equation and adjust the window and tick mark settings so that the intercepts can be clearly seen on both axes.

1. $y = -7.2x - 15$
2. $y - 2.13x = 27$
3. $5x + 6y = 84$
4. $2x - 7y = 150$
5. $3x + 2y = 50$
6. $y = 0.2x - 9$
7. $y = 1.3x - 15$
8. $25x - 20y = 1$

The following is a general procedure for graphing linear equations.

**GRAPHING LINEAR EQUATIONS**

1. If the equation is of the type $x = a$ or $y = b$, the graph will be a line parallel to an axis; $x = a$ is vertical and $y = b$ is horizontal.

   *Examples.*

   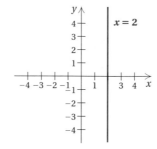
   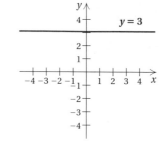

2. If the equation is of the type $y = mx$, both intercepts are the origin, $(0, 0)$. Plot $(0, 0)$ and two other points.

   *Example.*

   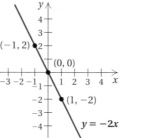

3. If the equation is of the type $y = mx + b$, plot the $y$-intercept $(0, b)$ and two other points.

   *Example.*

   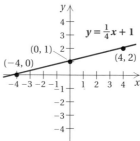

4. If the equation is of the type $Ax + By = C$, but not of the type $x = a$, $y = b$, $y = mx$, or $y = mx + b$, then either solve for $y$ and proceed as with the equation $y = mx + b$, or graph using intercepts. If the intercepts are too close together, choose another point or points farther from the origin.

   *Examples.*

   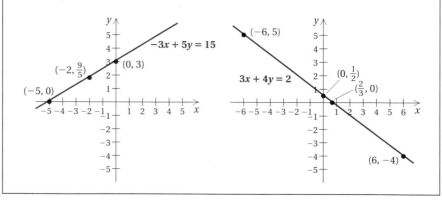

# Exercise Set 3.3

**a** For Exercises 1–4, find (a) the coordinates of the *y*-intercept and (b) the coordinates of the *x*-intercept.

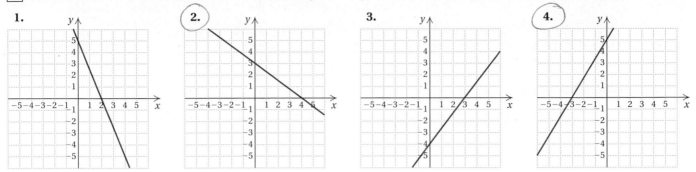

**1.**

**2.**

**3.**

**4.**

For Exercises 5–12, find (a) the coordinates of the *y*-intercept and (b) the coordinates of the *x*-intercept. Do not graph.

**5.** $3x + 5y = 15$

**6.** $5x + 2y = 20$

**7.** $7x - 2y = 28$

**8.** $3x - 4y = 24$

**9.** $-4x + 3y = 10$

**10.** $-2x + 3y = 7$

**11.** $6x - 3 = 9y$

**12.** $4y - 2 = 6x$

For each equation, find the intercepts. Then use the intercepts to graph the equation.

**13.** $x + 3y = 6$

**14.** $x + 2y = 2$

**15.** $-x + 2y = 4$

**16.** $-x + y = 5$

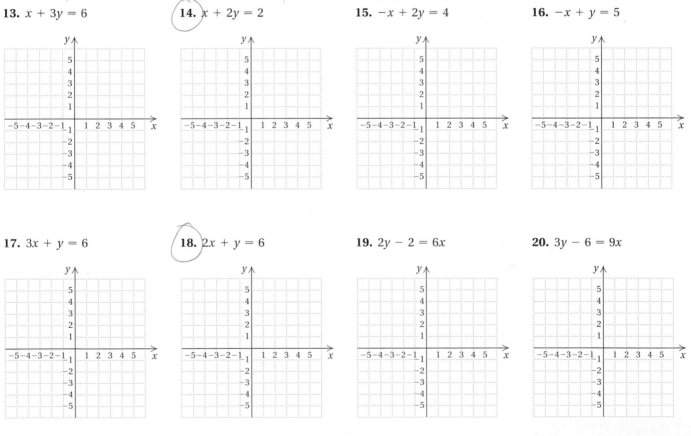

**17.** $3x + y = 6$

**18.** $2x + y = 6$

**19.** $2y - 2 = 6x$

**20.** $3y - 6 = 9x$

**21.** $3x - 9 = 3y$

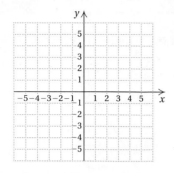

**22.** $5x - 10 = 5y$

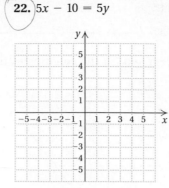

**23.** $2x - 3y = 6$

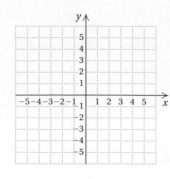

**24.** $2x - 5y = 10$

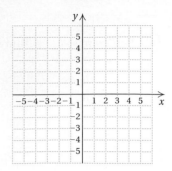

**25.** $4x + 5y = 20$

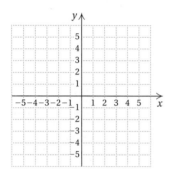

**26.** $2x + 6y = 12$

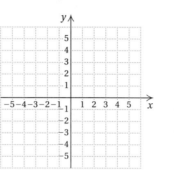

**27.** $2x + 3y = 8$

**28.** $x - 1 = y$

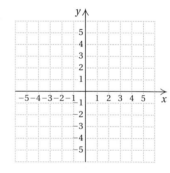

**29.** $x - 3 = y$

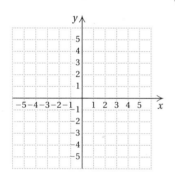

**30.** $2x - 1 = y$

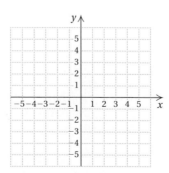

**31.** $3x - 2 = y$

**32.** $4x - 3y = 12$

**33.** $6x - 2y = 12$

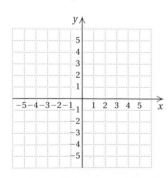

**34.** $7x + 2y = 6$

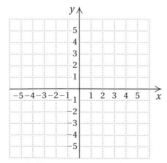

**35.** $3x + 4y = 5$

**36.** $y = -4 - 4x$

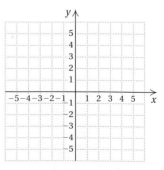

Copyright © 1999 Addison Wesley Longman

**37.** $y - -3 - 3x$

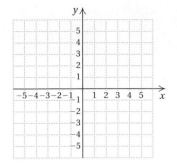

**38.** $-3x = 6y - 2$

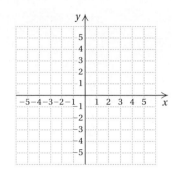

**39.** $y - 3x = 0$

**40.** $x + 2y = 0$

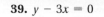

**b** Graph.

**41.** $x = -2$

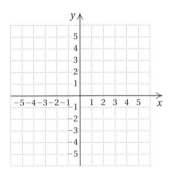

**42.** $x = 1$

**43.** $y = 2$

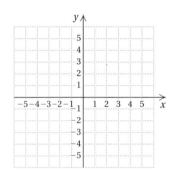

**44.** $y = -4$

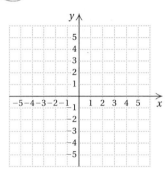

**45.** $x = 2$

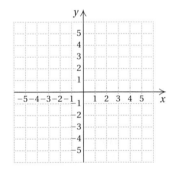

**46.** $x = 3$

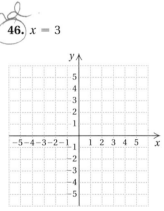

**47.** $y = 0$

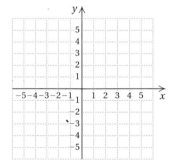

**48.** $y = -1$

**49.** $x = \dfrac{3}{2}$

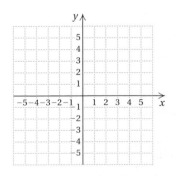

**50.** $x = -\dfrac{5}{2}$

**51.** $3y = -5$

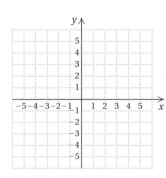

**52.** $12y = 45$

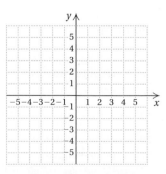

**53.** $4x + 3 = 0$

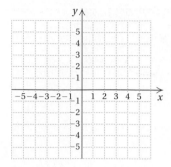

**54.** $-3x + 12 = 0$

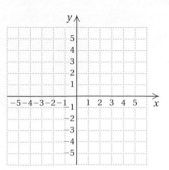

**55.** $48 - 3y = 0$

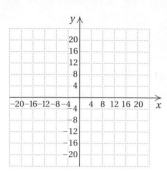

**56.** $63 + 7y = 0$

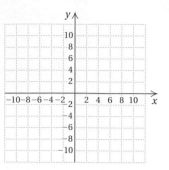

Write an equation for the graph shown.

**57.**

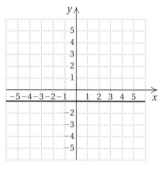

**58.**

**59.**

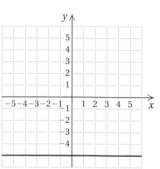

**60.**

---

### Skill Maintenance

Solve. [2.5a]

**61.** *Desserts.* If a restaurant sells 250 desserts in an evening, it is typical that 40 of them will be pie. What percent of the desserts sold will be pie?

**62.** *Desserts.* Of all desserts sold in restaurants, 20% of them are chocolate cake. One evening a restaurant sells 350 desserts. How many were chocolate cake?

**63.** Harry left a 20% tip of $6.50 for a meal. What was the cost of the meal before the tip?

**64.** Rambeau paid $27.60 for a taxi ride. This included a 20% tip. How much was the fare before the tip?

Solve. [2.7e]

**65.** $-1.6x < 64$

**66.** $-12x - 71 \geq 13$

**67.** $x + (x - 1) < (x + 2) - (x + 1)$

**68.** $6 - 18x \leq 4 - 12x - 5x$

---

### Synthesis

**69.** ◈ If the graph of the equation $Ax + By = C$ is a horizontal line, what can you conclude about $A$? Why?

**70.** ◈ Explain in your own words why the graph of $x = 7$ is a vertical line.

**71.** Write an equation for the $y$-axis.

**72.** Write an equation for the $x$-axis.

**73.** Write an equation of a line parallel to the $x$-axis and passing through $(-3, -4)$.

**74.** Find the value of $m$ such that the graph of $y = mx + 6$ has an $x$-intercept of $(2, 0)$.

**75.** Find the value of $k$ such that the graph of $3x + k = 5y$ has an $x$-intercept of $(-4, 0)$.

**76.** Find the value of $k$ such that the graph of $4x = k - 3y$ has a $y$-intercept of $(0, -8)$.

---

Copyright © 1999 Addison Wesley Longman

# 3.4 Applications and Data Analysis with Graphs

## a | Mean, Median, and Mode

One way to analyze data is to look for a single representative number, called a *center point* or *measure of central tendency*. Those most often used are the *mean* (or *average*), the *median*, and the *mode*.

### Mean

Let's first consider the *mean*, or average.

> The **mean**, or **average**, of a set of numbers is the sum of the numbers divided by the number of addends.

**Example 1** Consider the following data on revenue, in billions of dollars, at McDonald's restaurants in five recent years:

$12.5, \$13.2, \$14.2, \$14.9, \$15.9.

(**Source:** McDonalds Corporation). What is the mean of the numbers?

First, we add the numbers:

$$12.5 + 13.2 + 14.2 + 14.9 + 15.9 = 70.7.$$

Then we divide by the number of addends, 5:

$$\frac{(12.5 + 13.2 + 14.2 + 14.9 + 15.9)}{5} = \frac{70.7}{5} = 14.14.$$

The mean, or average, revenue of McDonald's for those five years is $14.14 billion.

Note that

$$14.14 + 14.14 + 14.14 + 14.14 + 14.14 = 70.7.$$

If we use this center point, 14.14, repeatedly as the addend, we get the same sum that we do when adding the individual data numbers.

*Do Exercises 1–3.*

### Median

The *median* is useful when we wish to de-emphasize extreme scores. For example, suppose five workers in a technology company manufactured the following numbers of computers during one day's work:

| | |
|---|---|
| Sarah: 88 | Jen: 94 |
| Matt: 92 | Mark: 91 |
| Pat: 66 | |

Let's first list the scores in order from smallest to largest:

66  88  **91**  92  94.
        ↑
    Middle number

The middle number—in this case, 91—is the **median.**

## Objectives

**a** Find the mean (average), the median, and the mode of a set of data and solve related applied problems.

**b** Compare two sets of data using their means.

**c** Make predictions from a set of data using interpolation or extrapolation.

**For Extra Help**

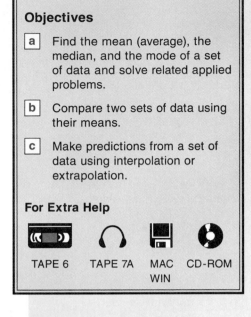

TAPE 6   TAPE 7A   MAC WIN   CD-ROM

Find the mean. Round to the nearest tenth.

**1.** 28, 103, 39

**2.** 85, 46, 105.7, 22.1

**3.** A student scored the following on five tests:

78, 95, 84, 100, 82.

What was the average score?

*Answers on page A-13*

Find the median.

**4.** 17, 13, 18, 14, 19

**5.** 17, 18, 16, 19, 13, 14

**6.** 122, 102, 103, 91, 83, 81, 78, 119, 88

Find any modes that exist.

**7.** 33, 55, 55, 88, 55

**8.** 90, 54, 88, 87, 87, 54

**9.** 23.7, 27.5, 54.9, 17.2, 20.1

**10.** In conducting laboratory tests, Carole discovers bacteria in different lab dishes grew to the following areas, in square millimeters:

25, 19, 29, 24, 28.

a) What is the mean?
b) What is the median?
c) What is the mode?

*Answers on page A-13*

▶ Once a set of data has been arranged from smallest to largest, the **median** of the set of data is the middle number if there is an odd number of data numbers. If there is an even number of data numbers, then there are two middle numbers and the median is the *average* of the two middle numbers.

**Example 2**   What is the median of the following set of yearly salaries?

$76,000,  $58,000,  $87,000,  $32,500,  $64,800,  $62,500

We first rearrange the numbers in order from smallest to largest.

$32,500   $58,000   $62,500   $64,800   $76,000   $87,000
↑
Median

There is an even number of numbers. We look for the middle two, which are $62,500 and $64,800. In this case, the median is the average of $62,₅ and $64,800:

$$\frac{\$62,500 + \$64,800}{2} = \$63,650.$$

*Do Exercises 4–6.*

## Mode

The last center point we consider is called the **mode**. A number that occurs most often in a set of data can be considered a representative number or center point.

▶ The **mode** of a set of data is the number or numbers that occur most often. If each number occurs the same number of times, there is *no* mode.

**Example 3**   Find the mode of the following data:

23, 24, 27, 18, 19, 27

The number that occurs most often is 27. Thus the mode is 27.

**Example 4**   Find the mode of the following data:

83,  84, 84, 84, 85, 86, 87, 87, 87, 88, 89, 90.

There are two numbers that occur most often, 84 and 87. Thus the modes are 84 and 87.

**Example 5**   Find the mode of the following data:

115, 117, 211, 213, 219.

Each number occurs the same number of times. The set of data has *no* mode.

*Do Exercises 7–10.*

## b Comparing Two Sets of Data

We have seen how data are displayed and interpreted using graphs, and we have calculated the mean, the median, and the mode from data. Now we look into using data analysis to solve applied problems.

One goal of analyzing two sets of data is to make a determination about which of two groups is "better." One way to do so is by comparing the means.

**Example 6** *Light-Bulb Testing.* An experiment is performed to compare the lives of two types of light bulb. Several bulbs of each type were tested and the results are listed in the following table. On the basis of this test, which bulb is better?

| Bulb A: HotLight Life Times (in hours) | | |
| --- | --- | --- |
| 983 | 964 | 1214 |
| 1417 | 1211 | 1521 |
| 1084 | 1075 | 892 |
| 1423 | 949 | |

| Bulb B: BrightBulb Life Times (in hours) | | |
| --- | --- | --- |
| 979 | 1083 | 1344 |
| 984 | 1445 | 975 |
| 1492 | 1325 | 1283 |
| 1325 | 1352 | 1432 |

Note that it is difficult to analyze the data at a glance because the numbers are close together and there is a different number of data points in each set. We need a way to compare the two groups. Let's compute the average of each set of data.

Bulb A: Average

$$= \frac{(983 + 964 + 1214 + 1417 + 1211 + 1521 + 1084 + 1075 + 892 + 1423 + 949)}{11}$$

$$= \frac{12{,}733}{11} \approx 1157.55;$$

Bulb B: Average

$$= \frac{(979 + 1083 + 1344 + 984 + 1445 + 975 + 1492 + 1325 + 1283 + 1325 + 1352 + 1432)}{12}$$

$$= \frac{15{,}019}{12} \approx 1251.58.$$

We see that the average life of bulb B is higher than that of bulb A and thus conclude that bulb B is "better." (It should be noted that statisticians might question whether these differences are what they call "significant." The answer to that question belongs to a later math course.)

*Do Exercise 11.*

11. *Quality of Baseballs.* Lauri experiments to see which of two kinds of baseball is better by determining which bounces highest. She drops balls of each kind from a height of 6 ft onto concrete and measures how high they bounce, in inches. Which kind of ball is better?

| Ball A Bouncing Heights (in inches) | | | |
| --- | --- | --- | --- |
| 16.2 | 22.3 | 19.5 | 15.7 |
| 19.6 | 18.0 | 15.6 | 21.7 |
| 19.8 | 16.4 | 18.4 | 16.6 |
| 21.5 | 18.7 | 22.0 | 18.3 |

| Ball B Bouncing Heights (in inches) | | | |
| --- | --- | --- | --- |
| 19.7 | 18.4 | 19.7 | 17.2 |
| 19.7 | 14.6 | 22.0 | 23.7 |
| 16.5 | 21.6 | 22.5 | 19.8 |
| 22.6 | 17.9 | 18.7 | |

*Answer on page A-13*

**12.** *Monthly Loan Payment.* The following table lists monthly payments on a loan of $110,000 at 9% interest. Note that there is no data point for a 35-yr loan. Use interpolation to estimate the missing value.

| Number of Years | Monthly Payment (in dollars) |
|---|---|
| 5 | $2283.42 |
| 10 | 1393.43 |
| 15 | 1115.69 |
| 20 | 989.70 |
| 25 | 923.12 |
| 30 | 885.08 |
| 35 | ? |
| 40 | 848.50 |

## **c** | **Making Predictions**

Sometimes we use data to make predictions or estimates of missing data points. One process for doing so is called *interpolation*. It uses graphs and/or averages to guess a missing data point between two data points.

**Example 7** *World Bicycle Production.* The following table shows how world bicycle production has grown in recent years. Note that there is no data for 1994. Use interpolation to estimate the missing value.

| Year | World Bicycle Production (in millions) |
|---|---|
| 1989 | 95 |
| 1990 | 90 |
| 1991 | 96 |
| 1992 | 103 |
| 1993 | 108 |
| 1994 | ? |
| 1995 | 114 |

*Source*: United Nations
Interbike Directory

First, we analyze the data and look for trends. Note that production decreases for the early years, but in more recent years it increases more like a straight line. It seems reasonable that we can draw a line between the points for 1993 and 1995. We zoom in on that portion of the graph, as shown below.

**World Bicycle Production**

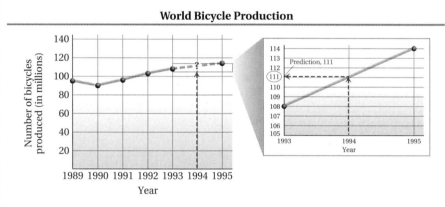

Then we visualize a vertical line up from the point for 1994 and see where the vertical line crosses the line between the data points (1993, 108) and (1995, 114). We move to the left and read off a value—about 111. We can also estimate this value by taking the average of the data values 108 and 114:

$$\frac{(108 + 114)}{2} = 111.$$

When we estimate in this way to find an "in-between value," we are using a process called **interpolation.** Real-world information about the data might tell us that an estimate found in this way is unreliable. For example, data from the stock market might be very erratic.

*Do Exercise 12.*

We often analyze data with the view of going "beyond" the data. One process for doing so is called *extrapolation*.

**Example 8**  *World Bicycle Production Extended.* Let's now consider that we know the world production of bicycles in 1994 to be 111 million and add it to the table of data. Use extrapolation to estimate world bicycle production in 1996.

| Year | World Bicycle Production (in millions) |
|------|----------------------------------------|
| 1989 | 95 |
| 1990 | 90 |
| 1991 | 96 |
| 1992 | 103 |
| 1993 | 108 |
| 1994 | 111 |
| 1995 | 114 |
| 1996 | ? |

*Source*: United Nations Interbike Directory

**World Bicycle Production**

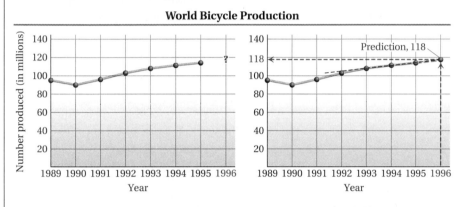

First, we analyze the data and note that they tend to follow a straight line past 1993. Keeping this trend in mind, we draw a "representative" line through the data and beyond. To estimate a value for 1996, we draw a vertical line up from 1996 until it hits the representative line. We go to the left and read off a value—about 118. When we estimate in this way to find a "go-beyond value," we are using a process called **extrapolation**. Answers found with this method can vary greatly depending on the points chosen to determine the "representative" line.

*Do Exercise 13.*

**13.** *Study Time and Test Scores.* A professor gathered the following data comparing study time and test scores. Use extrapolation to estimate the test score received when studying for 22 hr.

| Study Time (in hours) | Test Grade (in percent) |
|-----------------------|-------------------------|
| 18 | 84 |
| 19 | 86 |
| 20 | 89 |
| 21 | 92 |
| 22 | ? |

*Answer on page A-13*

# Calculator Spotlight

Trace and Table Features

**Trace Feature.** There are two ways in which we can determine the coordinates of points on a graph drawn by a grapher. One approach is to use a TRACE key.

Let's consider the equation for the FedEx mailing costs considered in Example 8 of Section 3.2.

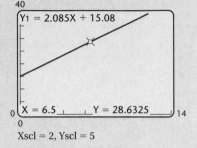
Xscl = 2, Yscl = 5

The cursor on the line means that the TRACE feature has been activated. The coordinates at the bottom indicate that the cursor is at the point with coordinates (6.5, 28.6325). By using the arrow keys, we can obtain coordinates of other points. For example, if we press the left arrow key $\triangleleft$ seven times, we move the cursor to the location shown below, obtaining a point on the graph with coordinates (5.5319149, 26.614043).

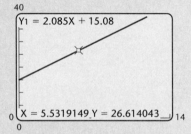

**Table Feature.** Another way to find the coordinates of solutions of equations makes use of the TABLE feature. We first press $\boxed{2nd}$ $\boxed{TBLSET}$ and set TblStart = 0 and $\triangle$Tbl = 10.

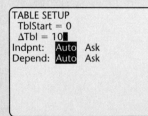
```
TABLE SETUP
  TblStart = 0
  △Tbl = 10
  Indpnt:  Auto   Ask
  Depend:  Auto   Ask
```

This means that the table's x-values will start at 0 and increase by 10. By setting Indpnt and Depend to Auto, we obtain the following when we press $\boxed{2nd}$ $\boxed{TABLE}$ : The arrow keys allow us to scroll up and down the table and extend it to other values not initially shown.

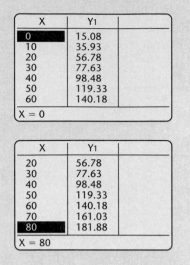

| X | Y1 |
|---|---|
| 0 | 15.08 |
| 10 | 35.93 |
| 20 | 56.78 |
| 30 | 77.63 |
| 40 | 98.48 |
| 50 | 119.33 |
| 60 | 140.18 |

X = 0

| X | Y1 |
|---|---|
| 20 | 56.78 |
| 30 | 77.63 |
| 40 | 98.48 |
| 50 | 119.33 |
| 60 | 140.18 |
| 70 | 161.03 |
| 80 | 181.88 |

X = 80

## Exercises

1. Use the TRACE feature to find five different ordered-pair solutions of the equation $y = 2.085x + 15.08$.

2. Use the TABLE feature to construct a table of solutions of the equation $y = 2.085x + 15.08$. Set TblStart = 100 with $\triangle$Tbl = 50. Find the value of $y$ when $x$ is 100. Then find the value of $y$ when $x$ is 150.

3. Again, use the equation $y = 2.085x + 15.08$ and adjust the table settings to Indpnt: Ask. How does the table change? Enter a number of your choice and see what happens. Use this setting to find the value of $y$ when $x$ is 187.

4. *Value of an Office Machine.* (Refer to Margin Exercise 14 in Section 3.2.) The value of Dupliographic's color copier is given by $v = -0.68t + 3.4$, where $v$ = the value, in thousands of dollars, $t$ years from the date of purchase.

   a) Graph the equation, choosing an appropriate viewing window that shows the intercepts.
   b) Does the equation have meaning for negative $x$-values?
   c) Find the $x$-intercept.
   d) For what $x$-values does the equation have meaning?
   e) Use the TRACE feature to find five ordered-pair solutions for values of $x$ between 0 and 5.
   f) After what amount of time is the value of the copier about $1500?
   g) Using the TABLE feature, find the value of the copier after 1.7 yr, 2.3 yr, 4.1 yr, and 5 yr.

# Exercise Set 3.4

**a** For each set of numbers, find the mean, the median, and any modes that exist. Round to the nearest tenth.

**1.** 15, 40, 30, 30, 45, 15, 25

**2.** 26, 28, 39, 24, 39, 29, 25

**3.** 81, 93, 96, 98, 102, 94

**4.** 23.4, 23.4, 22.6, 52.9

**5.** 23, 42, 35, 37, 23

**6.** 101.2, 104.3, 107.4, 105.7, 107.4

**7.** *Coffee Consumption.* The following lists the annual coffee consumption, in cups per person, for various countries.

| | |
|---|---|
| Germany | 1113 |
| United States | 610 |
| Switzerland | 1215 |
| France | 798 |
| Italy | 750 |

(*Source*: Beverage Marketing Corporation)

**8.** *Calories in Cereal.* The following lists the caloric content of a 2-cup bowl of certain cereals.

| | |
|---|---|
| Ralston Rice Chex | 240 |
| Kellogg's Complete Bran Flakes | 240 |
| Kellogg's Special K | 220 |
| Honey Nut Cheerios | 240 |
| Wheaties | 220 |

**9.** *NBA Tall Men.* The following is a list of the heights, in inches, of the tallest men in the NBA in a recent year.

| | |
|---|---|
| Shaquille O'Neal | 85 |
| Gheorghe Muresan | 91 |
| Shawn Bradley | 90 |
| Priest Lauderdale | 88 |
| Rik Smits | 88 |
| David Robinson | 85 |
| Arvydas Sabonis | 87 |

(*Source*: National Basketball Association)

**10.** *Movie Tickets Sold.* The following lists the number of movie tickets sold, in billions, in six recent years.

| | |
|---|---|
| 1990 | 1.19 |
| 1991 | 1.14 |
| 1992 | 1.17 |
| 1993 | 1.24 |
| 1994 | 1.29 |
| 1995 | 1.26 |

(*Source*: Motion Picture Association of America)

**b** Compare the set of data using their means.

11. *Battery Testing.* An experiment is performed to compare battery quality. Two kinds of battery were tested to see how long, in hours, they kept a portable CD player running. On the basis of this test, which battery is better?

| Battery A: EternReady Times (in hours) | | | Battery B: SturdyCell Times (in hours) | | |
|---|---|---|---|---|---|
| 27.9 | 28.3 | 27.4 | 28.3 | 27.6 | 27.8 |
| 27.6 | 27.9 | 28.0 | 27.4 | | 27.9 |
| 26.8 | 27.7 | 28.1 | 26.9 | 27.8 | 28.1 |
| 28.2 | 26.9 | 27.4 | 27.9 | 28.7 | 27.6 |

12. *Growth of Wheat.* A farmer experiments to see which of two kinds of wheat is better. (In this situation, the shorter wheat is considered "better.") He grows both kinds under similar conditions and measures stalk heights, in inches, as follows. Which kind is better?

| Wheat A Stalk Heights (in inches) | | | | Wheat B Stalk Heights (in inches) | | | |
|---|---|---|---|---|---|---|---|
| 16.2 | 42.3 | 19.5 | 25.7 | 19.7 | 18.4 | 32.0 | 25.7 |
| 25.6 | 18.0 | 15.6 | 41.7 | 19.7 | 21.6 | 42.5 | 32.6 |
| 22.6 | 26.4 | 18.4 | 12.6 | 14.0 | 10.9 | 26.7 | 22.8 |
| 41.5 | 13.7 | 42.0 | 21.6 | 22.6 | 19.7 | 17.2 | |

**c** Use interpolation or extrapolation to estimate the missing data values. Answers found using expolation can vary greatly.

13. *Height of Girls.*

| Age | Height (in centimeters) |
|---|---|
| 2 | 95.7 |
| 4 | 103.2 |
| 6 | 115.9 |
| 8 | 128.0 |
| 10 | 138.6 |
| 12 | 151.9 |
| 14 | 159.6 |
| 16 | 162.2 |
| 17 | ? |
| 18 | 162.5 |

*Source*: Kempe, C. Henry, M.D., et al., eds., *Current Pediatric Diagnosis & Treatment 1987.* Norwalk, CT: Appleton & Lange, 1987.

14. *Height of Boys.*

| Age | Height (in centimeters) |
|---|---|
| 2 | 96.2 |
| 4 | 103.4 |
| 6 | 117.5 |
| 8 | 130.0 |
| 10 | 140.3 |
| 12 | 149.6 |
| 14 | 162.7 |
| 15 | ? |
| 16 | 171.6 |
| 18 | 174.5 |

*Source*: Kempe, C. Henry, M.D., et al., eds., *Current Pediatric Diagnosis & Treatment 1987.* Norwalk, CT: Appleton & Lange, 1987.

**15.** *World Population.*

| Population (in billions) | Year in Which Population Is Reached |
|---|---|
| 1 | 1804 |
| 2 | 1927 |
| 3 | 1960 |
| 4 | 1974 |
| 5 | 1987 |
| 6 | 1998 |
| 7 | ? |

*Source*: U.S. Bureau of the Census

**16.** *SAT Scores.*

| Year | Average SAT Score, Math and Verbal |
|---|---|
| 1991 | 999 |
| 1992 | 1001 |
| 1993 | 1003 |
| 1994 | 1003 |
| 1995 | 1010 |
| 1996 | 1013 |
| 1997 | ? |

*Source*: The College Board

**17.** *Movie Tickets Sold.*

| Year | Tickets Sold (in billions) |
|---|---|
| 1991 | 1.19 |
| 1992 | 1.14 |
| 1993 | 1.17 |
| 1994 | 1.24 |
| 1995 | 1.29 |
| 1996 | 1.26 |
| 1997 | ? |
| 2000 | ? |

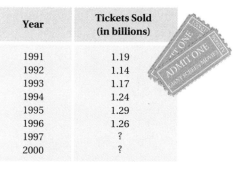

*Source*: Motion Picture Association of America

**18.** *McDonald's Restaurant Revenue in the United States.*

| Year | Revenue (in billions) |
|---|---|
| 1990 | $12.3 |
| 1991 | 12.5 |
| 1992 | 13.2 |
| 1993 | 14.2 |
| 1994 | 14.9 |
| 1995 | 15.9 |
| 1996 | ? |
| 2000 | ? |

*Source*: McDonald's Corporation

**19.** *Average Price of a 30-Second Super Bowl Commercial.*

| Year | Cost |
|---|---|
| 1991 | $  800,000 |
| 1992 | 850,000 |
| 1993 | 850,000 |
| 1994 | 900,000 |
| 1995 | 1,000,000 |
| 1996 | 1,300,000 |
| 1997 | ? |
| 2000 | ? |

*Source*: National Football League

**20.** *Retail Revenue from Lettuce.*

| Year | Revenue (in millions) |
|---|---|
| 1991 | $ 106 |
| 1992 | 168 |
| 1993 | 312 |
| 1994 | 577 |
| 1995 | 889 |
| 1996 | 1100 |
| 1997 | ? |
| 2000 | ? |

*Source*: International Fresh-Cut Produce Association, Information Resources

**21.** *Study Time vs. Grades.* A mathematics instructor asked her students to keep track of how much time each spent studying the chapter on percent notation in her basic mathematics course. She collected the information together with test scores from that chapter's test. The data are given in the following table. Estimate the missing data value.

| Study Time (in hours) | Test Grade (in percent) |
|:---:|:---:|
| 9 | 76 |
| 11 | 94 |
| 13 | 81 |
| 15 | 86 |
| 16 | 87 |
| 17 | 81 |
| 18 | ? |
| 19 | 87 |
| 20 | 92 |

**22.** *Maximum Heart Rate.* A person's maximum heart rate depends on his or her gender, age, and resting heart rate. The following table relates resting heart rate and maximum heart rate for a 20-yr-old man. Estimate the missing data value.

| Resting Heart Rate (in beats per minute) | Maximum Heart Rate (in beats per minute) |
|:---:|:---:|
| 50 | 166 |
| 60 | 168 |
| 70 | 170 |
| 75 | ? |
| 80 | 172 |

*Source*: American Heart Association

---

## Skill Maintenance

Convert to fractional notation.   [R.4b]

**23.** 16%

**24.** $33\frac{1}{3}\%$

**25.** 37.5%

**26.** 75%

Solve.   [2.5a]

**27.** Jennifer left an $8.50 tip for a meal that cost $42.50. What percent of the cost of the meal was the tip?

**28.** Kristen left an 18% tip of $3.24 for a meal. What was the cost of the meal before the tip?

**29.** Juan left a 15% tip for a meal. The total cost of the meal, including the tip, was $51.92. What was the cost of the meal before the tip was added?

**30.** After a 25% reduction, a sweater is on sale for $41.25. What was the original price?

---

## Synthesis

**31.** ◈ In a recent year, the average salary of all players in baseball's American League was $1.3 million and the median was $400,000 (***Source***: Major League Baseball). Discuss the merits of each value as a measure of central tendency.

**32.** ◈ Discuss how you might test the estimates that you found in Exercises 13–20.

⚏ Graph the equation using the standard viewing window. Then construct a table of *y*-values for *x*-values starting at $x = -10$ with $\triangle$Tbl = 0.1.

**33.** $y = 0.35x - 7$

**34.** $y = 5.6 - x^2$

**35.** $y = x^3 - 5$

**36.** $y = 4 + 3x - x^2$

---

Collaborative Learning Manual

Perform a statistical analysis of pulse rates.
Make predictions from a set of data.

# Summary and Review Exercises: Chapter 3

The objectives to be tested in addition to the material in this chapter are [R.3c], [1.2c], [1.2e], and [2.5a].

**1.** *Federal Spending.* The following pie chart shows how our federal income tax dollars are used. As a freelance graphic artist, Jennifer pays $3525 in taxes. How much of Jennifer's tax payment goes toward defense? toward social programs? [3.1a]

**Where Your Tax Dollars Are Spent**

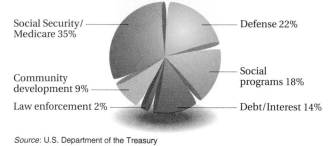

Social Security/Medicare 35%

Defense 22%

Community development 9%

Social programs 18%

Law enforcement 2%

Debt/Interest 14%

*Source*: U.S. Department of the Treasury

*Chicken Consumption.* The following line graph shows average chicken consumption from 1980 to 2000. (The value for the year 2000 is projected.) [3.1a]

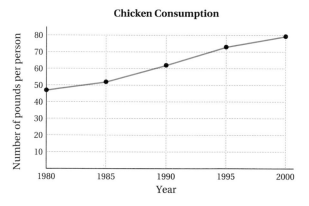

**Chicken Consumption**

**2.** About how many pounds of chicken were consumed per person in 1980?

**3.** About how many pounds of chicken will be consumed per person in 2000?

**4.** By what amount did chicken consumption increase from 1980 to 2000?

**5.** In what year did the consumption of chicken exceed 70 lb per person?

**6.** In what 5-yr period was the difference in consumption the greatest?

*Water Usage.* The following bar graph shows water usage, in gallons, for various tasks. [3.1a]

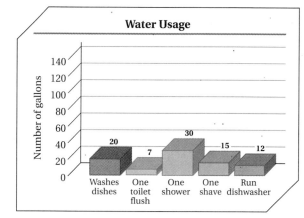

*Source*: American Water Works Association

**7.** Which task requires the most water?

**8.** Which task requires the least water?

**9.** Which tasks require 15 or more gallons?

**10.** Which task requires 7 gallons?

Find the coordinates of the point. [3.1d]

**11.** *A*          **12.** *B*          **13.** *C*

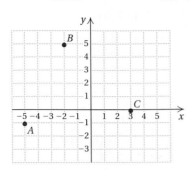

Plot the point. [3.1b]

**14.** $(2, 5)$      **15.** $(0, -3)$      **16.** $(-4, -2)$

In which quadrant is the point located? [3.1c]

**17.** $(3, -8)$      **18.** $(-20, -14)$      **19.** $(4.9, 1.3)$

Determine whether the point is a solution of $2y - x = 10$. [3.2a]

**20.** $(2, -6)$             **21.** $(0, 5)$

**22.** Show that the ordered pairs $(0, -3)$ and $(2, 1)$ are solutions of the equation $2x - y = 3$. Then use the graph of the two points to determine another solution. Answers may vary. [3.2a]

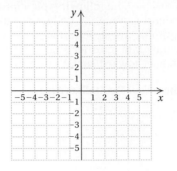

Graph the equation, identifying the *y*-intercept. [3.2b]

**23.** $y = 2x - 5$

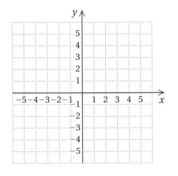

**24.** $y = -\dfrac{3}{4}x$

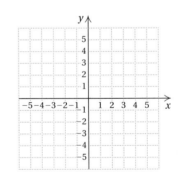

**25.** $y = -x + 4$

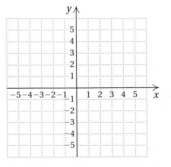

**26.** $y = 3 - 4x$

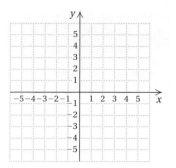

Graph the equation. [3.3b]

**27.** $y = 3$

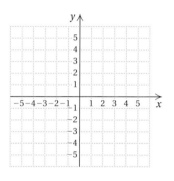

**28.** $5x - 4 = 0$

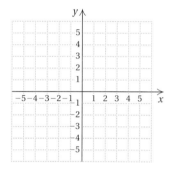

Find the intercepts of the equation. Then graph the equation. [3.3a]

**29.** $x - 2y = 6$

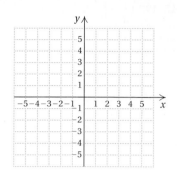

**30.** $5x - 2y = 10$

Solve. [3.2c]

**31.** *Kitchen Design.* Kitchen designers recommend that a refrigerator be selected on the basis of the number of people $n$ in the household. The appropriate size $S$, in cubic feet, is given by

$$S = \frac{3}{2}n + 13.$$

  **a)** Determine the recommended size of a refrigerator if the number of people is 1, 2, 5, and 10.
  **b)** Graph the equation and use the graph to estimate the recommended size of a refrigerator for 3 people sharing an apartment.
  **c)** A refrigerator is 22 ft³. For how many residents is it the recommended size?

Find the mean, the median, and the mode. [3.4a]

**32.** 27, 35, 44, 52

**33.** 8, 12, 15, 17, 19, 12, 17, 19

**34.** *Blood Alcohol Levels of Drivers in Fatal Accidents.* 0.18, 0.17, 0.21, 0.16

**35.** *Bowling Scores of the Author in a Recent Tournament.* 215, 259, 215, 223, 237

**36.** *Movie Production Costs (in millions of dollars) per Film in Five Recent Years.* $42.3, $44.0, $50.4, $54.1, $59.7

**37.** *Popcorn Testing.* An experiment is performed to compare the quality of two types of popcorn. The tester puts 200 kernels in a pan, pops them, and counts the number of unpopped kernels. The results are shown in the following table. Determine which type of popcorn is better by comparing their means. [3.4b]

| Popcorn A Unpopped Kernels | | |
|---|---|---|
| 20 | 23 | 35 |
| 10 | 12 | 18 |
| 18 | 24 | 11 |
| 19 | 21 | |

| Popcorn B Unpopped Kernels | | |
|---|---|---|
| 19 | 25 | 32 |
| 8 | 22 | 14 |
| 19 | 13 | 24 |
| 15 | 22 | 10 |

Estimate the missing data values using interpolation or extrapolation. [3.4c]

**38.** *Height of Girls.*

| Age | Height (in centimeters) |
|---|---|
| 2 | 95.7 |
| 4 | 103.2 |
| 5 | ? |
| 6 | 115.9 |
| 8 | 128.0 |
| 10 | 138.6 |
| 12 | 151.9 |
| 14 | 159.6 |
| 16 | 162.2 |
| 18 | 162.5 |

*Source*: Kempe, C. Henry, M.D., et al., eds., *Current Pediatric Diagnosis & Treatment 1987*. Norwalk, CT: Appleton & Lange, 1987.

**39.** *Movie Production Costs.*

| Year | Cost per Film (in millions) |
|---|---|
| 1992 | $42.3 |
| 1993 | 44.0 |
| 1994 | 50.4 |
| 1995 | 54.1 |
| 1996 | 59.7 |
| 1997 | ? |
| 2000 | ? |

*Source*: Motion Picture Association of America

**Skill Maintenance**

Convert to decimal notation. [1.2c]

**40.** $-\dfrac{11}{32}$

**41.** $\dfrac{8}{9}$

Find the absolute value. [1.2e]

**42.** $|-3.2|$

**43.** $\left|\dfrac{17}{19}\right|$

Round to the nearest hundredth. [R.3c]

**44.** 42.705

**45.** 112.5278

Solve. [2.5a]

**46.** An investment was made at 6% simple interest for 1 year. It grows to $10,340.40. How much was originally invested?

**47.** After a 20% reduction, a pair of slacks is on sale for $63.96. What was the original price (that is, the price before reduction)?

**Synthesis**

**48.** ◆ Describe two ways in which a small business might make use of graphs. [3.1a], [3.2c]

**49.** ◆ Explain why the first coordinate of the *y*-intercept is always 0. [3.2b]

**50.** Find the value of *m* in $y = mx + 3$ such that $(-2, 5)$ is on the graph. [3.2a]

**51.** Find the area and the perimeter of a rectangle for which $(-2, 2)$, $(7, 2)$, and $(7, -3)$ are three of the vertices. [3.1b]

# Test: Chapter 3

*Toothpaste Sales.* The following pie chart shows the percentages of sales of various toothpaste brands in the United States. In a recent year, total sales of toothpaste were $1,500,000,000.

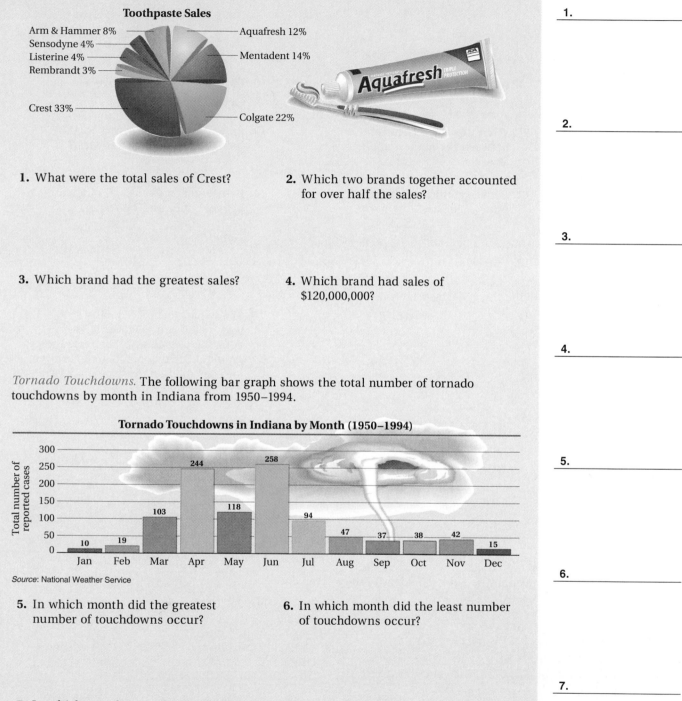

**Toothpaste Sales**

Arm & Hammer 8%
Sensodyne 4%
Listerine 4%
Rembrandt 3%
Crest 33%
Aquafresh 12%
Mentadent 14%
Colgate 22%

**1.** What were the total sales of Crest?

**2.** Which two brands together accounted for over half the sales?

**3.** Which brand had the greatest sales?

**4.** Which brand had sales of $120,000,000?

*Tornado Touchdowns.* The following bar graph shows the total number of tornado touchdowns by month in Indiana from 1950–1994.

**Tornado Touchdowns in Indiana by Month (1950–1994)**

Total number of reported cases

Jan 10
Feb 19
Mar 103
Apr 244
May 118
Jun 258
Jul 94
Aug 47
Sep 37
Oct 38
Nov 42
Dec 15

*Source*: National Weather Service

**5.** In which month did the greatest number of touchdowns occur?

**6.** In which month did the least number of touchdowns occur?

**7.** In which months was the number of touchdowns greater than 90?

**8.** In which month were there 47 touchdowns?

9. _____

10. _____

11. _____

12. _____

13. _____

14. _____

15. _____

16. _____

17. _____

18. _____

19. _____

20. _____

21. _____

_Average Salary of Major-League Baseball Players._ **The line graph at right shows the average salary of major-league baseball players over a recent seven-year period. Use the graph for Exercises 9–14.**

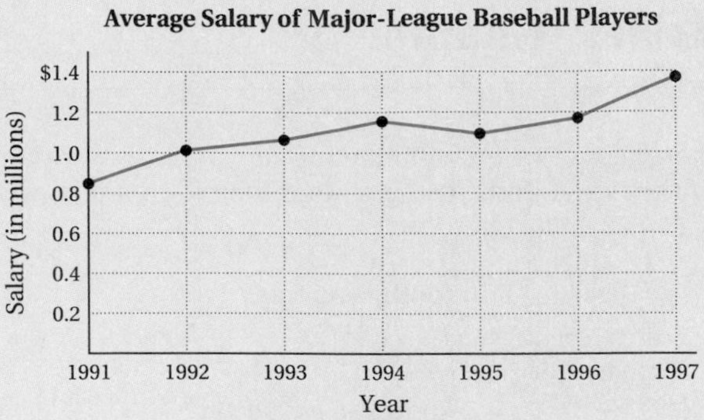

**Average Salary of Major-League Baseball Players**

**9.** In which year was the average salary the highest?

**10.** In which year was the average salary the lowest?

**11.** What was the difference in salary between the highest and lowest salaries?

**12.** Between which two years was the increase in salary the greatest?

**13.** Between which two years did the salary decrease?

**14.** By how much did salaries increase between 1991 and 1997?

In which quadrant is the point located?

**15.** $\left(-\frac{1}{2}, 7\right)$

**16.** $(-5, -6)$

Find the coordinates of the point.

**17.** $A$

**18.** $B$

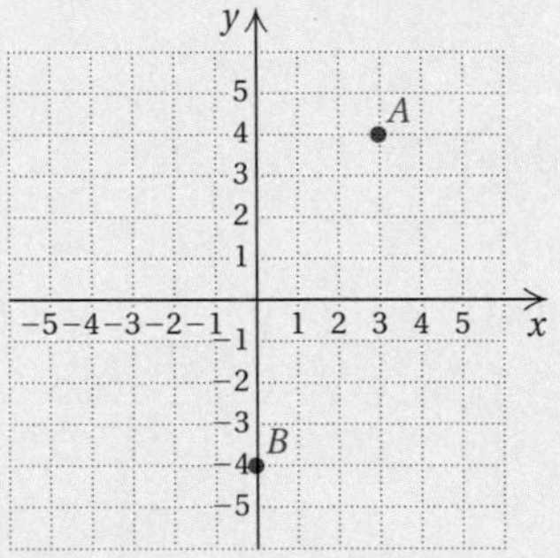

**19.** Show that the ordered pairs $(-4, -3)$ and $(-1, 3)$ are solutions of the equation $y - 2x = 5$. Then use the graph of the two points to determine another solution. Answers may vary.

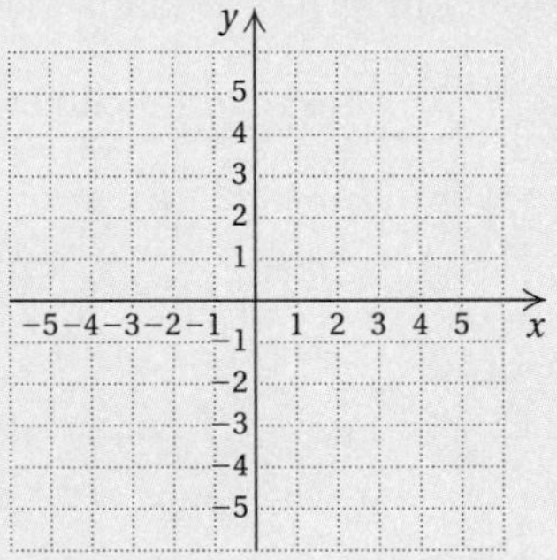

Graph the equation. Identify the $y$-intercept.

**20.** $y = 2x - 1$

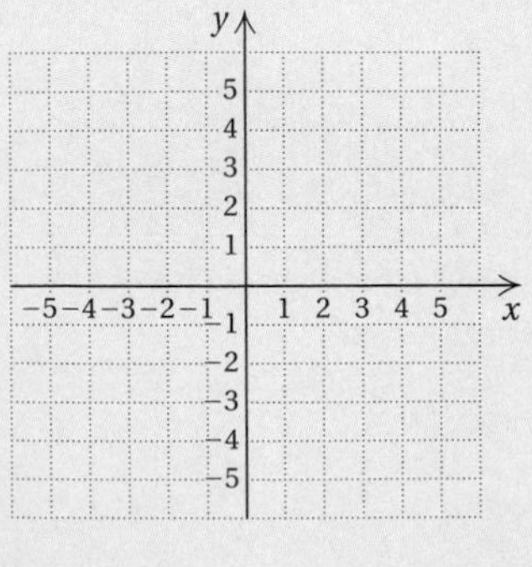

**21.** $y = -\frac{3}{2}x$

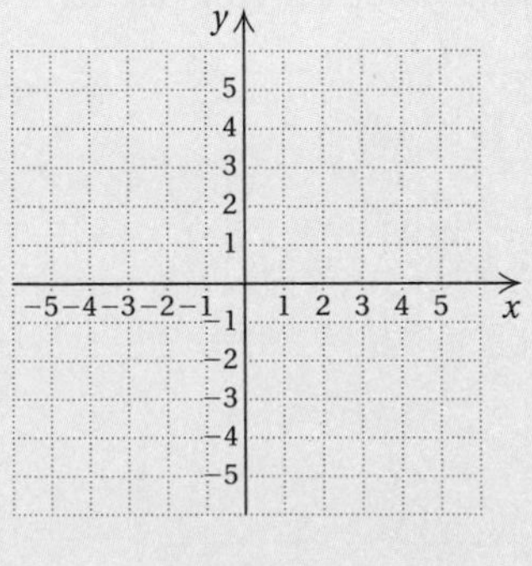

Copyright © 1999 Addison Wesley Longman

Graph the equation.

**22.** $2x + 8 = 0$

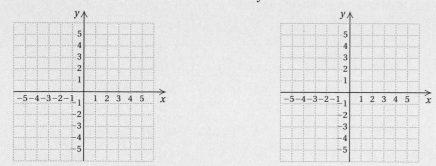

**23.** $y = 5$

Find the intercepts of the equation. Then graph the equation.

**24.** $2x - 4y = -8$

**25.** $2x - y = 3$

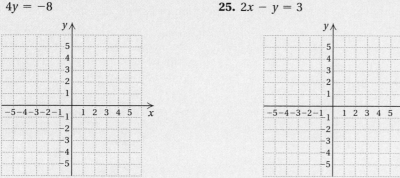

**26.** *Private-College Costs.* The cost $T$, in thousands of dollars, of tuition and fees at a private college (all expenses) can be approximated by

$$T = \frac{4}{5}n + 17,$$

where $n$ = the number of years since 1992 (**Source:** Statistical Abstract of the United States). That is, $n = 0$ corresponds to 1992, $n = 7$ corresponds to 1999, and so on.

**a)** Find the cost of tuition in 1992, 1995, 1999, and 2001.

**b)** Graph the equation and then use the graph to estimate the cost of tuition in 2005.

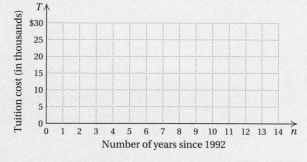

Number of years since 1992

**c)** Estimate the year in which the cost of tuition will be $25,000.

Find the mean, the median, and the mode.

**27.** 46, 50, 53, 55

**30.** *Animal Speeds.*

**28.** 2, 3, 4, 5, 6, 5

**29.** 4, 19, 20, 18, 19, 18

| Animal | Speed (in miles per hour) |
|---|---|
| Antelope | 61 |
| Bear | 30 |
| Cheetah | 70 |
| Fastest human | 28 |
| Greyhound | 39 |
| Lion | 50 |
| Zebra | 40 |

**Answers**

22. _____

23. _____

24. _____

25. _____

26. a) _____

b) _____

c) _____

27. _____

28. _____

29. _____

30. _____

31. _____

32. _____

33. _____

34. _____

35. _____

36. _____

37. _____

38. _____

39. _____

40. _____

41. _____

42. _____

43. _____

**31.** *Quality of Golf Balls.* A golf pro experiments to see which of two kinds of golf ball is better. He drops each type of ball from a height of 8 ft onto concrete and measures how high they bounce, in inches. Determine which type of ball is better by comparing their means.

| Ball A Bouncing Heights (in inches) | | | | Ball B Bouncing Heights (in inches) | | | |
|---|---|---|---|---|---|---|---|
| 59.7 | 58.4 | 59.7 | 57.2 | 56.2 | 62.3 | 59.5 | 65.7 |
| 59.7 | 64.6 | 62.0 | 63.7 | 59.6 | 58.0 | 61.6 | 62.7 |
| 66.5 | 61.6 | 62.5 | 59.8 | 59.8 | 56.4 | 58.4 | 66.6 |
| 61.6 | 57.9 | 58.7 | | 61.5 | 58.7 | 62.0 | 68.3 |

Estimate the missing data values using interpolation or extrapolation.

**32.** *Height of Boys.*

| Age | Height (in centimeters) |
|---|---|
| 2 | 96.2 |
| 4 | 103.4 |
| 6 | 117.5 |
| 8 | 130.0 |
| 9 | ? |
| 10 | 140.3 |
| 12 | 149.6 |
| 14 | 162.7 |
| 16 | 171.6 |
| 18 | 174.5 |

*Source*: Kempe, C. Henry, M.D., et al., eds., *Current Pediatric Diagnosis & Treatment 1987*. Norwalk, CT: Appleton & Lange, 1987.

**33.** *Deaths from Driving Incidents.* The following table lists for several years the number of driving incidents that resulted in death.

| Year | Incidents |
|---|---|
| 1990 | 1129 |
| 1991 | 1297 |
| 1992 | 1478 |
| 1993 | 1555 |
| 1994 | 1669 |
| 1995 | 1708 |
| 1996 | ? |
| 2000 | ? |

*Source*: AAA Foundation

---

**Skill Maintenance**

Convert to decimal notation.

**34.** $\dfrac{39}{40}$

**35.** $-\dfrac{13}{12}$

Find the absolute value.

**36.** $|71.2|$

**37.** $\left| -\dfrac{13}{47} \right|$

Round to the nearest thousandth.

**38.** 42.7047

**39.** 112.52702

Solve.

**40.** After a 24% reduction, a software game is on sale for $64.22. What was the original price (that is, the price before reduction)?

**41.** An investment was made at 7% simple interest for 1 year. It grows to $38,948. How much was originally invested?

---

**Synthesis**

**42.** A diagonal of a square connects the points $(-3, -1)$ and $(2, 4)$. Find the area and the perimeter of the square.

**43.** Write an equation of a line parallel to the *x*-axis and 3 units above it.

# Cumulative Review: Chapters 1–3

**1.** Evaluate $\dfrac{x}{2y}$ for $x = 10$ and $y = 2$.

**2.** Multiply: $3(4x - 5y + 7)$.

**3.** Factor: $15x - 9y + 3$.

**4.** Find the prime factorization of 42.

**5.** Find decimal notation: $\dfrac{9}{20}$.

**6.** Find the absolute value: $|-4|$.

**7.** Find the opposite of $-3.08$.

**8.** Find the reciprocal of $-\dfrac{8}{7}$.

**9.** Collect like terms: $2x - 5y + (-3x) + 4y$.

**10.** Find decimal notation: $78.5\%$.

Simplify.

**11.** $\dfrac{3}{4} - \dfrac{5}{12}$

**12.** $3.4 + (-0.8)$

**13.** $(-2)(-1.4)(2.6)$

**14.** $\dfrac{3}{8} \div \left(-\dfrac{9}{10}\right)$

**15.** $2 - [32 \div (4 + 2^2)]$

**16.** $-5 + 16 \div 2 \cdot 4$

**17.** $y - (3y + 7)$

**18.** $3(x - 1) - 2[x - (2x + 7)]$

Solve.

**19.** $1.5 = 2.7 + x$

**20.** $\dfrac{2}{7}x = -6$

**21.** $5x - 9 = 36$

**22.** $\dfrac{5}{2}y = \dfrac{2}{5}$

**23.** $5.4 - 1.9x = 0.8x$

**24.** $x - \dfrac{7}{8} = \dfrac{3}{4}$

**25.** $2(2 - 3x) = 3(5x + 7)$

**26.** $\dfrac{1}{4}x - \dfrac{2}{3} = \dfrac{3}{4} + \dfrac{1}{3}x$

**27.** $y + 5 - 3y = 5y - 9$

**28.** $x - 28 < 20 - 2x$

**29.** $2(x + 2) \geq 5(2x + 3)$

**30.** Solve $A = \frac{1}{2}h(b + c)$ for $h$.

**31.** In which quadrant is the point $(3, -1)$ located?

**32.** Graph on a number line: $-1 < x \leq 2$.

Graph.

**33.** $2x + 5y = 10$

**34.** $y = -2$

**35.** $y = -2x + 1$

**36.** $y = \dfrac{2}{3}x - 2$

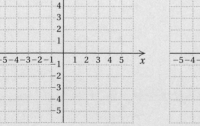

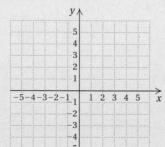

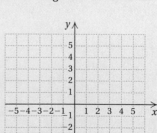

Find the intercepts. Do not graph.

**37.** $2x - 7y = 21$

**38.** $y = 4x + 5$

Solve.

**39.** *Blood Donors.* Each year, 8 million Americans donate blood. This is 5% of those healthy enough to do so. How many Americans are eligible to donate blood?

**40.** *Blood Types.* There are 117 million Americans with either O-positive or O-negative blood. Those with O-positive blood outnumber those with O-negative blood by 85.8 million. How many Americans have O-negative blood?

**41.** Tina paid $126 for a cordless drill. This included 5% for sales tax. How much did the drill cost before tax?

**42.** A 143-m wire is cut into three pieces. The second piece is 3 m longer than the first. The third is four-fifths as long as the first. How long is each piece?

**43.** Cory's contract stipulates that he cannot work more than 40 hr per week. For the first four days of one week, he worked 7, 10, 9, and 6 hr. Determine as an inequality the number of hours he can work on the fifth day without violating his contract.

**44.** *Telephone Line.* The cost $P$, in hundreds of dollars, of a telephone line for a business is given by $P = \frac{3}{4}n + 3$, where $n =$ the number of months that the line has been in service.

  a) Find the cost of the phone line for 1 month, 2 months, 3 months, and 7 months.
  b) Graph the equation and use the graph to estimate the cost of the phone line for 10 months.
  c) Estimate the number of months it will take for the cost to be $15,000.

**45.** Find the mean, the median, and the mode for these medical dosages:

  25.4 cc, 31.2 cc, 25.4 cc, 28.7 cc, 32.8 cc, 25.4 cc.

**46.** *Class Comparisons.* A math instructor conducts an experiment to compare two classes that took the same test. Determine which class performed better by comparing the means.

| Class A: 15 students | | | Class B: 14 students | | |
|---|---|---|---|---|---|
| 78 | 82 | 93 | 79 | 80 | 80 |
| 100 | 76 | 82 | 74 | 100 | 77 |
| 66 | 91 | 76 | 64 | 74 | 86 |
| 78 | 76 | 84 | 88 | 82 | 62 |
| 92 | 83 | 64 | 93 | 94 | |

**47.** *Height of Girls.* Estimate the missing data value.

| Age | Height (in centimeters) |
|---|---|
| 2 | 95.7 |
| 4 | 103.2 |
| 6 | 115.9 |
| 8 | 128.0 |
| 10 | 138.6 |
| 11 | ? |
| 12 | 151.9 |
| 14 | 159.6 |
| 16 | 162.2 |
| 18 | 162.5 |

## Synthesis

Solve.

**48.** $4|x| - 13 = 3$

**49.** $4(x + 2) = 4(x - 2) + 16$

**50.** $0(x + 3) + 4 = 0$

**51.** $\dfrac{2 + 5x}{4} = \dfrac{11}{28} + \dfrac{8x + 3}{7}$

**52.** $5(7 + x) = (x + 7)5$

**53.** $p = \dfrac{2}{m + Q}$, for $Q$

# 4

# Polynomials: Operations

## Introduction

Algebraic expressions like
$$-0.002d^2 + 0.8d + 6.6$$
and
$$x^3 - 5x^2 + 4x - 11$$
are called *polynomials*. One of the most important parts of introductory algebra is the study of polynomials. In this chapter, we learn to add, subtract, multiply, and divide polynomials.

Of particular importance here is the study of the quick ways to multiply certain polynomials. These *special products* will be helpful not only in this text but also in more advanced mathematics.

## An Application

The Olympic flame at the 1992 Summer Olympics was lit by a flaming arrow. As the arrow moved *d* feet horizontally from the archer, its height *h*, in feet, could be approximated by the polynomial

$$-0.002d^2 + 0.8d + 6.6.$$

Use the polynomial to approximate the height of the arrow after it has traveled 100 ft horizontally.

This problem appears as Exercise 19 in Exercise Set 4.3.

## The Mathematics

We substitute 100 for *d* and evaluate:

$$-0.002d^2 + 0.8d + 6.6 = -0.002(100)^2 + 0.8(100) + 6.6$$
$$= 66.6 \text{ ft.}$$

This is a polynomial.

World Wide Web   For more information, visit us at www.mathmax.com

**1.** Multiply: $x^{-3} \cdot x^5$.

**2.** Divide: $\dfrac{x^{-2}}{x^5}$.

**3.** Simplify: $(-4x^2y^{-3})^2$.

**4.** Express using a positive exponent: $p^{-3}$.

**5.** Convert to scientific notation: 0.000347.

**6.** Convert to decimal notation: $3.4 \times 10^6$.

**7.** Identify the degree of each term and the degree of the polynomial:

$$2x^3 - 4x^2 + 3x - 5.$$

**8.** Collect like terms:

$$2a^3b - a^2b^2 + ab^3 + 9 - 5a^3b - a^2b^2 + 12b^3.$$

**9.** Add:

$$(5x^2 - 7x + 8) + (6x^2 + 11x - 19).$$

**10.** Subtract:

$$(5x^2 - 7x + 8) - (6x^2 + 11x - 19).$$

Multiply.

**11.** $5x^2(3x^2 - 4x + 1)$

**12.** $(x + 5)^2$

**13.** $(x - 5)(x + 5)$

**14.** $(x^3 + 6)(4x^3 - 5)$

**15.** $(2x - 3y)(2x - 3y)$

**16.** Divide: $(x^3 - x^2 + x + 2) \div (x - 2)$.

## Objectives for Retesting

The objectives to be retested in addition to the material in this chapter are as follows.

[1.4a]   Subtract real numbers and simplify combinations of additions and subtractions.

[1.7d]   Use the distributive laws to factor expressions like $4x - 12 + 24y$.

[2.3b, c]   Solve equations in which like terms may need to be collected, and solve equations by first removing parentheses and collecting like terms.

[2.4a]   Solve applied problems by translating to equations.

# 4.1 Integers as Exponents

We introduced integer exponents of 2 or higher in Section R.5. Here we consider 0 and 1, as well as negative integers, as exponents.

## a Exponential Notation

An exponent of 2 or greater tells how many times the base is used as a factor. For example,

$$a \cdot a \cdot a \cdot a = a^4.$$

In this case, the **exponent** is 4 and the **base** is $a$. An expression for a power is called **exponential notation.**

$$a^n \leftarrow \text{This is the exponent.}$$
$$\uparrow$$
This is the base.

**Example 1**  What is the meaning of $3^5$? of $n^4$? of $(2n)^3$? of $50x^2$?

$3^5$ means $3 \cdot 3 \cdot 3 \cdot 3 \cdot 3$;      $n^4$ means $n \cdot n \cdot n \cdot n$;

$(2n)^3$ means $2n \cdot 2n \cdot 2n$;      $50x^2$ means $50 \cdot x \cdot x$

*Do Exercises 1–4.*

We read exponential notation as follows:

$a^n$ is read the **nth power of a,**   or simply **a to the nth,**   or **a to the n.**

We often read $x^2$ as "**x-squared.**" The reason for this is that the area of a square of side $x$ is $x \cdot x$, or $x^2$. We often read $x^3$ as "**x-cubed.**" The reason for this is that the volume of a cube with length, width, and height $x$ is $x \cdot x \cdot x$, or $x^3$.

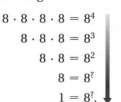

## b One and Zero as Exponents

Look for a pattern in the following:

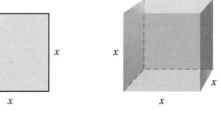

On each side, we divide by 8 at each step.

$8 \cdot 8 \cdot 8 \cdot 8 = 8^4$
$8 \cdot 8 \cdot 8 = 8^3$
$8 \cdot 8 = 8^2$
$8 = 8^?$
$1 = 8^?.$

On this side, the exponents decrease by 1.

To continue the pattern, we would say that

$$8 = 8^1$$

and   $1 = 8^0.$

## Objectives

a  Tell the meaning of exponential notation.

b  Evaluate exponential expressions with exponents of 0 and 1.

c  Evaluate algebraic expressions containing exponents.

d  Use the product rule to multiply exponential expressions with like bases.

e  Use the quotient rule to divide exponential expressions with like bases.

f  Express an exponential expression involving negative exponents with positive exponents.

### For Extra Help

TAPE 7      TAPE 7A      MAC WIN      CD-ROM

What is the meaning of each of the following?

**1.** $5^4$

**2.** $x^5$

**3.** $(3t)^2$

**4.** $3t^2$

*Answers on page A-16*

Evaluate.

**5.** $6^1$

**6.** $7^0$

**7.** $(8.4)^1$

**8.** $8654^0$

*Answers on page A-16*

We make the following definition.

> $a^1 = a$, for any number $a$;
> $a^0 = 1$, for any nonzero number $a$.

We consider $0^0$ to be undefined. We will explain why later in this section.

**Example 2**   Evaluate $5^1$, $8^1$, $3^0$, $(-7.3)^0$, and $(186,892,046)^0$.

$$5^1 = 5; \qquad 8^1 = 8; \qquad 3^0 = 1;$$
$$(-7.3)^0 = 1; \qquad (186,892,046)^0 = 1$$

*Do Exercises 5–8.*

## c   Evaluating Algebraic Expressions

Algebraic expressions can involve exponential notation. For example, the following are algebraic expressions:

$$x^4, \qquad (3x)^3 - 2, \qquad a^2 + 2ab + b^2.$$

We evaluate algebraic expressions by replacing variables with numbers and following the rules for order of operations.

**Example 3**   Evaluate $x^4$ for $x = 2$.

$$x^4 = 2^4 \qquad \text{\small Substituting}$$
$$= 2 \cdot 2 \cdot 2 \cdot 2 = 16$$

**Example 4**   *Area of a Compact Disc.* The standard compact disc used for software and music has a radius of 6 cm. Find the area of such a CD (ignoring the hole in the middle).

$$A = \pi r^2$$
$$= \pi \cdot (6 \text{ cm})^2$$
$$\approx 3.14 \times 36 \text{ cm}^2$$
$$\approx 113.04 \text{ cm}^2$$

$r = 6$ cm

In Example 4, "cm$^2$" means "square centimeters" and "$\approx$" means "is approximately equal to."

**Example 5**   Evaluate $(5x)^3$ for $x = -2$.

When we evaluate with a negative number, we often use extra parentheses to show the substitution.

$$(5x)^3 = [5 \cdot (-2)]^3 \qquad \text{\small Substituting}$$
$$= [-10]^3 \qquad \text{\small Multiplying within brackets first}$$
$$= -1000 \qquad \text{\small Evaluating the power}$$

**Example 6** Evaluate $5x^3$ for $x = -2$.

$$5x^3 = 5 \cdot (-2)^3 \qquad \text{Substituting}$$
$$= 5(-8) \qquad \text{Evaluating the power first}$$
$$= -40$$

Recall that two expressions are equivalent if they have the same value for all meaningful replacements. Note that Examples 5 and 6 show that $(5x)^3$ and $5x^3$ are *not* equivalent—that is, $(5x)^3 \neq 5x^3$.

*Do Exercises 9–13.*

## d | Multiplying Powers with Like Bases

There are several rules for manipulating exponential notation to obtain equivalent expressions. We first consider multiplying powers with like bases:

$$a^3 \cdot a^2 = \underbrace{(a \cdot a \cdot a)}_{\text{3 factors}}\underbrace{(a \cdot a)}_{\text{2 factors}} = \underbrace{a \cdot a \cdot a \cdot a \cdot a}_{\text{5 factors}} = a^5.$$

Since an integer exponent greater than 1 tells how many times we use a base as a factor, then $(a \cdot a \cdot a)(a \cdot a) = a \cdot a \cdot a \cdot a \cdot a = a^5$ by the associative law. Note that the exponent in $a^5$ is the sum of those in $a^3 \cdot a^2$. That is, $3 + 2 = 5$. Likewise,

$$b^4 \cdot b^3 = (b \cdot b \cdot b \cdot b)(b \cdot b \cdot b) = b^7, \quad \text{where} \quad 4 + 3 = 7.$$

Adding the exponents gives the correct result.

> **THE PRODUCT RULE**
>
> For any number $a$ and any positive integers $m$ and $n$,
> $$a^m \cdot a^n = a^{m+n}.$$
> (When multiplying with exponential notation, if the bases are the same, keep the base and add the exponents.)

**Examples** Multiply and simplify. By simplify, we mean write the expression as one number to a nonnegative power.

**7.** $8^4 \cdot 8^3 = 8^{4+3} \qquad$ Adding exponents: $a^m \cdot a^n = a^{m+n}$
$\qquad = 8^7$

**8.** $x^2 \cdot x^9 = x^{2+9}$
$\qquad = x^{11}$

**9.** $m^5 m^{10} m^3 = m^{5+10+3}$
$\qquad = m^{18}$

**10.** $x \cdot x^8 = x^1 \cdot x^8 = x^{1+8}$
$\qquad = x^9$

**11.** $(a^3 b^2)(a^3 b^5) = (a^3 a^3)(b^2 b^5)$
$\qquad = a^6 b^7$

*Do Exercises 14–18.*

---

**9.** Evaluate $t^3$ for $t = 5$.

**10.** Find the area of a circle when $r = 32$ cm. Use 3.14 for $\pi$.

**11.** Evaluate $200 - a^4$ for $a = 3$.

**12.** Evaluate $t^1 - 4$ and $t^0 - 4$ for $t = 7$.

**13. a)** Evaluate $(4t)^2$ for $t = -3$.

**b)** Evaluate $4t^2$ for $t = -3$.

**c)** Determine whether $(4t)^2$ and $4t^2$ are equivalent.

Multiply and simplify.
**14.** $3^5 \cdot 3^5$

**15.** $x^4 \cdot x^6$

**16.** $p^4 p^{12} p^8$

**17.** $x \cdot x^4$

**18.** $(a^2 b^3)(a^7 b^5)$

*Answers on page A-16*

Divide and simplify.

**19.** $\dfrac{4^5}{4^2}$

**20.** $\dfrac{y^6}{y^2}$

**21.** $\dfrac{p^{10}}{p}$

**22.** $\dfrac{a^7 b^6}{a^3 b^4}$

*Answers on page A-16*

### e | Dividing Powers with Like Bases

The following suggests a rule for dividing powers with like bases, such as $a^5/a^2$:

$$\frac{a^5}{a^2} = \frac{a \cdot a \cdot a \cdot a \cdot a}{a \cdot a} = \frac{a \cdot a \cdot a \cdot a \cdot a}{1 \cdot a \cdot a} = \frac{a \cdot a \cdot a}{1} \cdot \frac{a \cdot a}{a \cdot a} = \frac{a \cdot a \cdot a}{1} \cdot 1$$
$$= a \cdot a \cdot a = a^3.$$

Note that the exponent in $a^3$ is the difference of those in $a^5 \div a^2$. If we subtract exponents, we get $5 - 2$, which is 3.

> **THE QUOTIENT RULE**
>
> For any nonzero number $a$ and any positive integers $m$ and $n$,
>
> $$\frac{a^m}{a^n} = a^{m-n}.$$
>
> (When dividing with exponential notation, if the bases are the same, keep the base and subtract the exponent of the denominator from the exponent of the numerator.)

**Examples**  Divide and simplify. By simplify, we mean write the expression as one number to a nonnegative power.

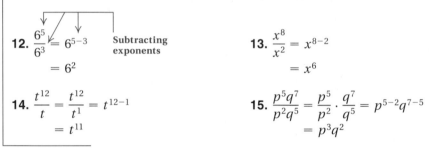

**12.** $\dfrac{6^5}{6^3} = 6^{5-3}$   Subtracting exponents

$\qquad = 6^2$

**13.** $\dfrac{x^8}{x^2} = x^{8-2}$

$\qquad = x^6$

**14.** $\dfrac{t^{12}}{t} = \dfrac{t^{12}}{t^1} = t^{12-1}$

$\qquad = t^{11}$

**15.** $\dfrac{p^5 q^7}{p^2 q^5} = \dfrac{p^5}{p^2} \cdot \dfrac{q^7}{q^5} = p^{5-2} q^{7-5}$

$\qquad = p^3 q^2$

The quotient rule can also be used to explain the definition of 0 as an exponent. Consider the expression $a^4/a^4$, where $a$ is nonzero:

$$\frac{a^4}{a^4} = \frac{a \cdot a \cdot a \cdot a}{a \cdot a \cdot a \cdot a} = 1.$$

This is true because the numerator and the denominator are the same. Now suppose we apply the rule for dividing powers with the same base:

$$\frac{a^4}{a^4} = a^{4-4} = a^0 = 1.$$

Since both expressions $a^4/a^4$ and $a^{4-4}$ are equivalent to 1, it follows that $a^0 = 1$, when $a \neq 0$.

We can explain why we do not define $0^0$ using the quotient rule. We know that $0^0$ is $0^{1-1}$. But $0^{1-1}$ is also equal to $0/0$. We have already seen that division by 0 is undefined, so $0^0$ is also undefined.

***Do Exercises 19–22.***

## f | Negative Integers as Exponents

We can use the rule for dividing powers with like bases to lead us to a definition of exponential notation when the exponent is a negative integer. Consider $5^3/5^7$ and first simplify it using procedures we have learned for working with fractions:

$$\frac{5^3}{5^7} = \frac{5 \cdot 5 \cdot 5}{5 \cdot 5 \cdot 5 \cdot 5 \cdot 5 \cdot 5 \cdot 5} = \frac{5 \cdot 5 \cdot 5 \cdot 1}{5 \cdot 5 \cdot 5 \cdot 5 \cdot 5 \cdot 5 \cdot 5}$$

$$= \frac{5 \cdot 5 \cdot 5}{5 \cdot 5 \cdot 5} \cdot \frac{1}{5 \cdot 5 \cdot 5 \cdot 5} = \frac{1}{5^4}.$$

Now we apply the rule for dividing powers with the same bases. Then

$$\frac{5^3}{5^7} = 5^{3-7} = 5^{-4}.$$

From these two expressions for $5^3/5^7$, it follows that

$$5^{-4} = \frac{1}{5^4}.$$

This leads to our definition of negative exponents:

> For any real number $a$ that is nonzero and any integer $n$,
>
> $$a^{-n} = \frac{1}{a^n}.$$

In fact, the numbers $a^n$ and $a^{-n}$ are reciprocals of each other because

$$a^n \cdot a^{-n} = a^n \cdot \frac{1}{a^n} = \frac{a^n}{a^n} = 1.$$

**Examples** Express using positive exponents. Then simplify.

**16.** $4^{-2} = \dfrac{1}{4^2} = \dfrac{1}{16}$

**17.** $(-3)^{-2} = \dfrac{1}{(-3)^2} = \dfrac{1}{(-3)(-3)} = \dfrac{1}{9}$

**18.** $m^{-3} = \dfrac{1}{m^3}$

**19.** $ab^{-1} = a\left(\dfrac{1}{b^1}\right) = a\left(\dfrac{1}{b}\right) = \dfrac{a}{b}$

**20.** $\dfrac{1}{x^{-3}} = x^{-(-3)} = x^3$

**21.** $3c^{-5} = 3\left(\dfrac{1}{c^5}\right) = \dfrac{3}{c^5}$

---

***CAUTION!*** Note in Example 16 that

$$4^{-2} \neq -16 \quad \text{and} \quad 4^{-2} \neq -\frac{1}{16}.$$

---

*Do Exercises 23–28.*

The rules for multiplying and dividing powers with like bases still hold when exponents are 0 or negative. We will state them in a summary at the end of this section.

Express with positive exponents. Then simplify.

**23.** $4^{-3}$

**24.** $5^{-2}$

**25.** $2^{-4}$

**26.** $(-2)^{-3}$

**27.** $4p^{-3}$

**28.** $\dfrac{1}{x^{-2}}$

*Answers on page A-16*

Simplify.

**29.** $5^{-2} \cdot 5^4$

**30.** $x^{-3} \cdot x^{-4}$

**31.** $\dfrac{7^{-2}}{7^3}$

**32.** $\dfrac{b^{-2}}{b^{-3}}$

**33.** $\dfrac{t}{t^{-5}}$

*Answers on page A-16*

**Examples**   Simplify. By simplify, we generally mean write the expression as one number to a nonnegative power.

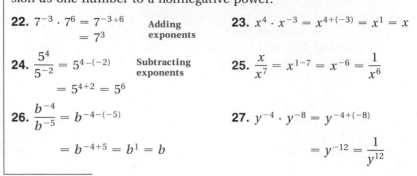

**22.** $7^{-3} \cdot 7^6 = 7^{-3+6}$   Adding
$= 7^3$   exponents

**23.** $x^4 \cdot x^{-3} = x^{4+(-3)} = x^1 = x$

**24.** $\dfrac{5^4}{5^{-2}} = 5^{4-(-2)}$   Subtracting
exponents
$= 5^{4+2} = 5^6$

**25.** $\dfrac{x}{x^7} = x^{1-7} = x^{-6} = \dfrac{1}{x^6}$

**26.** $\dfrac{b^{-4}}{b^{-5}} = b^{-4-(-5)}$
$= b^{-4+5} = b^1 = b$

**27.** $y^{-4} \cdot y^{-8} = y^{-4+(-8)}$
$= y^{-12} = \dfrac{1}{y^{12}}$

In Examples 24–26 (division with exponents), it may help to think as follows: After writing the base, write the top exponent. Then write a subtraction sign. Next write the bottom exponent. Then do the subtraction by adding the opposite. For example,

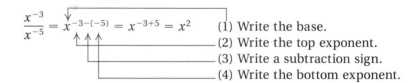

$$\frac{x^{-3}}{x^{-5}} = x^{-3-(-5)} = x^{-3+5} = x^2$$

(1) Write the base.
(2) Write the top exponent.
(3) Write a subtraction sign.
(4) Write the bottom exponent.

*Do Exercises 29–33.*

The following is another way to arrive at the definition of negative exponents.

On each side, we divide by 5 at each step.

$5 \cdot 5 \cdot 5 \cdot 5 = 5^4$
$5 \cdot 5 \cdot 5 = 5^3$
$5 \cdot 5 = 5^2$
$5 = 5^1$
$1 = 5^0$
$\dfrac{1}{5} = 5^?$
$\dfrac{1}{25} = 5^?$

On this side, the exponents decrease by 1.

To continue the pattern, it should follow that

$$\frac{1}{5} = \frac{1}{5^1} = 5^{-1} \quad \text{and} \quad \frac{1}{25} = \frac{1}{5^2} = 5^{-2}.$$

The following is a summary of the definitions and rules for exponents that we have considered in this section.

**DEFINITIONS AND RULES FOR EXPONENTS**

1 as an exponent:        $a^1 = a$;
0 as an exponent:        $a^0 = 1, a \neq 0$;

Negative integers as exponents:   $a^{-n} = \dfrac{1}{a^n}, \dfrac{1}{a^{-n}} = a^n; a \neq 0$

Product Rule:        $a^m \cdot a^n = a^{m+n}$;

Quotient Rule:        $\dfrac{a^m}{a^n} = a^{m-n}, a \neq 0$

# Calculator Spotlight

**Checking Equivalent Expressions.** Let's look at the expressions $x^2 \cdot x^3$ and $x^5$. We know from the product rule, $x^m \cdot x^n = x^{m+n}$, that these expressions are equivalent. In this case, $x^2 \cdot x^3 = x^{2+3} = x^5$ is true for any real-number substitution. This use of the product rule is an algebraic check of the correctness of the statement $x^2 \cdot x^3 = x^5$. How can we check the result using a grapher? We can do it *graphically* by looking at graphs and *numerically* by looking at a table of values.

**Graphical Check.** Let's first do a graphical check of $x^2 \cdot x^3 = x^5$. We consider each expression separately and form two equations to be graphed: $y_1 = x^2 \cdot x^3$ and $y_2 = x^5$. The "$y_1$" is read "y sub 1" and refers simply to a "first" equation. Similarly, "$y_2$" is read "y sub 2" and refers to a "second" equation. We enter these equations into the grapher using the $\boxed{y=}$ key and then graph them, as shown on the left below. Note that the graphs appear to coincide. This is a partial check that the expressions are equivalent. We say "partial check" because most graphs cannot be drawn completely so there is always an element of uncertainty.

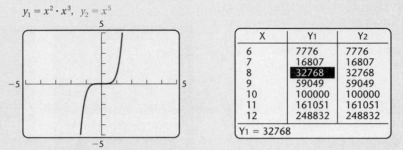

**Numerical Check.** Now let's use the TABLE feature to check $x^2 \cdot x^3 = x^5$. We already have the equations $y_1 = x^2 \cdot x^3$ and $y_2 = x^5$ entered. The TABLE feature allows us to compare $y$-values for various $x$-values. Note in the table on the right above that the $y_1$- and $y_2$-values agree. Thus we have a partial check that we have an identity. We say "partial check" because it is impossible to compute all possible $y$-values and there may be some that disagree.

Let's now consider the equation $x^2 \cdot x^3 = x^6$. Is this a correct result? It seems to violate the product rule, $x^m \cdot x^n = x^{m+n}$. Let's check the equation both graphically and numerically.

**Graphical Check.** We graph $y_1 = x^2 \cdot x^3$ and $y_2 = x^6$, as shown on the left below. On the TI-83, there is a way to choose a graphing style so that the graphs look different when graphed in the same window. See the window in the middle below. It is obvious that the graphs are different. Thus the equation is not correct.

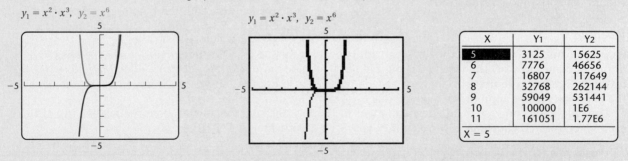

**Numerical Check.** Let's check a table of $y$-values. See the table on the right above. Here we note that the $y_1$- and $y_2$-values are not the same. Thus, $x^2 \cdot x^3 = x^6$ is not correct.

**Exercises** Determine whether each of the following equations is correct.

1. $x \cdot x^2 = x^3$

2. $x \cdot x^2 = x^2$

3. $\dfrac{x^3}{x^2} = x^5$

4. $\dfrac{x^5}{x^2} = x^3$

5. $\left(\dfrac{x}{3}\right)^2 = \dfrac{x^2}{9}$

6. $(5x)^2 = 25x^2$

7. $(x+2)^2 = x^2 + 4$

8. $(x+2)^2 = x^2 + 4x + 4$

9. $x + 3 = 3 + x$

10. $3(x-1) = 3x - 3$

11. $5x - 5 = 5(x-5)$

12. $10x + 20 = 5(2x + 4)$

13. $2 + (3 + x) = (2+3) + x$

14. $5(2x) = 5x$

# Improving Your Math Study Skills

## Classwork: During and After Class

### During Class

#### Asking Questions

Many students are afraid to ask questions in class. You will find that most instructors are not only willing to answer questions during class, but often encourage students to ask questions. In fact, some instructors would like more questions than are offered. Probably your question is one that other students in the class might have been afraid to ask!

Consider waiting for an appropriate time to ask questions. Some instructors will pause to ask the class if they have questions. Use this opportunity to get clarification on any concept you do not understand.

### After Class

#### Restudy Examples and Class Notes

As soon as possible after class, find some time to go over your notes. Read the appropriate sections from the textbook and try to correlate the text with your class notes. You may also want to restudy the examples in the textbook for added comprehension.

Often students make the mistake of doing the homework exercises without reading their notes or textbook. This is not a good idea, since you may lose the opportunity for a complete understanding of the concepts. Simply being able to work the exercises does not ensure that you know the material well enough to work problems on a test.

#### Videotapes

If you can find the time, visit the library, math lab, or media center to view the videotapes on the textbook. Look on the first page of each section in the textbook for the appropriate tape reference.

The videotapes provide detailed explanations of each objective and they may give you a different presentation than the one offered by your instructor. Being able to pause the tape while you take notes or work the examples or replay the tape as many times as you need are additional advantages to using the videos.

Also, consider studying the special tapes *Math Problem Solving in the Real World* and *Math Study Skills* prepared by the author. If these are not available in the media center, contact your instructor.

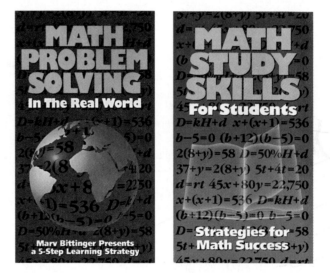

#### Software

If you would like additional practice on any section of the textbook, you can use the accompanying Interact Math Tutorial Software. This software can generate many different versions of the basic odd-numbered exercises for added practice. You can also ask the software to work out each problem step by step.

Ask your instructor about the availability of this software.

# Exercise Set 4.1

**a** What is the meaning of each of the following?

**1.** $3^4$

**2.** $4^3$   $4 \cdot 4 \cdot 4 = $ 64

**3.** $(1.1)^5$

**4.** $(87.2)^6$   $87.2 \cdot 87.2 \cdot 87.2 \cdot 87.2 \cdot 87.2 \cdot 87.2 = 50.42$

**5.** $\left(\dfrac{2}{3}\right)^4$

**6.** $\left(-\dfrac{5}{8}\right)^3$   $\left(\dfrac{5}{8}\right)\left(\dfrac{8}{6}\right) = \dfrac{40}{1}\left(\dfrac{5}{8}\right) = -25$

**7.** $(7p)^2$   $7p \cdot 7p$

**8.** $(11c)^3$

**9.** $8k^3$   $8 \cdot k \cdot k \cdot k$

**10.** $17x^2$

**b** Evaluate.

**11.** $a^0$, $a \neq 0$

**12.** $t^0$, $t \neq 0$

**13.** $b^1$   $b$

**14.** $c^1$

**15.** $\left(\dfrac{2}{3}\right)^0$

**16.** $\left(-\dfrac{5}{8}\right)^0$

**17.** $8.38^0$   $1$

**18.** $8.38^1$

**19.** $(ab)^1$   $ab$

**20.** $(ab)^0$, $a, b \neq 0$

**21.** $ab^1$

**22.** $ab^0$

**c** Evaluate.

**23.** $m^3$, for $m = 3$   $3 \cdot 3 \cdot 3 = 27$

**24.** $x^6$, for $x = 2$

**25.** $p^1$, for $p = 19$

**26.** $x^{19}$, for $x = 0$

**27.** $x^4$, for $x = 4$

**28.** $y^{15}$, for $y = 1$

**29.** $y^2 \div 7$, for $y = -10$   $-10 \cdot -10 = 100 - 7 = 93$

**30.** $z^5 + 5$, for $z = -2$

**31.** $x^1 + 3$ and $x^0 + 3$, for $x = 7$

**32.** $y^0 - 8$ and $y^1 - 8$, for $y = -3$

**33.** Find the area of a circle when $r = 34$ ft. Use 3.14 for $\pi$.

**34.** The area $A$ of a square with sides of length $s$ is given by $A = s^2$. Find the area of a square with sides of length 24 m.

**f** Express using positive exponents. Then simplify.

**35.** $3^{-2}$

$\frac{1}{3^2}$ or $\frac{1}{9}$

**36.** $2^{-3}$

**37.** $10^{-3}$

**38.** $5^{-4}$

**39.** $7^{-3}$

**40.** $5^{-2}$

**41.** $a^{-3}$

**42.** $x^{-2}$

**43.** $\frac{1}{8^{-2}}$

**44.** $\frac{1}{2^{-5}}$

**45.** $\frac{1}{y^{-4}}$

**46.** $\frac{1}{t^{-7}}$

**47.** $\frac{1}{z^{-n}}$

**48.** $\frac{1}{h^{-n}}$

Express using negative exponents.

**49.** $\frac{1}{4^3}$

**50.** $\frac{1}{5^2}$

**51.** $\frac{1}{x^3}$

**52.** $\frac{1}{y^2}$

**53.** $\frac{1}{a^5}$

**54.** $\frac{1}{b^7}$

**d** , **f** Multiply and simplify.

**55.** $2^4 \cdot 2^3$

**56.** $3^5 \cdot 3^2$

**57.** $8^5 \cdot 8^9$

**58.** $n^3 \cdot n^{20}$

**59.** $x^4 \cdot x^3$

**60.** $y^7 \cdot y^9$

**61.** $9^{17} \cdot 9^{21}$

**62.** $t^0 \cdot t^{16}$

**63.** $(3y)^4(3y)^8$

**64.** $(2t)^8(2t)^{17}$

**65.** $(7y)^1(7y)^{16}$

**66.** $(8x)^0(8x)^1$

**67.** $3^{-5} \cdot 3^8$

**68.** $5^{-8} \cdot 5^9$

**69.** $x^{-2} \cdot x$

**70.** $x \cdot x^{-1}$

Copyright © 1999 Addison Wesley Longman

**71.** $x^{14} \cdot x^3$

**72.** $x^9 \cdot x^4$

**73.** $x^{-7} \cdot x^{-6}$

**74.** $y^{-5} \cdot y^{-8}$

**75.** $a^{11} \cdot a^{-3} \cdot a^{-18}$

**76.** $a^{-11} \cdot a^{-3} \cdot a^{-7}$

**77.** $t^8 \cdot t^{-8}$

**78.** $m^{10} \cdot m^{-10}$

e , f  Divide and simplify.

**79.** $\dfrac{7^5}{7^2}$

**80.** $\dfrac{5^8}{5^6}$

**81.** $\dfrac{8^{12}}{8^6}$

**82.** $\dfrac{8^{13}}{8^2}$

**83.** $\dfrac{y^9}{y^5}$

**84.** $\dfrac{x^{11}}{x^9}$

**85.** $\dfrac{16^2}{16^8}$ $\dfrac{1}{16^6}$

**86.** $\dfrac{7^2}{7^9}$

**87.** $\dfrac{m^6}{m^{12}}$

**88.** $\dfrac{a^3}{a^4}$

**89.** $\dfrac{(8x)^6}{(8x)^{10}}$

**90.** $\dfrac{(8t)^4}{(8t)^{11}}$

**91.** $\dfrac{(2y)^9}{(2y)^9}$

**92.** $\dfrac{(6y)^7}{(6y)^7}$

**93.** $\dfrac{x}{x^{-1}} = x^2$

**94.** $\dfrac{y^8}{y}$

**95.** $\dfrac{x^7}{x^{-2}}$

**96.** $\dfrac{t^8}{t^{-3}}$

**97.** $\dfrac{z^{-6}}{z^{-2}}$

**98.** $\dfrac{x^{-9}}{x^{-3}}$

End

**99.** $\dfrac{x^{-5}}{x^{-8}}$

**100.** $\dfrac{y^{-2}}{y^{-9}}$

**101.** $\dfrac{m^{-9}}{m^{-9}}$

**102.** $\dfrac{x^{-7}}{x^{-7}}$

Simplify.

**103.** $5^2$, $5^{-2}$, $\left(\dfrac{1}{5}\right)^2$, $\left(\dfrac{1}{5}\right)^{-2}$, $-5^2$, and $(-5)^2$

**104.** $8^2$, $8^{-2}$, $\left(\dfrac{1}{8}\right)^2$, $\left(\dfrac{1}{8}\right)^{-2}$, $-8^2$, and $(-8)^2$

### Skill Maintenance

**105.** Translate to an algebraic expression: Sixty-four percent of $t$.  [1.1b]

**106.** Evaluate $3x/y$ for $x = 4$ and $y = 12$.  [1.1a]

**107.** Divide: $1555.2 \div 24.3$.  [1.6c]

**108.** Add: $1555.2 + 24.3$.  [1.3a]

**109.** Solve: $3x - 4 + 5x - 10x = x - 8$.  [2.3b]

**110.** Factor: $8x - 56$.  [1.7d]

Solve.  [2.4a]

**111.** A 12-in. submarine sandwich is cut into two pieces. One piece is twice as long as the other. How long are the pieces?

**112.** A book is opened. The sum of the page numbers on the facing pages is 457. Find the page numbers.

### Synthesis

**113.** ◈ Under what conditions does $a^n$ represent a negative number? Why?

**114.** ◈ Explain the errors in each of the following.

    **a)** $2^{-3} = \dfrac{1}{-8}$       **b)** $m^{-2}m^5 = m^{10}$

⌁ Determine whether each of the following is correct.

**115.** $(x + 1)^2 = x^2 + 1$

**116.** $(x - 1)^2 = x^2 - 2x + 1$

**117.** $(5x)^0 = 5x^0$

**118.** $\dfrac{x^3}{x^5} = x^2$

Simplify.

**119.** $(y^{2x})(y^{3x})$

**120.** $a^{5k} \div a^{3k}$

**121.** $\dfrac{a^{6t}(a^{7t})}{a^{9t}}$

**122.** $\dfrac{\left(\frac{1}{2}\right)^4}{\left(\frac{1}{2}\right)^5}$

**123.** $\dfrac{(0.8)^5}{(0.8)^3(0.8)^2}$

**124.** Determine whether $(a + b)^2$ and $a^2 + b^2$ are equivalent. (*Hint*: Choose values for $a$ and $b$ and evaluate.)

Use $>$, $<$, or $=$ for ▇ to write a true sentence.

**125.** $3^5$ ▇ $3^4$

**126.** $4^2$ ▇ $4^3$

**127.** $4^3$ ▇ $5^3$

**128.** $4^3$ ▇ $3^4$

Find a value of the variable that shows that the two expressions are *not* equivalent.

**129.** $3x^2$;  $(3x)^2$

**130.** $\dfrac{x + 2}{2}$;  $x$

Copyright © 1999 Addison Wesley Longman

# 4.2 Exponents and Scientific Notation

We now enhance our ability to manipulate exponential expressions by considering three more rules. The rules are also applied to a new way to name numbers called *scientific notation*.

## a | Raising Powers to Powers

Consider an expression like $(3^2)^4$. We are raising $3^2$ to the fourth power:

$$(3^2)^4 = (3^2)(3^2)(3^2)(3^2)$$
$$= (3 \cdot 3)(3 \cdot 3)(3 \cdot 3)(3 \cdot 3)$$
$$= 3 \cdot 3 \cdot 3 \cdot 3 \cdot 3 \cdot 3 \cdot 3 \cdot 3$$
$$= 3^8.$$

Note that in this case we could have multiplied the exponents:

$$(3^2)^4 = 3^{2 \cdot 4} = 3^8.$$

Likewise, $(y^8)^3 = (y^8)(y^8)(y^8) = y^{24}$. Once again, we get the same result if we multiply the exponents:

$$(y^8)^3 = y^{8 \cdot 3} = y^{24}.$$

---

> **THE POWER RULE**
>
> For any real number $a$ and any integers $m$ and $n$,
> $$(a^m)^n = a^{mn}.$$
> (To raise a power to a power, multiply the exponents.)

---

**Examples**  Simplify. Express the answers using positive exponents.

**1.** $(3^5)^4 = 3^{5 \cdot 4}$  **Multiplying exponents**
$= 3^{20}$

**2.** $(2^2)^5 = 2^{2 \cdot 5} = 2^{10}$

**3.** $(y^{-5})^7 = y^{-5 \cdot 7} = y^{-35} = \dfrac{1}{y^{35}}$

**4.** $(x^4)^{-2} = x^{4(-2)} = x^{-8} = \dfrac{1}{x^8}$

**5.** $(a^{-4})^{-6} = a^{(-4)(-6)} = a^{24}$

*Do Exercises 1–4.*

## b | Raising a Product or a Quotient to a Power

When an expression inside parentheses is raised to a power, the inside expression is the base. Let's compare $2a^3$ and $(2a)^3$:

$2a^3 = 2 \cdot a \cdot a \cdot a$;  The base is $a$.

$(2a)^3 = (2a)(2a)(2a)$  The base is $2a$.

$\quad\quad = (2 \cdot 2 \cdot 2)(a \cdot a \cdot a)$  Using the associative and commutative laws of multiplication to regroup the factors

$\quad\quad = 2^3 a^3$

$\quad\quad = 8a^3.$

We see that $2a^3$ and $(2a)^3$ are *not* equivalent. We also see that we can evaluate the power $(2a)^3$ by raising each factor to the power 3. This leads us to the following rule for raising a product to a power.

**Objectives**

a  Use the power rule to raise powers to powers.

b  Raise a product to a power and a quotient to a power.

c  Convert between scientific notation and decimal notation.

d  Multiply and divide using scientific notation.

e  Solve applied problems using scientific notation.

**For Extra Help**

TAPE 7   TAPE 7B   MAC WIN   CD-ROM

---

### Calculator Spotlight

 **Exercises**

Determine both graphically and numerically whether each of the following is correct. That is, use the GRAPH and TABLE features on your grapher.

**1.** $(x^2)^3 = x^6$

**2.** $(x^2)^3 = x^5$

---

Simplify. Express the answers using positive exponents.

**1.** $(3^4)^5$

**2.** $(x^{-3})^4$

**3.** $(y^{-5})^{-3}$

**4.** $(x^4)^{-8}$

*Answers on page A-16*

---

Simplify.

**5.** $(2x^5y^{-3})^4$

**6.** $(5x^5y^{-6}z^{-3})^2$

**7.** $[(-x)^{37}]^2$

**8.** $(3y^{-2}x^{-5}z^8)^3$

Simplify.

**9.** $\left(\dfrac{x^6}{5}\right)^2$

**10.** $\left(\dfrac{2t^5}{w^4}\right)^3$

**11.** $\left(\dfrac{x^4}{3}\right)^{-2}$

## Calculator Spotlight

### Exercises

Determine both graphically and numerically whether each of the following is correct. That is, use the GRAPH and TABLE features on your grapher.

**1.** $(3x)^2 = 9x^2$

**2.** $\left(\dfrac{x}{4}\right)^2 = \dfrac{x^2}{16}$

**3.** $(2x)^3 = 2x^3$

**4.** $\left(\dfrac{x}{5}\right)^2 = \dfrac{x^2}{5}$

*Answers on page A-16*

> **RAISING A PRODUCT TO A POWER**
>
> For any real numbers $a$ and $b$ and any integer $n$,
> $$(ab)^n = a^n b^n.$$
> (To raise a product to the $n$th power, raise each factor to the $n$th power.)

## Examples

**6.** $(4x^2)^3 = 4^3 \cdot (x^2)^3$    Raising each factor to the third power
$$= 64x^6$$

**7.** $(5x^3y^5z^2)^4 = 5^4(x^3)^4(y^5)^4(z^2)^4$    Raising each factor to the fourth power
$$= 625x^{12}y^{20}z^8$$

**8.** $(-5x^4y^3)^3 = (-5)^3(x^4)^3(y^3)^3$
$$= -125x^{12}y^9$$

**9.** $[(-x)^{25}]^2 = (-x)^{50}$    Using the power rule
$= (-1 \cdot x)^{50}$    Using the property of $-1$ (Section 1.8)
$= (-1)^{50}x^{50}$
$= 1 \cdot x^{50}$    The product of an even number of negative factors is positive.
$= x^{50}$

**10.** $(5x^2y^{-2})^3 = 5^3(x^2)^3(y^{-2})^3 = 125x^6y^{-6}$    Be sure to raise *each* factor to the third power.
$$= \dfrac{125x^6}{y^6}$$

**11.** $(3x^3y^{-5}z^2)^4 = 3^4(x^3)^4(y^{-5})^4(z^2)^4$
$$= 81x^{12}y^{-20}z^8 = \dfrac{81x^{12}z^8}{y^{20}}$$

*Do Exercises 5–8.*

There is a similar rule for raising a quotient to a power.

> **RAISING A QUOTIENT TO A POWER**
>
> For any real numbers $a$ and $b$, $b \neq 0$, and any integer $n$,
> $$\left(\dfrac{a}{b}\right)^n = \dfrac{a^n}{b^n}.$$
> (To raise a quotient to the $n$th power, raise both the numerator and the denominator to the $n$th power.)

## Examples  Simplify.

**12.** $\left(\dfrac{x^2}{4}\right)^3 = \dfrac{(x^2)^3}{4^3} = \dfrac{x^6}{64}$

**13.** $\left(\dfrac{3a^4}{b^3}\right)^2 = \dfrac{(3a^4)^2}{(b^3)^2} = \dfrac{3^2(a^4)^2}{b^{3 \cdot 2}} = \dfrac{9a^8}{b^6}$

**14.** $\left(\dfrac{y^3}{5}\right)^{-2} = \dfrac{(y^3)^{-2}}{5^{-2}} = \dfrac{y^{-6}}{5^{-2}} = \dfrac{\dfrac{1}{y^6}}{\dfrac{1}{5^2}} = \dfrac{1}{y^6} \div \dfrac{1}{5^2} = \dfrac{1}{y^6} \cdot \dfrac{5^2}{1} = \dfrac{25}{y^6}$

*Do Exercises 9–11.*

## c | Scientific Notation

There are many kinds of symbols, or notation, for numbers. You are already familiar with fractional notation, decimal notation, and percent notation. Now we study another, **scientific notation,** which is especially useful when calculations involve very large or very small numbers. The following are examples of scientific notation:

*Niagara Falls*: On the Canadian side, during the summer the amount of water that spills over the falls in 1 min is about

$1.3088 \times 10^8$ L = 130,880,000 L.  *move decimal 8 places to Right.*

*The mass of a hydrogen atom*:

$1.7 \times 10^{-24}$ g = 0.0000000000000000000000017 g.  *move decimal 23 places to left.*

> **Scientific notation** for a number is an expression of the type
>
> $$M \times 10^n,$$
>
> where $n$ is an integer, $M$ is greater than or equal to 1 and less than 10 ($1 \le M < 10$), and $M$ is expressed in decimal notation. $10^n$ is also considered to be scientific notation when $M = 1$.

You should try to make conversions to scientific notation mentally as much as possible. Here is a handy mental device.

> A positive exponent in scientific notation indicates a large number (greater than 1) and a negative exponent indicates a small number (less than 1).

**Examples**   Convert to scientific notation.

**15.** $78,000 = 7.8 \times 10^4$       7.8,000.
                                        4 places

Large number, so the exponent is positive.

**16.** $0.0000057 = 5.7 \times 10^{-6}$     0.000005.7
                                        6 places

Small number, so the exponent is negative.

Each of the following is *not* scientific notation.

$$\underbrace{12.46}_{} \times 10^7 \qquad\qquad \underbrace{0.347}_{} \times 10^{-5}$$

This number is greater than 10.   This number is less than 1.

*Do Exercises 12 and 13.*

**Examples**   Convert mentally to decimal notation.

**17.** $7.893 \times 10^5 = 789,300$      7.89300.
                                        5 places

Positive exponent, so the answer is a large number.

**18.** $4.7 \times 10^{-8} = 0.000000047$     0.00000004.7
*to left cuz (−) xponet*                   8 places
Negative exponent, so the answer is a small number.

Convert to scientific notation.
**12.** 0.000517

**13.** 523,000,000

Convert to decimal notation.
**14.** $6.893 \times 10^{11}$

**15.** $5.67 \times 10^{-5}$

*Answers on page A-16*

*Do Exercises 14 and 15 on the preceding page.*

## Calculator Spotlight

To enter a number in scientific notation into a graphing calculator, we first enter the decimal portion of the number and then press 2nd EE followed by the exponent.

For example, to enter $1.789 \times 10^{-11}$, we press

| 1 | . | 7 | 8 | 9 |

| 2nd | EE | (−) | 1 | 1 |

The display reads,

| 1.789 E −11 |

**Exercises**

Enter in scientific notation.

**1.** 260,000,000

**2.** 0.00000000006709

Multiply and write scientific notation for the result.

**16.** $(1.12 \times 10^{-8})(5 \times 10^{-7})$

**17.** $(9.1 \times 10^{-17})(8.2 \times 10^{3})$

---

## d | Multiplying and Dividing Using Scientific Notation

### Multiplying

Consider the product

$$400 \cdot 2000 = 800{,}000.$$

In scientific notation, this is

$$(4 \times 10^2) \cdot (2 \times 10^3) = (4 \cdot 2)(10^2 \cdot 10^3) = 8 \times 10^5.$$

By applying the commutative and associative laws, we can find this product by multiplying $4 \cdot 2$, to get 8, and $10^2 \cdot 10^3$, to get $10^5$ (we do this by adding the exponents).

**Example 19**  Multiply: $(1.8 \times 10^6) \cdot (2.3 \times 10^{-4})$.

We apply the commutative and associative laws to get

$$
\begin{aligned}
(1.8 \times 10^6) \cdot (2.3 \times 10^{-4}) &= (1.8 \cdot 2.3) \times (10^6 \cdot 10^{-4}) \\
&= 4.14 \times 10^{6+(-4)} \quad \text{**Adding exponents**} \\
&= 4.14 \times 10^2.
\end{aligned}
$$

**Example 20**  Multiply: $(3.1 \times 10^5) \cdot (4.5 \times 10^{-3})$.

We have

$$
\begin{aligned}
(3.1 \times 10^5) \cdot (4.5 \times 10^{-3}) &= (3.1 \times 4.5)(10^5 \cdot 10^{-3}) \\
&= 13.95 \times 10^2.
\end{aligned}
$$

The answer at this stage is $13.95 \times 10^2$, but this is *not* scientific notation, because 13.95 is not a number between 1 and 10. To find scientific notation for the product, we convert 13.95 to scientific notation and simplify:

$$
\begin{aligned}
13.95 \times 10^2 &= (1.395 \times 10^1) \times 10^2 \quad \text{**Substituting $1.395 \times 10^1$ for 13.95**} \\
&= 1.395 \times (10^1 \times 10^2) \quad \text{**Associative law**} \\
&= 1.395 \times 10^3. \quad \text{**Adding exponents**}
\end{aligned}
$$

The answer is

$$1.395 \times 10^3.$$

*Do Exercises 16 and 17.*

### Dividing

Consider the quotient

$$800{,}000 \div 400 = 2000.$$

In scientific notation, this is

$$(8 \times 10^5) \div (4 \times 10^2) = \frac{8 \times 10^5}{4 \times 10^2} = \frac{8}{4} \times \frac{10^5}{10^2} = 2 \times 10^3.$$

We can find this product by dividing 8 by 4, to get 2, and $10^5$ by $10^2$, to get $10^3$ (we do this by subtracting the exponents).

**Example 21** Divide: $(3.41 \times 10^5) \div (1.1 \times 10^{-3})$.

$$(3.41 \times 10^5) \div (1.1 \times 10^{-3}) = \frac{3.41 \times 10^5}{1.1 \times 10^{-3}}$$
$$= \frac{3.41}{1.1} \times \frac{10^5}{10^{-3}}$$
$$= 3.1 \times 10^{5-(-3)}$$
$$= 3.1 \times 10^8$$

**Example 22** Divide: $(6.4 \times 10^{-7}) \div (8.0 \times 10^6)$.

We have

$$(6.4 \times 10^{-7}) \div (8.0 \times 10^6) = \frac{6.4 \times 10^{-7}}{8.0 \times 10^6}$$
$$= \frac{6.4}{8.0} \times \frac{10^{-7}}{10^6}$$
$$= 0.8 \times 10^{-7-6}$$
$$= 0.8 \times 10^{-13}.$$

The answer at this stage is

$$0.8 \times 10^{-13},$$

but this is *not* scientific notation, because 0.8 is not a number between 1 and 10. To find scientific notation for the quotient, we convert 0.8 to scientific notation and simplify:

$$0.8 \times 10^{-13} = (8.0 \times 10^{-1}) \times 10^{-13} \qquad \text{Substituting } 8.0 \times 10^{-1} \text{ for } 0.8$$
$$= 8.0 \times (10^{-1} \times 10^{-13}) \qquad \text{Associative law}$$
$$= 8.0 \times 10^{-14}. \qquad \text{Adding exponents}$$

The answer is

$$8.0 \times 10^{-14}.$$

***Do Exercises 18 and 19.***

## e | Applications with Scientific Notation

**Example 23** *Distance from the Sun to Earth.* Light from the sun traveling at a rate of 300,000 kilometers per second (km/s) reaches Earth in 499 sec. Find the distance, expressed in scientific notation, from the sun to Earth.

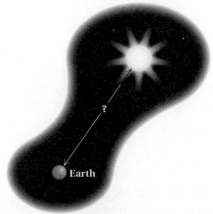
Earth

Divide and write scientific notation for the result.

**18.** $\dfrac{4.2 \times 10^5}{2.1 \times 10^2}$

**19.** $\dfrac{1.1 \times 10^{-4}}{2.0 \times 10^{-7}}$

*Answers on page A-16*

**20.** *Niagara Falls Water Flow.* On the Canadian side, during the summer the amount of water that spills over the falls in 1 min is about

$$1.3088 \times 10^8 \text{ L}.$$

How much water spills over the falls in one day? Express the answer in scientific notation.

**21.** *Earth vs. Saturn.* The mass of Earth is about $6 \times 10^{21}$ metric tons. The mass of Saturn is about $5.7 \times 10^{23}$ metric tons. About how many times the mass of Earth is the mass of Saturn? Express the answer in scientific notation.

Earth     Jupiter

*Answers on page A-16*

---

The time $t$ that it takes for light to reach Earth from the sun is $4.99 \times 10^2$ sec (s). The speed is $3.0 \times 10^5$ km/s. Recall that distance can be expressed in terms of speed and time as

$$\text{Distance} = \text{Speed} \cdot \text{Time}$$
$$d = rt.$$

We substitute $3.0 \times 10^5$ for $r$ and $4.99 \times 10^2$ for $t$:

$$
\begin{aligned}
d &= rt \\
&= (3.0 \times 10^5)(4.99 \times 10^2) \qquad \text{Substituting} \\
&= 14.97 \times 10^7 \\
&= 1.497 \times 10^8 \text{ km.} \qquad \text{Converting to scientific notation}
\end{aligned}
$$

Thus the distance from the sun to Earth is $1.497 \times 10^8$ km.

*Do Exercise 20.*

**Example 24**   *Earth vs. Jupiter.* The mass of Earth is about $6 \times 10^{21}$ metric tons. The mass of Jupiter is about $1.908 \times 10^{24}$ metric tons. About how many times the mass of Earth is the mass of Jupiter? Express the answer in scientific notation.

To determine how many times the mass of Jupiter is of the mass of Earth, we divide the mass of Jupiter by the mass of Earth:

$$
\begin{aligned}
\frac{1.908 \times 10^{24}}{6 \times 10^{21}} &= \frac{1.908}{6} \times \frac{10^{24}}{10^{21}} \\
&= 0.318 \times 10^3 \\
&= (3.18 \times 10^{-1}) \times 10^3 \\
&= 3.18 \times 10^2.
\end{aligned}
$$

Thus the mass of Jupiter is $3.18 \times 10^2$, or 318, times the mass of Earth.

*Do Exercise 21.*

The following is a summary of the definitions and rules for exponents that we have considered in this section and the preceding one.

| **DEFINITIONS AND RULES FOR EXPONENTS** | |
| --- | --- |
| Exponent of 1: | $a^1 = a$ |
| Exponent of 0: | $a^0 = 1, a \neq 0$ |
| Negative exponents: | $a^{-n} = \dfrac{1}{a^n}, a \neq 0$ |
| Product Rule: | $a^m \cdot a^n = a^{m+n}$ |
| Quotient Rule: | $\dfrac{a^m}{a^n} = a^{m-n}, a \neq 0$ |
| Power Rule: | $(a^m)^n = a^{mn}$ |
| Raising a product to a power: | $(ab)^n = a^n b^n$ |
| Raising a quotient to a power: | $\left(\dfrac{a}{b}\right)^n = \dfrac{a^n}{b^n}, b \neq 0$ |
| Scientific notation: | $M \times 10^n$, or $10^n$, where $1 \leq M < 10$ |

# Exercise Set 4.2

**a**, **b** Simplify.

**1.** $(2^3)^2$

**2.** $(5^2)^4$

**3.** $(5^2)^{-3}$

**4.** $(7^{-3})^5$

**5.** $(x^{-3})^{-4}$

**6.** $(a^{-5})^{-6}$

**7.** $(4x^3)^2$

**8.** $4(x^3)^2$

**9.** $(x^4y^5)^{-3}$

**10.** $(t^5x^3)^{-4}$

**11.** $(x^{-6}y^{-2})^{-4}$

**12.** $(x^{-2}y^{-7})^{-5}$

**13.** $(3x^3y^{-8}z^{-3})^2$

**14.** $(2a^2y^{-4}z^{-5})^3$

**15.** $\left(\dfrac{a^2}{b^3}\right)^4$

**16.** $\left(\dfrac{x^3}{y^4}\right)^5$

**17.** $\left(\dfrac{y^3}{2}\right)^2$

**18.** $\left(\dfrac{a^5}{3}\right)^3$

**19.** $\left(\dfrac{y^2}{2}\right)^{-3}$

**20.** $\left(\dfrac{a^4}{3}\right)^{-2}$

**21.** $\left(\dfrac{x^2y}{z}\right)^3$

**22.** $\left(\dfrac{m}{n^4p}\right)^3$

**23.** $\left(\dfrac{a^2b}{cd^3}\right)^{-2}$

**24.** $\left(\dfrac{2a^2}{3b^4}\right)^{-3}$

c Convert to scientific notation.

**25.** 28,000,000,000

**26.** 4,900,000,000,000

**27.** 907,000,000,000,000,000

**28.** 168,000,000,000,000

**29.** 0.00000304

**30.** 0.000000000865

**31.** 0.000000018

**32.** 0.00000000002

**33.** 100,000,000,000

**34.** 0.0000001

Convert the number in the sentence to scientific notation.

**35.** *Niagara Falls Water Flow.* On the American side, during the summer the amount of water that spills over the falls in 1 min is about 11.35 million L (1 million = $10^6$).

**36.** *Proctor & Gamble.* In a recent year, Proctor & Gamble led the nation's advertisers by spending $2.777 billion on advertising (***Source:*** *Advertising Age*) (1 billion = $10^9$).

Convert to decimal notation.

**37.** $8.74 \times 10^7$

**38.** $1.85 \times 10^8$

**39.** $5.704 \times 10^{-8}$

**40.** $8.043 \times 10^{-4}$

**41.** $10^7$

**42.** $10^6$

**43.** $10^{-5}$

**44.** $10^{-8}$

d Multiply or divide and write scientific notation for the result.

**45.** $(3 \times 10^4)(2 \times 10^5)$

**46.** $(3.9 \times 10^8)(8.4 \times 10^{-3})$

**47.** $(5.2 \times 10^5)(6.5 \times 10^{-2})$

**48.** $(7.1 \times 10^{-7})(8.6 \times 10^{-5})$

**49.** $(9.9 \times 10^{-6})(8.23 \times 10^{-8})$

**50.** $(1.123 \times 10^4) \times 10^{-9}$

**51.** $\dfrac{8.5 \times 10^8}{3.4 \times 10^{-5}}$

**52.** $\dfrac{5.6 \times 10^{-2}}{2.5 \times 10^5}$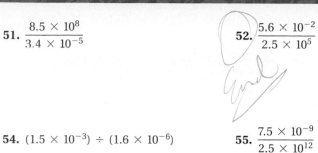

**53.** $(3.0 \times 10^6) \div (6.0 \times 10^9)$

**54.** $(1.5 \times 10^{-3}) \div (1.6 \times 10^{-6})$

**55.** $\dfrac{7.5 \times 10^{-9}}{2.5 \times 10^{12}}$

**56.** $\dfrac{4.0 \times 10^{-3}}{8.0 \times 10^{20}}$

**e** Solve.

**57.** *Total Income of Two-Person Households.* In 1993, there were about 31.2 million two-person households in the United States. The average income of these households was about $42,400. (**Source:** Statistical Abstract of the United States) Find the total income generated by two-person households in 1993. Express the answer in scientific notation.

**58.** *Niagara Falls Water Flow.* On the American side, during the summer the amount of water that spills over the falls in 1 min is about 11.35 million L (1 million = $10^6$). How much water spills over the falls in 1 yr? (Use 365 days for 1 yr.) Express the answer in scientific notation.

**59.** *Stars.* It is estimated that there are 10 billion trillion stars in the known universe. Express the number of stars in scientific notation.

**60.** *Closest Star.* Excluding the sun, the closest star to Earth is Proxima Centauri, which is 4.3 light-years away (one light-year = $5.88 \times 10^{12}$ mi). How far, in miles, is Proxima Centauri from Earth? Express the answer in scientific notation.

**61.** *Earth vs. Sun.* The mass of Earth is about $6 \times 10^{21}$ metric tons. The mass of the sun is about $1.998 \times 10^{27}$ metric tons. About how times the mass of Earth is the mass of the sun? Express the answer in scientific notation.

**62.** *Red Light.* The wavelength of light is given by the velocity divided by the frequency. The velocity of red light is 300,000,000 m/sec, and its frequency is 400,000,000,000,000 cycles per second. What is the wavelength of red light? Express the answer in scientific notation.

*Space Travel.* Use the following information for Exercises 63 and 64.

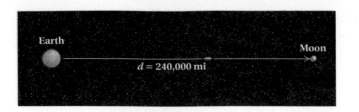

| Approximate Distance from Earth to: | |
|---|---|
| Moon | 240,000 mi |
| Mars | 35,000,000 mi |
| Pluto | 2,670,000,000 mi |

**63.** *Time to Reach Mars.* Suppose that it takes about 3 days for a space vehicle to travel from Earth to the moon. About how long would it take the same space vehicle traveling at the same speed to reach Mars? Express the answer in scientific notation.

**64.** *Time to Reach Pluto.* Suppose that it takes about 3 days for a space vehicle to travel from Earth to the moon. About how long would it take the same space vehicle traveling at the same speed to reach Pluto? Express the answer in scientific notation.

## Skill Maintenance

Factor.   [1.7d]

**65.** $9x - 36$

**66.** $4x - 2y + 16$

**67.** $3s + 3t + 24$

**68.** $-7x - 14$

Solve.   [2.3b]

**69.** $2x - 4 - 5x + 8 = x - 3$

**70.** $8x + 7 - 9x = 12 - 6x + 5$

Solve.   [2.3c]

**71.** $8(2x + 3) - 2(x - 5) = 10$

**72.** $4(x - 3) + 5 = 6(x + 2) - 8$

Graph.   [3.2b], [3.3a]

**73.** $y = x - 5$

**74.** $2x + y = 8$

## Synthesis

**75.** ◈ Using the quotient rule, explain why $9^0$ is defined to be 1.

**76.** ◈ Explain in your own words when exponents should be added and when they should be multiplied.

**77.** ▦ Carry out the indicated operations. Express the result in scientific notation.

$$\frac{(5.2 \times 10^6)(6.1 \times 10^{-11})}{1.28 \times 10^{-3}}$$

**78.** Find the reciprocal and express it in scientific notation.

$$6.25 \times 10^{-3}$$

Simplify.

**79.** $\dfrac{(5^{12})^2}{5^{25}}$

**80.** $\dfrac{a^{22}}{(a^2)^{11}}$

**81.** $\dfrac{(3^5)^4}{3^5 \cdot 3^4}$

**82.** $\dfrac{49^{18}}{7^{35}}$

**83.** $\left(\dfrac{1}{a}\right)^{-n}$

(*Hint*: Study Exercise 80.)

**84.** $\dfrac{(0.4)^5}{[(0.4)^3]^2}$

Determine whether each of the following is true for any pairs of integers $m$ and $n$ and any positive numbers $x$ and $y$.

**85.** $x^m \cdot y^n = (xy)^{mn}$

**86.** $x^m \cdot y^m = (xy)^{2m}$

**87.** $(x - y)^m = x^m - y^m$

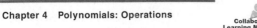

Use exponential and scientific notation to represent the salary for a job.

Collaborative Learning Manual

Copyright © 1999 Addison Wesley Longman

# 4.3 Introduction to Polynomials

We have already learned to evaluate and to manipulate certain kinds of algebraic expressions. We will now consider algebraic expressions called *polynomials*.

The following are examples of *monomials in one variable*:

$$3x^2, \quad 2x, \quad -5, \quad 37p^4, \quad 0.$$

Each expression is a constant or a constant times some variable to a non-negative integer power.

*real # constant*
$3x^2$ — *nonnegative integer*

> A **monomial** is an expression of the type $ax^n$, where $a$ is a real-number constant and $n$ is a nonnegative integer.

Algebraic expressions like the following are **polynomials:**

$$\tfrac{3}{4}y^5, \quad -2, \quad 5y + 3, \quad 3x^2 + 2x - 5, \quad -7a^3 + \tfrac{1}{2}a, \quad 6x, \quad 37p^4, \quad x, \quad 0.$$

> A **polynomial** is a monomial or a combination of sums and/or differences of monomials.

The following algebraic expressions are *not* polynomials:

$$\textbf{(1)} \ \frac{x + 3}{x - 4}, \qquad \textbf{(2)} \ 5x^3 - 2x^2 + \frac{1}{x}, \qquad \textbf{(3)} \ \frac{1}{x^3 - 2}.$$

Expressions (1) and (3) are not polynomials because they represent quotients, not sums. Expression (2) is not a polynomial because

$$\frac{1}{x} = x^{-1},$$

and this is not a monomial because the exponent is negative.

*Do Exercise 1.*

## a  Evaluating Polynomials and Applications

When we replace the variable in a polynomial with a number, the polynomial then represents a number called a **value** of the polynomial. Finding that number, or value, is called **evaluating the polynomial.** We evaluate a polynomial using the rules for order of operations (Section 1.8).

**Example 1**  Evaluate the polynomial for $x = 2$.

a) $3x + 5 = 3 \cdot 2 + 5$
$\qquad\quad = 6 + 5$
$\qquad\quad = 11$

b) $2x^2 - 7x + 3 = 2 \cdot 2^2 - 7 \cdot 2 + 3$
$\qquad\qquad\qquad = 2 \cdot 4 - 7 \cdot 2 + 3$
$\qquad\qquad\qquad = 8 - 14 + 3$
$\qquad\qquad\qquad = -3$

## Objectives

| | |
|---|---|
| **a** | Evaluate a polynomial for a given value of the variable. |
| **b** | Identify the terms of a polynomial. |
| **c** | Identify the like terms of a polynomial. |
| **d** | Identify the coefficients of a polynomial. |
| **e** | Collect the like terms of a polynomial. |
| **f** | Arrange a polynomial in descending order, or collect the like terms and then arrange in descending order. |
| **g** | Identify the degree of each term of a polynomial and the degree of the polynomial. |
| **h** | Identify the missing terms of a polynomial. |
| **i** | Classify a polynomial as a monomial, binomial, trinomial, or none of these. |

**For Extra Help**

| TAPE 7 | TAPE 7B | MAC WIN | CD-ROM |

1. Write three polynomials.

*Answer on page A-17*

Evaluate the polynomial for $x = 3$.

**2.** $-4x - 7$

**3.** $-5x^3 + 7x + 10$

Evaluate the polynomial for $x = -4$.

**4.** $5x + 7$

**5.** $2x^2 + 5x - 4$

**6.** Referring to Example 3, what is the total number of games to be played in a league of 12 teams?

**7.** *Perimeter of Baseball Diamond.* The perimeter of a square of side $x$ is given by the polynomial $4x$.

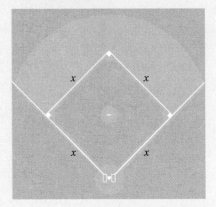

A baseball diamond is a square 90 ft on a side. Find the perimeter of a baseball diamond.

**8.** Use *only* the graph shown in Example 4 to evaluate the polynomial $2x - 2$ for $x = 4$ and for $x = -1$.

*Answers on page A-17*

**Example 2**   Evaluate the polynomial for $x = -4$.

a) $2 - x^3 = 2 - (-4)^3 = 2 - (-64)$
$= 2 + 64$
$= 66$

b) $-x^2 - 3x + 1 = -(-4)^2 - 3(-4) + 1$
$= -16 + 12 + 1$
$= -3$

*Do Exercises 2–5.*

Polynomials occur in many real-world situations.

**Example 3**   *Games in a Sports League.* In a sports league of $n$ teams in which each team plays every other team twice, the total number of games to be played is given by the polynomial

$$n^2 - n.$$

A women's slow-pitch softball league has 10 teams. What is the total number of games to be played?

We evaluate the polynomial for $n = 10$:

$$n^2 - n = 10^2 - 10 = 100 - 10 = 90.$$

The league plays 90 games.

*Do Exercises 6 and 7.*

**AG Algebraic–Graphical Connection**

An equation like $y = 2x - 2$, which has a polynomial on one side and $y$ on the other, is called a **polynomial equation.** We will here and in many places throughout the book connect graphs to related concepts.

Recall from Chapter 3 that in order to plot points before graphing an equation, we choose values for $x$ and compute the corresponding $y$-values. If the equation has $y$ on one side and a polynomial involving $x$ on the other, then determining $y$ is the same as evaluating the polynomial. Once the graph of such an equation has been drawn, we can evaluate the polynomial for a given $x$-value by finding the $y$-value that is paired with it on the graph.

**Example 4**   Use *only* the given graph of $y = 2x - 2$ to evaluate the polynomial $2x - 2$ for $x = 3$.

First, we locate 3 on the $x$-axis. From there we move vertically to the graph of the equation and then horizontally to the $y$-axis. There we locate the $y$-value that is paired with 3. Although our drawing may not be precise, it appears that the $y$-value 4 is paired with 3. Thus the value of $2x - 2$ is 4 when $x = 3$.

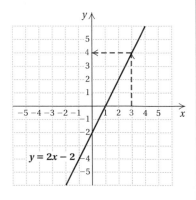

*Do Exercise 8.*

**Example 5**  *Medical Dosage.* The concentration $C$, in parts per million, of a certain antibiotic in the bloodstream after $t$ hours is given by the polynomial equation

$$C = -0.05t^2 + 2t + 2.$$

Find the concentration after 2 hr.

To find the concentration after 2 hr, we evaluate the polynomial for $t = 2$:

$$-0.05t^2 + 2t + 2 = -0.05(2)^2 + 2(2) + 2 \qquad \text{Carrying out the calculation using the rules for order of operations}$$

$$= -0.05(4) + 2(2) + 2$$

$$= -0.2 + 4 + 2$$

$$= -0.2 + 6$$

$$= 5.8.$$

The concentration after 2 hr is 5.8 parts per million.

*Do Exercise 9.*

**9.** Referring to Example 5, find the concentration after 3 hr.

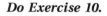 **Algebraic–Graphical Connection**

The polynomial equation in Example 5 can be graphed if we evaluate the polynomial for several values of $t$. We list the values in a table and show the graph below. Note that the concentration peaks at the 20-hr mark and after a bit more than 40 hr, the concentration is 0. Since neither time nor concentration can be negative, our graph uses only the first quadrant.

**10.** Use *only* the graph showing medical dosage to estimate the value of the polynomial for $t = 26$.

| $t$ | $-0.05t^2 + 2t + 2$ |
|-----|---------------------|
| 0   | 2                   |
| 2   | 5.8                 |
| 10  | 17                  |
| 20  | 22                  |
| 30  | 17                  |

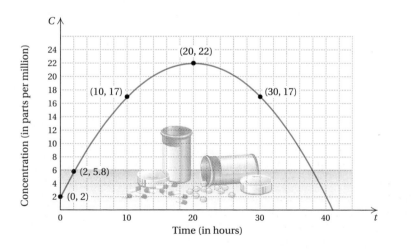

*Do Exercise 10.*

*Answers on page A-17*

# Calculator Spotlight

Evaluating Polynomials. One way to evaluate a polynomial like $-x^2 - 3x + 1$ for $x = -4$ (see Example 2b) is to graph $y_1 = -x^2 - 3x + 1$.

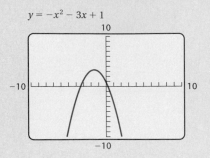

$y = -x^2 - 3x + 1$

We can also adjust the window to obtain a better view of the graph.

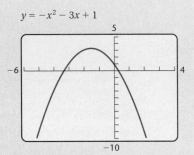

$y = -x^2 - 3x + 1$

To evaluate the polynomial, we then use the CALC feature and choose VALUE.

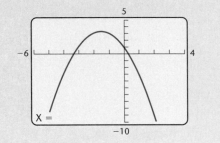

We enter $x = -4$. The $y$-value, $-3$, is shown together with a TRACE indicator showing the point $(-4, -3)$.

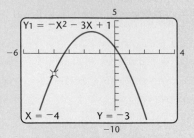

Polynomials can also be evaluated using the TABLE feature as described in Section 3.4.

## Exercises

1. Evaluate the polynomial $-x^2 - 3x + 1$ for $x = -1$, for $x = -0.3$, and for $x = 1.7$.
2. Evaluate the polynomial $-0.05x^2 + 2x + 2$ for $x = 0$, for $x = 10$, for $x = 23$, and for $x = 36.4$. Use the viewing window $[0, 41, 0, 25]$.
3. Evaluate the polynomial $2x^2 - x - 8$ for $x = -3$, for $x = -2$, for $x = 0$, for $x = 1.8$, and for $x = 3$.

## b | Identifying Terms

As we saw in Section 1.4, subtractions can be rewritten as additions. For any polynomial that has some subtractions, we can find an equivalent polynomial using only additions.

**Examples** Find an equivalent polynomial using only additions.

**6.** $-5x^2 - x = -5x^2 + (-x)$

**7.** $4x^5 - 2x^6 - 4x + 7 = 4x^5 + (-2x^6) + (-4x) + 7$

*Do Exercises 11 and 12.*

When a polynomial has only additions, the monomials being added are called **terms**. In Example 6, the terms are $-5x^2$ and $-x$. In Example 7, the terms are $4x^5$, $-2x^6$, $-4x$, and 7.

**Example 8** Identify the terms of the polynomial

$$4x^7 + 3x + 12 + 8x^3 + 5x.$$

Terms: $4x^7$, $3x$, 12, $8x^3$, and $5x$.

If there are subtractions, you can *think* of them as additions without rewriting.

**Example 9** Identify the terms of the polynomial

$$3t^4 - 5t^6 - 4t + 2.$$

Terms: $3t^4$, $-5t^6$, $-4t$, and 2.

*Do Exercises 13 and 14.*

## c | Like Terms

When terms have the same variable and the variable is raised to the same power, we say that they are **like terms,** or **similar terms.**

**Examples** Identify the like terms in the polynomials.

**10.** $4x^3 + 5x - 4x^2 + 2x^3 + x^2$

   Like terms: $4x^3$ and $2x^3$     Same variable and exponent
   Like terms: $-4x^2$ and $x^2$     Same variable and exponent

**11.** $6 - 3a^2 + 8 - a - 5a$

   Like terms: 6 and 8     Constant terms are like terms because $6 = 6x^0$ and $8 = 8x^0$.

   Like terms: $-a$ and $-5a$

*Do Exercises 15–17.*

## d | Coefficients

The coefficient of the term $5x^3$ is 5. In the following polynomial, the color numbers are the **coefficients**:

$$3x^5 - 2x^3 + 5x + 4.$$

---

Find an equivalent polynomial using only additions.

**11.** $-9x^3 - 4x^5$

**12.** $-2y^3 + 3y^7 - 7y$

Identify the terms of the polynomial.

**13.** $3x^2 + 6x + \dfrac{1}{2}$

**14.** $-4y^5 + 7y^2 - 3y - 2$

Identify the like terms in the polynomial.

**15.** $4x^3 - x^3 + 2$

**16.** $4t^4 - 9t^3 - 7t^4 + 10t^3$

**17.** $5x^2 + 3x - 10 + 7x^2 - 8x + 11$

*Answers on page A-17*

**18.** Identify the coefficient of each term in the polynomial

$$2x^4 - 7x^3 - 8.5x^2 + 10x - 4.$$

Collect like terms.

**19.** $3x^2 + 5x^2$

**20.** $4x^3 - 2x^3 + 2 + 5$

**21.** $\frac{1}{2}x^5 - \frac{3}{4}x^5 + 4x^2 - 2x^2$

**22.** $24 - 4x^3 - 24$

**23.** $5x^3 - 8x^5 + 8x^5$

**24.** $-2x^4 + 16 + 2x^4 + 9 - 3x^5$

Collect like terms.

**25.** $7x - x$

**26.** $5x^3 - x^3 + 4$

**27.** $\frac{3}{4}x^3 + 4x^2 - x^3 + 7$

**28.** $8x^2 - x^2 + x^3 - 1 - 4x^2 + 10$

Answers on page A-17

**Example 12**   Identify the coefficient of each term in the polynomial

$$3x^4 - 4x^3 + 7x^2 + x - 8.$$

The coefficient of the first term is 3.

The coefficient of the second term is $-4$.

The coefficient of the third term is 7.

The coefficient of the fourth term is 1.

The coefficient of the fifth term is $-8$.

*Do Exercise 18.*

## e   Collecting Like Terms

We can often simplify polynomials by **collecting like terms,** or **combining similar terms.** To do this, we use the distributive laws. We factor out the variable expression and add or subtract the coefficients. We try to do this mentally as much as possible.

**Examples**   Collect like terms.

**13.** $2x^3 - 6x^3 = (2 - 6)x^3 = -4x^3$   **Using a distributive law**

**14.** $5x^2 + 7 + 4x^4 + 2x^2 - 11 - 2x^4 = (5 + 2)x^2 + (4 - 2)x^4 + (7 - 11)$
$$= 7x^2 + 2x^4 - 4$$

Note that using the distributive laws in this manner allows us to collect like terms by adding or subtracting the coefficients. Often the middle step is omitted and we add or subtract mentally, writing just the answer. In collecting like terms, we may get 0.

**Examples**   Collect like terms.

**15.** $5x^3 - 5x^3 = (5 - 5)x^3 = 0x^3 = 0$

**16.** $3x^4 + 2x^2 - 3x^4 + 8 = (3 - 3)x^4 + 2x^2 + 8$
$$= 0x^4 + 2x^2 + 8 = 2x^2 + 8$$

*Do Exercises 19–24.*

Multiplying a term of a polynomial by 1 does not change the term, but it may make it easier to factor or collect like terms.

**Examples**   Collect like terms.

**17.** $5x^2 + x^2 = 5x^2 + 1x^2$   **Replacing $x^2$ with $1x^2$**
$$= (5 + 1)x^2$$   **Using a distributive law**
$$= 6x^2$$

**18.** $5x^4 - 6x^3 - x^4 = 5x^4 - 6x^3 - 1x^4$   $x^4 = 1x^4$
$$= (5 - 1)x^4 - 6x^3$$
$$= 4x^4 - 6x^3$$

**19.** $\frac{2}{3}x^4 - x^3 - \frac{1}{6}x^4 + \frac{2}{5}x^3 - \frac{3}{10}x^3 = \left(\frac{2}{3} - \frac{1}{6}\right)x^4 + \left(-1 + \frac{2}{5} - \frac{3}{10}\right)x^3$
$$= \left(\frac{4}{6} - \frac{1}{6}\right)x^4 + \left(-\frac{10}{10} + \frac{4}{10} - \frac{3}{10}\right)x^3$$
$$= \frac{3}{6}x^4 - \frac{9}{10}x^3 = \frac{1}{2}x^4 - \frac{9}{10}x^3$$

*Do Exercises 25–28.*

## f | Descending and Ascending Order

Note in the following polynomial that the exponents decrease from left to right. We say that the polynomial is arranged in **descending order:**

$$2x^4 - 8x^3 + 5x^2 - x + 3.$$

The term with the largest exponent is first. The term with the next largest exponent is second, and so on. The associative and commutative laws allow us to arrange the terms of a polynomial in descending order.

**Examples** Arrange the polynomial in descending order.

**20.** $6x^5 + 4x^7 + x^2 + 2x^3 = 4x^7 + 6x^5 + 2x^3 + x^2$

**21.** $\frac{2}{3} + 4x^5 - 8x^2 + 5x - 3x^3 = 4x^5 - 3x^3 - 8x^2 + 5x + \frac{2}{3}$

We usually arrange polynomials in descending order, but not always. The opposite order is called **ascending order.** Generally, if an exercise is written in a certain order, we give the answer in that same order.

*Do Exercises 29–31.*

**Example 22** Collect like terms and then arrange in descending order:

$$2x^2 - 4x^3 + 3 - x^2 - 2x^3.$$

We have

$$2x^2 - 4x^3 + 3 - x^2 - 2x^3 = x^2 - 6x^3 + 3 \qquad \text{Collecting like terms}$$
$$= -6x^3 + x^2 + 3 \qquad \text{Arranging in descending order}$$

*Do Exercises 32 and 33.*

## g | Degrees

The **degree** of a term is the exponent of the variable. The degree of the term $5x^3$ is 3.

**Example 23** Identify the degree of each term of $8x^4 + 3x + 7$.

The degree of $8x^4$ is 4.
The degree of $3x$ is 1.    **Recall that** $x = x^1$.
The degree of 7 is 0.    **Think of 7 as** $7x^0$. **Recall that** $x^0 = 1$.

The **degree of a polynomial** is the largest of the degrees of the terms, unless it is the polynomial 0. The polynomial 0 is a special case. We agree that it has *no* degree either as a term or as a polynomial. This is because we can express 0 as $0 = 0x^5 = 0x^7$, and so on, using any exponent we wish.

**Example 24** Identify the degree of the polynomial $5x^3 - 6x^4 + 7$.

We have

$$5x^3 - 6x^4 + 7. \qquad \text{The largest exponent is 4.}$$

The degree of the polynomial is 4.

*Do Exercise 34.*

Arrange the polynomial in descending order.

**29.** $x + 3x^5 + 4x^3 + 5x^2 + 6x^7 - 2x^4$

**30.** $4x^2 - 3 + 7x^5 + 2x^3 - 5x^4$

**31.** $-14 + 7t^2 - 10t^5 + 14t^7$

Collect like terms and then arrange in descending order.

**32.** $3x^2 - 2x + 3 - 5x^2 - 1 - x$

**33.** $-x + \frac{1}{2} + 14x^4 - 7x - 1 - 4x^4$

**34.** Identify the degree of each term and the degree of the polynomial

$$-6x^4 + 8x^2 - 2x + 9.$$

*Answers on page A-17*

Identify the missing terms in the polynomial.

**35.** $2x^3 + 4x^2 - 2$

**36.** $-3x^4$

**37.** $x^3 + 1$

**38.** $x^4 - x^2 + 3x + 0.25$

Classify the polynomial as a monomial, binomial, trinomial, or none of these.

**39.** $5x^4$

**40.** $4x^3 - 3x^2 + 4x + 2$

**41.** $3x^2 + x$

**42.** $3x^2 + 2x - 4$

*Answers on page A-17*

Let's summarize the terminology that we have learned, using the polynomial

$$3x^4 - 8x^3 + 5x^2 + 7x - 6.$$

| Term | Coefficient | Degree of the Term | Degree of the Polynomial |
|------|-------------|--------------------|--------------------------|
| $3x^4$ | 3 | 4 | |
| $-8x^3$ | $-8$ | 3 | |
| $5x^2$ | 5 | 2 | 4 |
| $7x$ | 7 | 1 | |
| $-6$ | $-6$ | 0 | |

## h | Missing Terms

If a coefficient is 0, we generally do not write the term. We say that we have a **missing term.**

**Example 25**   Identify the missing terms in the polynomial

$$8x^5 - 2x^3 + 5x^2 + 7x + 8.$$

There is no term with $x^4$. We say that the $x^4$-term (or the *fourth-degree term*) is missing.

For certain skills or manipulations, we can write missing terms with zero coefficients or leave space. For example, we can write the polynomial $3x^2 + 9$ as

$$3x^2 + 0x + 9 \quad \text{or} \quad 3x^2 + \qquad 9.$$

*Do Exercises 35–38.*

## i | Classifying Polynomials

Polynomials with just one term are called **monomials**. Polynomials with just two terms are called **binomials**. Those with just three terms are called **trinomials**. Those with more than three terms are generally not specified with a name.

**Example 26**

| Monomials | Binomials | Trinomials | None of These |
|-----------|-----------|------------|---------------|
| $4x^2$ | $2x + 4$ | $3x^3 + 4x + 7$ | $4x^3 - 5x^2 + x - 8$ |
| 9 | $3x^5 + 6x$ | $6x^7 - 7x^2 + 4$ | |
| $-23x^{19}$ | $-9x^7 - 6$ | $4x^2 - 6x - \frac{1}{2}$ | |

*Do Exercises 39–42.*

# Exercise Set 4.3

**a** Evaluate the polynomial for $x = 4$ and for $x = -1$.

**1.** $-5x + 2$

**2.** $-8x + 1$

**3.** $2x^2 - 5x + 7$

**4.** $3x^2 + x - 7$

**5.** $x^3 - 5x^2 + x$

**6.** $7 - x + 3x^2$

Evaluate the polynomial for $x = -2$ and for $x = 0$.

**7.** $3x + 5$

**8.** $8 - 4x$

**9.** $x^2 - 2x + 1$

**10.** $5x + 6 - x^2$

**11.** $-3x^3 + 7x^2 - 3x - 2$

**12.** $-2x^3 + 5x^2 - 4x + 3$

**13.** *Skydiving.* During the first 13 sec of a jump, the number of feet that a skydiver falls in $t$ seconds can be approximated by the polynomial

$$11.12t^2.$$

Approximately how far has a skydiver fallen 10 sec after having jumped from a plane?

**14.** *Skydiving.* For jumps that exceed 13 sec, the polynomial

$$173t - 369$$

can be used to approximate the distance, in feet, that a skydiver has fallen in $t$ seconds. Approximately how far has a skydiver fallen 20 sec after having jumped from a plane?

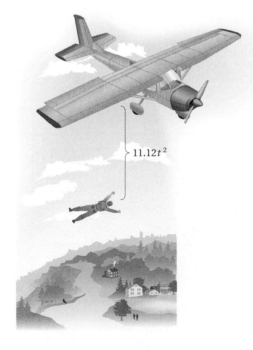

$11.12t^2$

**15.** *Total Revenue.* Hadley Electronics is marketing a new kind of high-density TV. The firm determines that when it sells $x$ TVs, its total revenue (the total amount of money taken in) will be

$$280x - 0.4x^2 \text{ dollars.}$$

What is the total revenue from the sale of 75 TVs? 100 TVs?

**16.** *Total Cost.* Hadley Electronics determines that the total cost of producing $x$ high-density TVs is given by

$$5000 + 0.6x^2 \text{ dollars.}$$

What is the total cost of producing 500 TVs? 650 TVs?

**17.** The graph of the polynomial equation $y = 5 - x^2$ is shown below. Use *only* the graph to estimate the value of the polynomial for $x = -3$, for $x = -1$, for $x = 0$, for $x = 1.5$, and for $x = 2$.

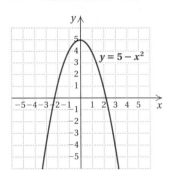

**18.** The graph of the polynomial equation $y = 6x^3 - 6x$ is shown below. Use *only* the graph to estimate the value of the polynomial for $x = -1$, for $x = -0.5$, for $x = 0.5$, for $x = 1$, and for $x = 1.1$.

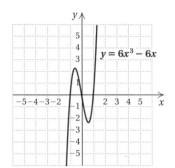

**19.** *Path of the Olympic Arrow.* The Olympic flame at the 1992 Summer Olympics was lit by a flaming arrow. As the arrow moved $d$ feet horizontally from the archer, its height $h$, in feet, could be approximated by the polynomial equation

$$h = -0.002d^2 + 0.8d + 6.6.$$

The graph of this equation is shown at right. Use either the graph or the polynomial to approximate the height of the arrow after it has traveled horizontally for 100 ft, 200 ft, 300 ft, and 350 ft.

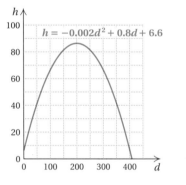

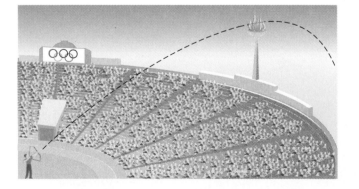

Copyright © 1999 Addison Wesley Longman

**20.** *Hearing-Impaired Americans.* The number $N$, in millions, of hearing-impaired Americans of age $x$ can be approximated by the polynomial equation

$$N = -0.00006x^3 + 0.006x^2 - 0.1x + 1.9$$

(***Source***: American Speech-Language Hearing Association). The graph of this equation is shown at right. Use either the graph or the polynomial to approximate the number of hearing-impaired Americans of ages 20, 40, 50, and 60.

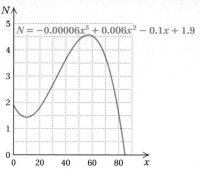

$N = -0.00006x^3 + 0.006x^2 - 0.1x + 1.9$

**b** Identify the terms of the polynomial.

**21.** $2 - 3x + x^2$

**22.** $2x^2 + 3x - 4$

**c** Identify the like terms in the polynomial.

**23.** $5x^3 + 6x^2 - 3x^2$

**24.** $3x^2 + 4x^3 - 2x^2$

**25.** $2x^4 + 5x - 7x - 3x^4$

**26.** $-3t + t^3 - 2t - 5t^3$

**27.** $3x^5 - 7x + 8 + 14x^5 - 2x - 9$

**28.** $8x^3 + 7x^2 - 11 - 4x^3 - 8x^2 - 29$

**d** Identify the coefficient of each term of the polynomial.

**29.** $-3x + 6$

**30.** $2x - 4$

**31.** $5x^2 + 3x + 3$

**32.** $3x^2 - 5x + 2$

**33.** $-5x^4 + 6x^3 - 3x^2 + 8x - 2$

**34.** $7x^3 - 4x^2 - 4x + 5$

e Collect like terms.

**35.** $2x - 5x$

**36.** $2x^2 + 8x^2$

**37.** $x - 9x$

**38.** $x - 5x$

**39.** $5x^3 + 6x^3 + 4$

**40.** $6x^4 - 2x^4 + 5$

**41.** $5x^3 + 6x - 4x^3 - 7x$

**42.** $3a^4 - 2a + 2a + a^4$

**43.** $6b^5 + 3b^2 - 2b^5 - 3b^2$

**44.** $2x^2 - 6x + 3x + 4x^2$

**45.** $\frac{1}{4}x^5 - 5 + \frac{1}{2}x^5 - 2x - 37$

**46.** $\frac{1}{3}x^3 + 2x - \frac{1}{6}x^3 + 4 - 16$

**47.** $6x^2 + 2x^4 - 2x^2 - x^4 - 4x^2$

**48.** $8x^2 + 2x^3 - 3x^3 - 4x^2 - 4x^2$

**49.** $\frac{1}{4}x^3 - x^2 - \frac{1}{6}x^2 + \frac{3}{8}x^3 + \frac{5}{16}x^3$

**50.** $\frac{1}{5}x^4 + \frac{1}{5} - 2x^2 + \frac{1}{10} - \frac{3}{15}x^4 + 2x^2 - \frac{3}{10}$

**f** Arrange the polynomial in descending order.

**51.** $x^5 + x + 6x^3 + 1 + 2x^2$

**52.** $3 + 2x^2 - 5x^6 - 2x^3 + 3x$

**53.** $5y^3 + 15y^9 + y - y^2 + 7y^8$

**54.** $9p - 5 + 6p^3 - 5p^4 + p^5$

Collect like terms and then arrange in descending order.

**55.** $3x^4 - 5x^6 - 2x^4 + 6x^6$

**56.** $-1 + 5x^3 - 3 - 7x^3 + x^4 + 5$

**57.** $-2x + 4x^3 - 7x + 9x^3 + 8$

**58.** $-6x^2 + x - 5x + 7x^2 + 1$

**59.** $3x + 3x + 3x - x^2 - 4x^2$

**60.** $-2x - 2x - 2x + x^3 - 5x^3$

**61.** $-x + \frac{3}{4} + 15x^4 - x - \frac{1}{2} - 3x^4$

**62.** $2x - \frac{5}{6} + 4x^3 + x + \frac{1}{3} - 2x$

Copyright © 1999 Addison Wesley Longman

g Identify the degree of each term of the polynomial and the degree of the polynomial.

**63.** $2x - 4$

**64.** $6 - 3x$

**65.** $3x^2 - 5x + 2$

**66.** $5x^3 - 2x^2 + 3$

**67.** $-7x^3 + 6x^2 + 3x + 7$

**68.** $5x^4 + x^2 - x + 2$

**69.** $x^2 - 3x + x^6 - 9x^4$

**70.** $8x - 3x^2 + 9 - 8x^3$

**71.** Complete the following table for the polynomial $-7x^4 + 6x^3 - 3x^2 + 8x - 2$.

| Term | Coefficient | Degree of the Term | Degree of the Polynomial |
|---|---|---|---|
|  |  |  |  |
| $6x^3$ | 6 |  |  |
|  |  | 2 |  |
| $8x$ |  | 1 |  |
|  | $-2$ |  |  |

**72.** Complete the following table for the polynomial $3x^2 + 8x^5 - 46x^3 + 6x - 2.4 - \frac{1}{2}x^4$.

| Term | Coefficient | Degree of the Term | Degree of the Polynomial |
|---|---|---|---|
|  |  | 5 |  |
| $-\frac{1}{2}x^4$ |  | 4 |  |
|  | $-46$ |  |  |
| $3x^2$ |  | 2 |  |
|  | 6 |  |  |
| $-2.4$ |  |  |  |

h Identify the missing terms in the polynomial.

**73.** $x^3 - 27$

**74.** $x^5 + x$

**75.** $x^4 - x$

**76.** $5x^4 - 7x + 2$

**77.** $2x^3 - 5x^2 + x - 3$

**78.** $-6x^3$

i Classify the polynomial as a monomial, binomial, trinomial, or none of these.

**79.** $x^2 - 10x + 25$

**80.** $-6x^4$

**81.** $x^3 - 7x^2 + 2x - 4$

**82.** $x^2 - 9$

**83.** $4x^2 - 25$

**84.** $2x^4 - 7x^3 + x^2 + x - 6$

**85.** $40x$

**86.** $4x^2 + 12x + 9$

**87.** Three tired campers stopped for the night. All they had to eat was a bag of apples. During the night, one awoke and ate one-third of the apples. Later, a second camper awoke and ate one-third of the apples that remained. Much later, the third camper awoke and ate one-third of those apples yet remaining after the other two had eaten. When they got up the next morning, 8 apples were left. How many apples did they begin with?   [2.4a]

Subtract.   [1.4a]

**88.** $1 - 20$

**89.** $\dfrac{1}{8} - \dfrac{5}{6}$

**90.** $\dfrac{3}{8} - \left(-\dfrac{1}{4}\right)$

**91.** $5.6 - 8.2$

**92.** Solve: $3(x + 2) = 5x - 9$.   [2.3c]

**93.** Solve $cx = ab - r$ for $b$.   [2.6a]

**94.** A nut dealer has 1800 lb of peanuts, 1500 lb of cashews, and 700 lb of almonds. What percent of the total is peanuts? cashews? almonds?   [2.5a]

**95.** Factor: $3x - 15y + 63$.   [1.7d]

**Synthesis**

**96.** ◆ Is it better to evaluate a polynomial before or after like terms have been collected? Why?

**97.** ◆ Explain why an understanding of the rules of order of operations is essential when evaluating polynomials.

Collect like terms.

**98.** $\dfrac{9}{2}x^8 + \dfrac{1}{9}x^2 + \dfrac{1}{2}x^9 + \dfrac{9}{2}x^1 + \dfrac{9}{2}x^9 + \dfrac{8}{9}x^2 + \dfrac{1}{2}x - \dfrac{1}{2}x^8$

**99.** $(3x^2)^3 + 4x^2 \cdot 4x^4 - x^4(2x)^2 + ((2x)^2)^3 - 100x^2(x^2)^2$

**100.** Construct a polynomial in $x$ (meaning that $x$ is the variable) of degree 5 with four terms and coefficients that are integers.

**101.** What is the degree of $(5m^5)^2$?

**102.** A polynomial in $x$ has degree 3. The coefficient of $x^2$ is 3 less than the coefficient of $x^3$. The coefficient of $x$ is three times the coefficient of $x^2$. The remaining coefficient is 2 more than the coefficient of $x^3$. The sum of the coefficients is $-4$. Find the polynomial.

Use the CALC feature and choose VALUE on your grapher to find the values in each of the following.

**103.** Exercise 17

**104.** Exercise 18

**105.** Exercise 19

**106.** Exercise 20

Copyright © 1999 Addison Wesley Longman

## 4.4 Addition and Subtraction of Polynomials

### a | Addition of Polynomials

To add two polynomials, we can write a plus sign between them and then collect like terms. Depending on the situation, you may see polynomials written in descending order, ascending order, or neither. Generally, if an exercise is written in a particular order, we write the answer in that same order.

**Example 1** Add: $(-3x^3 + 2x - 4) + (4x^3 + 3x^2 + 2)$.

$$(-3x^3 + 2x - 4) + (4x^3 + 3x^2 + 2)$$
$$= (-3 + 4)x^3 + 3x^2 + 2x + (-4 + 2) \quad \text{Collecting like terms}$$
$$\qquad\qquad\qquad\qquad\qquad\qquad (No \text{ signs are changed.})$$
$$= x^3 + 3x^2 + 2x - 2$$

**Example 2** Add:

$$\left(\tfrac{2}{3}x^4 + 3x^2 - 2x + \tfrac{1}{2}\right) + \left(-\tfrac{1}{3}x^4 + 5x^3 - 3x^2 + 3x - \tfrac{1}{2}\right).$$

We have

$$\left(\tfrac{2}{3}x^4 + 3x^2 - 2x + \tfrac{1}{2}\right) + \left(-\tfrac{1}{3}x^4 + 5x^3 - 3x^2 + 3x - \tfrac{1}{2}\right)$$
$$= \left(\tfrac{2}{3} - \tfrac{1}{3}\right)x^4 + 5x^3 + (3 - 3)x^2 + (-2 + 3)x + \left(\tfrac{1}{2} - \tfrac{1}{2}\right) \quad \begin{array}{l}\text{Collecting}\\\text{like terms}\end{array}$$
$$= \tfrac{1}{3}x^4 + 5x^3 + x.$$

We can add polynomials as we do because they represent numbers. After some practice, you will be able to add mentally.

*Do Exercises 1–4.*

**Example 3** Add: $(3x^2 - 2x + 2) + (5x^3 - 2x^2 + 3x - 4)$.

$$(3x^2 - 2x + 2) + (5x^3 - 2x^2 + 3x - 4)$$
$$= 5x^3 + (3 - 2)x^2 + (-2 + 3)x + (2 - 4) \quad \text{You might do this step mentally.}$$
$$= 5x^3 + x^2 + x - 2 \qquad\qquad\qquad\quad \text{Then you would write only this.}$$

*Do Exercises 5 and 6.*

We can also add polynomials by writing like terms in columns.

**Example 4** Add: $9x^5 - 2x^3 + 6x^2 + 3$ and $5x^4 - 7x^2 + 6$ and $3x^6 - 5x^5 + x^2 + 5$.

We arrange the polynomials with the like terms in columns.

$$\begin{array}{l}
9x^5 \qquad\quad -2x^3 + 6x^2 + \;\;3 \\
\qquad\quad 5x^4 \qquad\quad -7x^2 + \;\;6 \qquad \text{We leave spaces for missing terms.} \\
\underline{3x^6 - 5x^5 \qquad\qquad\quad + x^2 + \;\;5} \\
3x^6 + 4x^5 + 5x^4 - 2x^3 \qquad + 14 \qquad \text{Adding}
\end{array}$$

We write the answer as $3x^6 + 4x^5 + 5x^4 - 2x^3 + 14$ without the space.

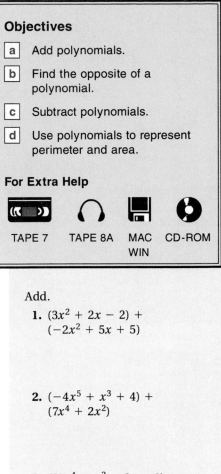

**Objectives**

a | Add polynomials.

b | Find the opposite of a polynomial.

c | Subtract polynomials.

d | Use polynomials to represent perimeter and area.

**For Extra Help**

TAPE 7    TAPE 8A    MAC WIN    CD-ROM

Add.

1. $(3x^2 + 2x - 2) + (-2x^2 + 5x + 5)$

2. $(-4x^5 + x^3 + 4) + (7x^4 + 2x^2)$

3. $(31x^4 + x^2 + 2x - 1) + (-7x^4 + 5x^3 - 2x + 2)$

4. $(17x^3 - x^2 + 3x + 4) + \left(-15x^3 + x^2 - 3x - \dfrac{2}{3}\right)$

Add mentally. Try to write just the answer.

5. $(4x^2 - 5x + 3) + (-2x^2 + 2x - 4)$

6. $(3x^3 - 4x^2 - 5x + 3) + \left(5x^3 + 2x^2 - 3x - \dfrac{1}{2}\right)$

*Answers on page A-17*

Add.

**7.**
$$-2x^3 + 5x^2 - 2x + 4$$
$$x^4 \qquad + 6x^2 + 7x - 10$$
$$-9x^4 + 6x^3 + x^2 \qquad - 2$$

**8.** $-3x^3 + 5x + 2$ and
$x^3 + x^2 + 5$ and
$x^3 - 2x - 4$

Find two equivalent expressions for the opposite of the polynomial.

**9.** $12x^4 - 3x^2 + 4x$

**10.** $-4x^4 + 3x^2 - 4x$

**11.** $-13x^6 + 2x^4 - 3x^2 + x - \frac{5}{13}$

**12.** $-7y^3 + 2y^2 - y + 3$

Simplify.

**13.** $-(4x^3 - 6x + 3)$

**14.** $-(5x^4 + 3x^2 + 7x - 5)$

**15.** $-\left(14x^{10} - \frac{1}{2}x^5 + 5x^3 - x^2 + 3x\right)$

Answers on page A-17

*Do Exercises 7 and 8.*

## b | Opposites of Polynomials

We now look at subtraction of polynomials. To do so, we first consider the opposite, or additive inverse, of a polynomial.

We know that two numbers are opposites of each other if their sum is zero. For example, 5 and $-5$ are opposites, since $5 + (-5) = 0$. The same definition holds for polynomials. Two polynomials are **opposites**, or **additive inverses,** of each other if their sum is zero.

To find a way to determine an opposite, look for a pattern in the following examples:

a) $2x + (-2x) = 0$;

b) $-6x^2 + 6x^2 = 0$;

c) $(5t^3 - 2) + (-5t^3 + 2) = 0$;

d) $(7x^3 - 6x^2 - x + 4) + (-7x^3 + 6x^2 + x - 4) = 0$.

Since $(5t^3 - 2) + (-5t^3 + 2) = 0$, we know that the opposite of $(5t^3 - 2)$ is $(-5t^3 + 2)$. To say the same thing with purely algebraic symbolism, consider

$$\underbrace{\text{The opposite of}} \quad \underbrace{(5t^3 - 2)} \quad \text{is} \quad \underbrace{-5t^3 + 2.}$$
$$- \qquad\qquad (5t^3 - 2) \quad = \quad -5t^3 + 2.$$

> We can find an equivalent polynomial for the opposite, or additive inverse, of a polynomial by replacing each term with its opposite—that is, *changing the sign of every term.*

**Example 5** Find two equivalent expressions for the opposite of
$$4x^5 - 7x^3 - 8x + \frac{5}{6}.$$

The opposite of $4x^5 - 7x^3 - 8x + \frac{5}{6}$ is

$$-\left(4x^5 - 7x^3 - 8x + \frac{5}{6}\right), \quad \text{or}$$
$$-4x^5 + 7x^3 + 8x - \frac{5}{6}. \qquad \text{\small Changing the sign of every term}$$

Thus, $-\left(4x^5 - 7x^3 - 8x + \frac{5}{6}\right)$ is equivalent to $-4x^5 + 7x^3 + 8x - \frac{5}{6}$, and each is the opposite of the original polynomial $4x^5 - 7x^3 - 8x + \frac{5}{6}$.

*Do Exercises 9–12.*

**Example 6** Simplify: $-\left(-7x^4 - \frac{5}{9}x^3 + 8x^2 - x + 67\right)$.

$$-\left(-7x^4 - \frac{5}{9}x^3 + 8x^2 - x + 67\right) = 7x^4 + \frac{5}{9}x^3 - 8x^2 + x - 67$$

*Do Exercises 13–15.*

## c Subtraction of Polynomials

Recall that we can subtract a real number by adding its opposite, or additive inverse: $a - b = a + (-b)$. This allows us to find an equivalent expression for the difference of two polynomials.

**Example 7** Subtract:

$$(9x^5 + x^3 - 2x^2 + 4) - (2x^5 + x^4 - 4x^3 - 3x^2).$$

We have

$(9x^5 + x^3 - 2x^2 + 4) - (2x^5 + x^4 - 4x^3 - 3x^2)$

$= 9x^5 + x^3 - 2x^2 + 4 + [-(2x^5 + x^4 - 4x^3 - 3x^2)]$    **Adding the opposite**

$= 9x^5 + x^3 - 2x^2 + 4 - 2x^5 - x^4 + 4x^3 + 3x^2$    **Finding the opposite by changing the sign of *each* term**

$= 7x^5 - x^4 + 5x^3 + x^2 + 4.$    **Collecting like terms**

*Do Exercises 16 and 17.*

As with similar work in Section 1.8, we combine steps by changing the sign of each term of the polynomial being subtracted and collecting like terms. Try to do this mentally as much as possible.

**Example 8** Subtract: $(9x^5 + x^3 - 2x) - (-2x^5 + 5x^3 + 6)$.

$(9x^5 + x^3 - 2x) - (-2x^5 + 5x^3 + 6)$

$= 9x^5 + x^3 - 2x + 2x^5 - 5x^3 - 6$    **Finding the opposite by changing the sign of each term**

$= 11x^5 - 4x^3 - 2x - 6$    **Collecting like terms**

*Do Exercises 18 and 19.*

We can use columns to subtract. We replace coefficients with their opposites, as shown in Example 7.

**Example 9** Write in columns and subtract:

$$(5x^2 - 3x + 6) - (9x^2 - 5x - 3).$$

a)    $5x^2 - 3x + 6$    **Writing similar terms in columns**
  $\underline{-(9x^2 - 5x - 3)}$

b)    $5x^2 - 3x + 6$
  $\underline{-9x^2 + 5x + 3}$    **Changing signs**

c)    $5x^2 - 3x + 6$
  $\underline{-9x^2 + 5x + 3}$
  $-4x^2 + 2x + 9$    **Adding**

If you can do so without error, you can arrange the polynomials in columns and write just the answer.

---

Subtract.

**16.** $(7x^3 + 2x + 4) - (5x^3 - 4)$

**17.** $(-3x^2 + 5x - 4) - (-4x^2 + 11x - 2)$

Subtract.

**18.** $(-6x^4 + 3x^2 + 6) - (2x^4 + 5x^3 - 5x^2 + 7)$

**19.** $\left(\dfrac{3}{2}x^3 - \dfrac{1}{2}x^2 + 0.3\right) - \left(\dfrac{1}{2}x^3 + \dfrac{1}{2}x^2 + \dfrac{4}{3}x + 1.2\right)$

*Answers on page A-17*

Write in columns and subtract.

**20.** $(4x^3 + 2x^2 - 2x - 3) -$
$(2x^3 - 3x^2 + 2)$

**21.** $(2x^3 + x^2 - 6x + 2) -$
$(x^5 + 4x^3 - 2x^2 - 4x)$

**22.** Find a polynomial for the sum of the perimeters and the areas of the rectangles.

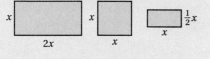

**23.** Find a polynomial for the shaded area.

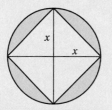

*Answers on page A-17*

**Example 10**  Write in columns and subtract:

$$(x^3 + x^2 + 2x - 12) - (-2x^3 + x^2 - 3x).$$

We have

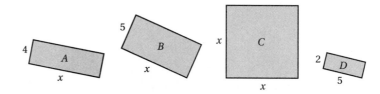

$$
\begin{array}{l}
\quad x^3 + x^2 + 2x - 12 \\
\underline{-2x^3 + x^2 - 3x} \qquad \text{Changing signs} \\
\quad 3x^3 \qquad\quad + 5x - 12. \qquad \text{Adding}
\end{array}
$$

*Do Exercises 20 and 21.*

## d | Polynomials and Geometry

**Example 11**  Find a polynomial for the sum of the areas of these rectangles.

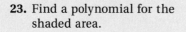

Recall that the area of a rectangle is the product of the length and the width. The sum of the areas is a sum of products. We find these products and then collect like terms.

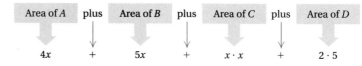

| Area of $A$ | plus | Area of $B$ | plus | Area of $C$ | plus | Area of $D$ |
|---|---|---|---|---|---|---|
| $4x$ | $+$ | $5x$ | $+$ | $x \cdot x$ | $+$ | $2 \cdot 5$ |

We collect like terms:

$$4x + 5x + x^2 + 10 = x^2 + 9x + 10.$$

*Do Exercise 22.*

**Example 12**  A water fountain with a 4-ft by 4-ft square base is placed on a square grassy park area that is $x$ ft on a side. To determine the amount of grass seed needed for the lawn, find a polynomial for the grassy area.

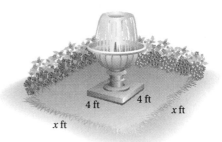

We draw a picture of the situation as shown here. We then reword the problem and write the polynomial as follows.

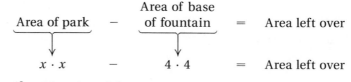

$$
\underbrace{\text{Area of park}} \quad - \quad \underbrace{\begin{array}{c}\text{Area of base}\\\text{of fountain}\end{array}} \quad = \quad \text{Area left over}
$$

$$x \cdot x \qquad - \qquad 4 \cdot 4 \qquad = \quad \text{Area left over}$$

Then $x^2 - 16 =$ Area left over.

*Do Exercise 23.*

# Exercise Set 4.4

**a** Add.

**1.** $(3x + 2) + (-4x + 3)$

**2.** $(6x + 1) + (-7x + 2)$

**3.** $(-6x + 2) + (x^2 + x - 3)$

**4.** $(x^2 - 5x + 4) + (8x - 9)$

**5.** $(x^2 - 9) + (x^2 + 9)$

**6.** $(x^3 + x^2) + (2x^3 - 5x^2)$

**7.** $(3x^2 - 5x + 10) + (2x^2 + 8x - 40)$

**8.** $(6x^4 + 3x^3 - 1) + (4x^2 - 3x + 3)$

**9.** $(1.2x^3 + 4.5x^2 - 3.8x) + (-3.4x^3 - 4.7x^2 + 23)$

**10.** $(0.5x^4 - 0.6x^2 + 0.7) + (2.3x^4 + 1.8x - 3.9)$

**11.** $(1 + 4x + 6x^2 + 7x^3) + (5 - 4x + 6x^2 - 7x^3)$

**12.** $(3x^4 - 6x - 5x^2 + 5) + (6x^2 - 4x^3 - 1 + 7x)$

**13.** $\left(\frac{1}{4}x^4 + \frac{2}{3}x^3 + \frac{5}{8}x^2 + 7\right) + \left(-\frac{3}{4}x^4 + \frac{3}{8}x^2 - 7\right)$

**14.** $\left(\frac{1}{3}x^9 + \frac{1}{5}x^5 - \frac{1}{2}x^2 + 7\right) +$
$\left(-\frac{1}{5}x^9 + \frac{1}{4}x^4 - \frac{3}{5}x^5 + \frac{3}{4}x^2 + \frac{1}{2}\right)$

**15.** $(0.02x^5 - 0.2x^3 + x + 0.08) +$
$(-0.01x^5 + x^4 - 0.8x - 0.02)$

**16.** $(0.03x^6 + 0.05x^3 + 0.22x + 0.05) +$
$\left(\frac{7}{100}x^6 - \frac{3}{100}x^3 + 0.5\right)$

**17.** $(9x^8 - 7x^4 + 2x^2 + 5) + (8x^7 + 4x^4 - 2x) +$
$(-3x^4 + 6x^2 + 2x - 1)$

**18.** $(4x^5 - 6x^3 - 9x + 1) + (6x^3 + 9x^2 + 9x) +$
$(-4x^3 + 8x^2 + 3x - 2)$

**19.**
$$
\begin{array}{l}
0.15x^4 + 0.10x^3 - 0.9x^2 \\
\qquad\quad - 0.01x^3 + 0.01x^2 + x \\
1.25x^4 \qquad\qquad + 0.11x^2 \qquad + 0.01 \\
\qquad\quad 0.27x^3 \qquad\qquad\quad + 0.99 \\
\underline{-0.35x^4 \qquad\qquad + \ 15x^2 \quad - 0.03}
\end{array}
$$

**20.**
$$
\begin{array}{l}
0.05x^4 + 0.12x^3 - 0.5x^2 \\
\qquad\quad - 0.02x^3 + 0.02x^2 + 2x \\
1.5x^4 \qquad\qquad + 0.01x^2 \qquad + 0.15 \\
\qquad\quad 0.25x^3 \qquad\qquad\quad + 0.85 \\
\underline{-0.25x^4 \qquad\qquad + \ 10x^2 \quad - 0.04}
\end{array}
$$

Copyright © 1999 Addison Wesley Longman

**b** Find two equivalent expressions for the opposite of the polynomial.

**21.** $-5x$

**22.** $x^2 - 3x$

**23.** $-x^2 + 10x - 2$

**24.** $-4x^3 - x^2 - x$

**25.** $12x^4 - 3x^3 + 3$

**26.** $4x^3 - 6x^2 - 8x + 1$

Simplify.

**27.** $-(3x - 7)$

**28.** $-(-2x + 4)$

**29.** $-(4x^2 - 3x + 2)$

**30.** $-(-6a^3 + 2a^2 - 9a + 1)$

**31.** $-(-4x^4 + 6x^2 + \frac{3}{4}x - 8)$

**32.** $-(-5x^4 + 4x^3 - x^2 + 0.9)$

**c** Subtract.

**33.** $(3x + 2) - (-4x + 3)$

**34.** $(6x + 1) - (-7x + 2)$

**35.** $(-6x + 2) - (x^2 + x - 3)$

**36.** $(x^2 - 5x + 4) - (8x - 9)$

**37.** $(x^2 - 9) - (x^2 + 9)$

**38.** $(x^3 + x^2) - (2x^3 - 5x^2)$

**39.** $(6x^4 + 3x^3 - 1) - (4x^2 - 3x + 3)$

**40.** $(-4x^2 + 2x) - (3x^3 - 5x^2 + 3)$

**41.** $(1.2x^3 + 4.5x^2 - 3.8x) - (-3.4x^3 - 4.7x^2 + 23)$

**42.** $(0.5x^4 - 0.6x^2 + 0.7) - (2.3x^4 + 1.8x - 3.9)$

**43.** $\left(\frac{5}{8}x^3 - \frac{1}{4}x - \frac{1}{3}\right) - \left(-\frac{1}{8}x^3 + \frac{1}{4}x - \frac{1}{3}\right)$

**44.** $\left(\frac{1}{5}x^3 + 2x^2 - 0.1\right) - \left(-\frac{2}{5}x^3 + 2x^2 + 0.01\right)$

**45.** $(0.08x^3 - 0.02x^2 + 0.01x) - (0.02x^3 + 0.03x^2 - 1)$

**46.** $(0.8x^4 + 0.2x - 1) - \left(\frac{7}{10}x^4 + \frac{1}{5}x - 0.1\right)$

Subtract.

**47.** $x^2 + 5x + 6$
   $\underline{x^2 + 2x\qquad}$

**48.** $x^3\qquad\ + 1$
   $\underline{x^3 + x^2\qquad}$

**49.** $\quad 5x^4 + 6x^3 - 9x^2$
   $\underline{-6x^4 - 6x^3\qquad\ + 8x + 9}$

**50.** $5x^4\qquad + 6x^2 - 3x + 6$
   $\underline{\qquad 6x^3 + 7x^2 - 8x - 9}$

**51.** $x^5\qquad\qquad\qquad - 1$
   $\underline{x^5 - x^4 + x^3 - x^2 + x - 1}$

**52.** $x^5 + x^4 - x^3 + x^2 - x + 2$
   $\underline{x^5 - x^4 + x^3 - x^2 - x + 2}$

---

**d** Solve.

**53.** Find a polynomial for the sum of the areas of these rectangles.

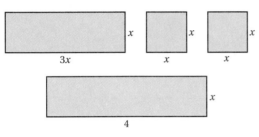

**54.** Find a polynomial for the sum of the areas of these circles.

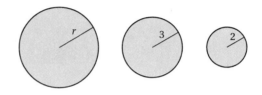

Find a polynomial for the perimeter of the figure.

**55.**

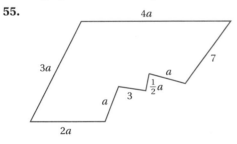

**56.**

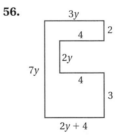

---

**Skill Maintenance**

Solve.  [2.3b]

**57.** $8x + 3x = 66$

**58.** $5x - 7x = 38$

**59.** $\frac{3}{8}x + \frac{1}{4} - \frac{3}{4}x = \frac{11}{16} + x$

**60.** $5x - 4 = 26 - x$

**61.** $1.5x - 2.7x = 22 - 5.6x$

**62.** $3x - 3 = -4x + 4$

Solve.  [2.3c]

**63.** $6(y - 3) - 8 = 4(y + 2) + 5$

**64.** $8(5x + 2) = 7(6x - 3)$

Solve.  [2.7e]

**65.** $3x - 7 \le 5x + 13$

**66.** $2(x - 4) > 5(x - 3) + 7$

## Synthesis

**67.** ◆ Is the sum of two binomials ever a trinomial? Why or why not?

**68.** ◆ Which, if any, of the commutative, associative, and distributive laws are needed for adding polynomials? Why?

Find two algebraic expressions for the area of the figure.

**69.**

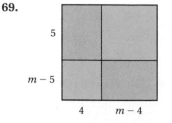

**70.**

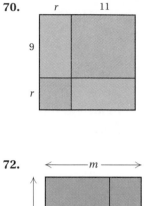

Find a polynomial for the shaded area of the figure.

**71.**

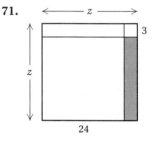

**72.**

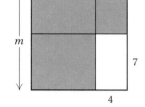

**73.** Find $(y - 2)^2$ using the four parts of this square.

**74.** Find a polynomial for the surface area of this right rectangular solid.

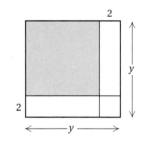

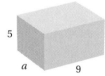

Simplify.

**75.** $(3x^2 - 4x + 6) - (-2x^2 + 4) + (-5x - 3)$

**76.** $(7y^2 - 5y + 6) - (3y^2 + 8y - 12) + (8y^2 - 10y + 3)$

**77.** $(-4 + x^2 + 2x^3) - (-6 - x + 3x^3) - (-x^2 - 5x^3)$

**78.** $(-y^4 - 7y^3 + y^2) + (-2y^4 + 5y - 2) - (-6y^3 + y^2)$

The TABLE feature can be used as a partial check that polynomials have been added or subtracted correctly. To check Example 3, we enter

$y_1 = (3x^2 - 2x + 2) + (5x^3 - 2x^2 + 3x - 4)$
and   $y_2 = 5x^3 + x^2 + x - 2.$

If our addition was correct, the $y_1$- and $y_2$-values should be the same, regardless of the table settings used.

**79.** Use the TABLE feature to check Exercise 7.

**80.** Use the TABLE feature to check Exercise 9.

| X | Y1 | Y2 |
|---|-----|-----|
| 4 | 338 | 338 |
| 5 | 653 | 653 |
| 6 | 1120 | 1120 |
| 7 | 1769 | 1769 |
| 8 | 2630 | 2630 |
| 9 | 3733 | 3733 |
| 10 | 5108 | 5108 |
| X = 4 | | |

Copyright © 1999 Addison Wesley Longman

# 4.5 Multiplication of Polynomials

We now multiply polynomials using techniques based, for the most part, on the distributive laws, but also on the associative and commutative laws. As we proceed in this chapter, we will develop special ways to find certain products.

## a | Multiplying Monomials

Consider $(3x)(4x)$. We multiply as follows:

$$
\begin{aligned}
(3x)(4x) &= 3 \cdot x \cdot 4 \cdot x && \text{By the associative law of multiplication} \\
&= 3 \cdot 4 \cdot x \cdot x && \text{By the commutative law of multiplication} \\
&= (3 \cdot 4)(x \cdot x) && \text{By the associative law} \\
&= 12x^2. && \text{Using the product rule for exponents}
\end{aligned}
$$

> To find an equivalent expression for the product of two monomials, multiply the coefficients and then multiply the variables using the product rule for exponents.

**Examples** Multiply.

**1.** $5x \cdot 6x = (5 \cdot 6)(x \cdot x)$    By the associative and commutative laws
$= 30x^2$    Multiplying the coefficients and multiplying the variables

**2.** $(3x)(-x) = (3x)(-1x)$
$= (3)(-1)(x \cdot x)$
$= -3x^2$

**3.** $(-7x^5)(4x^3) = (-7 \cdot 4)(x^5 \cdot x^3)$
$= -28x^{5+3}$    Adding the exponents
$= -28x^8$    Simplifying

After some practice, you can do this mentally. Multiply the coefficients and then the variables by keeping the base and adding the exponents. Write only the answer.

*Do Exercises 1–8.*

## b | Multiplying a Monomial and Any Polynomial

To find an equivalent expression for the product of a monomial, such as $2x$, and a binomial, such as $5x + 3$, we use a distributive law and multiply each term of $5x + 3$ by $2x$.

**Example 4** Multiply: $2x(5x + 3)$.

$$
\begin{aligned}
2x(5x + 3) &= (2x)(5x) + (2x)(3) && \text{Using a distributive law} \\
&= 10x^2 + 6x && \text{Multiplying the monomials}
\end{aligned}
$$

---

**Objectives**

a   Multiply monomials.

b   Multiply a monomial and any polynomial.

c   Multiply two binomials.

d   Multiply any two polynomials.

**For Extra Help**

TAPE 8    TAPE 8A    MAC WIN    CD-ROM

Multiply.

**1.** $(3x)(-5)$

**2.** $(-x) \cdot x$

**3.** $(-x)(-x)$

**4.** $(-x^2)(x^3)$

**5.** $3x^5 \cdot 4x^2$

**6.** $(4y^5)(-2y^6)$

**7.** $(-7y^4)(-y)$

**8.** $7x^5 \cdot 0$

*Answers on page A-18*

Multiply.

**9.** $4x(2x + 4)$

**10.** $3t^2(-5t + 2)$

**11.** $5x^3(x^3 + 5x^2 - 6x + 8)$

Multiply.

**12.** $(x + 8)(x + 5)$

**13.** $(x + 5)(x - 4)$

*Answers on page A-18*

**Example 5**   Multiply: $5x(2x^2 - 3x + 4)$.

$$5x(2x^2 - 3x + 4) = (5x)(2x^2) - (5x)(3x) + (5x)(4)$$
$$= 10x^3 - 15x^2 + 20x$$

> To multiply a monomial and a polynomial, multiply each term of the polynomial by the monomial.

**Example 6**   Multiply: $2x^2(x^3 - 7x^2 + 10x - 4)$.

$$2x^2(x^3 - 7x^2 + 10x - 4) = 2x^5 - 14x^4 + 20x^3 - 8x^2$$

*Do Exercises 9–11.*

## c | Multiplying Two Binomials

To find an equivalent expression for the product of two binomials, we use the distributive laws more than once. In Example 7, we use a distributive law three times.

**Example 7**   Multiply: $(x + 5)(x + 4)$.

$$(x + 5)(x + 4) = x(x + 4) + 5(x + 4) \qquad \text{Using a distributive law}$$
$$= x \cdot x + x \cdot 4 + 5 \cdot x + 5 \cdot 4 \qquad \text{Using a distributive law on each part}$$
$$= x^2 + 4x + 5x + 20 \qquad \text{Multiplying the monomials}$$
$$= x^2 + 9x + 20 \qquad \text{Collecting like terms}$$

To visualize the product in Example 7, consider a rectangle of length $x + 5$ and width $x + 4$.

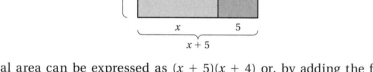

The total area can be expressed as $(x + 5)(x + 4)$ or, by adding the four smaller areas, $x^2 + 5x + 4x + 20$.

*Do Exercises 12 and 13.*

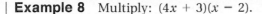

**Example 8**   Multiply: $(4x + 3)(x - 2)$.

$$(4x + 3)(x - 2) = 4x(x - 2) + 3(x - 2) \quad \text{Using a distributive law}$$
$$= 4x \cdot x - 4x \cdot 2 + 3 \cdot x - 3 \cdot 2 \quad \begin{array}{l}\text{Using a distributive law}\\\text{on each part}\end{array}$$
$$= 4x^2 - 8x + 3x - 6 \quad \text{Multiplying the monomials}$$
$$= 4x^2 - 5x - 6 \quad \text{Collecting like terms}$$

*Do Exercises 14 and 15.*

## d   Multiplying Any Two Polynomials

Let's consider the product of a binomial and a trinomial. We use a distributive law four times. You may see ways to skip some steps and do the work mentally.

**Example 9**   Multiply: $(x^2 + 2x - 3)(x^2 + 4)$.

$$(x^2 + 2x - 3)(x^2 + 4) = x^2(x^2 + 4) + 2x(x^2 + 4) - 3(x^2 + 4)$$
$$= x^2 \cdot x^2 + x^2 \cdot 4 + 2x \cdot x^2 + 2x \cdot 4 - 3 \cdot x^2 - 3 \cdot 4$$
$$= x^4 + 4x^2 + 2x^3 + 8x - 3x^2 - 12$$
$$= x^4 + 2x^3 + x^2 + 8x - 12$$

*Do Exercises 16 and 17.*

Perhaps you have discovered the following in the preceding examples.

> To multiply two polynomials $P$ and $Q$, select one of the polynomials—say, $P$. Then multiply each term of $P$ by every term of $Q$ and collect like terms.

We can use columns for long multiplications. We multiply each term at the top by every term at the bottom. We write like terms in columns, and then we add the results. Such multiplication is like multiplying with whole numbers:

$$
\begin{array}{r}
4\ 5\ 7 \\
\times\quad 6\ 3 \\
\hline
1\ 3\ 7\ 1 \\
2\ 7\ 4\ 2\ 0 \\
\hline
2\ 8\ 7\ 9\ 1
\end{array}
\qquad
\begin{array}{r}
4\ 5\ 7 \\
\times\quad 6\ 3 \\
\hline
1200 + 150 + 21 \\
24000 + 3000 + 420 \\
\hline
24000 + 4200 + 570 + 21
\end{array}
\qquad
\begin{array}{l}
= 400 + 50 + 7 \\
= 60 + 3 \\
= 3(457) = 3(400 + 50 + 7) \\
= 60(457) = 60(400 + 50 + 7) \\
= 28{,}791
\end{array}
$$

**Example 10**   Multiply: $(4x^2 - 2x + 3)(x + 2)$.

$$
\begin{array}{r}
4x^2 - 2x + 3 \\
x + 2 \\
\hline
8x^2 - 4x + 6 \qquad \text{Multiplying the top row by 2} \\
4x^3 - 2x^2 + 3x \phantom{ + 6} \qquad \text{Multiplying the top row by } x \\
\hline
4x^3 + 6x^2 - \phantom{0}x + 6 \qquad \text{Collecting like terms}
\end{array}
$$

Line up like terms in columns.

**Multiply.**

**14.** $(5x + 3)(x - 4)$

**15.** $(2x - 3)(3x - 5)$

**Multiply.**

**16.** $(x^2 + 3x - 4)(x^2 + 5)$

**17.** $(3y^2 - 7)(2y^3 - 2y + 5)$

*Answers on page A-18*

Multiply.

**18.** $3x^2 - 2x + 4$
$\phantom{3x^2 - 2x}\ x + 5$
$\overline{\phantom{3x^2 - 2x + 4x + 5}}$

**19.** $-5x^2 + 4x + 2$
$\phantom{-5x^2}\ -4x^2 - 8$
$\overline{\phantom{-5x^2 + 4x + 2 - 8}}$

**20.** Multiply.

$3x^2 - 2x - 5$
$2x^2 + \phantom{2}x - 2$
$\overline{\phantom{3x^2 - 2x - 5x - 2}}$

**Example 11** Multiply: $(5x^3 - 3x + 4)(-2x^2 - 3)$.

When missing terms occur, it helps to leave spaces for them and align like terms as we multiply.

$$\begin{array}{r}
5x^3 \phantom{000} - 3x + 4 \\
-2x^2 \phantom{0000} - 3 \\
\hline
-15x^3 \phantom{00} + 9x - 12 \qquad \text{Multiplying by } -3 \\
-10x^5 + 6x^3 - 8x^2 \phantom{0000000} \qquad \text{Multiplying by } -2x^2 \\
\hline
-10x^5 - 9x^3 - 8x^2 + 9x - 12 \qquad \text{Collecting like terms}
\end{array}$$

*Do Exercises 18 and 19.*

**Example 12** Multiply: $(2x^2 + 3x - 4)(2x^2 - x + 3)$.

$$\begin{array}{r}
2x^2 + 3x - 4 \\
2x^2 - \phantom{3}x + 3 \\
\hline
6x^2 + 9x - 12 \qquad \text{Multiplying by 3} \\
-2x^3 - 3x^2 + 4x \phantom{0000} \qquad \text{Multiplying by } -x \\
4x^4 + 6x^3 - 8x^2 \phantom{0000000} \qquad \text{Multiplying by } 2x^2 \\
\hline
4x^4 + 4x^3 - 5x^2 + 13x - 12 \qquad \text{Collecting like terms}
\end{array}$$

*Do Exercise 20.*

---

## Calculator Spotlight

Checking Multiplications with a Table or Graph

**Table.** The TABLE feature can be used as a partial check that polynomials have been multiplied correctly. To check whether

$$(x + 5)(x - 4) = x^2 + 9x - 20$$

is correct, we enter

$$y_1 = (x + 5)(x - 4) \quad \text{and} \quad y_2 = x^2 + 9x - 20.$$

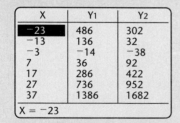

If our multiplication is correct, the $y_1$- and $y_2$-values should be the same, regardless of the table settings used. We see that $y_1$ and $y_2$ are not the same, so the multiplication is not correct.

**Graph.** Multiplication of polynomials can also be checked with the GRAPH feature. In this case, we see that the graphs differ, so the multiplication is not correct.

$$y_1 = (x + 5)(x - 4), \quad y_2 = x^2 + 9x - 20$$

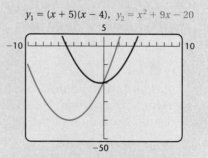

**Exercises**

Use the TABLE or GRAPH feature to check whether each of the following is correct.

**1.** $(x + 5)(x + 4) = x^2 + 9x + 20$ (Example 7)
**2.** $(4x + 3)(x - 2) = 4x^2 - 5x - 6$ (Example 8)
**3.** $(5x + 3)(x - 4) = 5x^2 + 17x - 12$
**4.** $(2x - 3)(3x - 5) = 6x^2 - 19x - 15$
**5.** $(x - 3)(x - 3) = x^2 - 9$
**6.** $(x - 3)(x + 3) = x^2 - 9$

---

*Answers on page A-18*

# Exercise Set 4.5

**a** Multiply.

**1.** $(8x^2)(5)$   **2.** $(4x^2)(-2)$   **3.** $(-x^2)(-x)$   **4.** $(-x^3)(x^2)$

**5.** $(8x^5)(4x^3)$   **6.** $(10a^2)(2a^2)$   **7.** $(0.1x^6)(0.3x^5)$   **8.** $(0.3x^4)(-0.8x^6)$

**9.** $\left(-\frac{1}{5}x^3\right)\left(-\frac{1}{3}x\right)$   **10.** $\left(-\frac{1}{4}x^4\right)\left(\frac{1}{5}x^8\right)$   **11.** $(-4x^2)(0)$   **12.** $(-4m^5)(-1)$

**13.** $(3x^2)(-4x^3)(2x^6)$   **14.** $(-2y^5)(10y^4)(-3y^3)$

**b** Multiply.

**15.** $2x(-x + 5)$   **16.** $3x(4x - 6)$   **17.** $-5x(x - 1)$

**18.** $-3x(-x - 1)$   **19.** $x^2(x^3 + 1)$   **20.** $-2x^3(x^2 - 1)$

**21.** $3x(2x^2 - 6x + 1)$   **22.** $-4x(2x^3 - 6x^2 - 5x + 1)$   **23.** $(-6x^2)(x^2 + x)$

**24.** $(-4x^2)(x^2 - x)$   **25.** $(3y^2)(6y^4 + 8y^3)$   **26.** $(4y^4)(y^3 - 6y^2)$

**c** Multiply.

**27.** $(x + 6)(x + 3)$   **28.** $(x + 5)(x + 2)$   **29.** $(x + 5)(x - 2)$   **30.** $(x + 6)(x - 2)$

**31.** $(x - 4)(x - 3)$   **32.** $(x - 7)(x - 3)$   **33.** $(x + 3)(x - 3)$   **34.** $(x + 6)(x - 6)$

**35.** $(5 - x)(5 - 2x)$   **36.** $(3 + x)(6 + 2x)$   **37.** $(2x + 5)(2x + 5)$   **38.** $(3x - 4)(3x - 4)$

**39.** $\left(x - \frac{5}{2}\right)\left(x + \frac{2}{5}\right)$   **40.** $\left(x + \frac{4}{3}\right)\left(x + \frac{3}{2}\right)$   **41.** $(x - 2.3)(x + 4.7)$   **42.** $(2x + 0.13)(2x - 0.13)$

**d** Multiply.

**43.** $(x^2 + x + 1)(x - 1)$   **44.** $(x^2 + x - 2)(x + 2)$   **45.** $(2x + 1)(2x^2 + 6x + 1)$

**46.** $(3x - 1)(4x^2 - 2x - 1)$   **47.** $(y^2 - 3)(3y^2 - 6y + 2)$   **48.** $(3y^2 - 3)(y^2 + 6y + 1)$

**49.** $(x^3 + x^2)(x^3 + x^2 - x)$

**50.** $(x^3 - x^2)(x^3 - x^2 + x)$

**51.** $(-5x^3 - 7x^2 + 1)(2x^2 - x)$

**52.** $(-4x^3 + 5x^2 - 2)(5x^2 + 1)$

**53.** $(1 + x + x^2)(-1 - x + x^2)$

**54.** $(1 - x + x^2)(1 - x + x^2)$

**55.** $(2t^2 - t - 4)(3t^2 + 2t - 1)$

**56.** $(3a^2 - 5a + 2)(2a^2 - 3a + 4)$

**57.** $(x - x^3 + x^5)(x^2 - 1 + x^4)$

**58.** $(x - x^3 + x^5)(3x^2 + 3x^6 + 3x^4)$

**59.** $(x^3 + x^2 + x + 1)(x - 1)$

**60.** $(x + 2)(x^3 - x^2 + x - 2)$

---

**Skill Maintenance**

Simplify.

**61.** $-\dfrac{1}{4} - \dfrac{1}{2}$  [1.4a]

**62.** $-3.8 - (-10.2)$  [1.4a]

**63.** $(10 - 2)(10 + 2)$  [1.8d]

**64.** $10 - 2 + (-6)^2 \div 3 \cdot 2$  [1.8d]

Factor.  [1.7d]

**65.** $15x - 18y + 12$

**66.** $16x - 24y + 36$

**67.** $-9x - 45y + 15$

**68.** $100x - 100y + 1000a$

**69.** Graph: $y = \dfrac{1}{2}x - 3$.  [3.2b]

**70.** Solve: $4(x - 3) = 5(2 - 3x) + 1$.  [2.3c]

---

**Synthesis**

**71.** ◆ Under what conditions will the product of two binomials be a trinomial?

**72.** ◆ Is it possible to understand polynomial multiplication without first understanding the distributive law? Why or why not?

**73.** Find a polynomial for the shaded area.

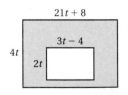

**74.** A box with a square bottom is to be made from a 12-in.-square piece of cardboard. Squares with side $x$ are cut out of the corners and the sides are folded up. Find polynomials for the volume and the outside surface area of the box.

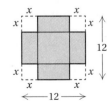

**75.** The height of a triangle is 4 ft longer than its base. Find a polynomial for the area.

Compute and simplify.

**76.** $(x + 3)(x + 6) + (x + 3)(x + 6)$

**77.** $(x - 2)(x - 7) - (x - 2)(x - 7)$

Visualize polynomial multiplication using rectangles.

# 4.6 Special Products

We encounter certain products so often that it is helpful to have faster methods of computing. We now consider special ways of multiplying any two binomials. Such techniques are called *special products*.

## a | Products of Two Binomials Using FOIL

To multiply two binomials, we can select one binomial and multiply each term of that binomial by every term of the other. Then we collect like terms. Consider the product $(x + 5)(x + 4)$:

$$(x + 5)(x + 4) = x \cdot x + 5 \cdot x + x \cdot 4 + 5 \cdot 4$$
$$= x^2 + 5x + 4x + 20$$
$$= x^2 + 9x + 20.$$

We can rewrite the first line of this product to show a special technique for finding the product of two binomials:

$$
\underbrace{\text{First terms}} \quad \underbrace{\text{Outside terms}} \quad \underbrace{\text{Inside terms}} \quad \underbrace{\text{Last terms}}
$$

$$(x + 5)(x + 4) = \quad x \cdot x \quad + \quad 4 \cdot x \quad + \quad 5 \cdot x \quad + \quad 5 \cdot 4.$$

To remember this method of multiplying, we use the initials **FOIL**.

---

**THE FOIL METHOD**

To multiply two binomials, $A + B$ and $C + D$, multiply the First terms $AC$, the Outside terms $AD$, the Inside terms $BC$, and then the Last terms $BD$. Then collect like terms, if possible.

$$(A + B)(C + D) = AC + AD + BC + BD$$

1. Multiply First terms: $AC$.
2. Multiply Outside terms: $AD$.
3. Multiply Inside terms: $BC$.
4. Multiply Last terms: $BD$.

↓

**FOIL**

$$(A + B)(C + D)$$

---

**Example 1** Multiply: $(x + 8)(x^2 - 5)$.

We have

$$(x + 8)(x^2 - 5) = \overset{F}{x^3} - \overset{O}{5x} + \overset{I}{8x^2} - \overset{L}{40}$$
$$= x^3 + 8x^2 - 5x - 40.$$

Since each of the original binomials is in descending order, we write the product in descending order, as is customary, but this is not a "must."

## Objectives

a | Multiply two binomials mentally using the FOIL method.

b | Multiply the sum and the difference of two terms mentally.

c | Square a binomial mentally.

d | Find special products when polynomial products are mixed together.

**For Extra Help**

TAPE 8    TAPE 8B    MAC WIN    CD-ROM

Multiply mentally, if possible. If you need extra steps, be sure to use them.

**1.** $(x + 3)(x + 4)$

**2.** $(x + 3)(x - 5)$

**3.** $(2x - 1)(x - 4)$

**4.** $(2x^2 - 3)(x - 2)$

**5.** $(6x^2 + 5)(2x^3 + 1)$

**6.** $(y^3 + 7)(y^3 - 7)$

**7.** $(t + 5)(t + 3)$

**8.** $(2x^4 + x^2)(-x^3 + x)$

Multiply.

**9.** $\left(x + \dfrac{4}{5}\right)\left(x - \dfrac{4}{5}\right)$

**10.** $(x^3 - 0.5)(x^2 + 0.5)$

**11.** $(2 + 3x^2)(4 - 5x^2)$

**12.** $(6x^3 - 3x^2)(5x^2 - 2x)$

*Answers on page A-18*

---

Often we can collect like terms after we have multiplied.

**Examples** Multiply.

**2.** $(x + 6)(x - 6) = x^2 - 6x + 6x - 36$    **Using FOIL**
$$= x^2 - 36 \qquad \text{Collecting like terms}$$

**3.** $(x + 7)(x + 4) = x^2 + 4x + 7x + 28$
$$= x^2 + 11x + 28$$

**4.** $(y - 3)(y - 2) = y^2 - 2y - 3y + 6$
$$= y^2 - 5y + 6$$

**5.** $(x^3 - 5)(x^3 + 5) = x^6 + 5x^3 - 5x^3 - 25$
$$= x^6 - 25$$

*Do Exercises 1–8.*

**Examples** Multiply.

**6.** $(4t^3 + 5)(3t^2 - 2) = 12t^5 - 8t^3 + 15t^2 - 10$

**7.** $\left(x - \dfrac{2}{3}\right)\left(x + \dfrac{2}{3}\right) = x^2 + \dfrac{2}{3}x - \dfrac{2}{3}x - \dfrac{4}{9}$
$$= x^2 - \dfrac{4}{9}$$

**8.** $(x^2 - 0.3)(x^2 - 0.3) = x^4 - 0.3x^2 - 0.3x^2 + 0.09$
$$= x^4 - 0.6x^2 + 0.09$$

**9.** $(3 - 4x)(7 - 5x^3) = 21 - 15x^3 - 28x + 20x^4$
$$= 21 - 28x - 15x^3 + 20x^4$$

(*Note*: If the original polynomials are in ascending order, it is natural to write the product in ascending order, but this is not a "must.")

**10.** $(5x^4 + 2x^3)(3x^2 - 7x) = 15x^6 - 35x^5 + 6x^5 - 14x^4$
$$= 15x^6 - 29x^5 - 14x^4$$

*Do Exercises 9–12.*

We can show the FOIL method geometrically as follows.

The area of the large rectangle is $(A + B)(C + D)$.

The area of rectangle ① is $AC$.

The area of rectangle ② is $AD$.

The area of rectangle ③ is $BC$.

The area of rectangle ④ is $BD$.

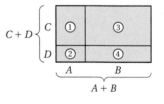

The area of the large rectangle is the sum of the areas of the smaller rectangles. Thus,

$$(A + B)(C + D) = AC + AD + BC + BD.$$

## b | Multiplying Sums and Differences of Two Terms

Consider the product of the sum and the difference of the same two terms, such as

$$(x + 2)(x - 2).$$

Since this is the product of two binomials, we can use FOIL. This type of product occurs so often, however, that it would be valuable if we could use an even faster method. To find a faster way to compute such a product, look for a pattern in the following:

a) $(x + 2)(x - 2) = x^2 - 2x + 2x - 4$
$$= x^2 - 4;$$

b) $(3x - 5)(3x + 5) = 9x^2 + 15x - 15x - 25$
$$= 9x^2 - 25.$$

**Do Exercises 13 and 14.**

Perhaps you discovered in each case that when you multiply the two binomials, two terms are opposites, or additive inverses, which add to 0 and "drop out."

> The product of the sum and the difference of the same two terms is the square of the first term minus the square of the second term:
> $$(A + B)(A - B) = A^2 - B^2.$$

It is helpful to memorize this rule in both words and symbols. (If you do forget it, you can, of course, use FOIL.)

**Examples** Multiply. (Carry out the rule and say the words as you go.)

$$(A + B) \, (A - B) = A^2 - B^2$$

**11.** $(x + 4) \, (x - 4) = x^2 - 4^2$   "The square of the first term, $x^2$, minus the square of the second, $4^2$"

$$= x^2 - 16 \quad \text{Simplifying}$$

**12.** $(5 + 2w)(5 - 2w) = 5^2 - (2w)^2$
$$= 25 - 4w^2$$

**13.** $(3x^2 - 7)(3x^2 + 7) = (3x^2)^2 - 7^2$
$$= 9x^4 - 49$$

**14.** $(-4x - 10)(-4x + 10) = (-4x)^2 - 10^2$
$$= 16x^2 - 100$$

**15.** $\left(x + \dfrac{3}{8}\right)\left(x - \dfrac{3}{8}\right) = x^2 - \left(\dfrac{3}{8}\right)^2 = x^2 - \dfrac{9}{64}$

**Do Exercises 15–19.**

## c | Squaring Binomials

Consider the square of a binomial, such as $(x + 3)^2$. This can be expressed as $(x + 3)(x + 3)$. Since this is the product of two binomials, we can again use FOIL. But again, this type of product occurs so often that we would like to use an even faster method. Look for a pattern in the following:

a) $(x + 3)^2 = (x + 3)(x + 3)$
$$= x^2 + 3x + 3x + 9$$
$$= x^2 + 6x + 9;$$

b) $(5 + 3p)^2 = (5 + 3p)(5 + 3p)$
$$= 25 + 15p + 15p + 9p^2$$
$$= 25 + 30p + 9p^2;$$

Multiply.

**13.** $(x + 5)(x - 5)$

**14.** $(2x - 3)(2x + 3)$

Multiply.

**15.** $(x + 2)(x - 2)$

**16.** $(x - 7)(x + 7)$

**17.** $(6 - 4y)(6 + 4y)$

**18.** $(2x^3 - 1)(2x^3 + 1)$

**19.** $\left(x - \dfrac{2}{5}\right)\left(x + \dfrac{2}{5}\right)$

*Answers on page A-18*

Multiply.

**20.** $(x + 8)(x + 8)$

**21.** $(x - 5)(x - 5)$

Multiply.

**22.** $(x + 2)^2$

**23.** $(a - 4)^2$

**24.** $(2x + 5)^2$

**25.** $(4x^2 - 3x)^2$

**26.** $(7.8 + 1.2y)(7.8 + 1.2y)$

**27.** $(3x^2 - 5)(3x^2 - 5)$

*Answers on page A-18*

**c)** $(x - 3)^2 = (x - 3)(x - 3)$
$= x^2 - 3x - 3x + 9$
$= x^2 - 6x + 9;$

**d)** $(3x - 5)^2 = (3x - 5)(3x - 5)$
$= 9x^2 - 15x - 15x + 25$
$= 9x^2 - 30x + 25.$

### Do Exercises 20 and 21.

When squaring a binomial, we multiply a binomial by itself. Perhaps you noticed that two terms are the same and when added give twice their product. The other two terms are squares.

> The square of a sum or a difference of two terms is the square of the first term, plus or minus twice the product of the two terms, plus the square of the last term:
> $$(A + B)^2 = A^2 + 2AB + B^2;$$
> $$(A - B)^2 = A^2 - 2AB + B^2.$$

It is helpful to memorize this rule in both words and symbols.

**Examples**   Multiply. (Carry out the rule and say the words as you go.)

$$(A + B)^2 = A^2 + 2 \cdot A \cdot B + B^2$$

**16.** $(x + 3)^2 = x^2 + 2 \cdot x \cdot 3 + 3^2$   "$x^2$ plus 2 times $x$ times 3 plus $3^2$"
$= x^2 + 6x + 9$

$$(A - B)^2 = A^2 - 2 \cdot A \cdot B + B^2$$

**17.** $(t - 5)^2 = t^2 - 2 \cdot t \cdot 5 + 5^2$   "$t^2$ minus 2 times $t$ times 5 plus $5^2$"
$= t^2 - 10t + 25$

**18.** $(2x + 7)^2 = (2x)^2 + 2 \cdot 2x \cdot 7 + 7^2$
$= 4x^2 + 28x + 49$

**19.** $(5x - 3x^2)^2 = (5x)^2 - 2 \cdot 5x \cdot 3x^2 + (3x^2)^2$
$= 25x^2 - 30x^3 + 9x^4$

**20.** $(2.3 - 5.4m)^2 = 2.3^2 - 2(2.3)(5.4m) + (5.4m)^2$
$= 5.29 - 24.84m + 29.16m^2$

### Do Exercises 22–27.

*CAUTION!*   Note carefully in these examples that the square of a sum is *not* the sum of the squares:

The middle term $2AB$ is missing.

$$(A + B)^2 \neq A^2 + B^2.$$

To see this, note that

$$(20 + 5)^2 = 25^2 = 625,$$

but

$$20^2 + 5^2 = 400 + 25 = 425 \quad \text{and} \quad 425 \neq 625.$$

However, $20^2 + 2(20)(5) + 5^2 = 625$, which illustrates that

$$(A + B)^2 = A^2 + 2AB + B^2.$$

We can look at the rule for finding $(A + B)^2$ geometrically as follows. The area of the large square is

$$(A + B)(A + B) = (A + B)^2.$$

This is equal to the sum of the areas of the smaller rectangles:

$$A^2 + AB + AB + B^2 = A^2 + 2AB + B^2.$$

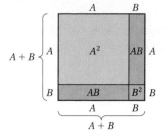

Thus,

$$(A + B)^2 = A^2 + 2AB + B^2.$$

## d | Multiplication of Various Types

We have considered how to quickly multiply certain kinds of polynomials. Let's now try several types of multiplications mixed together so that we can learn to sort them out. When you multiply, first see what kind of multiplication you have. Then use the best method. The formulas you should know and the questions you should ask yourself are as follows.

---

**MULTIPLYING TWO POLYNOMIALS**

1. Is the product the square of a binomial? If so, use the following:
   $$(A + B)(A + B) = (A + B)^2 = A^2 + 2AB + B^2,$$
   or  $(A - B)(A - B) = (A - B)^2 = A^2 - 2AB + B^2.$

   The square of a binomial is the square of the first term, plus or minus *twice* the product of the two terms, plus the square of the last term.

   [The answer has 3 terms.]

   *Example*: $(x + 7)(x + 7) = (x + 7)^2$
   $$= x^2 + 2 \cdot x \cdot 7 + 7^2 = x^2 + 14x + 49$$

2. Is it the product of the sum and the difference of the *same* two terms? If so, use the following:
   $$(A + B)(A - B) = A^2 - B^2.$$

   The product of the sum and the difference of the same two terms is the difference of the squares.

   [The answer has 2 terms.]

   *Example*: $(x + 7)(x - 7) = x^2 - 7^2 = x^2 - 49$

3. Is it the product of two binomials other than those above? If so, use FOIL.

   [The answer will have 3 or 4 terms.]

   *Example*: $(x + 7)(x - 4) = x^2 - 4x + 7x - 28 = x^2 + 3x - 28$

4. Is it the product of a monomial and a polynomial? If so, multiply each term of the polynomial by the monomial.

   *Example*: $5x(x + 7) = 5x \cdot x + 5x \cdot 7 = 5x^2 + 35x$

5. Is it the product of two polynomials other than those above? If so, multiply each term of one by every term of the other. Use columns if you wish.

   [The answer will have 2 or more terms, usually more than 2 terms.]

   *Example*: $(x^2 - 3x + 2)(x + 7) = x^2(x + 7) - 3x(x + 7) + 2(x + 7)$
   $$= x^2 \cdot x + x^2 \cdot 7 - 3x \cdot x - 3x \cdot 7$$
   $$+ 2 \cdot x + 2 \cdot 7$$
   $$= x^3 + 7x^2 - 3x^2 - 21x + 2x + 14$$
   $$= x^3 + 4x^2 - 19x + 14$$

---

Remember that FOIL will *always* work for two binomials. You can use it instead of either of the first two rules, but those rules will make your work go faster.

Multiply.

**28.** $(x + 5)(x + 6)$

**29.** $(t - 4)(t + 4)$

**30.** $4x^2(-2x^3 + 5x^2 + 10)$

**31.** $(9x^2 + 1)^2$

**32.** $(2a - 5)(2a + 8)$

**33.** $\left(5x + \dfrac{1}{2}\right)^2$

**34.** $\left(2x - \dfrac{1}{2}\right)^2$

**35.** $(x^2 - x + 4)(x - 2)$

---

**Example 21**    Multiply: $(x + 3)(x - 3)$.

$$(x + 3)(x - 3) = x^2 - 9 \qquad$$ Using method 2 (the product of the sum and the difference of two terms)

**Example 22**    Multiply: $(t + 7)(t - 5)$.

$$(t + 7)(t - 5) = t^2 + 2t - 35 \qquad$$ Using method 3 (the product of two binomials, but neither the square of a binomial nor the product of the sum and the difference of two terms)

**Example 23**    Multiply: $(x + 6)(x + 6)$.

$$(x + 6)(x + 6) = x^2 + 2(6)x + 36 \qquad$$ Using method 1 (the square of a binomial sum)

$$= x^2 + 12x + 36$$

**Example 24**    Multiply: $2x^3(9x^2 + x - 7)$.

$$2x^3(9x^2 + x - 7) = 18x^5 + 2x^4 - 14x^3 \qquad$$ Using method 4 (the product of a monomial and a trinomial; multiplying each term of the trinomial by the monomial)

**Example 25**    Multiply: $(5x^3 - 7x)^2$.

$$(5x^3 - 7x)^2 = 25x^6 - 2(5x^3)(7x) + 49x^2 \qquad$$ Using method 1 (the square of a binomial difference)

$$= 25x^6 - 70x^4 + 49x^2$$

**Example 26**    Multiply: $\left(3x + \dfrac{1}{4}\right)^2$.

$$\left(3x + \tfrac{1}{4}\right)^2 = 9x^2 + 2(3x)\left(\tfrac{1}{4}\right) + \tfrac{1}{16} \qquad$$ Using method 1 (the square of a binomial sum. To get the middle term, we multiply $3x$ by $\frac{1}{4}$ and double.)

$$= 9x^2 + \tfrac{3}{2}x + \tfrac{1}{16}$$

**Example 27**    Multiply: $\left(4x - \dfrac{3}{4}\right)^2$.

$$\left(4x - \tfrac{3}{4}\right)^2 = 16x^2 - 2(4x)\left(\tfrac{3}{4}\right) + \tfrac{9}{16} \qquad$$ Using method 1 (the square of a binomial difference)

$$= 16x^2 - 6x + \tfrac{9}{16}$$

**Example 28**    Multiply: $(p + 3)(p^2 + 2p - 1)$.

$$
\begin{array}{r}
p^2 + 2p - 1 \\
p + 3 \\
\hline
3p^2 + 6p - 3 \\
p^3 + 2p^2 - p \phantom{{} - 3} \\
\hline
p^3 + 5p^2 + 5p - 3
\end{array}
$$

Using method 5 (the product of two polynomials)

Multiplying by 3

Multiplying by $p$

*Do Exercises 28–35.*

---

*Answers on page A-18*

# Exercise Set 4.6

**a** Multiply. Try to write only the answer. If you need more steps, be sure to use them.

**1.** $(x + 1)(x^2 + 3)$

**2.** $(x^2 - 3)(x - 1)$

**3.** $(x^3 + 2)(x + 1)$

**4.** $(x^4 + 2)(x + 10)$

**5.** $(y + 2)(y - 3)$

**6.** $(a + 2)(a + 3)$

**7.** $(3x + 2)(3x + 2)$

**8.** $(4x + 1)(4x + 1)$

**9.** $(5x - 6)(x + 2)$

**10.** $(x - 8)(x + 8)$

**11.** $(3t - 1)(3t + 1)$

**12.** $(2m + 3)(2m + 3)$

**13.** $(4x - 2)(x - 1)$

**14.** $(2x - 1)(3x + 1)$

**15.** $\left(p - \frac{1}{4}\right)\left(p + \frac{1}{4}\right)$

**16.** $\left(q + \frac{3}{4}\right)\left(q + \frac{3}{4}\right)$

**17.** $(x - 0.1)(x + 0.1)$

**18.** $(x + 0.3)(x - 0.4)$

**19.** $(2x^2 + 6)(x + 1)$

**20.** $(2x^2 + 3)(2x - 1)$

**21.** $(-2x + 1)(x + 6)$

**22.** $(3x + 4)(2x - 4)$

**23.** $(a + 7)(a + 7)$

**24.** $(2y + 5)(2y + 5)$

**25.** $(1 + 2x)(1 - 3x)$

**26.** $(-3x - 2)(x + 1)$

**27.** $(x^2 + 3)(x^3 - 1)$

**28.** $(x^4 - 3)(2x + 1)$

**29.** $(3x^2 - 2)(x^4 - 2)$

**30.** $(x^{10} + 3)(x^{10} - 3)$

**31.** $(2.8x - 1.5)(4.7x + 9.3)$

**32.** $\left(x - \frac{3}{8}\right)\left(x + \frac{4}{7}\right)$

**33.** $(3x^5 + 2)(2x^2 + 6)$

**34.** $(1 - 2x)(1 + 3x^2)$

**35.** $(8x^3 + 1)(x^3 + 8)$

**36.** $(4 - 2x)(5 - 2x^2)$

**37.** $(4x^2 + 3)(x - 3)$

**38.** $(7x - 2)(2x - 7)$

**39.** $(4y^4 + y^2)(y^2 + y)$

**40.** $(5y^6 + 3y^3)(2y^6 + 2y^3)$

**b** Multiply mentally, if possible. If you need extra steps, be sure to use them.

**41.** $(x + 4)(x - 4)$

**42.** $(x + 1)(x - 1)$

**43.** $(2x + 1)(2x - 1)$

**44.** $(x^2 + 1)(x^2 - 1)$

**45.** $(5m - 2)(5m + 2)$

**46.** $(3x^4 + 2)(3x^4 - 2)$

**47.** $(2x^2 + 3)(2x^2 - 3)$

**48.** $(6x^5 - 5)(6x^5 + 5)$

**49.** $(3x^4 - 4)(3x^4 + 4)$

**50.** $(t^2 - 0.2)(t^2 + 0.2)$

**51.** $(x^6 - x^2)(x^6 + x^2)$

**52.** $(2x^3 - 0.3)(2x^3 + 0.3)$

**53.** $(x^4 + 3x)(x^4 - 3x)$

**54.** $\left(\frac{3}{4} + 2x^3\right)\left(\frac{3}{4} - 2x^3\right)$

**55.** $(x^{12} - 3)(x^{12} + 3)$

**56.** $(12 - 3x^2)(12 + 3x^2)$

**57.** $(2y^8 + 3)(2y^8 - 3)$

**58.** $\left(m - \frac{2}{3}\right)\left(m + \frac{2}{3}\right)$

**59.** $\left(\frac{5}{8}x - 4.3\right)\left(\frac{5}{8}x + 4.3\right)$

**60.** $(10.7 - x^3)(10.7 + x^3)$

**c** Multiply mentally, if possible. If you need extra steps, be sure to use them.

**61.** $(x + 2)^2$

**62.** $(2x - 1)^2$

**63.** $(3x^2 + 1)^2$

**64.** $\left(3x + \frac{3}{4}\right)^2$

**65.** $\left(a - \frac{1}{2}\right)^2$

**66.** $\left(2a - \frac{1}{5}\right)^2$

**67.** $(3 + x)^2$

**68.** $(x^3 - 1)^2$

**69.** $(x^2 + 1)^2$  **70.** $(8x - x^2)^2$  **71.** $(2 - 3x^4)^2$  **72.** $(6x^3 - 2)^2$

**73.** $(5 + 6t^2)^2$  **74.** $(3p^2 - p)^2$  **75.** $\left(x - \frac{5}{8}\right)^2$  **76.** $(0.3y + 2.4)^2$

d  Multiply mentally, if possible.

**77.** $(3 - 2x^3)^2$  **78.** $(x - 4x^3)^2$  **79.** $4x(x^2 + 6x - 3)$  **80.** $8x(-x^5 + 6x^2 + 9)$

**81.** $\left(2x^2 - \frac{1}{2}\right)\left(2x^2 - \frac{1}{2}\right)$  **82.** $(-x^2 + 1)^2$  **83.** $(-1 + 3p)(1 + 3p)$  **84.** $(-3q + 2)(3q + 2)$

**85.** $3t^2(5t^3 - t^2 + t)$  **86.** $-6x^2(x^3 + 8x - 9)$  **87.** $(6x^4 + 4)^2$  **88.** $(8a + 5)^2$

**89.** $(3x + 2)(4x^2 + 5)$  **90.** $(2x^2 - 7)(3x^2 + 9)$  **91.** $(8 - 6x^4)^2$  **92.** $\left(\frac{1}{5}x^2 + 9\right)\left(\frac{3}{5}x^2 - 7\right)$

**93.** $(t - 1)(t^2 + t + 1)$  **94.** $(y + 5)(y^2 - 5y + 25)$

Compute each of the following and compare.

**95.** $3^2 + 4^2$; $(3 + 4)^2$  **96.** $6^2 + 7^2$; $(6 + 7)^2$  **97.** $9^2 - 5^2$; $(9 - 5)^2$  **98.** $11^2 - 4^2$; $(11 - 4)^2$

### Skill Maintenance

**99.** In apartment 3B, lamps, an air conditioner, and a television set are all operating at the same time. The lamps use 10 times as many watts of electricity as the television set, and the air conditioner uses 40 times as many watts as the television set. The total wattage used in the apartment is 2550. How many watts are used by each appliance?  [2.4a]

Solve.  [2.3c]

**100.** $3x - 8x = 4(7 - 8x)$  **101.** $3(x - 2) = 5(2x + 7)$  **102.** $5(2x - 3) - 2(3x - 4) = 20$

Solve.  [2.6a]

**103.** $3x - 2y = 12$, for $y$  **104.** $ab - cd = 4$, for $a$

**105.** ◆ Under what conditions is the product of two binomials a binomial?

**106.** ◆ Todd feels that since the FOIL method can be used to find the product of any two binomials, he needn't study the other special products. What advice would you give him?

Multiply.

**107.** $5x(3x - 1)(2x + 3)$

**108.** $[(2x - 3)(2x + 3)](4x^2 + 9)$

**109.** $[(a - 5)(a + 5)]^2$

**110.** $(a - 3)^2(a + 3)^2$
(*Hint*: Examine Exercise 109.)

**111.** $(3t^4 - 2)^2(3t^4 + 2)^2$
(*Hint*: Examine Exercise 109.)

**112.** $[3a - (2a - 3)][3a + (2a - 3)]$

Solve.

**113.** $(x + 2)(x - 5) = (x + 1)(x - 3)$

**114.** $(2x + 5)(x - 4) = (x + 5)(2x - 4)$

**115.** *Factors and Sums.* To *factor* a number is to express it as a product. Since $12 = 4 \cdot 3$, we say that 12 is *factored* and that 4 and 3 are *factors* of 12. In the following table, the top number has been factored in such a way that the sum of the factors is the bottom number. For example, in the first column, 40 has been factored as $5 \cdot 8$, and $5 + 8 = 13$, the bottom number. Such thinking is important in algebra when we factor trinomials of the type $x^2 + bx + c$. Find the missing numbers in the table.

| Product | 40 | 63 | 36 | 72 | -140 | -96 | 48 | 168 | 110 | | | |
|---|---|---|---|---|---|---|---|---|---|---|---|---|
| Factor | 5 | | | | | | | | | -9 | -24 | -3 |
| Factor | 8 | | | | | | | | | -10 | 18 | |
| Sum | 13 | 16 | -20 | -38 | -4 | 4 | -14 | -29 | -21 | | | 18 |

Find the total shaded area.

**116.**

**117.**

**118.** A factored polynomial for the shaded area in this rectangle is $(A + B)(A - B)$.

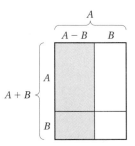

a) Find a polynomial for the area of the entire rectangle.
b) Find a polynomial for the sum of the areas of the two small unshaded rectangles.
c) Find a polynomial for the area in part (a) minus the area in part (b).
d) Find a polynomial for the area of the shaded region and compare this with the polynomial found in part (c).

〰️ Use the TABLE or GRAPH feature to check whether each of the following is correct.

**119.** $(x - 1)^2 = x^2 - 2x + 1$

**120.** $(x - 2)^2 = x^2 - 4x - 4$

**121.** $(x - 3)(x + 3) = x^2 - 6$

**122.** $(x - 3)(x + 2) = x^2 - x - 6$

Derive the special-product formulas.

# 4.7 Operations with Polynomials in Several Variables

The polynomials that we have been studying have only one variable. A **polynomial in several variables** is an expression like those you have already seen, but with more than one variable. Here are two examples:

$$3x + xy^2 + 5y + 4, \qquad 8xy^2z - 2x^3z - 13x^4y^2 + 15.$$

## a | Evaluating Polynomials

**Example 1** Evaluate the polynomial $4 + 3x + xy^2 + 8x^3y^3$ for $x = -2$ and $y = 5$.

We replace $x$ with $-2$ and $y$ with 5:

$$\begin{aligned} 4 + 3x + xy^2 + 8x^3y^3 &= 4 + 3(-2) + (-2) \cdot 5^2 + 8(-2)^3 \cdot 5^3 \\ &= 4 - 6 - 50 - 8000 \\ &= -8052. \end{aligned}$$

**Example 2** *Male Caloric Needs.* The number of calories needed each day by a moderately active man who weighs $w$ kilograms, is $h$ centimeters tall, and is $a$ years old can be estimated by the polynomial

$$19.18w + 7h - 9.52a + 92.4$$

(**Source:** Parker, M., *She Does Math*. Mathematical Association of America, p. 96). The author of this text is moderately active, weighs 97 kg, is 185 cm tall, and is 55 yr old. What are his daily caloric needs?

We evaluate the polynomial for $w = 97$, $h = 185$, and $a = 55$:

$$\begin{aligned} & 19.18w + 7h - 9.52a + 92.4 \\ &= 19.18(97) + 7(185) - 9.52(55) + 92.4 \qquad \textbf{Substituting} \\ &= 2724.26. \end{aligned}$$

His daily caloric need is about 2724 calories.

***Do Exercises 1–3.***

## Objectives

**a** Evaluate a polynomial in several variables for given values of the variables.

**b** Identify the coefficients and the degrees of the terms of a polynomial and the degree of a polynomial.

**c** Collect like terms of a polynomial.

**d** Add polynomials.

**e** Subtract polynomials.

**f** Multiply polynomials.

**For Extra Help**

TAPE 8    TAPE 8B    MAC WIN    CD-ROM

1. Evaluate the polynomial
$$4 + 3x + xy^2 + 8x^3y^3$$
for $x = 2$ and $y = -5$.

2. Evaluate the polynomial
$$8xy^2 - 2x^3z - 13x^4y^2 + 5$$
for $x = -1$, $y = 3$, and $z = 4$.

3. *Female Caloric Needs.* The number of calories needed each day by a moderately active woman who weighs $w$ pounds, is $h$ inches tall, and is $a$ years old can be estimated by the polynomial
$$917 + 6w + 6h - 6a$$
(**Source:** Parker, M., *She Does Math*. Mathematical Association of America, p. 96). Christine Poolos, a Project Manager for Addison Wesley Longman Publishing Co., is moderately active, weighs 125 lb, is 64 in. tall, and is 27 yr old. What are her daily caloric needs?

*Answers on page A-19*

**4.** Identify the coefficient of each term:

$$-3xy^2 + 3x^2y - 2y^3 + xy + 2.$$

**5.** Identify the degree of each term and the degree of the polynomial

$$4xy^2 + 7x^2y^3z^2 - 5x + 2y + 4.$$

Collect like terms.

**6.** $4x^2y + 3xy - 2x^2y$

**7.** $-3pq - 5pqr^3 - 12 + 8pq + 5pqr^3 + 4$

Answers on page A-19

## b | Coefficients and Degrees

The **degree** of a term is the sum of the exponents of the variables. The **degree of a polynomial** is the degree of the term of highest degree.

**Example 3**  Identify the coefficient and the degree of each term and the degree of the polynomial

$$9x^2y^3 - 14xy^2z^3 + xy + 4y + 5x^2 + 7.$$

| Term | Coefficient | Degree | Degree of the Polynomial |
|------|-------------|--------|--------------------------|
| $9x^2y^3$ | 9 | 5 | |
| $-14xy^2z^3$ | $-14$ | 6 | 6 |
| $xy$ | 1 | 2 | |
| $4y$ | 4 | 1 | |
| $5x^2$ | 5 | 2 | |
| 7 | 7 | 0 | |

Think: $4y = 4y^1$.

Think: $7 = 7x^0$, or $7x^0y^0z^0$.

*Do Exercises 4 and 5.*

## c | Collecting Like Terms

**Like terms** (or **similar terms**) have exactly the same variables with exactly the same exponents. For example,

$$3x^2y^3 \text{ and } -7x^2y^3 \text{ are like terms;}$$
$$9x^4z^7 \text{ and } 12x^4z^7 \text{ are like terms.}$$

But

$$13xy^5 \text{ and } -2x^2y^5 \text{ are } not \text{ like terms, because the } x\text{-factors have different exponents;}$$

and

$$3xyz^2 \text{ and } 4xy \text{ are } not \text{ like terms, because there is no factor of } z^2 \text{ in the second expression.}$$

Collecting like terms is based on the distributive laws.

**Examples**  Collect like terms.

**4.** $5x^2y + 3xy^2 - 5x^2y - xy^2 = (5 - 5)x^2y + (3 - 1)xy^2 = 2xy^2$

**5.** $8a^2 - 2ab + 7b^2 + 4a^2 - 9ab - 17b^2 = 12a^2 - 11ab - 10b^2$

**6.** $7xy - 5xy^2 + 3xy^2 - 7 + 6x^3 + 9xy - 11x^3 + y - 1$
    $= -2xy^2 + 16xy - 5x^3 + y - 8$

*Do Exercises 6 and 7.*

## d | Addition

We can find the sum of two polynomials in several variables by writing a plus sign between them and then collecting like terms.

**Example 7** Add: $(-5x^3 + 3y - 5y^2) + (8x^3 + 4x^2 + 7y^2)$.

$(-5x^3 + 3y - 5y^2) + (8x^3 + 4x^2 + 7y^2)$
$= (-5 + 8)x^3 + 4x^2 + 3y + (-5 + 7)y^2$
$= 3x^3 + 4x^2 + 3y + 2y^2$

**Example 8** Add:

$(5xy^2 - 4x^2y + 5x^3 + 2) + (3xy^2 - 2x^2y + 3x^3y - 5)$.

We first look for like terms. They are $5xy^2$ and $3xy^2$, $-4x^2y$ and $-2x^2y$, and 2 and $-5$. We collect these. Since there are no more like terms, the answer is

$8xy^2 - 6x^2y + 5x^3 + 3x^3y - 3$.

*Do Exercises 8–10.*

## e | Subtraction

We subtract a polynomial by adding its opposite, or additive inverse. The opposite of the polynomial

$4x^2y - 6x^3y^2 + x^2y^2 - 5y$

can be represented by

$-(4x^2y - 6x^3y^2 + x^2y^2 - 5y)$.

We find an equivalent expression for the opposite of a polynomial by replacing each coefficient with its opposite, or by changing the sign of each term. Thus,

$-(4x^2y - 6x^3y^2 + x^2y^2 - 5y) = -4x^2y + 6x^3y^2 - x^2y^2 + 5y$.

**Example 9** Subtract:

$(4x^2y + x^3y^2 + 3x^2y^3 + 6y + 10) - (4x^2y - 6x^3y^2 + x^2y^2 - 5y - 8)$.

We have

$(4x^2y + x^3y^2 + 3x^2y^3 + 6y + 10) - (4x^2y - 6x^3y^2 + x^2y^2 - 5y - 8)$
$= 4x^2y + x^3y^2 + 3x^2y^3 + 6y + 10 - 4x^2y + 6x^3y^2 - x^2y^2 + 5y + 8$

Finding the opposite by changing the sign of each term

$= 7x^3y^2 + 3x^2y^3 - x^2y^2 + 11y + 18$.    Collecting like terms. (Try to write just the answer!)

*Do Exercises 11 and 12.*

---

Add.

**8.** $(4x^3 + 4x^2 - 8y - 3) + (-8x^3 - 2x^2 + 4y + 5)$

**9.** $(13x^3y + 3x^2y - 5y) + (x^3y + 4x^2y - 3xy + 3y)$

**10.** $(-5p^2q^4 + 2p^2q^2 + 3q) + (6pq^2 + 3p^2q + 5)$

Subtract.

**11.** $(-4s^4t + s^3t^2 + 2s^2t^3) - (4s^4t - 5s^3t^2 + s^2t^2)$

**12.** $(-5p^4q + 5p^3q^2 - 3p^2q^3 - 7q^4 - 2) - (4p^4q - 4p^3q^2 + p^2q^3 + 2q^4 - 7)$

Answers on page A-19

Multiply.

**13.** $(x^2y^3 + 2x)(x^3y^2 + 3x)$

**14.** $(p^4q - 2p^3q^2 + 3q^3)(p + 2q)$

Multiply.

**15.** $(3xy + 2x)(x^2 + 2xy^2)$

**16.** $(x - 3y)(2x - 5y)$

**17.** $(4x + 5y)^2$

**18.** $(3x^2 - 2xy^2)^2$

**19.** $(2xy^2 + 3x)(2xy^2 - 3x)$

**20.** $(3xy^2 + 4y)(-3xy^2 + 4y)$

**21.** $(3y + 4 - 3x)(3y + 4 + 3x)$

**22.** $(2a + 5b + c)(2a - 5b - c)$

*Answers on page A-19*

## f  Multiplication

To multiply polynomials in several variables, we can multiply each term of one by every term of the other. We can use columns for long multiplications as with polynomials in one variable. We multiply each term at the top by every term at the bottom. We write like terms in columns, and then we add the results.

**Example 10**   Multiply:  $(3x^2y - 2xy + 3y)(xy + 2y)$.

$$
\begin{array}{r}
3x^2y - 2xy + 3y \\
xy + 2y \\
\hline
6x^2y^2 - 4xy^2 + 6y^2 \\
3x^3y^2 - 2x^2y^2 + 3xy^2 \\
\hline
3x^3y^2 + 4x^2y^2 - xy^2 + 6y^2
\end{array}
$$

Multiplying by $2y$
Multiplying by $xy$
Adding

*Do Exercises 13 and 14.*

Where appropriate, we use the special products that we have learned.

**Examples**   Multiply.

$$
\overset{\text{F} \qquad \text{O} \qquad \text{I} \qquad \text{L}}{}
$$
**11.** $(x^2y + 2x)(xy^2 + y^2) = x^3y^3 + x^2y^3 + 2x^2y^2 + 2xy^2$

**12.** $(p + 5q)(2p - 3q) = 2p^2 - 3pq + 10pq - 15q^2$
$$= 2p^2 + 7pq - 15q^2$$

$$(A + B)^2 = A^2 + 2 \cdot A \cdot B + B^2$$
**13.** $(3x + 2y)^2 = (3x)^2 + 2(3x)(2y) + (2y)^2$
$$= 9x^2 + 12xy + 4y^2$$

$$(A - B)^2 = A^2 - 2 \cdot A \cdot B + B^2$$
**14.** $(2y^2 - 5x^2y)^2 = (2y^2)^2 - 2(2y^2)(5x^2y) + (5x^2y)^2$
$$= 4y^4 - 20x^2y^3 + 25x^4y^2$$

$$(A + B)(A - B) = A^2 - B^2$$
**15.** $(3x^2y + 2y)(3x^2y - 2y) = (3x^2y)^2 - (2y)^2$
$$= 9x^4y^2 - 4y^2$$

**16.** $(-2x^3y^2 + 5t)(2x^3y^2 + 5t) = (5t - 2x^3y^2)(5t + 2x^3y^2)$
$$= (5t)^2 - (2x^3y^2)^2$$
$$= 25t^2 - 4x^6y^4$$

$$(A - B)(A + B) = A^2 - B^2$$
**17.** $( \boxed{2x + 3} - 2y)( \boxed{2x + 3} + 2y) = ( \boxed{2x + 3} )^2 - (2y)^2$
$$= 4x^2 + 12x + 9 - 4y^2$$

*Do Exercises 15–22.*

# Exercise Set 4.7

**a** Evaluate the polynomial for $x = 3$, $y = -2$, and $z = -5$.

**1.** $x^2 - y^2 + xy$

**2.** $x^2 + y^2 - xy$

**3.** $x^2 - 3y^2 + 2xy$

**4.** $x^2 - 4xy + 5y^2$

**5.** $8xyz$

**6.** $-3xyz^2$

**7.** $xyz^2 - z$

**8.** $xy - xz + yz$

*Lung Capacity.* The polynomial

$$0.041h - 0.018A - 2.69$$

can be used to estimate the lung capacity, in liters, of a female of height $h$, in centimeters, and age $A$, in years.

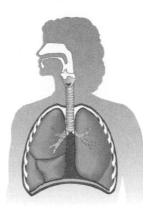

**9.** Find the lung capacity of a 20-yr-old woman who is 165 cm tall.

**10.** Find the lung capacity of a 50-yr-old woman who is 160 cm tall.

*Altitude of a Launched Object.* The altitude, in meters, of a launched object is given by the polynomial

$$h + vt - 4.9t^2,$$

where $h$ is the height, in meters, from which the launch occurs, $v$ is the initial upward speed (or velocity), in meters per second (m/s), and $t$ is the number of seconds for which the object is airborne.

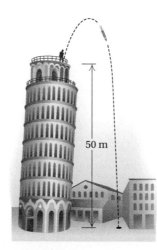

**11.** A model rocket is launched from the top of the Leaning Tower of Pisa, 50 m above the ground. The upward speed is 40 m/s. How high will the rocket be 2 sec after the blastoff?

**12.** A golf ball is thrown upward with an initial speed of 30 m/s by a golfer atop the Washington Monument, which is 160 m above the ground. How high above the ground will the ball be after 3 sec?

*Surface Area of a Right Circular Cylinder.* The area of a right circular cylinder is given by the polynomial

$$2\pi rh + 2\pi r^2,$$

where $h$ is the height and $r$ is the radius of the base.

**13.** A 16-oz beverage can has a height of 6.3 in. and a radius of 1.2 in. Evaluate the polynomial for $h = 6.3$ and $r = 1.2$ to find the area of the can. Use 3.14 for $\pi$.

**14.** A 26-oz coffee can has a height of 6.5 in. and a radius of 2.5 in. Evaluate the polynomial for $h = 6.5$ and $r = 2.5$ to find the area of the can. Use 3.14 for $\pi$.

**b** Identify the coefficient and the degree of each term of the polynomial. Then find the degree of the polynomial.

**15.** $x^3y - 2xy + 3x^2 - 5$

**16.** $5y^3 - y^2 + 15y + 1$

**17.** $17x^2y^3 - 3x^3yz - 7$

**18.** $6 - xy + 8x^2y^2 - y^5$

**c** Collect like terms.

**19.** $a + b - 2a - 3b$

**20.** $y^2 - 1 + y - 6 - y^2$

**21.** $3x^2y - 2xy^2 + x^2$

**22.** $m^3 + 2m^2n - 3m^2 + 3mn^2$

**23.** $6au + 3av + 14au + 7av$

**24.** $3x^2y - 2z^2y + 3xy^2 + 5z^2y$

**25.** $2u^2v - 3uv^2 + 6u^2v - 2uv^2$

**26.** $3x^2 + 6xy + 3y^2 - 5x^2 - 10xy - 5y^2$

**d** Add.

**27.** $(2x^2 - xy + y^2) + (-x^2 - 3xy + 2y^2)$

**28.** $(2z - z^2 + 5) + (z^2 - 3z + 1)$

**29.** $(r - 2s + 3) + (2r + s) + (s + 4)$

**30.** $(ab - 2a + 3b) + (5a - 4b) + (3a + 7ab - 8b)$

**31.** $(b^3a^2 - 2b^2a^3 + 3ba + 4) + (b^2a^3 - 4b^3a^2 + 2ba - 1)$

**32.** $(2x^2 - 3xy + y^2) + (-4x^2 - 6xy - y^2) + (x^2 + xy - y^2)$

**e** Subtract.

**33.** $(a^3 + b^3) - (a^2b - ab^2 + b^3 + a^3)$

**34.** $(x^3 - y^3) - (-2x^3 + x^2y - xy^2 + 2y^3)$

*Combine like terms, + change all signs.*

**35.** $(xy - ab - 8) - (xy - 3ab - 6)$

**36.** $(3y^4x^2 + 2y^3x - 3y - 7) - (2y^4x^2 + 2y^3x - 4y - 2x + 5)$

**37.** $(-2a + 7b - c) - (-3b + 4c - 8d)$

**38.** Find the sum of $2a + b$ and $3a - b$. Then subtract $5a + 2b$.

Copyright © 1999 Addison Wesley Longman

$\boxed{\text{f}}$ Multiply.

**39.** $(3z - u)(2z + 3u)$

**40.** $(a - b)(a^2 + b^2 + 2ab)$

**41.** $(a^2b - 2)(a^2b - 5)$

**42.** $(xy + 7)(xy - 4)$

**43.** $(a^3 + bc)(a^3 - bc)$

**44.** $(m^2 + n^2 - mn)(m^2 + mn + n^2)$

**45.** $(y^4x + y^2 + 1)(y^2 + 1)$

**46.** $(a - b)(a^2 + ab + b^2)$

**47.** $(3xy - 1)(4xy + 2)$

**48.** $(m^3n + 8)(m^3n - 6)$

**49.** $(3 - c^2d^2)(4 + c^2d^2)$

**50.** $(6x - 2y)(5x - 3y)$

**51.** $(m^2 - n^2)(m + n)$

**52.** $(pq + 0.2)(0.4pq - 0.1)$

**53.** $(xy + x^5y^5)(x^4y^4 - xy)$

**54.** $(x - y^3)(2y^3 + x)$

**55.** $(x + h)^2$

**56.** $(3a + 2b)^2$

**57.** $(r^3t^2 - 4)^2$

**58.** $(3a^2b - b^2)^2$

**59.** $(p^4 + m^2n^2)^2$

**60.** $(2ab - cd)^2$

**61.** $\left(2a^3 - \frac{1}{2}b^3\right)^2$

**62.** $-3x(x + 8y)^2$

**63.** $3a(a - 2b)^2$

**64.** $(a^2 + b + 2)^2$

**65.** $(2a - b)(2a + b)$

**66.** $(x - y)(x + y)$

**67.** $(c^2 - d)(c^2 + d)$

**68.** $(p^3 - 5q)(p^3 + 5q)$

**69.** $(ab + cd^2)(ab - cd^2)$

**70.** $(xy + pq)(xy - pq)$

**71.** $(x + y - 3)(x + y + 3)$

**72.** $(p + q + 4)(p + q - 4)$

**73.** $[x + y + z][x - (y + z)]$

**74.** $[a + b + c][a - (b + c)]$

**75.** $(a + b + c)(a - b - c)$

**76.** $(3x + 2 - 5y)(3x + 2 + 5y)$

---

**Skill Maintenance**

In which quadrant is the point located?   [3.1c]

**77.** $(2, -5)$

**78.** $(-8, -9)$

**79.** $(16, 23)$

**80.** $(-3, 2)$

Graph.   [3.3b]

**81.** $2x = -10$

**82.** $y = -4$

**83.** $8y - 16 = 0$

**84.** $x = 4$

Find the mean, the median, and the mode, if it exists.   [3.4a]

**85.** 23, 31, 24, 31, 25, 28, 31

**86.** 5.2, 5.6, 5.8, 6.1, 5.6, 5.2, 6.3

---

**Synthesis**

**87.** ◈ Is it possible for a polynomial in four variables to have a degree less than 4? Why or why not?

**88.** ◈ Can the sum of two trinomials in several variables be a trinomial in one variable? Why or why not?

Find a polynomial for the shaded area. (Leave results in terms of $\pi$ where appropriate.)

**89.**

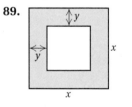

**90.**

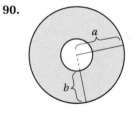

**91.**

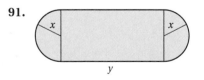

**92.**

Find a formula for the surface area of the solid object. Leave results in terms of $\pi$.

**93.**

**94.**

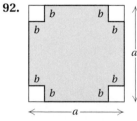

Copyright © 1999 Addison Wesley Longman

# 4.8 Division of Polynomials

In this section, we consider division of polynomials. You will see that such division is similar to what is done in arithmetic.

## a Divisor a Monomial

We first consider division by a monomial. When dividing a monomial by a monomial, we use the quotient rule of Section 4.1 to subtract exponents when the bases are the same. We also divide the coefficients.

**Examples** Divide.

1. $\dfrac{10x^2}{2x} = \dfrac{10}{2} \cdot \dfrac{x^2}{x} = 5x^{2-1} = 5x$

2. $\dfrac{x^9}{3x^2} = \dfrac{1x^9}{3x^2} = \dfrac{1}{3} \cdot \dfrac{x^9}{x^2} = \dfrac{1}{3}x^{9-2} = \dfrac{1}{3}x^7$

3. $\dfrac{-18x^{10}}{3x^3} = \dfrac{-18}{3} \cdot \dfrac{x^{10}}{x^3} = -6x^{10-3} = -6x^7$

4. $\dfrac{42a^2b^5}{-3ab^2} = \dfrac{42}{-3} \cdot \dfrac{a^2}{a} \cdot \dfrac{b^5}{b^2} = -14a^{2-1}b^{5-2} = -14ab^3$

*Do Exercises 1–4.*

To divide a polynomial by a monomial, we note that since

$$\frac{A}{C} + \frac{B}{C} = \frac{A+B}{C},$$

it follows that

$$\frac{A+B}{C} = \frac{A}{C} + \frac{B}{C}.$$

This is actually the procedure we use when performing divisions like $86 \div 2$. Although we might write

$$\frac{86}{2} = 43,$$

we could also calculate as follows:

$$\frac{86}{2} = \frac{80+6}{2} = \frac{80}{2} + \frac{6}{2} = 40 + 3 = 43.$$

Similarly, to divide a polynomial by a monomial, we divide each term by the monomial.

**Example 5** Divide: $(9x^8 + 12x^6) \div 3x^2$.

We have

$$(9x^8 + 12x^6) \div 3x^2 = \frac{9x^8 + 12x^6}{3x^2}$$

$$= \frac{9x^8}{3x^2} + \frac{12x^6}{3x^2}.$$

To see this, add and get the original expression.

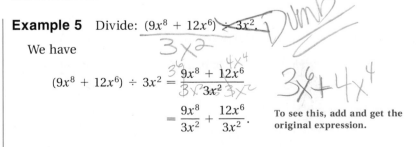

## Objectives

**a** Divide a polynomial by a monomial.

**b** Divide a polynomial by a divisor that is not a monomial.

### For Extra Help

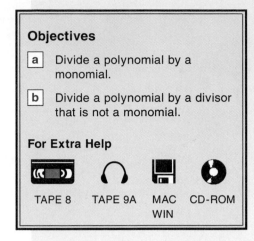

TAPE 8   TAPE 9A   MAC   CD-ROM
                    WIN

Divide.

1. $\dfrac{20x^3}{5x}$

2. $\dfrac{-28x^{14}}{4x^3}$

3. $\dfrac{-56p^5q^7}{2p^2q^6}$

4. $\dfrac{x^5}{4x}$

*Answers on page A-19*

**5.** Divide: $(28x^7 + 32x^5) \div 4x^3$.
Check the result.

We now perform the separate divisions:

$$\frac{9x^8}{3x^2} + \frac{12x^6}{3x^2} = \frac{9}{3} \cdot \frac{x^8}{x^2} + \frac{12}{3} \cdot \frac{x^6}{x^2}$$

$$= 3x^{8-2} + 4x^{6-2}$$

$$= 3x^6 + 4x^4.$$

> *Caution!* The coefficients are *divided*, but the exponents are *subtracted*.

To check, we multiply the quotient $3x^6 + 4x^4$ by the divisor $3x^2$:

$$(3x^6 + 4x^4)3x^2 = (3x^6)(3x^2) + (4x^4)(3x^2) = 9x^8 + 12x^6.$$

**6.** Divide: $(2x^3 + 6x^2 + 4x) \div 2x$.
Check the result.

*Do Exercises 5–7.*

**Example 6** Divide and check: $(10a^5b^4 - 2a^3b^2 + 6a^2b) \div (2a^2b)$.

$$\frac{10a^5b^4 - 2a^3b^2 + 6a^2b}{2a^2b} = \frac{10a^5b^4}{2a^2b} - \frac{2a^3b^2}{2a^2b} + \frac{6a^2b}{2a^2b}$$

$$= \frac{10}{2}a^{5-2}b^{4-1} - \frac{2}{2}a^{3-2}b^{2-1} + \frac{6}{2}$$

$$= 5a^3b^3 - ab + 3$$

**7.** Divide: $(6x^2 + 3x - 2) \div 3$.
Check the result.

CHECK:
$$\begin{array}{r} 5a^3b^3 - ab + 3 \\ 2a^2b \\ \hline 10a^5b^4 - 2a^3b^2 + 6a^2b \end{array}$$

We multiply.

The answer checks.

> To divide a polynomial by a monomial, divide each term by the monomial.

*Do Exercises 8 and 9.*

Divide and check.

**8.** $(8x^2 - 3x + 1) \div 2$

## b | Divisor Not a Monomial

Let's first consider long division as it is performed in arithmetic. When we divide, we repeat the following procedure.

> **LONG DIVISION**
>
> **1.** Divide,
> **2.** Multiply,
> **3.** Subtract, and
> **4.** Bring down the next term.

**9.** $\dfrac{2x^4y^6 - 3x^3y^4 + 5x^2y^3}{x^2y^2}$

We review this by considering the division $3711 \div 8$.

$$\begin{array}{r} 4 \\ 8 \overline{)\ 3\ 7\ 1\ 1} \\ 3\ 2 \\ \hline 5\ 1 \end{array}$$

① Divide: $37 \div 8 \approx 4$.

② Multiply: $4 \times 8 = 32$.

③ Subtract: $37 - 32 = 5$.

④ Bring down the 1.

$$\begin{array}{r} 4\ 6\ 3 \\ 8 \overline{)\ 3\ 7\ 1\ 1} \\ 3\ 2 \\ \hline 5\ 1 \\ 4\ 8 \\ \hline 3\ 1 \\ 2\ 4 \\ \hline 7 \end{array}$$

*Answers on page A-19*

Next, we repeat the process two more times. We obtain the complete division as shown on the right above. The quotient is 463. The remainder is 7, expressed as R = 7. We write the answer as

$$463 \text{ R } 7 \quad \text{or} \quad 463 + \frac{7}{8} = 463\frac{7}{8}.$$

We check by multiplying the quotient, 463, by the divisor, 8, and adding the remainder, 7:

$$8 \cdot 463 + 7 = 3704 + 7 = 3711.$$

Now let's look at long division with polynomials. We use this procedure when the divisor is not a monomial. We write polynomials in descending order and then write in missing terms.

**Example 7**  Divide $x^2 + 5x + 6$ by $x + 2$.

We have

$$
\begin{array}{r}
x \\
x + 2 \overline{)x^2 + 5x + 6} \\
x^2 + 2x \\
\hline
3x
\end{array}
$$

— Divide the first term by the first term: $x^2/x = x$.
Ignore the term 2.

— Multiply $x$ above by the divisor, $x + 2$.

— Subtract: $(x^2 + 5x) - (x^2 + 2x) = x^2 + 5x - x^2 - 2x$
$\quad = 3x$.

We now "bring down" the next term of the dividend—in this case, 6.

$$
\begin{array}{r}
x + 3 \\
x + 2 \overline{)x^2 + 5x + 6} \\
x^2 + 2x \\
\hline
3x + 6 \\
3x + 6 \\
\hline
0
\end{array}
$$

— Divide the first term by the first term: $3x/x = 3$.

— The 6 has been "brought down."

— Multiply 3 by the divisor, $x + 2$.

— Subtract: $(3x + 6) - (3x + 6) = 3x + 6 - 3x - 6 = 0$.

The quotient is $x + 3$. The remainder is 0, expressed as R = 0. A remainder of 0 is generally not listed in an answer.

To check, we multiply the quotient by the divisor and add the remainder, if any, to see if we get the dividend:

| Divisor | Quotient | Remainder | Dividend | |
|---------|----------|-----------|----------|--|
| $(x + 2) \cdot$ | $(x + 3)$ | $+ \quad 0$ | $= x^2 + 5x + 6$. | The division checks. |

*Do Exercise 10.*

**Example 8**  Divide and check: $(x^2 + 2x - 12) \div (x - 3)$.

We have

$$
\begin{array}{r}
x \\
x - 3 \overline{)x^2 + 2x - 12} \\
x^2 - 3x \\
\hline
5x
\end{array}
$$

— Divide the first term by the first term: $x^2/x = x$.

— Multiply $x$ above by the divisor, $x - 3$.

— Subtract: $(x^2 + 2x) - (x^2 - 3x) = x^2 + 2x - x^2 + 3x$
$\quad = 5x$.

We now "bring down" the next term of the dividend—in this case, $-12$.

$$
\begin{array}{r}
x + 5 \\
x - 3 \overline{)x^2 + 2x - 12} \\
x^2 - 3x \\
\hline
5x - 12 \\
5x - 15 \\
\hline
3
\end{array}
$$

— Divide the first term by the first term: $5x/x = 5$.

— Bring down the $-12$.

— Multiply 5 above by the divisor, $x - 3$.

— Subtract: $(5x - 12) - (5x - 15) = 5x - 12 - 5x + 15$
$\quad = 3$.

**10.** Divide and check:

$$(x^2 + x - 6) \div (x + 3).$$

*Answer on page A-19*

4.8  Division of Polynomials

**11.** Divide and check:

$$x - 2\overline{)x^2 + 2x - 8}.$$

Divide and check.

**12.** $x + 3\overline{)x^2 + 7x + 10}$

**13.** $(x^3 - 1) \div (x - 1)$

The answer is $x + 5$ with R = 3, or

Quotient    $\underbrace{x + 5}$ + $\dfrac{\overbrace{3}}{\underbrace{x - 3}}$ ⟶ Remainder

⟶ Divisor

(This is the way answers will be given at the back of the book.)

**CHECK:** We can check by multiplying the divisor by the quotient and adding the remainder, as follows:

$$(x - 3)(x + 5) + 3 = x^2 + 2x - 15 + 3$$
$$= x^2 + 2x - 12.$$

When dividing, an answer may "come out even" (that is, have a remainder of 0, as in Example 7), or it may not (as in Example 8). If a remainder is not 0, we continue dividing until the degree of the remainder is less than the degree of the divisor. Check this in each of Examples 7 and 8.

*Do Exercises 11 and 12.*

**Example 9**  Divide and check: $(x^3 + 1) \div (x + 1)$.

$$
\begin{array}{r}
x^2 - x + 1 \\
x + 1\overline{)x^3 + 0x^2 + 0x + 1} \\
\underline{x^3 + x^2} \\
-x^2 + 0x \\
\underline{-x^2 - x} \\
x + 1 \\
\underline{x + 1} \\
0
\end{array}
$$

⟵ Fill in the missing terms (see Section 4.3).

This subtraction is $x^3 - (x^3 + x^2)$.

This subtraction is $-x^2 - (-x^2 - x)$.

The answer is $x^2 - x + 1$. The check is left to the student.

**Example 10**  Divide and check: $(x^4 - 3x^2 + 1) \div (x - 4)$.

$$
\begin{array}{r}
x^3 + 4x^2 + 13x + 52 \\
x - 4\overline{)x^4 + 0x^3 - 3x^2 + 0x + 1} \\
\underline{x^4 - 4x^3} \\
4x^3 - 3x^2 \\
\underline{4x^3 - 16x^2} \\
13x^2 + 0x \\
\underline{13x^2 - 52x} \\
52x + 1 \\
\underline{52x - 208} \\
209
\end{array}
$$

⟵ Fill in the missing terms.

$x^4 - (x^4 - 4x^3)$

$(4x^3 - 3x^2) - (4x^3 - 16x^2)$

The answer is $x^3 + 4x^2 + 13x + 52$, with R = 209, or

$$x^3 + 4x^2 + 13x + 52 + \frac{209}{x - 4}.$$

**CHECK:**  $(x - 4)(x^3 + 4x^2 + 13x + 52) + 209$
$$= -4x^3 - 16x^2 - 52x - 208 + x^4 + 4x^3 + 13x^2 + 52x + 209$$
$$= x^4 - 3x^2 + 1$$

*Do Exercise 13.*

# Exercise Set 4.8

**a** Divide and check.

**1.** $\dfrac{24x^4}{8}$

**2.** $\dfrac{-2u^2}{u}$

**3.** $\dfrac{25x^3}{5x^2}$

**4.** $\dfrac{16x^7}{-2x^2}$

**5.** $\dfrac{-54x^{11}}{-3x^8}$

**6.** $\dfrac{-75a^{10}}{3a^2}$

**7.** $\dfrac{64a^5b^4}{16a^2b^3}$

**8.** $\dfrac{-34p^{10}q^{11}}{-17pq^9}$

**9.** $\dfrac{24x^4 - 4x^3 + x^2 - 16}{8}$

**10.** $\dfrac{12a^4 - 3a^2 + a - 6}{6}$

**11.** $\dfrac{u - 2u^2 - u^5}{u}$

**12.** $\dfrac{50x^5 - 7x^4 + x^2}{x}$

**13.** $(15t^3 + 24t^2 - 6t) \div (3t)$

**14.** $(25t^3 + 15t^2 - 30t) \div (5t)$

**15.** $(20x^6 - 20x^4 - 5x^2) \div (-5x^2)$

**16.** $(24x^6 + 32x^5 - 8x^2) \div (-8x^2)$

**17.** $(24x^5 - 40x^4 + 6x^3) \div (4x^3)$

**18.** $(18x^6 - 27x^5 - 3x^3) \div (9x^3)$

**19.** $\dfrac{18x^2 - 5x + 2}{2}$

**20.** $\dfrac{15x^2 - 30x + 6}{3}$

**21.** $\dfrac{12x^3 + 26x^2 + 8x}{2x}$

**22.** $\dfrac{2x^4 - 3x^3 + 5x^2}{x^2}$

**23.** $\dfrac{9r^2s^2 + 3r^2s - 6rs^2}{3rs}$

**24.** $\dfrac{4x^4y - 8x^6y^2 + 12x^8y^6}{4x^4y}$

**b** Divide.

**25.** $(x^2 + 4x + 4) \div (x + 2)$

**26.** $(x^2 - 6x + 9) \div (x - 3)$

**27.** $(x^2 - 10x - 25) \div (x - 5)$

**28.** $(x^2 + 8x - 16) \div (x + 4)$

**29.** $(x^2 + 4x - 14) \div (x + 6)$

**30.** $(x^2 + 5x - 9) \div (x - 2)$

**31.** $\dfrac{x^2 - 9}{x + 3}$

**32.** $\dfrac{x^2 - 25}{x - 5}$

**33.** $\dfrac{x^5 + 1}{x + 1}$

**34.** $\dfrac{x^5 - 1}{x - 1}$

**35.** $\dfrac{8x^3 - 22x^2 - 5x + 12}{4x + 3}$

*xtra credit*

**36.** $\dfrac{2x^3 - 9x^2 + 11x - 3}{2x - 3}$

**37.** $(x^6 - 13x^3 + 42) \div (x^3 - 7)$

**38.** $(x^6 + 5x^3 - 24) \div (x^3 - 3)$

**39.** $(x^4 - 16) \div (x - 2)$

**40.** $(x^4 - 81) \div (x - 3)$

**41.** $(t^3 - t^2 + t - 1) \div (t - 1)$

**42.** $(t^3 - t^2 + t - 1) \div (t + 1)$

---

**Skill Maintenance**

Subtract. [1.4a]

**43.** $17 - 45$

**44.** $-14 - 45$

**45.** $-2.3 - (-9.1)$

**46.** $-\dfrac{5}{8} - \dfrac{3}{4}$

Solve. [2.4a]

**47.** The perimeter of a rectangle is 640 ft. The length is 15 ft more than the width. Find the area of the rectangle.

**48.** The first angle of a triangle is 24° more than the second. The third angle is twice the first. Find the measures of the angles of the triangle.

Solve. [2.3c]

**49.** $-6(2 - x) + 10(5x - 7) = 10$

**50.** $-10(x - 4) = 5(2x + 5) - 7$

Factor. [1.7d]

**51.** $4x - 12 + 24y$

**52.** $256 - 2a - 4b$

---

**Synthesis**

**53.** ◈ Explain how the equation
$$(2x + 3)(3x - 1) = 6x^2 + 7x - 3$$
can be used to write two equations involving division.

**54.** ◈ Can the quotient of two binomials be a trinomial? Why or why not?

Divide.

**55.** $(x^4 + 9x^2 + 20) \div (x^2 + 4)$

**56.** $(y^4 + a^2) \div (y + a)$

**57.** $(5a^3 + 8a^2 - 23a - 1) \div (5a^2 - 7a - 2)$

**58.** $(15y^3 - 30y + 7 - 19y^2) \div (3y^2 - 2 - 5y)$

**59.** $(6x^5 - 13x^3 + 5x + 3 - 4x^2 + 3x^4) \div (3x^3 - 2x - 1)$

**60.** $(5x^7 - 3x^4 + 2x^2 - 10x + 2) \div (x^2 - x + 1)$

**61.** $(a^6 - b^6) \div (a - b)$

**62.** $(x^5 + y^5) \div (x + y)$

If the remainder is 0 when one polynomial is divided by another, the divisor is a *factor* of the dividend. Find the value(s) of $c$ for which $x - 1$ is a factor of the polynomial.

**63.** $x^2 + 4x + c$

**64.** $2x^2 + 3cx - 8$

**65.** $c^2x^2 - 2cx + 1$

Copyright © 1999 Addison Wesley Longman

# Summary and Review Exercises: Chapter 4

## Important Properties and Formulas

| | |
|---|---|
| FOIL: | $(A + B)(C + D) = AC + AD + BC + BD$ |
| *Square of a Sum:* | $(A + B)(A + B) = (A + B)^2 = A^2 + 2AB + B^2$ |
| *Square of a Difference:* | $(A - B)(A - B) = (A - B)^2 = A^2 - 2AB + B^2$ |
| *Product of a Sum and a Difference:* | $(A + B)(A - B) = A^2 - B^2$ |

**Definitions and Rules for Exponents**

See p. 242.

The objectives to be tested in addition to the material in this chapter are [1.4a], [1.7d], [2.3b, c], and [2.4a].

Multiply and simplify.  [4.1d, f]

**1.** $7^2 \cdot 7^{-4}$

**2.** $y^7 \cdot y^3 \cdot y$

**3.** $(3x)^5 \cdot (3x)^9$

**4.** $t^8 \cdot t^0$

Divide and simplify.  [4.1e, f]

**5.** $\dfrac{4^5}{4^2}$

**6.** $\dfrac{a^5}{a^8}$

**7.** $\dfrac{(7x)^4}{(7x)^4}$

Simplify.

**8.** $(3t^4)^2$  [4.2a, b]

**9.** $(2x^3)^2(-3x)^2$  [4.1d], [4.2a, b]

**10.** $\left(\dfrac{2x}{y}\right)^{-3}$  [4.2b]

**11.** Express using a negative exponent: $\dfrac{1}{t^5}$.  [4.1f]

**12.** Express using a positive exponent: $y^{-4}$.  [4.1f]

**13.** Convert to scientific notation: 0.0000328.  [4.2c]

**14.** Convert to decimal notation: $8.3 \times 10^6$.  [4.2c]

Multiply or divide and write scientific notation for the result.  [4.2d]

**15.** $(3.8 \times 10^4)(5.5 \times 10^{-1})$

**16.** $\dfrac{1.28 \times 10^{-8}}{2.5 \times 10^{-4}}$

**17.** *Diet-Drink Consumption.* It has been estimated that there will be 275 million people in the United States by the year 2000 and that on average, each of them will drink 15.3 gal of diet drinks that year (**Source:** U.S. Department of Agriculture). How many gallons of diet drinks will be consumed by the entire population in 2000? Express the answer in scientific notation.  [4.2e]

**18.** Evaluate the polynomial $x^2 - 3x + 6$ for $x = -1$.  [4.3a]

**19.** Identify the terms of the polynomial $-4y^5 + 7y^2 - 3y - 2$.  [4.3b]

**20.** Identify the missing terms in $x^3 + x$.  [4.3h]

**21.** Identify the degree of each term and the degree of the polynomial $4x^3 + 6x^2 - 5x + \frac{5}{3}$.  [4.3g]

Classify the polynomial as a monomial, binomial, trinomial, or none of these.  [4.3i]

**22.** $4x^3 - 1$

**23.** $4 - 9t^3 - 7t^4 + 10t^2$

**24.** $7y^2$

Collect like terms and then arrange in descending order.  [4.3f]

**25.** $3x^2 - 2x + 3 - 5x^2 - 1 - x$

**26.** $-x + \frac{1}{2} + 14x^4 - 7x^2 - 1 - 4x^4$

Add.  [4.4a]

**27.** $(3x^4 - x^3 + x - 4) + (x^5 + 7x^3 - 3x^2 - 5) + (-5x^4 + 6x^2 - x)$

**28.** $(3x^5 - 4x^4 + x^3 - 3) + (3x^4 - 5x^3 + 3x^2) + (-5x^5 - 5x^2) + (-5x^4 + 2x^3 + 5)$

Subtract.  [4.4c]

**29.** $(5x^2 - 4x + 1) - (3x^2 + 1)$

**30.** $(3x^5 - 4x^4 + 3x^2 + 3) - (2x^5 - 4x^4 + 3x^3 + 4x^2 - 5)$

**31.** Find a polynomial for the perimeter and for the area.   [4.4d], [4.5b]

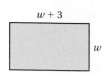

$w + 3$

$w$

Multiply.

**32.** $\left(x + \frac{2}{3}\right)\left(x + \frac{1}{2}\right)$   [4.6a]

**33.** $(7x + 1)^2$   [4.6c]

**34.** $(4x^2 - 5x + 1)(3x - 2)$   [4.5d]

**35.** $(3x^2 + 4)(3x^2 - 4)$   [4.6b]

**36.** $5x^4(3x^3 - 8x^2 + 10x + 2)$   [4.5b]

**37.** $(x + 4)(x - 7)$   [4.6a]

**38.** $(3y^2 - 2y)^2$   [4.6c]

**39.** $(2t^2 + 3)(t^2 - 7)$   [4.6a]

**40.** Evaluate the polynomial
$$2 - 5xy + y^2 - 4xy^3 + x^6$$
for $x = -1$ and $y = 2$.   [4.7a]

**41.** Identify the coefficient and the degree of each term of the polynomial
$$x^5y - 7xy + 9x^2 - 8.$$
Then find the degree of the polynomial.   [4.7b]

Collect like terms.   [4.7c]

**42.** $y + w - 2y + 8w - 5$

**43.** $m^6 - 2m^2n + m^2n^2 + n^2m - 6m^3 + m^2n^2 + 7n^2m$

**44.** Add:   [4.7d]
$(5x^2 - 7xy + y^2) + (-6x^2 - 3xy - y^2) +$
$(x^2 + xy - 2y^2)$.

**45.** Subtract:   [4.7e]
$(6x^3y^2 - 4x^2y - 6x) - (-5x^3y^2 + 4x^2y + 6x^2 - 6)$.

Multiply.   [4.7f]

**46.** $(p - q)(p^2 + pq + q^2)$     **47.** $\left(3a^4 - \frac{1}{3}b^3\right)^2$

Divide.

**48.** $(10x^3 - x^2 + 6x) \div (2x)$   [4.8a]

**49.** $(6x^3 - 5x^2 - 13x + 13) \div (2x + 3)$   [4.8b]

**50.** The graph of the polynomial equation $y = 10x^3 - 10x$ is shown below. Use *only* the graph to estimate the value of the polynomial for $x = -1$, for $x = -0.5$, for $x = 0.5$, for $x = 1$, and for $x = 1.1$.   [4.3a]

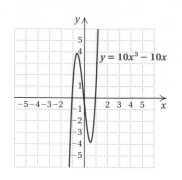

$y = 10x^3 - 10x$

**Skill Maintenance**

**51.** Factor: $25t - 50 + 100m$.   [1.7d]

**52.** Solve: $7x + 6 - 8x = 11 - 5x + 4$.   [2.3b]

**53.** Solve: $3(x - 2) + 6 = 5(x + 3) + 9$.   [2.3c]

**54.** Subtract: $-3.4 - 7.8$.   [1.4a]

**55.** The perimeter of a rectangle is 540 m. The width is 19 m less than the length. Find the width and the length.   [2.4a]

**Synthesis**

**56.** ◈ Explain why the expression $578.6 \times 10^{-7}$ is not in scientific notation.   [4.2c]

**57.** ◈ Write a short explanation of the difference between a monomial, a binomial, a trinomial, and a general polynomial.   [4.3i]

Find a polynomial for the shaded area. [4.4d], [4.6b]

**58.**      **59.**

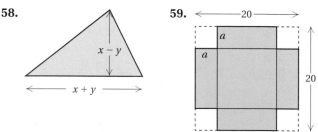

$x - y$

$x + y$

$20$

$a$

$a$

$20$

**60.** Collect like terms:   [4.1d], [4.2a], [4.3e]
$$-3x^5 \cdot 3x^3 - x^6(2x)^2 + (3x^4)^2 + (2x^2)^4 - 40x^2(x^3)^2.$$

**61.** Solve:   [4.6a]
$$(x - 7)(x + 10) = (x - 4)(x - 6).$$

**62.** The product of two polynomials is $x^5 - 1$. One of the polynomials is $x - 1$. Find the other.   [4.8b]

# Test: Chapter 4

Multiply and simplify.

**1.** $6^{-2} \cdot 6^{-3}$

**2.** $x^6 \cdot x^2 \cdot x$

**3.** $(4a)^3 \cdot (4a)^8$

Divide and simplify.

**4.** $\dfrac{3^5}{3^2}$

**5.** $\dfrac{x^3}{x^8}$

**6.** $\dfrac{(2x)^5}{(2x)^5}$

Simplify.

**7.** $(x^3)^2$

**8.** $(-3y^2)^3$

**9.** $(2a^3b)^4$

**10.** $\left(\dfrac{ab}{c}\right)^3$

**11.** $(3x^2)^3(-2x^5)^3$

**12.** $3(x^2)^3(-2x^5)^3$

**13.** $2x^2(-3x^2)^4$

**14.** $(2x)^2(-3x^2)^4$

**15.** Express using a positive exponent: $5^{-3}$.

**16.** Express using a negative exponent: $\dfrac{1}{y^8}$.

**17.** Convert to scientific notation: 3,900,000,000.

**18.** Convert to decimal notation: $5 \times 10^{-8}$.

Multiply or divide and write scientific notation for the answer.

**19.** $\dfrac{5.6 \times 10^6}{3.2 \times 10^{-11}}$

**20.** $(2.4 \times 10^5)(5.4 \times 10^{16})$

**21.** A CD-ROM can contain about 600 million pieces of information (bytes). How many sound files, each containing 40,000 bytes, can a CD-ROM hold? Express the answer in scientific notation.

**22.** Evaluate the polynomial $x^5 + 5x - 1$ for $x = -2$.

**23.** Identify the coefficient of each term of the polynomial $\frac{1}{3}x^5 - x + 7$.

**24.** Identify the degree of each term and the degree of the polynomial $2x^3 - 4 + 5x + 3x^6$.

**25.** Classify the polynomial $7 - x$ as a monomial, binomial, trinomial, or none of these.

Collect like terms.

**26.** $4a^2 - 6 + a^2$

**27.** $y^2 - 3y - y + \dfrac{3}{4}y^2$

## Answers

1. _____

2. _____

3. _____

4. _____

5. _____

6. _____

7. _____

8. _____

9. _____

10. _____

11. _____

12. _____

13. _____

14. _____

15. _____

16. _____

17. _____

18. _____

19. _____

20. _____

21. _____

22. _____

23. _____

24. _____

25. _____

26. _____

27. _____

28. _____

29. _____

30. _____

31. _____

32. _____

33. _____

34. _____

35. _____

36. _____

37. _____

38. _____

39. _____

40. _____

41. _____

42. _____

43. _____

44. _____

45. _____

46. _____

47. _____

48. _____

49. _____

50. _____

51. _____

52. _____

53. _____

**28.** Collect like terms and then arrange in descending order:

$$3 - x^2 + 2x^3 + 5x^2 - 6x - 2x + x^5.$$

Add.

**29.** $(3x^5 + 5x^3 - 5x^2 - 3) +$
$(x^5 + x^4 - 3x^3 - 3x^2 + 2x - 4)$

**30.** $\left(x^4 + \dfrac{2}{3}x + 5\right) + \left(4x^4 + 5x^2 + \dfrac{1}{3}x\right)$

Subtract.

**31.** $(2x^4 + x^3 - 8x^2 - 6x - 3) -$
$(6x^4 - 8x^2 + 2x)$

**32.** $(x^3 - 0.4x^2 - 12) -$
$(x^5 + 0.3x^3 + 0.4x^2 + 9)$

Multiply.

**33.** $-3x^2(4x^2 - 3x - 5)$  **34.** $\left(x - \dfrac{1}{3}\right)^2$  **35.** $(3x + 10)(3x - 10)$

**36.** $(3b + 5)(b - 3)$  **37.** $(x^6 - 4)(x^8 + 4)$  **38.** $(8 - y)(6 + 5y)$

**39.** $(2x + 1)(3x^2 - 5x - 3)$  **40.** $(5t + 2)^2$

**41.** Collect like terms: $x^3y - y^3 + xy^3 + 8 - 6x^3y - x^2y^2 + 11.$

**42.** Subtract: $(8a^2b^2 - ab + b^3) - (-6ab^2 - 7ab - ab^3 + 5b^3).$

**43.** Multiply: $(3x^5 - 4y^5)(3x^5 + 4y^5).$

Divide.

**44.** $(12x^4 + 9x^3 - 15x^2) \div (3x^2)$  **45.** $(6x^3 - 8x^2 - 14x + 13) \div (3x + 2)$

**46.** The graph of the polynomial equation $y = x^3 - 5x - 1$ is shown at right. Use _only_ the graph to estimate the value of the polynomial for $x = -1$, for $x = -0.5$, for $x = 0.5$, for $x = 1$, and for $x = 1.1$.

$y = x^3 - 5x - 1$

**Skill Maintenance**

**47.** Solve: $7x - 4x - 2 = 37.$  **48.** Solve: $4(x + 2) - 21 = 3(x - 6) + 2.$

**49.** Factor: $64t - 32m + 16.$  **50.** Subtract: $\dfrac{2}{5} - \left(-\dfrac{3}{4}\right).$

**51.** The first angle of a triangle is four times as large as the second. The measure of the third angle is 30° greater than that of the second. How large are the angles?

**Synthesis**

**52.** The height of a box is 1 less than its length, and the length is 2 more than its width. Find the volume in terms of the length.

**53.** Solve: $(x - 5)(x + 5) = (x + 6)^2.$

# Cumulative Review: Chapters 1–4

1. Evaluate $\dfrac{x}{2y}$ for $x = 10$ and $y = 2$.

2. Evaluate $2x^3 + x^2 - 3$ for $x = -1$.

3. Evaluate $x^3y^2 + xy + 2xy^2$ for $x = -1$ and $y = 2$.

4. Find the absolute value: $|-4|$.

5. Find the reciprocal of 5.

Compute and simplify.

6. $-\dfrac{3}{5} + \dfrac{5}{12}$

7. $3.4 - (-0.8)$

8. $(-2)(-1.4)(2.6)$

9. $\dfrac{3}{8} \div \left(-\dfrac{9}{10}\right)$

10. $(1.1 \times 10^{10})(2 \times 10^{12})$

11. $(3.2 \times 10^{-10}) \div (8 \times 10^{-6})$

Simplify.

12. $\dfrac{-9x}{3x}$

13. $y - (3y + 7)$

14. $3(x - 1) - 2[x - (2x + 7)]$

15. $2 - [32 \div (4 + 2^2)]$

Add.

16. $(x^4 + 3x^3 - x + 7) + (2x^5 - 3x^4 + x - 5)$

17. $(x^2 + 2xy) + (y^2 - xy) + (2x^2 - 3y^2)$

Subtract.

18. $(x^3 + 3x^2 - 4) - (-2x^2 + x + 3)$

19. $\left(\dfrac{1}{3}x^2 - \dfrac{1}{4}x - \dfrac{1}{5}\right) - \left(\dfrac{2}{3}x^2 + \dfrac{1}{2}x - \dfrac{1}{5}\right)$

Multiply.

20. $3(4x - 5y + 7)$

21. $(-2x^3)(-3x^5)$

22. $2x^2(x^3 - 2x^2 + 4x - 5)$

23. $(y^2 - 2)(3y^2 + 5y + 6)$

24. $(2p^3 + p^2q + pq^2)(p - pq + q)$

25. $(2x + 3)(3x + 2)$

26. $(3x^2 + 1)^2$

27. $\left(t + \dfrac{1}{2}\right)\left(t - \dfrac{1}{2}\right)$

28. $(2y^2 + 5)(2y^2 - 5)$

29. $(2x^4 - 3)(2x^2 + 3)$

30. $(t - 2t^2)^2$

31. $(3p + q)(5p - 2q)$

Divide.

32. $(18x^3 + 6x^2 - 9x) \div 3x$

33. $(3x^3 + 7x^2 - 13x - 21) \div (x + 3)$

Solve.

34. $1.5 = 2.7 + x$

35. $\dfrac{2}{7}x = -6$

36. $5x - 9 = 36$

37. $\dfrac{2}{3} = \dfrac{-m}{10}$

38. $5.4 - 1.9x = 0.8x$

39. $x - \dfrac{7}{8} = \dfrac{3}{4}$

40. $2(2 - 3x) = 3(5x + 7)$

41. $\dfrac{1}{4}x - \dfrac{2}{3} = \dfrac{3}{4} + \dfrac{1}{3}x$

42. $y + 5 - 3y = 5y - 9$

**43.** $\frac{1}{4}x - 7 < 5 - \frac{1}{2}x$      **44.** $2(x + 2) \geq 5(2x + 3)$      **45.** $A = 2\pi rh + \pi r^2$, for $h$

**46.** A 6-ft by 3-ft raft is floating in a swimming pool of radius $r$. Find a polynomial for the area of the surface of the pool not covered by the raft.

Solve.

**47.** The sum of the page numbers on the facing pages of a book is 37. What are the page numbers?

**48.** The perimeter of a room is 88 ft. The width is 4 ft less than the length. Find the width and the length.

**49.** The second angle of a triangle is five times as large as the first. The third angle is twice the sum of the other two angles. Find the measure of the first angle.

**50.** If you triple a number and then add 99, you get $\frac{4}{5}$ of the original number. What is the original number?

**51.** A bookstore sells books at a price that is 80% higher than the price the store pays for the books. A book is priced for sale at $6.30. How much did the store pay for the book?

**52.** *Coffee Consumption.* It has been estimated that there will be 275 million people in the United States by the year 2000 and that on average, each of them will drink 21.1 gal of coffee that year (*Source*: U.S. Department of Agriculture). How many gallons of coffee will be consumed by the entire population in 2000? Express the answer in scientific notation.

Simplify.

**53.** $y^2 \cdot y^{-6} \cdot y^8$      **54.** $\dfrac{x^6}{x^7}$      **55.** $(-3x^3y^{-2})^3$      **56.** $\dfrac{x^3x^{-4}}{x^{-5}x}$

**57.** Identify the coefficient of each term of the polynomial $\frac{2}{3}x^2 + 4x - 6$.

**58.** Identify the degree of each term and the degree of the polynomial $2x^4 + 3x^2 + 2x + 1$.

Classify the polynomial as a monomial, binomial, trinomial, or none of these.

**59.** $2x^2 + 1$

**60.** $2x^2 + x + 1$

**61.** Find the intercepts of $4x - 5y = 20$.

**62.** Graph: $4x - 5y = 20$.

**63.** *Height of Girls.* Estimate the missing data value.

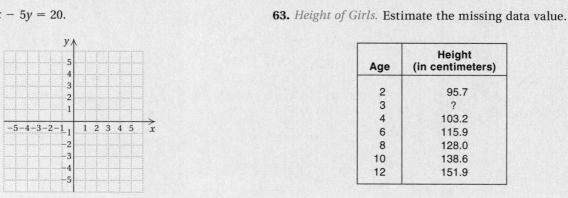

| Age | Height (in centimeters) |
|---|---|
| 2 | 95.7 |
| 3 | ? |
| 4 | 103.2 |
| 6 | 115.9 |
| 8 | 128.0 |
| 10 | 138.6 |
| 12 | 151.9 |

---

**Synthesis**

**64.** A picture frame is $x$ inches square. The picture that it frames is 2 in. shorter than the frame in both length and width. Find a polynomial for the area of the frame.

Add.

**65.** $[(2x)^2 - (3x)^3 + 2x^2x^3 + (x^2)^2] + [5x^2(2x^3) - ((2x)^2)^2]$

**66.** $(x - 3)^2 + (2x + 1)^2$

**67.** $[(3x^3 + 11x^2 + 11x + 15) \div (x + 3)] + [(2x^3 - 7x^2 + 2) \div (2x + 1)]$

Solve.

**68.** $(x + 3)(2x - 5) + (x - 1)^2 = (3x + 1)(x - 3)$

**69.** $(2x^2 + x - 6) \div (2x - 3) = (2x^2 - 9x - 5) \div (x - 5)$

**70.** $20 - 3|x| = 5$

**71.** $(x + 2)^2 = (x + 1)(x + 3)$

**72.** $(x - 3)(x + 4) = (x^3 - 4x^2 - 17x + 60) \div (x - 5)$

---

# 5

# Polynomials: Factoring

## An Application

A ladder of length 13 ft is placed against a building in such a way that the distance from the top of the ladder to the ground is 7 ft more than the distance from the bottom of the ladder to the building. Find both distances.

This problem appears as Example 6 in Section 5.8.

## The Mathematics

If we visualize this as a triangle, we can let $x =$ the length of the side (leg) across the bottom. Then $x + 7 =$ the length of the other side (leg). The hypotenuse has length 13 ft. Using the Pythagorean theorem, we translate the problem to

$$x^2 + (x + 7)^2 = 13^2$$
$$x^2 + x^2 + 14x + 49 = 169$$
$$\underline{2x^2 + 14x - 120 = 0.}$$

↑

This is a second-degree, or quadratic, equation.

 For more information, visit us at www.mathmax.com

# Pretest: Chapter 5

**1.** Find three factorizations of $-20x^6$.

Factor.

**2.** $2x^2 + 4x + 2$

**3.** $x^2 + 6x + 8$

**4.** $8a^5 + 4a^3 - 20a$

**5.** $-6 + 5x^2 - 13x$

**6.** $81 - z^4$

**7.** $y^6 - 4y^3 + 4$

**8.** $3x^3 + 2x^2 + 12x + 8$

**9.** $p^2 - p - 30$

**10.** $x^4y^2 - 64$

**11.** $2p^2 + 7pq - 4q^2$

Solve.

**12.** $x^2 - 5x = 0$

**13.** $(x - 4)(5x - 3) = 0$

**14.** $3x^2 + 10x - 8 = 0$

Solve.

**15.** Six less than the square of a number is five times the number. Find all such numbers.

**16.** The height of a triangle is 3 cm longer than the base. The area of the triangle is 44 cm$^2$. Find the base and the height.

## Objectives for Retesting

The objectives to be tested in addition to the material in this chapter are as follows.

[1.6c] Divide real numbers.

[2.7e] Solve inequalities using the addition and multiplication principles together.

[3.3a] Find the intercepts of a linear equation, and graph using intercepts.

[4.6d] Find special products when polynomial products are mixed together.

# 5.1 Introduction to Factoring

To solve certain types of algebraic equations involving polynomials of second degree, we must learn to factor polynomials.

Consider the product $15 = 3 \cdot 5$. We say that 3 and 5 are **factors** of 15 and that $3 \cdot 5$ is a **factorization** of 15. Since $15 = 15 \cdot 1$, we also know that 15 and 1 are factors of 15 and that $15 \cdot 1$ is a factorization of 15.

> To **factor** a polynomial is to express it as a product.
>
> A **factor** of a polynomial $P$ is a polynomial that can be used to express $P$ as a product.
>
> A **factorization** of a polynomial is an expression that names that polynomial as a product.

## a | Factoring Monomials

To factor a monomial, we find two monomials whose product is equivalent to the original monomial. Compare.

| *Multiplying* | *Factoring* |
|---|---|
| **a)** $(4x)(5x) = 20x^2$ | $20x^2 = (4x)(5x)$ |
| **b)** $(2x)(10x) = 20x^2$ | $20x^2 = (2x)(10x)$ |
| **c)** $(-4x)(-5x) = 20x^2$ | $20x^2 = (-4x)(-5x)$ |
| **d)** $(x)(20x) = 20x^2$ | $20x^2 = (x)(20x)$ |

You can see that the monomial $20x^2$ has many factorizations. There are still other ways to factor $20x^2$.

*Do Exercises 1 and 2.*

**Example 1** Find three factorizations of $15x^3$.

**a)** $15x^3 = (3 \cdot 5)(x \cdot x^2)$
$\quad = (3x)(5x^2)$
**b)** $15x^3 = (3 \cdot 5)(x^2 \cdot x)$
$\quad = (3x^2)(5x)$
**c)** $15x^3 = (-15)(-1)x^3$
$\quad = (-15)(-x^3)$

*Do Exercises 3–5.*

## b | Factoring When Terms Have a Common Factor

To factor polynomials quickly, we consider the special-product rules studied in Chapter 4, but we first factor out the largest common factor.

To multiply a monomial and a polynomial with more than one term, we multiply each term of the polynomial by the monomial using the distributive laws,

$$a(b + c) = ab + ac \quad \text{and} \quad a(b - c) = ab - ac.$$

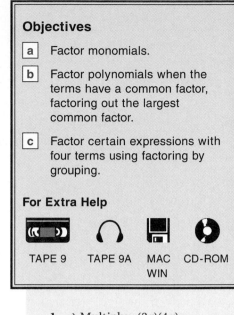

**Objectives**

a  Factor monomials.

b  Factor polynomials when the terms have a common factor, factoring out the largest common factor.

c  Factor certain expressions with four terms using factoring by grouping.

**For Extra Help**

TAPE 9     TAPE 9A     MAC WIN     CD-ROM

**1. a)** Multiply: $(3x)(4x)$.

**b)** Factor: $12x^2$.

**2. a)** Multiply: $(2x)(8x^2)$.

**b)** Factor: $16x^3$.

Find three factorizations of the monomial.

**3.** $8x^4$

**4.** $21x^2$

**5.** $6x^5$

*Answers on page A-21*

**6. a)** Multiply: $3(x + 2)$.

To factor, we do the reverse. We express a polynomial as a product using the distributive laws in reverse:

$$ab + ac = a(b + c) \quad \text{and} \quad ab - ac = a(b - c).$$

Compare.

*Multiply*

$3x(x^2 + 2x - 4)$

$\quad = 3x \cdot x^2 + 3x \cdot 2x - 3x \cdot 4$

$\quad = 3x^3 + 6x^2 - 12x$

*Factor*

$3x^3 + 6x^2 - 12x$

$\quad = 3x \cdot x^2 + 3x \cdot 2x - 3x \cdot 4$

$\quad = 3x(x^2 + 2x - 4)$

**b)** Factor: $3x + 6$.

*Do Exercises 6 and 7.*

---

**CAUTION!** Consider the following:

$$3x^3 + 6x^2 - 12x = 3 \cdot x \cdot x \cdot x + 2 \cdot 3 \cdot x \cdot x - 2 \cdot 2 \cdot 3 \cdot x.$$

The terms of the polynomial, $3x^3$, $6x^2$, and $-12x$, have been factored but the polynomial itself has not been factored. This is not what we mean by a factorization of the polynomial. The *factorization* is

$$3x(x^2 + 2x - 4).$$

The expressions $3x$ and $x^2 + 2x - 4$ are *factors* of $3x^3 + 6x^2 - 12x$.

---

To factor, we first try to find a factor common to all terms. There may not always be one other than 1. When there is, we generally use the factor with the largest possible coefficient and the largest possible exponent.

**Example 2**   Factor: $7x^2 + 14$.

We have

$\quad 7x^2 + 14 = 7 \cdot x^2 + 7 \cdot 2$     **Factoring each term**

$\qquad\qquad = 7(x^2 + 2).$     **Factoring out the common factor 7**

**CHECK:**   We multiply to check:

$\quad 7(x^2 + 2) = 7 \cdot x^2 + 7 \cdot 2 = 7x^2 + 14.$

**7. a)** Multiply: $2x(x^2 + 5x + 4)$.

**Example 3**   Factor: $16x^3 + 20x^2$.

$\quad 16x^3 + 20x^2 = (4x^2)(4x) + (4x^2)(5)$     **Factoring each term**

$\qquad\qquad = 4x^2(4x + 5)$     **Factoring out the common factor $4x^2$**

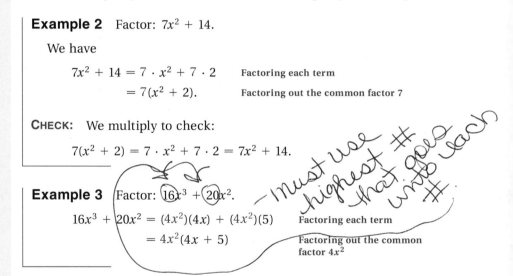

— must use highest # that goes into each #

**b)** Factor: $2x^3 + 10x^2 + 8x$.

Suppose in Example 3 that you had not recognized the largest common factor and removed only part of it, as follows:

$\quad 16x^3 + 20x^2 = (2x^2)(8x) + (2x^2)(10)$

$\qquad\qquad = 2x^2(8x + 10).$

Note that $8x + 10$ still has a common factor of 2. You need not begin again. Just continue factoring out common factors, as follows, until finished:

$\qquad\qquad = 2x^2[2(4x + 5)]$

$\qquad\qquad = 4x^2(4x + 5).$

*Answers on page A-21*

**Example 4**  Factor: $15x^5 - 12x^4 + 27x^3 - 3x^2$.

$$15x^5 - 12x^4 + 27x^3 - 3x^2 = (3x^2)(5x^3) - (3x^2)(4x^2) + (3x^2)(9x) - (3x^2)(1)$$
$$= 3x^2(5x^3 - 4x^2 + 9x - 1)$$

Factoring out $3x^2$

*crz lowest factor*

> CAUTION!  Don't forget the term $-1$.

**CHECK:**  We multiply to check:

$$3x^2(5x^3 - 4x^2 + 9x - 1)$$
$$= (3x^2)(5x^3) - (3x^2)(4x^2) + (3x^2)(9x) - (3x^2)(1)$$
$$= 15x^5 - 12x^4 + 27x^3 - 3x^2.$$

As you become more familiar with factoring, you will be able to spot the largest common factor without factoring each term. Then you can write just the answer.

**Examples**  Factor.

**5.** $8m^3 - 16m = 8m(m^2 - 2)$

**6.** $14p^2y^3 - 8py^2 + 2py = 2py(7py^2 - 4y + 1)$

**7.** $\frac{4}{5}x^2 + \frac{1}{5}x + \frac{2}{5} = \frac{1}{5}(4x^2 + x + 2)$

**8.** $2.4x^2 + 1.2x - 3.6 = 1.2(2x^2 + x - 3)$

**Do Exercises 8–13.**

There are two important points to keep in mind as we study this chapter.

---

- Before doing any other kind of factoring, first try to factor out the largest common factor.
- Always check the result of factoring by multiplying.

---

## c  Factoring by Grouping: Four Terms

Certain polynomials with four terms can be factored using a method called *factoring by grouping*.

**Example 9**  Factor: $x^2(x + 1) + 2(x + 1)$.

The binomial $x + 1$ is common to both terms:

$$x^2(x + 1) + 2(x + 1) = (x^2 + 2)(x + 1).$$

The factorization is $(x^2 + 2)(x + 1)$.

**Do Exercises 14 and 15.**

Factor. Check by multiplying.

**8.** $x^2 + 3x$

**9.** $3y^6 - 5y^3 + 2y^2$

**10.** $9x^4 - 15x^3 + 3x^2$

**11.** $\frac{3}{4}t^3 + \frac{5}{4}t^2 + \frac{7}{4}t + \frac{1}{4}$

**12.** $35x^7 - 49x^6 + 14x^5 - 63x^3$

**13.** $8.4x^2 - 5.6x + 2.8$

Factor.

**14.** $x^2(x + 7) + 3(x + 7)$

**15.** $x^2(a + b) + 2(a + b)$

*Answers on page A-21*

Factor by grouping.

**16.** $x^3 + 7x^2 + 3x + 21$

**17.** $8t^3 + 2t^2 + 12t + 3$

**18.** $3m^5 - 15m^3 + 2m^2 - 10$

**19.** $3x^3 - 6x^2 - x + 2$

**20.** $4x^3 - 6x^2 - 6x + 9$

**21.** $y^4 - 2y^3 - 2y - 10$

*Answers on page A-21*

Consider the four-term polynomial

$$x^3 + x^2 + 2x + 2.$$

There is no factor other than 1 that is common to all the terms. We can, however, factor $x^3 + x^2$ and $2x + 2$ separately:

$$x^3 + x^2 = x^2(x + 1); \qquad \text{Factoring } x^3 + x^2$$
$$2x + 2 = 2(x + 1). \qquad \text{Factoring } 2x + 2$$

We have grouped certain terms and factored each polynomial separately:

$$x^3 + x^2 + 2x + 2 = (x^3 + x^2) + (2x + 2)$$
$$= x^2(x + 1) + 2(x + 1)$$
$$= (x^2 + 2)(x + 1),$$

as in Example 9. This method is called **factoring by grouping.** We began with a polynomial with four terms. After grouping and removing common factors, we obtained a polynomial with two parts, each having a common factor $x + 1$. Not all polynomials with four terms can be factored by this procedure, but it does give us a method to try.

**Examples**   Factor by grouping.

**10.** $6x^3 - 9x^2 + 4x - 6$
$$= (6x^3 - 9x^2) + (4x - 6)$$
$$= 3x^2(2x - 3) + 2(2x - 3) \qquad \text{Factoring each binomial}$$
$$= (3x^2 + 2)(2x - 3) \qquad \text{Factoring out the common factor } 2x - 3$$

We think through this process as follows:

$$6x^3 - 9x^2 + 4x - 6 = 3x^2\underbrace{(2x - 3)}\ \ \underbrace{(2x - 3)}$$

(1) Factor the first two terms.

(2) This factor, $2x - 3$, gives us a hint to the factorization on the right.

(3) Now we ask ourselves, "What needs to be here to enable us to get $4x - 6$ when we multiply?"

**11.** $x^3 + x^2 + x + 1 = (x^3 + x^2) + (x + 1)$
$$= x^2(x + 1) + 1(x + 1) \qquad \text{Factoring each binomial}$$
$$= (x^2 + 1)(x + 1) \qquad \text{Factoring out the common factor } x + 1$$

**12.** $2x^3 - 6x^2 - x + 3$
$$= (2x^3 - 6x^2) + (-x + 3)$$
$$= 2x^2(x - 3) - 1(x - 3) \qquad \textit{Check: } -1(x - 3) = -x + 3.$$
$$= (2x^2 - 1)(x - 3) \qquad \text{Factoring out the common factor } x - 3$$

**13.** $12x^5 + 20x^2 - 21x^3 - 35 = 4x^2(3x^3 + 5) - 7(3x^3 + 5)$
$$= (4x^2 - 7)(3x^3 + 5)$$

**14.** $x^3 + x^2 + 2x - 2 = x^2(x + 1) + 2(x - 1)$

This polynomial is not factorable using factoring by grouping. It may be factorable, but not by methods that we will consider in this text.

*Do Exercises 16–21.*

# Exercise Set 5.1

**a** Find three factorizations for the monomial.

**1.** $8x^3$      **2.** $6x^4$      **3.** $-10a^6$      **4.** $-8y^5$      **5.** $24x^4$      **6.** $15x^5$

**b** Factor. Check by multiplying.

**7.** $x^2 - 6x$      **8.** $x^2 + 5x$      **9.** $2x^2 + 6x$

**10.** $8y^2 - 8y$      **11.** $x^3 + 6x^2$      **12.** $3x^4 - x^2$

**13.** $8x^4 - 24x^2$      **14.** $5x^5 + 10x^3$      **15.** $2x^2 + 2x - 8$

**16.** $8x^2 - 4x - 20$      **17.** $17x^5y^3 + 34x^3y^2 + 51xy$      **18.** $16p^6q^4 + 32p^5q^3 - 48pq^2$

**19.** $6x^4 - 10x^3 + 3x^2$      **20.** $5x^5 + 10x^2 - 8x$      **21.** $x^5y^5 + x^4y^3 + x^3y^3 - x^2y^2$

**22.** $x^9y^6 - x^7y^5 + x^4y^4 + x^3y^3$      **23.** $2x^7 - 2x^6 - 64x^5 + 4x^3$      **24.** $8y^3 - 20y^2 + 12y - 16$

**25.** $1.6x^4 - 2.4x^3 + 3.2x^2 + 6.4x$      **26.** $2.5x^6 - 0.5x^4 + 5x^3 + 10x^2$

**27.** $\dfrac{5}{3}x^6 + \dfrac{4}{3}x^5 + \dfrac{1}{3}x^4 + \dfrac{1}{3}x^3$      **28.** $\dfrac{5}{9}x^7 + \dfrac{2}{9}x^5 - \dfrac{4}{9}x^3 - \dfrac{1}{9}x$

**c** Factor.

**29.** $x^2(x + 3) + 2(x + 3)$      **30.** $3z^2(2z + 1) + (2z + 1)$

**31.** $5a^3(2a - 7) - (2a - 7)$      **32.** $m^4(8 - 3m) - 7(8 - 3m)$

Factor by grouping.

**33.** $x^3 + 3x^2 + 2x + 6$

**34.** $6z^3 + 3z^2 + 2z + 1$

*(handwritten: xtra credit, ext. 11 pg 322)*

**35.** $2x^3 + 6x^2 + x + 3$

**36.** $3x^3 + 2x^2 + 3x + 2$

**37.** $8x^3 - 12x^2 + 6x - 9$

**38.** $10x^3 - 25x^2 + 4x - 10$

**39.** $12x^3 - 16x^2 + 3x - 4$

**40.** $18x^3 - 21x^2 + 30x - 35$

**41.** $5x^3 - 5x^2 - x + 1$

**42.** $7x^3 - 14x^2 - x + 2$

**43.** $x^3 + 8x^2 - 3x - 24$

**44.** $2x^3 + 12x^2 - 5x - 30$

**45.** $2x^3 - 8x^2 - 9x + 36$

**46.** $20g^3 - 4g^2 - 25g + 5$

---

**Skill Maintenance**

Solve.

**47.** $-2x < 48$  [2.7d]

**48.** $4x - 8x + 16 \geq 6(x - 2)$  [2.7e]

**49.** Divide: $\dfrac{-108}{-4}$.  [1.6a]

**50.** Solve $A = \dfrac{p + q}{2}$ for $p$.  [2.6a]

Multiply.  [4.6d]

**51.** $(y + 5)(y + 7)$

**52.** $(y + 7)^2$

**53.** $(y + 7)(y - 7)$

**54.** $(y - 7)^2$

Find the intercepts of the equation. Then graph the equation.  [3.3a]

**55.** $x + y = 4$

**56.** $x - y = 3$

**57.** $5x - 3y = 15$

**58.** $y - 3x = 6$

---

**Synthesis**

**59.** ◆ Josh says that there is no need to print answers for Exercises 1–46 at the back of the book. Is he correct in saying this? Why or why not?

**60.** ◆ Explain how one could construct a polynomial with four terms that can be factored by grouping.

Factor.

**61.** $4x^5 + 6x^3 + 6x^2 + 9$

**62.** $x^6 + x^4 + x^2 + 1$

**63.** $x^{12} + x^7 + x^5 + 1$

**64.** $x^3 - x^2 - 2x + 5$

**65.** $p^3 + p^2 - 3p + 10$

# 5.2 Factoring Trinomials of the Type $x^2 + bx + c$

## a

We now begin a study of the factoring of trinomials. We first factor trinomials like

$$x^2 + 5x + 6 \quad \text{and} \quad x^2 + 3x - 10$$

by a refined *trial-and-error* process. In this section, we restrict our attention to trinomials of the type $ax^2 + bx + c$, where $a = 1$. The coefficient $a$ is often called the **leading coefficient.**

### Constant Term Positive

Recall the FOIL method of multiplying two binomials:

$$
\begin{array}{ccccc}
 & \text{F} & \text{O} & \text{I} & \text{L} \\
(x + 2)(x + 5) = & x^2 & + 5x & + 2x & + 10
\end{array}
$$

$$= x^2 \quad + 7x \quad + 10.$$

The product above is a trinomial. The term of highest degree, $x^2$, called the leading term, has a coefficient of 1. The constant term, 10, is positive. To factor $x^2 + 7x + 10$, we think of FOIL in reverse. We multiplied $x$ times $x$ to get the first term of the trinomial, so we know that the first term of each binomial factor is $x$. Next, we look for numbers $p$ and $q$ such that

$$x^2 + 7x + 10 = (x + p)(x + q).$$

To get the middle term and the last term of the trinomial, we look for two numbers $p$ and $q$ whose product is 10 and whose sum is 7. Those numbers arc 2 and 5. Thus the factorization is

$$(x + 2)(x + 5).$$

**Example 1** Factor: $x^2 + 5x + 6$.

Think of FOIL in reverse. The first term of each factor is $x$: $(x + 2)(x + 3)$. Next, we look for two numbers whose product is 6 and whose sum is 5. All the pairs of factors of 6 are shown in the table on the left below. Since both the product, 6, and the sum, 5, of the pair of numbers must be positive, we need consider only the positive factors, listed in the table on the right.

| Pairs of Factors | Sums of Factors |
|:---:|:---:|
| 1,  6 | 7 |
| −1, −6 | −7 |
| 2,  3 | 5 |
| −2, −3 | −5 |

| Pairs of Factors | Sums of Factors |
|:---:|:---:|
| 1, 6 | 7 |
| 2, 3 | 5 |

↑
The numbers we need are 2 and 3.

The factorization is $(x + 2)(x + 3)$. We can check by multiplying to see whether we get the original trinomial.

**CHECK:** $(x + 2)(x + 3) = x^2 + 3x + 2x + 6 = x^2 + 5x + 6.$

*Do Exercises 1 and 2.*

---

**Objective**

a Factor trinomials of the type $x^2 + bx + c$ by examining the constant term $c$.

**For Extra Help**

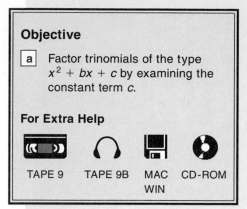

TAPE 9    TAPE 9B    MAC    CD-ROM
                     WIN

1. Consider the trinomial $x^2 + 7x + 12$.

a) Complete the following table.

| Pairs of Factors | Sums of Factors |
|:---:|:---:|
| 1,   12 | 13 |
| −1, −12 | |
| 2,   6 | |
| −2, −6 | |
| 3,   4 | |
| −3,  −4 | |

b) Explain why you need to consider only positive factors, as in the following table.

| Pairs of Factors | Sums of Factors |
|:---:|:---:|
| 1, 12 | |
| 2,  6 | |
| 3,  4 | |

c) Factor: $x^2 + 7x + 12$.

2. Factor: $x^2 + 13x + 36$.

*Answers on page A-21*

**3.** Explain why you would not consider the pairs of factors listed below in factoring $y^2 - 8y + 12$.

| Pairs of Factors | Sums of Factors |
|---|---|
| 1, 12 | |
| 2, 6 | |
| 3, 4 | |

Factor.

**4.** $x^2 - 8x + 15$

**5.** $t^2 - 9t + 20$

Consider this multiplication:

$$(x - 2)(x - 5) = x^2 \overset{F}{\phantom{=}} \overset{O}{-5x} \overset{I}{-2x} \overset{L}{+10}$$

$$= x^2 \quad - 7x \quad + 10.$$

> When the constant term of a trinomial is positive, we look for two numbers with the same sign (both negative or both positive). The sign is that of the middle term:
>
> $x^2 - 7x + 10 = (x - 2)(x - 5)$, or $x^2 + 7x + 10 = (x + 2)(x + 5)$.

**Example 2** Factor: $y^2 \ominus 8y \oplus 12$. *means*

Since the constant term, 12, is positive and the coefficient of the middle term, $-8$, is negative, we look for a factorization of 12 in which both factors are negative. Their sum must be $-8$.

| Pairs of Factors | Sums of Factors |
|---|---|
| $-1, -12$ | $-13$ |
| $-2, -6$ | $-8$ ← The numbers we need are $-2$ and $-6$. |
| $-3, -4$ | $-7$ |

The factorization is $(y \ominus 2)(y \ominus 6)$.

*Do Exercises 3–5.*

### Constant Term Negative

Sometimes when we use FOIL, the product has a negative constant term. Consider these multiplications:

$$\text{a) } (x - 5)(x + 2) = x^2 \overset{F}{\phantom{=}} \overset{O}{+2x} \overset{I}{-5x} \overset{L}{-10}$$

$$= x^2 \quad - 3x \quad - 10 ;$$

$$\text{b) } (x + 5)(x - 2) = x^2 \overset{F}{\phantom{=}} \overset{O}{-2x} \overset{I}{+5x} \overset{L}{-10}$$

$$= x^2 \quad + 3x \quad - 10.$$

Reversing the signs of the factors changes the sign of the middle term.

> When the constant term of a trinomial is negative, we look for two factors whose product is negative. One of them must be positive and the other negative. Their sum must be the coefficient of the middle term:
>
> $x^2 - 3x - 10 = (x - 5)(x + 2)$, or $x^2 + 3x - 10 = (x + 5)(x - 2)$.

**Example 3** Factor: $x^3 - 8x^2 - 20x$.

*Always* look first for a common factor. This time there is one, $x$. We first factor it out: $x^3 - 8x^2 - 20x = x(x^2 - 8x - 20)$. Now consider the expression $x^2 - 8x - 20$. Since the constant term, $-20$, is negative, we look for a factorization of $-20$ in which one factor is positive and one factor is negative. The sum must be $-8$, so the negative factor must have the larger absolute value. Thus we consider only pairs of factors in which the negative factor has the larger absolute value.

| Pairs of Factors | Sums of Factors | |
|:---:|:---:|:---|
| 1, −20 | −19 | |
| 2, −10 | −8 ← | The numbers we need are |
| 4, −5 | −1 | 2 and −10. |

The factorization of $x^2 - 8x - 20$ is $(x + 2)(x - 10)$. *means* But we must also remember to include the common factor. The factorization of the original polynomial is

$$x(x + 2)(x - 10). \quad \text{different signs}$$

*Do Exercise 6.*

**Example 4** Factor: $t^2 - 24 + 5t$.

It helps to first write the trinomial in descending order: $t^2 + 5t - 24$. Since the constant term, $-24$, is negative, we look for a factorization of $-24$ in which one factor is positive and one factor is negative. Their sum must be 5, so the positive factor must have the larger absolute value. Thus we consider only pairs of factors in which the positive term has the larger absolute value.

| Pairs of Factors | Sums of Factors | |
|:---:|:---:|:---|
| −1, 24 | 23 | |
| −2, 12 | 10 | |
| −3, 8 | 5 ← | The numbers we need are |
| −4, 6 | 2 | −3 and 8. |

The factorization is $(t - 3)(t + 8)$.

*Do Exercise 7.*

**Example 5** Factor: $x^4 - x^2 - 110$.

Consider this trinomial as $(x^2)^2 - x^2 - 110$. We look for numbers $p$ and $q$ such that

$$x^4 - x^2 - 110 = (x^2 + p)(x^2 + q).$$

Since the constant term, $-110$, is negative, we look for a factorization of $-110$ in which one factor is positive and one factor is negative. Their sum must be $-1$. The middle-term coefficient, $-1$, is small compared to $-110$. This tells us that the desired factors are close to each other in absolute value. The numbers we want are 10 and $-11$. The factorization is

$$(x^2 + 10)(x^2 - 11).$$

6. Explain why you would not consider the pairs of factors listed below in factoring $x^2 - 8x - 20$.

| Pairs of Factors | Sums of Factors |
|:---:|:---:|
| −1, 20 | |
| −2, 10 | |
| −4, 5 | |

7. Explain why you would not consider the pairs of factors listed below in factoring $t^2 + 5t - 24$.

| Pairs of Factors | Sums of Factors |
|:---:|:---:|
| 1, −24 | |
| 2, −12 | |
| 3, −8 | |
| 4, −6 | |

*Answers on page A-21*

Factor.

**8.** $x^3 + 4x^2 - 12x$

**9.** $y^2 - 12 - 4y$

**10.** $t^4 + 5t^2 - 14$

**11.** $p^2 - pq - 3pq^2$

**12.** $x^2 + 2x + 7$

**13.** Factor: $x^2 + 8x + 16$.

*Answers on page A-21*

**Example 6** Factor: $a^2 + 4ab - 21b^2$.

We consider the trinomial in the equivalent form

$a^2 + 4ba - 21b^2$. *means*

We think of $4b$ as a "coefficient" of $a$. Then we look for factors of $-21b^2$ whose sum is $4b$. Those factors are $-3b$ and $7b$. The factorization is

$(a - 3b)(a + 7b)$. *different signs*

There are polynomials that are not factorable.

**Example 7** Factor: $x^2 - x + 5$. *— Primed.*

Since 5 has very few factors, we can easily check all possibilities.

| Pairs of Factors | Sums of Factors |
|:---:|:---:|
| 5, 1 | 6 |
| −5, −1 | −6 |

There are no factors whose sum is $-1$. Thus the polynomial is *not* factorable into binomials.

*Do Exercises 8–12.*

We can factor a trinomial that is a perfect square using this method.

**Example 8** Factor: $x^2 - 10x + 25$.

Since the constant term, 25, is positive and the coefficient of the middle term, $-10$, is negative, we look for a factorization of 25 in which both factors are negative. Their sum must be $-10$.

| Pairs of Factors | Sums of Factors | |
|:---:|:---:|:---|
| −25, −1 | −26 | |
| −5, −5 | −10 ← | The numbers we need are −5 and −5. |

The factorization is $(x - 5)(x - 5)$, or $(x - 5)^2$.

*Do Exercise 13.*

The following is a summary of our procedure for factoring $x^2 + bx + c$.

To factor $x^2 + bx + c$:

1. First arrange in descending order.
2. Use a trial-and-error process that looks for factors of $c$ whose sum is $b$.
3. If $c$ is positive, the signs of the factors are the same as the sign of $b$.
4. If $c$ is negative, one factor is positive and the other is negative. If the sum of two factors is the opposite of $b$, changing the sign of each factor will give the desired factors whose sum is $b$.
5. Check by multiplying.

# Exercise Set 5.2

**a** Factor. Remember that you can check by multiplying.

**1.** $x^2 + 8x + 15$      **2.** $x^2 + 5x + 6$      **3.** $x^2 + 7x + 12$      **4.** $x^2 + 9x + 8$

**5.** $x^2 - 6x + 9$      **6.** $y^2 - 11y + 28$      **7.** $x^2 + 9x + 14$      **8.** $a^2 + 11a + 30$

**9.** $b^2 + 5b + 4$      **10.** $z^2 - 8z + 7$      **11.** $x^2 + \dfrac{2}{3}x + \dfrac{1}{9}$      **12.** $x^2 - \dfrac{2}{5}x + \dfrac{1}{25}$

**13.** $d^2 - 7d + 10$      **14.** $t^2 - 12t + 35$      **15.** $y^2 - 11y + 10$      **16.** $x^2 - 4x - 21$

**17.** $x^2 + x - 42$      **18.** $x^2 + 2x - 15$      **19.** $x^2 - 7x - 18$      **20.** $y^2 - 3y - 28$

**21.** $x^3 - 6x^2 - 16x$      **22.** $x^3 - x^2 - 42x$      **23.** $y^3 - 4y^2 - 45y$      **24.** $x^3 - 7x^2 - 60x$

**25.** $-2x - 99 + x^2$      **26.** $x^2 - 72 + 6x$      **27.** $c^4 + c^2 - 56$      **28.** $b^4 + 5b^2 - 24$

**29.** $a^4 + 2a^2 - 35$      **30.** $x^4 - x^2 - 6$      **31.** $x^2 + x + 1$      **32.** $x^2 + 5x + 3$

**33.** $7 - 2p + p^2$      **34.** $11 - 3w + w^2$      **35.** $x^2 + 20x + 100$      **36.** $a^2 + 19a + 88$

**37.** $x^4 - 21x^3 - 100x^2$      **38.** $x^4 - 20x^3 + 96x^2$      **39.** $x^2 - 21x - 72$      **40.** $4x^2 + 40x + 100$

**41.** $x^2 - 25x + 144$      **42.** $y^2 - 21y + 108$      **43.** $a^2 + a - 132$      **44.** $a^2 + 9a - 90$

**45.** $120 - 23x + x^2$      **46.** $96 + 22d + d^2$      **47.** $108 - 3x - x^2$      **48.** $112 + 9y - y^2$

**49.** $y^2 - 0.2y - 0.08$    **50.** $t^2 - 0.3t - 0.10$    **51.** $p^2 + 3pq - 10q^2$    **52.** $a^2 + 2ab - 3b^2$

**53.** $m^2 + 5mn + 4n^2$    **54.** $x^2 + 11xy + 24y^2$    **55.** $s^2 - 2st - 15t^2$    **56.** $p^2 + 5pq - 24q^2$

---

### Skill Maintenance

Multiply.  [4.6d]

**57.** $8x(2x^2 - 6x + 1)$    **58.** $(7w + 6)(4w - 11)$    **59.** $(7w + 6)^2$

**60.** $(4w - 11)^2$    **61.** $(4w - 11)(4w + 11)$

**62.** Simplify: $(3x^4)^3$.  [4.2b]

Solve.  [2.3a]

**63.** $3x - 8 = 0$    **64.** $2x + 7 = 0$

Solve.

**65.** *Arrests for Counterfeiting.* In a recent year, 29,200 people were arrested for counterfeiting. This number was down 1.2% from the preceding year. How many people were arrested the preceding year?  [2.5a]

**66.** The first angle of a triangle is four times as large as the second. The measure of the third angle is 30° greater than that of the second. Find the angle measures.  [2.4a]

---

### Synthesis

**67.** ◈ Without doing the multiplication $(x - 17)(x - 18)$, explain why it cannot possibly be a factorization of $x^2 + 35x + 306$.

**68.** ◈ When searching for a factorization of $x^2 + bx + c$, why do we list pairs of numbers with the specified product $c$ instead of pairs of numbers with the specified sum $b$?

**69.** Find all integers $m$ for which $y^2 + my + 50$ can be factored.

**70.** Find all integers $b$ for which $a^2 + ba - 50$ can be factored.

Factor completely.

**71.** $x^2 - \frac{1}{2}x - \frac{3}{16}$    **72.** $x^2 - \frac{1}{4}x - \frac{1}{8}$    **73.** $x^2 + \frac{30}{7}x - \frac{25}{7}$

**74.** $\frac{1}{3}x^3 + \frac{1}{3}x^2 - 2x$    **75.** $b^{2n} + 7b^n + 10$    **76.** $a^{2m} - 11a^m + 28$

Find a polynomial in factored form for the shaded area. (Leave answers in terms of $\pi$.)

**77.**
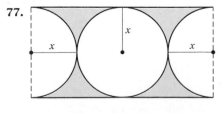

**78.**

Copyright © 1999 Addison Wesley Longman

# 5.3 Factoring $ax^2 + bx + c$, $a \neq 1$, Using FOIL

In Section 5.2, we learned a trial-and-error method to factor trinomials of the type $x^2 + bx + c$. In this section, we factor trinomials in which the coefficient of the leading term $x^2$ is not 1. The procedure we learn is a refined trial-and-error method. (In Section 5.4, we will consider an alternative method for the same kind of factoring. It involves *factoring by grouping*.)

**Objective**

**a** Factor trinomials of the type $ax^2 + bx + c$, $a \neq 1$.

**For Extra Help**

TAPE 9    TAPE 9B    MAC WIN    CD-ROM

**a** We want to factor trinomials of the type $ax^2 + bx + c$. Consider the following multiplication:

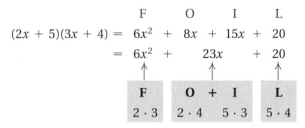

$$
\begin{array}{ccccccc}
 & \text{F} & \text{O} & \text{I} & \text{L} \\
(2x + 5)(3x + 4) = & 6x^2 & + & 8x & + & 15x & + & 20 \\
 = & 6x^2 & + & & 23x & & + & 20
\end{array}
$$

| F | O + I | L |
|---|-------|---|
| $2 \cdot 3$ | $2 \cdot 4 \quad 5 \cdot 3$ | $5 \cdot 4$ |

To factor $6x^2 + 23x + 20$, we reverse the above multiplication, using what we might call an "unFOIL" process. We look for two binomials $rx + p$ and $sx + q$ whose product is this trinomial. The product of the First terms must be $6x^2$. The product of the Outside terms plus the product of the Inside terms must be $23x$. The product of the Last terms must be 20. We know from the preceding discussion that the answer is $(2x + 5)(3x + 4)$. Generally, however, finding such an answer is a refined trial-and-error process. It turns out that $(-2x - 5)(-3x - 4)$ is also a correct answer, but we usually choose an answer in which the first coefficients are positive.

We will use the following trial-and-error method.

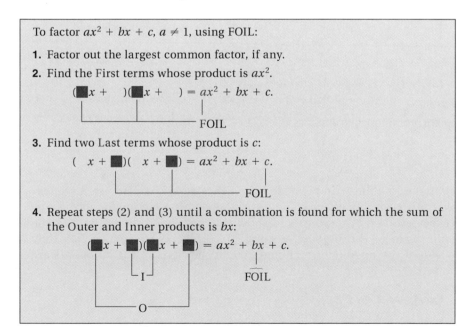

To factor $ax^2 + bx + c$, $a \neq 1$, using FOIL:

**1.** Factor out the largest common factor, if any.

**2.** Find the First terms whose product is $ax^2$.

$$(\blacksquare x + \phantom{)}\phantom{)})(\blacksquare x + \phantom{)}\phantom{)}) = ax^2 + bx + c.$$

FOIL

**3.** Find two Last terms whose product is $c$:

$$(\phantom{)} x + \blacksquare)(\phantom{)} x + \blacksquare) = ax^2 + bx + c.$$

FOIL

**4.** Repeat steps (2) and (3) until a combination is found for which the sum of the Outer and Inner products is $bx$:

$$(\blacksquare x + \blacksquare)(\blacksquare x + \blacksquare) = ax^2 + bx + c.$$

I
FOIL
O

Factor.

**1.** $2x^2 - x - 15$

**2.** $12x^2 - 17x - 5$

**Example 1**   Factor: $3x^2 - 10x - 8$.

1) First, we check for a common factor. Here there is none (other than 1 or $-1$).

2) Find two **F**irst terms whose product is $3x^2$.

   The only possibilities for the **F**irst terms are $3x$ and $x$, so any factorization must be of the form

   $$(3x + \phantom{x})(x + \phantom{x}).$$

3) Find two **L**ast terms whose product is $-8$.

   Possible factorizations of $-8$ are

   $$(-8) \cdot 1, \quad 8 \cdot (-1), \quad (-2) \cdot 4, \quad \text{and} \quad 2 \cdot (-4).$$

   Since the **F**irst terms are not identical, we must also consider

   $$1 \cdot (-8), \quad (-1) \cdot 8, \quad 4 \cdot (-2), \quad \text{and} \quad (-4) \cdot 2.$$

4) Inspect the **O**uter and **I**nner products resulting from steps (2) and (3). Look for a combination in which the sum of the products is the middle term, $-10x$:

| Trial | Product | |
|---|---|---|
| $(3x - 8)(x + 1)$ | $3x^2 + 3x - 8x - 8$ | |
| | $= 3x^2 - 5x - 8$ | ← Wrong middle term |
| $(3x + 8)(x - 1)$ | $3x^2 - 3x + 8x - 8$ | |
| | $= 3x^2 + 5x - 8$ | ← Wrong middle term |
| $(3x - 2)(x + 4)$ | $3x^2 + 12x - 2x - 8$ | |
| | $= 3x^2 + 10x - 8$ | ← Wrong middle term |
| $(3x + 2)(x - 4)$ | $3x^2 - 12x + 2x - 8$ | |
| | $= 3x^2 - 10x - 8$ | ← Correct middle term! |
| $(3x + 1)(x - 8)$ | $3x^2 - 24x + x - 8$ | |
| | $= 3x^2 - 23x - 8$ | ← Wrong middle term |
| $(3x - 1)(x + 8)$ | $3x^2 + 24x - x - 8$ | |
| | $= 3x^2 + 23x - 8$ | ← Wrong middle term |
| $(3x + 4)(x - 2)$ | $3x^2 - 6x + 4x - 8$ | |
| | $= 3x^2 - 2x - 8$ | ← Wrong middle term |
| $(3x - 4)(x + 2)$ | $3x^2 + 6x - 4x - 8$ | |
| | $= 3x^2 + 2x - 8$ | ← Wrong middle term |

The correct factorization is $(3x + 2)(x - 4)$.

**CHECK:**   $(3x + 2)(x - 4) = 3x^2 - 10x - 8$.

Two observations can be made from Example 1. First, we listed all possible trials even though we could have stopped after having found the correct factorization. We did this to show that each trial differs only in the middle term of the product. Second, note that as in Section 5.2, only the sign of the middle term changes when the signs in the binomials are reversed.

***Do Exercises 1 and 2.***

*Answers on page A-22*

**Example 2** Factor: $24x^2 - 76x + 40$.

**1)** First, we factor out the largest common factor, 4:

$$4(6x^2 - 19x + 10).$$

Now we factor the trinomial $6x^2 - 19x + 10$.

**2)** Because $6x^2$ can be factored as $3x \cdot 2x$ or $6x \cdot x$, we have these possibilities for factorizations:

$$(3x + \quad)(2x + \quad) \quad \text{or} \quad (6x + \quad)(x + \quad).$$

**3)** There are four pairs of factors of 10 and they each can be listed in two ways:

$$10, 1 \qquad -10, -1 \qquad 5, 2 \qquad -5, -2$$

and

$$1, 10 \qquad -1, -10 \qquad 2, 5 \qquad -2, -5.$$

**4)** The two possibilities from step (2) and the eight possibilities from step (3) give $2 \cdot 8$, or 16 possibilities for factorizations. We look for **O**uter and **I**nner products resulting from steps (2) and (3) for which the sum is the middle term, $-19x$. Since the sign of the middle term is negative, but the sign of the last term, 10, is positive, the two factors of 10 must both be negative. This means only four pairings from step (3) need be considered. We first try these factors with $(3x + \quad)(2x + \quad)$. If none gives the correct factorization, we will consider $(6x + \quad)(x + \quad)$.

| Trial | Product | |
|---|---|---|
| $(3x - 10)(2x - 1)$ | $6x^2 - 3x - 20x + 10$ | |
| | $= 6x^2 - 23x + 10$ | ← **Wrong middle term** |
| $(3x - 1)(2x - 10)$ | $6x^2 - 30x - 2x + 10$ | |
| | $= 6x^2 - 32x + 10$ | ← **Wrong middle term** |
| $(3x - 5)(2x - 2)$ | $6x^2 - 6x - 10x + 10$ | |
| | $= 6x^2 - 16x + 10$ | ← **Wrong middle term** |
| $(3x - 2)(2x - 5)$ | $6x^2 - 15x - 4x + 10$ | |
| | $= 6x^2 - 19x + 10$ | ← **Correct middle term!** |

Since we have a correct factorization, we need not consider

$$(6x + \quad)(x + \quad).$$

The factorization of $6x^2 - 19x + 10$ is $(3x - 2)(2x - 5)$, but *do not forget the common factor*! We must include it in order to factor the original trinomial:

$$24x^2 - 76x + 40 = 4(6x^2 - 19x + 10)$$
$$= 4(3x - 2)(2x - 5).$$

---

*CAUTION!* When factoring any polynomial, always look for a common factor. Failure to do so is such a common error that this caution bears repeating.

---

Factor.

**3.** $3x^2 - 19x + 20$

**4.** $20x^2 - 46x + 24$

**5.** Factor: $6x^2 + 7x + 2$.

*Answers on page A-22*

In Example 2, look again at the possibility $(3x - 5)(2x - 2)$. Without multiplying, we can reject such a possibility. To see why, consider the following:

$$(3x - 5)(2x - 2) = 2(3x - 5)(x - 1).$$

The expression $2x - 2$ has a common factor, 2. But we removed the *largest* common factor in the first step. If $2x - 2$ were one of the factors, then 2 would have to be a common factor in addition to the original 4. Thus, $(2x - 2)$ cannot be part of the factorization of the original trinomial.

> Given that the largest common factor is factored out at the outset, we need not consider factorizations that have a common factor.

*Do Exercises 3 and 4.*

**Example 3**   Factor: $10x^2 + 37x + 7$.

**1)** There is no common factor (other than 1 or $-1$).

**2)** Because $10x^2$ factors as $10x \cdot x$ or $5x \cdot 2x$, we have these possibilities for factorizations:

$$(10x + \;\;)(x + \;\;) \quad \text{or} \quad (5x + \;\;)(2x + \;\;).$$

**3)** There are two pairs of factors of 7 and they each can be listed in two ways:

$$1, 7 \quad -1, -7 \qquad \text{and} \qquad 7, 1 \quad -7, -1.$$

**4)** From steps (2) and (3), we see that there are 8 possibilities for factorizations. Look for **O**uter and **I**nner products for which the sum is the middle term. Because all coefficients in $10x^2 + 37x + 7$ are positive, we need consider only positive factors of 7. The possibilities are

$$(10x + 1)(x + 7) = 10x^2 + 71x + 7,$$
$$(10x + 7)(x + 1) = 10x^2 + 17x + 7,$$
$$(5x + 7)(2x + 1) = 10x^2 + 19x + 7,$$
$$(5x + 1)(2x + 7) = 10x^2 + 37x + 7.$$

The factorization is $(5x + 1)(2x + 7)$.

Keep in mind that this method of factoring trinomials of the type $ax^2 + bx + c$ involves *trial and error*. As you practice, you will find that you can make better and better guesses.

*Do Exercise 5.*

> **TIPS FOR FACTORING $ax^2 + bx + c$, $a \neq 1$**
>
> **1.** Always factor out the largest common factor, if one exists. Once the common factor has been factored out of the original trinomial, no binomial factor can contain a common factor (other than 1 or $-1$).
>
> **2.** If $c$ is positive, then the signs in both binomial factors must match the sign of $b$. (This assumes that $a > 0$.)
>
> **3.** Reversing the signs in the binomials reverses the sign of the middle term of their product.
>
> **4.** Be systematic about your trials. Keep track of those pairs you have tried and those you have not.
>
> **5.** Always check by multiplying.

**Example 4** Factor: $6p^2 - 13pq - 28q^2$.

**1)** Factor out a common factor, if any. There is none (other than 1 or $-1$).

**2)** Factor the first term, $6p^2$. Possibilities are $2p$, $3p$ and $6p$, $p$. We have these as possibilities for factorizations:

$$(2p + \ \ )(3p + \ \ ) \quad \text{or} \quad (6p + \ \ )(p + \ \ ).$$

**3)** Factor the last term, $-28q^2$, which has a negative coefficient. The possibilities are $-14q$, $2q$ and $14q$, $-2q$; $-28q$, $q$ and $28q$, $-q$; and $-7q$, $4q$ and $7q$, $-4q$.

**4)** The coefficient of the middle term is negative, so we look for combinations of factors from steps (2) and (3) such that the sum of their products has a negative coefficient. We try some possibilities:

$$(2p + q)(3p - 28q) = 6p^2 - 53pq - 28q^2,$$
$$(2p - 7q)(3p + 4q) = 6p^2 - 13pq - 28q^2.$$

The factorization of $6p^2 - 13pq - 28q^2$ is $(2p - 7q)(3p + 4q)$.

*Do Exercises 6 and 7.*

Factor.

**6.** $6a^2 - 5ab + b^2$

**7.** $6x^2 + 15xy + 9y^2$

---

## Calculator Spotlight

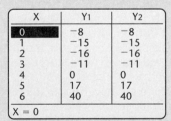 Checking Factorizations with a Table or a Graph

**Table.** The TABLE feature can be used as a partial check that polynomials have been factored correctly. To check whether the factoring of Example 1,

$$3x^2 - 10x - 8 = (3x + 2)(x - 4),$$

is correct, we enter

$$y_1 = 3x^2 - 10x - 8 \quad \text{and} \quad y_2 = (3x + 2)(x - 4).$$

If our factoring is correct, the $y_1$- and $y_2$-values should be the same, regardless of the table settings used.

| X | Y1 | Y2 |
|---|-----|-----|
| 0 | −8 | −8 |
| 1 | −15 | −15 |
| 2 | −16 | −16 |
| 3 | −11 | −11 |
| 4 | 0 | 0 |
| 5 | 17 | 17 |
| 6 | 40 | 40 |

X = 0

We see that $y_1$ and $y_2$ are the same, so the factoring seems to be correct. Remember, though, that this is only a partial check.

**Graph.** This factorization of Example 1 can also be checked with the GRAPH feature. We see that the graphs

of $y_1$ and $y_2$ are the same, so the factoring seems to be correct.

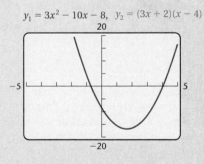
$y_1 = 3x^2 - 10x - 8, \quad y_2 = (3x + 2)(x - 4)$

**Exercises**

Use the TABLE or the GRAPH feature to check whether the factorization is correct.

**1.** $24x^2 - 76x + 40 = 4(3x - 2)(2x - 5)$ (Example 2)

**2.** $4x^2 - 5x - 6 = (4x + 3)(x - 2)$

**3.** $5x^2 + 17x - 12 = (5x + 3)(x - 4)$

**4.** $10x^2 + 37x + 7 = (5x - 1)(2x + 7)$

**5.** $12x^2 - 17x - 5 = (6x + 1)(2x - 5)$

**6.** $12x^2 - 17x - 5 = (4x + 1)(3x - 5)$

**7.** $x^2 - 4 = (x - 2)(x - 2)$

**8.** $x^2 - 4 = (x + 2)(x - 2)$

---

*Answers on page A-22*

# Improving Your Math Study Skills

## How Many Women Have Won the Ultimate Math Contest?

*Although this Study Skill feature does not contain specific tips on studying mathematics, we hope that you will find this article both challenging and encouraging.*

Every year on college campuses across the United States and Canada, the most brilliant math students face the ultimate challenge. For six hours, they struggle with problems from the merely intractable to the seemingly impossible.

Every spring, five are chosen winners of the William Lowell Putnam Mathematical Competition, the Olympics of college mathematics. Every year for 56 years, all have been men.

Until this year.

This spring, Ioana Dumitriu (pronounced yo-AHN-na doo-mee-TREE-oo), 20, a New York University sophomore from Romania, became the first woman to win the award.

Ms. Dumitriu, the daughter of two electrical engineering professors in Romania, who as a girl solved math puzzles for fun, was identified as a math talent early in her schooling in Bucharest. At 11, Ms. Dumitriu was steered into years of math training camps as preparation for the Romanian entry in the International Mathematics Olympiad.

It was this training, and a handsome young coach, that led her to New York City. He was several years older. They fell in love. He chose N.Y.U. for its graduate school in mathematics, and at 19 she joined him in New York.

The test Ms. Dumitriu won is dauntingly difficult, even for math majors. About half of the 2,407 test-takers scored 2 or less of a possible 120, and a third scored 0. Some students simply walk out after staring at the questions for a while.

Ms. Dumitriu said that in the six hours allotted, she had time to do 8 of the 12 problems, each worth a maximum of 10 points. The last one she did in 10 minutes. This year, Ms. Dumitriu and her five co-winners (there was a tie for fifth place) scored between 76 and 98. She does not know her exact score or rank because the organizers do not announce them.

"I didn't ever tell myself that I was unlikely to win, that no woman before had ever won and therefore I couldn't," she said. "It is not that I forget that I'm a woman. It's just that I don't see it as an obstacle or a ——."

Her English is near-perfect, but she paused because she could not find the right word. "The mathematics community is made up of persons, and that is what I am primarily."

Prof. Joel Spencer, who was a Putnam winner himself, said her work for his class in problem solving last year was remarkable. "What really got me was her fearlessness," he said. "To be good at math, you have to go right at it and start playing around with it, and she had that from the start."

In the graduate lounge in the Courant Institute of Mathematical Sciences at N.Y.U., Ms. Dumitriu, a tall, striking redhead, stands out. Instead of jeans and T-shirts, she wears gray pin-striped slacks and a rust-colored turtleneck and vest.

"There is a social perception of women and math, a stereotype," Ms. Dumitriu said during an interview. "What's happening right now is that the stereotype is defied. It starts breaking."

Still, even as women began to flock to sciences, math has remained largely a male bastion.

"Math remains the bottom line of sex differences for many," said Sheila Tobias, author of "Overcoming Math Anxiety" (W.W. Norton & Company, 1994). "It's one thing for women to write books, negotiate bills through Congress, litigate, fire missiles; quite another for them to do math."

Besides collecting the $1,000 awarded to each Putnam fellow, Ms. Dumitriu also won the $500 Elizabeth Lowell Putman prize for the top woman finisher for the second year in a row, a prize created five years ago to encourage women to take the test. This year 414 did.

In her view, there are never too many problems, never too much practice.

Besides, each new problem holds its own allure: "When you have all the pieces and you put them together and you see the puzzle, that moment always amazes me."

Copyright © 1997 by The New York Times Co. Reprinted by permission. Article by Karen W. Arenson.

# Exercise Set 5.3

**a** Factor.

1. $2x^2 - 7x - 4$

2. $3x^2 - x - 4$

3. $5x^2 - x - 18$

4. $4x^2 - 17x + 15$

5. $6x^2 + 23x + 7$

6. $6x^2 - 23x + 7$

7. $3x^2 + 4x + 1$

8. $7x^2 + 15x + 2$

9. $4x^2 + 4x - 15$

10. $9x^2 + 6x - 8$

11. $2x^2 - x - 1$

12. $15x^2 - 19x - 10$

13. $9x^2 + 18x - 16$

14. $2x^2 + 5x + 2$

15. $3x^2 - 5x - 2$

16. $18x^2 - 3x - 10$

17. $12x^2 + 31x + 20$

18. $15x^2 + 19x - 10$

19. $14x^2 + 19x - 3$

20. $35x^2 + 34x + 8$

21. $9x^2 + 18x + 8$

22. $6 - 13x + 6x^2$

23. $49 - 42x + 9x^2$

24. $16 + 36x^2 + 48x$

25. $24x^2 + 47x - 2$

26. $16p^2 - 78p + 27$

27. $35x^2 - 57x - 44$

28. $9a^2 + 12a - 5$

29. $20 + 6x - 2x^2$

30. $15 + x - 2x^2$

31. $12x^2 + 28x - 24$

32. $6x^2 + 33x + 15$

33. $30x^2 - 24x - 54$

34. $18t^2 - 24t + 6$

35. $4y + 6y^2 - 10$

36. $-9 + 18x^2 - 21x$

37. $3x^2 - 4x + 1$

38. $6t^2 + 13t + 6$

39. $12x^2 - 28x - 24$

40. $6x^2 - 33x + 15$

41. $-1 + 2x^2 - x$

42. $-19x + 15x^2 + 6$

43. $9x^2 - 18x - 16$

44. $14y^2 + 35y + 14$

45. $15x^2 - 25x - 10$

46. $18x^2 + 3x - 10$

47. $12p^3 + 31p^2 + 20p$

48. $15x^3 + 19x^2 - 10x$

**49.** $14x^4 + 19x^3 - 3x^2$     **50.** $70x^4 + 68x^3 + 16x^2$     **51.** $168x^3 - 45x^2 + 3x$     **52.** $144x^5 + 168x^4 + 48x^3$

**53.** $15x^4 - 19x^2 + 6$     **54.** $9x^4 + 18x^2 + 8$     **55.** $25t^2 + 80t + 64$     **56.** $9x^2 - 42x + 49$

**57.** $6x^3 + 4x^2 - 10x$     **58.** $18x^3 - 21x^2 - 9x$     **59.** $25x^2 + 79x + 64$     **60.** $9y^2 + 42y + 47$

**61.** $6x^2 - 19x - 5$     **62.** $2x^2 + 11x - 9$     **63.** $12m^2 - mn - 20n^2$     **64.** $12a^2 - 17ab + 6b^2$

**65.** $6a^2 - ab - 15b^2$     **66.** $3p^2 - 16pq - 12q^2$     **67.** $9a^2 + 18ab + 8b^2$     **68.** $10s^2 + 4st - 6t^2$

**69.** $35p^2 + 34pq + 8q^2$     **70.** $30a^2 + 87ab + 30b^2$     **71.** $18x^2 - 6xy - 24y^2$     **72.** $15a^2 - 5ab - 20b^2$

---

**Skill Maintenance**

Solve.   [2.6a]

**73.** $A = pq - 7$, for $q$     **74.** $y = mx + b$, for $x$     **75.** $3x + 2y = 6$, for $y$     **76.** $p - q + r = 2$, for $q$

Solve.   [2.7e]

**77.** $5 - 4x < -11$                 **78.** $2x - 4(x + 3x) \geq 6x - 8 - 9x$

**79.** Graph: $y = \dfrac{2}{5}x - 1$.   [3.2b]         **80.** Divide: $\dfrac{y^{12}}{y^4}$.   [4.1e]

Multiply.   [4.6d]

**81.** $(3x - 5)(3x + 5)$                 **82.** $(4a - 3)^2$

---

**Synthesis**

**83.** ◈ Explain how the factoring in Exercise 21 can be used to aid the factoring in Exercise 67.

**84.** ◈ A student presents the following work:
$$4x^2 + 28x + 48 = (2x + 6)(2x + 8)$$
$$= 2(x + 3)(x + 4).$$
Is it correct? Explain.

Factor.

**85.** $20x^{2n} + 16x^n + 3$     **86.** $-15x^{2m} + 26x^m - 8$     **87.** $3x^{6a} - 2x^{3a} - 1$     **88.** $x^{2n+1} - 2x^{n+1} + x$

Copyright © 1999 Addison Wesley Longman

# 5.4 Factoring $ax^2 + bx + c$, $a \neq 1$, Using Grouping

**a**    Another method of factoring trinomials of the type $ax^2 + bx + c$, $a \neq 1$, is known as the **grouping method.** It involves factoring by grouping. We know how to factor the trinomial $x^2 + 5x + 6$. We look for factors of the constant term, 6, whose sum is the coefficient of the middle term, 5:

$$x^2 + 5x + 6.$$
$$\text{(1) Factor: } 6 = 2 \cdot 3$$
$$\text{(2) Sum: } 2 + 3 = 5$$

What happens when the leading coefficient is not 1? To factor a trinomial like $3x^2 - 10x - 8$, we can use a method similar to what we used for the preceding trinomial, but we need two more steps. That method is outlined as follows.

---

To factor $ax^2 + bx + c$, $a \neq 1$, using the grouping method:

1. Factor out a common factor, if any.
2. Multiply the leading coefficient $a$ and the constant $c$.
3. Try to factor the product $ac$ so that the sum of the factors is $b$. That is, find integers $p$ and $q$ such that $pq = ac$ and $p + q = b$.
4. Split the middle term. That is, write it as a sum using the factors found in step (3).
5. Then factor by grouping.

---

**Example 1**    Factor: $3x^2 - 10x - 8$.

1) First, we factor out a common factor, if any. There is none (other than 1 or $-1$).

2) We multiply the leading coefficient, 3, and the constant, $-8$:

$$3(-8) = -24.$$

3) Then we look for a factorization of $-24$ in which the sum of the factors is the coefficient of the middle term, $-10$.

| Pairs of Factors | Sums of Factors |
|---|---|
| $-1$,   24 | 23 |
| 1, $-24$ | $-23$ |
| $-2$,   12 | 10 |
| 2, $-12$ | $-10$   ←   $2 + (-12) = -10$ |
| $-3$,    8 | 5 |
| 3,  $-8$ | $-5$ |
| $-4$,    6 | 2 |
| 4,  $-6$ | $-2$ |

4) Next, we split the middle term as a sum or a difference using the factors found in step (3):

$$10x = 2x - 12x.$$

**Objective**

**a**   Factor trinomials of the type $ax^2 + bx + c$, $a \neq 1$, by splitting the middle term and using grouping.

**For Extra Help**

TAPE 9    TAPE 10A    MAC WIN    CD-ROM

Factor.

**1.** $6x^2 + 7x + 2$

**2.** $12x^2 - 17x - 5$

Factor.

**3.** $6x^2 + 15x + 9$

**4.** $20x^2 - 46x + 24$

**5)** Finally, we factor by grouping, as follows:

$$3x^2 - 10x - 8 = 3x^2 + 2x - 12x - 8 \qquad \text{Substituting } 2x - 12x \text{ for } -10x$$

$$= x(3x + 2) - 4(3x + 2) \qquad \text{Factoring by grouping; see Section 5.1}$$

$$= (x - 4)(3x + 2).$$

We can also split the middle term as $-12x + 2x$. We still get the same factorization, although the factors may be in a different order. Note the following:

$$3x^2 - 10x - 8 = 3x^2 - 12x + 2x - 8 \qquad \text{Substituting } -12x + 2x \text{ for } -10x$$

$$= 3x(x - 4) + 2(x - 4) \qquad \text{Factoring by grouping; see Section 5.1}$$

$$= (3x + 2)(x - 4).$$

*Check by multiplying:* $\qquad (3x + 2)(x - 4) = 3x^2 - 10x - 8.$

*Do Exercises 1 and 2.*

**Example 2**   Factor: $8x^2 + 8x - 6$.

**1)** First, we factor out a common factor, if any. The number 2 is common to all three terms, so we factor it out:

$$2(4x^2 + 4x - 3).$$

**2)** Next, we factor the trinomial $4x^2 + 4x - 3$. We multiply the leading coefficient and the constant, 4 and $-3$:

$$4(-3) = -12.$$

**3)** We try to factor $-12$ so that the sum of the factors is 4.

| Pairs of Factors | Sums of Factors | |
|---|---|---|
| −1,  12 | 11 | |
| 1, −12 | −11 | |
| −2,   6 | 4 | ← ─── −2 + 6 = 4 |
| 2,  −6 | −4 | |
| −3,   4 | 1 | |
| 3,  −4 | −1 | |

**4)** Then we split the middle term, $4x$, as follows:

$$4x = -2x + 6x.$$

**5)** Finally, we factor by grouping:

$$4x^2 + 4x - 3 = 4x^2 - 2x + 6x - 3 \qquad \text{Substituting } -2x + 6x \text{ for } 4x$$

$$= 2x(2x - 1) + 3(2x - 1) \qquad \text{Factoring by grouping}$$

$$= (2x + 3)(2x - 1).$$

The factorization of $4x^2 + 4x - 3$ is $(2x + 3)(2x - 1)$. But don't forget the common factor! We must include it to get a factorization of the original trinomial:

$$8x^2 + 8x - 6 = 2(2x + 3)(2x - 1).$$

*Do Exercises 3 and 4.*

# Exercise Set 5.4

**a** Factor. Note that the middle term has already been split.

**1.** $x^2 + 2x + 7x + 14$

**2.** $x^2 + 3x + x + 3$

**3.** $x^2 - 4x - x + 4$

**4.** $a^2 + 5a - 2a - 10$

**5.** $6x^2 + 4x + 9x + 6$

**6.** $3x^2 - 2x + 3x - 2$

**7.** $3x^2 - 4x - 12x + 16$

**8.** $24 - 18y - 20y + 15y^2$

**9.** $35x^2 - 40x + 21x - 24$

**10.** $8x^2 - 6x - 28x + 21$

**11.** $4x^2 + 6x - 6x - 9$

**12.** $2x^4 - 6x^2 - 5x^2 + 15$

**13.** $2x^4 + 6x^2 + 5x^2 + 15$

**14.** $9x^4 - 6x^2 - 6x^2 + 4$

Factor by grouping.

**15.** $2x^2 - 7x - 4$

**16.** $5x^2 - x - 18$

**17.** $3x^2 + 4x - 15$

**18.** $3x^2 + x - 4$

**19.** $6x^2 + 23x + 7$

**20.** $6x^2 + 13x + 6$

**21.** $3x^2 + 4x + 1$

**22.** $7x^2 + 15x + 2$

**23.** $4x^2 + 4x - 15$

**24.** $9x^2 + 6x - 8$

**25.** $2x^2 + x - 1$

**26.** $15x^2 + 19x - 10$

**27.** $9x^2 - 18x - 16$

**28.** $2x^2 - 5x + 2$

**29.** $3x^2 + 5x - 2$

**30.** $18x^2 + 3x - 10$

**31.** $12x^2 - 31x + 20$      **32.** $15x^2 - 19x - 10$      **33.** $14x^2 + 19x - 3$      **34.** $35x^2 + 34x + 8$

**35.** $9x^2 + 18x + 8$      **36.** $6 - 13x + 6x^2$      **37.** $49 - 42x + 9x^2$      **38.** $25x^2 + 40x + 16$

**39.** $24x^2 + 47x - 2$      **40.** $16a^2 + 78a + 27$      **41.** $35x^5 - 57x^4 - 44x^3$      **42.** $18a^3 + 24a^2 - 10a$

**43.** $60x + 18x^2 - 6x^3$      **44.** $60x + 4x^2 - 8x^3$      **45.** $15x^3 + 33x^4 + 6x^5$      **46.** $8x^2 + 2x + 6x^3$

---

**Skill Maintenance**

Solve.    [2.7d, e]

**47.** $-10x > 1000$      **48.** $-3.8x \leq -824.6$      **49.** $6 - 3x \geq -18$

**50.** $3 - 2x - 4x > -9$      **51.** $\frac{1}{2}x - 6x + 10 \leq x - 5x$      **52.** $-2(x + 7) > -4(x - 5)$

**53.** $3x - 6x + 2(x - 4) > 2(9 - 4x)$      **54.** $-6(x - 4) + 8(4 - x) \leq 3(x - 7)$

---

**Synthesis**

**55.** ◈ If you have studied both the FOIL and the grouping methods of factoring $ax^2 + bx + c$, $a \neq 1$, decide which method you think is better and explain why.

**56.** ◈ Explain factoring $ax^2 + bx + c$, $a \neq 1$, by grouping as though you were teaching a fellow student.

Factor.

**57.** $9x^{10} - 12x^5 + 4$      **58.** $24x^{2n} + 22x^n + 3$      **59.** $16x^{10} + 8x^5 + 1$      **60.** $(a + 4)^2 - 2(a + 4) + 1$

**61.–70.** ⌁ Use the TABLE feature to check the factoring in Exercises 15–24.

Copyright © 1999 Addison Wesley Longman

# 5.5 Factoring Trinomial Squares and Differences of Squares

In this section, we first learn to factor trinomials that are squares of binomials. Then we factor binomials that are differences of squares.

## a Recognizing Trinomial Squares

Some trinomials are squares of binomials. For example, the trinomial $x^2 + 10x + 25$ is the square of the binomial $x + 5$. To see this, we can calculate $(x + 5)^2$. It is $x^2 + 2 \cdot x \cdot 5 + 5^2$, or $x^2 + 10x + 25$. A trinomial that is the square of a binomial is called a **trinomial square.**

In Chapter 4, we considered squaring binomials as special-product rules:

$$(A + B)^2 = A^2 + 2AB + B^2;$$
$$(A - B)^2 = A^2 - 2AB + B^2.$$

We can use these equations in reverse to factor trinomial squares.

▶ $A^2 + 2AB + B^2 = (A + B)^2;$
$A^2 - 2AB + B^2 = (A - B)^2$

How can we recognize when an expression to be factored is a trinomial square? Look at $A^2 + 2AB + B^2$ and $A^2 - 2AB + B^2$. In order for an expression to be a trinomial square:

**a)** Two terms, $A^2$ and $B^2$, must be squares, such as

   $4, \quad x^2, \quad 25x^4, \quad 16t^2.$

   When the coefficient is a perfect square and the power(s) of the variable(s) is (are) even, then the expression is a perfect square.

**b)** There must be no minus sign before $A^2$ or $B^2$.

**c)** If we multiply $A$ and $B$ (expressions whose squares are $A^2$ and $B^2$) and double the result, we get either the remaining term $2 \cdot A \cdot B$, or its opposite, $-2 \cdot A \cdot B$.

**Example 1** Determine whether $x^2 + 6x + 9$ is a trinomial square.

**a)** We know that $x^2$ and 9 are squares.

**b)** There is no minus sign before $x^2$ or 9.

**c)** If we multiply the square roots, $x$ and 3, and double the product, we get the remaining term: $2 \cdot x \cdot 3 = 6x$.

Thus, $x^2 + 6x + 9$ is the square of a binomial. In fact, $x^2 + 6x + 9 = (x + 3)^2$.

**Example 2** Determine whether $x^2 + 6x + 11$ is a trinomial square.

The answer is no, because only one term is a square.

## Objectives

**a** Recognize trinomial squares.

**b** Factor trinomial squares.

**c** Recognize differences of squares.

**d** Factor differences of squares, being careful to factor completely.

## For Extra Help

TAPE 10   TAPE 10A   MAC WIN   CD-ROM

---

*Handwritten notes:*

① 1st + last term must be perfect square #

② last term must be (+)

③ Sq. root of 1st + last #'s must be multiplied & doubled. it should = the middle #.

Determine whether each is a tri-nomial square. Write "yes" or "no."

**1.** $x^2 + 8x + 16$

**2.** $25 - x^2 + 10x$

**3.** $t^2 - 12t + 4$

**4.** $25 + 20y + 4y^2$

**5.** $5x^2 + 16 - 14x$

**6.** $16x^2 + 40x + 25$

**7.** $p^2 + 6p - 9$

**8.** $25a^2 + 9 - 30a$

Factor.

**9.** $x^2 + 2x + 1$

**10.** $1 - 2x + x^2$

**11.** $4 + t^2 + 4t$

**12.** $25x^2 - 70x + 49$

**13.** $49 - 56y + 16y^2$

*Answers on page A-22*

**Example 3**   Determine whether $16x^2 + 49 - 56x$ is a trinomial square.

It helps to first write the trinomial in descending order:

$$16x^2 - 56x + 49.$$

a) We know that $16x^2$ and 49 are squares.

b) There is no minus sign before $16x^2$ or 49.

c) If we multiply the square roots, $4x$ and 7, and double the product, we get the opposite of the remaining term: $2 \cdot 4x \cdot 7 = 56x$; $56x$ is the oppo-site of $-56x$.

Thus, $16x^2 + 49 - 56x$ is a trinomial square. In fact, $16x^2 - 56x + 49 = (4x - 7)^2$. or $(4x-7)(4x-7)$

*Do Exercises 1–8.*

## b   Factoring Trinomial Squares

We can use the trial-and-error or grouping methods from Sections 5.2–5.4 to factor such trinomial squares, but there is a faster method using the fol-lowing equations:

$$A^2 + 2AB + B^2 = (A + B)^2;$$
$$A^2 - 2AB + B^2 = (A - B)^2.$$

We consider 3 to be a square root of 9 because $3^2 = 9$. Similarly, $A$ is a square root of $A^2$. We use square roots of the squared terms and the sign of the remaining term to factor a trinomial square.

**Example 4**   Factor: $x^2 + 6x + 9$.

$$x^2 + 6x + 9 = x^2 + 2 \cdot x \cdot 3 + 3^2 = (x + 3)^2$$
$$\quad\quad\quad\quad A^2 + 2 \quad A \quad B + B^2 = (A + B)^2$$

The sign of the middle term is positive.

**Example 5**   Factor: $x^2 + 49 - 14x$.

$$x^2 + 49 - 14x = x^2 - 14x + 49 \quad\quad \text{Changing order}$$
$$= x^2 - 2 \cdot x \cdot 7 + 7^2 \quad\quad \text{The sign of the middle term is negative.}$$
$$= (x - 7)^2$$

**Example 6**   Factor: $16x^2 - 40x + 25$.

$$16x^2 - 40x + 25 = (4x)^2 - 2 \cdot 4x \cdot 5 + 5^2 = (4x - 5)^2$$
$$\quad\quad\quad\quad A^2 \quad - 2 \quad A \quad B + B^2 = (A - B)^2$$

*Do Exercises 9–13.*

**Example 7**  Factor: $t^4 + 20t^2 + 100$.

$$t^4 + 20t^2 + 100 = (t^2)^2 + 2(t^2)(10) + 10^2$$
$$= (t^2 + 10)^2$$

**Example 8**  Factor: $75m^3 + 210m^2 + 147m$.

*Always* look first for a common factor. This time there is one, $3m$:

$$75m^3 + 210m^2 + 147m = 3m[25m^2 + 70m + 49]$$
$$= 3m[(5m)^2 + 2(5m)(7) + 7^2]$$
$$= 3m(5m + 7)^2.$$

**Example 9**  Factor: $4p^2 - 12pq + 9q^2$.

$$4p^2 - 12pq + 9q^2 = (2p)^2 - 2(2p)(3q) + (3q)^2$$
$$= (2p - 3q)^2$$

*Do Exercises 14–17.*

**c** **Recognizing Differences of Squares**

The following polynomials are *differences of squares*:

$$x^2 - 9, \qquad 4t^2 - 49, \qquad a^2 - 25b^2.$$

To factor a difference of squares such as $x^2 - 9$, think about the formula we used in Chapter 4:

$$(A + B)(A - B) = A^2 - B^2.$$

Equations are reversible, so we also know that

> $A^2 - B^2 = (A + B)(A - B).$

Thus,

$$x^2 - 9 = (x + 3)(x - 3).$$

To use this formula, we must be able to recognize when it applies. A **difference of squares** is an expression like the following:

$$A^2 - B^2.$$

How can we recognize such expressions? Look at $A^2 - B^2$. In order for a binomial to be a difference of squares:

**a)** There must be two expressions, both squares, such as

$$4x^2, \quad 9, \quad 25t^4, \quad 1, \quad x^6, \quad 49y^8.$$

**b)** The terms must have different signs.

*(handwritten notes in margin:)*

$(x+4)(x-4)=x^2-16$

$(2x-3)(2x+3)=4x^2-9$

$(5x-7)(5x+7)=25x^2-49$

1st term must be squared. Must be (−) sign + must be difference of 2 squares.

Determine whether each is a difference of squares. Write "yes" or "no."

**18.** $x^2 - 25$

**19.** $t^2 - 24$

**20.** $y^2 + 36$

**21.** $4x^2 - 15$

**22.** $16x^4 - 49$

**23.** $9w^6 - 1$

**24.** $-49 + 25t^2$

*Answers on page A-22*

**Example 10**   Is $9x^2 - 64$ a difference of squares?

a) The first expression is a square: $9x^2 = (3x)^2$.
   The second expression is a square: $64 = 8^2$.

b) The terms have different signs.

Thus we have a difference of squares, $(3x)^2 - 8^2$.

**Example 11**   Is $25 - t^3$ a difference of squares?

a) The expression $t^3$ is not a square.

The expression is not a difference of squares.

**Example 12**   Is $-4x^2 + 16$ a difference of squares?

a) The expressions $4x^2$ and $16$ are squares: $4x^2 = (2x)^2$ and $16 = 4^2$.

b) The terms have different signs.

Thus we have a difference of squares. We can also see this by rewriting in the equivalent form: $16 - 4x^2$.

*Do Exercises 18–24.*

## d | Factoring Differences of Squares

To factor a difference of squares, we use the following equation:

$$A^2 - B^2 = (A + B)(A - B).$$

To factor a difference of squares $A^2 - B^2$, we find $A$ and $B$, which are square roots of the expressions $A^2$ and $B^2$. We then use $A$ and $B$ to form two factors. One is the sum $A + B$, and the other is the difference $A - B$.

**Example 13**   Factor: $x^2 - 4$.

$$x^2 - 4 = x^2 - 2^2 = (x + 2)(x - 2)$$
$$A^2 - B^2 = (A + B)(A - B)$$

**Example 14**   Factor: $9 - 16t^4$.

$$9 - 16t^4 = 3^2 - (4t^2)^2 = (3 + 4t^2)(3 - 4t^2)$$
$$A^2 - B^2 = (A + B) (A - B)$$

**Example 15**   Factor: $m^2 - 4p^2$.

$$m^2 - 4p^2 = m^2 - (2p)^2 = (m + 2p)(m - 2p)$$

**Example 16**   Factor: $x^2 - \dfrac{1}{9}$.

$$x^2 - \frac{1}{9} = x^2 - \left(\frac{1}{3}\right)^2 = \left(x + \frac{1}{3}\right)\left(x - \frac{1}{3}\right)$$

**Example 17** Factor: $18x^2 - 50x^6$.

*Always* look first for a factor common to all terms. This time there is one, $2x^2$.

$$18x^2 - 50x^6 = 2x^2(9 - 25x^4)$$
$$= 2x^2[3^2 - (5x^2)^2]$$
$$= 2x^2(3 + 5x^2)(3 - 5x^2)$$

**Example 18** Factor: $49x^4 - 9x^6$.

$$49x^4 - 9x^6 = x^4(49 - 9x^2) = x^4(7 + 3x)(7 - 3x)$$

*Do Exercises 25–29.*

---

*CAUTION!* Note carefully in these examples that a difference of squares is *not* the square of the difference; that is,

$$A^2 - B^2 \neq (A - B)^2 = A^2 - 2AB + B^2.$$

For example,

$$(45 - 5)^2 = 40^2 = 1600,$$

but

$$45^2 - 5^2 = 2025 - 25 = 2000.$$

---

## Factoring Completely

If a factor with more than one term can still be factored, you should do so. When no factor can be factored further, you have **factored completely.** Always factor completely whenever told to factor.

**Example 19** Factor: $p^4 - 16$.

$$p^4 - 16 = (p^2)^2 - 4^2$$
$$= (p^2 + 4)(p^2 - 4) \qquad \text{Factoring a difference of squares}$$
$$= (p^2 + 4)(p + 2)(p - 2) \qquad \begin{array}{l}\text{Factoring further. The factor}\\ p^2 - 4 \text{ is a difference of squares.}\end{array}$$

The polynomial $p^2 + 4$ cannot be factored further into polynomials with real coefficients.

---

*CAUTION!* If the greatest common factor has been removed, then you cannot factor a sum of squares further. In particular,

$$(A + B)^2 \neq A^2 + B^2.$$

Consider $25x^2 + 100$. This is a case in which we have a sum of squares, but there is a common factor, 25. Factoring, we get $25(x^2 + 4)$. Now $x^2 + 4$ cannot be factored further.

---

**Example 20** Factor: $y^4 - 16x^{12}$.

$$y^4 - 16x^{12} = (y^2 + 4x^6)(y^2 - 4x^6) \qquad \begin{array}{l}\text{Factoring a difference}\\ \text{of squares}\end{array}$$

$$= (y^2 + 4x^6)(y + 2x^3)(y - 2x^3) \qquad \begin{array}{l}\text{Factoring further. The}\\ \text{factor } y^2 - 4x^6 \text{ is a}\\ \text{difference of squares.}\end{array}$$

Factor.

**25.** $x^2 - 9$

**26.** $64 - 4t^2$

**27.** $a^2 - 25b^2$

**28.** $64x^4 - 25x^6$

**29.** $5 - 20t^6$
[*Hint:* $1 = 1^2$, $t^6 = (t^3)^2$.]

*Answers on page A-22*

Factor completely.

**30.** $81x^4 - 1$

**31.** $49p^4 - 25q^6$

**FACTORING HINTS**

**1.** Always look first for a common factor. If there is one, factor out the largest common factor.

**2.** Always factor completely.

**3.** Check by multiplying.

*Do Exercises 30 and 31.*

# Improving Your Math Study Skills

## A Checklist of Your Study Skills

You are now about halfway through this textbook as well as the course. How are you doing? If you are struggling, we might ask if you are making full use of the study skills that we have suggested in these inserts. To determine this, review the following list of all study skill suggestions made so far and answer the questions "yes" or "no."

| Study Skill Questions | Yes | No |
|---|---|---|
| **1.** Are you doing a thorough job of reading the book? | | |
| **2.** Are you stopping and working the margin exercises when directed to do so? | | |
| **3.** Are you doing your homework as soon as possible after class? | | |
| **4.** Are you doing your homework at a specified time and in a quiet setting? | | |
| **5.** Have you found a study group in which to work? | | |
| **6.** Are you consistently trying to apply the five-step problem-solving strategy when working applied problems? | | |
| **7.** Are you asking questions in class and in tutoring sessions? | | |
| **8.** Are you doing lots of even-numbered exercises for which answers are not available? | | |

| Study Skill Questions | Yes | No |
|---|---|---|
| **9.** Are you keeping one section ahead of your syllabus? | | |
| **10.** Are you using the book supplements, such as the *Student's Solutions Manual* and the *InterAct Math Tutorial Software*? | | |
| **11.** When you study the book, are you marking the points that you do not understand as a source for in-class questions? | | |
| **12.** Are you reading and studying each step of each example? | | |
| **13.** Are you using the objective code symbols ( $\boxed{a}$ , $\boxed{b}$ , $\boxed{c}$ , etc.) that appear at the beginning of each section, throughout the section, and in the exercise sets, the summary–reviews, and the answers for the chapter tests? | | |

If you have answered "no" seven or more times and are struggling in the course, you need to improve your study skills by following more of these suggestions.

A consultation with your instructor regarding your situation is strongly advised.

*Answers on page A-22*

# Exercise Set 5.5

**a** Determine whether each of the following is a trinomial square.

**1.** $x^2 - 14x + 49$

**2.** $x^2 - 16x + 64$
Yes

**3.** $x^2 + 16x - 64$

**4.** $x^2 - 14x - 49$

**5.** $x^2 - 2x + 4$

**6.** $x^2 + 3x + 9$
No

**7.** $9x^2 - 36x + 24$

**8.** $36x^2 - 24x + 16$

**b** Factor completely. Remember to look first for a common factor and to check by multiplying.

**9.** $x^2 - 14x + 49$
$(x-7)^2$

**10.** $x^2 - 20x + 100$

**11.** $x^2 + 16x + 64$
$(x+8)^2$

**12.** $x^2 + 20x + 100$

**13.** $x^2 - 2x + 1$

**14.** $x^2 + 2x + 1$
$(x+1)^2$
$(x+1)(x+1)$

**15.** $4 + 4x + x^2$
$(x+2)^2$

**16.** $4 + x^2 - 4x$

**17.** $q^4 - 6q^2 + 9$

**18.** $64 + 16a^2 + a^4$

**19.** $49 + 56y + 16y^2$

**20.** $75 + 48a^2 - 120a$

**21.** $2x^2 - 4x + 2$
$2(x-1)^2$

**22.** $2x^2 - 40x + 200$

**23.** $x^3 - 18x^2 + 81x$
$x(x-9)^2$

**24.** $x^3 + 24x^2 + 144x$

**25.** $12q^2 - 36q + 27$

**26.** $20p^2 + 100p + 125$

**27.** $49 - 42x + 9x^2$

**28.** $64 - 112x + 49x^2$

**29.** $5y^4 + 10y^2 + 5$
$5(y^2+1)^2$

**30.** $a^4 + 14a^2 + 49$

**31.** $1 + 4x^4 + 4x^2$

**32.** $1 - 2a^5 + a^{10}$

**33.** $4p^2 + 12pq + 9q^2$

**34.** $25m^2 + 20mn + 4n^2$

**35.** $a^2 - 6ab + 9b^2$

**36.** $x^2 - 14xy + 49y^2$

**37.** $81a^2 - 18ab + b^2$

**38.** $64p^2 + 16pq + q^2$

**39.** $36a^2 + 96ab + 64b^2$

**40.** $16m^2 - 40mn + 25n^2$

[c] Determine whether each of the following is a difference of squares.

**41.** $x^2 - 4$

**42.** $x^2 - 36$

**43.** $x^2 + 25$

**44.** $x^2 + 9$

**45.** $x^2 - 45$

**46.** $x^2 - 80y^2$

**47.** $16x^2 - 25y^2$

**48.** $-1 + 36x^2$

[d] Factor completely. Remember to look first for a common factor.

**49.** $y^2 - 4$

**50.** $q^2 - 1$

**51.** $p^2 - 9$

**52.** $x^2 - 36$

**53.** $-49 + t^2$

**54.** $-64 + m^2$

**55.** $a^2 - b^2$

**56.** $p^2 - q^2$

**57.** $25t^2 - m^2$

**58.** $w^2 - 49z^2$

**59.** $100 - k^2$

**60.** $81 - w^2$

**61.** $16a^2 - 9$

**62.** $25x^2 - 4$

**63.** $4x^2 - 25y^2$

**64.** $9a^2 - 16b^2$

**65.** $8x^2 - 98$  **66.** $24x^2 - 54$  **67.** $36x - 49x^3$  **68.** $16x - 81x^3$

**69.** $49a^4 - 81$  **70.** $25a^4 - 9$  **71.** $a^4 - 16$  **72.** $y^4 - 1$

**73.** $5x^4 - 405$  **74.** $4x^4 - 64$  **75.** $1 - y^8$  **76.** $x^8 - 1$

**77.** $x^{12} - 16$  **78.** $x^8 - 81$  **79.** $y^2 - \dfrac{1}{16}$  **80.** $x^2 - \dfrac{1}{25}$

**81.** $25 - \dfrac{1}{49}x^2$  **82.** $\dfrac{1}{4} - 9q^2$  **83.** $16m^4 - t^4$  **84.** $p^4q^4 - 1$

---

**Skill Maintenance**

Divide.  [1.6c]

**85.** $(-110) \div 10$  **86.** $-1000 \div (-2.5)$  **87.** $\left(-\dfrac{2}{3}\right) \div \dfrac{4}{5}$

**88.** $8.1 \div (-9)$  **89.** $-64 \div (-32)$  **90.** $-256 \div 1.6$

Find a polynomial for the shaded area. (Leave results in terms of $\pi$ where appropriate.)  [4.4d]

**91.**

**92.**

Simplify.

**93.** $y^5 \cdot y^7$ [4.1d]

**94.** $(5a^2b^3)^2$ [4.2a]

Find the intercepts. Then graph the equation. [3.3a]

**95.** $y - 6x = 6$

**96.** $3x - 5y = 15$

---

**Synthesis**

**97.** ◆ Explain in your own words how to determine whether a polynomial is a trinomial square.

**98.** ◆ A student concludes that since $x^2 - 9 = (x - 3)(x + 3)$, it must follow that $x^2 + 9 = (x + 3)(x + 3)$. What mistake is the student making? How would you go about correcting the misunderstanding?

Factor completely, if possible.

**99.** $49x^2 - 216$

**100.** $27x^3 - 13x$

**101.** $x^2 + 22x + 121$

**102.** $x^2 - 5x + 25$

**103.** $18x^3 + 12x^2 + 2x$

**104.** $162x^2 - 82$

**105.** $x^8 - 2^8$

**106.** $4x^4 - 4x^2$

**107.** $3x^5 - 12x^3$

**108.** $3x^2 - \frac{1}{3}$

**109.** $18x^3 - \frac{8}{25}x$

**110.** $x^2 - 2.25$

**111.** $0.49p - p^3$

**112.** $3.24x^2 - 0.81$

**113.** $0.64x^2 - 1.21$

**114.** $1.28x^2 - 2$

**115.** $(x + 3)^2 - 9$

**116.** $(y - 5)^2 - 36q^2$

**117.** $x^2 - \left(\dfrac{1}{x}\right)^2$

**118.** $a^{2n} - 49b^{2n}$

**119.** $81 - b^{4k}$

**120.** $9x^{18} + 48x^9 + 64$

**121.** $9b^{2n} + 12b^n + 4$

**122.** $(x + 7)^2 - 4x - 24$

**123.** $(y + 3)^2 + 2(y + 3) + 1$

**124.** $49(x + 1)^2 - 42(x + 1) + 9$

Find $c$ such that the polynomial is the square of a binomial.

**125.** $cy^2 + 6y + 1$

**126.** $cy^2 - 24y + 9$

⌁ Use the TABLE feature to determine whether the factorization is correct.

**127.** $x^2 + 9 = (x + 3)(x + 3)$

**128.** $x^2 - 49 = (x - 7)(x + 7)$

**129.** $x^2 + 9 = (x + 3)^2$

**130.** $x^2 - 49 = (x - 7)^2$

# 5.6 Factoring: A General Strategy

**a** We now combine all of our factoring techniques and consider a general strategy for factoring polynomials. Here we will encounter polynomials of all the types we have considered, in random order, so you will have the opportunity to determine which method to use.

**Objective**

**a** Factor polynomials completely using any of the methods considered in this chapter.

**For Extra Help**

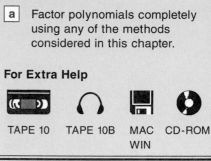

TAPE 10    TAPE 10B    MAC    CD-ROM
                       WIN

> To factor a polynomial:
>
> **a)** Always <u>look first for a common factor</u>. If there is one, factor out the largest common factor.
>
> **b)** Then look at the number of terms.
>
> *Two terms*: Determine whether you have a difference of squares. Do not try to factor a sum of squares: $A^2 + B^2$.
>
> *Three terms*: Determine whether the trinomial is a square. If it is, you know how to factor. If not, try trial and error, using FOIL or grouping.
>
> *Four terms*: Try factoring by grouping.
>
> **c)** *Always factor completely.* If a factor with more than one term can still be factored, you should factor it. When no factor can be factored further, you have finished.

**Example 1**   Factor: $5t^4 - 80$.

**a)** We look for a common factor:

$$5t^4 - 80 = 5(t^4 - 16).$$

**b)** The factor $t^4 - 16$ has only two terms. It is a difference of squares: $(t^2)^2 - 4^2$. We factor it, being careful to include the common factor:

$$5(t^2 + 4)(t^2 - 4).$$

**c)** We see that one of the factors is again a difference of squares. We factor it:

$$5(t^2 + 4)(t + 2)(t - 2).$$
$\uparrow$

This is a sum of squares. It cannot be factored!

We have factored completely because no factor with more than one term can be factored further.

**Example 2**   Factor: $2x^3 + 10x^2 + x + 5$.

**a)** We look for a common factor. There isn't one.

**b)** There are four terms. We try factoring by grouping:

$$2x^3 + 10x^2 + x + 5$$
$$= (2x^3 + 10x^2) + (x + 5) \qquad \text{Separating into two binomials}$$
$$= 2x^2(x + 5) + 1(x + 5) \qquad \text{Factoring each binomial}$$
$$= (2x^2 + 1)(x + 5). \qquad \text{Factoring out the common factor } x + 5$$

**c)** None of these factors can be factored further, so we have factored completely.

Factor.

**1.** $3m^4 - 3$

**2.** $x^6 + 8x^3 + 16$

**3.** $2x^4 + 8x^3 + 6x^2$

**4.** $3x^3 + 12x^2 - 2x - 8$

**5.** $8x^3 - 200x$

*Answers on page A-23*

**Example 3** Factor: $x^5 - 2x^4 - 35x^3$.

a) We look first for a common factor. This time there is one, $x^3$:

$$x^5 - 2x^4 - 35x^3 = x^3(x^2 - 2x - 35).$$

b) The factor $x^2 - 2x - 35$ has three terms, but it is not a trinomial square. We factor it using trial and error (FOIL or grouping):

$$x^5 - 2x^4 - 35x^3 = x^3(x^2 - 2x - 35) = x^3(x - 7)(x + 5).$$

> Don't forget to include the common factor in the final answer!

c) No factor with more than one term can be factored further, so we have factored completely.

**Example 4** Factor: $x^4 - 10x^2 + 25$.

a) We look first for a common factor. There isn't one.

b) There are three terms. We see that this polynomial is a trinomial square. We factor it:

$$x^4 - 10x^2 + 25 = (x^2)^2 - 2 \cdot x^2 \cdot 5 + 5^2 = (x^2 - 5)^2.$$

c) Since $x^2 - 5$ cannot be factored further, we have factored completely.

*Do Exercises 1–5.*

**Example 5** Factor: $6x^2y^4 - 21x^3y^5 + 3x^2y^6$.

a) We look first for a common factor:

$$6x^2y^4 - 21x^3y^5 + 3x^2y^6 = 3x^2y^4(2 - 7xy + y^2).$$

b) There are three terms in $2 - 7xy + y^2$. We determine whether the trinomial is a square. Since only $y^2$ is a square, we do not have a trinomial square. Can the trinomial be factored by trial and error? A key to the answer is that $x$ is only in the term $-7xy$. The polynomial might be in a form like $(1 - y)(2 + y)$, but there would be no $x$ in the middle term. Thus, $2 - 7xy + y^2$ cannot be factored.

c) Have we factored completely? Yes, because no factor with more than one term can be factored further.

**Example 6** Factor: $(p + q)(x + 2) + (p + q)(x + y)$.

a) We look for a common factor:

$$(p + q)(x + 2) + (p + q)(x + y) = (p + q)[(x + 2) + (x + y)]$$
$$= (p + q)(2x + y + 2).$$

b) There are three terms in $2x + y + 2$, but this trinomial cannot be factored further.

c) Neither factor can be factored further, so we have factored completely.

**Example 7**   Factor: $px + py + qx + qy$.

a) We look first for a common factor. There isn't one.

b) There are four terms. We try factoring by grouping:

$$px + py + qx + qy = p(x + y) + q(x + y)$$
$$= (p + q)(x + y).$$

c) Have we factored completely? Since neither factor can be factored further, we have factored completely.

**Example 8**   Factor: $25x^2 + 20xy + 4y^2$.

a) We look first for a common factor. There isn't one.

b) There are three terms. We determine whether the trinomial is a square. The first term and the last term are squares:

$$25x^2 = (5x)^2 \quad \text{and} \quad 4y^2 = (2y)^2.$$

Since twice the product of $5x$ and $2y$ is the other term,

$$2 \cdot 5x \cdot 2y = 20xy,$$

the trinomial is a perfect square.

   We factor by writing the square roots of the square terms and the sign of the middle term:

$$25x^2 + 20xy + 4y^2 = (5x + 2y)^2.$$

We can check by squaring $5x + 2y$.

c) Since $5x + 2y$ cannot be factored further, we have factored completely.

**Example 9**   Factor: $p^2q^2 + 7pq + 12$.

a) We look first for a common factor. There isn't one.

b) There are three terms. We determine whether the trinomial is a square. The first term is a square, but neither of the other terms is a square, so we do not have a trinomial square. We use the trial-and-error or grouping method, thinking of the product $pq$ as a single variable. We consider this possibility for factorization:

$$(pq + \ \ )(pq + \ \ ).$$

We factor the last term, 12. All the signs are positive, so we consider only positive factors. Possibilities are 1, 12 and 2, 6 and 3, 4. The pair 3, 4 gives a sum of 7 for the coefficient of the middle term. Thus,

$$p^2q^2 + 7pq + 12 = (pq + 3)(pq + 4).$$

c) No factor with more than one term can be factored further, so we have factored completely.

Factor.

**6.** $x^4y^2 + 2x^3y + 3x^2y$

**7.** $10p^6q^2 + 4p^5q^3 + 2p^4q^4$

**8.** $(a - b)(x + 5) + (a - b)(x + y^2)$

**9.** $ax^2 + ay + bx^2 + by$

**10.** $x^4 + 2x^2y^2 + y^4$

**11.** $x^2y^2 + 5xy + 4$

**12.** $p^4 - 81q^4$

*Answers on page A-23*

**Example 10**  Factor: $8x^4 - 20x^2y - 12y^2$.

a) We look first for a common factor:

$$8x^4 - 20x^2y - 12y^2 = 4(2x^4 - 5x^2y - 3y^2).$$

b) There are three terms in $2x^4 - 5x^2y - 3y^2$. We determine whether the trinomial is a square. Since none of the terms is a square, we do not have a trinomial square. We factor $2x^4$. Possibilities are $2x^2$, $x^2$ and $2x$, $x^3$ and others. We also factor the last term, $-3y^2$. Possibilities are $3y$, $-y$ and $-3y$, $y$ and others. We look for factors such that the sum of their products is the middle term. We try some possibilities:

$$(2x - y)(x^3 + 3y) = 2x^4 + 6xy - x^3y - 3y^2,$$
$$(2x^2 - y)(x^2 + 3y) = 2x^4 + 5x^2y - 3y^2,$$
$$(2x^2 + y)(x^2 - 3y) = 2x^4 - 5x^2y - 3y^2.$$

c) No factor with more than one term can be factored further, so we have factored completely. The factorization, including the common factor, is

$$4(2x^2 + y)(x^2 - 3y).$$

**Example 11**  Factor: $a^4 - 16b^4$.

a) We look first for a common factor. There isn't one.

b) There are two terms. Since $a^4 = (a^2)^2$ and $16b^4 = (4b^2)^2$, we see that we do have a difference of squares. Thus,

$$a^4 - 16b^4 = (a^2 + 4b^2)(a^2 - 4b^2).$$

c) The last factor can be factored further. It is also a difference of squares. Thus,

$$a^4 - 16b^4 = (a^2 + 4b^2)(a + 2b)(a - 2b).$$

*Do Exercises 6–12.*

# Exercise Set 5.6

**a** Factor completely.

**1.** $3x^2 - 192$

$3(x+8)(x-8)$

**2.** $2t^2 - 18$

$2(t^2+9)$

**3.** $a^2 + 25 - 10a$

$(a-5)^2$

**4.** $y^2 + 49 + 14y$

$(y+2)(y+7)$

**5.** $2x^2 - 11x + 12$

$(2x-3)(x-4)$

**6.** $8y^2 - 18y - 5$

$(4y+1)(4y-5)$

**7.** $x^3 + 24x^2 + 144x$

$x(x+12)^2$

**8.** $x^3 - 18x^2 + 81x$

**9.** $x^3 + 3x^2 - 4x - 12$

$(x+2)(x-2)(x+3)$

**10.** $x^3 - 5x^2 - 25x + 125$

**11.** $48x^2 - 3$

$3(4x+1)(4x-1)$

**12.** $50x^2 - 32$

**13.** $9x^3 + 12x^2 - 45x$

$3x(3x-5)(x+3)$

**14.** $20x^3 - 4x^2 - 72x$

**15.** $x^2 + 4$

**16.** $t^2 + 25$

**17.** $x^4 + 7x^2 - 3x^3 - 21x$

**18.** $m^4 + 8m^3 + 8m^2 + 64m$

**19.** $x^5 - 14x^4 + 49x^3$

**20.** $2x^6 + 8x^5 + 8x^4$

**21.** $20 - 6x - 2x^2$

**22.** $45 - 3x - 6x^2$

**23.** $x^2 - 6x + 1$

**24.** $x^2 + 8x + 5$

**25.** $4x^4 - 64$

**26.** $5x^5 - 80x$

**27.** $1 - y^8$

**28.** $t^8 - 1$

**29.** $x^5 - 4x^4 + 3x^3$

**30.** $x^6 - 2x^5 + 7x^4$

**31.** $\dfrac{1}{81}x^6 - \dfrac{8}{27}x^3 + \dfrac{16}{9}$

**32.** $36a^2 - 15a + \dfrac{25}{16}$

**33.** $mx^2 + my^2$

**34.** $12p^2 + 24q^3$

**35.** $9x^2y^2 - 36xy$

**36.** $x^2y - xy^2$

**37.** $2\pi rh + 2\pi r^2$

**38.** $10p^4q^4 + 35p^3q^3 + 10p^2q^2$

**39.** $(a + b)(x - 3) + (a + b)(x + 4)$

**40.** $5c(a^3 + b) - (a^3 + b)$

**41.** $(x - 1)(x + 1) - y(x + 1)$

**42.** $3(p - q) - q^2(p - q)$

**43.** $n^2 + 2n + np + 2p$

**44.** $a^2 - 3a + ay - 3y$

**45.** $6q^2 - 3q + 2pq - p$

**46.** $2x^2 - 4x + xy - 2y$

**47.** $4b^2 + a^2 - 4ab$

**48.** $x^2 + y^2 - 2xy$

**49.** $16x^2 + 24xy + 9y^2$

**50.** $9c^2 + 6cd + d^2$

**51.** $49m^4 - 112m^2n + 64n^2$

**52.** $4x^2y^2 + 12xyz + 9z^2$

**53.** $y^4 + 10y^2z^2 + 25z^4$

**54.** $0.01x^4 - 0.1x^2y^2 + 0.25y^4$

**55.** $\dfrac{1}{4}a^2 + \dfrac{1}{3}ab + \dfrac{1}{9}b^2$

**56.** $4p^2q + pq^2 + 4p^3$

**57.** $a^2 - ab - 2b^2$

**58.** $3b^2 - 17ab - 6a^2$

**59.** $2mn - 360n^2 + m^2$

**60.** $15 + x^2y^2 + 8xy$

**61.** $m^2n^2 - 4mn - 32$

**62.** $p^2q^2 + 7pq + 6$

**63.** $a^2b^6 + 4ab^5 - 32b^4$

**64.** $p^5q^2 + 3p^4q - 10p^3$

**65.** $a^5 + 4a^4b - 5a^3b^2$

**66.** $2s^6t^2 + 10s^3t^3 + 12t^4$

**67.** $a^2 - \dfrac{1}{25}b^2$

**68.** $p^2 - \dfrac{1}{49}b^2$

**69.** $x^2 - y^2$

**70.** $p^2q^2 - r^2$

**71.** $16 - p^4q^4$

**72.** $15a^4 - 15b^4$

**73.** $1 - 16x^{12}y^{12}$

**74.** $81a^4 - b^4$

**75.** $q^3 + 8q^2 - q - 8$

**76.** $m^3 - 7m^2 - 4m + 28$

**77.** $112xy + 49x^2 + 64y^2$

**78.** $4ab^5 - 32b^4 + a^2b^6$

*Sports-Car Sales.* The sales of sports cars rise and fall over the years due often to new or redesigned models, such as the 1997 Corvette. Sales for recent years are shown in the following line graph. Use the graph for Exercises 79–84. [3.1a]

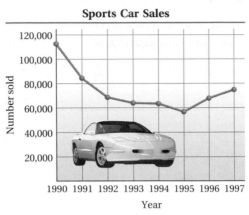

**Sports Car Sales**

*Source*: Autodata

**79.** In which year were sports-car sales the highest?

**80.** In which year were sports-car sales the lowest?

**81.** In which year were sports-car sales about 68,000?

**82.** What were sports-car sales in 1997?

**83.** By how much did sales increase from 1995 to 1997?

**84.** By how much did sales decrease from 1990 to 1995?

**85.** Divide: $\dfrac{7}{5} \div \left(-\dfrac{11}{10}\right)$. [1.6c]

**86.** Multiply: $(5x - t)^2$. [4.6d]

**87.** Solve $A = aX + bX - 7$ for $X$. [2.6a]

**88.** Solve: $4(x - 9) - 2(x + 7) < 14$. [2.7e]

---

**Synthesis**

**89.** ◈ Kelly factored $16 - 8x + x^2$ as $(x - 4)^2$, while Tony factored it as $(4 - x)^2$. Evaluate each expression for several values of $x$. Then explain why both answers are correct.

**90.** ◈ Describe in your own words a strategy that can be used to factor polynomials.

Factor completely.

**91.** $a^4 - 2a^2 + 1$

**92.** $x^4 + 9$

**93.** $12.25x^2 - 7x + 1$

**94.** $\dfrac{1}{5}x^2 - x + \dfrac{4}{5}$

**95.** $5x^2 + 13x + 7.2$

**96.** $x^3 - (x - 3x^2) - 3$

**97.** $18 + y^3 - 9y - 2y^2$

**98.** $-(x^4 - 7x^2 - 18)$

**99.** $a^3 + 4a^2 + a + 4$

**100.** $x^3 + x^2 - (4x + 4)$

**101.** $x^4 - 7x^2 - 18$

**102.** $3x^4 - 15x^2 + 12$

**103.** $x^3 - x^2 - 4x + 4$

**104.** $y^2(y + 1) - 4y(y + 1) - 21(y + 1)$

**105.** $y^2(y - 1) - 2y(y - 1) + (y - 1)$

**106.** $6(x - 1)^2 + 7y(x - 1) - 3y^2$

**107.** $(y + 4)^2 + 2x(y + 4) + x^2$

**108.** $a^4 - 81$

---

Create polynomials for factoring.

# 5.7 Solving Quadratic Equations by Factoring

*[handwritten: 2nd degree equation]*

*[handwritten: Has 2 answers.]*

In this section, we introduce a new equation-solving method and use it along with factoring to solve certain equations like $x^2 + x - 156 = 0$.

> A **quadratic equation** is an equation equivalent to an equation of the type
>
> $$ax^2 + bx + c = 0, \quad \text{where } a > 0.$$
>
> The trinomial on the left is of second degree.

## a  The Principle of Zero Products

The product of two numbers is 0 if one or both of the numbers is 0. Furthermore, *if any product is 0, then a factor must be 0*. For example:

If $7x = 0$, then we know that $x = 0$.

If $x(2x - 9) = 0$, then we know that $x = 0$ or $2x - 9 = 0$.

If $(x + 3)(x - 2) = 0$, then we know that $x + 3 = 0$ or $x - 2 = 0$.

In a product such as $ab = 24$, we cannot conclude with certainty that $a$ is 24 or that $b$ is 24, but if $ab = 0$, we can conclude that $a = 0$ or $b = 0$.

**Example 1**  Solve: $(x + 3)(x - 2) = 0$.

We have a product of 0. This equation will be true when either factor is 0. Thus it is true when

$$x + 3 = 0 \quad \text{or} \quad x - 2 = 0.$$

Here we have two simple equations that we know how to solve:

$$x = -3 \quad \text{or} \quad x = 2.$$

Each of the numbers $-3$ and 2 is a solution of the original equation, as we can see in the following checks.

CHECK:    For $-3$:

$$\frac{(x + 3)(x - 2) = 0}{(-3 + 3)(-3 - 2) \; ? \; 0}$$
$$0(-5) \quad$$
$$0 \quad \text{TRUE}$$

For 2:

$$\frac{(x + 3)(x - 2) = 0}{(2 + 3)(2 - 2) \; ? \; 0}$$
$$5(0) \quad$$
$$0 \quad \text{TRUE}$$

We now have a principle to help in solving quadratic equations.

> **THE PRINCIPLE OF ZERO PRODUCTS**
>
> An equation $ab = 0$ is true if and only if $a = 0$ is true or $b = 0$ is true, or both are true. (A product is 0 if and only if one or both of the factors is 0.)

*[handwritten: $x^2 = 16 \quad \{-4, 4\}$]*

## Objectives

**a**  Solve equations (already factored) using the principle of zero products.

**b**  Solve quadratic equations by factoring and then using the principle of zero products.

### For Extra Help

TAPE 10    TAPE 10B    MAC WIN    CD-ROM

Solve using the principle of zero products.

**1.** $(x - 3)(x + 4) = 0$

**2.** $(x - 7)(x - 3) = 0$

**3.** $(4t + 1)(3t - 2) = 0$

**4.** Solve: $y(3y - 17) = 0$.

*Answers on page A-23*

## Calculator Spotlight

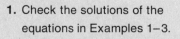 Checking Solutions on a Grapher. We can check solutions using the TABLE or the CALC feature. Consider the equation $(x + 3)(x - 2) = 0$ of Example 1. We enter $y_1 = (x + 3)(x - 2)$ into the grapher.

**Table.** First, we set up the table with TblStart $= -4$ and $\Delta$Tbl $= 1$. By scrolling through the table, we see that when $x = -3$, $y = 0$ and when $x = 2$, $y = 0$.

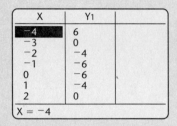

| X | Y1 | |
|---|---|---|
| -4 | 6 | |
| -3 | 0 | |
| -2 | -4 | |
| -1 | -6 | |
| 0 | -6 | |
| 1 | -4 | |
| 2 | 0 | |
| X = -4 | | |

**Calc.** We graph the equation $y = (x + 3)(x - 2)$ and use the CALC and VALUE features to check the solutions to the equation $(x + 3)(x - 2) = 0$. By entering $-3$ for $x$, we see that when $x = -3$, $y = 0$ (that is, when $x = -3$, the graph crosses the $x$-axis). It can also be shown that when $x = 2$, $y = 0$.

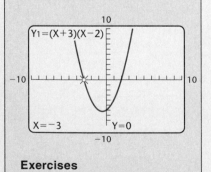

**Exercises**

1. Check the solutions of the equations in Examples 1–3.

**Example 2** Solve: $(5x + 1)(x - 7) = 0$.

$$(5x + 1)(x - 7) = 0$$

$5x + 1 = 0 \quad or \quad x - 7 = 0$  Using the principle of zero products

$5x = -1 \quad or \quad x = 7$  Solving the two equations separately

$x = -\frac{1}{5} \quad or \quad x = 7$

**CHECK:** For $-\frac{1}{5}$:

$$\frac{(5x + 1)(x - 7) = 0}{\left(5\left(-\frac{1}{5}\right) + 1\right)\left(-\frac{1}{5} - 7\right) \ ? \ 0}$$

$$(-1 + 1)\left(-7\frac{1}{5}\right)$$

$$0\left(-7\frac{1}{5}\right)$$

$$0 \ \Big| \quad \text{TRUE}$$

For 7:

$$\frac{(5x + 1)(x - 7) = 0}{(5(7) + 1)(7 - 7) \ ? \ 0}$$

$$(35 + 1) \cdot 0$$

$$36 \cdot 0$$

$$0 \ \Big| \quad \text{TRUE}$$

The solutions are $-\frac{1}{5}$ and 7.

When you solve an equation using the principle of zero products, you may wish to check by substitution, as in Examples 1 and 2. Such a check will detect errors in solving.

*Do Exercises 1–3 on the preceding page.*

When some factors have only one term, you can still use the principle of zero products.

**Example 3** Solve: $x(2x - 9) = 0$.

$$x(2x - 9) = 0$$

$x = 0 \quad or \quad 2x - 9 = 0$  Using the principle of zero products

$x = 0 \quad or \quad 2x = 9$

$x = 0 \quad or \quad x = \dfrac{9}{2}$

The solutions are 0 and $\frac{9}{2}$. The check is left to the student.

*Do Exercise 4 on the preceding page.*

## b | Using Factoring to Solve Equations

Using factoring and the principle of zero products, we can solve some new kinds of equations. Thus we have extended our equation-solving abilities.

**Example 4** Solve: $x^2 + 5x + 6 = 0$.

Compare this equation to those that we know how to solve from Chapter 2. There are no like terms to collect, and we have a squared term. We first factor the polynomial. Then we use the principle of zero products.

$$x^2 + 5x + 6 = 0$$

$(x + 2)(x + 3) = 0$  Factoring

$x + 2 = 0 \quad or \quad x + 3 = 0$  Using the principle of zero products

$x = -2 \quad or \quad x = -3$

CHECK:     For $-2$:              For $-3$:

$$\frac{x^2 + 5x + 6 = 0}{(-2)^2 + 5(-2) + 6 \; ? \; 0}$$
$$4 - 10 + 6$$
$$-6 + 6$$
$$0 \quad | \quad \text{TRUE}$$

$$\frac{x^2 + 5x + 6 = 0}{(-3)^2 + 5(-3) + 6 \; ? \; 0}$$
$$9 - 15 + 6$$
$$-6 + 6$$
$$0 \quad | \quad \text{TRUE}$$

The solutions are $-2$ and $-3$.

---

*CAUTION!*   Keep in mind that you *must* have 0 on one side of the equation before you can use the principle of zero products. Get all nonzero terms on one side and 0 on the other.

---

*Do Exercise 5.*

**Example 5**   Solve: $x^2 - 8x = -16$.

We first add 16 to get a 0 on one side:

$$x^2 - 8x = -16$$
$$x^2 - 8x + 16 = 0 \qquad \text{Adding 16}$$
$$(x - 4)(x - 4) = 0 \qquad \text{Factoring}$$
$$x - 4 = 0 \quad or \quad x - 4 = 0 \qquad \text{Using the principle of zero products}$$
$$x = 4 \quad or \qquad x = 4$$

There is only one solution, 4. The check is left to the student.

*Do Exercises 6 and 7.*

**Example 6**   Solve: $x^2 + 5x = 0$.

$$x^2 + 5x = 0$$
$$x(x + 5) = 0 \qquad \text{Factoring out a common factor}$$
$$x = 0 \quad or \quad x + 5 = 0 \qquad \text{Using the principle of zero products}$$
$$x = 0 \quad or \qquad x = -5$$

The solutions are 0 and $-5$. The check is left to the student.

**Example 7**   Solve: $4x^2 = 25$.

$$4x^2 = 25$$
$$4x^2 - 25 = 0 \qquad \text{Subtracting 25 on both sides to get 0 on one side}$$
$$(2x - 5)(2x + 5) = 0 \qquad \text{Factoring a difference of squares}$$
$$2x - 5 = 0 \quad or \quad 2x + 5 = 0$$
$$2x = 5 \quad or \qquad 2x = -5$$
$$x = \frac{5}{2} \quad or \qquad x = -\frac{5}{2}$$

The solutions are $\frac{5}{2}$ and $-\frac{5}{2}$.

*Do Exercises 8 and 9.*

**5.** Solve: $x^2 - x - 6 = 0$.

Solve.

**6.** $x^2 - 3x = 28$

**7.** $x^2 = 6x - 9$

Solve.

**8.** $x^2 - 4x = 0$

**9.** $9x^2 = 16$

*Answers on page A-23*

**10.** Solve: $(x + 1)(x - 1) = 8$.

**11.** Find the $x$-intercepts of the graph shown below.

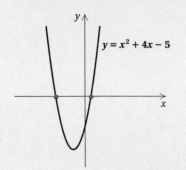

**12.** Use *only* the graph shown below to solve $3x - x^2 = 0$.

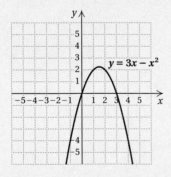

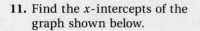

**Example 8** Solve: $(x + 2)(x - 2) = 5$.

Be careful with an equation like this one! It might be tempting to set each factor equal to 5. Remember: We must have a 0 on one side. We first carry out the product on the left. Then we subtract 5 on both sides to get 0 on one side. Then we proceed with the principle of zero products.

$$(x + 2)(x - 2) = 5$$
$$x^2 - 4 = 5 \qquad \text{Multiplying on the left}$$
$$x^2 - 4 - 5 = 5 - 5 \qquad \text{Subtracting 5}$$
$$x^2 - 9 = 0 \qquad \text{Simplifying}$$
$$(x + 3)(x - 3) = 0 \qquad \text{Factoring}$$
$$x + 3 = 0 \quad or \quad x - 3 = 0 \qquad \text{Using the principle of zero products}$$
$$x = -3 \quad or \qquad x = 3$$

The solutions are $-3$ and $3$. The check is left to the student.

*Do Exercise 10.*

### Algebraic–Graphical Connection

In Chapter 3, we graphed linear equations of the type $y = mx + b$ and $Ax + By = C$. Recall that to find the $x$-intercept, we replaced $y$ with 0 and solved for $x$. This procedure can also be used to find the $x$-intercepts when an equation of the form $y = ax^2 + bx + c$, $a \neq 0$, is to be graphed. Although the details of creating such graphs will be left to Chapter 10, we consider them briefly here from the standpoint of finding the $x$-intercepts. The graphs are shaped like the following curves. Note that each $x$-intercept represents a solution of $ax^2 + bx + c = 0$.

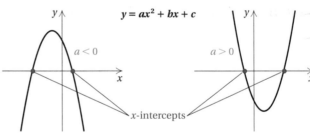

**Example 9** Find the $x$-intercepts of the graph of $y = x^2 - 4x - 5$ shown at right.

To find the $x$-intercepts, we let $y = 0$ and solve for $x$:

$$0 = x^2 - 4x - 5 \qquad \text{Substituting 0 for } y$$
$$0 = (x - 5)(x + 1) \qquad \text{Factoring}$$
$$x - 5 = 0 \quad or \quad x + 1 = 0 \qquad \text{Using the principle of zero products}$$
$$x = 5 \quad or \qquad x = -1.$$

The $x$-intercepts are $(5, 0)$ and $(-1, 0)$.

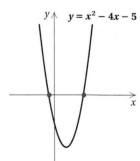

*Do Exercises 11 and 12.*

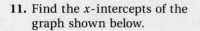

*Answer on page A-23*

# Exercise Set 5.7

**a** Solve using the principle of zero products. $-2 \quad +7$

**1.** $(x + 4)(x + 9) = 0$

**2.** $(x + 2)(x - 7) = 0$

**3.** $(x + 3)(x - 8) = 0$

**4.** $(x + 6)(x - 8) = 0$

**5.** $(x + 12)(x - 11) = 0$

**6.** $(x - 13)(x + 53) = 0$

**7.** $x(x + 3) = 0$

$0, -3$

**8.** $y(y + 5) = 0$

**9.** $0 = y(y + 18)$

**10.** $0 = x(x - 19)$

**11.** $(2x + 5)(x + 4) = 0$

**12.** $(2x + 9)(x + 8) = 0$

**13.** $(5x + 1)(4x - 12) = 0$

$-\frac{1}{5}, 3$

**14.** $(4x + 9)(14x - 7) = 0$

**15.** $(7x - 28)(28x - 7) = 0$

**16.** $(13x + 14)(6x - 5) = 0$

**17.** $2x(3x - 2) = 0$

**18.** $55x(8x - 9) = 0$

**19.** $\left(\frac{1}{5} + 2x\right)\left(\frac{1}{9} - 3x\right) = 0$

**20.** $\left(\frac{7}{4}x - \frac{1}{16}\right)\left(\frac{2}{3}x - \frac{16}{15}\right) = 0$

**21.** $(0.3x - 0.1)(0.05x + 1) = 0$

**22.** $(0.1x + 0.3)(0.4x - 20) = 0$

**23.** $9x(3x - 2)(2x - 1) = 0$

**24.** $(x + 5)(x - 75)(5x - 1) = 0$

**b** Solve by factoring and using the principle of zero products. Remember to check.

**25.** $x^2 + 6x + 5 = 0$

**26.** $x^2 + 7x + 6 = 0$

**27.** $x^2 + 7x - 18 = 0$

**28.** $x^2 + 4x - 21 = 0$

**29.** $x^2 - 8x + 15 = 0$

**30.** $x^2 - 9x + 14 = 0$

**31.** $x^2 - 8x = 0$

**32.** $x^2 - 3x = 0$

**33.** $x^2 + 18x = 0$

**34.** $x^2 + 16x = 0$

**35.** $x^2 = 16$

**36.** $100 = x^2$

**37.** $9x^2 - 4 = 0$

**38.** $4x^2 - 9 = 0$

**39.** $0 = 6x + x^2 + 9$

**40.** $0 = 25 + x^2 + 10x$

**41.** $x^2 + 16 = 8x$

**42.** $1 + x^2 = 2x$

**43.** $5x^2 = 6x$

**44.** $7x^2 = 8x$

**45.** $6x^2 - 4x = 10$

**46.** $3x^2 - 7x = 20$

**47.** $12y^2 - 5y = 2$

**48.** $2y^2 + 12y = -10$

**49.** $t(3t + 1) = 2$  **50.** $x(x - 5) = 14$  **51.** $100y^2 = 49$  **52.** $64a^2 = 81$

**53.** $x^2 - 5x = 18 + 2x$  **54.** $3x^2 + 8x = 9 + 2x$  **55.** $10x^2 - 23x + 12 = 0$  **56.** $12x^2 + 17x - 5 = 0$

Find the $x$-intercepts for the graph of the equation.

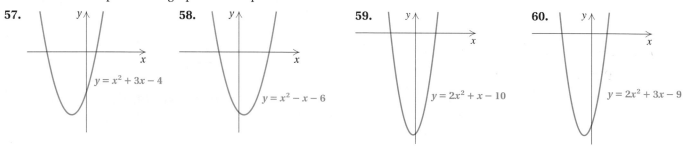

**57.** $y = x^2 + 3x - 4$

**58.** $y = x^2 - x - 6$

**59.** $y = 2x^2 + x - 10$

**60.** $y = 2x^2 + 3x - 9$

---

**Skill Maintenance**

Translate to an algebraic expression.  [1.1b]

**61.** The square of the sum of $a$ and $b$

**62.** The sum of the squares of $a$ and $b$

Divide.  [1.6c]

**63.** $144 \div (-9)$  **64.** $-24.3 \div 5.4$  **65.** $-\frac{5}{8} \div \frac{3}{16}$  **66.** $-\frac{3}{16} \div \left(-\frac{5}{8}\right)$

---

**Synthesis**

**67.** ◈ What is wrong with the following? Explain the correct method of solution.

$(x - 3)(x + 4) = 8$

$x - 3 = 8$   or  $x + 4 = 8$

$x = 11$   or      $x = 4$

**68.** ◈ What is incorrect about solving $x^2 = 3x$ by dividing by $x$ on both sides?

Solve.

**69.** $b(b + 9) = 4(5 + 2b)$  **70.** $y(y + 8) = 16(y - 1)$  **71.** $(t - 3)^2 = 36$  **72.** $(t - 5)^2 = 2(5 - t)$

**73.** $x^2 - \frac{1}{64} = 0$  **74.** $x^2 - \frac{25}{36} = 0$  **75.** $\frac{5}{16}x^2 = 5$  **76.** $\frac{27}{25}x^2 = \frac{1}{3}$

**77.** Find an equation that has the given numbers as solutions. For example, 3 and $-2$ are solutions to $x^2 - x - 6 = 0$.

a) $-3, 4$   b) $-3, -4$   c) $\frac{1}{2}, \frac{1}{2}$   d) $5, -5$   e) $0, 0.1, \frac{1}{4}$

**Create and solve quadratic equations.**

Collaborative
Learning Manual

Copyright © 1999 Addison Wesley Longman

# 5.8 Applications and Problem Solving

**a** We can now use our new method for solving quadratic equations and the five steps for solving problems.

**Example 1** One more than a number times one less than the number is 8. Find all such numbers.

1. **Familiarize.** Let's make a guess. Try 5. One more than 5 is 6. One less than the number is 4. The product of one more than the number and one less than the number is 6(4), or 24, which is too large. We could continue to guess, but let's use our algebraic skills to find the numbers. Let $x =$ the number (there could be more than one).

2. **Translate.** From the familiarization, we can translate as follows:

$$\underbrace{\text{One more than a number}}_{(x + 1)} \quad \underbrace{\text{times}}_{\cdot} \quad \underbrace{\text{One less than that number}}_{(x - 1)} \quad \underbrace{\text{is}}_{=} \quad \underbrace{8.}_{8}$$

3. **Solve.** We solve the equation as follows:

$$(x + 1)(x - 1) = 8$$
$$x^2 - 1 = 8 \qquad \text{Multiplying}$$
$$x^2 - 1 - 8 = 8 - 8 \qquad \text{Subtracting 8 on both sides to get 0 on one side}$$
$$x^2 - 9 = 0 \qquad \text{Simplifying}$$
$$(x - 3)(x + 3) = 0 \qquad \text{Factoring}$$
$$x - 3 = 0 \quad or \quad x + 3 = 0 \qquad \text{Using the principle of zero products}$$
$$x = 3 \quad or \quad x = -3.$$

4. **Check.** One more than 3 (this is 4) times one less than 3 (this is 2) is 8. Thus, 3 checks. One more than $-3$ (this is $-2$) times one less than $-3$ (this is $-4$) is 8. Thus, $-3$ also checks.

5. **State.** There are two such numbers, 3 and $-3$.

*Do Exercises 1 and 2.*

**Example 2** The square of a number minus twice the number is 48. Find all such numbers.

1. **Familiarize.** Let's make a guess to help understand the problem and ease the translation. Try 6. The square of 6 is 36, and twice the number is 12. Then $36 - 12 = 24$, so 6 is not a number we want. We find the numbers using our algebraic skills. Let $x =$ the number, or numbers.

2. **Translate.** We translate as follows:

$$\underbrace{\text{The square of a number}}_{x^2} \quad \underbrace{\text{minus}}_{-} \quad \underbrace{\text{Twice the number}}_{2x} \quad \underbrace{\text{is}}_{=} \quad \underbrace{48.}_{48}$$

3. **Solve.** We solve the equation as follows:

$$x^2 - 2x = 48$$
$$x^2 - 2x - 48 = 48 - 48 \qquad \text{Subtracting 48 to get 0 on one side}$$
$$x^2 - 2x - 48 = 0 \qquad \text{Simplifying}$$
$$(x - 8)(x + 6) = 0 \qquad \text{Factoring}$$
$$x - 8 = 0 \quad or \quad x + 6 = 0 \qquad \text{Using the principle of zero products}$$
$$x = 8 \quad or \quad x = -6.$$

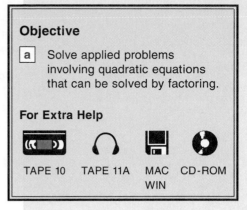

**Objective**

**a** Solve applied problems involving quadratic equations that can be solved by factoring.

**For Extra Help**

TAPE 10    TAPE 11A    MAC    CD-ROM
WIN

1. One more than a number times one less than the number is 24. Find all such numbers.

2. Seven less than a number times eight less than the number is 0. Find all such numbers.

*Answers on page A-23*

**3.** The square of a number minus the number is 20. Find all such numbers.

**4.** The width of a rectangle is 2 cm less than the length. The area is 15 cm². Find the length and the width.

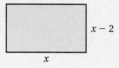

*x − 2*

*x*

---

**4. Check.** The square of 8 is 64, and twice the number 8 is 16. Then 64 − 16 is 48, so 8 checks. The square of −6 is $(-6)^2$, or 36, and twice −6 is −12. Then 36 − (−12) is 48, so −6 checks.

**5. State.** There are two such numbers, 8 and −6.

*Do Exercise 3.*

**Example 3** *Sailing.* The height of a triangular foresail on a racing yacht is 7 ft more than the base. The area of the triangle is 30 ft². Find the height and the base.

**1. Familiarize.** We first make a drawing. If you don't remember the formula for the area of a triangle, look it up on the inside front cover of this book or in a geometry book. The area is $\frac{1}{2}$ (base)(height).

$b + 7$

$b$

We let $b =$ the base of the triangle. Then $b + 7 =$ the height.

**2. Translate.** It helps to reword this problem before translating:

| $\frac{1}{2}$ | times | Base | times | Height | is | 30. | Rewording |
|---|---|---|---|---|---|---|---|
| ↓ | ↓ | ↓ | ↓ | ↓ | ↓ | ↓ | |
| $\frac{1}{2}$ | $\cdot$ | $b$ | $\cdot$ | $(b+7)$ | $=$ | 30 | Translating |

**3. Solve.** We solve the equation as follows:

$$\frac{1}{2} \cdot b \cdot (b+7) = 30$$

$$\frac{1}{2}(b^2 + 7b) = 30 \qquad \text{Multiplying}$$

$$2 \cdot \frac{1}{2}(b^2 + 7b) = 2 \cdot 30 \qquad \text{Multiplying by 2}$$

$$b^2 + 7b = 60 \qquad \text{Simplifying}$$

$$b^2 + 7b - 60 = 60 - 60 \qquad \text{Subtracting 60 to get 0 on one side}$$

$$b^2 + 7b - 60 = 0$$

$$(b + 12)(b - 5) = 0 \qquad \text{Factoring}$$

$$b + 12 = 0 \quad \text{or} \quad b - 5 = 0 \qquad \text{Using the principle of zero products}$$

$$b = -12 \quad \text{or} \qquad b = 5.$$

**4. Check.** The base of a triangle cannot have a negative length, so −12 cannot be a solution. Suppose the base is 5 ft. Then the height is 7 ft more than the base, so the height is 12 ft and the area is $\frac{1}{2}(5)(12)$, or 30 ft². These numbers check in the original problem.

**5. State.** The height is 12 ft and the base is 5 ft.

*Do Exercise 4.*

**Example 4** *Games in a Sports League.* In a sports league of $n$ teams in which each team plays every other team twice, the total number $N$ of games to be played is given by

$$n^2 - n = N.$$

If a basketball league plays a total of 240 games, how many teams are in the league?

**1., 2. Familiarize and Translate.** We are given that $n$ = the number of teams in a league and $N$ = the number of games. To familiarize yourself with this problem, reread Example 3 in Section 4.3 where we first considered it. To find the number of teams $n$ in a league in which 240 games are played, we substitute 240 for $N$ in the equation:

$$n^2 - n = 240. \qquad \text{Substituting 240 for } N$$

**3. Solve.** We solve the equation as follows:

$$n^2 - n = 240$$
$$n^2 - n - 240 = 240 - 240 \qquad \text{Subtracting 240 to get 0 on one side}$$
$$n^2 - n - 240 = 0$$
$$(n - 16)(n + 15) = 0 \qquad \text{Factoring}$$
$$n - 16 = 0 \quad or \quad n + 15 = 0 \qquad \text{Using the principle of zero products}$$
$$n = 16 \quad or \qquad n = -15.$$

**4. Check.** The solutions of the equation are 16 and $-15$. Since the number of teams cannot be negative, $-15$ cannot be a solution. But 16 checks, since $16^2 - 16 = 256 - 16 = 240$.

**5. State.** There are 16 teams in the league.

*Do Exercise 5.*

**Example 5** The product of the numbers of two consecutive entrants in a marathon race is 156. Find the numbers.

**1. Familiarize.** The numbers are consecutive integers. Recall that consecutive integers are next to each other, such as 49 and 50, or $-6$ and $-5$. Let $x$ = the smaller integer; then $x + 1$ = the larger integer.

**2. Translate.** It helps to reword the problem before translating:

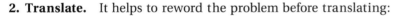

| First integer | times | Second integer | is | 156. | Rewording |
|:---:|:---:|:---:|:---:|:---:|:---|
| $x$ | $\cdot$ | $(x + 1)$ | $=$ | 156 | Translating |

**3. Solve.** We solve the equation as follows:

$$x(x + 1) = 156$$
$$x^2 + x = 156 \qquad \text{Multiplying}$$
$$x^2 + x - 156 = 156 - 156 \qquad \text{Subtracting 156 to get 0 on one side}$$
$$x^2 + x - 156 = 0 \qquad \text{Simplifying}$$
$$(x - 12)(x + 13) = 0 \qquad \text{Factoring}$$
$$x - 12 = 0 \quad or \quad x + 13 = 0 \qquad \text{Using the principle of zero products}$$
$$x = 12 \quad or \qquad x = -13.$$

**4. Check.** The solutions of the equation are 12 and $-13$. When $x$ is 12, then $x + 1$ is 13, and $12 \cdot 13 = 156$. The numbers 12 and 13 are consecutive integers that are solutions to the problem. When $x$ is $-13$, then $x + 1$ is $-12$, and $(-13)(-12) = 156$. The numbers $-13$ and $-12$ are also consecutive integers, but they are not solutions of the problem because negative numbers are not used as entry numbers.

**5. State.** The entry numbers are 12 and 13.

**5.** Use $N = n^2 - n$ for the following.

a) *Volleyball League.* A women's volleyball league has 19 teams. What is the total number of games to be played?

b) *Softball League.* A slow-pitch softball league plays a total of 72 games. How many teams are in the league?

*Answers on page A-23*

**6.** The product of the page numbers on two facing pages of a book is 506. Find the page numbers.

**Do Exercise 6.**

The following example involves the Pythagorean theorem, which relates the lengths of the sides of a right triangle. A **right triangle** has a 90° angle. The side opposite the 90° angle is called the **hypotenuse**. The other sides are called **legs**.

> **THE PYTHAGOREAN THEOREM**
>
> The sum of the squares of the legs of a right triangle is equal to the square of the hypotenuse:
> $$a^2 + b^2 = c^2.$$

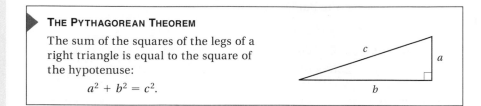

**Example 6** *Ladder Settings.* A ladder of length 13 ft is placed against a building in such a way that the distance from the top of the ladder to the ground is 7 ft more than the distance from the bottom of the ladder to the building. Find both distances.

1. **Familiarize.** We first make a drawing. The ladder and the missing distances form the hypotenuse and legs of a right triangle. We let $x$ = the length of the side (leg) across the bottom. Then $x + 7$ = the length of the other side (leg). The hypotenuse has length 13 ft.

13 ft · · · $x + 7$ · · · $x$

**7.** The length of one leg of a right triangle is 1 m longer than the other. The length of the hypotenuse is 5 m. Find the lengths of the legs.

2. **Translate.** Since a right triangle is formed, we can use the Pythagorean theorem:
$$a^2 + b^2 = c^2$$
$$x^2 + (x + 7)^2 = 13^2. \quad \text{Substituting}$$

3. **Solve.** We solve the equation as follows:

| | |
|---|---|
| $x^2 + (x^2 + 14x + 49) = 169$ | Squaring the binomial and 13 |
| $2x^2 + 14x + 49 = 169$ | Collecting like terms |
| $2x^2 + 14x + 49 - 169 = 169 - 169$ | Subtracting 169 to get 0 on one side |
| $2x^2 + 14x - 120 = 0$ | Simplifying |
| $2(x^2 + 7x - 60) = 0$ | Factoring out a common factor |
| $x^2 + 7x - 60 = 0$ | Dividing by 2 |
| $(x + 12)(x - 5) = 0$ | Factoring |

$$x + 12 = 0 \quad or \quad x - 5 = 0$$
$$x = -12 \quad or \quad x = 5.$$

4. **Check.** The negative integer $-12$ cannot be the length of a side. When $x = 5$, $x + 7 = 12$, and $5^2 + 12^2 = 13^2$. So 5 and 12 check.

5. **State.** The distance from the top of the ladder to the ground is 12 ft. The distance from the bottom of the ladder to the building is 5 ft.

*Answers on page A-23*

**Do Exercise 7.**

# Exercise Set 5.8

**a** Solve.

1. If 7 is added to the square of a number, the result is 32. Find all such numbers.

2. If you subtract a number from four times its square, the result is 3. Find all such numbers.

3. Fifteen more than the square of a number is eight times the number. Find all such numbers.

4. Eight more than the square of a number is six times the number. Find all such numbers.

5. *Calculator Dimensions.* The length of a rectangular calculator is 5 cm greater than the width. The area of the calculator is 84 cm². Find the length and the width.

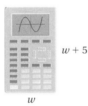

$w + 5$

$w$

6. *Garden Dimensions.* The length of a rectangular garden is 4 m greater than the width. The area of the garden is 96 m². Find the length and the width.

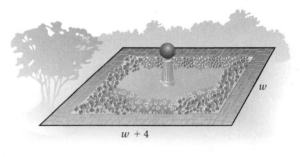

$w$

$w + 4$

7. *Consecutive Page Numbers.* The product of the page numbers on two facing pages of a book is 210. Find the page numbers.

$x$   $x + 1$

8. *Consecutive Page Numbers.* The product of the page numbers on two facing pages of a book is 420. Find the page numbers.

**9.** The product of two consecutive even integers is 168. Find the integers.

**10.** The product of two consecutive even integers is 224. Find the integers.

**11.** The product of two consecutive odd integers is 255. Find the integers.

**12.** The product of two consecutive odd integers is 143. Find the integers.

**13.** The area of a square bookcase is 5 more than the perimeter. Find the length of a side.

**14.** The perimeter of a square porch is 3 more than the area. Find the length of a side.

**15.** *Sharks' Teeth.* Sharks' teeth are shaped like triangles. The height of a tooth of a great white shark is 1 cm longer than the base. The area is 15 cm². Find the height and the base.

**16.** The base of a triangle is 6 cm greater than twice the height. The area is 28 cm². Find the height and the base.

**17.** If the sides of a square are lengthened by 3 km, the area becomes 81 km². Find the length of a side of the original square.

**18.** The base and the height of a triangle are the same length. If the length of the base is increased by 4 in., the area becomes 96 in². Find the length of the base of the original triangle.

Copyright © 1999 Addison Wesley Longman

*Rocket Launch.* A model water rocket is launched with an initial velocity of 180 ft/sec. Its height $h$, in feet, after $t$ seconds is given by the formula

$$h = 180t - 16t^2.$$

Use this formula for Exercises 19 and 20.

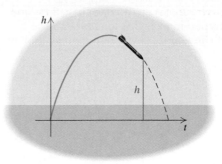

**19.** After how many seconds will the rocket first reach a height of 464 ft?

**20.** After how many seconds will the rocket again be at that same height of 464 ft? (See Exercise 19.)

**21.** The sum of the squares of two consecutive odd positive integers is 74. Find the integers.

**22.** The sum of the squares of two consecutive odd positive integers is 130. Find the integers.

*Games in a League.* Use $n^2 - n = N$ for Exercises 23–26.

**23.** A chess league has 14 teams. What is the total number of games to be played?

**24.** A women's volleyball league has 23 teams. What is the total number of games to be played?

**25.** A slow-pitch softball league plays a total of 132 games. How many teams are in the league?

**26.** A basketball league plays a total of 90 games. How many teams are in the league?

*Handshakes.* A researcher wants to investigate the potential spread of germs by contact. She knows that the number of possible handshakes within a group of $n$ people is given by

$$N = \tfrac{1}{2}(n^2 - n).$$

**27.** There are 100 people at a party. How many handshakes are possible?

**28.** There are 40 people at a meeting. How many handshakes are possible?

**29.** Everyone at a meeting shook hands. There were 300 handshakes in all. How many people were at the meeting?

**30.** Everyone at a party shook hands. There were 190 handshakes in all. How many people were at the party?

**31.** The length of one leg of a right triangle is 8 ft. The length of the hypotenuse is 2 ft longer than the other leg. Find the length of the hypotenuse and the other leg.

**32.** The length of one leg of a right triangle is 24 ft. The length of the other leg is 16 ft shorter than the hypotenuse. Find the length of the hypotenuse and the other leg.

---

### Skill Maintenance

Multiply. [4.6d], [4.7f]

**33.** $(3x - 5y)(3x + 5y)$      **34.** $(3x - 5y)^2$      **35.** $(3x + 5y)^2$      **36.** $(3x - 5y)(2x + 7y)$

Find the intercepts of the equation. [3.3a]

**37.** $4x - 16y = 64$      **38.** $4x + 16y = 64$      **39.** $x - 1.3y = 6.5$      **40.** $\frac{2}{3}x + \frac{5}{8}y = \frac{5}{12}$

---

### Synthesis

**41.** ◈ Write a problem in which a quadratic equation must be solved.

**42.** ◈ Write a problem for a classmate to solve such that only one of the two solutions of a quadratic equation can be used as an answer.

**43.** A cement walk of constant width is built around a 20-ft by 40-ft rectangular pool. The total area of the pool and the walk is 1500 ft². Find the width of the walk.

**44.** An open rectangular gutter is made by turning up the sides of a piece of metal 20 in. wide. The area of the cross-section of the gutter is 50 in². Find the depth of the gutter.

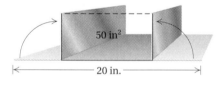

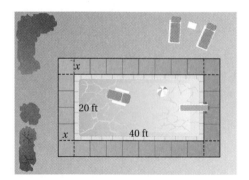

**45.** The ones digit of a number less than 100 is 4 greater than the tens digit. The sum of the number and the product of the digits is 58. Find the number.

**46.** The total surface area of a closed box is 350 m². The box is 9 m high and has a square base and lid. Find the length of the side of the base.

**47.** A rectangular piece of cardboard is twice as long as it is wide. A 4-cm square is cut out of each corner, and the sides are turned up to make a box with an open top. The volume of the box is 616 cm³. Find the original dimensions of the cardboard.

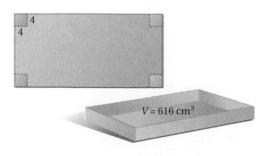

Copyright © 1999 Addison Wesley Longman

# Summary and Review Exercises: Chapter 5

## Important Properties and Formulas

*Factoring Formulas:*     $A^2 - B^2 = (A + B)(A - B)$,
$A^2 + 2AB + B^2 = (A + B)^2$,
$A^2 - 2AB + B^2 = (A - B)^2$

*The Principle of Zero Products:*     An equation $ab = 0$ is true if and only if $a = 0$ is true or $b = 0$ is true, or both are true.

The objectives to be tested in addition to the material in this chapter are [1.6c], [2.7e], [3.3a], and [4.6d].

Find three factorizations of the monomial.   [5.1a]

**1.** $-10x^2$

**2.** $36x^5$

Factor completely.   [5.6a]

**3.** $5 - 20x^6$

**4.** $x^2 - 3x$

**5.** $9x^2 - 4$

**6.** $x^2 + 4x - 12$

**7.** $x^2 + 14x + 49$

**8.** $6x^3 + 12x^2 + 3x$

**9.** $x^3 + x^2 + 3x + 3$

**10.** $6x^2 - 5x + 1$

**11.** $x^4 - 81$

**12.** $9x^3 + 12x^2 - 45x$

**13.** $2x^2 - 50$

**14.** $x^4 + 4x^3 - 2x - 8$

**15.** $16x^4 - 1$

**16.** $8x^6 - 32x^5 + 4x^4$

**17.** $75 + 12x^2 + 60x$

**18.** $x^2 + 9$

**19.** $x^3 - x^2 - 30x$

**20.** $4x^2 - 25$

**21.** $9x^2 + 25 - 30x$

**22.** $6x^2 - 28x - 48$

**23.** $x^2 - 6x + 9$

**24.** $2x^2 - 7x - 4$

**25.** $18x^2 - 12x + 2$

**26.** $3x^2 - 27$

**27.** $15 - 8x + x^2$

**28.** $25x^2 - 20x + 4$

**29.** $49b^{10} + 4a^8 - 28a^4b^5$

**30.** $x^2y^2 + xy - 12$

**31.** $12a^2 + 84ab + 147b^2$

**32.** $m^2 + 5m + mt + 5t$

**33.** $32x^4 - 128y^4z^4$

Solve.   [5.7a], [5.7b]

**34.** $(x - 1)(x + 3) = 0$

**35.** $x^2 + 2x - 35 = 0$

**36.** $x^2 + x - 12 = 0$

**37.** $3x^2 + 2 = 5x$

**38.** $2x^2 + 5x = 12$

**39.** $16 = x(x - 6)$

Solve.   [5.8a]

**40.** The square of a number is 6 more than the number. Find all such numbers.

**41.** The product of two consecutive even integers is 288. Find the integers.

**42.** Twice the square of a number is 10 more than the number. Find all such numbers.

**43.** The product of two consecutive odd integers is 323. Find the integers.

**44.** *House Plan.* An architect has allocated a rectangular space of 264 ft$^2$ for a square dining room and a 10-ft wide kitchen. Find the dimensions of each room.

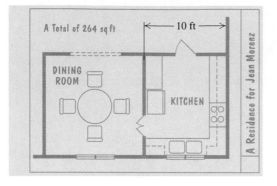

**45.** *Antenna Guy Wire.* The guy wires for a television antenna are 1 m longer than the height of the antenna. The guy wires are anchored 3 m from the foot of the antenna. How tall is the antenna?

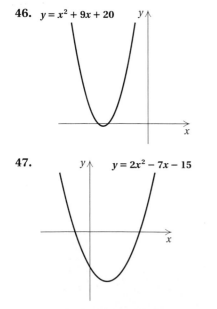

Find the $x$-intercepts for the graph of the equation. [5.7b]

**46.** $y = x^2 + 9x + 20$

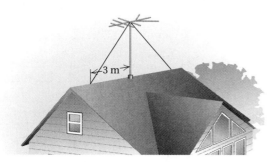

**47.** $y = 2x^2 - 7x - 15$

Skill Maintenance

**48.** Divide: $-\dfrac{12}{25} \div \left(-\dfrac{21}{10}\right)$. [1.6c]

**49.** Solve: $20 - (3x + 2) \geq 2(x + 5) + x$. [2.7e]

**50.** Multiply: $(2a - 3)(2a + 3)$. [4.6d]

**51.** Find the intercepts. Then graph the equation. [3.3a]

$$3y - 4x = -12$$

**Synthesis**

**52.** ◆ Compare the types of equations that we are able to solve after having studied this chapter with those we have studied earlier. [2.3a, b, c], [5.7a, b]

**53.** ◆ Describe as many procedures as you can for checking the result of factoring a polynomial. [5.3a], [5.7a]

Solve. [5.8a]

**54.** The pages of a book measure 15 cm by 20 cm. Margins of equal width surround the printing on each page and constitute one-half of the area of the page. Find the width of the margins.

**55.** The cube of a number is the same as twice the square of the number. Find all such numbers.

**56.** The length of a rectangle is two times its width. When the length is increased by 20 and the width decreased by 1, the area is 160. Find the original length and width.

Solve. [5.7b]

**57.** $x^2 + 25 = 0$

**58.** $(x - 2)(x + 3)(2x - 5) = 0$

**59.** For each equation in group A, find an equivalent equation in group B. [5.6a]

**A. a)** $3x^2 - 4x + 8 = 0$
**b)** $(x - 6)(x + 3) = 0$
**c)** $x^2 + 2x + 9 = 0$
**d)** $(2x - 5)(x + 4) = 0$
**e)** $5x^2 - 5 = 0$
**f)** $x^2 + 10x - 2 = 0$

**B. g)** $4x^2 + 8x + 36 = 0$
**h)** $(2x + 8)(2x - 5) = 0$
**i)** $9x^2 - 12x + 24 = 0$
**j)** $(x + 1)(5x - 5) = 0$
**k)** $x^2 - 3x - 18 = 0$
**l)** $2x^2 + 20x - 4 = 0$

**60.** Which is greater, $2^{90} + 2^{90}$, or $2^{100}$? Why? [5.1b]

# Test: Chapter 5

**1.** Find three factorizations of $4x^3$.

Factor completely.

**2.** $x^2 - 7x + 10$

**3.** $x^2 + 25 - 10x$

**4.** $6y^2 - 8y^3 + 4y^4$

**5.** $x^3 + x^2 + 2x + 2$

**6.** $x^2 - 5x$

**7.** $x^3 + 2x^2 - 3x$

**8.** $28x - 48 + 10x^2$

**9.** $4x^2 - 9$

**10.** $x^2 - x - 12$

**11.** $6m^3 + 9m^2 + 3m$

**12.** $3w^2 - 75$

**13.** $60x + 45x^2 + 20$

**14.** $3x^4 - 48$

**15.** $49x^2 - 84x + 36$

**16.** $5x^2 - 26x + 5$

**17.** $x^4 + 2x^3 - 3x - 6$

**18.** $80 - 5x^4$

**19.** $4x^2 - 4x - 15$

**20.** $6t^3 + 9t^2 - 15t$

**21.** $3m^2 - 9mn - 30n^2$

## Answers

1. _____

2. _____

3. _____

4. _____

5. _____

6. _____

7. _____

8. _____

9. _____

10. _____

11. _____

12. _____

13. _____

14. _____

15. _____

16. _____

17. _____

18. _____

19. _____

20. _____

21. _____

Solve.

**22.** $x^2 - x - 20 = 0$   **23.** $2x^2 + 7x = 15$   **24.** $x(x - 3) = 28$

Solve.

**25.** The square of a number is 24 more than five times the number. Find all such numbers.

**26.** *Dimensions of a Sail.* The height of the jib sail on a Lightning sailboat is 5 ft greater than the length of its "foot." If the area of the sail is 42 ft², find the length of the foot and the height of the sail.

$x + 5$

$x$

Find the $x$-intercepts for the graph of the equation.

**27.**

$y = x^2 - 2x - 35$

**28.**

$y = 3x^2 - 5x + 2$

## Skill Maintenance

**29.** Divide: $\dfrac{5}{8} \div \left(-\dfrac{11}{16}\right)$.

**30.** Solve: $10(x - 3) < 4(x + 2)$.

**31.** Find the intercepts. Then graph the equation.

$2y - 5x = 10$

**32.** Multiply: $(5x^2 - 7)^2$.

## Synthesis

**33.** The length of a rectangle is five times its width. When the length is decreased by 3 and the width is increased by 2, the area of the new rectangle is 60. Find the original length and width.

**34.** Factor: $(a + 3)^2 - 2(a + 3) - 35$.

**35.** If $x^2 - 4 = (14)(18)$, then one possibility for $x$ is which of the following?

a) 12   b) 14
c) 16   d) 18

**36.** If $x + y = 4$ and $x - y = 6$, then $x^2 - y^2 = $ ?

a) 2   b) 10
c) 34   d) 24

**Answers**

22. _____

23. _____

24. _____

25. _____

26. _____

27. _____

28. _____

29. _____

30. _____

31. _____

32. _____

33. _____

34. _____

35. _____

36. _____

# Cumulative Review: Chapters 1–5

Use either $<$ or $>$ for ▆ to write a true sentence.

**1.** $\dfrac{2}{3}$ ▆ $\dfrac{5}{7}$

**2.** $-\dfrac{4}{7}$ ▆ $-\dfrac{8}{11}$

Compute and simplify.

**3.** $2.06 + (-4.79) - (-3.08)$

**4.** $5.652 \div (-3.6)$

**5.** $\left(\dfrac{2}{9}\right)\left(-\dfrac{3}{8}\right)\left(\dfrac{6}{7}\right)$

**6.** $\dfrac{21}{5} \div \left(-\dfrac{7}{2}\right)$

Simplify.

**7.** $[3x + 2(x - 1)] - [2x - (x + 3)]$

**8.** $1 - [14 + 28 \div 7 - (6 + 9 \div 3)]$

**9.** $(2x^2 y^{-1})^3$

**10.** $\dfrac{3x^5}{4x^3} \cdot \dfrac{-2x^{-3}}{9x^2}$

**11.** Add:
$(2x^2 - 3x^3 + x - 4) + (x^4 - x - 5x^2).$

**12.** Subtract:
$(2x^2 y^2 + xy - 2xy^2) - (2xy - 2xy^2 + x^2 y).$

**13.** Divide: $(x^3 + 2x^2 - x + 1) \div (x - 1).$

Multiply.

**14.** $(2t - 3)^2$

**15.** $(x^2 - 3)(x^2 + 3)$

**16.** $(2x + 4)(3x - 4)$

**17.** $2x(x^3 + 3x^2 + 4x)$

**18.** $(2y - 1)(2y^2 + 3y + 4)$

**19.** $\left(x + \dfrac{2}{3}\right)\left(x - \dfrac{2}{3}\right)$

Factor.

**20.** $x^2 + 2x - 8$

**21.** $4x^2 - 25$

**22.** $3x^3 - 4x^2 + 3x - 4$

**23.** $x^2 - 26x + 169$

**24.** $75x^2 - 108y^2$

**25.** $6x^2 - 13x - 63$

**26.** $x^4 - 2x^2 - 3$

**27.** $4y^3 - 6y^2 - 4y + 6$

**28.** $6p^2 + pq - q^2$

**29.** $10x^3 + 52x^2 + 10x$

**30.** $49x^3 - 42x^2 + 9x$

**31.** $3x^2 + 5x - 4$

**32.** $75x^3 + 27x$

**33.** $3x^8 - 48y^8$

**34.** $14x^2 + 28 + 42x$

**35.** $x^5 - x^3 + x^2 - 1$

Solve.

**36.** $3x - 5 = 2x + 10$

**37.** $3y + 4 > 5y - 8$

**38.** $(x - 15)\left(x + \dfrac{1}{4}\right) = 0$

**39.** $-98x(x + 37) = 0$

**40.** $x^3 + x^2 = 25x + 25$

**41.** $2x^2 = 72$

**42.** $9x^2 + 1 = 6x$

**43.** $x^2 + 17x + 70 = 0$

**44.** $14y^2 = 21y$

**45.** $1.6 - 3.5x = 0.9$

**46.** $(x + 3)(x - 4) = 8$

**47.** $1.5x - 3.6 \le 1.3x + 0.4$

**48.** $2x - [3x - (2x + 3)] = 3x + [4 - (2x + 1)]$

**49.** $y = mx + b$, for $m$

Solve.

**50.** The sum of two consecutive even integers is 102. Find the integers.

**51.** The product of two consecutive even integers is 360. Find the integers.

**52.** The length of a rectangular window is 3 ft longer than the height. The area of the window is 18 ft². Find the length and the height.

**53.** The length of a rectangular lot is 200 m longer than the width. The perimeter of the lot is 1000 m. Find the dimensions of the lot.

**54.** Money is borrowed at 12% simple interest. After 1 year, $7280 pays off the loan. How much was originally borrowed?

**55.** The length of one leg of a right triangle is 15 m. The length of the other leg is 9 m shorter than the length of the hypotenuse. Find the length of the hypotenuse.

**56.** A 100-m wire is cut into three pieces. The second piece is twice as long as the first piece. The third piece is one-third as long as the first piece. How long is each piece?

**57.** After a 25% price reduction, a pair of shoes is on sale for $21.75. What was the price before reduction?

**58.** The height of a triangle is 2 cm more than the base. The area of the triangle is 144 cm². Find the height and the base.

**59.** Find the intercepts. Then graph the equation.
$$3x + 4y = -12$$

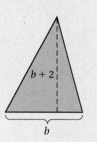

$b + 2$

$b$

---

### Synthesis

Solve.

**60.** $(x + 3)(x - 5) \le (x + 2)(x - 1)$

**61.** $\dfrac{x - 3}{2} - \dfrac{2x + 5}{26} = \dfrac{4x + 11}{13}$

**62.** $(x + 1)^2 = 25$

Factor.

**63.** $x^2(x - 3) - x(x - 3) - 2(x - 3)$

**64.** $4a^2 - 4a + 1 - 9b^2 - 24b - 16$

Solve.

**65.** Find $c$ such that the polynomial will be the square of a binomial: $cx^2 - 40x + 16$.

**66.** The length of the radius of a circle is increased by 2 cm to form a new circle. The area of the new circle is four times the area of the original circle. Find the length of the radius of the original circle.

# 6

# Rational Expressions and Equations

---

**An Application**

**The Mathematics**

Erin and Nick work as volunteers at a community recycling depot. Erin can sort a morning's accumulation of recyclables in 4 hr, while Nick requires 6 hr to do the same job. How long would it take them, working together, to sort the recyclables?

This problem appears as Example 3 in Section 6.7.

We let $t =$ the time it takes them, working together, to sort the recyclables. The problem then translates to the equation

These are *rational expressions.*

$$\underbrace{\frac{t}{4} + \frac{t}{6}}_{} = 1.$$

This is a *rational equation.*

# Pretest: Chapter 6

1. Find the LCM of $x^2 + 5x + 6$ and $x^2 + 6x + 9$.

Perform the indicated operations and simplify.

2. $\dfrac{b-1}{2-b} + \dfrac{b^2-3}{b^2-4}$

3. $\dfrac{4y-4}{y^2-y-2} - \dfrac{3y-5}{y^2-y-2}$

4. $\dfrac{4}{a+2} + \dfrac{3}{a}$

5. $\dfrac{x}{x+1} - \dfrac{x}{x-1} + \dfrac{2x^2}{x^2-1}$

6. $\dfrac{4x+8}{x+1} \cdot \dfrac{x^2-2x-3}{2x^2-8}$

7. $\dfrac{x+3}{x^2-9} \div \dfrac{x+3}{x^2-6x+9}$

8. Simplify: $\dfrac{\dfrac{1}{x} + \dfrac{1}{y}}{\dfrac{1}{x} - \dfrac{1}{y}}$.

Solve.

9. $\dfrac{1}{x+4} = \dfrac{5}{x}$

10. $\dfrac{3}{x-2} + \dfrac{x}{2} = \dfrac{6}{2x-4}$

11. Solve $R = \dfrac{1}{3}M(a-b)$ for $M$.

12. It takes 6 hr for a paper carrier to deliver 200 papers. At this rate, how long would it take to deliver 350 papers?

13. One data-entry clerk can key in a report in 6 hr. Another can key in the same report in 5 hr. How long would it take them, working together, to key in the same report?

14. One car travels 20 mph faster than another. While one car travels 300 mi, the other travels 400 mi. Find their speeds.

## Objectives for Retesting

The objectives to be tested in addition to the material in this chapter are as follows.

[4.2b]  Raise a product to a power and a quotient to a power.
[4.4c]  Subtract polynomials.
[5.6a]  Factor polynomials completely.
[5.8a]  Solve applied problems involving quadratic equations that can be solved by factoring.

# 6.1 Multiplying and Simplifying Rational Expressions

## a Rational Expressions and Replacements

Rational numbers are quotients of integers. Some examples are

$$\frac{2}{3}, \quad \frac{4}{-5}, \quad \frac{-8}{17}, \quad \frac{563}{1}.$$

The following are called **rational expressions** or **fractional expressions.** They are quotients, or ratios, of polynomials:

$$\frac{3}{4}, \quad \frac{z}{6}, \quad \frac{5}{x+2}, \quad \frac{t^2 + 3t - 10}{7t^2 - 4}.$$

A rational expression is also a division. For example,

$$\frac{3}{4} \quad \text{means} \quad 3 \div 4 \quad \text{and} \quad \frac{x-8}{x+2} \quad \text{means} \quad (x-8) \div (x+2).$$

Because rational expressions indicate division, we must be careful to avoid denominators of zero. When a variable is replaced with a number that produces a denominator equal to zero, the rational expression is undefined. For example, in the expression

$$\frac{x-8}{x+2},$$

when $x$ is replaced with $-2$, the denominator is 0, and the expression is undefined:

$$\frac{x-8}{x+2} = \frac{-2-8}{-2+2} = \frac{-10}{0} \quad \leftarrow \text{Undefined}$$

When $x$ is replaced with a number other than $-2$, such as 3, the expression *is* defined because the denominator is nonzero:

$$\frac{x-8}{x+2} = \frac{3-8}{3+2} = \frac{-5}{5} = -1.$$

**Example 1** Find all numbers for which the rational expression

$$\frac{x+4}{x^2 - 3x - 10} \leftarrow \text{only work denominator. don't look @ numer}$$

is undefined.

To determine which numbers make the rational expression undefined, we set the denominator equal to 0 and solve:

$$x^2 - 3x - 10 = 0$$
$$(x - 5)(x + 2) = 0 \qquad \text{Factoring}$$
$$x - 5 = 0 \quad or \quad x + 2 = 0 \qquad \text{Using the principle of zero products}$$
$$x = 5 \quad or \quad x = -2.$$

The expression is undefined for the replacement numbers 5 and $-2$.

*Do Exercises 1–3.*

Answers on page A-25

---

### Objectives

**a** Find all numbers for which a rational expression is undefined.

**b** Multiply a rational expression by 1, using an expression such as $A/A$.

**c** Simplify rational expressions by factoring the numerator and the denominator and removing factors of 1.

**d** Multiply rational expressions and simplify.

### For Extra Help

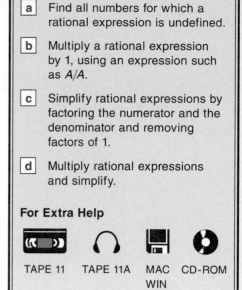

TAPE 11    TAPE 11A    MAC    CD-ROM
                       WIN

Find all numbers for which the rational expression is undefined.

1. $\dfrac{16}{x-3}$

2. $\dfrac{2x-7}{x^2 + 5x - 24}$

3. $\dfrac{x+5}{8}$

Multiply.

4. $\dfrac{2x + 1}{3x - 2} \cdot \dfrac{x}{x}$

5. $\dfrac{x + 1}{x - 2} \cdot \dfrac{x + 2}{x + 2}$

6. $\dfrac{x - 8}{x - y} \cdot \dfrac{-1}{-1}$

## b | Multiplying by 1

We multiply rational expressions in the same way that we multiply fractional notation in arithmetic. For a review, see Section R.2. We saw there that

$$\frac{3}{7} \cdot \frac{2}{5} = \frac{3 \cdot 2}{7 \cdot 5} = \frac{6}{35}.$$

> To multiply rational expressions, multiply the numerators and multiply the denominators.

For example,

$$\frac{x - 2}{3} \cdot \frac{x + 2}{x + 7} = \frac{(x - 2)(x + 2)}{3(x + 7)}. \qquad \text{Multiplying the numerators and the denominators}$$

Note that we leave the numerator, $(x - 2)(x + 2)$, and the denominator, $3(x + 7)$, in factored form because it is easier to simplify if we do not multiply. In order to learn to simplify, we first need to consider multiplying the rational expression by 1.

Any rational expression with the same numerator and denominator is a symbol for 1:

$$\frac{19}{19} = 1, \qquad \frac{x + 8}{x + 8} = 1, \qquad \frac{3x^2 - 4}{3x^2 - 4} = 1, \qquad \frac{-1}{-1} = 1.$$

> Expressions that have the same value for all allowable (or meaningful) replacements are called **equivalent expressions.**

We can multiply by 1 to obtain an equivalent expression. At this point, we select expressions for 1 arbitrarily. Later, we will have a system for our choices when we add and subtract.

**Examples**   Multiply.

**2.** $\dfrac{3x + 2}{x + 1} \cdot 1 = \dfrac{3x + 2}{x + 1} \cdot \dfrac{2x}{2x} = \dfrac{(3x + 2)2x}{(x + 1)2x}$   Identity property of 1

**3.** $\dfrac{x + 2}{x - 7} \cdot \dfrac{x + 3}{x + 3} = \dfrac{(x + 2)(x + 3)}{(x - 7)(x + 3)}$

**4.** $\dfrac{2 + x}{2 - x} \cdot \dfrac{-1}{-1} = \dfrac{(2 + x)(-1)}{(2 - x)(-1)}$

*Do Exercises 4–6.*

## c | Simplifying Rational Expressions

Simplify by removing a factor of 1.

**7.** $\dfrac{5y}{y}$

Simplifying rational expressions is similar to simplifying fractional expressions in arithmetic. For a review, see Section R.2. We saw there, for example, that an expression like $\frac{15}{40}$ can be simplified as follows:

$$\frac{15}{40} = \frac{3 \cdot 5}{8 \cdot 5} \qquad \text{Factoring the numerator and the denominator. Note the common factor of 5.}$$

$$= \frac{3}{8} \cdot \frac{5}{5} \qquad \text{Factoring the fractional expression}$$

$$= \frac{3}{8} \cdot 1 \qquad \frac{5}{5} = 1$$

$$= \frac{3}{8} \cdot \qquad \text{Using the identity property of 1, or "removing a factor of 1"}$$

In algebra, instead of simplifying

$$\frac{15}{40},$$

we may need to simplify an expression like

$$\frac{x^2 - 16}{x + 4}.$$

Just as factoring is important in simplifying in arithmetic, so too is it important in simplifying rational expressions. The factoring we use most is the factoring of polynomials, which we studied in Chapter 5.

To simplify, we can do the reverse of multiplying. We factor the numerator and the denominator and "remove" a factor of 1.

**Example 5** Simplify by removing a factor of 1: $\dfrac{8x^2}{24x}$.

$$\frac{8x^2}{24x} = \frac{8 \cdot x \cdot x}{3 \cdot 8 \cdot x} \qquad \text{Factoring the numerator and the denominator}$$

$$= \frac{8x}{8x} \cdot \frac{x}{3} \qquad \text{Factoring the rational expression}$$

$$= 1 \cdot \frac{x}{3} \qquad \frac{8x}{8x} = 1$$

$$= \frac{x}{3} \qquad \text{We removed a factor of 1.}$$

**8.** $\dfrac{9x^2}{36x}$

*Do Exercises 7 and 8.*

**Examples** Simplify by removing a factor of 1.

**6.** $\dfrac{5a + 15}{10} = \dfrac{5(a + 3)}{5 \cdot 2} \qquad \text{Factoring the numerator and the denominator}$

$$= \frac{5}{5} \cdot \frac{a + 3}{2} \qquad \text{Factoring the rational expression}$$

$$= 1 \cdot \frac{a + 3}{2} \qquad \frac{5}{5} = 1$$

$$= \frac{a + 3}{2} \qquad \text{Removing a factor of 1}$$

*Answers on page A-25*

Simplify by removing a factor of 1.

9. $\dfrac{2x^2 + x}{3x^2 + 2x}$

10. $\dfrac{x^2 - 1}{2x^2 - x - 1}$

11. $\dfrac{6x + 14}{7}$   $x + 14; x + 2$

12. $\dfrac{12y + 24}{48}$

7. $\dfrac{6a + 12}{7a + 14} = \dfrac{6(a + 2)}{7(a + 2)}$   Factoring the numerator and the denominator

$= \dfrac{6}{7} \cdot \dfrac{a + 2}{a + 2}$   Factoring the rational expression

$= \dfrac{6}{7} \cdot 1$   $\dfrac{a + 2}{a + 2} = 1$

$= \dfrac{6}{7}$   Removing a factor of 1

8. $\dfrac{6x^2 + 4x}{2x^2 + 2x} = \dfrac{2x(3x + 2)}{2x(x + 1)}$   Factoring the numerator and the denominator

$= \dfrac{2x}{2x} \cdot \dfrac{3x + 2}{x + 1}$   Factoring the rational expression

$= 1 \cdot \dfrac{3x + 2}{x + 1}$   $\dfrac{2x}{2x} = 1$

$= \dfrac{3x + 2}{x + 1}$   Removing a factor of 1. Note in this step that you *cannot* remove the $x$'s because $x$ is not a factor of the entire numerator and the entire denominator.

9. $\dfrac{x^2 + 3x + 2}{x^2 - 1} = \dfrac{(x + 2)(x + 1)}{(x + 1)(x - 1)}$

$= \dfrac{x + 1}{x + 1} \cdot \dfrac{x + 2}{x - 1}$

$= 1 \cdot \dfrac{x + 2}{x - 1}$

$= \dfrac{x + 2}{x - 1}$

## Canceling

You may have encountered canceling when working with rational expressions. With great concern, we mention it as a possible way to speed up your work. Our concern is that canceling be done with care and understanding. Example 9 might have been done faster as follows:

$\dfrac{x^2 + 3x + 2}{x^2 - 1} = \dfrac{(x + 2)(x + 1)}{(x + 1)(x - 1)}$   Factoring the numerator and the denominator

$= \dfrac{(x + 2)\cancel{(x + 1)}}{\cancel{(x + 1)}(x - 1)}$   When a factor of 1 is noted, it is canceled, as shown: $\dfrac{x + 1}{x + 1} = 1$.

$= \dfrac{x + 2}{x - 1}$.   Simplifying

---

*CAUTION!*  The difficulty with canceling is that it is often applied incorrectly, as in the following situations:

$\dfrac{\cancel{x} + 3}{\cancel{x}} = 3$;   $\dfrac{\cancel{4} + 1}{\cancel{4} + 2} = \dfrac{1}{2}$;   $\dfrac{1\cancel{5}}{\cancel{5}4} = \dfrac{1}{4}$.

Wrong!   Wrong!   Wrong!

In each of these situations, the expressions canceled were *not* factors of 1. Factors are parts of products. For example, in $2 \cdot 3$, 2 and 3 are factors, but in $2 + 3$, 2 and 3 are *not* factors. **If you can't factor, you can't cancel. If in doubt, don't cancel!**

---

*Do Exercises 9–12.*

## Factors That Are Opposites

Consider

$$\frac{x - 4}{4 - x}.$$

At first glance, the numerator and the denominator do not appear to have any common factors other than 1. But $x - 4$ and $4 - x$ are opposites, or additive inverses, of each other. Thus we can rewrite one as the opposite of the other by factoring out a $-1$.

**Example 10**  Simplify: $\dfrac{x - 4}{4 - x}$.

$$\frac{x - 4}{4 - x} = \frac{x - 4}{-(-4 + x)} = \frac{x - 4}{-1(x - 4)}$$

$$= -1 \cdot \frac{x - 4}{x - 4}$$

$$= -1 \cdot 1$$

$$= -1$$

*Do Exercises 13–15.*

## d | Multiplying and Simplifying

We try to simplify after we multiply. That is why we leave the numerator and the denominator in factored form.

**Example 11**  Multiply and simplify: $\dfrac{5a^3}{4} \cdot \dfrac{2}{5a}$.

$$\frac{5a^3}{4} \cdot \frac{2}{5a} = \frac{5a^3(2)}{4(5a)} \qquad \text{Multiplying the numerators and the denominators}$$

$$= \frac{2 \cdot 5 \cdot a \cdot a \cdot a}{2 \cdot 2 \cdot 5 \cdot a} \qquad \text{Factoring the numerator and the denominator}$$

$$= \frac{\cancel{2} \cdot \cancel{5} \cdot \cancel{a} \cdot a \cdot a}{\cancel{2} \cdot 2 \cdot \cancel{5} \cdot \cancel{a}} \qquad \text{Removing a factor of 1: } \frac{2 \cdot 5 \cdot a}{2 \cdot 5 \cdot a} = 1$$

$$= \frac{a^2}{2} \qquad \text{Simplifying}$$

**Example 12**  Multiply and simplify: $\dfrac{x^2 + 6x + 9}{x^2 - 4} \cdot \dfrac{x - 2}{x + 3}$.

$$\frac{x^2 + 6x + 9}{x^2 - 4} \cdot \frac{x - 2}{x + 3} = \frac{(x^2 + 6x + 9)(x - 2)}{(x^2 - 4)(x + 3)} \qquad \text{Multiplying the numerators and the denominators}$$

$$= \frac{(x + 3)(x + 3)(x - 2)}{(x + 2)(x - 2)(x + 3)} \qquad \text{Factoring the numerator and the denominator}$$

$$= \frac{\cancel{(x + 3)}(x + 3)\cancel{(x - 2)}}{(x + 2)\cancel{(x - 2)}\cancel{(x + 3)}} \qquad \text{Removing a factor of 1: } \frac{(x + 3)(x - 2)}{(x + 3)(x - 2)} = 1$$

$$= \frac{x + 3}{x + 2} \qquad \text{Simplifying}$$

*Do Exercise 16.*

Simplify.

**13.** $\dfrac{x - 8}{8 - x}$  $-1$

**14.** $\dfrac{c - d}{d - c}$  $-1$

**15.** $\dfrac{-x - 7}{x + 7}$

**16.** Multiply and simplify:

$$\frac{a^2 - 4a + 4}{a^2 - 9} \cdot \frac{a + 3}{a - 2}.$$

*Answers on page A-25*

**17.** Multiply and simplify:

$$\frac{x^2 - 25}{6} \cdot \frac{3}{x + 5}.$$

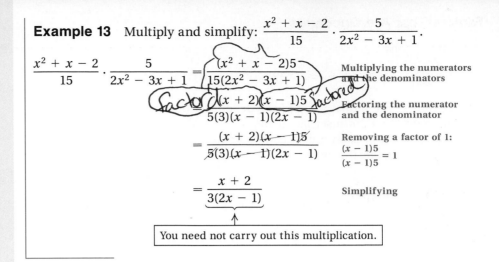

**Example 13**   Multiply and simplify: $\dfrac{x^2 + x - 2}{15} \cdot \dfrac{5}{2x^2 - 3x + 1}$.

$$\frac{x^2 + x - 2}{15} \cdot \frac{5}{2x^2 - 3x + 1} = \frac{(x^2 + x - 2)5}{15(2x^2 - 3x + 1)}$$

Multiplying the numerators and the denominators

$$= \frac{(x + 2)(x - 1)5}{5(3)(x - 1)(2x - 1)}$$

Factoring the numerator and the denominator

$$= \frac{(x + 2)(x - 1)5}{5(3)(x - 1)(2x - 1)}$$

Removing a factor of 1: $\dfrac{(x - 1)5}{(x - 1)5} = 1$

$$= \frac{x + 2}{3(2x - 1)}$$

Simplifying

↑
You need not carry out this multiplication.

*Do Exercise 17.*

---

## Calculator Spotlight

Checking Simplifications Using a Table.   We can use the TABLE feature as a partial check that rational expressions have been multiplied and/or simplified correctly. To check whether the simplifying of Example 9,

$$\frac{x^2 + 3x + 2}{x^2 - 1} = \frac{x + 2}{x - 1},$$

is correct, we enter

$$y_1 = \frac{x^2 + 3x + 2}{x^2 - 1} \quad \text{as} \quad y_1 = (x^2 + 3x + 2)/(x^2 - 1)$$

and

$$y_2 = \frac{x + 2}{x - 1} \quad \text{as} \quad y_2 = (x + 2)/(x - 1).$$

If our simplifying is correct, the $y_1$- and $y_2$-values should be the same for all allowable replacements, regardless of the table settings used. Let TblStart = −4 and ΔTbl = 1.

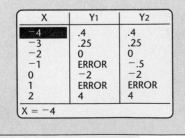

Note that −1 and 1 are not allowable replacements in the first expression, and 1 is not an allowable replacement in the second. These facts are indicated by the ERROR messages. For all other numbers, we see that $y_1$ and $y_2$ are the same, so the simplifying seems to be correct. Remember, this is only a partial check.

**Exercises**

Use the TABLE feature to check whether each of the following is correct.

1. $\dfrac{8x^2}{24x} = \dfrac{x}{3}$

2. $\dfrac{5a + 15}{10} = \dfrac{a + 3}{2}$

3. $\dfrac{x + 3}{x} = 3$

4. $\dfrac{x^2 - 3x + 2}{x^2 - 1} = \dfrac{x + 2}{x - 1}$

5. $\dfrac{x^2 - 9}{x^2 - 3} = 3$

6. $\dfrac{x^2 + 6x + 9}{x^2 - 4} \cdot \dfrac{x - 2}{x + 3} = \dfrac{x + 3}{x + 2}$

7. $\dfrac{x^2 - 25}{6} \cdot \dfrac{3}{x + 5} = \dfrac{x - 5}{3}$

---

*Answer on page A-25*

# Exercise Set 6.1

[a] Find all numbers for which the rational expression is undefined.

1. $\dfrac{-3}{2x}$

2. $\dfrac{24}{-8y}$

3. $\dfrac{5}{x-8}$

4. $\dfrac{y-4}{y+6}$

5. $\dfrac{3}{2y+5}$

6. $\dfrac{x^2-9}{4x-12}$

7. $\dfrac{x^2+11}{x^2-3x-28}$

8. $\dfrac{p^2-9}{p^2-7p+10}$

9. $\dfrac{m^3-2m}{m^2-25}$

10. $\dfrac{7-3x+x^2}{49-x^2}$

11. $\dfrac{x-4}{3}$

12. $\dfrac{x^2-25}{14}$

[b] Multiply. Do not simplify. Note that in each case you are multiplying by 1.

13. $\dfrac{4x}{4x} \cdot \dfrac{3x^2}{5y}$

14. $\dfrac{5x^2}{5x^2} \cdot \dfrac{6y^3}{3z^4}$

15. $\dfrac{2x}{2x} \cdot \dfrac{x-1}{x+4}$

16. $\dfrac{2a-3}{5a+2} \cdot \dfrac{a}{a}$

17. $\dfrac{3-x}{4-x} \cdot \dfrac{-1}{-1}$

18. $\dfrac{x-5}{5-x} \cdot \dfrac{-1}{-1}$

19. $\dfrac{y+6}{y+6} \cdot \dfrac{y-7}{y+2}$

20. $\dfrac{x^2+1}{x^3-2} \cdot \dfrac{x-4}{x-4}$

[c] Simplify.

21. $\dfrac{8x^3}{32x}$

22. $\dfrac{4x^2}{20x}$

23. $\dfrac{48p^7q^5}{18p^5q^4}$

24. $\dfrac{-76x^8y^3}{-24x^4y^3}$

25. $\dfrac{4x-12}{4x}$

26. $\dfrac{5a-40}{5}$

**27.** $\dfrac{3m^2 + 3m}{6m^2 + 9m}$

**28.** $\dfrac{4y^2 - 2y}{5y^2 - 5y}$

**29.** $\dfrac{a^2 - 9}{a^2 + 5a + 6}$

**30.** $\dfrac{t^2 - 25}{t^2 + t - 20}$

**31.** $\dfrac{a^2 - 10a + 21}{a^2 - 11a + 28}$

**32.** $\dfrac{x^2 - 2x - 8}{x^2 - x - 6}$

**33.** $\dfrac{x^2 - 25}{x^2 - 10x + 25}$

**34.** $\dfrac{x^2 + 8x + 16}{x^2 - 16}$

**35.** $\dfrac{a^2 - 1}{a - 1}$

**36.** $\dfrac{t^2 - 1}{t + 1}$

**37.** $\dfrac{x^2 + 1}{x + 1}$

**38.** $\dfrac{m^2 + 9}{m + 3}$

**39.** $\dfrac{6x^2 - 54}{4x^2 - 36}$

**40.** $\dfrac{8x^2 - 32}{4x^2 - 16}$

**41.** $\dfrac{6t + 12}{t^2 - t - 6}$

**42.** $\dfrac{4x + 32}{x^2 + 9x + 8}$

**43.** $\dfrac{2t^2 + 6t + 4}{4t^2 - 12t - 16}$

**44.** $\dfrac{3a^2 - 9a - 12}{6a^2 + 30a + 24}$

**45.** $\dfrac{t^2 - 4}{(t + 2)^2}$

**46.** $\dfrac{m^2 - 10m + 25}{m^2 - 25}$

Copyright © 1990 Addison Wesley Longman

**47.** $\dfrac{6 - x}{x - 6}$

**48.** $\dfrac{t - 3}{3 - t}$

**49.** $\dfrac{a - b}{b - a}$

**50.** $\dfrac{y - x}{-x + y}$

**51.** $\dfrac{6t - 12}{2 - t}$

**52.** $\dfrac{5a - 15}{3 - a}$

**53.** $\dfrac{x^2 - 1}{1 - x}$

**54.** $\dfrac{a^2 - b^2}{b^2 - a^2}$

d Multiply and simplify.

**55.** $\dfrac{4x^3}{3x} \cdot \dfrac{14}{x}$

**56.** $\dfrac{18}{x^3} \cdot \dfrac{5x^2}{6}$

**57.** $\dfrac{3c}{d^2} \cdot \dfrac{4d}{6c^3}$

**58.** $\dfrac{3x^2y}{2} \cdot \dfrac{4}{xy^3}$

**59.** $\dfrac{x^2 - 3x - 10}{x^2 - 4x + 4} \cdot \dfrac{x - 2}{x - 5}$

**60.** $\dfrac{t^2}{t^2 - 4} \cdot \dfrac{t^2 - 5t + 6}{t^2 - 3t}$

**61.** $\dfrac{a^2 - 9}{a^2} \cdot \dfrac{a^2 - 3a}{a^2 + a - 12}$

**62.** $\dfrac{x^2 + 10x - 11}{x^2 - 1} \cdot \dfrac{x + 1}{x + 11}$

**63.** $\dfrac{4a^2}{3a^2 - 12a + 12} \cdot \dfrac{3a - 6}{2a}$

**64.** $\dfrac{5v + 5}{v - 2} \cdot \dfrac{v^2 - 4v + 4}{v^2 - 1}$

**65.** $\dfrac{t^4 - 16}{t^4 - 1} \cdot \dfrac{t^2 + 1}{t^2 + 4}$

**66.** $\dfrac{x^4 - 1}{x^4 - 81} \cdot \dfrac{x^2 + 9}{x^2 + 1}$

**67.** $\dfrac{(x+4)^3}{(x+2)^3} \cdot \dfrac{x^2+4x+4}{x^2+8x+16}$

**68.** $\dfrac{(t-2)^3}{(t-1)^3} \cdot \dfrac{t^2-2t+1}{t^2-4t+4}$

**69.** $\dfrac{5a^2-180}{10a^2-10} \cdot \dfrac{20a+20}{2a-12}$

**70.** $\dfrac{2t^2-98}{4t^2-4} \cdot \dfrac{8t+8}{16t-112}$

---

**Skill Maintenance**

Solve.

**71.** The product of two consecutive even integers is 360. Find the integers.   [5.8a]

**72.** About 5 L of oxygen can be dissolved in 100 L of water at 0°C. This is 1.6 times the amount that can be dissolved in the same volume of water at 20°C. How much oxygen can be dissolved in 100 L at 20°C?   [2.4a]

Factor.   [5.6a]

**73.** $x^2 - x - 56$

**74.** $a^2 - 16a + 64$

**75.** $x^5 - 2x^4 - 35x^3$

**76.** $2y^3 - 10y^2 + y - 5$

**77.** $16 - t^4$

**78.** $10x^2 + 80x + 70$

**79.** $x^2 - 9x + 14$

**80.** $x^2 + x + 7$

**81.** $16x^2 - 40xy + 25y^2$

**82.** $a^2 - 9ab + 14b^2$

---

**Synthesis**

**83.** ◆ How is the process of canceling related to the identity property of 1?

**84.** ◆ Explain how a rational expression can be formed for which $-3$ and $4$ are not allowable replacements.

Simplify.

**85.** $\dfrac{x^4 - 16y^4}{(x^2+4y^2)(x-2y)}$

**86.** $\dfrac{(a-b)^2}{b^2-a^2}$

**87.** $\dfrac{t^4-1}{t^4-81} \cdot \dfrac{t^2-9}{t^2+1} \cdot \dfrac{(t-9)^2}{(t+1)^2}$

**88.** $\dfrac{(t+2)^3}{(t+1)^3} \cdot \dfrac{t^2+2t+1}{t^2+4t+4} \cdot \dfrac{t+1}{t+2}$

**89.** $\dfrac{x^2-y^2}{(x-y)^2} \cdot \dfrac{x^2-2xy+y^2}{x^2-4xy-5y^2}$

**90.** $\dfrac{x-1}{x^2+1} \cdot \dfrac{x^4-1}{(x-1)^2} \cdot \dfrac{x^2-1}{x^4-2x^2+1}$

# 6.2 Division and Reciprocals

There is a similarity between what we do with rational expressions and what we do with rational numbers. In fact, after variables have been replaced with rational numbers, a rational expression represents a rational number.

## a Finding Reciprocals

Two expressions are reciprocals of each other if their product is 1. The reciprocal of a rational expression is found by interchanging the numerator and the denominator.

### Examples

**1.** The reciprocal of $\frac{2}{5}$ is $\frac{5}{2}$. $\left(\text{This is because } \frac{2}{5} \cdot \frac{5}{2} = \frac{10}{10} = 1.\right)$

**2.** The reciprocal of $\frac{2x^2 - 3}{x + 4}$ is $\frac{x + 4}{2x^2 - 3}$.

**3.** The reciprocal of $x + 2$ is $\frac{1}{x + 2}$. $\left(\text{Think of } x + 2 \text{ as } \frac{x + 2}{1}.\right)$

*Do Exercises 1–4.*

## b Division

We divide rational expressions in the same way that we divide fractional notation in arithmetic. For a review, see Section R.2.

> To divide rational expressions, multiply by the reciprocal of the divisor. Then factor and simplify the result.

### Examples   Divide.

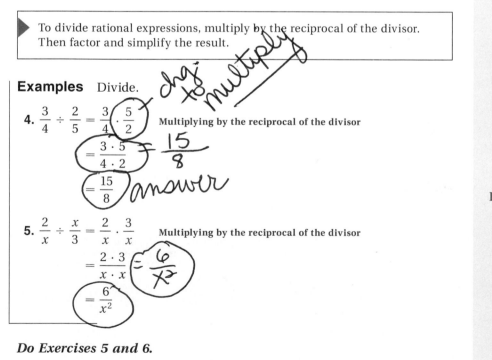

**4.** $\frac{3}{4} \div \frac{2}{5} = \frac{3}{4} \cdot \frac{5}{2}$   **Multiplying by the reciprocal of the divisor**

$= \frac{3 \cdot 5}{4 \cdot 2} = \frac{15}{8}$

$= \frac{15}{8}$   *answer*

**5.** $\frac{2}{x} \div \frac{x}{3} = \frac{2}{x} \cdot \frac{3}{x}$   **Multiplying by the reciprocal of the divisor**

$= \frac{2 \cdot 3}{x \cdot x} = \frac{6}{x^2}$

$= \frac{6}{x^2}$

*Do Exercises 5 and 6.*

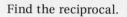

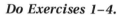

---

### Objectives

**a** Find the reciprocal of a rational expression.

**b** Divide rational expressions and simplify.

### For Extra Help

TAPE 11    TAPE 11B    MAC    CD-ROM
                        WIN

---

Find the reciprocal.

**1.** $\frac{7}{2}$

**2.** $\frac{x^2 + 5}{2x^3 - 1}$

**3.** $x - 5$

**4.** $\frac{1}{x^2 - 3}$

Divide.

**5.** $\frac{3}{5} \div \frac{7}{2}$

**6.** $\frac{x}{8} \div \frac{5}{x}$

*Answers on page A-25*

---

**7.** Divide:

$$\frac{x-3}{x+5} \div \frac{x+5}{x-2}.$$

Divide and simplify.

**8.** $\dfrac{x-3}{x+5} \div \dfrac{x+2}{x+5}$

**9.** $\dfrac{x^2-5x+6}{x+5} \div \dfrac{x+2}{x+5}$

**10.** $\dfrac{y^2-1}{y+1} \div \dfrac{y^2-2y+1}{y+1}$

## Calculator Spotlight

Use the TABLE feature to check the divisions in Examples 5–8.

*Answers on page A-25*

**Example 6**   Divide: $\dfrac{x+1}{x+2} \div \dfrac{x-1}{x+3}$

$$\frac{x+1}{x+2} \div \frac{x-1}{x+3} = \frac{x+1}{x+2} \cdot \frac{x+3}{x-1} \qquad \text{Multiplying by the reciprocal of the divisor}$$

$$= \frac{(x+1)(x+3)}{(x+2)(x-1)}$$

> We usually do not carry out the multiplication in the numerator or the denominator. It is not wrong to do so, but the factored form is often more useful.

*Do Exercise 7.*

**Example 7**   Divide and simplify: $\dfrac{x+1}{x^2-1} \div \dfrac{x+1}{x^2-2x+1}$.

$$\frac{x+1}{x^2-1} \div \frac{x+1}{x^2-2x+1}$$

$$= \frac{x+1}{x^2-1} \cdot \frac{x^2-2x+1}{x+1} \qquad \text{Multiplying by the reciprocal}$$

$$= \frac{(x+1)(x^2-2x+1)}{(x^2-1)(x+1)}$$

$$= \frac{(x+1)(x-1)(x-1)}{(x-1)(x+1)(x+1)} \qquad \text{Factoring the numerator and the denominator}$$

$$= \frac{(x+1)(x-1)(x-1)}{(x-1)(x+1)(x+1)} \qquad \text{Removing a factor of 1: } \frac{(x+1)(x-1)}{(x+1)(x-1)} = 1$$

$$= \frac{x-1}{x+1}$$

**Example 8**   Divide and simplify: $\dfrac{x^2-2x-3}{x^2-4} \div \dfrac{x+1}{x+5}$.

$$\frac{x^2-2x-3}{x^2-4} \div \frac{x+1}{x+5}$$

$$= \frac{x^2-2x-3}{x^2-4} \cdot \frac{x+5}{x+1} \qquad \text{Multiplying by the reciprocal}$$

$$= \frac{(x^2-2x-3)(x+5)}{(x^2-4)(x+1)}$$

$$= \frac{(x-3)(x+1)(x+5)}{(x-2)(x+2)(x+1)} \qquad \text{Factoring the numerator and the denominator}$$

$$= \frac{(x-3)(x+1)(x+5)}{(x-2)(x+2)(x+1)} \qquad \text{Removing a factor of 1: } \frac{x+1}{x+1} = 1$$

$$= \frac{(x-3)(x+5)}{(x-2)(x+2)}$$

> You need not carry out the multiplications in the numerator and the denominator.

*Do Exercises 8–10.*

# Exercise Set 6.2

**a** Find the reciprocal.

1. $\dfrac{4}{x}$

2. $\dfrac{a+3}{a-1}$

3. $x^2 - y^2$

4. $x^2 - 5x + 7$

5. $\dfrac{1}{a+b}$

6. $\dfrac{x^2}{x^2 - 3}$

7. $\dfrac{x^2 + 2x - 5}{x^2 - 4x + 7}$

8. $\dfrac{(a-b)(a+b)}{(a+4)(a-5)}$

**b** Divide and simplify.

9. $\dfrac{2}{5} \div \dfrac{4}{3}$

10. $\dfrac{3}{10} \div \dfrac{3}{2}$

11. $\dfrac{2}{x} \div \dfrac{8}{x}$

12. $\dfrac{t}{3} \div \dfrac{t}{15}$

13. $\dfrac{a}{b^2} \div \dfrac{a^2}{b^3} = \dfrac{b}{a}$

14. $\dfrac{x^2}{y} \div \dfrac{x^3}{y^3}$

15. $\dfrac{a+2}{a-3} \div \dfrac{a-1}{a+3}$

16. $\dfrac{x-8}{x+9} \div \dfrac{x+2}{x-1}$

17. $\dfrac{x^2 - 1}{x} \div \dfrac{x+1}{x-1}$

18. $\dfrac{4y-8}{y+2} \div \dfrac{y-2}{y^2 - 4}$

19. $\dfrac{x+1}{6} \div \dfrac{x+1}{3}$

20. $\dfrac{a}{a-b} \div \dfrac{b}{a-b}$

21. $\dfrac{5x-5}{16} \div \dfrac{x-1}{6}$

22. $\dfrac{4y-12}{12} \div \dfrac{y-3}{3}$

23. $\dfrac{-6+3x}{5} \div \dfrac{4x-8}{25}$

24. $\dfrac{-12+4x}{4} \div \dfrac{-6+2x}{6}$

25. $\dfrac{a+2}{a-1} \div \dfrac{3a+6}{a-5}$

26. $\dfrac{t-3}{t+2} \div \dfrac{4t-12}{t+1}$

27. $\dfrac{x^2-4}{x} \div \dfrac{x-2}{x+2}$

28. $\dfrac{x+y}{x-y} \div \dfrac{x^2+y}{x^2-y^2}$

29. $\dfrac{x^2-9}{4x+12} \div \dfrac{x-3}{6}$

30. $\dfrac{a-b}{2a} \div \dfrac{a^2-b^2}{8a^3}$

31. $\dfrac{c^2+3c}{c^2+2c-3} \div \dfrac{c}{c+1}$

32. $\dfrac{y+5}{2y} \div \dfrac{y^2-25}{4y^2}$

**33.** $\dfrac{2y^2 - 7y + 3}{2y^2 + 3y - 2} \div \dfrac{6y^2 - 5y + 1}{3y^2 + 5y - 2}$

**34.** $\dfrac{x^2 + x - 20}{x^2 - 7x + 12} \div \dfrac{x^2 + 10x + 25}{x^2 - 6x + 9}$

**35.** $\dfrac{x^2 - 1}{4x + 4} \div \dfrac{2x^2 - 4x + 2}{8x + 8}$

**36.** $\dfrac{5t^2 + 5t - 30}{10t + 30} \div \dfrac{2t^2 - 8}{6t^2 + 36t + 54}$

---

## Skill Maintenance

Solve.

**37.** Bonnie is taking an astronomy course. In order to receive an A, she must average at least 90 after four exams. Bonnie scored 96, 98, and 89 on the first three tests. Determine (in terms of an inequality) what scores on the last test will earn her an A. [2.8b]

**38.** Sixteen more than the square of a number is eight times the number. Find the number. [5.8a]

Subtract. [4.4c]

**39.** $(8x^3 - 3x^2 + 7) - (8x^2 + 3x - 5)$

**40.** $(3p^2 - 6pq + 7q^2) - (5p^2 - 10pq + 11q^2)$

Simplify. [4.2b]

**41.** $(2x^{-3}y^4)^2$

**42.** $(5x^6y^{-4})^3$

**43.** $\left(\dfrac{2x^3}{y^5}\right)^2$

**44.** $\left(\dfrac{a^{-3}}{b^4}\right)^5$

---

## Synthesis

**45.** ◆ Explain why 5, −1, and 7 are *not* allowable replacements in the division

$$\dfrac{x + 3}{x - 5} \div \dfrac{x - 7}{x + 1}.$$

**46.** ◆ Is the reciprocal of a product the product of the reciprocals? Why or why not?

Simplify.

**47.** $\dfrac{3a^2 - 5ab - 12b^2}{3ab + 4b^2} \div (3b^2 - ab)$

**48.** $\dfrac{3x + 3y + 3}{9x} \div \left(\dfrac{x^2 + 2xy + y^2 - 1}{x^4 + x^2}\right)$

**49.** The volume of this rectangular solid is $x - 3$. What is its height?

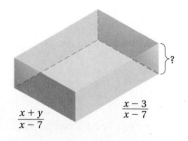

$\dfrac{x + y}{x - 7}$

$\dfrac{x - 3}{x - 7}$

---

Copyright © 1999 Addison Wesley Longman

# 6.3 Least Common Multiples and Denominators

## a Least Common Multiples

To add when denominators are different, we first find a common denominator. For a review, see Sections R.1 and R.2. We saw there, for example, that to add $\frac{5}{12}$ and $\frac{7}{30}$, we first look for the **least common multiple, LCM,** of both 12 and 30. That number becomes the **least common denominator, LCD.** To find the LCM of 12 and 30, we factor:

$$12 = 2 \cdot 2 \cdot 3;$$
$$30 = 2 \cdot 3 \cdot 5.$$

The LCM is the number that has 2 as a factor twice, 3 as a factor once, and 5 as a factor once:

$$LCM = 2 \cdot 2 \cdot 3 \cdot 5, \text{ or } 60.$$

> To find the LCM, use each factor the greatest number of times that it appears in any one factorization.

**Example 1** Find the LCM of 24 and 36.

$$\left. \begin{array}{l} 24 = 2 \cdot 2 \cdot 2 \cdot 3 \\ 36 = 2 \cdot 2 \cdot 3 \cdot 3 \end{array} \right\} \quad LCM = 2 \cdot 2 \cdot 2 \cdot 3 \cdot 3, \text{ or } 72$$

*Do Exercises 1–4.*

## b Adding Using the LCD

Let's finish adding $\frac{5}{12}$ and $\frac{7}{30}$:

$$\frac{5}{12} + \frac{7}{30} = \frac{5}{2 \cdot 2 \cdot 3} + \frac{7}{2 \cdot 3 \cdot 5}.$$

The least common denominator, LCD, is $2 \cdot 2 \cdot 3 \cdot 5$. To get the LCD in the first denominator, we need a 5. To get the LCD in the second denominator, we need another 2. We get these numbers by multiplying by 1:

$$\frac{5}{12} + \frac{7}{30} = \frac{5}{2 \cdot 2 \cdot 3} \cdot \frac{5}{5} + \frac{7}{2 \cdot 3 \cdot 5} \cdot \frac{2}{2} \qquad \text{Multiplying by 1}$$

$$= \frac{25}{2 \cdot 2 \cdot 3 \cdot 5} + \frac{14}{2 \cdot 3 \cdot 5 \cdot 2} \qquad \begin{array}{l}\text{The denominators are} \\ \text{now the LCD.}\end{array}$$

$$= \frac{39}{2 \cdot 2 \cdot 3 \cdot 5} \qquad \begin{array}{l}\text{Adding the numerators} \\ \text{and keeping the LCD}\end{array}$$

$$= \frac{3 \cdot 13}{2 \cdot 2 \cdot 3 \cdot 5} \qquad \begin{array}{l}\text{Factoring the numerator and} \\ \text{removing a factor of 1: } \frac{3}{3} = 1\end{array}$$

$$= \frac{13}{20}. \qquad \text{Simplifying}$$

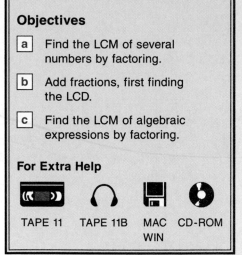

**Objectives**

a Find the LCM of several numbers by factoring.

b Add fractions, first finding the LCD.

c Find the LCM of algebraic expressions by factoring.

**For Extra Help**

TAPE 11    TAPE 11B    MAC WIN    CD-ROM

Find the LCM by factoring.

1. 16, 18

2. 6, 12

3. 2, 5

4. 24, 30, 20

*Answers on page A-25*

Add, first finding the LCD.
Simplify, if possible.

**5.** $\dfrac{3}{16} + \dfrac{1}{18}$

**6.** $\dfrac{1}{6} + \dfrac{1}{12}$

**7.** $\dfrac{1}{2} + \dfrac{3}{5}$

**8.** $\dfrac{1}{24} + \dfrac{1}{30} + \dfrac{3}{20}$

Find the LCM.

**9.** $12xy^2, \quad 15x^3y$

**10.** $y^2 + 5y + 4, \quad y^2 + 2y + 1$

**11.** $t^2 + 16, \quad t - 2, \quad 7$

**12.** $x^2 + 2x + 1, \quad 3x^2 - 3x, \quad x^2 - 1$

*Answers on page A-26*

**Example 2**  Add: $\dfrac{5}{12} + \dfrac{11}{18}$.

$$\left.\begin{array}{l} 12 = 2 \cdot 2 \cdot 3 \\ 18 = 2 \cdot 3 \cdot 3 \end{array}\right\} \quad \text{LCD} = 2 \cdot 2 \cdot 3 \cdot 3, \text{ or } 36$$

$$\frac{5}{12} + \frac{11}{18} = \frac{5}{2 \cdot 2 \cdot 3} \cdot \frac{3}{3} + \frac{11}{2 \cdot 3 \cdot 3} \cdot \frac{2}{2} = \frac{15 + 22}{2 \cdot 2 \cdot 3 \cdot 3} = \frac{37}{36}$$

*Do Exercises 5–8.*

## c | LCMs of Algebraic Expressions

To find the LCM of two or more algebraic expressions, we factor them. Then we use each factor the greatest number of times that it occurs in any one expression.

**Example 3**  Find the LCM of $12x$, $16y$, and $8xyz$.

$$\left.\begin{array}{l} 12x = 2 \cdot 2 \cdot 3 \cdot x \\ 16y = 2 \cdot 2 \cdot 2 \cdot 2 \cdot y \\ 8xyz = 2 \cdot 2 \cdot 2 \cdot x \cdot y \cdot z \end{array}\right\} \quad \begin{array}{l} \text{LCM} = 2 \cdot 2 \cdot 2 \cdot 2 \cdot 3 \cdot x \cdot y \cdot z \\ \phantom{\text{LCM}} = 48xyz \end{array}$$

**Example 4**  Find the LCM of $x^2 + 5x - 6$ and $x^2 - 1$.

$$\left.\begin{array}{l} x^2 + 5x - 6 = (x + 6)(x - 1) \\ x^2 - 1 = (x + 1)(x - 1) \end{array}\right\} \quad \text{LCM} = (x + 6)(x - 1)(x + 1)$$

**Example 5**  Find the LCM of $x^2 + 4$, $x + 1$, and 5.

These expressions do not share a common factor other than 1, so the LCM is their product:

$$5(x^2 + 4)(x + 1).$$

**Example 6**  Find the LCM of $x^2 - 25$ and $2x - 10$.

$$\left.\begin{array}{l} x^2 - 25 = (x + 5)(x - 5) \\ 2x - 10 = 2(x - 5) \end{array}\right\} \quad \text{LCM} = 2(x + 5)(x - 5)$$

**Example 7**  Find the LCM of $x^2 - 4y^2$, $x^2 - 4xy + 4y^2$, and $x - 2y$.

$$\left.\begin{array}{l} x^2 - 4y^2 = (x - 2y)(x + 2y) \\ x^2 - 4xy + 4y^2 = (x - 2y)(x - 2y) \\ x - 2y = x - 2y \end{array}\right\} \quad \begin{array}{l} \text{LCM} = (x + 2y)(x - 2y)(x - 2y) \\ \phantom{\text{LCM}} = (x + 2y)(x - 2y)^2 \end{array}$$

*Do Exercises 9–12.*

# Exercise Set 6.3

**a** Find the LCM.

**1.** 12, 27      **2.** 10, 15      **3.** 8, 9      **4.** 12, 18      **5.** 6, 9, 21

**6.** 8, 36, 40      **7.** 24, 36, 40      **8.** 4, 5, 20      **9.** 10, 100, 500      **10.** 28, 42, 60

**b** Add, first finding the LCD. Simplify, if possible.

**11.** $\dfrac{7}{24} + \dfrac{11}{18}$      **12.** $\dfrac{7}{60} + \dfrac{2}{25}$      **13.** $\dfrac{1}{6} + \dfrac{3}{40}$

**14.** $\dfrac{5}{24} + \dfrac{3}{20}$      **15.** $\dfrac{1}{20} + \dfrac{1}{30} + \dfrac{2}{45}$      **16.** $\dfrac{2}{15} + \dfrac{5}{9} + \dfrac{3}{20}$

**c** Find the LCM.

**17.** $6x^2, \ 12x^3$      **18.** $2a^2b, \ 8ab^3$      **19.** $2x^2, \ 6xy, \ 18y^2$      **20.** $p^3q, \ p^2q, \ pq^2$

**21.** $2(y-3), \ 6(y-3)$      **22.** $5(m+2), \ 15(m+2)$      **23.** $t, \ t+2, \ t-2$      **24.** $y, \ y-5, \ y+5$

**25.** $x^2 - 4, \ x^2 + 5x + 6$      **26.** $x^2 - 4, \ x^2 - x - 2$      **27.** $t^3 + 4t^2 + 4t, \ t^2 - 4t$      **28.** $m^4 - m^2, \ m^3 - m^2$

**29.** $a + 1, \ (a-1)^2, \ a^2 - 1$      **30.** $a^2 - 2ab + b^2, \ a^2 - b^2, \ 3a + 3b$

**31.** $m^2 - 5m + 6, \ m^2 - 4m + 4$      **32.** $2x^2 + 5x + 2, \ 2x^2 - x - 1$

**33.** $2 + 3x, \ 4 - 9x^2, \ 2 - 3x$      **34.** $9 - 4x^2, \ 3 + 2x, \ 3 - 2x$

**35.** $10v^2 + 30v, \quad 5v^2 + 35v + 60$

**36.** $12a^2 + 24a, \quad 4a^2 + 20a + 24$

**37.** $9x^3 - 9x^2 - 18x, \quad 6x^5 - 24x^4 + 24x^3$

**38.** $x^5 - 4x^3, \quad x^3 + 4x^2 + 4x$

**39.** $x^5 + 4x^4 + 4x^3, \quad 3x^2 - 12, \quad 2x + 4$

**40.** $x^5 + 2x^4 + x^3, \quad 2x^3 - 2x, \quad 5x - 5$

---

**Skill Maintenance**

Factor.  [5.6a]

**41.** $x^2 - 6x + 9$

**42.** $6x^2 + 4x$

**43.** $x^2 - 9$

**44.** $x^2 + 4x - 21$

**45.** $x^2 + 6x + 9$

**46.** $x^2 - 4x - 21$

*Divorce Rate.* The graph at right is that of the equation

$$D = 0.00509x^2 - 19.17x + 18,065.305$$

for values of $x$ ranging from 1900 to 2010. It shows the percentage of couples who are married in a given year, $x$, whose marriages, it is predicted, will end in divorce. Use *only* the graph to answer the questions in Exercises 47–52.  [3.1a], [4.3a]

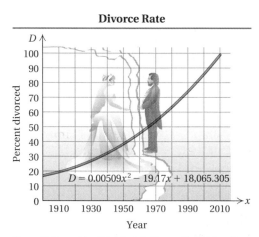

**Divorce Rate**

$D = 0.00509x^2 - 19.17x + 18,065.305$

*Source*: Gottman, John, *What Predicts Divorce: The Relationship Between Marital Processes and Marital Outcomes.* New Jersey: Lawrence Erlbaum Associates, 1993.

**47.** Estimate the divorce percentage of those married in 1970.

**48.** Estimate the divorce percentage of those married in 1980.

**49.** Estimate the divorce percentage of those married in 1990.

**50.** Estimate the divorce percentage of those married in 2010.

**51.** In what year was the divorce percentage about 50%?

**52.** In what year was the divorce percentage about 84%?

---

**Synthesis**

**53.** ◈ If the LCM of a binomial and a trinomial is the trinomial, what relationship exists between the two expressions?

**54.** ◈ Explain how you might find the LCD of these two expressions:

$$\frac{x+1}{x^2-4}, \quad \frac{x-2}{x^2+5x+6}.$$

Copyright © 1999 Addison Wesley Longman

# 6.4 Adding Rational Expressions

**a** We add rational expressions as we do rational numbers.

> To add when the denominators are the same, add the numerators and keep the same denominator.

**Objective**

**a** Add rational expressions.

**For Extra Help**

TAPE 11    TAPE 12A    MAC WIN    CD-ROM

**Examples** Add.

**1.** $\dfrac{x}{x+1} + \dfrac{2}{x+1} = \dfrac{x+2}{x+1}$

*Never Cancel / Not multiplying / Were adding*

**2.** $\dfrac{2x^2 + 3x - 7}{2x + 1} + \dfrac{x^2 + x - 8}{2x + 1} = \dfrac{(2x^2 + 3x - 7) + (x^2 + x - 8)}{2x + 1}$

$\qquad = \dfrac{3x^2 + 4x - 15}{2x + 1}$

**3.** $\dfrac{x-5}{x^2 - 9} + \dfrac{2}{x^2 - 9} = \dfrac{(x - 5) + 2}{x^2 - 9} = \dfrac{x-3}{x^2 - 9}$

$\qquad = \dfrac{x - 3}{(x - 3)(x + 3)}$  **Factoring**

$\qquad = \dfrac{x - 3}{(x - 3)(x + 3)}$  **Removing a factor of 1:** $\dfrac{x-3}{x-3} = 1$

$\qquad = \dfrac{1}{x + 3}$  **Simplifying**

*Cancel because multiplying / Must put 1 for Numerator.*

As in Example 3, simplifying should be done if possible after adding.

*Do Exercises 1–3.*

When denominators are not the same, we multiply by 1 to obtain equivalent expressions with the same denominator. When one denominator is the opposite of the other, we can first multiply either expression by 1 using $-1/-1$.

**Examples**

**4.** $\dfrac{x}{2} + \dfrac{3}{-2} = \dfrac{x}{2} + \dfrac{3}{-2} \cdot \dfrac{-1}{-1}$  **Multiplying by 1 using** $\dfrac{-1}{-1}$

$\qquad = \dfrac{x}{2} + \dfrac{-3}{2}$  **The denominators are now the same.**

$\qquad = \dfrac{x + (-3)}{2} = \dfrac{x - 3}{2}$

*Which would make -3, and +2 - for same denominator. You put to Numerator. / You for same denominator.*

**5.** $\dfrac{3x + 4}{x - 2} + \dfrac{x - 7}{2 - x} = \dfrac{3x + 4}{x - 2} + \dfrac{x - 7}{2 - x} \cdot \dfrac{-1}{-1}$

We could have chosen to multiply this expression by $-1/-1$. We multiply only one expression, *not* both.

$\qquad = \dfrac{3x + 4}{x - 2} + \dfrac{-x + 7}{x - 2}$  **Note:** $(2 - x)(-1) = -2 + x$
$= x - 2.$

$\qquad = \dfrac{(3x + 4) + (-x + 7)}{x - 2} = \dfrac{2x + 11}{x - 2}$

*Switch but must also switch numerator also. / Want to do this to have common denominator.*

*Do Exercises 4 and 5.*

Add.

**1.** $\dfrac{5}{9} + \dfrac{2}{9}$

**2.** $\dfrac{3}{x - 2} + \dfrac{x}{x - 2}$

**3.** $\dfrac{4x + 5}{x - 1} + \dfrac{2x - 1}{x - 1}$

Add.

**4.** $\dfrac{x}{4} + \dfrac{5}{-4}$

**5.** $\dfrac{2x + 1}{x - 3} + \dfrac{x + 2}{3 - x}$

*Answers on page A-26*

Add.

**6.** $\dfrac{3x}{16} + \dfrac{5x^2}{24}$

**7.** $\dfrac{3}{16x} + \dfrac{5}{24x^2}$

When denominators are different, we find the least common denominator, LCD. The procedure we will use is as follows.

> To add rational expressions with different denominators:
>
> **1.** Find the LCM of the denominators. This is the least common denominator (LCD).
> **2.** For each rational expression, find an equivalent expression with the LCD. To do so, multiply by 1 using an expression for 1 made up of factors of the LCD that are missing from the original denominator.
> **3.** Add the numerators. Write the sum over the LCD.
> **4.** Simplify, if possible.

**Example 6**   Add: $\dfrac{5x^2}{8} + \dfrac{7x}{12}$.

First, we find the LCD:

$$\left. \begin{array}{l} 8 = 2 \cdot 2 \cdot 2 \\ 12 = 2 \cdot 2 \cdot 3 \end{array} \right\} \quad \text{LCD} = 2 \cdot 2 \cdot 2 \cdot 3, \text{ or } 24.$$

Compare the factorization $8 = 2 \cdot 2 \cdot 2$ with the factorization of the LCD, $24 = 2 \cdot 2 \cdot 2 \cdot 3$. The factor of the LCD missing from 8 is 3. Compare $12 = 2 \cdot 2 \cdot 3$ and $24 = 2 \cdot 2 \cdot 2 \cdot 3$. The factor of the LCD missing from 12 is 2. We multiply by 1 to get the LCD in each expression, and then add and simplify, if possible:

$$\begin{aligned} \frac{5x^2}{8} + \frac{7x}{12} &= \frac{5x^2}{2 \cdot 2 \cdot 2} + \frac{7x}{2 \cdot 2 \cdot 3} \\[2mm] &= \frac{5x^2}{2 \cdot 2 \cdot 2} \cdot \frac{3}{3} + \frac{7x}{2 \cdot 2 \cdot 3} \cdot \frac{2}{2} \qquad \text{\small Multiplying by 1 to get} \\ &\qquad\qquad\qquad\qquad\qquad\qquad\qquad\quad\text{\small the same denominators} \\[2mm] &= \frac{15x^2}{24} + \frac{14x}{24} \\[2mm] &= \frac{15x^2 + 14x}{24}. \end{aligned}$$

**Example 7**   Add: $\dfrac{3}{8x} + \dfrac{5}{12x^2}$.

First, we find the LCD:

$$\left. \begin{array}{l} 8x = 2 \cdot 2 \cdot 2 \cdot x \\ 12x^2 = 2 \cdot 2 \cdot 3 \cdot x \cdot x \end{array} \right\} \quad \text{LCD} = 2 \cdot 2 \cdot 2 \cdot 3 \cdot x \cdot x, \text{ or } 24x^2.$$

The factors of the LCD missing from $8x$ are 3 and $x$. The factor of the LCD missing from $12x^2$ is 2. We multiply by 1 to get the LCD in each expression, and then add and simplify, if possible:

$$\begin{aligned} \frac{3}{8x} + \frac{5}{12x^2} &= \frac{3}{8x} \cdot \frac{3 \cdot x}{3 \cdot x} + \frac{5}{12x^2} \cdot \frac{2}{2} \\[2mm] &= \frac{9x}{24x^2} + \frac{10}{24x^2} \\[2mm] &= \frac{9x + 10}{24x^2}. \end{aligned}$$

*Do Exercises 6 and 7.*

**Example 8**  Add: $\dfrac{2a}{a^2 - 1} + \dfrac{1}{a^2 + a}$.

First, we find the LCD:

$$\left.\begin{array}{l} a^2 - 1 = (a - 1)(a + 1) \\ a^2 + a = a(a + 1) \end{array}\right\} \quad \text{LCD} = a(a - 1)(a + 1).$$

We multiply by 1 to get the LCD in each expression, and then add and simplify:

$$\frac{2a}{(a - 1)(a + 1)} \cdot \frac{a}{a} + \frac{1}{a(a + 1)} \cdot \frac{a - 1}{a - 1}$$

$$= \frac{2a^2}{a(a - 1)(a + 1)} + \frac{a - 1}{a(a - 1)(a + 1)}$$

$$= \frac{2a^2 + a - 1}{a(a - 1)(a + 1)}$$

$$= \frac{(a + 1)(2a - 1)}{a(a - 1)(a + 1)} \qquad \text{Factoring the numerator in order to simplify}$$

$$= \frac{\cancel{(a + 1)}(2a - 1)}{a(a - 1)\cancel{(a + 1)}} \qquad \text{Removing a factor of 1: } \frac{a + 1}{a + 1} = 1$$

$$= \frac{2a - 1}{a(a - 1)}.$$

*Do Exercise 8.*

**Example 9**  Add: $\dfrac{x + 4}{x - 2} + \dfrac{x - 7}{x + 5}$.

First, we find the LCD. It is just the product of the denominators:

$$\text{LCD} = (x - 2)(x + 5).$$

We multiply by 1 to get the LCD in each expression, and then add and simplify:

$$\frac{x + 4}{x - 2} \cdot \frac{x + 5}{x + 5} + \frac{x - 7}{x + 5} \cdot \frac{x - 2}{x - 2} = \frac{(x + 4)(x + 5)}{(x - 2)(x + 5)} + \frac{(x - 7)(x - 2)}{(x - 2)(x + 5)}$$

$$= \frac{x^2 + 9x + 20}{(x - 2)(x + 5)} + \frac{x^2 - 9x + 14}{(x - 2)(x + 5)}$$

$$= \frac{x^2 + 9x + 20 + x^2 - 9x + 14}{(x - 2)(x + 5)}$$

$$= \frac{2x^2 + 34}{(x - 2)(x + 5)}.$$

*Do Exercise 9.*

**8.** Add:

$$\frac{3}{x^3 - x} + \frac{4}{x^2 + 2x + 1}.$$

**9.** Add:

$$\frac{x - 2}{x + 3} + \frac{x + 7}{x + 8}.$$

## Calculator Spotlight

Use the TABLE feature to check the additions in Examples 7–9. Then check your answers to Margin Exercises 7–9.

*Answers on page A-26*

**10.** Add:

$$\frac{5}{x^2 + 17x + 16} + \frac{3}{x^2 + 9x + 8}.$$

**11.** Add:

$$\frac{x + 3}{x^2 - 16} + \frac{5}{12 - 3x}.$$

**Example 10**   Add: $\dfrac{x}{x^2 + 11x + 30} + \dfrac{-5}{x^2 + 9x + 20}.$

$$\frac{x}{x^2 + 11x + 30} + \frac{-5}{x^2 + 9x + 20}$$

$$= \frac{x}{(x + 5)(x + 6)} + \frac{-5}{(x + 5)(x + 4)} \qquad \begin{array}{l}\text{Factoring the denominators in}\\\text{order to find the LCD. The}\\\text{LCD is } (x + 4)(x + 5)(x + 6).\end{array}$$

$$= \frac{x}{(x + 5)(x + 6)} \cdot \frac{x + 4}{x + 4} + \frac{-5}{(x + 5)(x + 4)} \cdot \frac{x + 6}{x + 6} \qquad \textbf{Multiplying by 1}$$

$$= \frac{x(x + 4) + (-5)(x + 6)}{(x + 4)(x + 5)(x + 6)} = \frac{x^2 + 4x - 5x - 30}{(x + 4)(x + 5)(x + 6)}$$

$$= \frac{x^2 - x - 30}{(x + 4)(x + 5)(x + 6)}$$

$$\left.\begin{array}{l}= \dfrac{(x - 6)\cancel{(x + 5)}}{(x + 4)\cancel{(x + 5)}(x + 6)}\\[2em]= \dfrac{(x - 6)}{(x + 4)(x + 6)}\end{array}\right\} \rightarrow \text{Always simplify at the end if possible: } \dfrac{x + 5}{x + 5} = 1.$$

*Do Exercise 10.*

Suppose that after we factor to find the LCD, we find factors that are opposites. There are several ways to handle this, but the easiest is to first go back and multiply by $-1/-1$ appropriately to change factors so that they are not opposites.

**Example 11**   Add: $\dfrac{x}{x^2 - 25} + \dfrac{3}{10 - 2x}.$

First, we factor as though we are going to find the LCD:

$$x^2 - 25 = (x - 5)(x + 5);$$
$$10 - 2x = 2(5 - x).$$

We note that there is an $x - 5$ as one factor and a $5 - x$ as another factor. If the denominator of the second expression were $2x - 10$, this situation would not occur. To rewrite the second expression with a denominator of $2x - 10$, we multiply by 1 using $-1/-1$, and then continue as before:

$$\frac{x}{x^2 - 25} + \frac{3}{10 - 2x} = \frac{x}{(x - 5)(x + 5)} + \frac{3}{10 - 2x} \cdot \frac{-1}{-1}$$

$$= \frac{x}{(x - 5)(x + 5)} + \frac{-3}{2x - 10}$$

$$= \frac{x}{(x - 5)(x + 5)} + \frac{-3}{2(x - 5)} \qquad \text{LCD} = 2(x - 5)(x + 5)$$

$$= \frac{x}{(x - 5)(x + 5)} \cdot \frac{2}{2} + \frac{-3}{2(x - 5)} \cdot \frac{x + 5}{x + 5}$$

$$= \frac{2x - 3(x + 5)}{2(x - 5)(x + 5)} = \frac{2x - 3x - 15}{2(x - 5)(x + 5)}$$

$$= \frac{-x - 15}{2(x - 5)(x + 5)}. \qquad \begin{array}{l}\text{Collecting like}\\\text{terms}\end{array}$$

*Do Exercise 11.*

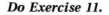

*Answers on page A-26*

# Exercise Set 6.4

**a** Add. Simplify, if possible.

**1.** $\dfrac{5}{8} + \dfrac{3}{8}$

**2.** $\dfrac{3}{16} + \dfrac{5}{16} = \dfrac{8}{16} = \dfrac{1}{2}$

**3.** $\dfrac{1}{3 + x} + \dfrac{5}{3 + x}$

**4.** $\dfrac{4x + 6}{2x - 1} + \dfrac{5 - 8x}{-1 + 2x} = \dfrac{-4x + 11}{2x - 1}$

$2x - 1$

**5.** $\dfrac{x^2 + 7x}{x^2 - 5x} + \dfrac{x^2 - 4x}{x^2 - 5x}$

**6.** $\dfrac{4}{x + y} + \dfrac{9}{y + x} = \dfrac{13}{x + y}$

$x + y$

**7.** $\dfrac{7}{8} + \dfrac{5}{-8}$

**8.** $\dfrac{5}{-3} + \dfrac{11}{3}$

**9.** $\dfrac{3}{t} + \dfrac{4}{-t}$

**10.** $\dfrac{5}{-a} + \dfrac{8}{a}$

**11.** $\dfrac{2x + 7}{x - 6} + \dfrac{3x}{6 - x}$

**12.** $\dfrac{2x - 7}{5x - 8} + \dfrac{6 + 10x}{8 - 5x}$

**13.** $\dfrac{y^2}{y - 3} + \dfrac{9}{3 - y}$

**14.** $\dfrac{t^2}{t - 2} + \dfrac{4}{2 - t}$

**15.** $\dfrac{b - 7}{b^2 - 16} + \dfrac{7 - b}{16 - b^2}$

**16.** $\dfrac{a - 3}{a^2 - 25} + \dfrac{a - 3}{25 - a^2}$

**17.** $\dfrac{a^2}{a - b} + \dfrac{b^2}{b - a}$

**18.** $\dfrac{x^2}{x - 7} + \dfrac{49}{7 - x}$

**19.** $\dfrac{x + 3}{x - 5} + \dfrac{2x - 1}{5 - x} + \dfrac{2(3x - 1)}{x - 5}$

**20.** $\dfrac{3(x - 2)}{2x - 3} + \dfrac{5(2x + 1)}{2x - 3} + \dfrac{3(x + 1)}{3 - 2x}$

**21.** $\dfrac{2(4x + 1)}{5x - 7} + \dfrac{3(x - 2)}{7 - 5x} + \dfrac{-10x - 1}{5x - 7}$

**22.** $\dfrac{5(x - 2)}{3x - 4} + \dfrac{2(x - 3)}{4 - 3x} + \dfrac{3(5x + 1)}{4 - 3x}$

**23.** $\dfrac{x + 1}{(x + 3)(x - 3)} + \dfrac{4(x - 3)}{(x - 3)(x + 3)} + \dfrac{(x - 1)(x - 3)}{(3 - x)(x + 3)}$

**24.** $\dfrac{2(x + 5)}{(2x - 3)(x - 1)} + \dfrac{3x + 4}{(2x - 3)(1 - x)} + \dfrac{x - 5}{(3 - 2x)(x - 1)}$

**25.** $\dfrac{2}{x} + \dfrac{5}{x^2}$

**26.** $\dfrac{3}{y^2} + \dfrac{6}{y}$

**27.** $\dfrac{5}{6r} + \dfrac{7}{8r}$

**28.** $\dfrac{13}{18x} + \dfrac{7}{24x}$

**29.** $\dfrac{4}{xy^2} + \dfrac{6}{x^2 y}$

**30.** $\dfrac{8}{ab^3} + \dfrac{3}{a^2 b}$

**31.** $\dfrac{2}{9t^3} + \dfrac{1}{6t^2}$

**32.** $\dfrac{5}{c^2 d^3} + \dfrac{-4}{7cd^2}$

**33.** $\dfrac{x + y}{xy^2} + \dfrac{3x + y}{x^2 y}$

**34.** $\dfrac{2c - d}{c^2 d} + \dfrac{c + d}{cd^2}$

**35.** $\dfrac{3}{x - 2} + \dfrac{3}{x + 2}$

**36.** $\dfrac{2}{y + 1} + \dfrac{2}{y - 1}$

**37.** $\dfrac{3}{x + 1} + \dfrac{2}{3x}$

**38.** $\dfrac{4}{5y} + \dfrac{7}{y - 2}$

**39.** $\dfrac{2x}{x^2 - 16} + \dfrac{x}{x - 4}$

Copyright © 1999 Addison Wesley Longman

**40.** $\dfrac{4x}{x^2 - 25} + \dfrac{x}{x + 5}$

**41.** $\dfrac{5}{z + 4} + \dfrac{3}{3z + 12}$

**42.** $\dfrac{t}{t - 3} + \dfrac{5}{4t - 12}$

**43.** $\dfrac{3}{x - 1} + \dfrac{2}{(x - 1)^2}$

**44.** $\dfrac{8}{(y + 3)^2} + \dfrac{5}{y + 3}$

**45.** $\dfrac{4a}{5a - 10} + \dfrac{3a}{10a - 20}$

**46.** $\dfrac{9x}{6x - 30} + \dfrac{3x}{4x - 20}$

**47.** $\dfrac{x + 4}{x} + \dfrac{x}{x + 4}$

**48.** $\dfrac{a}{a - 3} + \dfrac{a - 3}{a}$

**49.** $\dfrac{4}{a^2 - a - 2} + \dfrac{3}{a^2 + 4a + 3}$

**50.** $\dfrac{a}{a^2 - 2a + 1} + \dfrac{1}{a^2 - 5a + 4}$

**51.** $\dfrac{x + 3}{x - 5} + \dfrac{x - 5}{x + 3}$

**52.** $\dfrac{3x}{2y - 3} + \dfrac{2x}{3y - 2}$

**53.** $\dfrac{a}{a^2 - 1} + \dfrac{2a}{a^2 - a}$

**54.** $\dfrac{3x + 2}{3x + 6} + \dfrac{x - 2}{x^2 - 4}$

**55.** $\dfrac{6}{x - y} + \dfrac{4x}{y^2 - x^2}$

**56.** $\dfrac{a - 2}{3 - a} + \dfrac{4 - a^2}{a^2 - 9}$

**57.** $\dfrac{4 - a}{25 - a^2} + \dfrac{a + 1}{a - 5}$

**58.** $\dfrac{x + 2}{x - 7} + \dfrac{3 - x}{49 - x^2}$

**59.** $\dfrac{2}{t^2 + t - 6} + \dfrac{3}{t^2 - 9}$

**60.** $\dfrac{10}{a^2 - a - 6} + \dfrac{3a}{a^2 + 4a + 4}$

## Skill Maintenance

Subtract.   [4.4c]

**61.** $(x^2 + x) - (x + 1)$

**62.** $(4y^3 - 5y^2 + 7y - 24) - (-9y^3 + 9y^2 - 5y + 49)$

Simplify.   [4.2b]

**63.** $(2x^4y^3)^{-3}$

**64.** $\left(\dfrac{x^3}{5y}\right)^2$

**65.** $\left(\dfrac{x^{-4}}{y^7}\right)^3$

**66.** $(5x^{-2}y^{-3})^2$

Graph.

**67.** $y = \dfrac{1}{2}x - 5$

[3.2b], [3.3a]

**68.** $2y + x + 10 = 0$

[3.2b], [3.3a]

**69.** $y = 3$   [3.3b]

**70.** $x = -5$   [3.3b]

Solve.

**71.** $3x - 7 = 5x + 9$   [2.3b]

**72.** $2a + 8 = 13 - 4a$   [2.3b]

**73.** $x^2 - 8x + 15 = 0$   [5.7b]

**74.** $x^2 - 7x = 18$   [5.7b]

## Synthesis

**75.** ◈ Explain why the expressions

$$\frac{1}{3 - x} \quad \text{and} \quad \frac{1}{x - 3}$$

are opposites.

**76.** ◈ Why is it better to use the *least* common denominator, rather than *any* common denominator, when adding rational expressions?

Find the perimeter and the area of the figure.

**77.**

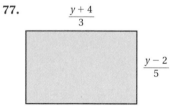

$$\frac{y + 4}{3}$$

$$\frac{y - 2}{5}$$

**78.**

$$\frac{3}{x + 4}$$

$$\frac{2}{x - 5}$$

Add. Simplify, if possible.

**79.** $\dfrac{5}{z + 2} + \dfrac{4z}{z^2 - 4} + 2$

**80.** $\dfrac{-2}{y^2 - 9} + \dfrac{4y}{(y - 3)^2} + \dfrac{6}{3 - y}$

**81.** $\dfrac{3z^2}{z^4 - 4} + \dfrac{5z^2 - 3}{2z^4 + z^2 - 6}$

**82.** Find an expression equivalent to

$$\frac{a - 3b}{a - b}$$

that is a sum of two fractional expressions. Answers may vary.

**83.–88.** 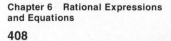 Use the TABLE feature to check the additions in Exercises 47–52.

Copyright © 1999 Addison Wesley Longman

# 6.5 Subtracting Rational Expressions

**a** We subtract rational expressions as we do rational numbers.

> To subtract when the denominators are the same, subtract the numerators and keep the same denominator.

**Example 1** Subtract: $\dfrac{8}{x} - \dfrac{3}{x}$.

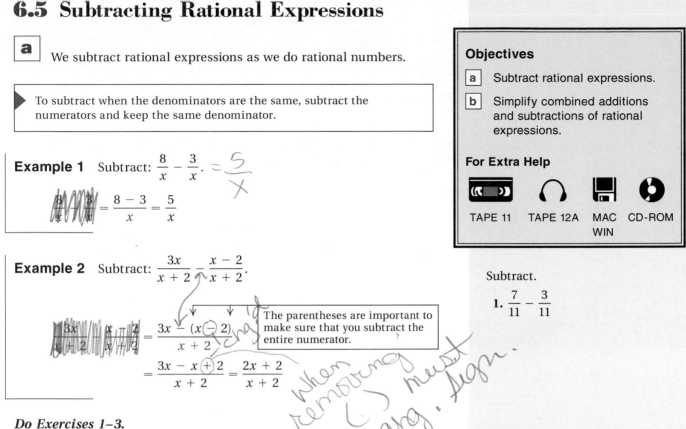

$$\dfrac{8}{x} - \dfrac{3}{x} = \dfrac{8 - 3}{x} = \dfrac{5}{x}$$

**Example 2** Subtract: $\dfrac{3x}{x + 2} - \dfrac{x - 2}{x + 2}$.

$$\dfrac{3x}{x + 2} - \dfrac{x - 2}{x + 2} = \dfrac{3x - (x - 2)}{x + 2}$$

> The parentheses are important to make sure that you subtract the entire numerator.

$$= \dfrac{3x - x + 2}{x + 2} = \dfrac{2x + 2}{x + 2}$$

*Do Exercises 1–3.*

When one denominator is the opposite of the other, we can first multiply one expression by $-1/-1$ to obtain a common denominator.

**Example 3** Subtract: $\dfrac{x}{5} - \dfrac{3x - 4}{-5}$.

$$\dfrac{x}{5} - \dfrac{3x - 4}{-5} = \dfrac{x}{5} - \dfrac{3x - 4}{-5} \cdot \dfrac{-1}{-1}$$

Multiplying by 1 using $\dfrac{-1}{-1}$

This is equal to 1 (not $-1$).

$$= \dfrac{x}{5} - \dfrac{(3x - 4)(-1)}{(-5)(-1)}$$

$$= \dfrac{x}{5} - \dfrac{4 - 3x}{5}$$

Remember the parentheses!

$$= \dfrac{x - (4 - 3x)}{5}$$

$$= \dfrac{x - 4 + 3x}{5} = \dfrac{4x - 4}{5}$$

**Example 4** Subtract: $\dfrac{5y}{y - 5} - \dfrac{2y - 3}{5 - y}$.

$$\dfrac{5y}{y - 5} - \dfrac{2y - 3}{5 - y} = \dfrac{5y}{y - 5} - \dfrac{2y - 3}{5 - y} \cdot \dfrac{-1}{-1}$$

$$= \dfrac{5y}{y - 5} - \dfrac{(2y - 3)(-1)}{(5 - y)(-1)} = \dfrac{5y}{y - 5} - \dfrac{3 - 2y}{y - 5}$$

Remember the parentheses!

$$= \dfrac{5y - (3 - 2y)}{y - 5}$$

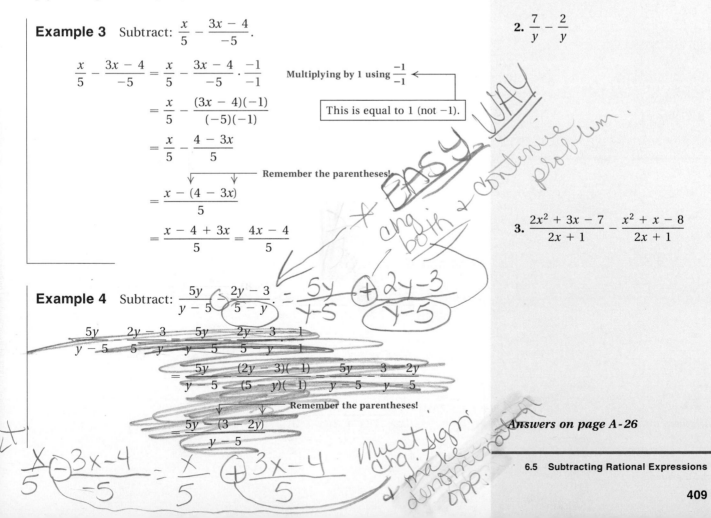

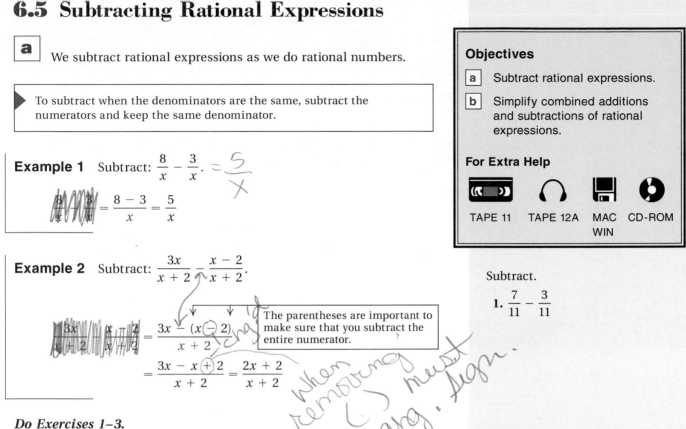

**Objectives**

**a** Subtract rational expressions.

**b** Simplify combined additions and subtractions of rational expressions.

**For Extra Help**

TAPE 11    TAPE 12A    MAC WIN    CD-ROM

Subtract.

**1.** $\dfrac{7}{11} - \dfrac{3}{11}$

**2.** $\dfrac{7}{y} - \dfrac{2}{y}$

**3.** $\dfrac{2x^2 + 3x - 7}{2x + 1} - \dfrac{x^2 + x - 8}{2x + 1}$

*Answers on page A-26*

6.5 Subtracting Rational Expressions

**409**

Subtract.

**4.** $\dfrac{x}{3} - \dfrac{2x-1}{-3}$

**5.** $\dfrac{3x}{x-2} - \dfrac{x-3}{2-x}$

**6.** Subtract:

$\dfrac{x-2}{3x} - \dfrac{2x-1}{5x}$.

Then
$$= \frac{5y - 3 + 2y}{y - 5}$$
$$= \frac{7y - 3}{y - 5}.$$

### Do Exercises 4 and 5.

To subtract rational expressions with different denominators, we use a procedure similar to what we used for addition, except that we subtract numerators and write the difference over the LCD.

> To subtract rational expressions with different denominators:
>
> **1.** Find the LCM of the denominators. This is the least common denominator (LCD).
>
> **2.** For each rational expression, find an equivalent expression with the LCD. To do so, multiply by 1 using a symbol for 1 made up of factors of the LCD that are missing from the original denominator.
>
> **3.** Subtract the numerators. Write the difference over the LCD.
>
> **4.** Simplify, if possible.

**Example 5**  Subtract: $\dfrac{x+2}{x-4} - \dfrac{x+1}{x+4}$.

The LCD $= (x-4)(x+4)$.

$$\frac{x+2}{x-4} \cdot \frac{x+4}{x+4} - \frac{x+1}{x+4} \cdot \frac{x-4}{x-4} \qquad \text{Multiplying by 1}$$

$$= \frac{(x+2)(x+4)}{(x-4)(x+4)} - \frac{(x+1)(x-4)}{(x-4)(x+4)}$$

$$= \frac{x^2 + 6x + 8}{(x-4)(x+4)} - \frac{x^2 - 3x - 4}{(x-4)(x+4)}$$

$$= \frac{x^2 + 6x + 8 - (x^2 - 3x - 4)}{(x-4)(x+4)} \qquad \begin{array}{l}\text{Subtracting this numerator.}\\\text{Don't forget the parentheses.}\end{array}$$

$$= \frac{x^2 + 6x + 8 - x^2 + 3x + 4}{(x-4)(x+4)}$$

$$= \frac{9x + 12}{(x-4)(x+4)}$$

### Do Exercise 6.

**Example 6**  Subtract: $\dfrac{x}{x^2 + 5x + 6} - \dfrac{2}{x^2 + 3x + 2}$.

$$\frac{x}{x^2 + 5x + 6} - \frac{2}{x^2 + 3x + 2}$$

$$= \frac{x}{(x+2)(x+3)} - \frac{2}{(x+2)(x+1)} \qquad \text{LCD} = (x+1)(x+2)(x+3)$$

$$= \frac{x}{(x+2)(x+3)} \cdot \frac{x+1}{x+1} - \frac{2}{(x+2)(x+1)} \cdot \frac{x+3}{x+3}$$

$$= \frac{x^2 + x}{(x+1)(x+2)(x+3)} - \frac{2x + 6}{(x+1)(x+2)(x+3)}$$

Then

$$= \frac{x^2 + x - (2x + 6)}{(x + 1)(x + 2)(x + 3)}$$

Subtracting this numerator.
Don't forget the parentheses.

$$= \frac{x^2 + x - 2x - 6}{(x + 1)(x + 2)(x + 3)}$$

$$= \frac{x^2 - x - 6}{(x + 1)(x + 2)(x + 3)}$$

$$= \frac{(x + 2)(x - 3)}{(x + 1)(x + 2)(x + 3)}$$

$$= \frac{(x + 2)(x - 3)}{(x + 1)(x + 2)(x + 3)}$$

Simplifying by removing
a factor of 1: $\dfrac{x + 2}{x + 2} = 1$

$$= \frac{x - 3}{(x + 1)(x + 3)}.$$

*Do Exercise 7.*

Suppose that after we factor to find the LCD, we find factors that are opposites. Then we multiply by $-1/-1$ appropriately to change factors so that they are not opposites.

**Example 7**   Subtract: $\dfrac{p}{64 - p^2} - \dfrac{5}{p - 8}$.

Factoring $64 - p^2$, we get $(8 - p)(8 + p)$. Note that the factors $8 - p$ in the first denominator and $p - 8$ in the second denominator are opposites. We multiply the first expression by $-1/-1$ to avoid this situation. Then we proceed as before.

$$\frac{p}{64 - p^2} - \frac{5}{p - 8} = \frac{p}{64 - p^2} \cdot \frac{-1}{-1} - \frac{5}{p - 8}$$

$$= \frac{-p}{p^2 - 64} - \frac{5}{p - 8}$$

$$= \frac{-p}{(p - 8)(p + 8)} - \frac{5}{p - 8} \qquad \text{LCD} = (p - 8)(p + 8)$$

$$= \frac{-p}{(p - 8)(p + 8)} - \frac{5}{p - 8} \cdot \frac{p + 8}{p + 8}$$

$$= \frac{-p}{(p - 8)(p + 8)} - \frac{5p + 40}{(p - 8)(p + 8)}$$

$$= \frac{-p - (5p + 40)}{(p - 8)(p + 8)}$$

Subtracting this numerator.
Don't forget the parentheses.

$$= \frac{-p - 5p - 40}{(p - 8)(p + 8)}$$

$$= \frac{-6p - 40}{(p - 8)(p + 8)}$$

*Do Exercise 8.*

**7.** Subtract:

$$\frac{x}{x^2 + 15x + 56} - \frac{6}{x^2 + 13x + 42}.$$

**8.** Subtract:

$$\frac{y}{16 - y^2} - \frac{7}{y - 4}.$$

---

## Calculator Spotlight

Use the TABLE feature to check the subtractions in Examples 5–7. Then check your answers to Margin Exercises 6–8.

---

*Answers on page A-26*

6.5   Subtracting Rational Expressions

**9.** Perform the indicated operations and simplify:

$$\frac{x+2}{x^2-9} - \frac{x-7}{9-x^2} + \frac{-8-x}{x^2-9}.$$

**b** | **Combined Additions and Subtractions**

Now let's look at some combined additions and subtractions.

**Example 8** Perform the indicated operations and simplify:

$$\frac{x+9}{x^2-4} + \frac{5-x}{4-x^2} - \frac{2+x}{x^2-4}.$$

We have

$$\frac{x+9}{x^2-4} + \frac{5-x}{4-x^2} - \frac{2+x}{x^2-4} = \frac{x+9}{x^2-4} + \frac{5-x}{4-x^2} \cdot \frac{-1}{-1} - \frac{2+x}{x^2-4}$$

$$= \frac{x+9}{x^2-4} + \frac{x-5}{x^2-4} - \frac{2+x}{x^2-4}$$

$$= \frac{(x+9)+(x-5)-(2+x)}{x^2-4}$$

$$= \frac{x+9+x-5-2-x}{x^2-4}$$

$$= \frac{x+2}{x^2-4}$$

$$= \frac{(x+2) \cdot 1}{(x+2)(x-2)} \qquad \frac{x+2}{x+2} = 1$$

$$= \frac{1}{x-2}.$$

*Do Exercise 9.*

**10.** Perform the indicated operations and simplify:

$$\frac{1}{x} - \frac{5}{3x} + \frac{2x}{x+1}.$$

**Example 9** Perform the indicated operations and simplify:

$$\frac{1}{x} - \frac{1}{x^2} + \frac{2}{x+1}.$$

The LCD $= x \cdot x(x+1)$, or $x^2(x+1)$.

$$\frac{1}{x} \cdot \frac{x(x+1)}{x(x+1)} - \frac{1}{x^2} \cdot \frac{(x+1)}{(x+1)} + \frac{2}{x+1} \cdot \frac{x^2}{x^2}$$

$$= \frac{x(x+1)}{x^2(x+1)} - \frac{x+1}{x^2(x+1)} + \frac{2x^2}{x^2(x+1)}$$

$$= \frac{x(x+1) - (x+1) + 2x^2}{x^2(x+1)} \qquad \text{Subtracting this numerator. Don't forget the parentheses.}$$

$$= \frac{x^2 + x - x - 1 + 2x^2}{x^2(x+1)}$$

$$= \frac{3x^2 - 1}{x^2(x+1)}$$

*Do Exercise 10.*

*Answers on page A-26*

# Exercise Set 6.5

**a** Subtract. Simplify, if possible.

**1.** $\dfrac{7}{x} - \dfrac{3}{x}$

**2.** $\dfrac{5}{a} - \dfrac{8}{a}$

**3.** $\dfrac{y}{y-4} - \dfrac{4}{y-4}$

**4.** $\dfrac{t^2}{t+5} - \dfrac{25}{t+5}$

**5.** $\dfrac{2x-3}{x^2+3x-4} - \dfrac{x-7}{x^2+3x-4}$

**6.** $\dfrac{x+1}{x^2-2x+1} - \dfrac{5-3x}{x^2-2x+1}$

**7.** $\dfrac{11}{6} - \dfrac{5}{-6}$

**8.** $\dfrac{5}{9} - \dfrac{7}{-9}$

**9.** $\dfrac{5}{a} - \dfrac{8}{-a}$

**10.** $\dfrac{8}{x} - \dfrac{3}{-x}$

**11.** $\dfrac{4}{y-1} - \dfrac{4}{1-y}$

**12.** $\dfrac{5}{a-2} - \dfrac{3}{2-a}$

**13.** $\dfrac{3-x}{x-7} - \dfrac{2x-5}{7-x}$

**14.** $\dfrac{t^2}{t-2} - \dfrac{4}{2-t}$

**15.** $\dfrac{a-2}{a^2-25} - \dfrac{6-a}{25-a^2}$

**16.** $\dfrac{x-8}{x^2-16} - \dfrac{x-8}{16-x^2}$

**17.** $\dfrac{4-x}{x-9} - \dfrac{3x-8}{9-x}$

**18.** $\dfrac{4x-6}{x-5} - \dfrac{7-2x}{5-x}$

**19.** $\dfrac{2(x-1)}{2x-3} - \dfrac{3(x+2)}{2x-3} - \dfrac{x-1}{3-2x}$

**20.** $\dfrac{5(2y+1)}{2y-3} - \dfrac{3(y-1)}{3-2y} - \dfrac{3(y-2)}{2y-3}$

**21.** $\dfrac{a-2}{10} - \dfrac{a+1}{5}$

**22.** $\dfrac{y+3}{2} - \dfrac{y-4}{4}$

**23.** $\dfrac{4z-9}{3z} - \dfrac{3z-8}{4z}$

**24.** $\dfrac{a-1}{4a} - \dfrac{2a+3}{a}$

**25.** $\dfrac{4x+2t}{3xt^2} - \dfrac{5x-3t}{x^2t}$

**26.** $\dfrac{5x+3y}{2x^2y} - \dfrac{3x+4y}{xy^2}$

**27.** $\dfrac{5}{x+5} - \dfrac{3}{x-5}$

**28.** $\dfrac{3t}{t-1} - \dfrac{8t}{t+1}$

**29.** $\dfrac{3}{2t^2-2t} - \dfrac{5}{2t-2}$

**30.** $\dfrac{11}{x^2-4} - \dfrac{8}{x+2}$

**31.** $\dfrac{2s}{t^2-s^2} - \dfrac{s}{t-s}$

**32.** $\dfrac{3}{12+x-x^2} - \dfrac{2}{x^2-9}$

**33.** $\dfrac{y-5}{y} - \dfrac{3y-1}{4y}$

**34.** $\dfrac{3x-2}{4x} - \dfrac{3x+1}{6x}$

**35.** $\dfrac{a}{x+a} - \dfrac{a}{x-a}$

**36.** $\dfrac{a}{a-b} - \dfrac{a}{a+b}$

**37.** $\dfrac{5x}{x^2 - 9} - \dfrac{4}{3 - x}$

**38.** $\dfrac{8x}{16 - x^2} - \dfrac{5}{x - 4}$

**39.** $\dfrac{t^2}{2t^2 - 2t} - \dfrac{1}{2t - 2}$

**40.** $\dfrac{4}{5a^2 - 5a} - \dfrac{2}{5a - 5}$

**41.** $\dfrac{x}{x^2 + 5x + 6} - \dfrac{2}{x^2 + 3x + 2}$

**42.** $\dfrac{a}{a^2 + 11a + 30} - \dfrac{5}{a^2 + 9a + 20}$

**b** Perform the indicated operations and simplify.

**43.** $\dfrac{3(2x + 5)}{x - 1} - \dfrac{3(2x - 3)}{1 - x} + \dfrac{6x - 1}{x - 1}$

**44.** $\dfrac{a - 2b}{b - a} - \dfrac{3a - 3b}{a - b} + \dfrac{2a - b}{a - b}$

**45.** $\dfrac{x - y}{x^2 - y^2} + \dfrac{x + y}{x^2 - y^2} - \dfrac{2x}{x^2 - y^2}$

**46.** $\dfrac{x - 3y}{2(y - x)} + \dfrac{x + y}{2(x - y)} - \dfrac{2x - 2y}{2(x - y)}$

**47.** $\dfrac{10}{2y - 1} - \dfrac{6}{1 - 2y} + \dfrac{y}{2y - 1} + \dfrac{y - 4}{1 - 2y}$

**48.** $\dfrac{(x + 1)(2x - 1)}{(2x - 3)(x - 3)} - \dfrac{(x - 3)(x + 1)}{(3 - x)(3 - 2x)} + \dfrac{(2x + 1)(x + 3)}{(3 - 2x)(x - 3)}$

**49.** $\dfrac{a + 6}{4 - a^2} - \dfrac{a + 3}{a + 2} + \dfrac{a - 3}{2 - a}$

**50.** $\dfrac{4t}{t^2 - 1} - \dfrac{2}{t} - \dfrac{2}{t + 1}$

**51.** $\dfrac{2z}{1-2z} + \dfrac{3z}{2z+1} - \dfrac{3}{4z^2-1}$

**52.** $\dfrac{1}{x-y} - \dfrac{2x}{x^2-y^2} + \dfrac{1}{x+y}$

**53.** $\dfrac{1}{x+y} - \dfrac{1}{x-y} + \dfrac{2x}{x^2-y^2}$

**54.** $\dfrac{2b}{a^2-b^2} - \dfrac{1}{a+b} + \dfrac{1}{a-b}$

---

## Skill Maintenance

Simplify.

**55.** $\dfrac{x^8}{x^3}$  [4.1e]

**56.** $3x^4 \cdot 10x^8$  [4.1d]

**57.** $(a^2b^{-5})^{-4}$  [4.2b]

**58.** $\dfrac{54x^{10}}{3x^7}$  [4.1e]

**59.** $\dfrac{66x^2}{11x^5}$  [4.1e]

**60.** $5x^{-7} \cdot 2x^4$  [4.1d]

Find a polynomial for the shaded area of the figure.  [4.4d]

**61.**

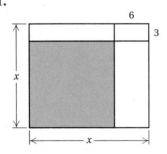

**62.**

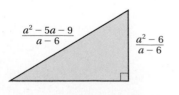

---

## Synthesis

**63.** ◆ Are parentheses as important when adding rational expressions as they are when subtracting? Why or why not?

**64.** ◆ Is it possible to add or subtract rational expressions without knowing how to factor? Why or why not?

Perform the indicated operations and simplify.

**65.** $\dfrac{2x+11}{x-3} \cdot \dfrac{3}{x+4} + \dfrac{2x+1}{4+x} \cdot \dfrac{3}{3-x}$

**66.** $\dfrac{x^2}{3x^2-5x-2} - \dfrac{2x}{3x+1} \cdot \dfrac{1}{x-2}$

**67.** $\dfrac{x}{x^4-y^4} - \left(\dfrac{1}{x+y}\right)^2$

**68.** $\left(\dfrac{a}{a-b} + \dfrac{b}{a+b}\right)\left(\dfrac{1}{3a+b} + \dfrac{2a+6b}{9a^2-b^2}\right)$

**69.** The perimeter of the following right triangle is $2a + 5$. Find the length of the missing side and the area.

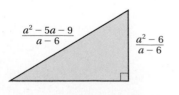

$\dfrac{a^2-5a-9}{a-6}$

$\dfrac{a^2-6}{a-6}$

**70.–75.** 📉 Use the TABLE feature to check the subtractions in Exercises 29–34.

# 6.6 Solving Rational Equations

##  Rational Equations

In Sections 6.1–6.5, we studied operations with *rational expressions*. These expressions have no equals signs. We can perform the operations and simplify, but we cannot solve if there are no equals signs—as, for example, in

$$\frac{x^2 + 6x + 9}{x^2 - 4} \cdot \frac{x - 2}{x + 3}, \qquad \frac{x + y}{x - y} \div \frac{x^2 + y}{x^2 - y^2}, \quad \text{and} \quad \frac{a + 3}{a^2 - 16} + \frac{5}{12 - 3a}.$$

Operation signs occur. There are no equals signs!

Most often, the result of our calculation is another rational expression that has not been cleared of fractions.

Equations *do have* equals signs, and we can clear them of fractions as we did in Section 2.3. A **rational**, or **fractional, equation** is an equation containing one or more rational expressions. Here are some examples:

$$\frac{2}{3} + \frac{5}{6} = \frac{x}{9}, \qquad x + \frac{6}{x} = -5, \quad \text{and} \quad \frac{x^2}{x - 1} = \frac{1}{x - 1}.$$

There are equals signs as well as operation signs.

> To solve a rational equation, the first step is to clear the equation of fractions. To do this, multiply both sides of the equation by the LCM of all the denominators. Then carry out the equation-solving process as we learned it in Chapter 2.

When clearing an equation of fractions, we use the terminology LCM instead of LCD because we are *not* adding or subtracting rational expressions.

**Example 1**  Solve: $\frac{2}{3} + \frac{5}{6} = \frac{x}{9}$.

The LCM of all denominators is $2 \cdot 3 \cdot 3$, or 18. We multiply by 18 on both sides:

$$18\left(\frac{2}{3} + \frac{5}{6}\right) = 18 \cdot \frac{x}{9} \qquad \text{Multiplying by the LCM on both sides}$$

$$18 \cdot \frac{2}{3} + 18 \cdot \frac{5}{6} = 18 \cdot \frac{x}{9} \qquad \text{Multiplying to remove parentheses}$$

When clearing an equation of fractions, be sure to multiply *each* term by the LCM.

$$12 + 15 = 2x \qquad \text{Simplifying. Note that we have now cleared fractions.}$$

$$27 = 2x$$

$$\frac{27}{2} = x.$$

The solution is $\frac{27}{2}$.

*Do Exercise 1.*

**Objective**

**a**  Solve rational equations.

**For Extra Help**

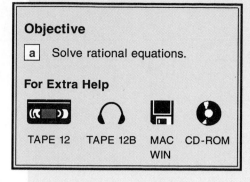

TAPE 12      TAPE 12B      MAC WIN      CD-ROM

1. Solve: $\frac{3}{4} + \frac{5}{8} = \frac{x}{12}$.

*Answer on page A-27*

**2.** Solve: $\dfrac{1}{x} = \dfrac{1}{6-x}$.

**Example 2** Solve: $\dfrac{1}{x} = \dfrac{1}{4-x}$.

The LCM is $x(4-x)$. We multiply by $x(4-x)$ on both sides:

$$\frac{1}{x} = \frac{1}{4-x}$$

$$x(4-x) \cdot \frac{1}{x} = x(4-x) \cdot \frac{1}{4-x} \qquad \text{Multiplying by the LCM on both sides}$$

$$4 - x = x \qquad\qquad\qquad \text{Simplifying}$$

$$4 = 2x$$

$$x = 2.$$

**CHECK:**

$$\frac{1}{x} = \frac{1}{4-x}$$

| $\dfrac{1}{2}$ | $\dfrac{1}{4-2}$ |
|---|---|
| | $\dfrac{1}{2}$  TRUE |

This checks, so the solution is 2.

*Do Exercise 2.*

**3.** Solve: $\dfrac{x}{4} - \dfrac{x}{6} = \dfrac{1}{8}$.

**Example 3** Solve: $\dfrac{x}{6} - \dfrac{x}{8} = \dfrac{1}{12}$.

The LCM is 24. We multiply by 24 on both sides:

$$\frac{x}{6} - \frac{x}{8} = \frac{1}{12}$$

$$24\left(\frac{x}{6} - \frac{x}{8}\right) = 24 \cdot \frac{1}{12} \qquad \text{Multiplying by the LCM on both sides}$$

$$24 \cdot \frac{x}{6} - 24 \cdot \frac{x}{8} = 24 \cdot \frac{1}{12} \qquad \text{Multiplying to remove parentheses}$$

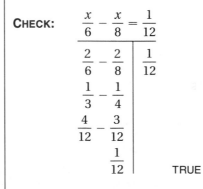

> Be sure to multiply *each* term by the LCM.

$$4x - 3x = 2 \qquad \text{Simplifying}$$

$$x = 2.$$

**CHECK:**

$$\frac{x}{6} - \frac{x}{8} = \frac{1}{12}$$

| $\dfrac{2}{6} - \dfrac{2}{8}$ | $\dfrac{1}{12}$ |
|---|---|
| $\dfrac{1}{3} - \dfrac{1}{4}$ | |
| $\dfrac{4}{12} - \dfrac{3}{12}$ | |
| $\dfrac{1}{12}$ | TRUE |

This checks, so the solution is 2.

*Do Exercise 3.*

*Answers on page A-27*

## Algebraic–Graphical Connection

We can obtain a visual check of the solutions of a rational equation by graphing. For example, consider the equation

$$\frac{x}{4} + \frac{x}{2} = 6.$$

We can examine the solution by graphing the equations

$$y = \frac{x}{4} + \frac{x}{2} \quad \text{and} \quad y = 6$$

using the same set of axes.

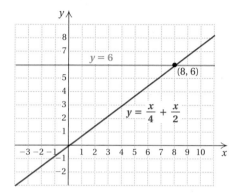

The $y$-values for each equation will be the same where the graphs intersect. The $x$-value of that point will yield that value, so it will be the solution of the equation. It appears from the graph that when $x = 8$, the value of $x/4 + x/2$ is 6. We can check by substitution:

$$\frac{x}{4} + \frac{x}{2} = \frac{8}{4} + \frac{8}{2} = 2 + 4 = 6.$$

Thus the solution is 8.

**Example 4**  Solve: $\dfrac{2}{3x} + \dfrac{1}{x} = 10$.

The LCM is $3x$. We multiply by $3x$ on both sides:

$$\frac{2}{3x} + \frac{1}{x} = 10$$

$$3x\left(\frac{2}{3x} + \frac{1}{x}\right) = 3x \cdot 10 \qquad \text{Multiplying by the LCM on both sides}$$

$$3x \cdot \frac{2}{3x} + 3x \cdot \frac{1}{x} = 3x \cdot 10 \qquad \text{Multiplying to remove parentheses}$$

$$2 + 3 = 30x \qquad \text{Simplifying}$$

$$5 = 30x$$

$$\frac{5}{30} = x$$

$$\frac{1}{6} = x.$$

We leave the check to the student. The solution is $\frac{1}{6}$.

*Do Exercise 4.*

*Answer on page A-27*

**5.** Solve: $x + \dfrac{1}{x} = 2$.

**Example 5** Solve: $x + \dfrac{6}{x} = -5$.

The LCM is $x$. We multiply by $x$ on both sides:

$$x + \frac{6}{x} = -5$$

$$x\left(x + \frac{6}{x}\right) = -5x \qquad \text{Multiplying by } x \text{ on both sides}$$

$$x \cdot x + x \cdot \frac{6}{x} = -5x \qquad \text{Note that each rational expression on the left is now multiplied by } x.$$

$$x^2 + 6 = -5x \qquad \text{Simplifying}$$

$$x^2 + 5x + 6 = 0 \qquad \text{Adding } 5x \text{ to get a 0 on one side}$$

$$(x + 3)(x + 2) = 0 \qquad \text{Factoring}$$

$$x + 3 = 0 \quad or \quad x + 2 = 0 \qquad \text{Using the principle of zero products}$$

$$x = -3 \quad or \qquad x = -2.$$

**CHECK:** For $-3$:

$$\begin{array}{c|c} x + \dfrac{6}{x} = -5 & \\ \hline -3 + \dfrac{6}{-3} & -5 \\ -3 - 2 & \\ -5 & \text{TRUE} \end{array}$$

For $-2$:

$$\begin{array}{c|c} x + \dfrac{6}{x} = -5 & \\ \hline -2 + \dfrac{6}{-2} & -5 \\ -2 - 3 & \\ -5 & \text{TRUE} \end{array}$$

Both of these check, so there are two solutions, $-3$ and $-2$.

*Answer on page A-27*

***Do Exercise 5.***

---

## Calculator Spotlight

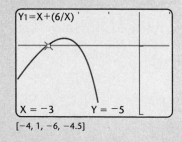

Checking Solutions Graphically. A grapher can be used to check the solutions of the equation in Example 5:

$$x + \frac{6}{x} = -5.$$

To do so, we graph

$$y_1 = x + \frac{6}{x} \quad \text{and} \quad y_2 = -5.$$

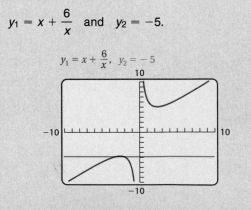

We see that the graphs appear to cross each other in the third quadrant. To get a better look, we change window settings and obtain the following graph.

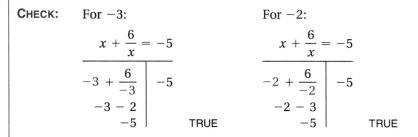

Next, we can use the CALC and VALUE features to confirm that the points of intersection occur at $x = -3$ and $x = -2$.

### Exercises

Use a grapher to check the solutions in each of the following.

1. Example 1
2. Margin Exercise 1
3. Example 3
4. Margin Exercise 5

When we multiply on both sides of an equation by the LCM, the resulting equation might have solutions that are *not* solutions of the original equation. Thus we must *always* check possible solutions in the original equation.

1. If you have carried out all algebraic procedures correctly, you need only check to see if a number makes a denominator 0 in the original equation. If it does make a denominator 0, it is *not* a solution.

2. To be sure that no computational errors have been made and that you indeed have a solution, a complete check is necessary, as we did in Chapter 2.

The next example illustrates the importance of checking all possible solutions.

**6.** Solve: $\dfrac{x^2}{x+2} = \dfrac{4}{x+2}$.

**Example 6**   Solve: $\dfrac{x^2}{x-1} = \dfrac{1}{x-1}$.

The LCM is $x - 1$. We multiply by $x - 1$ on both sides:

$$\frac{x^2}{x-1} = \frac{1}{x-1}$$

$$(x-1) \cdot \frac{x^2}{x-1} = (x-1) \cdot \frac{1}{x-1} \qquad \text{Multiplying by } x-1 \text{ on both sides}$$

$$x^2 = 1 \qquad \text{Simplifying}$$

$$x^2 - 1 = 0 \qquad \text{Subtracting 1 to get a 0 on one side}$$

$$(x-1)(x+1) = 0 \qquad \text{Factoring}$$

$$x - 1 = 0 \quad or \quad x + 1 = 0 \qquad \text{Using the principle of zero products}$$

$$x = 1 \quad or \qquad x = -1.$$

The numbers 1 and $-1$ are possible solutions. We look at the original equation and see that 1 makes a denominator 0 and is therefore not a solution. The number $-1$ checks and is a solution.

*Do Exercise 6.*

**7.** Solve: $\dfrac{4}{x-2} + \dfrac{1}{x+2} = \dfrac{26}{x^2-4}$.

**Calculator Spotlight**

Use a grapher to check the solution to Example 6.

**Example 7**   Solve: $\dfrac{3}{x-5} + \dfrac{1}{x+5} = \dfrac{2}{x^2-25}$.

The LCM is $(x-5)(x+5)$. We multiply by $(x-5)(x+5)$ on both sides:

$$(x-5)(x+5)\left(\frac{3}{x-5} + \frac{1}{x+5}\right) = (x-5)(x+5)\left(\frac{2}{x^2-25}\right)$$

Multiplying on both sides by the LCM

$$(x-5)(x+5)\cdot\frac{3}{x-5} + (x-5)(x+5)\cdot\frac{1}{x+5} = (x-5)(x+5)\cdot\frac{2}{x^2-25}$$

$$3(x+5) + (x-5) = 2 \qquad \text{Simplifying}$$

$$3x + 15 + x - 5 = 2 \qquad \text{Removing parentheses}$$

$$4x + 10 = 2$$

$$4x = -8$$

$$x = -2.$$

The check is left to the student. The number $-2$ checks and is the solution.

*CAUTION!*  We have introduced a new use of the LCM in this section. We previously used the LCM in adding or subtracting rational expressions. *Now* we have equations with equals signs. We clear fractions by multiplying on both sides of the equation by the LCM. This eliminates the denominators. Do *not* make the mistake of trying to clear fractions when you do not have an equation.

*Do Exercise 7.*

*Answers on page A-27*

# Improving Your Math Study Skills

## Are You Calculating or Solving?

At the beginning of this section, we noted that one of the common difficulties with this chapter is knowing for sure the task at hand. Are you combining expressions using operations to get another *rational expression,* or are you solving equations for which the results are numbers that are *solutions* of an equation? To learn to make these decisions, complete the following list by writing in the blank the type of answer you should get: "Rational expression" or "Solutions." You do not need to complete the mathematical operations.

| Task | Answer (Just write "Rational expression" or "Solutions.") |
|---|---|
| **1.** Add: $\dfrac{4}{x-2} + \dfrac{1}{x+2}$. | |
| **2.** Solve: $\dfrac{4}{x-2} = \dfrac{1}{x+2}$. | |
| **3.** Subtract: $\dfrac{4}{x-2} - \dfrac{1}{x+2}$. | |
| **4.** Multiply: $\dfrac{4}{x-2} \cdot \dfrac{1}{x+2}$. | |
| **5.** Divide: $\dfrac{4}{x-2} \div \dfrac{1}{x+2}$. | |
| **6.** Solve: $\dfrac{4}{x-2} + \dfrac{1}{x+2} = \dfrac{26}{x^2-4}$. | |
| **7.** Perform the indicated operations and simplify: $\dfrac{4}{x-2} + \dfrac{1}{x+2} - \dfrac{26}{x^2-4}$. | |
| **8.** Solve: $\dfrac{x^2}{x-1} = \dfrac{1}{x-1}$. | |
| **9.** Solve: $\dfrac{2}{y^2-25} = \dfrac{3}{y-5} + \dfrac{1}{y-5}$. | |
| **10.** Solve: $\dfrac{x}{x+4} - \dfrac{4}{x-4} = \dfrac{x^2+16}{x^2-16}$. | |
| **11.** Perform the indicated operations and simplify: $\dfrac{x}{x+4} - \dfrac{4}{x-4} - \dfrac{x^2+16}{x^2-16}$. | |
| **12.** Solve: $\dfrac{5}{y-3} - \dfrac{30}{y^2-9} = 1$. | |
| **13.** Add: $\dfrac{5}{y-3} + \dfrac{30}{y^2-9} + 1$. | |

# Exercise Set 6.6

**a** Solve. Don't forget to check!

**1.** $\dfrac{4}{5} - \dfrac{2}{3} = \dfrac{x}{9}$

**2.** $\dfrac{x}{20} = \dfrac{3}{8} - \dfrac{4}{5}$

**3.** $\dfrac{3}{5} + \dfrac{1}{8} = \dfrac{1}{x}$

**4.** $\dfrac{2}{3} + \dfrac{5}{6} = \dfrac{1}{x}$

**5.** $\dfrac{3}{8} + \dfrac{4}{5} = \dfrac{x}{20}$

**6.** $\dfrac{3}{5} + \dfrac{2}{3} = \dfrac{x}{9}$

**7.** $\dfrac{1}{x} = \dfrac{2}{3} - \dfrac{5}{6}$

**8.** $\dfrac{1}{x} = \dfrac{1}{8} - \dfrac{3}{5}$

**9.** $\dfrac{1}{6} + \dfrac{1}{8} = \dfrac{1}{t}$

**10.** $\dfrac{1}{8} + \dfrac{1}{12} = \dfrac{1}{t}$

**11.** $x + \dfrac{4}{x} = -5$

**12.** $\dfrac{10}{x} - x = 3$

**13.** $\dfrac{x}{4} - \dfrac{4}{x} = 0$

**14.** $\dfrac{x}{5} - \dfrac{5}{x} = 0$

**15.** $\dfrac{5}{x} = \dfrac{6}{x} - \dfrac{1}{3}$

**16.** $\dfrac{4}{x} = \dfrac{5}{x} - \dfrac{1}{2}$

**17.** $\dfrac{5}{3x} + \dfrac{3}{x} = 1$

**18.** $\dfrac{5}{2y} + \dfrac{8}{y} = 1$

**19.** $\dfrac{t-2}{t+3} = \dfrac{3}{8}$

**20.** $\dfrac{x-7}{x+2} = \dfrac{1}{4}$

**21.** $\dfrac{2}{x+1} = \dfrac{1}{x-2}$

**22.** $\dfrac{8}{y-3} = \dfrac{6}{y+4}$

**23.** $\dfrac{x}{6} - \dfrac{x}{10} = \dfrac{1}{6}$

**24.** $\dfrac{x}{8} - \dfrac{x}{12} = \dfrac{1}{8}$

**25.** $\dfrac{t+2}{5} - \dfrac{t-2}{4} = 1$

**26.** $\dfrac{x+1}{3} - \dfrac{x-1}{2} = 1$

**27.** $\dfrac{5}{x-1} = \dfrac{3}{x+2}$

**28.** $\dfrac{x-7}{x-9} = \dfrac{2}{x-9}$

**29.** $\dfrac{a-3}{3a+2} = \dfrac{1}{5}$

**30.** $\dfrac{x+7}{8x-5} = \dfrac{2}{3}$

**31.** $\dfrac{x-1}{x-5} = \dfrac{4}{x-5}$

**32.** $\dfrac{y+11}{y+8} = \dfrac{3}{y+8}$

**33.** $\dfrac{2}{x+3} = \dfrac{5}{x}$

**34.** $\dfrac{6}{y} = \dfrac{5}{y-8}$

**35.** $\dfrac{x-2}{x-3} = \dfrac{x-1}{x+1}$

**36.** $\dfrac{t+5}{t-2} = \dfrac{t-2}{t+4}$

**37.** $\dfrac{1}{x+3} + \dfrac{1}{x-3} = \dfrac{1}{x^2-9}$

**38.** $\dfrac{4}{x-3} + \dfrac{2x}{x^2-9} = \dfrac{1}{x+3}$

**39.** $\dfrac{x}{x+4} - \dfrac{4}{x-4} = \dfrac{x^2+16}{x^2-16}$

**40.** $\dfrac{5}{y-3} - \dfrac{30}{y^2-9} = 1$

**41.** $\dfrac{4-a}{8-a} = \dfrac{4}{a-8}$

**42.** $\dfrac{3}{x-7} = \dfrac{x+10}{x-7}$

**43.** $2 - \dfrac{a-2}{a+3} = \dfrac{a^2-4}{a+3}$

**44.** $\dfrac{5}{x-1} + x + 1 = \dfrac{5x+4}{x-1}$

---

**Skill Maintenance**

Simplify.

**45.** $(a^2b^5)^{-3}$  [4.2b]

**46.** $(x^{-2}y^{-3})^{-4}$  [4.2b]

**47.** $\left(\dfrac{2x}{t^2}\right)^4$  [4.2b]

**48.** $\left(\dfrac{y^3}{w^2}\right)^{-2}$  [4.2b]

**49.** $4x^{-5} \cdot 8x^{11}$  [4.1d]

**50.** $(8x^5y^{-4})^2$  [4.2b]

Find the intercepts. Then graph the equation.  [3.3a]

**51.** $5x + 10y = 20$

**52.** $2x - 4y = 8$

**53.** $10y - 4x = -20$

**54.** $y - 5x = 5$

---

**Synthesis**

**55.** ◈ Why is it especially important to check the possible solutions to a rational equation?

**56.** ◈ How can a graph be used to determine how many solutions an equation has?

Solve.

**57.** $\dfrac{4}{y-2} - \dfrac{2y-3}{y^2-4} = \dfrac{5}{y+2}$

**58.** $\dfrac{x}{x^2+3x-4} + \dfrac{x+1}{x^2+6x+8} = \dfrac{2x}{x^2+x-2}$

**59.** $\dfrac{x+1}{x+2} = \dfrac{x+3}{x+4}$

**60.** $\dfrac{x^2}{x^2-4} = \dfrac{x}{x+2} - \dfrac{2x}{2-x}$

**61.** $4a - 3 = \dfrac{a+13}{a+1}$

**62.** $\dfrac{3x-9}{x-3} = \dfrac{5x-4}{2}$

**63.** $\dfrac{y^2-4}{y+3} = 2 - \dfrac{y-2}{y+3}$

**64.** $\dfrac{3a-5}{a^2+4a+3} + \dfrac{2a+2}{a+3} = \dfrac{a-3}{a+1}$

**65.** 〰 Use a grapher to check the solutions to Exercises 1–4.

**66.** 〰 Use a grapher to check the solutions to Exercises 13, 15, and 25.

Copyright © 1999 Addison Wesley Longman

# 6.7 Applications, Proportions, and Problem Solving

## a Solving Applied Problems

**Example 1** If 2 is subtracted from a number and then the reciprocal is found, the result is twice the reciprocal of the number itself. What is the number?

1. **Familiarize.** Let's try to guess such a number. Try 10: $10 - 2$ is 8, and the reciprocal of 8 is $\frac{1}{8}$. Two times the reciprocal of 10 is $2\left(\frac{1}{10}\right)$, or $\frac{1}{5}$. Since $\frac{1}{8} \neq \frac{1}{5}$, the number 10 does not check, but the process helps us understand the translation. Let $x =$ the number.

2. **Translate.** From the *Familiarize* step, we get the following translation. Subtracting 2 from the number gives us $x - 2$. Twice the reciprocal of the original number is $2(1/x)$.

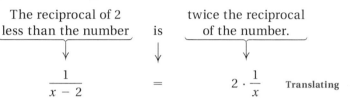

| The reciprocal of 2 less than the number | is | twice the reciprocal of the number. | |
| :---: | :---: | :---: | :--- |
| $\dfrac{1}{x-2}$ | $=$ | $2 \cdot \dfrac{1}{x}$ | **Translating** |

3. **Solve.** We solve the equation. The LCM is $x(x - 2)$.

$$x(x - 2) \cdot \frac{1}{x - 2} = x(x - 2) \cdot \frac{2}{x} \qquad \text{\textbf{Multiplying by the LCM}}$$

$$x = 2(x - 2) \qquad \text{\textbf{Simplifying}}$$

$$x = 2x - 4$$

$$-x = -4$$

$$x = 4$$

4. **Check.** We go back to the original problem. The number to be checked is 4. Two from 4 is 2. The reciprocal of 2 is $\frac{1}{2}$. The reciprocal of the number itself is $\frac{1}{4}$. Since $\frac{1}{2}$ is twice $\frac{1}{4}$, the conditions are satisfied.

5. **State.** The number is 4.

*Do Exercise 1.*

**Example 2** *Animal Speeds.* A cheetah can run 20 mph faster than a lion. A cheetah can run 7 mi in the same time that a lion can run 5 mi. Find the speed of each animal.

1. **Familiarize.** We first make a drawing. Let $r =$ the speed of the lion. Then $r + 20 =$ the speed of the cheetah.

5 mi, $r$ mph

7 mi, $r + 20$ mph

---

**Objectives**

**a** Solve applied problems using rational equations.

**b** Solve proportion problems.

**For Extra Help**

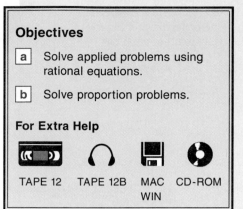

TAPE 12    TAPE 12B    MAC WIN    CD-ROM

1. The reciprocal of 2 more than a number is three times the reciprocal of the number. Find the number.

*Answer on page A-27*

**2.** *Car Speeds.* One car travels 20 km/h faster than another. While one car travels 240 km, the other travels 160 km. Find the speed of each car.

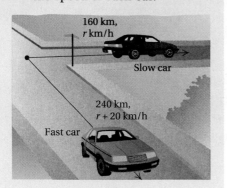

160 km,
r km/h

Slow car

240 km,
r + 20 km/h

Fast car

Recall that sometimes we need to find a formula in order to solve an application. A formula that relates the notions of distance, speed, and time is $d = rt$, or

$$Distance = Speed \cdot Time.$$

(Indeed, you may need to look up such a formula.)

Since each animal travels the same length of time, we can use just $t$ for time. We organize the information in a chart, as follows.

$$d \;=\; r \;\cdot\; t$$

|  | Distance | Speed | Time |  |
|---|---|---|---|---|
| Lion | 5 | $r$ | $t$ | $\rightarrow 5 = rt$ |
| Cheetah | 7 | $r + 20$ | $t$ | $\rightarrow 7 = (r + 20)t$ |

**2. Translate.** We can apply the formula $d = rt$ along the rows of the table to obtain two equations:

$$5 = rt, \qquad \textbf{(1)}$$
$$7 = (r + 20)t. \qquad \textbf{(2)}$$

We know that the animals travel for the same length of time. Thus if we solve each equation for $t$ and set the results equal to each other, we get an equation in terms of $r$.

Solving $5 = rt$ for $t$: $\qquad t = \dfrac{5}{r}$

Solving $7 = (r + 20)t$ for $t$: $\quad t = \dfrac{7}{r + 20}$

Since the times are the same, we have the following equation:

$$\frac{5}{r} = \frac{7}{r + 20}.$$

**3. Solve.** To solve the equation, we first multiply on both sides by the LCM, which is $r(r + 20)$:

$$r(r + 20) \cdot \frac{5}{r} = r(r + 20) \cdot \frac{7}{r + 20} \qquad \text{Multiplying on both sides by the LCM, which is } r(r + 20)$$

$$5(r + 20) = 7r \qquad \text{Simplifying}$$

$$5r + 100 = 7r \qquad \text{Removing parentheses}$$

$$100 = 2r$$

$$50 = r.$$

We now have a possible solution. The speed of the lion is 50 mph, and the speed of the cheetah is $r = 50 + 20$, or 70 mph.

**4. Check.** We first reread the problem to see what we were to find. We check the speeds of 50 for the lion and 70 for the cheetah. The cheetah does travel 20 mph faster than the lion and will travel farther than the lion, which runs at a slower speed. If the cheetah runs 7 mi at 70 mph, the time it has traveled is $\frac{7}{70}$, or $\frac{1}{10}$ hr. If the lion runs 5 mi at 50 mph, the time it has traveled is $\frac{5}{50}$, or $\frac{1}{10}$ hr. Since the times are the same, the speeds check.

**5. State.** The speed of the lion is 50 mph and the speed of the cheetah is 70 mph.

*Do Exercise 2.*

*Answer on page A-27*

**Example 3** *Recyclable Work.* Erin and Nick work as volunteers at a community recycling depot. Erin can sort a morning's accumulation of recyclables in 4 hr, while Nick requires 6 hr to do the same job. How long would it take them, working together, to sort the recyclables?

1. **Familiarize.** We familiarize ourselves with the problem by considering two *incorrect* ways of translating the problem to mathematical language.

   a) A common *incorrect* way to translate the problem is to add the two times: 4 hr + 6 hr = 10 hr. Let's think about this. Erin can do the job alone in 4 hr. If Erin and Nick work together, whatever time it takes them should be *less* than 4 hr. Thus we reject 10 hr as a solution, but we do have a partial check on any answer we get. The answer should be less than 4 hr.

   b) Another *incorrect* way to translate the problem is as follows. Suppose the two people split up the sorting job in such a way that Erin does half the sorting and Nick does the other half. Then

   $$\text{Erin sorts } \frac{1}{2} \text{ the recyclables in } \frac{1}{2}(4 \text{ hr}), \text{ or 2 hr,}$$

   and    $$\text{Nick sorts } \frac{1}{2} \text{ the recyclables in } \frac{1}{2}(6 \text{ hr}), \text{ or 3 hr.}$$

But time is wasted since Erin would finish 1 hr earlier than Nick. In effect, they have not worked together to get the job done as fast as possible. If Erin helps Nick after completing her half, the entire job could be done in a time somewhere between 2 hr and 3 hr.

We proceed to a translation by considering how much of the job is finished in 1 hr, 2 hr, 3 hr, and so on. It takes Erin 4 hr to do the sorting job alone. Then, in 1 hr, she can do $\frac{1}{4}$ of the job. It takes Nick 6 hr to do the job alone. Then, in 1 hr, he can do $\frac{1}{6}$ of the job. Working together, they can do

$$\frac{1}{4} + \frac{1}{6}, \text{ or } \frac{5}{12} \text{ of the job in 1 hr.}$$

In 2 hr, Erin can do $2\left(\frac{1}{4}\right)$ of the job and Nick can do $2\left(\frac{1}{6}\right)$ of the job. Working together, they can do

$$2\left(\frac{1}{4}\right) + 2\left(\frac{1}{6}\right), \text{ or } \frac{5}{6} \text{ of the job in 2 hr.}$$

Continuing this reasoning, we can create a table like the following one.

| Time | Fraction of the Job Completed | | |
|------|------|------|------|
| | Erin | Nick | Together |
| 1 hr | $\dfrac{1}{4}$ | $\dfrac{1}{6}$ | $\dfrac{1}{4} + \dfrac{1}{6}$, or $\dfrac{5}{12}$ |
| 2 hr | $2\left(\dfrac{1}{4}\right)$ | $2\left(\dfrac{1}{6}\right)$ | $2\left(\dfrac{1}{4}\right) + 2\left(\dfrac{1}{6}\right)$, or $\dfrac{5}{6}$ |
| 3 hr | $3\left(\dfrac{1}{4}\right)$ | $3\left(\dfrac{1}{6}\right)$ | $3\left(\dfrac{1}{4}\right) + 3\left(\dfrac{1}{6}\right)$, or $1\dfrac{1}{4}$ |
| $t$ hr | $t\left(\dfrac{1}{4}\right)$ | $t\left(\dfrac{1}{6}\right)$ | $t\left(\dfrac{1}{4}\right) + t\left(\dfrac{1}{6}\right)$ |

3. By checking work records, a contractor finds that it takes Eduardo 6 hr to construct a wall of a certain size. It takes Yolanda 8 hr to construct the same wall. How long would it take if they worked together?

*Answer on page A-27*

## Calculator Spotlight

〜 Work Problems on a Calculator.
Using a reciprocal key $\boxed{x^{-1}}$ on a calculator can make it easier to solve problems when using the work formula.

**Example.** It takes Bianca 6.75 hr, working alone, to keyboard a financial statement. It takes Arturo 7 hr, working alone, and Alani 8 hr, working alone, to produce the same statement. How long does it take if they all work together?

Let $t =$ the time it takes them to do the job working together. Then

$$\frac{t}{6.75} + \frac{t}{7} + \frac{t}{8} = 1, \quad \text{or}$$

$$\frac{1}{6.75} + \frac{1}{7} + \frac{1}{8} = \frac{1}{t}.$$

We calculate $1/t$ using the reciprocal key:

$$\frac{1}{t} = 6.75^{-1} + 7^{-1} + 8^{-1}$$

$$\frac{1}{t} = 0.416005291.$$

This is $1/t$, so we press the reciprocal key $\boxed{x^{-1}}$ again to obtain

$$t = 2.40381558.$$

The financial statement will take about 2.4 hr to complete if they all work together.

**Exercises**

Solve.

1. Irwin can edge a lawn in 45 min if he works alone. It takes Hannah 38 min to edge the same lawn if she works alone. How long would it take if they work together?

2. Five sorority sisters work on invitations to a dance. Working alone, it would take them, respectively, 2.8, 3.1, 2.2, 3.4, and 2.4 hr to do the job. How long would it take if they work together?

---

From the table, we see that if they work 3 hr, the fraction of the job completed is $1\frac{1}{4}$, which is more of the job than needs to be done. We see again that the answer is somewhere between 2 hr and 3 hr. What we want is a number $t$ such that the fraction of the job that gets completed is 1; that is, the job is just completed.

2. **Translate.** From the table, we see that the time we want is some number $t$ for which

$$t\left(\frac{1}{4}\right) + t\left(\frac{1}{6}\right) = 1, \quad \text{or} \quad \frac{t}{4} + \frac{t}{6} = 1,$$

where 1 represents the idea that the entire job is completed in time $t$.

3. **Solve.** We solve the equation:

$$12\left(\frac{t}{4} + \frac{t}{6}\right) = 12 \cdot 1 \qquad \begin{array}{l}\text{Multiplying by the LCM,} \\ \text{which is } 2 \cdot 2 \cdot 3, \text{ or } 12\end{array}$$

$$12 \cdot \frac{t}{4} + 12 \cdot \frac{t}{6} = 12$$

$$3t + 2t = 12$$

$$5t = 12$$

$$t = \frac{12}{5}, \text{ or } 2\frac{2}{5} \text{ hr.}$$

4. **Check.** The check can be done by recalculating:

$$\frac{12}{5}\left(\frac{1}{4}\right) + \frac{12}{5}\left(\frac{1}{6}\right) = \frac{3}{5} + \frac{2}{5} = \frac{5}{5} = 1.$$

We also have another check in what we learned from the *Familiarize* step. The answer, $2\frac{2}{5}$ hr, is between 2 hr and 3 hr (see the table), and it is less than 4 hr, the time it takes Erin working alone.

5. **State.** It takes $2\frac{2}{5}$ hr for them to do the sorting, working together.

---

▶ **THE WORK PRINCIPLE**

Suppose $a =$ the time it takes A to do a job, $b =$ the time it takes B to do the same job, and $t =$ the time it takes them to do the same job working together. Then

$$\frac{t}{a} + \frac{t}{b} = 1, \quad \text{or} \quad \frac{1}{a} + \frac{1}{b} = \frac{1}{t}.$$

---

*Do Exercise 3 on the preceding page.*

## b ▏ Applications Involving Proportions

We now consider applications with proportions. A **proportion** involves ratios. A **ratio** of two quantities is their quotient. For example, 73% is the ratio of 73 to 100, $\frac{73}{100}$. The ratio of two different kinds of measure is called a **rate**. Suppose an animal travels 720 ft in 2.5 hr. Its **rate**, or **speed**, is then

$$\frac{720 \text{ ft}}{2.5 \text{ hr}} = 288 \frac{\text{ft}}{\text{hr}}.$$

*Do Exercises 4–7 on the following page.*

> An equality of ratios, $A/B = C/D$, is called a **proportion**. The numbers named in a proportion are said to be **proportional**.

Proportions can be used to solve applications by expressing a single ratio in two ways.

**Example 4**   *Gas Mileage.* A Ford Taurus can travel 135 mi of city driving on 6 gal of gas (*Source:* Ford Motor Company). How much gas would be required for 360 mi of city driving?

1. **Familiarize.** We know that the Taurus can travel 135 mi on 6 gal of gas. Thus we can set up ratios, letting $x =$ the amount of gas required to drive 360 mi.

2. **Translate.** We assume that the car uses gas at the same rate throughout the 360 miles. Thus the ratios are the same and we can write a proportion. Note that the units of *mileage* are in the numerators and the units of *gasoline* are in the denominators.

$$\text{Miles} \longrightarrow \frac{135}{6} = \frac{360}{x} \longleftarrow \text{Miles}$$
$$\text{Gas} \longrightarrow \phantom{\frac{135}{6}} \phantom{=} \phantom{\frac{360}{x}} \longleftarrow \text{Gas}$$

3. **Solve.** To solve for $x$, we multiply on both sides by the LCM, which is $6x$:

$$6x \cdot \frac{135}{6} = 6x \cdot \frac{360}{x}$$

$$135x = 2160 \qquad \text{Simplifying}$$

$$\frac{135x}{135} = \frac{2160}{135} \qquad \text{Dividing by 135}$$

$$x = 16. \qquad \text{Simplifying}$$

We can also use **cross products** to solve the proportion:

$$\frac{135}{6} = \frac{360}{x} \qquad \text{135}x \text{ and } 6 \cdot 360 \text{ are called cross products.}$$

$$135x = 6 \cdot 360 \qquad \text{Equating the cross products}$$

$$\frac{135x}{135} = \frac{6 \cdot 360}{135} \qquad \text{Dividing by 135}$$

$$x = 16.$$

4. **Check.** We leave the check to the student.

5. **State.** The Taurus will require 16 gal of gas for 360 mi of city driving.

*Do Exercise 8.*

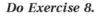

4. Find the ratio of 145 km to 2.5 liters (L).

5. *Batting Average.* Recently, a baseball player got 7 hits in 25 times at bat. What was the rate, or batting average, in hits per times at bat?

6. Impulses in nerve fibers travel 310 km in 2.5 hr. What is the rate, or speed, in kilometers per hour?

7. A lake of area 550 yd² contains 1320 fish. What is the population density of the lake in fish per square yard?

8. *Gas Mileage.* An Oldsmobile Achieva can travel 576 mi of interstate driving on 18 gal of gas (*Source:* General Motors Corporation). How much gas would be required for 2592 mi of interstate driving?

*Answers on page A-27*

**9.** In 1997, Mark McGwire of the Oakland Athletics (and later with the St. Louis Cardinals) had 27 home runs after 77 games.

   **a)** At this rate, how many home runs could McGwire hit in 162 games?

   **b)** Could it be predicted that he would break Maris's record? (McGwire actually completed the season hitting a major-league high of 58 home runs.) (**Source:** Major League Baseball)

**10.** A sample of 184 light bulbs contained 6 defective bulbs. How many would you expect to find in a sample of 1288 bulbs?

---

Proportions can be used in many types of applications. In the following example, we predict whether an important home-run record can be broken.

**Example 5** *Home-Run Record.* Baseball fans enjoy speculating about records being broken. Roger Maris hit 61 home runs in 1961 to claim the major-league season home-run record. In 1997, Ken Griffey, Jr., had 20 home runs after 44 games. The season consists of 162 games. At this rate, could it be predicted that Griffey would break Maris's record? (**Source:** Major League Baseball)

**1. Familiarize.** Let's assume that Griffey's rate of hitting 20 home runs in 44 games will continue for the 162-game season. We let $H =$ the number of home runs that Griffey can hit in 162 games.

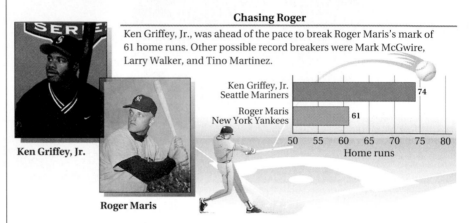

**Chasing Roger**

Ken Griffey, Jr., was ahead of the pace to break Roger Maris's mark of 61 home runs. Other possible record breakers were Mark McGwire, Larry Walker, and Tino Martinez.

Ken Griffey, Jr. (Seattle Mariners): 74
Roger Maris (New York Yankees): 61

Ken Griffey, Jr.

Roger Maris

**2. Translate.** Assuming the rate of hitting home runs continues, the ratios are the same, and we have the proportion

$$\text{Number of home runs} \rightarrow \frac{H}{162} = \frac{20}{44} \leftarrow \text{Number of home runs} \atop \leftarrow \text{Number of games}$$

**3. Solve.** We solve the equation:

$$\frac{H}{162} = \frac{20}{44}$$

$$44H = 162 \cdot 20 \qquad \text{Equating cross products}$$

$$\frac{44H}{44} = \frac{162 \cdot 20}{44} \qquad \text{Dividing by 44}$$

$$H \approx 73.64.$$

**4. Check.** We leave the check to the student.

**5. State.** We can indeed predict that Griffey, Jr., will hit about 74 home runs and break Maris's record. (Griffey actually completed the season with 56 home runs, having hit only 8 home runs in June and July.)

*Do Exercises 9 and 10.*

---

*Answers on page A-27*

**Example 6** *Estimating Wildlife Populations.* To determine the number of fish in a lake, a park ranger catches 225 fish, tags them, and throws them back into the lake. Later, 108 fish are caught, and 15 of them are found to be tagged. Estimate how many fish are in the lake.

1. **Familiarize.** The ratio of fish tagged to the total number of fish in the lake, $F$, is $\frac{225}{F}$. Of the 108 fish caught later, 15 fish were tagged. The ratio of fish tagged to fish caught is $\frac{15}{108}$.

2. **Translate.** Assuming that the two ratios are the same, we can translate to a proportion.

$$\text{Fish tagged originally} \rightarrow \frac{225}{F} = \frac{15}{108} \leftarrow \text{Tagged fish caught later}$$
$$\text{Fish in lake} \rightarrow \phantom{\frac{225}{F}} \phantom{=} \phantom{\frac{15}{}} \leftarrow \text{Fish caught later}$$

3. **Solve.** We solve the proportion. We multiply by the LCM, which is $108F$:

$$108F \cdot \frac{225}{F} = 108F \cdot \frac{15}{108} \qquad \text{Multiplying by } 108F$$

$$108 \cdot 225 = F \cdot 15$$

$$\frac{108 \cdot 225}{15} = F \qquad \text{Dividing by 15}$$

$$1620 = F.$$

4. **Check.** We leave the check to the student.

5. **State.** We estimate that there are about 1620 fish in the lake.

*Do Exercise 11.*

## Similar Triangles

Proportions also occur geometrically with *similar triangles.* Although similar triangles have the same shape, their sizes may be different.

$$\frac{a}{r} = \frac{b}{s} = \frac{c}{t}$$

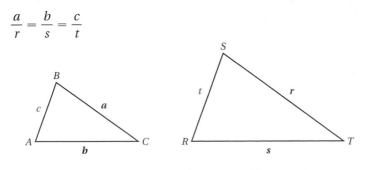

> In **similar triangles,** corresponding angles have the same measure and the lengths of corresponding sides are proportional.

**11.** To determine the number of deer in a forest, a conservationist catches 612 deer, tags them, and lets them loose. Later, 244 deer are caught, 72 of which are tagged. Estimate how many deer are in the forest.

*Answer on page A-27*

**12.** *Height of a Flagpole.* How high is a flagpole that casts a 45-ft shadow at the same time that a 5.5-ft woman casts a 10-ft shadow?

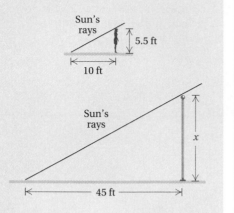

**13.** *F-106 Blueprint.* Referring to Example 8, find the length $x$ on the plane.

---

**Example 7** Triangles $ABC$ and $XYZ$ below are similar triangles. Solve for $z$ if $x = 10$, $a = 8$, and $c = 5$.

We make a sketch, write a proportion, and then solve. Note that side $a$ is always opposite angle $A$, side $x$ is always opposite angle $X$, and so on.

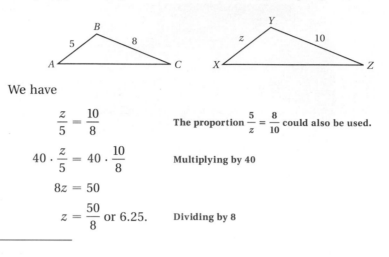

We have

$$\frac{z}{5} = \frac{10}{8} \qquad \text{The proportion } \frac{5}{z} = \frac{8}{10} \text{ could also be used.}$$

$$40 \cdot \frac{z}{5} = 40 \cdot \frac{10}{8} \qquad \text{Multiplying by 40}$$

$$8z = 50$$

$$z = \frac{50}{8} \text{ or } 6.25. \qquad \text{Dividing by 8}$$

**Example 8** *F-106 Blueprint.* A blueprint for an F-106 Delta Dart fighter plane is a scale drawing, as shown below. Each wing has a triangular shape. The blueprint shows similar triangles. Find the length of side $a$ of the wing.

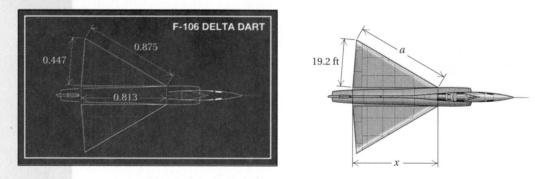

We let $a$ = the length of the wing. Thus we have the proportion

Length on the blueprint → $\dfrac{0.447}{19.2} = \dfrac{0.875}{a}$ ← Length on the blueprint
Length of the wing →　　　　　　　　 ← Length of the wing

*Solve*: $0.447 \cdot a = 19.2 \cdot 0.875$ 　 **Equating cross products**

$$a = \frac{19.2 \cdot 0.875}{0.447} \qquad \text{Dividing by 0.447}$$

$$a \approx 37.6 \text{ ft.}$$

The length of side $a$ of the wing is about 37.6 ft.

***Do Exercises 12 and 13.***

*Answers on page A-27*

# Exercise Set 6.7

**a** Solve.

**1.** The reciprocal of 6 plus the reciprocal of 8 is the reciprocal of what number?

**2.** The reciprocal of 5 plus the reciprocal of 4 is the reciprocal of what number?

**3.** One number is 5 more than another. The quotient of the larger divided by the smaller is $\frac{4}{3}$. Find the numbers.

**4.** One number is 4 more than another. The quotient of the larger divided by the smaller is $\frac{5}{2}$. Find the numbers.

**5.** *Car Speeds.* Rick drives his four-wheel-drive truck 40 km/h faster than Sarah drives her Saturn. While Sarah travels 150 km, Rick travels 350 km. Find their speeds.

Complete this table and the equations as part of the *Familiarize* step.

| $d$ | $=$ | $r$ | $\cdot$ | $t$ | |
|-----|-----------|-------|------|--------|
| | **Distance** | **Speed** | **Time** | |
| **Car** | 150 | $r$ | | $\rightarrow 150 = r(\quad)$ |
| **Truck** | 350 | | $t$ | $\rightarrow 350 = (\quad)t$ |

150 km, $r$ km/h

Sarah's car

Rick's truck

350 km, $r + 40$ km/h

**6.** *Car Speeds.* A passenger car travels 30 km/h faster than a delivery truck. While the car goes 400 km, the truck goes 250 km. Find their speeds.

**7.** *Train Speeds.* The speed of a freight train is 14 mph slower than the speed of a passenger train. The freight train travels 330 mi in the same time that it takes the passenger train to travel 400 mi. Find the speed of each train.

Complete this table and the equations as part of the *Familiarize* step.

| $d$ | $=$ | $r$ | $\cdot$ | $t$ | |
|-----|-----------|-------|------|--------|
| | **Distance** | **Speed** | **Time** | |
| **Freight** | 330 | | $t$ | $\rightarrow 330 = (\quad)t$ |
| **Passenger** | 400 | $r$ | | $\rightarrow 400 = r(\quad)$ |

**8.** *Train Speeds.* The speed of a freight train is 15 mph slower than the speed of a passenger train. The freight train travels 390 mi in the same time that it takes the passenger train to travel 480 mi. Find the speed of each train.

**9.** A long-distance trucker traveled 120 mi in one direction during a snowstorm. The return trip in rainy weather was accomplished at double the speed and took 3 hr less time. Find the speed going.

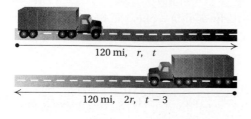

120 mi, $r$, $t$

120 mi, $2r$, $t - 3$

**10.** After making a trip of 126 mi, a person found that the trip would have taken 1 hr less time by increasing the speed by 8 mph. What was the actual speed?

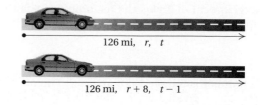

126 mi, $r$, $t$

126 mi, $r + 8$, $t - 1$

**11.** The Brother MFC4500 can fax a year-end report in 10 min while the Xerox 850 can fax the same report in 8 min. How long would it take the two machines, working together, to fax the report? (Assume that the recipient has two machines for incoming faxes.)

**12.** Zack mows the backyard in 40 min, while Angela can mow the same yard in 50 min. How long would it take them, working together with two mowers, to mow the yard?

**13.** By checking work records, a plumber finds that Rory can fit a kitchen in 12 hr. Mira can do the same job in 9 hr. How long would it take if they worked together?

**14.** Morgan can proofread 25 pages in 40 min. Shelby can proofread the same 25 pages in 30 min. How long would it take them, working together, to proofread 25 pages?

**b** Find the ratio of the following. Simplify, if possible.

**15.** 54 days, 6 days

**16.** 800 mi, 50 gal

**17.** A black racer snake travels 4.6 km in 2 hr. What is the speed in kilometers per hour?

**18.** *Speed of Light.* Light travels 558,000 mi in 3 sec. What is the speed in miles per second?

Solve.

**19.** A 120-lb person should eat a minimum of 44 g of protein each day. How much protein should a 180-lb person eat each day?

**20.** *Coffee Beans.* The coffee beans from 14 trees are required to produce 7.7 kg of coffee (this is the average amount that each person in the United States drinks each year). How many trees are required to produce 320 kg of coffee?

Copyright © 1999 Addison Wesley Longman

**21.** A student traveled 234 km in 14 days. At this same rate, how far would the student travel in 42 days?

**22.** In a potato bread recipe, the ratio of milk to flour is $\frac{3}{13}$. If 5 cups of milk are used, how many cups of flour are used?

**23.** A sample of 144 firecrackers contained 9 "duds." How many duds would you expect in a sample of 3200 firecrackers?

**24.** *Grass Seed.* It takes 60 oz of grass seed to seed 3000 ft$^2$ of lawn. At this rate, how much would be needed to seed 5000 ft$^2$ of lawn?

**25.** *Home Runs.* In 1997, Tino Martinez of the New York Yankees had 17 home runs after 44 games (***Source:*** Major League Baseball).

  **a)** At this rate, how many home runs could Martinez hit in 162 games?
  **b)** Could it be predicted that Martinez would break Maris's record of 61 home runs in a season?

**26.** *Home Runs.* In 1997, Larry Walker of the Colorado Rockies had 14 home runs after 40 games (***Source:*** Major League Baseball).

  **a)** At this rate, how many home runs could Walker hit in 162 games?
  **b)** Could it be predicted that Walker would break Maris's record of 61 home runs in a season?

**27.** *Estimating Whale Population.* To determine the number of blue whales in the world's oceans, marine biologists tag 500 blue whales in various parts of the world. Later, 400 blue whales are checked, and it is found that 20 of them are tagged. Estimate the blue whale population.

**28.** *Estimating Trout Population.* To determine the number of trout in a lake, a conservationist catches 112 trout, tags them, and throws them back into the lake. Later, 82 trout are caught; 32 of them are tagged. Estimate the number of trout in the lake.

**29.** *Weight on Mars.* The ratio of the weight of an object on Mars to the weight of an object on Earth is 0.4 to 1.

  **a)** How much would a 12-ton rocket weigh on Mars?
  **b)** How much would a 120-lb astronaut weigh on Mars?

**30.** *Weight on Moon.* The ratio of the weight of an object on the moon to the weight of an object on Earth is 0.16 to 1.

  **a)** How much would a 12-ton rocket weigh on the moon?
  **b)** How much would a 180-lb astronaut weigh on the moon?

**31.** A basketball team has 12 more games to play. They have won 25 of the 36 games they have played. How many more games must they win in order to finish with a 0.750 record?

**32.** Simplest fractional notation for a rational number is $\frac{9}{17}$. Find an equal ratio in which the sum of the numerator and the denominator is 104.

For each pair of similar triangles, find the length of the indicated letter.

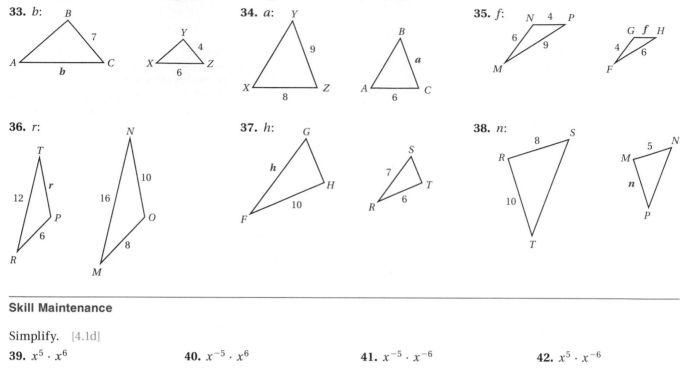

**33.** $b$:

**34.** $a$:

**35.** $f$:

**36.** $r$:

**37.** $h$:

**38.** $n$:

---

**Skill Maintenance**

Simplify.   [4.1d]

**39.** $x^5 \cdot x^6$

**40.** $x^{-5} \cdot x^6$

**41.** $x^{-5} \cdot x^{-6}$

**42.** $x^5 \cdot x^{-6}$

---

**Synthesis**

**43.** ◈ Explain why it is incorrect to assume that two workers can complete a task twice as quickly as one person working alone.

**44.** ◈ Write a problem similar to Example 3 or Margin Exercise 3 for a classmate to solve. Design the problem so that the translation step is

$$\frac{t}{7} + \frac{t}{5} = 1.$$

**45.** Larry, Moe, and Curly are accountants who can complete a financial report together in 3 days. Larry can do the job in 8 days and Moe can do it in 10 days. How many days will it take Curly to complete the job?

**46.** Ann and Betty work together and complete a sales report in 4 hr. It would take Betty 6 hr longer, working alone, to do the job than it would Ann. How long would it take each of them to do the job working alone?

**47.** The denominator of a fraction is 1 more than the numerator. If 2 is subtracted from both the numerator and the denominator, the resulting fraction is $\frac{1}{2}$. Find the original fraction.

**48.** Express 100 as the sum of two numbers for which the ratio of one number, increased by 5, to the other number, decreased by 5, is 4.

**49.** How soon after 5 o'clock will the hands on a clock first be together?

**50.** Rachel allows herself 1 hr to reach a sales appointment 50 mi away. After she has driven 30 mi, she realizes that she must increase her speed by 15 mph in order to get there on time. What was her speed for the first 30 mi?

Copyright © 1999 Addison Wesley Longman

# 6.8 Formulas and Applications

**a** The use of formulas is important in many applications of mathematics. We use the following procedure to solve a rational formula for a letter.

---

To solve a rational formula for a given letter, identify the letter, and:

1. Multiply on both sides to clear fractions or decimals, if that is needed.
2. Multiply to remove parentheses, if necessary.
3. Get all terms with the letter to be solved for on one side of the equation and all other terms on the other side, using the addition principle.
4. Factor out the unknown.
5. Solve for the letter in question, using the multiplication principle.

---

**Example 1** *Gravitational Force.*
The gravitational force $f$ between planets of mass $M$ and $m$, at a distance $d$ from each other, is given by

$$f = \frac{kMm}{d^2},$$

where $k$ represents a fixed number constant. Solve for $m$.

We have

$$f \cdot d^2 = \frac{kMm}{d^2} \cdot d^2 \qquad \text{Multiplying by the LCM, } d^2$$

$$fd^2 = kMm \qquad \text{Simplifying}$$

$$\frac{fd^2}{kM} = m. \qquad \text{Dividing by } kM$$

*Do Exercise 1.*

**Example 2** *Area of a Trapezoid.* The area $A$ of a trapezoid is half the product of the height $h$ and the sum of the lengths $b_1$ and $b_2$ of the parallel sides. Solve for $b_2$.

$$A = \frac{1}{2}h(b_1 + b_2)$$

We consider $b_1$ and $b_2$ to be different variables (or constants). The letter $b_1$ represents the length of the first parallel side and $b_2$ represents the length of the second parallel side. The small numbers 1 and 2 are called **subscripts**. Subscripts are used to identify different variables with related meanings.

$$2 \cdot A = 2 \cdot \frac{1}{2}h(b_1 + b_2) \qquad \text{Multiplying by 2 to clear fractions}$$

$$2A = h(b_1 + b_2) \qquad \text{Simplifying}$$

**Objective**

**a** Solve a formula for a letter.

**For Extra Help**

TAPE 12    TAPE 13A    MAC WIN    CD-ROM

1. Solve for $M$: $f = \dfrac{kMm}{d^2}$.

*Answer on page A-27*

**2.** Solve for $b_1$: $A = \dfrac{1}{2}h(b_1 + b_2)$.

Then

$$2A = hb_1 + hb_2 \qquad \text{Using a distributive law to remove parentheses}$$

$$2A - hb_1 = hb_2 \qquad \text{Subtracting } hb_1$$

$$\dfrac{2A - hb_1}{h} = b_2. \qquad \text{Dividing by } h$$

**Do Exercise 2.**

**Example 3** *A Work Formula.* The following work formula was considered in Section 6.7. Solve it for $t$.

$$\frac{t}{a} + \frac{t}{b} = 1$$

We multiply by the LCM, which is $ab$:

$$ab \cdot \left( \frac{t}{a} + \frac{t}{b} \right) = ab \cdot 1 \qquad \text{Multiplying by } ab$$

**3.** Solve for $f$: $\dfrac{1}{p} + \dfrac{1}{q} = \dfrac{1}{f}$.
(This is an optics formula.)

$$ab \cdot \frac{t}{a} + ab \cdot \frac{t}{b} = ab \qquad \text{Using a distributive law to remove parentheses}$$

$$bt + at = ab \qquad \text{Simplifying}$$

$$(b + a)t = ab \qquad \text{Factoring out } t$$

$$t = \frac{ab}{b + a}. \qquad \text{Dividing by } b + a$$

**Do Exercise 3.**

In Examples 1 and 2, the letter for which we solved was on the right side of the equation. In Example 3, the letter was on the left. Since all equations are reversible, the location of the letter is a matter of choice.

---

**TIP FOR FORMULA SOLVING**

The variable to be solved for should be alone on one side of the equation, with *no* occurrence of that variable on the other side.

---

**4.** Solve for $b$: $Q = \dfrac{a - b}{2b}$.

**Example 4** Solve for $b$: $S = \dfrac{a + b}{3b}$.

We multiply by the LCM, which is $3b$:

$$3b \cdot S = 3b \cdot \frac{a + b}{3b} \qquad \text{Multiplying by } 3b$$

$$3bS = a + b \qquad \text{Simplifying}$$

If we divide by $3S$, we will have $b$ alone on the left, but we will still have a term with $b$ on the right.

$$3bS - b = a \qquad \text{Subtracting } b \text{ to get all terms involving } b \text{ on one side}$$

$$b(3S - 1) = a$$

$$b = \frac{a}{3S - 1}. \qquad \text{Dividing by } 3S - 1$$

**Do Exercise 4.**

*Answers on page A-27*

# Exercise Set 6.8

**a** Solve.

**1.** $S = 2\pi rh$, for $r$

**2.** $A = P(1 + rt)$, for $t$
(An interest formula)

**3.** $A = \frac{1}{2}bh$, for $b$
(The area of a triangle)

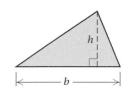

**4.** $s = \frac{1}{2}gt^2$, for $g$

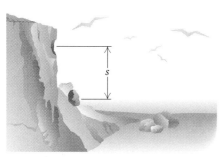

**5.** $S = 180(n - 2)$, for $n$

**6.** $S = \frac{n}{2}(a + l)$, for $a$

**7.** $V = \frac{1}{3}k(B + b + 4M)$, for $b$

**8.** $A = P + Prt$, for $P$
(*Hint*: Factor the right-hand side.)

**9.** $S(r - 1) = rl - a$, for $r$

**10.** $T = mg - mf$, for $m$
(*Hint*: Factor the right-hand side.)

**11.** $A = \frac{1}{2}h(b_1 + b_2)$, for $h$

**12.** $S = 2\pi r(r + h)$, for $h$
(The surface area of a right circular cylinder)

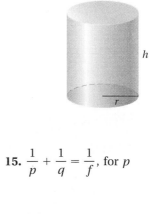

**13.** $\frac{A - B}{AB} = Q$, for $B$

**14.** $L = \frac{Mt + g}{t}$, for $t$

**15.** $\frac{1}{p} + \frac{1}{q} = \frac{1}{f}$, for $p$

**16.** $\frac{1}{a} + \frac{1}{b} = \frac{1}{t}$, for $b$

**17.** $\frac{A}{P} = 1 + r$, for $A$

**18.** $\frac{2A}{h} = a + b$, for $h$

**19.** $\dfrac{1}{R} = \dfrac{1}{r_1} + \dfrac{1}{r_2}$, for $R$
(An electricity formula)

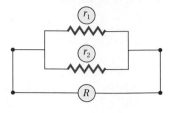

**20.** $\dfrac{1}{R} = \dfrac{1}{r_1} + \dfrac{1}{r_2}$, for $r_1$

**21.** $\dfrac{A}{B} = \dfrac{C}{D}$, for $D$

**22.** $q = \dfrac{VQ}{I}$, for $I$
(An engineering formula)

**23.** $h_1 = q\left(1 + \dfrac{h_2}{p}\right)$, for $h_2$

**24.** $S = \dfrac{a - ar^n}{1 - r}$, for $a$

**25.** $C = \dfrac{Ka - b}{a}$, for $a$

**26.** $Q = \dfrac{Pt - h}{t}$, for $t$

---

### Skill Maintenance

Subtract. [4.4c]

**27.** $(5x^3 - 7x^2 + 9) - (8x^3 - 2x^2 + 4)$

**28.** $(5x^4 - 6x^3 + 23x^2 - 79x + 24) -$
$(-18x^4 - 56x^3 + 84x - 17)$

Factor. [5.6a]

**29.** $x^2 - 4$

**30.** $30y^4 + 9y^2 - 12$

**31.** $49m^2 - 112mn + 64n^2$

**32.** $y^2 + 2y - 35$

**33.** $y^4 - 1$

**34.** $a^2 - 100b^2$

Divide and check. [4.8b]

**35.** $(x^3 + 4x - 4) \div (x - 2)$

**36.** $(x^4 - 6x^2 + 9) \div (x^2 - 3)$

---

### Synthesis

**37.** ◈ Describe a situation in which the result of Example 3,
$$t = \dfrac{ab}{a + b},$$
would be especially useful.

**38.** ◈ Which of the following is easier to solve for $x$?
$$\dfrac{1}{23} + \dfrac{1}{25} = \dfrac{1}{x} \quad \text{or} \quad \dfrac{1}{a} + \dfrac{1}{b} = \dfrac{1}{x}$$
Explain the reasons for your choice.

Solve.

**39.** $u = -F\left(E - \dfrac{P}{T}\right)$, for $T$

**40.** $l = a + (n - 1)d$, for $d$

**41.** The formula
$$N = \dfrac{(b + d)f_1 - v}{(b - v)f_2}$$
is used when monitoring the water in fisheries. Solve for $v$.

**42.** In
$$N = \dfrac{a}{c},$$
what is the effect on $N$ when $c$ increases? when $c$ decreases? Assume that $a$, $c$, and $N$ are positive.

Collaborative Learning Manual

Develop a formula for calculating the time required to complete a task when two or more people are working together.

# 6.9 Complex Rational Expressions

**a** A **complex rational expression,** or **complex fractional expression,** is a rational expression that has one or more rational expressions within its numerator or denominator. Here are some examples:

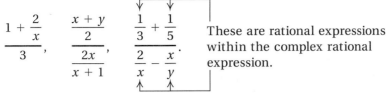

$$\frac{1 + \dfrac{2}{x}}{3}, \quad \frac{\dfrac{x + y}{2}}{\dfrac{2x}{x + 1}}, \quad \frac{\dfrac{1}{3} + \dfrac{1}{5}}{\dfrac{2}{x} - \dfrac{x}{y}}.$$

These are rational expressions within the complex rational expression.

### Objective

**a** Simplify complex rational expressions.

**For Extra Help**

TAPE 12   TAPE 13A   MAC WIN   CD-ROM

There are two methods to simplify complex rational expressions. We will consider them both. Use the one that works best for you or the one that your instructor directs you to use.

## Multiplying by the LCM of All the Denominators: Method 1

> **METHOD 1**
>
> To simplify a complex rational expression:
>
> 1. First, find the LCM of all the denominators of all the rational expressions occurring *within* both the numerator and the denominator of the complex rational expression.
> 2. Then multiply by 1 using LCM/LCM.
> 3. If possible, simplify by removing a factor of 1.

**Example 1** Simplify: $\dfrac{\dfrac{1}{2} + \dfrac{3}{4}}{\dfrac{5}{6} - \dfrac{3}{8}}$.

We have

$$\frac{\dfrac{1}{2} + \dfrac{3}{4}}{\dfrac{5}{6} - \dfrac{3}{8}}$$

The denominators *within* the complex rational expression are 2, 4, 6, and 8. The LCM of these denominators is 24. We multiply by 1 using $\frac{24}{24}$.

$$= \frac{\dfrac{1}{2} + \dfrac{3}{4}}{\dfrac{5}{6} - \dfrac{3}{8}} \cdot \frac{24}{24} \qquad \text{Multiplying by 1}$$

$$= \frac{\left(\dfrac{1}{2} + \dfrac{3}{4}\right)24}{\left(\dfrac{5}{6} - \dfrac{3}{8}\right)24} \begin{matrix} \leftarrow \text{Multiplying the numerator by 24} \\ \\ \leftarrow \text{Multiplying the denominator by 24} \end{matrix}$$

**1.** Simplify. Use method 1.

$$\dfrac{\dfrac{1}{3} + \dfrac{4}{5}}{\dfrac{7}{8} - \dfrac{5}{6}}$$

**2.** Simplify. Use method 1.

$$\dfrac{\dfrac{x}{2} + \dfrac{2x}{3}}{\dfrac{1}{x} - \dfrac{x}{2}}$$

**3.** Simplify. Use method 1.

$$\dfrac{1 + \dfrac{1}{x}}{1 - \dfrac{1}{x^2}}$$

Answers on page A-27

Using the distributive laws, we carry out the multiplications:

$$= \dfrac{\dfrac{1}{2}(24) + \dfrac{3}{4}(24)}{\dfrac{5}{6}(24) - \dfrac{3}{8}(24)}$$

$$= \dfrac{12 + 18}{20 - 9} \quad \textbf{Simplifying}$$

$$= \dfrac{30}{11}.$$

Multiplying in this manner has the effect of clearing fractions in both the top and the bottom of the complex rational expression.

*Do Exercise 1.*

**Example 2**  Simplify: $\dfrac{\dfrac{3}{x} + \dfrac{1}{2x}}{\dfrac{1}{3x} - \dfrac{3}{4x}}.$

The denominators within the complex expression are $x$, $2x$, $3x$, and $4x$. The LCM of these denominators is $12x$. We multiply by 1 using $12x/12x$.

$$\dfrac{\dfrac{3}{x} + \dfrac{1}{2x}}{\dfrac{1}{3x} - \dfrac{3}{4x}} \cdot \dfrac{12x}{12x} = \dfrac{\left(\dfrac{3}{x} + \dfrac{1}{2x}\right)12x}{\left(\dfrac{1}{3x} - \dfrac{3}{4x}\right)12x} = \dfrac{\dfrac{3}{x}(12x) + \dfrac{1}{2x}(12x)}{\dfrac{1}{3x}(12x) - \dfrac{3}{4x}(12x)}$$

$$= \dfrac{36 + 6}{4 - 9} = -\dfrac{42}{5}$$

*Do Exercise 2.*

**Example 3**  Simplify: $\dfrac{1 - \dfrac{1}{x}}{1 - \dfrac{1}{x^2}}.$

The denominators within the complex expression are $x$ and $x^2$. The LCM of these denominators is $x^2$. We multiply by 1 using $x^2/x^2$. Then, after obtaining a single rational expression, we simplify:

$$\dfrac{1 - \dfrac{1}{x}}{1 - \dfrac{1}{x^2}} \cdot \dfrac{x^2}{x^2} = \dfrac{\left(1 - \dfrac{1}{x}\right)x^2}{\left(1 - \dfrac{1}{x^2}\right)x^2} = \dfrac{1(x^2) - \dfrac{1}{x}(x^2)}{1(x^2) - \dfrac{1}{x^2}(x^2)} = \dfrac{x^2 - x}{x^2 - 1}$$

$$= \dfrac{x(x - 1)}{(x + 1)(x - 1)} = \dfrac{x}{x + 1}.$$

*Do Exercise 3.*

## Adding in the Numerator and the Denominator: Method 2

**METHOD 2**

To simplify a complex rational expression:

**1.** Add or subtract, as necessary, to get a single rational expression in the numerator.

**2.** Add or subtract, as necessary, to get a single rational expression in the denominator.

**3.** Divide the numerator by the denominator.

**4.** If possible, simplify by removing a factor of 1.

We will redo Examples 1–3 using this method.

**Example 4** Simplify: $\dfrac{\frac{1}{2} + \frac{3}{4}}{\frac{5}{6} - \frac{3}{8}}$.

We have

$$\frac{\frac{1}{2} + \frac{3}{4}}{\frac{5}{6} - \frac{3}{8}} = \frac{\frac{1}{2} \cdot \frac{2}{2} + \frac{3}{4}}{\frac{5}{6} \cdot \frac{4}{4} - \frac{3}{8} \cdot \frac{3}{3}} \quad \begin{array}{l} \leftarrow \text{Multiplying the } \frac{1}{2} \text{ by 1 to get} \\ \text{a common denominator} \\ \leftarrow \text{Multiplying the } \frac{5}{6} \text{ and the } \frac{3}{8} \text{ by 1 to get} \\ \text{a common denominator} \end{array}$$

$$= \frac{\frac{2}{4} + \frac{3}{4}}{\frac{20}{24} - \frac{9}{24}}$$

$$= \frac{\frac{5}{4}}{\frac{11}{24}} \qquad \begin{array}{l} \text{Adding in the numerator;} \\ \text{subtracting in the denominator} \end{array}$$

$$= \frac{5}{4} \cdot \frac{24}{11} \qquad \begin{array}{l} \text{Multiplying by the reciprocal} \\ \text{of the divisor} \end{array}$$

$$= \frac{5 \cdot 3 \cdot 2 \cdot 2 \cdot 2}{2 \cdot 2 \cdot 11} \qquad \text{Factoring}$$

$$= \frac{5 \cdot 3 \cdot 2 \cdot \cancel{2} \cdot \cancel{2}}{\cancel{2} \cdot \cancel{2} \cdot 11} \qquad \text{Removing a factor of 1: } \frac{2 \cdot 2}{2 \cdot 2} = 1$$

$$= \frac{30}{11}.$$

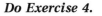

*Do Exercise 4.*

**4.** Simplify. Use method 2.

$$\frac{\frac{1}{3} + \frac{4}{5}}{\frac{7}{8} - \frac{5}{6}}$$

*Answer on page A-27*

**5.** Simplify. Use method 2.

$$\frac{\dfrac{x}{2}+\dfrac{2x}{3}}{\dfrac{1}{x}-\dfrac{x}{2}}$$

**Example 5** Simplify: $\dfrac{\dfrac{3}{x}+\dfrac{1}{2x}}{\dfrac{1}{3x}-\dfrac{3}{4x}}$.

We have

$$\frac{\dfrac{3}{x}+\dfrac{1}{2x}}{\dfrac{1}{3x}-\dfrac{3}{4x}}=\frac{\dfrac{3}{x}\cdot\dfrac{2}{2}+\dfrac{1}{2x}}{\dfrac{1}{3x}\cdot\dfrac{4}{4}-\dfrac{3}{4x}\cdot\dfrac{3}{3}}\quad\left.\begin{array}{l}\phantom{x}\end{array}\right\}$$

$\leftarrow$ Finding the LCD, $2x$, and multiplying by 1 in the numerator

$\leftarrow$ Finding the LCD, $12x$, and multiplying by 1 in the denominator

$$=\frac{\dfrac{6}{2x}+\dfrac{1}{2x}}{\dfrac{4}{12x}-\dfrac{9}{12x}}=\frac{\dfrac{7}{2x}}{\dfrac{-5}{12x}}$$

Adding in the numerator and subtracting in the denominator

$$=\frac{7}{2x}\cdot\frac{12x}{-5}$$

Multiplying by the reciprocal of the divisor

$$=\frac{7}{2x}\cdot\frac{6(2x)}{-5}$$

Factoring

$$=\frac{7}{2\!\!\!/x}\cdot\frac{6(2\!\!\!/x)}{-5}$$

Removing a factor of 1: $\dfrac{2x}{2x}=1$

$$=\frac{42}{-5}=-\frac{42}{5}.$$

*Do Exercise 5.*

**6.** Simplify. Use method 2.

$$\frac{1+\dfrac{1}{x}}{1-\dfrac{1}{x^2}}$$

**Example 6** Simplify: $\dfrac{1-\dfrac{1}{x}}{1-\dfrac{1}{x^2}}$.

We have

$$\frac{1-\dfrac{1}{x}}{1-\dfrac{1}{x^2}}=\frac{\dfrac{x}{x}-\dfrac{1}{x}}{\dfrac{x^2}{x^2}-\dfrac{1}{x^2}}\quad\left.\begin{array}{l}\phantom{x}\end{array}\right\}$$

$\leftarrow$ Finding the LCD, $x$, and multiplying by 1 in the numerator

$\leftarrow$ Finding the LCD, $x^2$, and multiplying by 1 in the denominator

$$=\frac{\dfrac{x-1}{x}}{\dfrac{x^2-1}{x^2}}$$

Subtracting in the numerator and subtracting in the denominator

$$=\frac{x-1}{x}\cdot\frac{x^2}{x^2-1}$$

Multiplying by the reciprocal of the divisor

$$=\frac{(x-1)x\cdot x}{x(x-1)(x+1)}$$

Factoring

$$=\frac{(x-1)\!\!\!/\,x\cdot x}{x\!\!\!/(x-1)\!\!\!/(x+1)}$$

Removing a factor of 1: $\dfrac{x(x-1)}{x(x-1)}=1$

$$=\frac{x}{x+1}.$$

*Do Exercise 6.*

*Answers on page A-27*

# Exercise Set 6.9

**a** Simplify.

**1.** $\dfrac{1 + \dfrac{9}{16}}{1 - \dfrac{3}{4}}$

**2.** $\dfrac{6 - \dfrac{3}{8}}{4 + \dfrac{5}{6}}$

**3.** $\dfrac{1 - \dfrac{3}{5}}{1 + \dfrac{1}{5}}$

**4.** $\dfrac{2 + \dfrac{2}{3}}{2 - \dfrac{2}{3}}$

**5.** $\dfrac{\dfrac{1}{2} + \dfrac{3}{4}}{\dfrac{5}{8} - \dfrac{5}{6}}$

**6.** $\dfrac{\dfrac{3}{4} + \dfrac{7}{8}}{\dfrac{2}{3} - \dfrac{5}{6}}$

**7.** $\dfrac{\dfrac{1}{x} + 3}{\dfrac{1}{x} - 5}$

**8.** $\dfrac{2 - \dfrac{1}{a}}{4 + \dfrac{1}{a}}$

**9.** $\dfrac{4 - \dfrac{1}{x^2}}{2 - \dfrac{1}{x}}$

**10.** $\dfrac{\dfrac{2}{y} + \dfrac{1}{2y}}{y + \dfrac{y}{2}}$

**11.** $\dfrac{8 + \dfrac{8}{d}}{1 + \dfrac{1}{d}}$

**12.** $\dfrac{3 + \dfrac{2}{t}}{3 - \dfrac{2}{t}}$

**13.** $\dfrac{\dfrac{x}{8} - \dfrac{8}{x}}{\dfrac{1}{8} + \dfrac{1}{x}}$

**14.** $\dfrac{\dfrac{2}{m} + \dfrac{m}{2}}{\dfrac{m}{3} - \dfrac{3}{m}}$

**15.** $\dfrac{1 + \dfrac{1}{y}}{1 - \dfrac{1}{y^2}}$

**16.** $\dfrac{\dfrac{1}{q^2} - 1}{\dfrac{1}{q} + 1}$

**17.** $\dfrac{\dfrac{1}{5} - \dfrac{1}{a}}{\dfrac{5 - a}{5}}$

**18.** $\dfrac{\dfrac{4}{t}}{4 + \dfrac{1}{t}}$

**19.** $\dfrac{\dfrac{1}{a}+\dfrac{1}{b}}{\dfrac{1}{a^2}-\dfrac{1}{b^2}}$

**20.** $\dfrac{\dfrac{1}{x^2}-\dfrac{1}{y^2}}{\dfrac{2}{x}-\dfrac{2}{y}}$

**21.** $\dfrac{\dfrac{p}{q}+\dfrac{q}{p}}{\dfrac{1}{p}+\dfrac{1}{q}}$

**22.** $\dfrac{x-3+\dfrac{2}{x}}{x-4+\dfrac{3}{x}}$

---

**Skill Maintenance**

Add.   [4.4a]

**23.** $(2x^3 - 4x^2 + x - 7) + (4x^4 + x^3 + 4x^2 + x)$

**24.** $(2x^3 - 4x^2 + x - 7) + (-2x^3 + 4x^2 - x + 7)$

Factor.   [5.6a]

**25.** $p^2 - 10p + 25$

**26.** $p^2 + 10p + 25$

**27.** $50p^2 - 100$

**28.** $5p^2 - 40p - 100$

Solve.   [5.8a]

**29.** The length of a rectangle is 3 yd greater than the width. The area of the rectangle is 10 yd². Find the perimeter.

**30.** A ladder of length 13 ft is placed against a building in such a way that the distance from the top of the ladder to the ground is 7 ft more than the distance from the bottom of the ladder to the building. Find these distances.

---

**Synthesis**

**31.** ◈ Why is factoring an important skill when simplifying complex rational expressions?

**32.** ◈ Why is the distributive law especially important when using method 1 of this section?

**33.** Find the reciprocal of $\dfrac{2}{x-1} - \dfrac{1}{3x-2}$.

Simplify.

**34.** $\dfrac{\dfrac{a}{b}+\dfrac{c}{d}}{\dfrac{b}{a}+\dfrac{d}{c}}$

**35.** $\dfrac{\dfrac{a}{b}-\dfrac{c}{d}}{\dfrac{b}{a}-\dfrac{d}{c}}$

**36.** $\left[\dfrac{\dfrac{x+1}{x-1}+1}{\dfrac{x+1}{x-1}-1}\right]^5$

**37.** $1 + \dfrac{1}{1+\dfrac{1}{1+\dfrac{1}{1+\dfrac{1}{x}}}}$

**38.** $\dfrac{\dfrac{z}{1-\dfrac{z}{2+2z}}-2z}{\dfrac{2z}{5z-2}-3}$

# Summary and Review Exercises: Chapter 6

The objectives to be tested in addition to the material in this chapter are [4.2b], [4.4c], [5.6a], and [5.8a].

Find all numbers for which the rational expression is undefined.  [6.1a]

**1.** $\dfrac{3}{x}$

**2.** $\dfrac{4}{x - 6}$

**3.** $\dfrac{x + 5}{x^2 - 36}$

**4.** $\dfrac{x^2 - 3x + 2}{x^2 + x - 30}$

**5.** $\dfrac{-4}{(x + 2)^2}$

**6.** $\dfrac{x - 5}{x^3 - 8x^2 + 15x}$

Simplify.  [6.1c]

**7.** $\dfrac{4x^2 - 8x}{4x^2 + 4x}$

**8.** $\dfrac{14x^2 - x - 3}{2x^2 - 7x + 3}$

**9.** $\dfrac{(y - 5)^2}{y^2 - 25}$

Multiply and simplify.  [6.1d]

**10.** $\dfrac{a^2 - 36}{10a} \cdot \dfrac{2a}{a + 6}$

**11.** $\dfrac{6t - 6}{2t^2 + t - 1} \cdot \dfrac{t^2 - 1}{t^2 - 2t + 1}$

Divide and simplify.  [6.2b]

**12.** $\dfrac{10 - 5t}{3} \div \dfrac{t - 2}{12t}$

**13.** $\dfrac{4x^4}{x^2 - 1} \div \dfrac{2x^3}{x^2 - 2x + 1}$

Find the LCM.  [6.3c]

**14.** $3x^2, \quad 10xy, \quad 15y^2$

**15.** $a - 2, \quad 4a - 8$

**16.** $y^2 - y - 2, \quad y^2 - 4$

Add and simplify.  [6.4a]

**17.** $\dfrac{x + 8}{x + 7} + \dfrac{10 - 4x}{x + 7}$

**18.** $\dfrac{3}{3x - 9} + \dfrac{x - 2}{3 - x}$

**19.** $\dfrac{2a}{a + 1} + \dfrac{4a}{a^2 - 1}$

**20.** $\dfrac{d^2}{d - c} + \dfrac{c^2}{c - d}$

Subtract and simplify.  [6.5a]

**21.** $\dfrac{6x - 3}{x^2 - x - 12} - \dfrac{2x - 15}{x^2 - x - 12}$

**22.** $\dfrac{3x - 1}{2x} - \dfrac{x - 3}{x}$

**23.** $\dfrac{x + 3}{x - 2} - \dfrac{x}{2 - x}$

**24.** $\dfrac{1}{x^2 - 25} - \dfrac{x - 5}{x^2 - 4x - 5}$

**25.** Perform the indicated operations and simplify:
[6.5b]
$$\dfrac{3x}{x + 2} - \dfrac{x}{x - 2} + \dfrac{8}{x^2 - 4}.$$

Simplify.  [6.9a]

**26.** $\dfrac{\dfrac{1}{z} + 1}{\dfrac{1}{z^2} - 1}$

**27.** $\dfrac{\dfrac{c}{d} - \dfrac{d}{c}}{\dfrac{1}{c} + \dfrac{1}{d}}$

Solve.  [6.6a]

**28.** $\dfrac{3}{y} - \dfrac{1}{4} = \dfrac{1}{y}$

**29.** $\dfrac{15}{x} - \dfrac{15}{x + 2} = 2$

Solve.  [6.7a]

**30.** In checking records, a contractor finds that crew A can pave a certain length of highway in 9 hr, while crew B can do the same job in 12 hr. How long would it take if they worked together?

**31.** *Train Speeds.* A manufacturer is testing two high-speed trains. One train travels 40 km/h faster than the other. While one train travels 70 km, the other travels 60 km. Find the speed of each train.

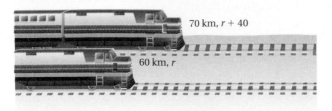

70 km, $r + 40$

60 km, $r$

**32.** The reciprocal of 1 more than a number is twice the reciprocal of the number itself. What is the number?

**33.** *Airplane Speeds.* One plane travels 80 mph faster than another. While one travels 1750 mi, the other travels 950 mi. Find the speed of each plane.

Solve.  [6.7b]

**34.** A sample of 250 calculators contained 8 defective calculators. How many defective calculators would you expect to find in a sample of 5000?

**35.** It is known that 10 cm³ of a normal specimen of human blood contains 1.2 g of hemoglobin. How many grams of hemoglobin would 16 cm³ of the same blood contain?

**36.** Triangles *ABC* and *XYZ* below are similar. Find the value of *x*.

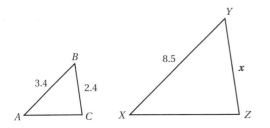

Solve for the letter indicated.  [6.8a]

**37.** $\dfrac{1}{r} + \dfrac{1}{s} = \dfrac{1}{t}$, for $s$

**38.** $F = \dfrac{9C + 160}{5}$, for $C$

**39.** $V = \dfrac{4}{3}\pi r^3$, for $r^3$

(The volume of a sphere)

**Skill Maintenance**

**40.** Factor: $5x^3 + 20x^2 - 3x - 12$.  [5.6a]

**41.** Simplify: $(5x^3y^2)^{-3}$.  [4.2b]

**42.** Subtract:  [4.4c]
$(5x^3 - 4x^2 + 3x - 4) - (7x^3 - 7x^2 - 9x + 14)$.

**43.** The width of a rectangle is 2 cm less than the length. The area is 15 cm². Find the dimensions and the perimeter of the rectangle.  [5.8a]

**Synthesis**

◆ Carry out the direction for each of the following. Explain the use of the LCM in each case.

**44.** Add: $\dfrac{4}{x - 2} + \dfrac{1}{x + 2}$.  [6.4a]

**45.** Subtract: $\dfrac{4}{x - 2} - \dfrac{1}{x + 2}$.  [6.5a]

**46.** Solve: $\dfrac{4}{x - 2} + \dfrac{1}{x + 2} = \dfrac{26}{x^2 - 4}$.  [6.6a]

**47.** Simplify: $\dfrac{1 - \dfrac{2}{x}}{1 + \dfrac{x}{4}}$.  [6.9a]

Simplify.

**48.** $\dfrac{2a^2 + 5a - 3}{a^2} \cdot \dfrac{5a^3 + 30a^2}{2a^2 + 7a - 4} \div \dfrac{a^2 + 6a}{a^2 + 7a + 12}$

[6.1d], [6.2b]

**49.** $\dfrac{12a}{(a - b)(b - c)} - \dfrac{2a}{(b - a)(c - b)}$  [6.5a]

**50.** Compare

$$\dfrac{A + B}{B} = \dfrac{C + D}{D}$$

with the proportion

$$\dfrac{A}{B} = \dfrac{C}{D}.$$

[6.7b]

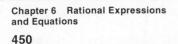

# Test:  Chapter 6

Find all numbers for which the rational expression is undefined.

**1.** $\dfrac{8}{2x}$

**2.** $\dfrac{5}{x + 8}$

**3.** $\dfrac{x - 7}{x^2 - 49}$

**4.** $\dfrac{x^2 + x - 30}{x^2 - 3x + 2}$

**5.** $\dfrac{11}{(x - 1)^2}$

**6.** $\dfrac{x + 2}{x^3 + 8x^2 + 15x}$

**7.** Simplify:

$$\frac{6x^2 + 17x + 7}{2x^2 + 7x + 3}.$$

**8.** Multiply and simplify:

$$\frac{a^2 - 25}{6a} \cdot \frac{3a}{a - 5}.$$

**9.** Divide and simplify:

$$\frac{25x^2 - 1}{9x^2 - 6x} \div \frac{5x^2 + 9x - 2}{3x^2 + x - 2}.$$

**10.** Find the LCM:

$$y^2 - 9,\ y^2 + 10y + 21,\ y^2 + 4y - 21.$$

Add or subtract. Simplify, if possible.

**11.** $\dfrac{16 + x}{x^3} + \dfrac{7 - 4x}{x^3}$

**12.** $\dfrac{5 - t}{t^2 + 1} - \dfrac{t - 3}{t^2 + 1}$

**13.** $\dfrac{x - 4}{x - 3} + \dfrac{x - 1}{3 - x}$

**14.** $\dfrac{x - 4}{x - 3} - \dfrac{x - 1}{3 - x}$

**15.** $\dfrac{5}{t - 1} + \dfrac{3}{t}$

**16.** $\dfrac{1}{x^2 - 16} - \dfrac{x + 4}{x^2 - 3x - 4}$

**17.** $\dfrac{1}{x - 1} + \dfrac{4}{x^2 - 1} - \dfrac{2}{x^2 - 2x + 1}$

**18.** Simplify: $\dfrac{9 - \dfrac{1}{y^2}}{3 - \dfrac{1}{y}}.$

**Answers**

1. _____

2. _____

3. _____

4. _____

5. _____

6. _____

7. _____

8. _____

9. _____

10. _____

11. _____

12. _____

13. _____

14. _____

15. _____

16. _____

17. _____

18. _____

Solve.

**19.** $\dfrac{7}{y} - \dfrac{1}{3} = \dfrac{1}{4}$

**20.** $\dfrac{15}{x} - \dfrac{15}{x-2} = -2$

19. _____

20. _____

Solve.

**21.** The reciprocal of 3 less than a number is four times the reciprocal of the number itself. What is the number?

**22.** A sample of 125 spark plugs contained 4 defective spark plugs. How many defective spark plugs would you expect to find in a sample of 500?

21. _____

22. _____

**23.** One car travels 20 mph faster than another on a freeway. While one goes 225 mi, the other goes 325 mi. Find the speed of each car.

23. _____

24. _____

**24.** Solve $L = \dfrac{Mt - g}{t}$ for $t$.

**25.** This pair of triangles is similar. Find the missing length $x$.

25. _____

26. _____

27. _____

**Skill Maintenance**

28. _____

**26.** Factor: $16a^2 - 49$.

**27.** Simplify: $\left(\dfrac{3x^2}{y^3}\right)^{-4}$.

**28.** Subtract:
$(5x^2 - 19x + 34) - (-8x^2 + 10x - 42)$.

**29.** The product of two consecutive integers is 462. Find the integers.

29. _____

**Synthesis**

30. _____

**30.** Team A and team B work together and complete a job in $2\frac{6}{7}$ hr. It would take team B 6 hr longer, working alone, to do the job than it would team A. How long would it take each of them to do the job working alone?

**31.** Simplify: $1 + \dfrac{1}{1 + \dfrac{1}{1 + \dfrac{1}{a}}}$.

31. _____

Copyright © 1990 Addison-Wesley Longman

# Cumulative Review: Chapters 1–6

Evaluate.

**1.** $\dfrac{2x + 5}{y - 10}$, for $x = 2$ and $y = 5$

**2.** $4 - x^3$, for $x = -2$

Simplify.

**3.** $x - [x - 2(x + 3)]$

**4.** $(2x^{-2})^{-2}(3x)^3$

**5.** $\dfrac{24x^8}{18x^{-2}}$

**6.** $\dfrac{2t^2 + 8t - 42}{2t^2 + 13t - 7}$

**7.** $\dfrac{\dfrac{2}{x} + 1}{\dfrac{x}{x + 2}}$

**8.** $\dfrac{a^2 - 16}{a^2 - 8a + 16}$

Add. Simplify, if possible.

**9.** $\dfrac{9}{14} + \left(-\dfrac{5}{21}\right)$

**10.** $\dfrac{2x + y}{x^2 y} + \dfrac{x + 2y}{xy^2}$

**11.** $\dfrac{z}{z^2 - 1} + \dfrac{2}{z + 1}$

**12.** $(2x^4 + 5x^3 + 4) + (3x^3 - 2x + 5)$

Subtract. Simplify, if possible.

**13.** $1.53 - (-0.8)$

**14.** $(x^2 - xy - y^2) - (x^2 - y^2)$

**15.** $\dfrac{3}{x^2 - 9} - \dfrac{x}{9 - x^2}$

**16.** $\dfrac{2x}{x^2 - x - 20} - \dfrac{4}{x^2 - 10x + 25}$

Multiply. Simplify, if possible.

**17.** $(1.3)(-0.5)(2)$

**18.** $3x^2(2x^2 + 4x - 5)$

**19.** $\left(3t + \dfrac{1}{2}\right)\left(3t - \dfrac{1}{2}\right)$

**20.** $(2p - q)^2$

**21.** $(3x + 5)(x - 4)$

**22.** $(2x^2 + 1)(2x^2 - 1)$

**23.** $\dfrac{6t + 6}{t^3 - 2t^2} \cdot \dfrac{t^3 - 3t^2 + 2t}{3t + 3}$

**24.** $\dfrac{a^2 - 1}{a^2} \cdot \dfrac{2a}{1 - a}$

Divide. Simplify, if possible.

**25.** $(3x^3 - 7x^2 + 9x - 5) \div (x - 1)$

**26.** $-\dfrac{21}{25} \div \dfrac{28}{15}$

**27.** $\dfrac{x^2 - x - 2}{4x^3 + 8x^2} \div \dfrac{x^2 - 2x - 3}{2x^2 + 4x}$

**28.** $\dfrac{3 - 3x}{x^2} \div \dfrac{x - 1}{4x}$

Factor completely.

**29.** $4x^3 + 12x^2 - 9x - 27$

**30.** $x^2 + 7x - 8$

**31.** $3x^2 - 14x - 5$

**32.** $16y^2 + 40xy + 25x^2$

**33.** $3x^3 + 24x^2 + 45x$

**34.** $2x^2 - 2$

**35.** $x^2 - 28x + 196$

**36.** $4y^3 + 10y^2 + 12y + 30$

Solve.

**37.** $2(x - 3) = 5(x + 3)$

**38.** $2x(3x + 4) = 0$

**39.** $x^2 = 8x$

**40.** $x^2 + 16 = 8x$

**41.** $x - 5 \leq 2x + 4$

**42.** $3x^2 = 27$

**43.** $\dfrac{1}{3}x - \dfrac{2}{5} = \dfrac{4}{5}x + \dfrac{1}{3}$

**44.** $\dfrac{x}{3} = \dfrac{3}{x}$

**45.** $\dfrac{x + 5}{2x + 1} = \dfrac{x - 7}{2x - 1}$

**46.** $\dfrac{1}{3}x\left(2x - \dfrac{1}{5}\right) = 0$

**47.** $\dfrac{3 - x}{x - 1} = \dfrac{2}{x - 1}$

**48.** $\dfrac{3}{2x + 5} = \dfrac{2}{5 - x}$

**49.** $\dfrac{1}{x} + \dfrac{1}{y} = \dfrac{1}{z}$, for $z$

**50.** $\dfrac{3N}{T} = D$, for $N$

**51.** Find the intercepts. Then graph the equation.
$$5y - 2x = -10$$

Solve.

**52.** The sum of three consecutive integers is 99. What are the integers?

**53.** The speed of one bicyclist is 2 km/h faster than the speed of another bicyclist. The first bicyclist travels 60 km in the same time that it takes the second to travel 50 km. Find the speed of each bicyclist.

**54.** A swimming pool can be filled in 5 hr by hose A alone and in 6 hr by hose B alone. How long would it take to fill the tank if both hoses were working?

**55.** The sum of the page numbers on the facing pages of a book is 69. What are the page numbers?

**56.** The product of the page numbers on two facing pages of a book is 272. Find the page numbers.

**57.** In 1997, Mark McGwire of the Oakland Athletics (and later with the St. Louis Cardinals) had 30 home runs after 85 games (**Source**: Major League Baseball).

    a) At this rate, how many home runs could McGwire hit in 162 games?

    b) At this rate, could it be predicted that he would break Maris's record of 61 home runs in a season?

**58.** The area of a circle is $35\pi$ more than the circumference. Find the length of the radius.

**59.** The sum of the squares of two consecutive odd positive integers is 202. Find the integers.

---

**Synthesis**

Solve.

**60.** $(2x - 1)^2 = (x + 3)^2$

**61.** $\dfrac{x + 2}{3x + 2} = \dfrac{1}{x}$

**62.** $\dfrac{2 + \dfrac{2}{x}}{x + 2 + \dfrac{1}{x}} = \dfrac{x + 2}{3}$

**63.** $\dfrac{x^6 x^4}{x^9 x^{-1}} = \dfrac{5^{14}}{25^6}$

**64.** Find the reciprocal of $\dfrac{1 - x}{x + 3} + \dfrac{x + 1}{2 - x}$.

**65.** Find the reciprocal of $2.0 \times 10^{-8}$ and express in scientific notation.

# 7

# Graphs, Slope, and Applications

## Introduction

We began our study of graphs in Chapter 3, where we focused on linear equations and intercepts. Here we expand our study of linear equations to consider the concept of *slope*. Slope is a number that describes the way in which a line slants. We will also consider applications such as variation and the graphing of inequalities in two variables.

**7.1** Slope and Applications

**7.2** Equations of Lines

**7.3** Parallel and Perpendicular Lines

**7.4** Graphing Inequalities in Two Variables

**7.5** Direct and Inverse Variation

| An Application | The Mathematics |
|---|---|

In order to meet federal standards, a wheelchair ramp must not rise more than 1 ft over a horizontal distance of 12 ft. Express this slope as a grade.

This problem appears as Exercise 40 in Exercise Set 7.1.

The slope, or grade, is the vertical change in distance divided by the horizontal change in distance, or

$$m = \frac{1}{12} \approx 0.083 \approx 8.3\%.$$

**World Wide Web** For more information, visit us at www.mathmax.com

Find the slope, if it exists, of the line.

**1.** $-4x + y = 6$

**2.** $y = 3$

**3.** Find the slope and the $y$-intercept of the line $x - 3y = 7$.

**4.** Find the slope, if it exists, of the line containing the points $(3, 0)$ and $(3, 6)$.

**5.** Find an equation of the line containing the points $(3, -1)$ and $(1, -3)$.

**6.** Find an equation of the line containing the point $(-1, 3)$ and having slope 4.

**7.** Find an equation of variation in which $y$ varies directly as $x$ and $y = 10$ when $x = 4$.

**8.** Find an equation of variation in which $y$ varies inversely as $x$ and $y = 10$ when $x = 4$.

Graph on a plane.

**9.** $y < x + 2$

**10.** $2y - 3x \geq 6$

Determine whether the graphs of the equations are parallel, perpendicular, or neither.

**11.** $y - 3x = 9$, $y - 3x = 7$

**12.** $-x + 2y = 7$, $2x + y = 4$

**13.** $y = \dfrac{2}{3}x - 5$, $y = -\dfrac{3}{2}x + 4$

**14.** Determine whether the ordered pair $(-3, 4)$ is a solution of $2x + 5y < 17$.

**15.** *Consumer Spending on Software.* The line graph at right describes the amount of consumer spending $S$, in billions, on software in recent years.

a) Find an equation of the line.
b) What is the rate of change in software spending?
c) Use the equation to predict consumer spending on software in 2002.

*Source:* Veronis, Suhler & Associates, PC Data

## Objectives for Retesting

The objectives to be tested in addition to the material in this chapter are as follows.

[1.8d]  Simplify expressions using rules for order of operations.
[5.7b]  Solve quadratic equations by factoring and then using the principle of zero products.
[6.6a]  Solve rational equations.
[6.7a]  Solve applied problems using rational equations.

# 7.1 Slope and Applications

## a | Slope

In Chapter 3, we considered two forms of a linear equation,

$$Ax + By = C \quad \text{and} \quad y = mx + b.$$

We found that from the form of the equation $y = mx + b$, we know certain information, namely, that the y-intercept of the line is $(0, b)$.

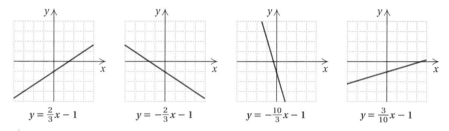

What about the constant $m$? Does it give us certain information about the line? Look at the following graphs and see if you can make any connection between the constant $m$ and the "slant" of the line.

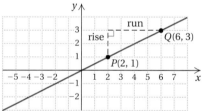

$y = \frac{2}{3}x - 1$    $y = -\frac{2}{3}x - 1$    $y = -\frac{10}{3}x - 1$    $y = \frac{3}{10}x - 1$

The graphs of some linear equations slant upward from left to right. Others slant downward. Some are vertical and some are horizontal. Some slant more steeply than others. We now look for a way to describe such possibilities with numbers.

Consider a line with two points marked $P$ and $Q$. As we move from $P$ to $Q$, the y-coordinate changes from 1 to 3 and the x-coordinate changes from 2 to 6. The change in $y$ is $3 - 1$, or $2$. The change in $x$ is $6 - 2$, or $4$.

We call the change in $y$ the **rise** and the change in $x$ the **run**. The ratio rise/run is the same for any two points on a line. We call this ratio the **slope**. Slope describes the slant of a line. The slope of the line in the graph above is given by

$$\frac{\text{rise}}{\text{run}} = \frac{\text{the change in } y}{\text{the change in } x}, \text{ or } \frac{2}{4}, \text{ or } \frac{1}{2}.$$

> The **slope** of a line containing points $(x_1, y_1)$ and $(x_2, y_2)$ is given by
> $$m = \frac{\text{rise}}{\text{run}} = \frac{\text{the change in } y}{\text{the change in } x} = \frac{y_2 - y_1}{x_2 - x_1}.$$

In the definition above, $(x_1, y_1)$ and $(x_2, y_2)$—read "$x$ sub-one, $y$ sub-one and $x$ sub-two, $y$ sub-two"—represent two different points on a line. It does not matter which point is considered $(x_1, y_1)$ and which is considered $(x_2, y_2)$ so long as coordinates are subtracted in the same order in both the numerator and the denominator.

## Objectives

**a** Given the coordinates of two points on a line, find the slope of the line.

**b** Find the slope of a line from an equation.

**c** Find the slope or rate of change in an applied problem involving slope.

**For Extra Help**

TAPE 13    TAPE 13B    MAC WIN    CD-ROM

Graph the line containing the points and find the slope in two different ways.

**1.** $(-2, 3)$ and $(3, 5)$

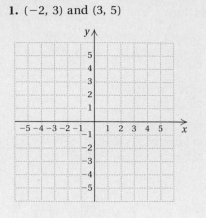

**2.** $(0, -3)$ and $(-3, 2)$

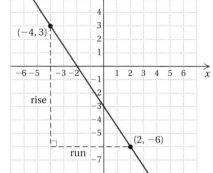

**Example 1** Graph the line containing the points $(-4, 3)$ and $(2, -6)$ and find the slope.

The graph is shown below. From $(-4, 3)$ and $(2, -6)$, we see that the change in $y$, or the rise, is $-6 - 3$, or $-9$. The change in $x$, or the run, is $2 - (-4)$, or 6. We consider $(x_1, y_1)$ to be $(-4, 3)$ and $(x_2, y_2)$ to be $(2, -6)$:

$$\text{Slope} = \frac{\text{rise}}{\text{run}} = \frac{\text{change in } y}{\text{change in } x}$$

$$= \frac{y_2 - y_1}{x_2 - x_1}$$

$$= \frac{-6 - 3}{2 - (-4)}$$

$$= \frac{-9}{6} = -\frac{9}{6}, \text{ or } -\frac{3}{2}.$$

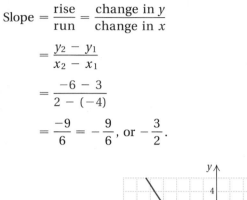

When we use the formula

$$m = \frac{y_2 - y_1}{x_2 - x_1},$$

we can subtract in two ways. We must remember, however, to subtract the $y$-coordinates in the same order that we subtract the $x$-coordinates. Let's redo Example 1, where we consider $(x_1, y_1)$ to be $(2, -6)$ and $(x_2, y_2)$ to be $(-4, 3)$:

$$\text{Slope} = \frac{\text{change in } y}{\text{change in } x} = \frac{3 - (-6)}{-4 - 2} = \frac{9}{-6} = -\frac{3}{2}.$$

The slope of a line tells how it slants. A line with positive slope slants up from left to right. The larger the slope, the steeper the slant. A line with negative slope slants downward from left to right.

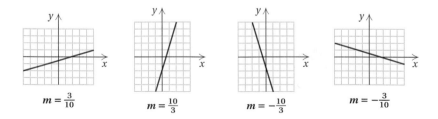

*Do Exercises 1 and 2.*

*Answers on page A-29*

## b | Finding the Slope from an Equation

It is possible to find the slope of a line from its equation. Let's consider the equation $y = 2x + 3$, which is in the form $y = mx + b$. We can find two points by choosing convenient values for $x$—say, 0 and 1—and substituting to find the corresponding $y$-values. We find the two points on the line to be $(0, 3)$ and $(1, 5)$. The slope of the line is found using the definition of slope:

$$m = \frac{\text{change in } y}{\text{change in } x} = \frac{5 - 3}{1 - 0} = \frac{2}{1} = 2.$$

The slope is 2. Note that this is also the coefficient of the $x$-term in the equation $y = 2x + 3$.

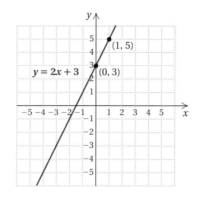

> The slope of the line $y = mx + b$ is $m$. To find the slope of a nonvertical line, solve the linear equation in $x$ and $y$ for $y$ and get the resulting equation in the form $y = mx + b$. The coefficient of the $x$-term, $m$, is the slope of the line.

**Examples**  Find the slope of the line.

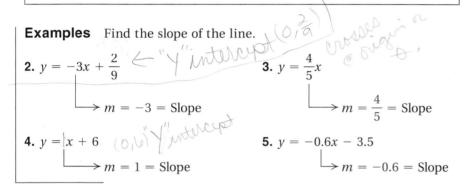

**2.** $y = -3x + \dfrac{2}{9}$  ← "y" intercept $(0, \frac{2}{9})$

$\qquad \longrightarrow m = -3 = \text{Slope}$

**3.** $y = \dfrac{4}{5}x$  crosses @ origin

$\qquad \longrightarrow m = \dfrac{4}{5} = \text{Slope}$

**4.** $y = x + 6$  $(0, 6)$ "y" intercept

$\qquad \longrightarrow m = 1 = \text{Slope}$

**5.** $y = -0.6x - 3.5$

$\qquad \longrightarrow m = -0.6 = \text{Slope}$

*Do Exercises 3–6.*

To find slope from an equation, we may have to first find an equivalent form of the equation.

**Example 6**  Find the slope of the line $2x + 3y = 7$.

We solve for $y$ to get the equation in the form $y = mx + b$:

$$2x + 3y = 7 - 2x$$
$$-2x$$
$$3y = -2x + 7$$
$$y = \frac{-2x + 7}{3}$$
$$y = -\frac{2}{3}x + \frac{7}{3}. \quad \text{This is } y = mx + b.$$

The slope is $-\frac{2}{3}$. (Slope)

*Do Exercises 7 and 8.*

*Answers on page A-29*

---

### Calculator Spotlight

Visualizing Slope

**Exercises**

Graph each of the following sets of equations using the window settings $[-6, 6, -4, 4]$, Xscl = 1, Yscl = 1.

**1.** $y = x$, $y = 2x$,
$y = 5x$, $y = 10x$

What do you think the graph of $y = 123x$ will look like?

**2.** $y = x$, $y = \dfrac{3}{4}x$,
$y = 0.38x$, $y = \dfrac{5}{32}x$

What do you think the graph of $y = 0.000043x$ will look like?

Find the slope of the line.

**3.** $y = 4x + 11$

**4.** $y = -17x + 8$

**5.** $y = -x + \dfrac{1}{2}$

**6.** $y = \dfrac{2}{3}x - 1$

Find the slope of the line.

**7.** $4x + 4y = 7$

**8.** $5x - 4y = 8$

## Calculator Spotlight

 Visualizing Slope

**Exercises**

Graph each of the following sets of equations using the window settings $[-6, 6, -4, 4]$, Xscl = 1, Yscl = 1.

**1.** $y = -x$, $y = -2x$, $y = -5x$, $y = -10x$

What do you think the graph of $y = -123x$ will look like?

**2.** $y = -x$, $y = -\frac{3}{4}x$, $y = -0.38x$, $y = -\frac{5}{32}x$

What do you think the graph of $y = -0.000043x$ will look like?

Find the slope, if it exists, of the line.

**9.** $x = 7$

**10.** $y = -5$

What about the slope of a horizontal or a vertical line?

**Example 7** Find the slope of the line $y = 5$.

We can think of $y = 5$ as $y = 0x + 5$. Then from this equation, we see that $m = 0$. Consider the points $(-3, 5)$ and $(4, 5)$, which are on the line. The change in $y = 5 - 5$, or 0. The change in $x = -3 - 4$, or $-7$. We have

$$m = \frac{5 - 5}{-3 - 4}$$

$$= \frac{0}{-7}$$

$$= 0.$$

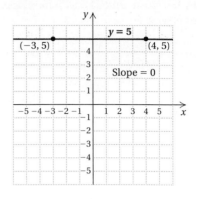

Any two points on a horizontal line have the same $y$-coordinate. Thus the change in $y$ is 0.

**Example 8** Find the slope of the line $x = -4$.

Consider the points $(-4, 3)$ and $(-4, -2)$, which are on the line. The change in $y = 3 - (-2)$, or 5. The change in $x = -4 - (-4)$, or 0. We have

$$m = \frac{3 - (-2)}{-4 - (-4)}$$

$$= \frac{5}{0}. \quad \textbf{Undefined}$$

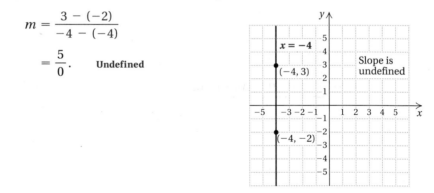

Since division by 0 is undefined, the slope of this line is undefined. The answer in this example is "The slope of this line is undefined."

> A horizontal line has slope 0. The slope of a vertical line is undefined.

*Do Exercises 9 and 10.*

## c  Applications of Slope

Slope has many real-world applications. For example, numbers like 2%, 3%, and 6% are often used to represent the *grade* of a road, a measure of how steep a road on a hill or mountain is. For example, a 3% grade $\left(3\% = \frac{3}{100}\right)$ means that for every horizontal distance of 100 ft, the road rises 3 ft, and a −3% grade means that for every horizontal distance of 100 ft, the road drops 3 ft. The concept of grade also occurs in skiing or snowboarding, where a 4% grade is considered very tame, but a 40% grade is considered extremely steep. And in cardiology, a physician may change the grade of a treadmill to measure its effect on heartbeat.

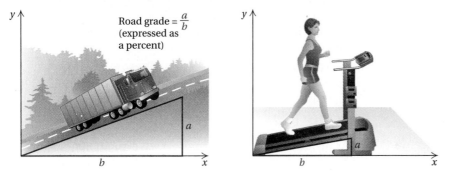

Road grade $= \dfrac{a}{b}$
(expressed as a percent)

Architects and carpenters use slope when designing and building stairs, ramps, or roof pitches. Another application occurs in hydrology. When a river flows, the strength or force of the river depends on how far the river falls vertically compared to how far it flows horizontally. Slope can also be considered as a **rate of change.**

### Algebraic–Graphical Connection

**Example 9**  *Cost of a Formal Wedding.* The cost of a formal wedding has increased over the years, as shown in the following graph. Find the rate of change of the cost.

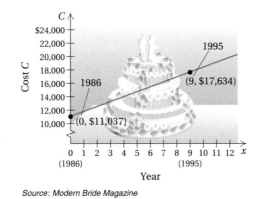

Source: Modern Bride Magazine

We determine the coordinates of two points on the graph. In this case, they are given as (0, $11,037) and (9, $17,634). Then we compute the slope, or rate of change:

$$\text{Slope} = \text{Rate of change} = \frac{\text{change in } y}{\text{change in } x}$$

$$= \frac{\$17,634 - \$11,037}{9 - 0} = \frac{\$6597}{9} = 733\,\frac{\$}{\text{yr}}.$$

What this means is that each year the cost of a formal wedding is $733 more than it was the preceding year.

*Do Exercise 11.*

---

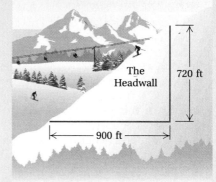

**11.** *Headwall Ski Slope.* One of the steepest ski slopes in North America, the Headwall on Mount Washington in New Hampshire, drops 720 ft over a horizontal distance of 900 ft. Find the grade of the Headwall.

Mt. Washington

The Headwall

720 ft

900 ft

*Answer on page A-29*

# Improving Your Math Study Skills

## On Reading and Writing Mathematics

*Mike Rosenborg is a former student of Marv Bittinger. He went on to receive a Master's Degree in mathematics and is now a math teacher. Here are some of his study tips regarding the reading and writing of mathematics.*

Why read your math text? This is a legitimate question when you consider that the instructor usually covers most of the material in the text. I have a reason: I don't want to be spoon-fed the material; I want to learn on my own. It's a lot more fun, and it builds my self-confidence to know that I can learn the material without the need for a teacher or a classroom.

It's a good idea to read your math text regularly for several reasons. When you read mathematics, you rely exclusively on the written word, and this is where mathematics derives much of its power. Mathematics is very precise, and it depends on writing to maintain and communicate this precision. Definitions and theorems in mathematics are stated in precise terms, and mathematical manipulations (such as solving equations) are performed by writing in a precise way.

In general, math texts develop the concepts in mathematics in a clear, tightly reasoned format, showing many examples along the way. Remember: The authors are mathematicians, and the way they write reflects their extensive mathematical training and thought processes. If you carefully read through the text, you will experience what it is like to think in a mathematical, rigorous, and precise way. This will not happen if you rely exclusively on oral lectures, because oral presentations are intrinsically "loose."

Reading your math text has other benefits. You will often find how to solve a difficult problem in the exercise set by looking at the text; in fact, there may be an example developed for you in the text that is much like your problem. Often your instructor will not have the time to cover everything in the text, or may want to cover something a little different. In these cases, reading your text will fill in the gaps.

But how do you read a math text? There is, of course, a difference between reading mathematics and a novel. Mathematics is like a chain with each link being developed in sequence and in order, and each link demands careful thought and attention before one can proceed to the next link—each link depends on the link before it. This is why you will experience troubles throughout an entire math course if you miss a single concept. Here, then, is a list of math reading tips and techniques:

- **Always read with a pencil and a piece of paper nearby.** When you find a section in your text that is difficult to understand, stop and work it out on your scratch paper.

- **Make notes in the margins of your text.** For instance, if you come across a word you don't understand, look it up and write its definition in the margin. Also, if the book refers to something covered previously that you have forgotten, find where it was originally covered, write the page number down in the margin, and go back and review the word.

- **Proceed slowly and carefully, making sure you understand what you read before continuing.** If there are worked-out examples in the book, read one and then try the next ones on your own. If you make a mistake, the details on the worked examples in the text will enable you to find your mistake quickly.

- **Do the Thinking and Writing Exercises.** The benefits of writing mathematics are that writing forces you to think through what you are writing about in a step-by-step manner, the act of writing itself helps to reinforce the concepts in your mind, and you have a ready, easy-to-read reference for further study or review. For instance, when studying for a test, if you have written up all your homework problems in a complete way, it will be easy for you to study for the test directly from your homework.

# Exercise Set 7.1

**a** Find the slope, if it exists, of the line.

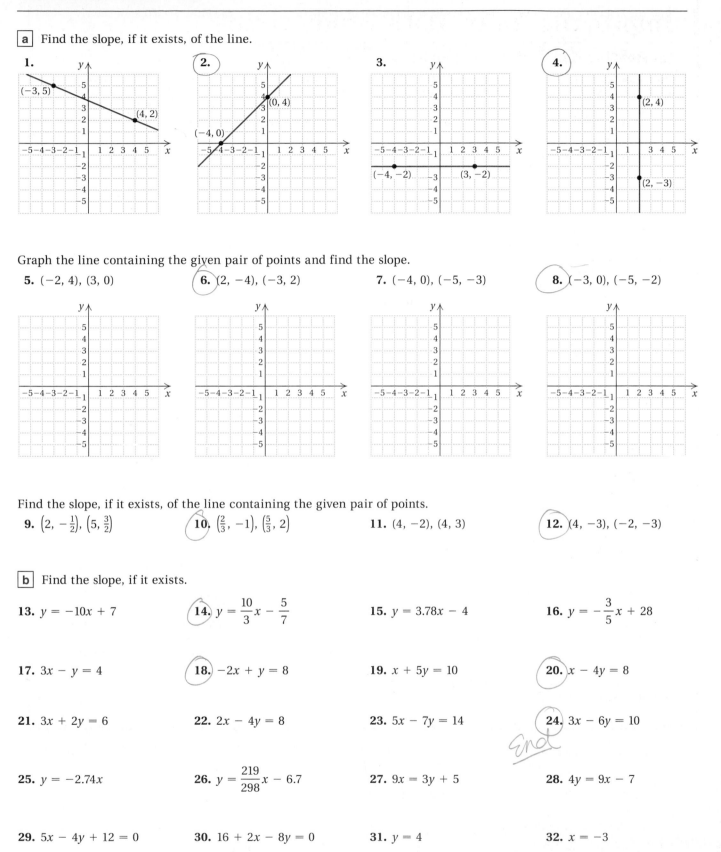

**1.**

**2.**

**3.**

**4.**

Graph the line containing the given pair of points and find the slope.

**5.** $(-2, 4)$, $(3, 0)$　　　**6.** $(2, -4)$, $(-3, 2)$　　　**7.** $(-4, 0)$, $(-5, -3)$　　　**8.** $(-3, 0)$, $(-5, -2)$

Find the slope, if it exists, of the line containing the given pair of points.

**9.** $\left(2, -\frac{1}{2}\right)$, $\left(5, \frac{3}{2}\right)$　　　**10.** $\left(\frac{2}{3}, -1\right)$, $\left(\frac{5}{3}, 2\right)$　　　**11.** $(4, -2)$, $(4, 3)$　　　**12.** $(4, -3)$, $(-2, -3)$

**b** Find the slope, if it exists.

**13.** $y = -10x + 7$　　　**14.** $y = \dfrac{10}{3}x - \dfrac{5}{7}$　　　**15.** $y = 3.78x - 4$　　　**16.** $y = -\dfrac{3}{5}x + 28$

**17.** $3x - y = 4$　　　**18.** $-2x + y = 8$　　　**19.** $x + 5y = 10$　　　**20.** $x - 4y = 8$

**21.** $3x + 2y = 6$　　　**22.** $2x - 4y = 8$　　　**23.** $5x - 7y = 14$　　　**24.** $3x - 6y = 10$

**25.** $y = -2.74x$　　　**26.** $y = \dfrac{219}{298}x - 6.7$　　　**27.** $9x = 3y + 5$　　　**28.** $4y = 9x - 7$

**29.** $5x - 4y + 12 = 0$　　　**30.** $16 + 2x - 8y = 0$　　　**31.** $y = 4$　　　**32.** $x = -3$

**c** Find the slope (or rate of change) in each exercise.

**33.** Find the slope (or pitch) of the roof.

2.4 ft

8.2 ft

**34.** Find the grade of the road.

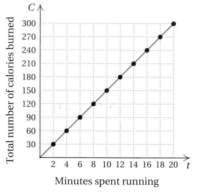

920.58 m

13,740 m

**35.** Find the slope of the river.

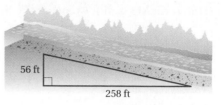

56 ft

258 ft

**36.** Find the slope (or grade) of the treadmill.

0.4 ft

5 ft

**37.** Find the rate at which a runner burns calories.

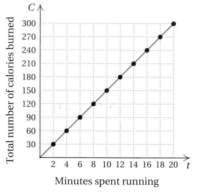

**38.** Find the rate of change in the number of U.S. farms.

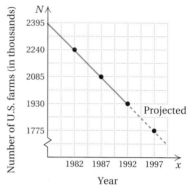

*Source*: Statistical Abstract of the United States

**39.** *Slope of Long's Peak.* From a base elevation of 9600 ft, Long's Peak in Colorado rises to a summit elevation of 14,255 ft over a horizontal distance of 15,840 ft. Find the grade of Long's Peak.

**40.** *Ramps for the Disabled.* In order to meet federal standards, a wheelchair ramp must not rise more than 1 ft over a horizontal distance of 12 ft. Express this slope as a grade.

---

**Skill Maintenance**

Simplify. [1.8d]

**41.** $11 \cdot 6 \div 3 \cdot 2 \div 7$

**42.** $2^4 - 2^4 \div 2^2 - 2$

**43.** $10 - 3(7 - 2)$

**44.** $5^3 - 4^2 + 6(5 \cdot 7 + 4 \cdot 3)$

**45.** $\dfrac{4^3 + 2^2}{5^3 - 4^2}$

**46.** $(4^3 + 2^2) \cdot (5^3 - 4^2)$

**47.** $1000 \div 100 \div 10 \div 2$

**48.** $3^{10} \div 3^2 \div 3^4 \div 9$

---

**Synthesis**

**49.** ◆ If one line has a slope of $-3$ and another has a slope of 2, which line is steeper? Why?

**50.** ◆ Graph several equations that have the same slope. How are they related?

Collaborative Learning Manual

Verify that *m* and *b* represent the slope and the *y*-intercept of the line $y = mx + b$.

Copyright © 1999 Addison Wesley Longman

# 7.2 Equations of Lines

## a Finding an Equation of a Line When the Slope and the *y*-Intercept Are Given

We know that in the equation $y = mx + b$ the slope is $m$ and the *y*-intercept is $(0, b)$. Thus we call the equation $y = mx + b$ the **slope–intercept equation.**

> **THE SLOPE–INTERCEPT EQUATION: $y = mx + b$**
>
> The equation $y = mx + b$ is called the **slope–intercept equation.** The slope is $m$ and the *y*-intercept is $(0, b)$.

**Example 1**  Find the slope and the *y*-intercept of $2x - 3y = 8$.

We first solve for *y*:

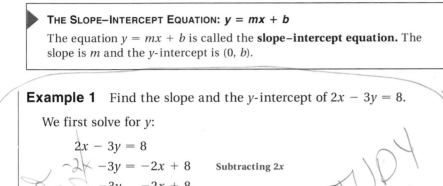

$$2x - 3y = 8$$
$$-3y = -2x + 8 \qquad \text{Subtracting } 2x$$
$$\frac{-3y}{-3} = \frac{-2x + 8}{-3} \qquad \text{Dividing by } -3$$
$$y = \frac{-2x}{-3} + \frac{8}{-3}$$
$$y = \frac{2}{3}x - \frac{8}{3}$$

The slope is $\dfrac{2}{3}$.  The *y*-intercept is $\left(0, -\dfrac{8}{3}\right)$.

*Do Exercises 1–5.*

**Example 2**  A line has slope $-2.4$ and *y*-intercept $(0, 11)$. Find an equation of the line.

We use the slope–intercept equation and substitute $-2.4$ for *m* and 11 for *b*:

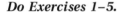

$$y = mx + b$$
$$y = -2.4x + 11. \qquad \text{Substituting}$$

*Do Exercise 6.*

## b Finding an Equation of a Line When the Slope and a Point Are Given

Suppose we know the slope of a line and a certain point on that line. We can use the slope–intercept equation $y = mx + b$ to find an equation of the line. To write an equation in this form, we need to know the slope ($m$) and the *y*-intercept ($b$).

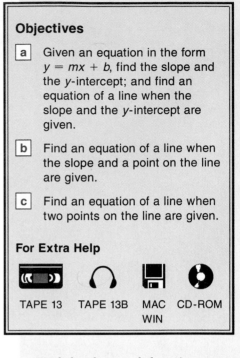

**Objectives**

**a**  Given an equation in the form $y = mx + b$, find the slope and the *y*-intercept; and find an equation of a line when the slope and the *y*-intercept are given.

**b**  Find an equation of a line when the slope and a point on the line are given.

**c**  Find an equation of a line when two points on the line are given.

**For Extra Help**

TAPE 13   TAPE 13B   MAC   CD-ROM
WIN

Find the slope and the *y*-intercept.

**1.** $y = 5x$

**2.** $y = -\dfrac{3}{2}x - 6$

**3.** $3x + 4y = 15$

**4.** $2y = 4x - 17$

**5.** $-7x - 5y = 22$

**6.** A line has slope 3.5 and *y*-intercept $(0, -23)$. Find an equation of the line.

*Answers on page A-29*

Find an equation of the line that contains the given point and has the given slope.

**7.** $(4, 2)$, $m = 5$

**8.** $(-2, 1)$, $m = -3$

**9.** $(3, 5)$, $m = 6$

**10.** $(1, 4)$, $m = -\dfrac{2}{3}$

Find an equation of the line containing the given points.

**11.** $(2, 4)$ and $(3, 5)$

**12.** $(-1, 2)$ and $(-3, -2)$

*Answers on page A-30*

**Example 3**   Find the equation of the line with slope 3 that contains the point $(4, 1)$.

We know that the slope is 3, so the equation is $y = 3x + b$. Using the point $(4, 1)$, we substitute 4 for $x$ and 1 for $y$ in $y = 3x + b$. Then we solve for $b$:

$$y = 3x + b$$
$$1 = 3(4) + b \quad \text{Substituting}$$
$$-11 = b. \quad \text{Solving for } b, \text{ the } y\text{-intercept}$$

We use the equation $y = mx + b$ and substitute 3 for $m$ and $-11$ for $b$:

$$y = 3x - 11.$$

**Example 4**   Find an equation of the line with slope $-5$ that contains the point $(-2, 3)$.

We know that the slope is $-5$, so the equation is $y = -5x + b$. Using the point $(-2, 3)$, we substitute $-2$ for $x$ and 3 for $y$ in $y = -5x + b$. Then we solve for $b$:

$$y = -5x + b$$
$$3 = -5(-2) + b \quad \text{Substituting}$$
$$3 = 10 + b$$
$$-7 = b. \quad \text{Solving for } b$$

We use the equation $y = mx + b$ and substitute $-5$ for $m$ and $-7$ for $b$:

$$y = -5x - 7.$$

*Do Exercises 7–10.*

## c   Finding an Equation of a Line When Two Points Are Given

We can also use the slope–intercept equation to find an equation of a line when two points are given.

**Example 5**   Find an equation of the line containing the points $(2, 3)$ and $(-6, 1)$.

First, we find the slope:

$$m = \frac{3 - 1}{2 - (-6)} = \frac{2}{8}, \text{ or } \frac{1}{4}.$$

Thus, $y = \frac{1}{4}x + b$. We then proceed as we did in Example 4, using either point to find $b$. We choose $(2, 3)$ and substitute 2 for $x$ and 3 for $y$:

$$y = \frac{1}{4}x + b$$
$$3 = \frac{1}{4} \cdot 2 + b \quad \text{Substituting}$$
$$3 = \frac{1}{2} + b$$
$$\frac{5}{2} = b. \quad \text{Solving for } b$$

We use the equation $y = mx + b$ and substitute $\frac{1}{4}$ for $m$ and $\frac{5}{2}$ for $b$:

$$y = \frac{1}{4}x + \frac{5}{2}.$$

*Do Exercises 11 and 12.*

# Exercise Set 7.2

**a** Find the slope and the $y$-intercept.

**1.** $y = -4x - 9$    **2.** $y = -2x + 3$    **3.** $y = 1.8x$    **4.** $y = -27.4x$

**5.** $-8x - 7y = 21$    **6.** $-2x - 8y = 16$    **7.** $4x = 9y + 7$    **8.** $5x + 4y = 12$

**9.** $-6x = 4y + 2$    **10.** $4.8x - 1.2y = 36$    **11.** $y = -17$    **12.** $y = 28$

Find an equation of the line with the given slope and $y$-intercept.

**13.** Slope $= -7$,
$y$-intercept $= (0, -13)$

**14.** Slope $= 73$,
$y$-intercept $= (0, 54)$

**15.** Slope $= 1.01$,
$y$-intercept $= (0, -2.6)$

**16.** Slope $= -\frac{3}{8}$,
$y$-intercept $= \left(0, \frac{7}{11}\right)$

**b** Find an equation of the line containing the given point and having the given slope.

**17.** $(-3, 0)$,   $m = -2$    **18.** $(2, 5)$,   $m = 5$    **19.** $(2, 4)$,   $m = \frac{3}{4}$    **20.** $\left(\frac{1}{2}, 2\right)$,   $m = -1$

**21.** $(2, -6)$,   $m = 1$    **22.** $(4, -2)$,   $m = 6$    **23.** $(0, 3)$,   $m = -3$    **24.** $(-2, -4)$,   $m = 0$

**c** Find an equation of the line that contains the given pair of points.

**25.** $(12, 16)$ and $(1, 5)$    **26.** $(-6, 1)$ and $(2, 3)$    **27.** $(0, 4)$ and $(4, 2)$    **28.** $(0, 0)$ and $(4, 2)$

**29.** (3, 2) and (1, 5)      **30.** (−4, 1) and (−1, 4)      **31.** (−4, 5) and (−2, −3)      **32.** (−2, −4) and (2, −1)

**33.** *Aerobic Exercise.* The line graph below describes the *target heart rate*, T, in beats per minute, of a person of age *a*, who is exercising. The goal is to get the number of heart beats per minute to this target level.

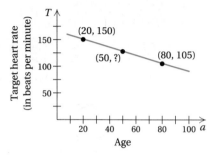

a) Find an equation of the line.
b) What is the rate of change in the target heart rate?
c) Use the equation to calculate the target heart rate of a person of age 50.

**34.** *Diabetes Cases.* The line graph below describes the number N, in millions, of cases of diabetes in this country in years *x* since 1983.

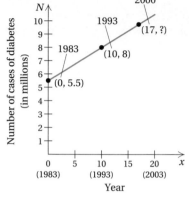

*Source*: U.S. National Center for Health Statistics

a) Find an equation of the line.
b) What is the rate of change of the number of cases of diabetes?
c) Use the equation to predict the number of cases of diabetes in 2000.

---

**Skill Maintenance**

Solve.   [5.7b]

**35.** $2x^2 + 6x = 0$      **36.** $x^2 − 49 = 0$      **37.** $x^2 − x − 6 = 0$      **38.** $x^2 + 4x − 5 = 0$

**39.** $2x^2 + 11x = 21$      **40.** $5x^2 = 14x + 24$      **41.** $x^2 + 5x − 14 = 0$      **42.** $12x^2 + 16x − 16 = 0$

Solve.   [2.3c]

**43.** $3x − 4(9 − x) = 17$      **44.** $2(5 + 2y) + 4y = 13$

**45.** $40(2x − 7) = 50(4 − 6x)$      **46.** $\frac{2}{3}(x − 5) = \frac{3}{8}(x + 5)$

---

**Synthesis**

**47.** ◈ Do all graphs of linear equations have *y*-intercepts? Why or why not?

**48.** ◈ Do all graphs of linear equations have *x*-intercepts? Why or why not?

**49.** Find an equation of the line that contains the point (2, −3) and has the same slope as the line $3x − y + 4 = 0$.

**50.** Find an equation of the line that has the same *y*-intercept as the line $x − 3y = 6$ and contains the point (5, −1).

**51.** Find an equation of the line with the same slope as the line $3x − 2y = 8$ and the same *y*-intercept as the line $2y + 3x = −4$.

---

Copyright © 1999 Addison Wesley Longman

# 7.3 Parallel and Perpendicular Lines

When we graph a pair of linear equations, there are three possibilities:

1. The graphs are the same.
2. The graphs intersect at exactly one point.
3. The graphs are parallel (they do not intersect).

## a Parallel Lines

The graphs shown at right are of the linear equations

$$y = 2x + 5$$

and $y = 2x - 3$.

The slope of each line is 2. The $y$-intercepts are (0, 5) and (0, −3) and are different. The lines do not intersect and are parallel.

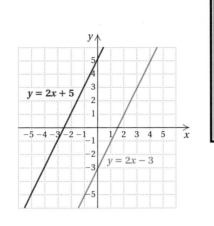

> **PARALLEL LINES**
>
> - Parallel nonvertical lines have the *same* slope, $m_1 = m_2$, and *different* $y$-intercepts, $b_1 \neq b_2$.
> - Parallel horizontal lines have equations $y = p$ and $y = q$, where $p \neq q$.
> - Parallel vertical lines have equations $x = p$ and $x = q$, where $p \neq q$.

By simply graphing, we may find it difficult to determine whether lines are parallel. Sometimes they may intersect only very far from the origin. We can use the preceding statements about slopes, $y$-intercepts, and parallel lines to determine for certain whether lines are parallel.

**Example 1** Determine whether the graphs of the lines $y = -3x + 4$ and $6x + 2y = -10$ are parallel.

The graphs of these equations are shown below, but they are not necessary in order to determine whether the lines are parallel.

We first solve each equation for $y$. In this case, the first equation is already solved for $y$.

*Slope intercept form*

a) $y = -3x + 4$

b) $6x + 2y = -10$  *chg. to slope int. form here*

$$2y = -6x - 10$$

$$y = \frac{1}{2}(-6x - 10)$$

$$y = -3x - 5$$

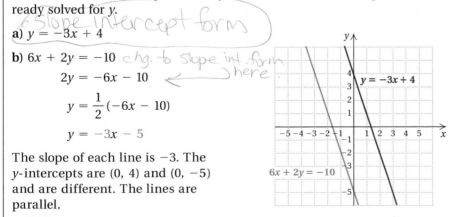

The slope of each line is −3. The $y$-intercepts are (0, 4) and (0, −5) and are different. The lines are parallel.

*Do Exercises 1 and 2.*

*Answers on page A-30*

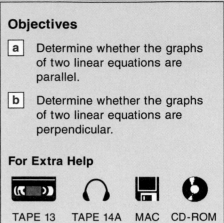

**Objectives**

a   Determine whether the graphs of two linear equations are parallel.

b   Determine whether the graphs of two linear equations are perpendicular.

**For Extra Help**

TAPE 13    TAPE 14A    MAC WIN    CD-ROM

Determine whether the graphs of the pair of equations are parallel.

1. $y - 3x = 1,$
   $-2y = 3x + 2$

2. $3x - y = -5,$
   $y - 3x = -2$

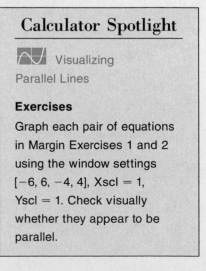

**Calculator Spotlight**

   Visualizing Parallel Lines

**Exercises**

Graph each pair of equations in Margin Exercises 1 and 2 using the window settings [−6, 6, −4, 4], Xscl = 1, Yscl = 1. Check visually whether they appear to be parallel.

Determine whether the graphs of the pair of equations are perpendicular.

3. $y = -\dfrac{3}{4}x + 7,$

$y = \dfrac{4}{3}x - 9$

4. $4x - 5y = 8,$
$6x + 9y = -12$

Answers on page A-30

## Calculator Spotlight

~~ Visualizing Perpendicular Lines

**Exercises**

Graph each pair of equations in Margin Exercises 3 and 4 using the window settings $[-9, 9, -6, 6]$, Xscl = 1, Yscl = 1. Check visually whether they appear to be perpendicular.

## b | Perpendicular Lines

Perpendicular lines in a plane are lines that intersect at a right angle. The measure of a right angle is 90°. The lines whose graphs are shown below are perpendicular. You can check this approximately by using a protractor or placing a rectangular piece of paper at the intersection.

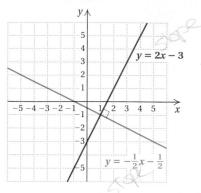

The slopes of the lines are 2 and $-\frac{1}{2}$. Note that $2\left(-\frac{1}{2}\right) = -1$. That is, the product of the slopes is $-1$.

> **PERPENDICULAR LINES**
>
> • Two nonvertical lines are perpendicular if the product of their slopes is $-1$, $m_1 \cdot m_2 = -1$. (If one line has slope $m$, the slope of the line perpendicular to it is $-1/m$.)
> • If one equation in a pair of perpendicular lines is vertical, then the other is horizontal. These equations are of the form $x = a$ and $y = b$.

**Example 2** Determine whether the graphs of the lines $3y = 9x + 3$ and $6y + 2x = 6$ are perpendicular.

The graphs are shown below, but they are not necessary in order to determine whether the lines are perpendicular.

We first solve each equation for $y$ in order to determine the slopes:

a) $3y = 9x + 3$

$y = \frac{1}{3}(9x + 3)$

$y = 3x + 1;$

b) $6y + 2x = 6$

$6y = -2x + 6$

$y = \frac{1}{6}(-2x + 6)$

$y = -\frac{1}{3}x + 1.$

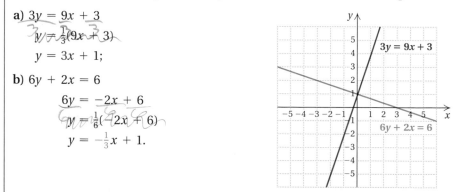

The slopes are 3 and $-\frac{1}{3}$. The product of the slopes is $3\left(-\frac{1}{3}\right) = -1$. The lines are perpendicular.

*Do Exercises 3 and 4.*

# Exercise Set 7.3

**a** Determine whether the graphs of the equations are parallel lines.

**1.** $x + 4 = y$,
$y - x = -3$

**2.** $3x - 4 = y$,
$y - 3x = 8$

**3.** $y + 3 = 6x$,
$-6x - y = 2$

**4.** $y = -4x + 2$,
$-5 = -2y + 8x$

**5.** $10y + 32x = 16.4$,
$y + 3.5 = 0.3125x$

**6.** $y = 6.4x + 8.9$,
$5y - 32x = 5$

**7.** $y = 2x + 7$,
$5y + 10x = 20$

**8.** $y + 5x = -6$,
$3y + 5x = -15$

**9.** $3x - y = -9$,
$2y - 6x = -2$

**10.** $y - 6 = -6x$,
$-2x + y = 5$

**11.** $x = 3$,
$x = 4$

**12.** $y = 1$,
$y = -2$

**b** Determine whether the graphs of the equations are perpendicular lines.

**13.** $y = -4x + 3$,
$4y + x = -1$

**14.** $y = -\dfrac{2}{3}x + 4$,
$3x + 2y = 1$

**15.** $x + y = 6$,
$4y - 4x = 12$

**16.** $2x - 5y = -3$,
$5x + 2y = 6$

**17.** $y = -0.3125x + 11$,
$y - 3.2x = -14$

**18.** $y = -6.4x - 7$,
$64y - 5x = 32$

**19.** $y = -x + 8$,
$x - y = -1$

**20.** $2x + 6y = -3$,
$12y = 4x + 20$

**21.** $\dfrac{3}{8}x - \dfrac{y}{2} = 1$,
$\dfrac{4}{3}x - y + 1 = 0$

**22.** $\dfrac{1}{2}x + \dfrac{3}{4}y = 6$,
$-\dfrac{3}{2}x + y = 4$

**23.** $x = 0$,
$y = -2$

**24.** $x = -3$,
$y = 5$

**Skill Maintenance**

Solve.  [6.7a]

**25.** A train leaves a station and travels west at 70 km/h. Two hours later, a second train leaves on a parallel track and travels west at 90 km/h. When will it overtake the first train?

**26.** One car travels 10 km/h faster than another. While one car travels 130 km, the other travels 140 km. What is the speed of each car?

Solve.  [6.6a]

**27.** $\dfrac{x^2}{x+4} = \dfrac{16}{x+4}$

**28.** $\dfrac{2}{3} - \dfrac{5}{6} = \dfrac{1}{x}$

**29.** $\dfrac{t}{3} + \dfrac{t}{10} = 1$

**30.** $\dfrac{5}{x-4} = \dfrac{3}{x+2}$

**31.** $\dfrac{4}{x-2} + \dfrac{7}{x-3} = \dfrac{10}{x^2-5x+6}$

**32.** $\dfrac{3}{x-5} + \dfrac{4}{x+5} = \dfrac{2}{x^2-25}$

---

**Synthesis**

**33.** ◈ Consider two equations of the type $Ax + By = C$. Explain how you would go about showing that their graphs are perpendicular.

**34.** ◈ Consider two equations of the type $Ax + By = C$. Explain how you would go about showing that their graphs are parallel.

**35.–40.** 📈 Check the results of Exercises 1–6 by graphing each pair of equations using the window settings $[-6, 6, -4, 4]$, Xscl = 1, Yscl = 1.

**41.–46.** 📈 Check the results of Exercises 13–18 by graphing each pair of equations using the window settings $[-24, 24, -16, 16]$, Xscl = 1, Yscl = 1.

**47.** Find an equation of a line that contains the point $(0, 6)$ and is parallel to $y - 3x = 4$.

**48.** Find an equation of the line that contains the point $(-2, 4)$ and is parallel to $y = 2x - 3$.

**49.** Find an equation of the line that contains the point $(0, 2)$ and is perpendicular to $3y - x = 0$.

**50.** Find an equation of the line that contains the point $(1, 0)$ and is perpendicular to $2x + y = -4$.

**51.** Find an equation of the line that has $x$-intercept $(-2, 0)$ and is parallel to $4x - 8y = 12$.

**52.** Find the value of $k$ such that $4y = kx - 6$ and $5x + 20y = 12$ are parallel.

**53.** Find the value of $k$ such that $4y = kx - 6$ and $5x + 20y = 12$ are perpendicular.

The lines in the graphs in Exercises 54 and 55 are perpendicular and the lines in the graph in Exercise 56 are parallel. Find an equation of each line.

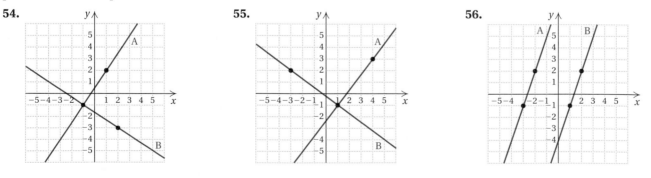

**54.** **55.** **56.**

Copyright © 1999 Addison Wesley Longman

# 7.4 Graphing Inequalities in Two Variables

A graph of an inequality is a drawing that represents its solutions. An inequality in one variable can be graphed on a number line. An inequality in two variables can be graphed on a coordinate plane.

## a | Solutions of Inequalities in Two Variables

The solutions of inequalities in two variables are ordered pairs.

**Example 1** Determine whether $(-3, 2)$ is a solution of $5x + 4y < 13$.

We use alphabetical order to replace $x$ with $-3$ and $y$ with 2.

$$\frac{5x + 4y < 13}{\begin{array}{c|c} 5(-3) + 4 \cdot 2 \; ? \; 13 & \\ -15 + 8 & \\ -7 & \text{TRUE} \end{array}}$$

Since $-7 < 13$ is true, $(-3, 2)$ is a solution.

**Example 2** Determine whether $(6, 8)$ is a solution of $5x + 4y < 13$.

We use alphabetical order to replace $x$ with 6 and $y$ with 8.

$$\frac{5x + 4y < 13}{\begin{array}{c|c} 5(6) + 4(8) \; ? \; 13 & \\ 30 + 32 & \\ 62 & \text{FALSE} \end{array}}$$

Since $62 < 13$ is false, $(6, 8)$ is not a solution.

*Do Exercises 1 and 2.*

## b | Graphing Inequalities in Two Variables

**Example 3** Graph: $y > x$.

We first graph the line $y = x$. Every solution of $y = x$ is an ordered pair like $(3, 3)$. The first and second coordinates are the same. We draw the line $y = x$ dashed because its points (as shown on the left below) are *not* solutions of $y > x$.

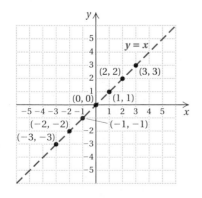
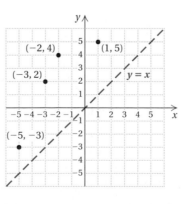

### Objectives

**a** Determine whether an ordered pair of numbers is a solution of an inequality in two variables.

**b** Graph linear inequalities.

### For Extra Help

TAPE 13    TAPE 14A    MAC WIN    CD-ROM

1. Determine whether $(4, 3)$ is a solution of $3x - 2y < 1$.

2. Determine whether $(2, -5)$ is a solution of $4x + 7y \geq 12$.

*Answers on page A-30*

**3.** Graph: $y < x$.

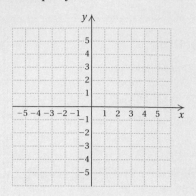

Now look at the graph on the right on the preceding page. Several ordered pairs are plotted in the half-plane above the line $y = x$. Each is a solution of $y > x$.

We can check a pair such as $(-2, 4)$ as follows:

$$\frac{y > x}{4 \ ? \ -2} \quad \text{TRUE}$$

It turns out that any point on the same side of $y = x$ as $(-2, 4)$ is also a solution. *If we know that one point in a half-plane is a solution, then all points in that half-plane are solutions.* The graph of $y > x$ is shown below. (Solutions are indicated by color shading throughout.) We shade the half-plane above $y = x$.

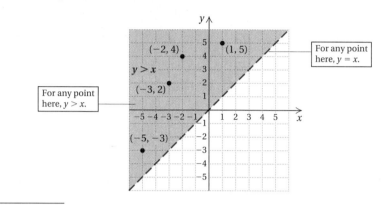

*Do Exercise 3.*

A **linear inequality** is one that we can get from a linear equation by changing the equals symbol to an inequality symbol. Every linear equation has a graph that is a straight line. The graph of a linear inequality is a half-plane, sometimes including the line along the edge.

---

To graph an inequality in two variables:

1. Replace the inequality symbol with an equals sign and graph this related equation.

2. If the inequality symbol is $<$ or $>$, draw the line dashed. If the inequality symbol is $\leq$ or $\geq$, draw the line solid.

3. The graph consists of a half-plane, either above or below or left or right of the line, and, if the line is solid, the line as well. To determine which half-plane to shade, choose a point not on the line as a test point. Substitute to find whether that point is a solution of the inequality. If it is, shade the half-plane containing that point. If it is not, shade the half-plane on the opposite side of the line.

---

**Example 4**   Graph: $5x - 2y < 10$.

1. We first graph the line $5x - 2y = 10$. The intercepts are $(0, -5)$ and $(2, 0)$. This line forms the boundary of the solutions of the inequality.

2. Since the inequality contains the $<$ symbol, points on the line are not solutions of the inequality, so we draw a dashed line.

*Answer on page A-30*

**3.** To determine which half-plane to shade, we consider a test point *not* on the line. We try $(3, -2)$ and substitute:

$$\frac{5x - 2y < 10}{\begin{array}{c|c} 5(3) - 2(-2) \ ? \ 10 \\ 15 + 4 \\ 19 \end{array}} \quad \text{FALSE}$$

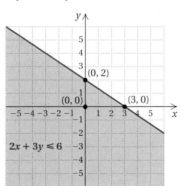

Since this inequality is false, the point $(3, -2)$ is *not* a solution; no point in the half-plane containing $(3, -2)$ is a solution. Thus the points in the opposite half-plane are solutions. The graph is shown above.

*Do Exercise 4.*

**Example 5** Graph: $2x + 3y \le 6$.

First, we graph the line $2x + 3y = 6$. The intercepts are $(0, 2)$ and $(3, 0)$. Since the inequality contains the $\le$ symbol, we draw the line solid to indicate that any pair on the line is a solution. Next, we choose a test point that does not belong to the line. We substitute to determine whether this point is a solution. The origin $(0, 0)$ is generally an easy one to use:

$$\frac{2x + 3y \le 6}{\begin{array}{c|c} 2 \cdot 0 + 3 \cdot 0 \ ? \ 6 \\ 0 \end{array}} \quad \text{TRUE}$$

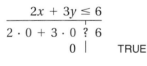

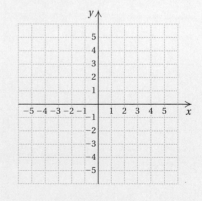

We see that $(0, 0)$ is a solution, so we shade the lower half-plane. Had the substitution given us a false inequality, we would have shaded the other half-plane.

*Do Exercises 5 and 6.*

**Example 6** Graph $x < 3$ on a plane.

There is a missing variable in this inequality. Thus we rewrite this inequality as $x + 0y < 3$. We use the same technique that we have used with the other examples. First, we graph the related equation $x = 3$ on the plane and draw the graph with a dashed line since the inequality symbol is $<$.

**4.** Graph: $2x + 4y < 8$.

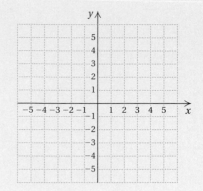

Graph.

**5.** $3x - 5y < 15$

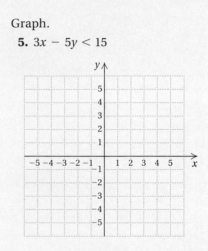

**6.** $2x + 3y \ge 12$

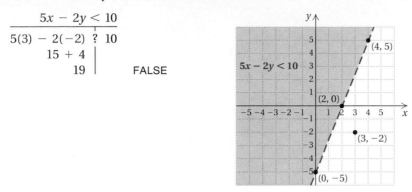

*Answers on page A-30*

Graph.

**7.** $x > -3$

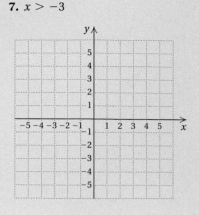

**8.** $y \leq 4$

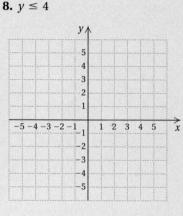

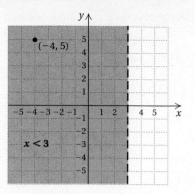

The graph is a half-plane either to the left or to the right of the line $x = 3$. To determine which, we consider a test point, $(-4, 5)$:

$$\frac{x + 0y < 3}{-4 + 0(5) \;?\; 3}$$
$$-4 \;|\; \qquad \text{TRUE}$$

We see that $(-4, 5)$ is a solution, so all the pairs in the half-plane containing $(-4, 5)$ are solutions. We shade that half-plane.

We see from the graph that the solutions of $x < 3$ are all those ordered pairs whose first coordinates are less than 3.

If we graph the inequality in Example 6 on a line rather than on a plane, its graph is as follows:

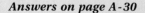

**Example 7**  Graph $y \geq -4$ on a plane.

We first graph $y = -4$ using a solid line to indicate that all points on the line are solutions. We then use $(2, 3)$ as a test point and substitute:

$$\frac{0x + y \geq -4}{0(2) + 3 \;?\; -4}$$
$$3 \;|\; \qquad \text{TRUE}$$

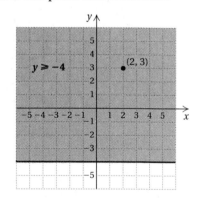

Since $(2, 3)$ is a solution, all points in the half-plane containing $(2, 3)$ are solutions. Note that this half-plane consists of all ordered pairs whose second coordinate is greater than or equal to $-4$.

*Do Exercises 7 and 8.*

*Answers on page A-30*

# Exercise Set 7.4

**a**

**1.** Determine whether $(-3, -5)$ is a solution of

$$-x - 3y < 18.$$

**2.** Determine whether $(2, -3)$ is a solution of

$$5x - 4y \geq 1.$$

**3.** Determine whether $\left(\frac{1}{2}, -\frac{1}{4}\right)$ is a solution of

$$7y - 9x \leq -3.$$

**4.** Determine whether $(-8, 5)$ is a solution of

$$x + 0 \cdot y > 4.$$

**b** Graph on a plane.

**5.** $x > 2y$

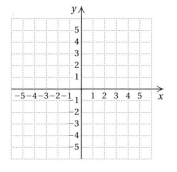

**6.** $x > 3y$

**7.** $y \leq x - 3$

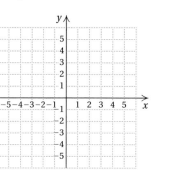

**8.** $y \leq x - 5$

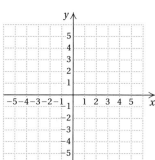

**9.** $y < x + 1$

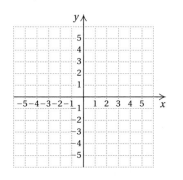

**10.** $y < x + 4$

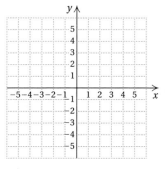

**11.** $y \geq x - 2$

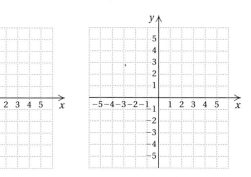

**12.** $y \geq x - 1$

**13.** $y \leq 2x - 1$

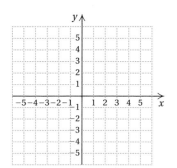

**14.** $y \leq 3x + 2$

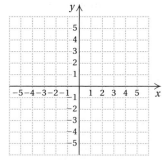

**15.** $x + y \leq 3$

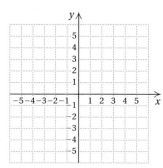

**16.** $x + y \leq 4$

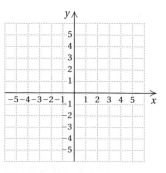

**17.** $x - y > 7$

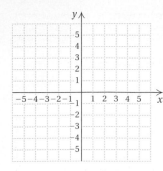

**18.** $x - y > -2$

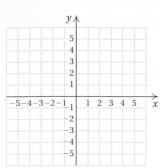

**19.** $2x + 3y \leq 12$

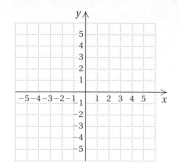

**20.** $5x + 4y \geq 20$

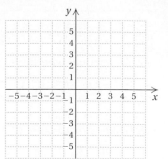

**21.** $y \geq 1 - 2x$

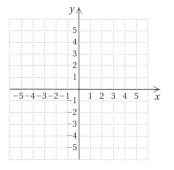

**22.** $y - 2x \leq -1$

**23.** $2x - 3y > 6$

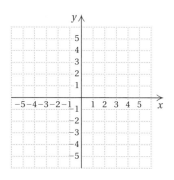

**24.** $5y - 2x \leq 10$

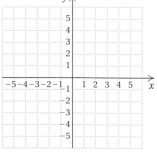

**25.** $y \leq 3$

**26.** $y > -1$

**27.** $x \geq -1$

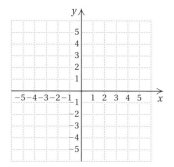

**28.** $x < 0$

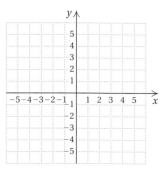

## Skill Maintenance

Solve.  [6.6a]

**29.** $\dfrac{12}{x} = \dfrac{48}{x + 9}$

**30.** $x + 5 = -\dfrac{6}{x}$

Solve.  [5.7b]

**31.** $x^2 + 16 = 8x$

**32.** $12x^2 + 17x = 5$

## Synthesis

**33.** ◈ Why is (0, 0) such a "convenient" test point?

**34.** ◈ Is the graph of any inequality in the form $y > mx + b$ shaded *above* the line $y = mx + b$? Why or why not?

**35.** *Elevators.* Many elevators have a capacity of 1 metric ton (1000 kg). Suppose $c$ children, each weighing 35 kg, and $a$ adults, each weighing 75 kg, are on an elevator. Find and graph an inequality that asserts that the elevator is overloaded.

**36.** *Hockey Wins and Losses.* A hockey team determines that it needs at least 60 points for the season in order to make the playoffs. A win $w$ is worth 2 points and a tie $t$ is worth 1 point. Find and graph an inequality that describes the situation.

Collaborative
Learning Manual

Practice graphing linear inequalities by playing a variation of the game Battleship.

# 7.5 Direct and Inverse Variation

## a | Equations of Direct Variation

A bicycle is traveling at a speed of 15 km/h. In 1 hr, it goes 15 km; in 2 hr, it goes 30 km; in 3 hr, it goes 45 km; and so on. We can form a set of ordered pairs using the number of hours as the first coordinate and the number of kilometers traveled as the second coordinate. These determine a set of ordered pairs:

$$(1, 15), \ (2, 30), \ (3, 45), \ (4, 60), \ \text{and so on.}$$

Note that the ratio of the second coordinate to the first is the same number:

$$\frac{15}{1} = 15, \quad \frac{30}{2} = 15, \quad \frac{45}{3} = 15, \quad \frac{60}{4} = 15, \quad \text{and so on.}$$

Whenever a situation produces pairs of numbers in which the *ratio is constant*, we say that there is **direct variation.** Here the distance varies directly as the time:

$$\frac{d}{t} = 15 \ \text{(a constant)}, \quad \text{or} \quad d = 15t.$$

The equation is an **equation of direct variation.** The coefficient—in this case, 15—is called the **variation constant.** The graph of $d = 15t$ is shown at right.

---

> **DIRECT VARIATION**
>
> If a situation translates to an equation described by $y = kx$, where $k$ is a positive constant, $y = kx$ is called an **equation of direct variation,** and $k$ is called the **variation constant.** We say that $y$ varies directly as $x$.

---

The terminologies "$y$ varies as $x$," "$y$ is directly proportional to $x$," and "$y$ is proportional to $x$" also imply direct variation and are used in many situations. The constant $k$ is often referred to as a **constant of proportionality.**

When there is direct variation $y = kx$, the variation constant can be found if one pair of values of $x$ and $y$ is known. Then other values can be found.

**Example 1**  Find an equation of variation in which $y$ varies directly as $x$ and $y = 7$ when $x = 25$.

We first substitute to find $k$:

$$y = kx$$
$$7 = k \cdot 25$$
$$\frac{7}{25} = k, \quad \text{or} \quad k = 0.28.$$

Then the equation of variation is

$$y = 0.28x.$$

Note that the answer is an *equation.*

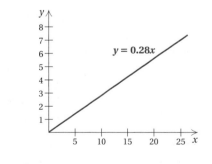

*Do Exercises 1 and 2 on the following page.*

**Objectives**

**a** Find an equation of direct variation given a pair of values of the variables.

**b** Solve applied problems involving direct variation.

**c** Find an equation of inverse variation given a pair of values of the variables.

**d** Solve applied problems involving inverse variation.

**For Extra Help**

TAPE 13    TAPE 14B    MAC WIN    CD-ROM

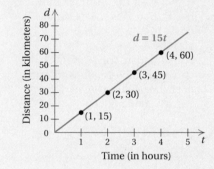

Find an equation of variation in which $y$ varies directly as $x$ and the following is true.

**1.** $y = 84$ when $x = 12$

**2.** $y = 50$ when $x = 80$

**3.** *Electricity Costs.* The cost $C$ of operating a television varies directly as the number $n$ of hours that it is in operation. It costs $14.00 to operate a standard-size color TV continuously for 30 days. At this rate, how much would it cost to operate the TV for 1 day? for 1 hour?

**4.** *Weight on Venus.* The weight $V$ of an object on Venus varies directly as its weight $E$ on Earth. A person weighing 165 lb on Earth would weigh 145.2 lb on Venus. How much would a person weighing 198 lb on Earth weigh on Venus?

Answers on page A-31

## b | Applications of Direct Variation

**Example 2**   *Karat Ratings of Gold Objects.* It is known that the karat rating $K$ of a gold object varies directly as the actual percentage $P$ of gold in the object. A 14-karat gold ring is 58.25% gold. What is the percentage of gold in a 10-karat chain?

**1., 2.   Familiarize** and **Translate.**   The problem states that we have direct variation between the variables $K$ and $P$. Thus an equation $K = kP$, $k > 0$, applies. As the percentage of gold increases, the karat rating increases. The letters $K$ and $k$ represent different quantities.

**3. Solve.**   The mathematical manipulation has two parts. First, we determine the equation of variation by substituting known values for $K$ and $P$ to find the variation constant $k$. Second, we compute the percentage of gold in a 10-karat chain.

a) First, we find an equation of variation:

$$K = kP$$
$$14 = k(0.5825) \qquad \text{Substituting 14 for } K \text{ and 58.25\%, or 0.5825, for } P$$
$$\frac{14}{0.5825} = k$$
$$24.03 \approx k. \qquad \text{Dividing and rounding to the nearest hundredth}$$

The equation of variation is $K = 24.03P$.

b) We then use the equation to find the percentage of gold in a 10-karat chain:

$$K = 24.03P$$
$$10 = 24.03P \qquad \text{Substituting 10 for } K$$
$$\frac{10}{24.03} = P$$
$$0.416 \approx P$$
$$41.6\% \approx P.$$

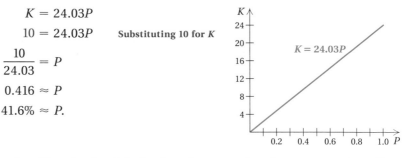

**4. Check.**   The check might be done by repeating the computations. You might also do some reasoning about the answer. The karat rating decreased from 14 to 10. Similarly, the percentage decreased from 58.25% to 41.6%.

**5. State.**   A 10-karat chain is 41.6% gold.

*Do Exercises 3 and 4.*

Let's consider direct variation from the standpoint of a graph. The graph of $y = kx$, $k > 0$, always goes through the origin and rises from left to right. Note that as $x$ increases, $y$ increases; and as $x$ decreases, $y$ decreases. This is why the terminology "direct" is used. What one variable does, the other does as well.

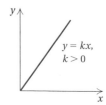

## c Equations of Inverse Variation

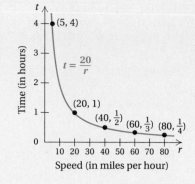

A car is traveling a distance of 20 mi. At a speed of 5 mph, it will take 4 hr; at 20 mph, it will take 1 hr; at 40 mph, it will take $\frac{1}{2}$ hr; and so on. We use speed as the first coordinate and the time as the second coordinate. These determine a set of ordered pairs:

$$(5, 4), \quad (20, 1), \quad \left(40, \tfrac{1}{2}\right), \quad \left(60, \tfrac{1}{3}\right), \quad \text{and so on.}$$

Note that the products of the coordinates in each ordered pair are all the same number:

$$5 \cdot 4 = 20, \quad 20 \cdot 1 = 20, \quad 40 \cdot \tfrac{1}{2} = 20, \quad 60 \cdot \tfrac{1}{3} = 20, \quad \text{and so on.}$$

Whenever a situation produces pairs of numbers in which the *product is constant,* we say that there is **inverse variation.** Here the time varies inversely as the speed:

$$rt = 20 \text{ (a constant)}, \quad \text{or} \quad t = \frac{20}{r}.$$

The equation is an **equation of inverse variation.** The coefficient—in this case, 20—is called the **variation constant.** Note that as the first number gets larger, the second number gets smaller.

> ▶ **INVERSE VARIATION**
>
> If a situation translates to an equation described by $y = k/x$, where $k$ is a positive constant, $y = k/x$ is called an **equation of inverse variation.** We say that $y$ varies inversely as $x$.

The terminology "$y$ is inversely proportional to $x$" also implies inverse variation and is used in some situations.

**Example 3** Find an equation of variation in which $y$ varies inversely as $x$ and $y = 145$ when $x = 0.8$.

We first substitute to find $k$:

$$y = \frac{k}{x}$$
$$145 = \frac{k}{0.8}$$
$$(0.8)145 = k$$
$$116 = k.$$

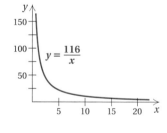

The equation of variation is $y = \dfrac{116}{x}$.

*Do Exercises 5 and 6.*

The graph of $y = k/x$, $k > 0$, is shaped like the figure at right for positive values of $x$. (You need not know how to graph such equations at this time.) Note that as $x$ increases, $y$ decreases; and as $x$ decreases, $y$ increases. This is why the terminology "inverse" is used. One variable does the opposite of what the other does.

---

Find an equation of variation in which $y$ varies inversely as $x$ and the following is true.

**5.** $y = 105$ when $x = 0.6$

**6.** $y = 45$ when $x = 20$

*Answers on page A-31*

**7.** Referring to Example 4, determine how long it would take 10 people to do the job.

**8.** The time required to drive a fixed distance varies inversely as the speed $r$. It takes 5 hr at 60 km/h to drive a fixed distance. How long would it take at 40 km/h?

---

## d | Applications of Inverse Variation

Often in an applied situation we must decide which kind of variation, if any, might apply to the problem.

**Example 4**  *Work Time.* Molly is a maintenance supervisor. She notes that it takes 4 hr for 20 people to wash and wax the floors in a building. How long would it then take 25 people to do the job?

1. **Familiarize.** Think about the problem situation. What kind of variation would be used? It seems reasonable that the more people there are working on the job, the less time it will take to finish. (One might argue that too many people in a crowded area would be counterproductive, but we will disregard that possibility.) Thus inverse variation might apply. We let $T$ = the time to do the job, in hours, and $N$ = the number of people. Assuming inverse variation, we know that an equation $T = k/N$, $k > 0$, applies. As the number of people increases, the time it takes to do the job decreases.

2. **Translate.** We write an equation of variation:

$$T = \frac{k}{N}.$$

Time varies inversely as the number of people involved.

3. **Solve.** The mathematical manipulation has two parts. First, we find the equation of variation by substituting known values for $T$ and $N$ to find $k$. Second, we compute the amount of time it would take 25 people to do the job.

a) First, we find an equation of variation:

$$T = \frac{k}{N}$$

$$4 = \frac{k}{20} \qquad \text{Substituting 4 for } T \text{ and 20 for } N$$

$$20 \cdot 4 = k$$

$$80 = k.$$

The equation of variation is $T = \dfrac{80}{N}$.

b) We then use the equation to find the amount of time that it takes 25 people to do the job:

$$T = \frac{80}{N}$$

$$= \frac{80}{25} \qquad \text{Substituting 25 for } N$$

$$= 3.2.$$

4. **Check.** The check might be done by repeating the computations. We might also analyze the results. The number of people increased from 20 to 25. Did the time decrease? It did, and this confirms what we expect with inverse variation.

5. **State.** It should take 3.2 hr for 25 people to complete the job.

---

*Do Exercises 7 and 8.*

# Exercise Set 7.5

**a** Find an equation of variation in which *y* varies directly as *x* and the following are true.

**1.** $y = 36$ when $x = 9$   **2.** $y = 60$ when $x = 16$   **3.** $y = 0.8$ when $x = 0.5$   **4.** $y = 0.7$ when $x = 0.4$

**5.** $y = 630$ when $x = 175$   **6.** $y = 400$ when $x = 125$   **7.** $y = 500$ when $x = 60$   **8.** $y = 200$ when $x = 300$

**b** Solve.

**9.** A person's paycheck *P* varies directly as the number *H* of hours worked. For working 15 hr, the pay is $84. Find the pay for 35 hr of work.

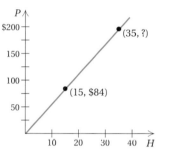

**10.** The interest *I* earned in 1 yr on a fixed principal varies directly as the interest rate *r*. An investment earns $53.55 at an interest rate of 4.25%. How much will the investment earn at a rate of 5.75%?

**11.** The cost *C* to fill a sandbox varies directly as the depth *S* of the sand. Lucinda checks at her local hardware store and finds that it would cost $75 to fill the box with 6 in. of sand. She decides to fill the sandbox to a depth of 8 in. How much will the sand cost Lucinda?

$C = k \cdot S$

**12.** The cost *C* of cement needed to pave a driveway varies directly as the depth *D* of the driveway. John checks at his local building materials store and finds that it costs $500 to install his driveway with a depth of 8 in. He decides to build a stronger driveway at a depth of 12 in. How much will it cost?

$C = k \cdot D$

**13.** A chef is planning meals in a refreshment tent at a golf tournament. The number of servings $S$ of meat that can be obtained from a turkey varies directly as its weight $W$. From a turkey weighing 14 kg, one can get 40 servings of meat. How many servings can be obtained from an 8-kg turkey?

**14.** The number of servings $S$ of meat that can be obtained from round steak varies directly as the weight $W$. From 9 kg of round steak, one can get 70 servings of meat. How many servings can one get from 12 kg of round steak?

**15.** *Weight on Moon.* The weight $M$ of an object on the moon varies directly as its weight $E$ on Earth. A person who weighs 171.6 lb on Earth weighs 28.6 lb on the moon. How much would a 220-lb person weigh on the moon?

**16.** *Weight on Mars.* The weight $M$ of an object on Mars varies directly as its weight $E$ on Earth. A person who weighs 209 lb on Earth weighs 79.42 lb on Mars. How much would a 176-lb person weigh on Mars?

**17.** *Computer Megahertz.* The number of computer instructions $N$ per second varies directly as the speed $S$ of its internal processor. A processor with a speed of 25 megahertz can perform 2,000,000 instructions per second. How many instructions will the same processor perform if it is running at a speed of 200 megahertz?

**18.** *Water in Human Body.* The number of kilograms $W$ of water in a human body varies directly as the total body weight $B$. A person who weighs 75 kg contains 54 kg of water. How many kilograms of water are in a person who weighs 95 kg?

---

**c** Find an equation of variation in which $y$ varies inversely as $x$ and the following are true.

**19.** $y = 3$ when $x = 25$

**20.** $y = 2$ when $x = 45$

**21.** $y = 10$ when $x = 8$

**22.** $y = 10$ when $x = 7$

**23.** $y = 6.25$ when $x = 0.16$

**24.** $y = 0.125$ when $x = 8$

**25.** $y = 50$ when $x = 42$

**26.** $y = 25$ when $x = 42$

**27.** $y = 0.2$ when $x = 0.3$

**28.** $y = 0.4$ when $x = 0.6$

**29.** A production line produces 15 compact disc players every 8 hr. How many players can it produce in 37 hr?

    **a)** What kind of variation might apply to this situation?

    **b)** Solve the problem.

**30.** A person works for 15 hr and makes $93.75. How much will the person make by working 35 hr?

    **a)** What kind of variation might apply to this situation?

    **b)** Solve the problem.

**31.** It takes 4 hr for 9 cooks to prepare the food for a wedding rehearsal dinner. How long will it take 8 cooks to prepare the dinner?

    **a)** What kind of variation might apply to this situation?

    **b)** Solve the problem.

**32.** It takes 16 hr for 2 people to resurface a tennis court. How long will it take 6 people to do the job?

    **a)** What kind of variation might apply to this situation?

    **b)** Solve the problem.

**33.** *Miles per Gallon.* To travel a fixed distance, the number of gallons $N$ of gasoline needed is inversely proportional to the miles-per-gallon rating $P$ of the car. A car that gets 20 miles per gallon (mpg) needs 14 gal to travel the distance. How much gas will be needed for a car that gets 28 mpg?

**34.** *Miles per Gallon.* To travel a fixed distance, the number of gallons $N$ of gasoline needed is inversely proportional to the miles-per-gallon rating $P$ of the car. A car that gets 25 miles per gallon (mpg) needs 12 gal to travel the distance. How much gas will be needed for a car that gets 20 mpg?

**35.** *Electrical Current.* The current $I$ in an electrical conductor varies inversely as the resistance $R$ of the conductor. The current is 96 amperes when the resistance is 20 ohms. What is the current when the resistance is 60 ohms?

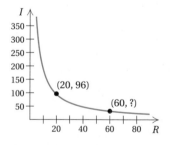

**36.** *Gas Volume.* The volume $V$ of a gas varies inversely as the pressure $P$ on it. The volume of a gas is 200 cm$^3$ under a pressure of 32 kg/cm$^2$. What will be its volume under a pressure of 20 kg/cm$^2$?

**37.** The number of files $N$ of the same size that a computer's hard drive will hold varies inversely as the size of the files. Loretta's hard drive will hold 1600 files if each is 50,000 bytes long. How many files will the drive hold if each is 125,000 bytes long?

**38.** The time $t$ required to empty a tank varies inversely as the rate $r$ of pumping. A pump can empty a tank in 90 min at a rate of 1200 L/min. How long will it take the pump to empty the tank at a rate of 2000 L/min?

**39.** The apparent size $A$ of an object varies inversely as the distance $d$ of the object from the eye. A flagpole 30 ft from an observer appears to be 27.5 ft tall. How tall will the same flagpole appear to be if it is 100 ft from the eye?

**40.** The time $t$ required to drive a fixed distance varies inversely as the speed $r$. It takes 5 hr at 55 mph to drive a fixed distance. How long would it take at 40 mph?

---

**Skill Maintenance**

Solve. [6.6a]

**41.** $\dfrac{x + 2}{x + 5} = \dfrac{x - 4}{x - 6}$

**42.** $\dfrac{x - 3}{x - 5} = \dfrac{x + 5}{x + 1}$

Solve. [5.7b]

**43.** $x^2 - 25x + 144 = 0$

**44.** $t^2 + 21t + 108 = 0$

**45.** $35x^2 + 8 = 34x$

**46.** $14x^2 - 19x - 3 = 0$

Calculate. [1.8d]

**47.** $3^7 \div 3^4 \div 3^3 \div 3$

**48.** $\dfrac{37 - 5(4 - 6)}{2 \cdot 6 + 8}$

---

**Synthesis**

In Exercises 49–52, determine whether the situation represents direct variation, inverse variation, or neither. Give a reason for your answer.

**49.** ◈ The cost of mailing a first-class letter in the United States and the distance that it travels

**50.** ◈ The number of hours that a student watches TV per week and the student's grade point average

**51.** ◈ The weight of a turkey and the cooking time

**52.** ◈ The number of plays that it takes to go 80 yd for a touchdown and the average gain per play

**53.** 🖩 Graph the equation that corresponds to Exercise 12. Then use the TABLE feature to create a table with TblStart = 1 and ΔTbl = 1. What happens to the $y$-values as the $x$-values become larger?

**54.** 🖩 Graph the equation that corresponds to Exercise 13. Then use the TABLE feature to create a table with TblStart = 1 and ΔTbl = 1. What happens to the $y$-values as the $x$-values become larger?

Write an equation of variation for the situation.

**55.** The square of the pitch $P$ of a vibrating string varies directly as the tension $t$ on the string.

**56.** In a stream, the amount $S$ of salt carried varies directly as the sixth power of the speed $V$ of the stream.

**57.** The power $P$ in a windmill varies directly as the cube of the wind speed $V$.

**58.** The volume $V$ of a sphere varies directly as the cube of the radius $r$.

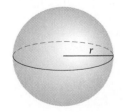

Copyright © 1999 Addison Wesley Longman

# Summary and Review Exercises: Chapter 7

## Important Properties and Formulas

Slope $= m = \dfrac{y_2 - y_1}{x_2 - x_1}$

Slope–Intercept Equation: $\quad y = mx + b$
Parallel Lines: $\qquad\qquad$ Slopes equal, $y$-intercepts different
Perpendicular Lines: $\qquad$ Product of slopes $= -1$
Equation of Direct Variation: $\quad y = kx$
Equation of Inverse Variation: $\quad y = \dfrac{k}{x}$

The objectives to be tested in addition to the material in this chapter are [1.8d], [5.7b], [6.6a], and [6.7a].

Find the slope, if it exists, of the line.  [7.1a]

1.

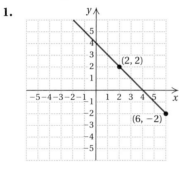

2.

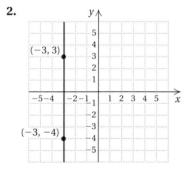

Graph the line containing the given pair of points and find the slope.  [7.1a]

3. $(5, -2)$ and $(-3, 4)$

4. $(0, -3)$ and $(5, -3)$

Find the slope, if it exists, of the line containing the given pair of points.  [7.1a]

5. $(6, 8)$ and $(-2, -4)$

6. $(-8.3, 4.6)$ and $(-9.9, 1.4)$

Find the slope of the line, if it exists.  [7.1b]

7. $y = -6$

8. $x = 90$

9. $4x + 3y = -20$

Find the slope and the $y$-intercept.  [7.2a]

10. $y = -9x + 46$

11. $x + y = 9$

12. $3x - 5y = 4$

Find an equation of the line with the given slope and $y$-intercept.  [7.2a]

13. Slope $= -2.8$; $y$-intercept: $(0, 19)$

14. Slope $= \frac{5}{8}$; $y$-intercept: $\left(0, -\frac{7}{8}\right)$

Find an equation of the line containing the given point and with the given slope.  [7.2b]

15. $(1, 2)$, $m = 3$

16. $(-2, -5)$, $m = \frac{2}{3}$

17. $(0, -4)$, $m = -2$

Find an equation of the line containing the given pair of points.  [7.2c]

18. $(5, 7)$ and $(-1, 1)$

19. $(2, 0)$ and $(-4, -3)$

Solve.  [7.2c]

**20.** *Median Age of Cars.* People are driving cars for longer periods of time. The line graph below describes the *median age of cars A,* in years, for years since 1990.

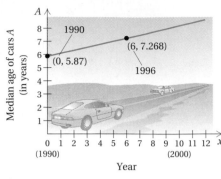

Source: The Polk Co.

a) Find an equation of the line.
b) What is the rate of change in the median age of cars?
c) Use the equation to predict the median age of cars in 2000.

Determine whether the graphs of the equations are parallel, perpendicular, or neither.  [7.3a, b]

**21.** $4x + y = 6,$
$4x + y = 8$

**22.** $2x + y = 10,$
$y = \frac{1}{2}x - 4$

**23.** $x + 4y = 8,$
$x = -4y - 10$

**24.** $3x - y = 6,$
$3x + y = 8$

Determine whether the given point is a solution of the inequality $x - 2y > 1.$  [7.4a]

**25.** $(0, 0)$

**26.** $(1, 3)$

**27.** $(4, -1)$

Graph on a plane.  [7.4b]

**28.** $x < y$

**29.** $x + 2y \geq 4$

**30.** $x > -2$

Find an equation of variation in which $y$ varies directly as $x$ and the following are true.  [7.5a]

**31.** $y = 12$ when $x = 4$

**32.** $y = 4$ when $x = 8$

**33.** $y = 0.4$ when $x = 0.5$

Find an equation of variation in which $y$ varies inversely as $x$ and the following are true.  [7.5c]

**34.** $y = 5$ when $x = 6$

**35.** $y = 0.5$ when $x = 2$

**36.** $y = 1.3$ when $x = 0.5$

Solve.

**37.** A person's paycheck $P$ varies directly as the number $H$ of hours worked. The pay is $165.00 for working 20 hr. Find the pay for 35 hr of work. [7.5b]

**38.** It takes 5 hr for 2 washing machines to wash a fixed amount of laundry. How long would it take 10 washing machines to do the same job? (The number of hours varies inversely as the number of washing machines.)  [7.5d]

---

### Skill Maintenance

**39.** Judd can paint a shed alone in 5 hr. Bud can paint the same shed in 10 hr. How long would it take both of them, working together, to paint the fence? [6.7a]

**40.** Compute: $13 \cdot 6 \div 3 \cdot 26 \div 13.$  [1.8d]

Solve.

**41.** $\dfrac{x^2}{x - 4} = \dfrac{16}{x - 4}$  [6.6a]

**42.** $a^2 + 6a - 55 = 0$  [5.7b]

---

### Synthesis

**43.** ◈ Briefly describe the concept of slope.  [7.1a]

**44.** ◈ Graph $x < 1$ on both a number line and a plane, and explain the difference between the graphs. [7.4b]

**45.** In chess, the knight can move to any of the eight squares shown by $a, b, c, d, e, f, g,$ and $h$ below. If lines are drawn from the beginning to the end of the move, what slopes are possible for these lines?  [7.1a]

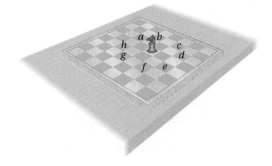

# Test: Chapter 7

Find the slope, if it exists, of the line.

**1.**

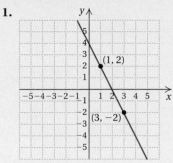

**2.**

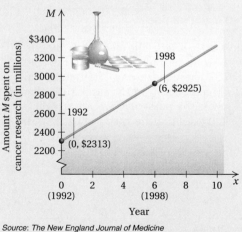

**3.** Graph the line containing the given pair of points and find the slope.

$(5, -4)$ and $(-2, -2)$

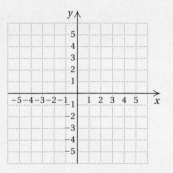

Find the slope, if it exists, of the line containing the given pair of points.

**4.** $(4, 7)$ and $(4, -1)$

**5.** $(9, 2)$ and $(-3, -5)$

Find the slope of the line, if it exists.

**6.** $y = -7$

**7.** $x = 6$

Find the slope and the $y$-intercept.

**8.** $y = 2x - \frac{1}{4}$

**9.** $-4x + 3y = -6$

Find an equation of the line with the given slope and $y$-intercept.

**10.** Slope $= 1.8$; $y$-intercept: $(0, -7)$

**11.** Slope $= -\frac{3}{8}$; $y$-intercept: $\left(0, -\frac{1}{8}\right)$

Find an equation of the line containing the given point and with the given slope.

**12.** $(3, 5)$, $m = 1$

**13.** $(-2, 0)$, $m = -3$

Find an equation of the line containing the given pair of points.

**14.** $(1, 1)$ and $(2, -2)$

**15.** $(4, -1)$ and $(-4, -3)$

**16.** *Cancer Research.* Increasing amounts of money are being spent each year on cancer research. The line graph at right describes the amount spent on cancer research $M$, in millions of dollars, for years since 1992.

    **a)** Find an equation of the line.
    **b)** What is the rate of change in the amount spent on cancer research?
    **c)** Use the equation to predict the amount spent on cancer research in 2000.

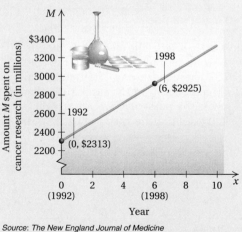

Source: The New England Journal of Medicine

Determine whether the graphs of the equations are parallel, perpendicular, or neither.

**17.** $2x + y = 8$,
    $2x + y = 4$

**18.** $2x + 5y = 2$,
    $y = 2x + 4$

**19.** $x + 2y = 8$,
    $-2x + y = 8$

## Answers

1. _____

2. _____

3. _____

4. _____

5. _____

6. _____

7. _____

8. _____

9. _____

10. _____

11. _____

12. _____

13. _____

14. _____

15. _____

16. **a)** _____

    **b)** _____

    **c)** _____

17. _____

18. _____

19. _____

Determine whether the given point is a solution of the inequality $3y - 2x < -2$.

**20.** $(0, 0)$

**21.** $(-4, -10)$

Graph on a plane.

**22.** $y > x - 1$

**23.** $2x - y \leq 4$

Find an equation of variation in which $y$ varies directly as $x$ and the following are true.

**24.** $y = 6$ when $x = 3$

**25.** $y = 1.5$ when $x = 3$

Find an equation of variation in which $y$ varies inversely as $x$ and the following are true.

**26.** $y = 6$ when $x = 3$

**27.** $y = 11$ when $x = 2$

Solve.

**28.** The distance $d$ traveled by a train varies directly as the time $t$ that it travels. The train travels 60 km in $\frac{1}{2}$ hr. How far will it travel in 2 hr?

**29.** It takes 3 hr for 2 concrete mixers to mix a fixed amount of concrete. The number of hours varies inversely as the number of concrete mixers used. How long would it take 5 concrete mixers to do the same job?

**Skill Maintenance**

**30.** *Train Speeds.* The speed of a freight train is 15 mph slower than the speed of a passenger train. The freight train travels 360 mi in the same time that it takes the passenger train to travel 420 mi. Find the speed of each train.

**31.** Compute: $\dfrac{3^2 - 2^3}{2^2 + 3 - 12 \div 2}$.

Solve.

**32.** $\dfrac{x^2}{x + 10} = \dfrac{100}{x + 10}$

**33.** $a^2 + 3a - 28 = 0$

**Synthesis**

**34.** Find the value of $k$ such that $3x + 7y = 14$ and $ky - 7x = -3$ are perpendicular.

**35.** Find the slope–intercept equation of the line that contains the point $(-4, 1)$ and has the same slope as the line $2x - 3y = -6$.

Answers

20. _____

21. _____

22. _____

23. _____

24. _____

25. _____

26. _____

27. _____

28. _____

29. _____

30. _____

31. _____

32. _____

33. _____

34. _____

35. _____

Copyright © 1999 Addison Wesley Longman

# Cumulative Review: Chapters 1–7

**1.** Find the absolute value: $|3.5|$.

**2.** Identify the coefficient of each term of the polynomial
$$x^3 - 2x^2 + x - 1.$$

**3.** Identify the degree of each term and the degree of the polynomial
$$x^3 - 2x^2 + x - 1.$$

**4.** Classify this polynomial as a monomial, binomial, trinomial, or none of these:
$$x^3 - 2x^2 + x - 1.$$

**5.** Collect like terms: $x^2 - 3x^3 - 4x^2 + 5x^3 - 2$.

Simplify.

**6.** $\dfrac{1}{2}x - \left[\dfrac{3}{8}x - \left(\dfrac{2}{3} + \dfrac{1}{4}x\right) - \dfrac{1}{3}\right]$

**7.** $\left(\dfrac{2x^3}{3x^{-1}}\right)^{-2}$

**8.**

Perform the indicated operations. Simplify, if possible.

**9.** $(5xy^2 - 6x^2y^2 - 3xy^3) - (-4xy^3 + 7xy^2 - 2x^2y^2)$

**10.** $(4x^4 + 6x^3 - 6x^2 - 4) + (2x^5 + 2x^4 - 4x^3 - 4x^2 + 3x - 5)$

**11.** $\dfrac{2y + 4}{21} \cdot \dfrac{7}{y^2 + 4y + 4}$

**12.** $\dfrac{x^2 - 9}{x^2 + 8x + 15} \div \dfrac{x - 3}{2x + 10}$

**13.** $\dfrac{x^2}{x - 4} + \dfrac{16}{4 - x}$

**14.** $\dfrac{5x}{x^2 - 4} - \dfrac{-3}{2 - x}$

Multiply.

**15.** $(2.5a + 7.5)(0.4a - 1.2)$

**16.** $(2x^2 - 1)(x^3 + x - 3)$

**17.** $(2x^3 + 1)(2x^3 - 1)$

**18.** $(6x - 5)^2$

**19.** $4x(3x^3 + 4x^2 + x)$

**20.** $(2x^5 + 3)(3x^2 - 6)$

Solve.

**21.** $x - [x - (x - 1)] = 2$

**22.** $2x^2 + 7x = 4$

**23.** $x^2 + x - 20 = 0$

**24.** $3(x - 2) \le 4(x + 5)$

**25.** $x(x - 4) = 0$

**26.** $x^2 = 10x$

**27.** $2x^2 = 800$

**28.** $t = ax + ay$, for $a$

**29.** $\dfrac{5x - 2}{4} - \dfrac{4x - 5}{3} = 1$

**30.** $\dfrac{2x}{x - 3} - \dfrac{6}{x} = \dfrac{18}{x^2 - 3x}$

Factor.

**31.** $-6 - 2x - 12y$

**32.** $x^2 - 10x + 24$

**33.** $2x^2 - 18$

**34.** $m^4 + 2m^3 - 3m - 6$

**35.** $16x^2 + 40x + 25$

**36.** $8x^2 + 10x + 3$

Solve.

**37.** The product of a number and 1 more than the number is 20. Find the number.

**38.** A person's salary varies directly as the number of hours worked. For working 9 hr, the salary is $117. Find the salary for working 6 hr.

**39.** Money is borrowed at 6% simple interest. After 1 yr, $2650 pays off the loan. How much was originally borrowed?

**40.** One car travels 105 mi in the same time that a car traveling 10 mph slower travels 75 mi. Find the speed of each car.

**41.** If the sides of a square are increased by 2 ft, the sum of the areas of the two squares is 452 ft$^2$. Find the length of a side of the original square.

**42.** One number is 7 more than another number. The quotient of the larger divided by the smaller is $\frac{5}{4}$. Find the numbers.

Graph on a plane.

**43.** $y = \dfrac{1}{2}x$

**44.** $3x - 5y = 15$

**45.** $y = 1$

**46.** $y < -x - 2$

**47.** $x \le -3$

**48.** Find an equation of variation in which $y$ varies directly as $x$ and $y = 8$ when $x = 12$.

**49.** Find an equation of variation in which $y$ varies inversely as $x$ and $y = 20$ when $x = 0.5$.

Find the slope, if it exists, of the line containing the given pair of points.

**50.** $(-2, 6)$ and $(-2, -1)$

**51.** $(-4, 1)$ and $(3, -2)$

**52.** Find the slope and the $y$-intercept of $4x - 3y = 6$.

**53.** Find an equation for the line containing the point $(2, -3)$ and having slope $m = -4$.

**54.** Find an equation of the line containing the points $(-1, -3)$ and $(5, -2)$.

Determine whether the graphs of the equations are parallel, perpendicular, or neither.

**55.** $2x = 7 - 3y$,
$7 + 2x = 3y$

**56.** $x - y = 4$,
$y = x + 5$

---

**Synthesis**

**57.** Simplify: $(x + 7)(x - 4) - (x + 8)(x - 5)$.

**58.** Multiply: $[4y^3 - (y^2 - 3)][4y^3 + (y^2 - 3)]$.

**59.** Factor: $2a^{32} - 13{,}122b^{40}$.

**60.** Solve: $(x - 4)(x + 7)(x - 12) = 0$.

**61.** Find an equation of the line that contains the point $(-3, -2)$ and is parallel to the line $2x - 3y = -12$.

**62.** Find all numbers for which the following complex rational expression is undefined:

$$\dfrac{\dfrac{1}{x} + x}{2 + \dfrac{1}{x - 3}}.$$

# Systems of Equations

## Introduction

We now consider how the graphs of two linear equations might intersect. Such a point of intersection is a solution of what is called a *system of equations*. Many applications and problems involve two facts about two quantities and are easier to solve by translating to a system of two equations in two variables. Systems of equations have extensive applications in many fields such as psychology, sociology, business, education, engineering, and science.

## An Application

In most areas of the United States, gas stations offer three grades of gasoline, indicated by octane ratings on the pumps, such as 87, 89, and 93. When a tanker delivers gas, it brings only two grades of gasoline, the highest and lowest, filling two large underground tanks. If you purchase the middle grade, the pump's computer mixes the other two grades appropriately. How much 87-octane gas and 93-octane gas should be blended in order to make 18 gal of 89-octane gas?

This problem appears as Exercise 37 in Exercise Set 8.4.

## The Mathematics

We let $x =$ the number of gallons of the 87-octane gasoline and $y =$ the number of gallons of the 93-octane gasoline. Then we can translate the problem to this pair of equations:

$$\left.\begin{array}{l} x + y = 18, \\ 87x + 93y = 89 \cdot 18. \end{array}\right\}$$

This is a *system of equations.*

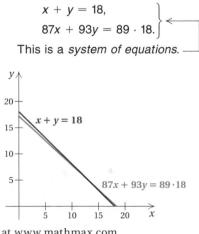

# Pretest: Chapter 8

1. Determine whether the ordered pair $(-1, 1)$ is a solution of the system of equations

    $2x + y = -1,$
    $3x - 2y = -5.$

2. Solve this system by graphing.

    $2x = y + 1,$
    $2x - y = 5$

Solve by the substitution method.

3. $x + y = 7,$
   $x = 2y + 1$

4. $2x - 3y = 7,$
   $x + y = 1$

Solve by the elimination method.

5. $2x - y = 1,$
   $2x + y = 2$

6. $2x - 3y = -4,$
   $3x - 4y = -7$

7. $\dfrac{3}{5}x - \dfrac{1}{4}y = 4,$

   $\dfrac{1}{5}x + \dfrac{3}{4}y = 8$

8. Find two numbers whose sum is 74 and whose difference is 26.

9. Two angles are complementary. (Complementary angles are angles whose sum is 90°.) One angle is 15° more than twice the other. Find the angles.

10. A train leaves a station and travels north at 96 mph. Two hours later, a second train leaves on a parallel track and travels north at 120 mph. When will it overtake the first train?

## Objectives for Retesting

The objectives to be tested in addition to the material in this chapter are as follows.

[3.3a]     Find the intercepts of a linear equation, and graph using intercepts.

[4.1d, e, f]   Use the product and quotient rules to multiply and divide exponential expressions with like bases, and express exponential expressions involving negative exponents with positive exponents.

[6.1c]     Simplify rational expressions by factoring the numerator and the denominator and removing factors of 1.

[6.5a]     Subtract rational expressions.

# 8.1 Systems of Equations in Two Variables

## a | Systems of Equations and Solutions

Many problems can be solved more easily by translating to two equations in two variables. The following is such a **system of equations:**

$$x + y = 8,$$
$$2x - y = 1.$$

> A **solution** of a system of two equations is an ordered pair that makes both equations true.

Look at the graphs shown at right. Recall that a graph of an equation is a drawing that represents its solution set. Each point on the graph corresponds to a solution of that equation. Which points (ordered pairs) are solutions of *both* equations?

The graph shows that there is only one. It is the point $P$ where the graphs cross. This point looks as if its coordinates are $(3, 5)$. We check to see if $(3, 5)$ is a solution of *both* equations, substituting 3 for $x$ and 5 for $y$.

CHECK:

$$\begin{array}{c|c} x + y = 8 \\ \hline 3 + 5 \ ? \ 8 \\ 8 \ | \qquad \text{TRUE} \end{array} \qquad \begin{array}{c|c} 2x - y = 1 \\ \hline 2 \cdot 3 - 5 \ ? \ 1 \\ 6 - 5 \ | \\ 1 \ | \qquad \text{TRUE} \end{array}$$

There is just one solution of the system of equations. It is $(3, 5)$. In other words, $x = 3$ and $y = 5$.

**Example 1** Determine whether $(1, 2)$ is a solution of the system

$$y = x + 1,$$
$$2x + y = 4.$$

We check by substituting alphabetically 1 for $x$ and 2 for $y$.

CHECK:

$$\begin{array}{c|c} y = x + 1 \\ \hline 2 \ ? \ 1 + 1 \\ | \ 2 \qquad \text{TRUE} \end{array} \qquad \begin{array}{c|c} 2x + y = 4 \\ \hline 2 \cdot 1 + 2 \ ? \ 4 \\ 2 + 2 \ | \\ 4 \ | \qquad \text{TRUE} \end{array}$$

This checks, so $(1, 2)$ is a solution of the system.

**Example 2** Determine whether $(-3, 2)$ is a solution of the system

$$p + q = -1,$$
$$q + 3p = 4.$$

*don't intersect*

We check by substituting alphabetically $-3$ for $p$ and 2 for $q$.

CHECK:

$$\begin{array}{c|c} p + q = -1 \\ \hline -3 + 2 \ ? \ -1 \\ -1 \ | \qquad \text{TRUE} \end{array} \qquad \begin{array}{c|c} q + 3p = 4 \\ \hline 2 + 3(-3) \ ? \ 4 \\ 2 - 9 \ | \\ -7 \ | \qquad \text{FALSE} \end{array}$$

The point $(-3, 2)$ is not a solution of $q + 3p = 4$. Thus it is not a solution of the system.

## Objectives

a | Determine whether an ordered pair is a solution of a system of equations.

b | Solve systems of two linear equations in two variables by graphing.

### For Extra Help

TAPE 14    TAPE 14B    MAC    CD-ROM
                        WIN

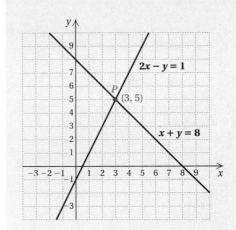

Determine whether the given ordered pair is a solution of the system of equations.

1. $(2, -3)$; $\quad x = 2y + 8$,
$\qquad\qquad 2x + y = 1$

2. $(20, 40)$; $\quad a = \dfrac{1}{2}b$,
$\qquad\qquad b - a = 60$

3. Solve this system by graphing:

$\qquad 2x + y = 1$,
$\qquad x = 2y + 8$.

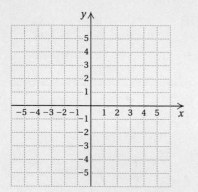

### Calculator Spotlight

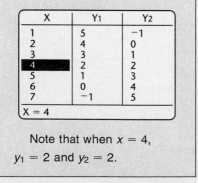

Checking with the TABLE feature. We can use the TABLE feature to check a possible solution of a system. Consider the system in Example 3. We enter each equation in the "y =" form:

$\qquad y_1 = 6 - x$,
$\qquad y_2 = x - 2$.

We set TblStart = 1 and ΔTbl = 1 and obtain the following table.

| X | Y₁ | Y₂ |
|---|---|---|
| 1 | 5 | −1 |
| 2 | 4 | 0 |
| 3 | 3 | 1 |
| 4 | 2 | 2 |
| 5 | 1 | 3 |
| 6 | 0 | 4 |
| 7 | −1 | 5 |
| X = 4 | | |

Note that when $x = 4$, $y_1 = 2$ and $y_2 = 2$.

*Answers on page A-33*

Example 2 illustrates that an ordered pair may be a solution of one equation but *not both*. If that is the case, it is *not* a solution of the system.

*Do Exercises 1 and 2.*

## b | Graphing Systems of Equations

Recall that the **graph** of an equation is a drawing that represents its solution set. If the graph of an equation is a line, then every point on the line corresponds to an ordered pair that is a solution of the equation. If we graph a **system** of two linear equations, we graph both equations and find the coordinates of the points of intersection, if any exist.

**Example 3** Solve this system of equations by graphing:

$\qquad x + y = 6$,
$\qquad x = y + 2$.

We graph the equations using any of the methods studied in Chapter 3. Point $P$ with coordinates $(4, 2)$ looks as if it is the solution.

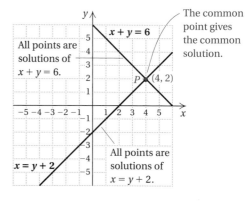

We check the pair as follows.

CHECK:

$\dfrac{x + y = 6}{4 + 2 \;?\; 6}$ 　　　　$\dfrac{x = y + 2}{4 \;?\; 2 + 2}$
$\qquad\quad 6 \mid$ 　TRUE 　　　$\qquad\quad \mid 4$ 　TRUE

The solution is $(4, 2)$.

*Do Exercise 3.*

**Example 4** Solve this system of equations by graphing:

$\qquad x = 2$,
$\qquad y = -3$.

The graph of $x = 2$ is a vertical line, and the graph of $y = -3$ is a horizontal line. They intersect at the point $(2, -3)$. The solution is $(2, -3)$.

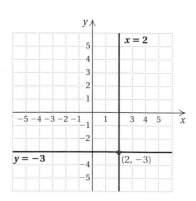

*Do Exercise 4 on the following page.*

Sometimes the equations in a system have graphs that are parallel lines.

**Example 5**  Solve this system of equations by graphing:

$$y = 3x + 4,$$
$$y = 3x - 3.$$

We graph the equations, again using any of the methods studied in Chapter 3. The lines have the same slope, 3, and different $y$-intercepts, $(0, 4)$ and $(0, -3)$, so they are parallel.

There is no point at which the lines cross, so the system has no solution. The solution set is the empty set, denoted $\varnothing$, or $\{\ \}$.

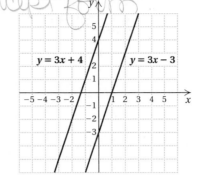

*Do Exercise 5.*

Sometimes the equations in a system have the same graph.

**Example 6**  Solve this system of equations by graphing:

$$2x + 3y = 6,$$
$$-8x - 12y = -24.$$

We graph the equations and see that the graphs are the same. Thus any solution of one of the equations is a solution of the other. Each equation has an infinite number of solutions, some of which are indicated on the graph.

We check one such solution, $(0, 2)$: the $y$-intercept of each equation.

CHECK:

| $2x + 3y = 6$ | |
|---|---|
| $2(0) + 3(2)\ ?\ 6$ | |
| $0 + 6$ | |
| $6$ | TRUE |

| $-8x - 12y = -24$ | |
|---|---|
| $-8(0) - 12(2)\ ?\ -24$ | |
| $0 - 24$ | |
| $-24$ | TRUE |

We leave it to the student to check that $(-3, 4)$ is also a solution of the system. If $(0, 2)$ and $(-3, 4)$ are solutions, then all points on the line containing them are solutions. The system has an infinite number of solutions.

*Do Exercise 6.*

When we graph a system of two equations in two variables, we obtain one of the following three results.

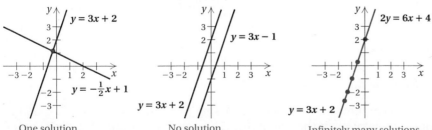

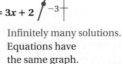

One solution.
Graphs intersect.

No solution.
Graphs are parallel.

Infinitely many solutions.
Equations have
the same graph.

**4.** Solve this system by graphing:

$$x = -4,$$
$$y = 3.$$

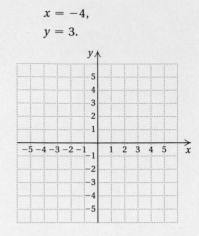

**5.** Solve this system by graphing:

$$y + 4 = x,$$
$$x - y = -2.$$

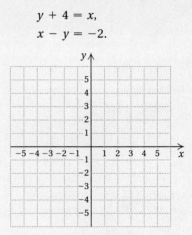

**6.** Solve this system by graphing:

$$2x + y = 4,$$
$$-6x - 3y = -12.$$

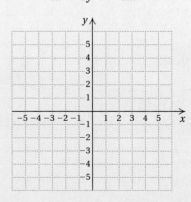

*Answers on page A-33*

**7. a)** Solve $2x - 1 = 8 - x$ algebraically.

**b)** Solve $2x - 1 = 8 - x$ graphically using method 1.

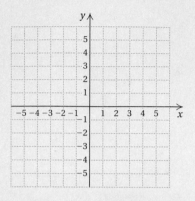

**c)** Compare your answers to parts (a) and (b).

**8. a)** Solve $2x - 1 = 8 - x$ graphically using method 2.

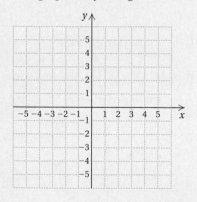

**b)** Compare your answers to 7(a), 7(b), and 8(a).

*Answers on page A-33*

**AG** **Algebraic–Graphical Connection**

To bring together the concepts of Chapters 1–8, let's take an algebraic–graphical look at equation solving. Such interpretation is useful when using a graphing calculator or computer graphing software.

Consider the equation $6 - x = x - 2$. Let's solve it algebraically as we did in Chapter 2:

$$6 - x = x - 2$$

$6 = 2x - 2$     Adding $x$

$8 = 2x$     Adding 2

$4 = x.$     Dividing by 2

Can we also solve the equation graphically? We can, as we see in the following two methods.

**METHOD 1**   Solve $6 - x = x - 2$ graphically.

We let $y = 6 - x$ and $y = x - 2$. Graphing the system of equations as we did in Example 3 gives us the following. The point of intersection is $(4, 2)$. Note that the $x$-coordinate of the intersection is 4. This value for $x$ is *also* the solution of the equation $6 - x = x - 2$.

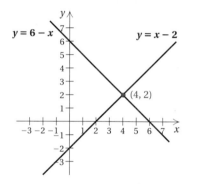

*Do Exercise 7.*

**METHOD 2**   Solve $6 - x = x - 2$ graphically.

Adding $x$ and $-6$ on both sides, we obtain the form $0 = 2x - 8$. In this case, we let $y = 0$ and $y = 2x - 8$. Since $y = 0$ is the $x$-axis, we need only graph $y = 2x - 8$ and see where it crosses the $x$-axis. Note that the $x$-intercept of $y = 2x - 8$ is $(4, 0)$, or just 4. This $x$-value is *also* the solution of the equation $6 - x = x - 2$.

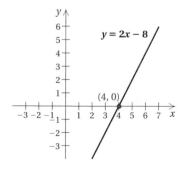

*Do Exercise 8.*

Let's compare the two methods. Using method 1, we graph two equations. The solution of the original equation is the $x$-coordinate of the point of intersection. Using method 2, we find that the solution of the original equation is the $x$-intercept of the graph.

# Exercise Set 8.1

**a** Determine whether the given ordered pair is a solution of the system of equations. Use alphabetical order of the variables.

**1.** (1, 5); $5x - 2y = -5$,
$3x - 7y = -32$

**2.** (3, 2); $2x + 3y = 12$,
$x - 4y = -5$

**3.** (4, 2); $3b - 2a = -2$,
$b + 2a = 8$

**4.** (6, −6); $t + 2s = 6$,
$t - s = -12$

**5.** (15, 20); $3x - 2y = 5$,
$6x - 5y = -10$

$(-1) \ (-5)$

**6.** (−1, −5); $4r + s = -9$,
$3r = 2 + s$

**7.** (−1, 1); $x = -1$,
$x - y = -2$

**8.** (−3, 4); $2x = -y - 2$,
$y = -4$

**9.** (18, 3); $y = \dfrac{1}{6}x$,
$2x - y = 33$

**10.** (−3, 1); $y = -\dfrac{1}{3}x$,
$3y = -5x - 12$

**b** Solve the system of equations by graphing.

**11.** $x - y = 2$,
$x + y = 6$

**12.** $x + y = 3$,
$x - y = 1$

**13.** $8x - y = 29$,
$2x + y = 11$

**14.** $4x - y = 10$,
$3x + 5y = 19$

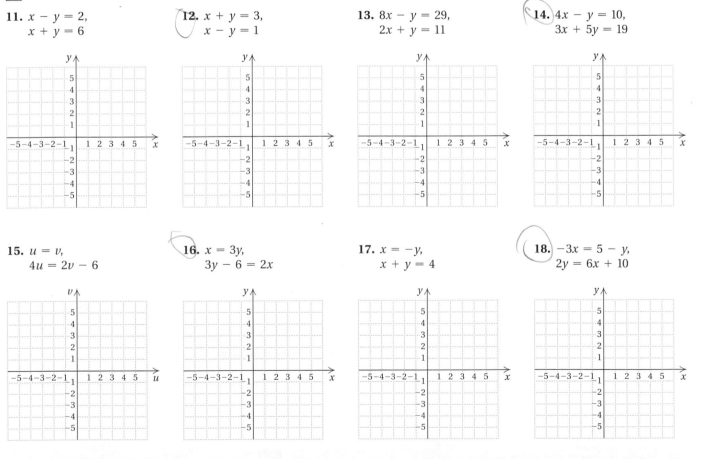

**15.** $u = v$,
$4u = 2v - 6$

**16.** $x = 3y$,
$3y - 6 = 2x$

**17.** $x = -y$,
$x + y = 4$

**18.** $-3x = 5 - y$,
$2y = 6x + 10$

**19.** $a = \dfrac{1}{2}b + 1,$

$\quad a - 2b = -2$

**20.** $x = \dfrac{1}{3}y + 2,$

$\quad -2x - y = 1$

**21.** $y - 2x = 0,$

$\quad y = 6x - 2$

**22.** $y = 3x,$

$\quad y = -3x + 2$

**23.** $x + y = 9,$

$\quad 3x + 3y = 27$

**24.** $x + y = 4,$

$\quad x + y = -4$

**25.** $x = 5,$

$\quad y = -3$

**26.** $y = 2,$

$\quad y = -4$

---

**Skill Maintenance**

**27.** Multiply: $(9x^{-5})(12x^{-8})$.   [4.1d, f].

**28.** Divide: $\dfrac{9x^{-5}}{3x^{-8}}$.   [4.1e, f]

Simplify.

**29.** $\dfrac{1}{x} - \dfrac{1}{x^2} + \dfrac{1}{x+1}$   [6.5b]

**30.** $\dfrac{3-x}{x-2} - \dfrac{x-7}{2-x}$   [6.5a]

**31.** $\dfrac{x+2}{x-4} - \dfrac{x+1}{x+4}$   [6.5a]

**32.** $\dfrac{2x^2 - x - 15}{x^2 - 9}$   [6.1c]

Classify the polynomial as a monomial, binomial, trinomial, or none of these.   [4.3i]

**33.** $5x^2 - 3x + 7$

**34.** $4x^3 - 2x^2$

**35.** $1.8x^5$

**36.** $x^3 + 2x^2 - 3x + 1$

---

**Synthesis**

**37.** ◆ Suppose you have shown that the solution of the equation $3x - 1 = 9 - 2x$ is 2. How can this result be used to determine where the graphs of $y = 3x - 1$ and $y = 9 - 2x$ intersect?

**38.** ◆ Graph this system of equations. What happens when you try to determine a solution from the graph?

$\quad x - 2y = 6,$

$\quad 3x + 2y = 4$

**39.** The solution of the following system is $(2, -3)$. Find $A$ and $B$.

$\quad Ax - 3y = 13,$

$\quad x - By = 8$

**40.** Find an equation to go with $5x + 2y = 11$ such that the solution of the system is $(3, -2)$. Answers may vary.

**41.** Find a system of equations with $(6, -2)$ as a solution. Answers may vary.

**42.–49.** 〰 Use the TABLE feature to check your answers to Exercises 11–18.

Copyright © 1999 Addison Wesley Longman

# 8.2 The Substitution Method

Consider the following system of equations:

$$3x + 7y = 5,$$
$$6x - 7y = 1.$$

Suppose we try to solve this system graphically. We obtain this graph.

**Objectives**

**a** Solve a system of two equations in two variables by the substitution method when one of the equations has a variable alone on one side.

**b** Solve a system of two equations in two variables by the substitution method when neither equation has a variable alone on one side.

**c** Solve applied problems by translating to a system of two equations and then solving using the substitution method.

**For Extra Help**

TAPE 14     TAPE 15A     MAC WIN     CD-ROM

What is the solution? It is rather difficult to tell exactly. It would appear that fractions are involved. It turns out that the solution is $\left(\frac{2}{3}, \frac{3}{7}\right)$. We need techniques involving algebra to determine the solution exactly. Graphing helps us picture the solution of a system of equations, but solving by graphing, though practical in many applications, is not always fast or accurate in cases where solutions are not integers. We now learn other methods using algebra. Because they use algebra, they are called **algebraic**.

## a   Solving by the Substitution Method

One nongraphical method for solving systems is known as the **substitution method.** In Example 1, we use the substitution method to solve a system we graphed in Example 3 of Section 8.1.

**Example 1**   Solve the system

$$x + y = 6, \qquad \textbf{(1)}$$
$$x = y + 2. \qquad \textbf{(2)}$$

Equation (2) says that $x$ and $y + 2$ name the same thing. Thus in equation (1), we can substitute $y + 2$ for $x$:

$$x + y = 6 \qquad \text{Equation (1)}$$
$$(y + 2) + y = 6. \qquad \text{Substituting } y + 2 \text{ for } x$$

This last equation has only one variable. We solve it:

$$y + 2 + y = 6 \qquad \text{Removing parentheses}$$
$$2y + 2 = 6 \qquad \text{Collecting like terms}$$
$$2y + 2 - 2 = 6 - 2 \qquad \text{Subtracting 2 on both sides}$$
$$2y = 4 \qquad \text{Simplifying}$$
$$\frac{2y}{2} = \frac{4}{2} \qquad \text{Dividing by 2}$$
$$y = 2. \qquad \text{Simplifying}$$

We have found the $y$-value of the solution. To find the $x$-value, we return to the original pair of equations. Substituting into either equation will give us the $x$-value.

1. Solve by the substitution method. Do not graph.

$$x + y = 5,$$
$$x = y + 1$$

We choose equation (2) because it has $x$ alone on one side:

$$x = y + 2 \qquad \text{Equation (2)}$$
$$\phantom{x} = 2 + 2 \qquad \text{Substituting 2 for } y$$
$$\phantom{x} = 4.$$

The ordered pair (4, 2) may be a solution. We check.

CHECK:

$$\begin{array}{c} x + y = 6 \\ \hline 4 + 2 \ ? \ 6 \\ 6 \ | \qquad \text{TRUE} \end{array} \qquad \begin{array}{c} x = y + 2 \\ \hline 4 \ ? \ 2 + 2 \\ | \ 4 \qquad \text{TRUE} \end{array}$$

Since (4, 2) checks, we have the solution. We could also express the answer as $x = 4$, $y = 2$.

Note in Example 1 that substituting 2 for $y$ in equation (1) will also give us the $x$-value of the solution:

$$x + y = 6$$
$$x + 2 = 6$$
$$x = 4.$$

Note also that we are using alphabetical order in listing the coordinates in an ordered pair. That is, since $x$ precedes $y$, we list 4 before 2 in the pair (4, 2).

*Do Exercise 1.*

**Example 2** Solve the system

$$t = 1 - 3s, \qquad \textbf{(1)}$$
$$s - t = 11. \qquad \textbf{(2)}$$

We substitute $1 - 3s$ for $t$ in equation (2):

$$s - t = 11 \qquad \text{Equation (2)}$$
$$s - (1 - 3s) = 11. \qquad \text{Substituting } 1 - 3s \text{ for } t$$

Remember to use parentheses when you substitute.

Now we solve for $s$:

$$s - 1 + 3s = 11 \qquad \text{Removing parentheses}$$
$$4s - 1 = 11 \qquad \text{Collecting like terms}$$
$$4s = 12 \qquad \text{Adding 1}$$
$$s = 3. \qquad \text{Dividing by 4}$$

Next, we substitute 3 for $s$ in equation (1) of the original system:

$$t = 1 - 3s \qquad \text{Equation (1)}$$
$$\phantom{t} = 1 - 3 \cdot 3 \qquad \text{Substituting 3 for } s$$
$$\phantom{t} = -8.$$

The pair $(3, -8)$ checks and is the solution. Remember: We list the answer in alphabetical order, $(s, t)$. That is, since $s$ comes before $t$ in the alphabet, 3 is listed first and $-8$ second.

*Do Exercise 2.*

Answers on page A-33

---

**Calculator Spotlight**

Use the TABLE feature to check the possible solutions of the systems in Example 1 and Margin Exercise 1.

---

2. Solve by the substitution method:

$$a - b = 4,$$
$$b = 2 - a.$$

## b | Solving for the Variable First

Sometimes neither equation of a pair has a variable alone on one side. Then we solve one equation for one of the variables and proceed as before, substituting into the *other* equation. If possible, we solve in either equation for a variable that has a coefficient of 1.

**Example 3** Solve the system

$$x - 2y = 6, \quad \textbf{(1)}$$
$$3x + 2y = 4. \quad \textbf{(2)}$$

We solve one equation for one variable. Since the coefficient of $x$ is 1 in equation (1), it is easier to solve that equation for $x$:

$$x - 2y = 6 \qquad \text{Equation (1)}$$
$$x = 6 + 2y. \qquad \text{Adding } 2y \quad \textbf{(3)}$$

We substitute $6 + 2y$ for $x$ in equation (2) of the original pair and solve for $y$:

$$3x + 2y = 4 \qquad \text{Equation (2)}$$
$$3(6 + 2y) + 2y = 4 \qquad \text{Substituting } 6 + 2y \text{ for } x$$
$$18 + 6y + 2y = 4 \qquad \text{Removing parentheses}$$
$$18 + 8y = 4 \qquad \text{Collecting like terms}$$
$$8y = -14 \qquad \text{Subtracting 18}$$
$$y = \frac{-14}{8}, \text{ or } -\frac{7}{4}. \qquad \text{Dividing by 8}$$

To find $x$, we go back to either of the original equations (1) or (2) or to equation (3), which we solved for $x$. It is generally easier to use an equation like equation (3) where we have solved for a specific variable. We substitute $-\frac{7}{4}$ for $y$ in equation (3) and compute $x$:

$$x = 6 + 2y \qquad \text{Equation (3)}$$
$$= 6 + 2\left(-\frac{7}{4}\right) \qquad \text{Substituting } -\frac{7}{4} \text{ for } y$$
$$= 6 - \frac{7}{2} = \frac{5}{2}.$$

We check the ordered pair $\left(\frac{5}{2}, -\frac{7}{4}\right)$.

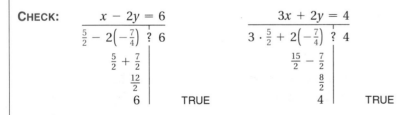

CHECK:

$$\begin{array}{c|c}
x - 2y = 6 & 3x + 2y = 4 \\
\hline
\frac{5}{2} - 2\left(-\frac{7}{4}\right) \; ? \; 6 & 3 \cdot \frac{5}{2} + 2\left(-\frac{7}{4}\right) \; ? \; 4 \\
\frac{5}{2} + \frac{7}{2} & \frac{15}{2} - \frac{7}{2} \\
\frac{12}{2} & \frac{8}{2} \\
6 \;\; | \quad \text{TRUE} & 4 \;\; | \quad \text{TRUE}
\end{array}$$

Since $\left(\frac{5}{2}, -\frac{7}{4}\right)$ checks, it is the solution.

This solution would have been difficult to find graphically because it involves fractions.

***Do Exercise 3.***

## c | Solving Applied Problems

Now let's solve an applied problem using systems of equations and the substitution method.

---

**3.** Solve:

$$x - 2y = 8,$$
$$2x + y = 8.$$

### Calculator Spotlight

 Solving Systems. Graphers can be used to solve systems of equations, provided each equation has been solved first for $y$. Consider the system in Example 3. We put each equation in the "$y =$" form by solving for $y$:

$$y_1 = (1/2)x - 3, \qquad \textbf{(1)}$$
$$y_2 = (-3/2)x + 2. \qquad \textbf{(2)}$$

We then enter the equations.

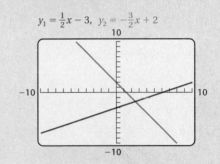

To find the solution, we use the CALC-INTERSECT feature.

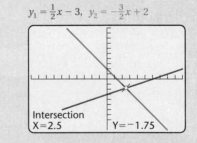

The solution is listed at the bottom of the screen.

**Exercises**

Use a grapher to solve each of the following systems.

**1.** Margin Exercise 3

**2.** $6x - 7y = 1,$
$\quad 3x + 7y = 5$

*Answers on page A-33*

**4.** *Perimeter of High School Court.* The perimeter of a standard high school basketball court is 268 ft. The length is 34 ft longer than the width. Find the dimensions of the court.

**Example 4**  *Perimeter of NBA Court.* The perimeter of an NBA-sized basketball court is 288 ft. The length is 44 ft longer than the width. (*Source*: National Basketball Association) Find the dimensions of the court.

1. **Familiarize.**  We make a drawing and label it, calling the length $l$ and the width $w$.

2. **Translate.**  The perimeter of the rectangle is $2l + 2w$. We translate the first statement.

$$\underbrace{\text{The perimeter}}_{\downarrow} \quad \underbrace{\text{is}}_{\downarrow} \quad \underbrace{\text{288 ft.}}_{\downarrow}$$

$$2l + 2w \quad = \quad 288$$

We translate the second statement:

$$\underbrace{\text{The length}}_{\downarrow} \quad \underbrace{\text{is}}_{\downarrow} \quad \underbrace{\text{44 ft longer than the width.}}_{\downarrow}$$

$$l \quad = \quad 44 + w$$

We now have a system of equations:

$$2l + 2w = 288, \quad \textbf{(1)}$$
$$l = 44 + w. \quad \textbf{(2)}$$

3. **Solve.**  We solve the system. To begin, we substitute $44 + w$ for $l$ in the first equation and solve:

| | |
|---|---|
| $2(44 + w) + 2w = 288$ | Substituting $44 + w$ for $l$ in equation (1) |
| $2 \cdot 44 + 2 \cdot w + 2w = 288$ | Using a distributive law |
| $4w + 88 = 288$ | Collecting like terms |
| $4w + 88 - 88 = 288 - 88$ | Subtracting 88 |
| $4w = 200$ | |
| $w = 50.$ | Dividing by 4 |

We go back to the original equations and substitute 50 for $w$:

$$l = 44 + w = 44 + 50 = 94. \quad \text{Substituting in equation (2)}$$

4. **Check.**  If the width is 50 ft and the length is 94 ft, then the length is 44 ft more than the width and the perimeter is 2(50 ft) + 2(94 ft), or 288 ft. This checks.

5. **State.**  The width is 50 ft and the length is 94 ft.

The problem in Example 4 illustrates that many problems that can be solved by translating to *one* equation in *one* variable are actually easier to solve by translating to *two* equations in *two* variables.

*Answer on page A-33*

*Do Exercise 4.*

# Exercise Set 8.2

**a** Solve using the substitution method.

**1.** $x + y = 10,$
$\quad y = x + 8$

**2.** $x + y = 4,$
$\quad y = 2x + 1$

**3.** $y = x - 6,$
$\quad x + y = -2$

**4.** $y = x + 1,$
$\quad 2x + y = 4$

**5.** $y = 2x - 5,$
$\quad 3y - x = 5$

**6.** $y = 2x + 1,$
$\quad x + y = -2$

**7.** $x = -2y,$
$\quad x + 4y = 2$

**8.** $r = -3s,$
$\quad r + 4s = 10$

**b** Solve using the substitution method. First, solve one equation for one variable.

**9.** $x - y = 6,$
$\quad x + y = -2$

**10.** $s + t = -4,$
$\quad s - t = 2$

**11.** $y - 2x = -6,$
$\quad 2y - x = 5$

**12.** $x - y = 5,$
$\quad x + 2y = 7$

**13.** $2x + 3y = -2,$
$\quad 2x - y = 9$

**14.** $x + 2y = 10,$
$\quad 3x + 4y = 8$

**15.** $x - y = -3,$
$\quad 2x + 3y = -6$

**16.** $3b + 2a = 2,$
$\quad -2b + a = 8$

**17.** $r - 2s = 0,$
$\quad 4r - 3s = 15$

**18.** $y - 2x = 0,$
$\quad 3x + 7y = 17$

**c** Solve.

**19.** The sum of two numbers is 37. One number is 5 more than the other. Find the numbers.

**20.** The sum of two numbers is 26. One number is 12 more than the other. Find the numbers.

**21.** Find two numbers whose sum is 52 and whose difference is 28.

**22.** Find two numbers whose sum is 63 and whose difference is 5.

**23.** The difference between two numbers is 12. Two times the larger is five times the smaller. What are the numbers?

**24.** The difference between two numbers is 18. Twice the smaller number plus three times the larger is 74. What are the numbers?

**25.** *Dimensions of Colorado.* The state of Colorado is roughly in the shape of a rectangle whose perimeter is 1300 mi. The width is 110 mi less than the length. Find the length and the width.

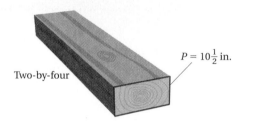

**26.** *Standard Billboard.* A standard rectangular highway billboard has a perimeter of 124 ft. The length is 6 ft more than three times the width. Find the length and the width.

**27.** *Two-by-Four.* The perimeter of a cross section of a "two-by-four" piece of lumber is $10\frac{1}{2}$ in. The length is twice the width. Find the actual dimensions of the cross section of a two-by-four.

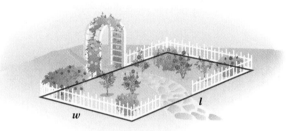

$P = 10\frac{1}{2}$ in.

Two-by-four

**28.** The perimeter of a rectangular rose garden is 400 m. The length is 3 m more than twice the width. Find the length and the width.

## Skill Maintenance

Graph.   [3.3a, b]

**29.** $2x - 3y = 6$

**30.** $2x + 3y = 6$

**31.** $y = 2x - 5$

**32.** $y = 4$

Factor completely.   [5.6a]

**33.** $6x^2 - 13x + 6$

**34.** $4p^2 - p - 3$

**35.** $4x^2 + 3x + 2$

**36.** $9a^2 - 25$

## Synthesis

**37.** ◈ Joel solves every system of two equations (in $x$ and $y$) by first solving for $y$ in the first equation and then substituting into the second equation. Is he using the best approach, given that he always uses substitution? Why or why not?

**38.** ◈ Compare the solution of Example 4 in this section with the solution of the same problem in Example 5 of Section 2.4. Discuss the merits of using a system of equations to solve the problem.

Solve using a grapher and its CALC-INTERSECT feature.

**39.** $x - y = 5$,
$\quad x + 2y = 7$

**40.** $y - 2x = -6$,
$\quad 2y - x = 5$

**41.** $y - 2.35x = -5.97$,
$\quad 2.14y - x = 4.88$

**42.** $y = 1.2x - 32.7$,
$\quad y = -0.7x + 46.15$

# 8.3 The Elimination Method

## a | Solving by the Elimination Method

The **elimination method** for solving systems of equations makes use of the *addition principle*. Some systems are much easier to solve using this method. Trying to solve the system in Example 1 by substitution would necessitate the use of fractions and extra steps because no variable has a coefficient of 1. Instead we use the elimination method.

**Example 1**  Solve the system

$$2x + 3y = 13, \quad \textbf{(1)}$$
$$4x - 3y = 17. \quad \textbf{(2)}$$

The key to the advantage of the elimination method for solving this system involves the $3y$ in one equation and the $-3y$ in the other. The terms are opposites. If we add the terms on the sides of the equations, the $y$-terms will add to 0, and in effect, the variable $y$ will be eliminated.

We will use the addition principle for equations. According to equation (2), $4x - 3y$ and 17 are the same number. Thus we can use a vertical form and add $4x - 3y$ to the left side of equation (1) and 17 to the right side—in effect, adding the same number on both sides of equation (1):

$$\begin{array}{ll} 2x + 3y = 13 & \textbf{(1)} \\ \underline{4x - 3y = 17} & \textbf{(2)} \\ 6x + 0y = 30. & \textbf{Adding} \end{array}$$

We have "eliminated" one variable. This is why we call this the **elimination method.** We now have an equation with just one variable that can be solved for $x$:

$$6x = 30$$
$$x = 5.$$

Next, we substitute 5 for $x$ in either of the original equations:

$$\begin{array}{ll} 2x + 3y = 13 & \textbf{Equation (1)} \\ 2(5) + 3y = 13 & \textbf{Substituting 5 for } x \\ 10 + 3y = 13 & \\ 3y = 3 & \\ y = 1. & \textbf{Solving for } y \end{array}$$

We check the ordered pair $(5, 1)$.

CHECK:

$$\begin{array}{c|c} 2x + 3y = 13 & 4x - 3y = 17 \\ \hline 2(5) + 3(1) \ \overset{?}{} \ 13 & 4(5) - 3(1) \ \overset{?}{} \ 17 \\ 10 + 3 \ \big| & 20 - 3 \ \big| \\ 13 \ \big| \quad \text{TRUE} & 17 \ \big| \quad \text{TRUE} \end{array}$$

Since $(5, 1)$ checks, it is the solution.

*Do Exercises 1 and 2.*

### Objectives

**a** | Solve a system of two equations in two variables using the elimination method when no multiplication is necessary.

**b** | Solve a system of two equations in two variables using the elimination method when multiplication is necessary.

**c** | Solve applied problems by translating to a system of two equations and then solving using the elimination method.

### For Extra Help

TAPE 14    TAPE 15A    MAC    CD-ROM
WIN

Solve using the elimination method.

**1.** $x + y = 5,$
    $2x - y = 4$

**2.** $-2x + y = -4,$
    $2x - 5y = 12$

---

**Calculator Spotlight**

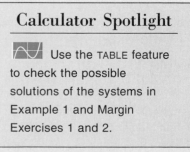 Use the TABLE feature to check the possible solutions of the systems in Example 1 and Margin Exercises 1 and 2.

---

*Answers on page A-33*

**3.** Solve. Multiply one equation by −1 first.

$$5x + 3y = 17,$$
$$5x - 2y = -3$$

**4.** Solve the system

$$3x - 2y = -30,$$
$$5x - 2y = -46.$$

## Calculator Spotlight

Use the CALC-INTERSECT feature to solve the systems in Examples 2 and 3 and Margin Exercises 3 and 4.

*Answers on page A-33*

## b | Using the Multiplication Principle First

The elimination method allows us to eliminate a variable. We may need to multiply by certain numbers first, however, so that terms become opposites.

**Example 2** Solve the system

$$2x + 3y = 8, \quad \textbf{(1)}$$
$$x + 3y = 7. \quad \textbf{(2)}$$

If we add, we will not eliminate a variable. However, if the $3y$ were $-3y$ in one equation, we could eliminate $y$. Thus we multiply by $-1$ on both sides of equation (2) and then add, using a vertical form:

$$
\begin{array}{ll}
2x + 3y = \phantom{-}8 & \text{Equation (1)} \\
\underline{-x - 3y = -7} & \text{Multiplying equation (2) by } -1 \\
x \phantom{- 3y} = \phantom{-}1. & \text{Adding}
\end{array}
$$

Next, we substitute 1 for $x$ in one of the original equations:

$$
\begin{array}{ll}
x + 3y = 7 & \text{Equation (2)} \\
1 + 3y = 7 & \text{Substituting 1 for } x \\
3y = 6 \\
y = 2. & \text{Solving for } y
\end{array}
$$

We check the ordered pair (1, 2).

CHECK:

$$
\begin{array}{c|c}
\dfrac{2x + 3y = 8}{2 \cdot 1 + 3 \cdot 2 \ ? \ 8} \\
2 + 6 \quad | \\
8 \quad | \quad \text{TRUE}
\end{array}
\qquad
\begin{array}{c|c}
\dfrac{x + 3y = 7}{1 + 3 \cdot 2 \ ? \ 7} \\
1 + 6 \quad | \\
7 \quad | \quad \text{TRUE}
\end{array}
$$

Since (1, 2) checks, it is the solution.

*Do Exercises 3 and 4.*

In Example 2, we used the multiplication principle, multiplying by $-1$. However, we often need to multiply by something other than $-1$.

**Example 3** Solve the system

$$3x + 6y = -6, \quad \textbf{(1)}$$
$$5x - 2y = 14. \quad \textbf{(2)}$$

Looking at the terms with variables, we see that if $-2y$ were $-6y$, we would have terms that are opposites. We can achieve this by multiplying by 3 on both sides of equation (2). Then we add and solve for $x$:

$$
\begin{array}{ll}
3x + 6y = -6 & \text{Equation (1)} \\
\underline{15x - 6y = \phantom{-}42} & \text{Multiplying equation (2) by 3} \\
18x \phantom{- 6y} = \phantom{-}36 & \text{Adding} \\
x = 2. & \text{Solving for } x
\end{array}
$$

Next, we go back to equation (1) and substitute 2 for $x$:

$$3 \cdot 2 + 6y = -6 \qquad \text{Substituting}$$
$$6 + 6y = -6$$
$$6y = -12$$
$$y = -2. \qquad \text{Solving for } y$$

We check the ordered pair $(2, -2)$.

CHECK:

$$\frac{3x + 6y = -6}{3 \cdot 2 + 6 \cdot (-2) \;?\; -6}$$
$$6 + (-12)$$
$$-6 \;\bigm|\; \text{TRUE}$$

$$\frac{5x - 2y = 14}{5 \cdot 2 - 2 \cdot (-2) \;?\; 14}$$
$$10 - (-4)$$
$$14 \;\bigm|\; \text{TRUE}$$

Since $(2, -2)$ checks, it is the solution.

*Do Exercises 5 and 6.*

Part of the strategy in using the elimination method is making a decision about which variable to eliminate. So long as the algebra has been carried out correctly, the solution can be found by eliminating *either* variable. We multiply so that terms involving the variable to be eliminated are opposites. It is helpful to first get each equation in a form equivalent to $Ax + By = C$.

**Example 4**   Solve the system

$$3y + 1 + 2x = 0, \qquad \textbf{(1)}$$
$$5x = 7 - 4y. \qquad \textbf{(2)}$$

We first rewrite each equation in a form equivalent to $Ax + By = C$:

$$2x + 3y = -1, \qquad \textbf{(1)} \qquad \substack{\text{Subtracting 1 on both sides} \\ \text{and rearranging terms}}$$
$$5x + 4y = 7. \qquad \textbf{(2)} \qquad \text{Adding } 4y \text{ on both sides}$$

We decide to eliminate the $x$-term. We do this by multiplying by 5 on both sides of equation (1) and by $-2$ on both sides of equation (2). Then we add and solve for $y$:

$$10x + 15y = -5 \qquad \text{Multiplying by 5 on both sides of equation (1)}$$
$$\underline{-10x - \;\; 8y = -14} \qquad \text{Multiplying by } -2 \text{ on both sides of equation (2)}$$
$$7y = -19 \qquad \text{Adding}$$
$$y = \frac{-19}{7}, \text{ or } -\frac{19}{7}. \qquad \text{Dividing by 7}$$

Next, we substitute $-\frac{19}{7}$ for $y$ in one of the original equations:

$$2x + 3y = -1 \qquad \text{Equation (1)}$$
$$2x + 3\left(-\tfrac{19}{7}\right) = -1 \qquad \text{Substituting } -\tfrac{19}{7} \text{ for } y$$
$$2x - \tfrac{57}{7} = -1$$
$$2x = -1 + \tfrac{57}{7}$$
$$2x = -\tfrac{7}{7} + \tfrac{57}{7}$$
$$2x = \tfrac{50}{7}$$
$$x = \tfrac{50}{7} \cdot \tfrac{1}{2}, \text{ or } \tfrac{25}{7}. \qquad \text{Solving for } x$$

We check the ordered pair $\left(\tfrac{25}{7}, -\tfrac{19}{7}\right)$.

Solve the system.

**5.** $4a + 7b = 11,$
   $2a + 3b = 5$

**6.** $3x - 8y = 2,$
   $5x + 2y = -12$

---

*CAUTION!*   Solving a *system* of equations in two variables requires finding an ordered *pair* of numbers. Once you have solved for one variable, don't forget the other, and remember to list the ordered-pair solution using alphabetical order.

---

*Answers on page A-33*

**7.** Solve the system

$$3x = 5 + 2y,$$
$$2x + 3y - 1 = 0.$$

**8.** Solve the system

$$2x + \ y = 15,$$
$$4x + 2y = 23.$$

**9.** Solve the system

$$5x - 2y = 3,$$
$$-15x + 6y = -9.$$

---

## Calculator Spotlight

### ⌐⌐ Exercises

Use a viewing window of $[-6, 6, -4, 4]$ for the following.

**1.** Check the results of Example 5 on a grapher. What happens when you solve each equation for $y$?

**2.** Check the results of Example 6 on a grapher. What happens when you solve each equation for $y$?

---

*Answers on page A-33*

**CHECK:**

$$\begin{array}{c|c} 3y + 1 + 2x = 0 & 5x = 7 - 4y \\ \hline 3\left(-\frac{19}{7}\right) + 1 + 2\left(\frac{25}{7}\right) \;?\; 0 & 5\left(\frac{25}{7}\right) \;?\; 7 - 4\left(-\frac{19}{7}\right) \\ -\frac{57}{7} + \frac{7}{7} + \frac{50}{7} & \frac{125}{7} \;\bigg|\; \frac{49}{7} + \frac{76}{7} \\ 0 \;\bigg|\; \text{TRUE} & \frac{125}{7} \;\bigg|\; \text{TRUE} \end{array}$$

The solution is $\left(\frac{25}{7}, -\frac{19}{7}\right)$.

*Do Exercise 7.*

Let's consider a system with no solution and see what happens when we apply the elimination method.

**Example 5**   Solve the system

$$y - 3x = 2, \qquad \textbf{(1)}$$
$$y - 3x = 1. \qquad \textbf{(2)}$$

We multiply by $-1$ on both sides of equation (2) and then add:

$$\begin{array}{ll} y - 3x = 2 & \\ \underline{-y + 3x = -1} & \text{Multiplying by } -1 \\ \phantom{-y +} 0 = 1. & \text{Adding} \end{array}$$

We obtain a false equation, $0 = 1$, so there is *no solution*. The slope–intercept forms of these equations are

$$y = 3x + 2,$$
$$y = 3x + 1.$$

The slopes are the same and the $y$-intercepts are different. Thus the lines are parallel. They do not intersect.

*Do Exercise 8.*

Sometimes there is an infinite number of solutions. Let's look at a system that we graphed in Example 6 of Section 8.1.

**Example 6**   Solve the system

$$2x + \ 3y = 6, \qquad \textbf{(1)}$$
$$-8x - 12y = -24. \qquad \textbf{(2)}$$

We multiply by 4 on both sides of equation (1) and then add the two equations:

$$\begin{array}{ll} 8x + 12y = 24 & \text{Multiplying by 4} \\ \underline{-8x - 12y = -24} & \\ \phantom{-8x -1} 0 = 0. & \text{Adding} \end{array}$$

We have eliminated both variables, and what remains, $0 = 0$, is an equation easily seen to be true. If this happens when we use the elimination method, we have an infinite number of solutions.

*Do Exercise 9.*

When decimals or fractions appear, we first multiply to clear them. Then we proceed as before.

**Example 7** Solve the system

$$\frac{1}{3}x + \frac{1}{2}y = -\frac{1}{6}, \quad \textbf{(1)}$$

$$\frac{1}{2}x + \frac{2}{5}y = \frac{7}{10}. \quad \textbf{(2)}$$

The number 6 is a multiple of all the denominators of equation (1). The number 10 is a multiple of all the denominators of equation (2). We multiply by 6 on both sides of equation (1) and by 10 on both sides of equation (2):

$$6\left(\frac{1}{3}x + \frac{1}{2}y\right) = 6\left(-\frac{1}{6}\right) \qquad 10\left(\frac{1}{2}x + \frac{2}{5}y\right) = 10\left(\frac{7}{10}\right)$$

$$6 \cdot \frac{1}{3}x + 6 \cdot \frac{1}{2}y = -1 \qquad 10 \cdot \frac{1}{2}x + 10 \cdot \frac{2}{5}y = 7$$

$$2x + 3y = -1; \qquad\qquad 5x + 4y = 7.$$

The resulting system is

$$2x + 3y = -1,$$
$$5x + 4y = 7.$$

As we saw in Example 4, the solution of this system is $\left(\frac{25}{7}, -\frac{19}{7}\right)$.

**Do Exercises 10 and 11.**

The following is a summary that compares the graphical, substitution, and elimination methods for solving systems of equations.

| Method | Strengths | Weaknesses |
|---|---|---|
| Graphical | Can "see" solution. | Inexact when solution involves numbers that are not integers or are very large and off the graph. |
| Substitution | Works well when solutions are not integers.<br>Easy to use when a variable is alone on one side. | Introduces extensive computations with fractions for more complicated systems where coefficients are not 1 or −1.<br>Cannot "see" solution. |
| Elimination | Works well when solutions are not integers, when coefficients are not 1 or −1, and when coefficients involve decimals or fractions. | Cannot "see" solution. |

When deciding which method to use, consider the preceding chart and directions from your instructor. The situation is like having a piece of wood to cut and three saws with which to cut it. The saw you use depends on the type of wood, the type of cut you are making, and how you want the wood to turn out.

Solve the system.

**10.** $\dfrac{1}{2}x + \dfrac{3}{10}y = \dfrac{1}{5}$,

$\quad\ \dfrac{3}{5}x + \quad y = -\dfrac{2}{5}$

**11.** $3.3x + 6.6y = -6.6$,

$\quad 0.1x - 0.04y = 0.28$

*Answers on page A-33*

**12.** *Car Rental.* Budget Rent-A-Car rents a car at a daily rate of $41.95 plus 43 cents per mile. Speedo Rentzit rents a car for $44.95 plus 39 cents per mile. For what mileage are the costs the same?

## c | Solving Applied Problems

We now use the elimination method to solve an applied problem.

**Example 8** *Truck Rental.* At one time, Value Rent-A-Car rented pickup trucks at a daily rate of $43.95 plus 40 cents per mile. Thrifty Rent-A-Car rented the same type of pickup trucks at a daily rate of $42.95 plus 42 cents per mile. For what mileage are the costs the same?

1. **Familiarize.** To become familiar with the problem, we make a guess. Suppose a person rents a pickup truck from each rental agency and drives it 100 mi. The total cost at Value is $43.95 + $0.40(100) = $43.95 + $40.00, or $83.95. The total cost at Thrifty is $42.95 + $0.42(100) = $42.95 + $42.00, or $84.95. Note that we converted all of the money units to dollars. The resulting costs are very nearly the same, so our guess is close. We can, of course, refine our guess. Instead, we will use algebra to solve the problem. We let $M =$ the number of miles driven and $C =$ the total cost of the truck rental.

2. **Translate.** We translate the first statement, using $0.40 for 40 cents. It helps to reword the problem before translating.

$\underbrace{\$43.95}\ \underbrace{plus}\ \underbrace{40\ cents}\ \underbrace{times}\ \underbrace{Number\ of\ miles\ driven}\ \underbrace{is}\ \underbrace{Cost.}$  Rewording

$\quad\ \$43.95\qquad +\qquad \$0.40\qquad\cdot\qquad\qquad\qquad M\qquad\qquad = \quad C$  Translating

We translate the second statement, but again it helps to reword it first.

$\underbrace{\$42.95}\ \underbrace{plus}\ \underbrace{42\ cents}\ \underbrace{times}\ \underbrace{Number\ of\ miles\ driven}\ \underbrace{is}\ \underbrace{Cost.}$  Rewording

$\quad\ \$42.95\qquad +\qquad \$0.42\qquad\cdot\qquad\qquad\qquad M\qquad\qquad = \quad C$  Translating

We have now translated to a system of equations:

$$43.95 + 0.40M = C,$$
$$42.95 + 0.42M = C.$$

3. **Solve.** We solve the system of equations. We clear the system of decimals by multiplying by 100 on both sides. Then we multiply the second equation by $-1$ and add:

$$4395 + 40M = 100C$$
$$\underline{-4295 - 42M = -100C}$$
$$100 - 2M = 0$$
$$100 = 2M$$
$$50 = M.$$

4. **Check.** For 50 mi, the cost of the Value truck is $43.95 + $0.40(50), or $43.95 + $20, or $63.95, and the cost of the Thrifty truck is $42.95 + $0.42(50), or $42.95 + $21, or $63.95. Thus the costs are the same when the mileage is 50.

5. **State.** When the trucks are driven 50 mi, the costs will be the same.

*Do Exercise 12.*

*Answer on page A-33*

# Exercise Set 8.3

**a** Solve using the elimination method.

**1.** $x - y = 7$,
$x + y = 5$

**2.** $x + y = 11$,
$x - y = 7$

**3.** $x + y = 8$,
$-x + 2y = 7$

**4.** $x + y = 6$,
$-x + 3y = -2$

**5.** $5x - y = 5$,
$3x + y = 11$

**6.** $2x - y = 8$,
$3x + y = 12$

**7.** $4a + 3b = 7$,
$-4a + b = 5$

**8.** $7c + 5d = 18$,
$c - 5d = -2$

**9.** $8x - 5y = -9$,
$3x + 5y = -2$

**10.** $3a - 3b = -15$,
$-3a - 3b = -3$

**11.** $4x - 5y = 7$,
$-4x + 5y = 7$

**12.** $2x + 3y = 4$,
$-2x - 3y = -4$

**b** Solve using the multiplication principle first. Then add.

**13.** $x + y = -7$,
$3x + y = -9$

**14.** $-x - y = 8$,
$2x - y = -1$

**15.** $3x - y = 8$,
$x + 2y = 5$

**16.** $x + 3y = 19$,
$x - y = -1$

**17.** $x - y = 5,$
$4x - 5y = 17$

**18.** $x + y = 4,$
$5x - 3y = 12$

**19.** $2w - 3z = -1,$
$3w + 4z = 24$

**20.** $7p + 5q = 2,$
$8p - 9q = 17$

**21.** $2a + 3b = -1,$
$3a + 5b = -2$

**22.** $3x - 4y = 16,$
$5x + 6y = 14$

**23.** $x = 3y,$
$5x + 14 = y$

**24.** $5a = 2b,$
$2a + 11 = 3b$

**25.** $2x + 5y = 16,$
$3x - 2y = 5$

**26.** $3p - 2q = 8,$
$5p + 3q = 7$

**27.** $p = 32 + q,$
$3p = 8q + 6$

**28.** $3x = 8y + 11,$
$x + 6y - 8 = 0$

**29.** $3x - 2y = 10,$
$-6x + 4y = -20$

**30.** $2x + y = 13,$
$4x + 2y = 23$

**31.** $0.06x + 0.05y = 0.07,$
$0.4x - 0.3y = 1.1$

**32.** $1.8x - 2y = 0.9,$
$0.04x + 0.18y = 0.15$

**33.** $\frac{1}{3}x + \frac{3}{2}y = \frac{5}{4},$
$\frac{3}{4}x - \frac{5}{6}y = \frac{3}{8}$

**34.** $x - \frac{3}{2}y = 13,$
$\frac{3}{2}x - y = 17$

**35.** $-4.5x + 7.5y = 6,$
$-x + 1.5y = 5$

**36.** $0.75x + 0.6y = -0.3,$
$3.9x + 5.2y = 96.2$

Copyright © 1999 Addison Wesley Longman

Solve.

**37.** *Van Rental.* A family plans to rent a van to move a daughter to college. Quick-Haul rents a 10-ft moving van at a daily rate of $19.95 plus 39 cents per mile. Another company rents the same size van for $39.95 plus 29 cents per mile. For what mileage are the costs the same?

**38.** *Car Rental.* Elite Rent-A-Car rents a basic car at a daily rate of $45.95 plus 40 cents per mile. Another company rents a basic car for $46.95 plus 20 cents per mile. For what mileage are the costs the same?

**39.** Two angles are supplementary. (**Supplementary angles** are angles whose sum is 180°.) One is 30° more than two times the other. Find the angles.

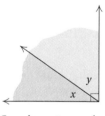

Supplementary angles
$x + y = 180°$

**40.** Two angles are supplementary. One is 8° less than three times the other. Find the angles.

**41.** Two angles are complementary. (**Complementary angles** are angles whose sum is 90°.) Their difference is 34°. Find the angles.

Complementary angles
$x + y = 90°$

**42.** Two angles are complementary. One angle is 42° more than one-half the other. Find the angles.

**43.** The Rolling Velvet Horse Farm allots 650 hectares to plant hay and oats. The owners know that their needs are best met if they plant 180 hectares more of hay than of oats. How many hectares of each should they plant?

**44.** In a vineyard, a vintner uses 820 hectares to plant Chardonnay and Riesling grapes. The vintner knows that the profits will be greatest by planting 140 hectares more of Chardonnay than of Riesling. How many hectares of each grape should be planted?

## Skill Maintenance

Simplify. [4.1d, e, f]

**45.** $x^{-2} \cdot x^{-5}$

**46.** $x^{-2} \cdot x^5$

**47.** $x^2 \cdot x^{-5}$

**48.** $x^2 \cdot x^5$

**49.** $\dfrac{x^{-2}}{x^{-5}}$

**50.** $\dfrac{x^2}{x^{-5}}$

**51.** $(a^2 b^{-3})(a^5 b^{-6})$

**52.** $\dfrac{a^2 b^{-3}}{a^5 b^{-6}}$

Simplify. [6.1c]

**53.** $\dfrac{x^2 - 5x + 6}{x^2 - 4}$

**54.** $\dfrac{x^2 - 25}{x^2 - 10x + 25}$

Subtract. [6.5a]

**55.** $\dfrac{x - 2}{x + 3} - \dfrac{2x - 5}{x - 4}$

**56.** $\dfrac{x + 7}{x^2 - 1} - \dfrac{3}{x + 1}$

---

## SYNTHESIS

**57.** ◈ The following lists the steps a student uses to solve a system of equations, but an error occurs. Find and describe the error and correct the answer.

$$
\begin{aligned}
3x - y &= 4 \\
\underline{2x + y} &= \underline{16} \\
5x &= 20 \\
x &= 4
\end{aligned}
$$

$$
\begin{aligned}
3x - y &= 4 \\
3(4) - y &= 4 \\
y &= 4 - 12 \\
y &= -8
\end{aligned}
$$

The solution is $(4, -8)$.

**58.** ◈ Explain how the addition and multiplication principles are used in this section. Then count the number of times that these principles are used in Example 4.

**59.–68.** ⊠ Use the TABLE feature to check the possible solutions to Exercises 1–10.

**69.–78.** ⊠ Use a grapher and the CALC-INTERSECT feature to solve the systems in Exercises 21–30.

**79.** Will's age is 20% of his father's age. Twenty years from now, Will's age will be 52% of his father's age. How old are Will and his father now?

**80.** If 5 is added to a woman's age and the total is divided by 5, the result will be her daughter's age. Five years ago, the woman's age was eight times her daughter's age. Find their present ages.

Solve using either the substitution or the elimination method.

**81.** $3(x - y) = 9,$
$x + y = 7$

**82.** $2(x - y) = 3 + x,$
$x = 3y + 4$

**83.** $2(5a - 5b) = 10,$
$-5(6a + 2b) = 10$

**84.** $\dfrac{x}{3} + \dfrac{y}{2} = 1\dfrac{1}{3},$
$x + 0.05y = 4$

**85.** Several ancient Chinese books included problems that can be solved by translating to systems of equations. *Arithmetical Rules in Nine Sections* is a book of 246 problems compiled by a Chinese mathematician, Chang Tsang, who died in 152 B.C. One of the problems is: Suppose there are a number of rabbits and pheasants confined in a cage. In all, there are 35 heads and 94 feet. How many rabbits and how many pheasants are there? Solve the problem.

---

Collaborative
Learning Manual

Compare the three methods for solving systems of equations in two variables.

# 8.4 Applications and Problem Solving

### a
We now use systems of equations to solve applied problems that involve two equations in two variables.

**Example 1**  *Pizza and Soda Prices.* A campus vendor charges $3.50 for one slice of pizza and one medium soda and $9.15 for three slices of pizza and two medium sodas. Determine the price of one medium soda and the price of one slice of pizza.

1. **Familiarize.** We let $p$ = the price of one slice of pizza and $s$ = the price of one medium soda.

2. **Translate.** The price of one slice of pizza and one medium soda is $3.50. This gives us one equation:

   $$p + s = 3.50.$$

   The price of three slices of pizza and two medium sodas is $9.15. This gives us another equation:

   $$3p + 2s = 9.15.$$

3. **Solve.** We solve the system of equations

   $$p + s = 3.50, \qquad \textbf{(1)}$$
   $$3p + 2s = 9.15. \qquad \textbf{(2)}$$

   Which method should we use? As we discussed in Section 8.3, any method can be used. Each has its advantages and disadvantages. We decide to proceed with the elimination method, because we see that if we multiply each side of equation (1) by $-2$ and add, the $s$-terms can be eliminated:

   | | |
   |---|---|
   | $-2p - 2s = -7.00$ | Multiplying equation (1) by $-2$ |
   | $\underline{3p + 2s = \phantom{0}9.15}$ | Equation (2) |
   | $p \phantom{+ 2s} = \phantom{0}2.15.$ | Adding |

   Next, we substitute 2.15 for $p$ in equation (1) and solve for $s$:

   $$p + s = 3.50$$
   $$2.15 + s = 3.50$$
   $$s = 1.35.$$

4. **Check.** The sum of the prices for one slice of pizza and one medium soda is

   $$\$2.15 + \$1.35, \quad \text{or} \quad \$3.50.$$

   Three times the price of one slice of pizza plus twice the price of a medium soda is

   $$3(\$2.15) + 2(\$1.35), \quad \text{or} \quad \$9.15.$$

   The prices check.

5. **State.** The price of one slice of pizza is $2.15, and the price of one medium soda is $1.35.

*Do Exercise 1.*

*Answer on page A-34*

---

### Objective

**a** Solve applied problems by translating to a system of two equations in two variables.

**For Extra Help**

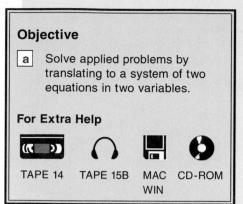

TAPE 14   TAPE 15B   MAC        CD-ROM
                      WIN

1. *Chicken and Hamburger Prices.* Fast Rick's Burger restaurant decides to include chicken on its menu. It offers a special two-and-one promotion. The price of one hamburger and two pieces of chicken is $5.39, and the price of two hamburgers and one piece of chicken is $5.68. Find the price of one hamburger and the price of one piece of chicken.

**2.** *Ages.* Sarah is 26 yr older than Malcolm. In 5 yr, Sarah will be twice as old as Malcolm. How old are they now?

Complete the following table to aid with the familiarization.

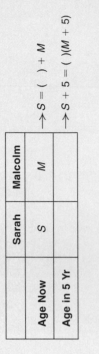

**Example 2** *Ages.* Caleb is 21 yr older than Tanya. In 6 yr, Caleb will be twice as old as Tanya. How old are they now?

1. **Familiarize.** Let's consider some conditions of the problem. We let $C$ = Caleb's age now and $T$ = Tanya's age now. How will the ages relate in 6 yr? In 6 yr, Tanya will be $T + 6$ and Caleb will be $C + 6$. We make a table to organize our information.

|  | Caleb | Tanya |  |
|---|---|---|---|
| **Age Now** | $C$ | $T$ | $\rightarrow C = 21 + T$ |
| **Age in 6 Yr** | $C + 6$ | $T + 6$ | $\rightarrow C + 6 = 2(T + 6)$ |

2. **Translate.** From the present ages, we get the following rewording and translation.

| Caleb's age | is | 21 | more than | Tanya's age. | Rewording |
|---|---|---|---|---|---|
| $C$ | $=$ | 21 | $+$ | $T$ | Translating |

From their ages in 6 yr, we get the following rewording and translation.

| Caleb's age in 6 yr | will be | twice | Tanya's age in 6 yr | Rewording |
|---|---|---|---|---|
| $C + 6$ | $=$ | $2 \cdot$ | $(T + 6)$ | Translating |

The problem has been translated to the following system of equations:

$$C = 21 + T, \qquad \textbf{(1)}$$
$$C + 6 = 2(T + 6). \qquad \textbf{(2)}$$

3. **Solve.** We solve the system. This time we use the substitution method since there is a variable alone on one side. We substitute $21 + T$ for $C$ in equation (2) and solve for $T$:

$$C + 6 = 2(T + 6)$$
$$(21 + T) + 6 = 2(T + 6)$$
$$T + 27 = 2T + 12$$
$$15 = T.$$

We find $C$ by substituting 15 for $T$ in the first equation:

$$C = 21 + T$$
$$= 21 + 15$$
$$= 36.$$

4. **Check.** Caleb's age is 36, which is 21 more than 15, Tanya's age. In 6 yr, when Caleb will be 42 and Tanya 21, Caleb's age will be twice Tanya's age.

5. **State.** Caleb is now 36 and Tanya is 15.

*Do Exercise 2.*

*Answer on page A-34*

**Example 3** *Imax Movie Prices.* There were 270 people at a recent showing of the IMAX 3D movie *Antarctica*. Admission was $8.00 each for adults and $4.75 each for children, and receipts totaled $2088.50. How many adults and how many children attended?

1. **Familiarize.** There are many ways in which to familiarize ourselves with a problem situation. This time, let's make a guess and do some calculations. The total number of people at the movie was 270, so we choose numbers that total 270. Let's try

   220 adults and
   50 children.

   How much money was taken in? The problem says that adults paid $8.00 each, so the total amount of money collected from the adults was

   220($8), or $1760.

   Children paid $4.75 each, so the total amount of money collected from the children was

   50($4.75), or $237.50.

   This makes the total receipts $1760 + $237.50, or $1997.50.

   Our guess is not the answer to the problem because the total taken in, according to the problem, was $2088.50. If we were to continue guessing, we would need to add more adults and fewer children, since our first guess gave us an amount of total receipts that was lower than $2088.50. The steps we have used to see if our guesses are correct help us to understand the actual steps involved in solving the problem.

   Let's list the information in a table. That usually helps in the familiarization process. We let $a$ = the number of adults and $c$ = the number of children.

|  | Adults | Children | Total |  |
|---|---|---|---|---|
| **Admission** | $8.00 | $4.75 |  |  |
| **Number Attending** | $a$ | $c$ | 270 | $\rightarrow$ $a + c = 270$ |
| **Money Taken in** | 8.00$a$ | 4.75$c$ | $2088.50 | $\rightarrow$ $8.00a + 4.75c = 2088.50$ |

2. **Translate.** The total number of people attending was 270, so

   $a + c = 270$.

   The amount taken in from the adults was 8.00$a$, and the amount taken in from the children was 4.75$c$. These amounts are in dollars. The total was $2088.50, so we have

   $8.00a + 4.75c = 2088.50$.

   We can multiply by 100 on both sides to clear decimals. Thus we have a translation to a system of equations:

$$a + c = 270, \qquad \textbf{(1)}$$
$$800a + 475c = 208{,}850. \qquad \textbf{(2)}$$

Answer on page A-34

**3.** *Game Admissions.* There were 166 paid admissions to a game. The price was $3.10 each for adults and $1.75 each for children. The amount taken in was $459.25. How many adults and how many children attended?

Complete the following table to aid with the familiarization.

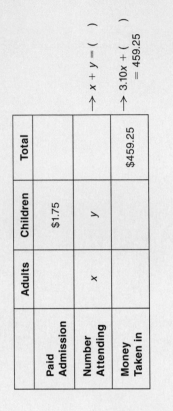

| | Adults | Children | Total | |
|---|---|---|---|---|
| **Paid Admission** | | $1.75 | | |
| **Number Attending** | $x$ | $y$ | | $\rightarrow x + y = (\ \ )$ |
| **Money Taken in** | | | $459.25 | $\rightarrow 3.10x + (\ \ ) = 459.25$ |

**3. Solve.** We solve the system. We use the elimination method since the equations are both in the form $Ax + By = C$. (A case can certainly be made for using the substitution method since we can solve for one of the variables quite easily in the first equation. Very often a decision is just a matter of choice.) We multiply by $-475$ on both sides of equation (1) and then add and solve for $a$:

$$
\begin{array}{rll}
-475a - 475c = -128{,}250 & \text{Multiplying by } -475 \\
\underline{800a + 475c = \phantom{-}208{,}850} & \\
325a \phantom{+ 475c} = \phantom{-}80{,}600 & \text{Adding} \\
a = \dfrac{80{,}600}{325} & \text{Dividing by 325} \\
a = 248.
\end{array}
$$

Next, we go back to equation (1), substituting 248 for $a$, and solve for $c$:

$$
\begin{aligned}
a + c &= 270 \\
248 + c &= 270 \\
c &= 22.
\end{aligned}
$$

**4. Check.** We leave the check to the student. It is similar to what we did in the *Familiarize* step.

**5. State.** Attending the showing were 248 adults and 22 children.

*Do Exercise 3.*

**Example 4** *Mixture of Solutions.* A chemist has one solution that is 80% acid (that is, 8 parts are acid and 2 parts are water) and another solution that is 30% acid. What is needed is 200 L of a solution that is 62% acid. The chemist will prepare it by mixing the two solutions. How much of each should be used?

**1. Familiarize.** We can make a drawing of the situation. The chemist uses $x$ liters of the first solution and $y$ liters of the second solution.

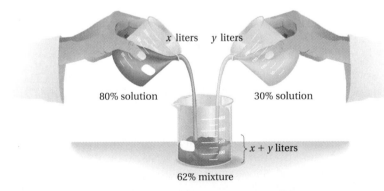

*x* liters  *y* liters

80% solution  30% solution

$x + y$ liters

62% mixture

We can also arrange the information in a table.

| Type of Solution | First | Second | Mixture | |
|---|---|---|---|---|
| **Amount of Solution** | $x$ | $y$ | 200 L | $\rightarrow x + y = 200$ |
| **Percent of Acid** | 80% | 30% | 62% | |
| **Amount of Acid in Solution** | 80%$x$ | 30%$y$ | 62% × 200, or 124 L | $\rightarrow 80\%x + 30\%y = 124$ |

**2. Translate.** The chemist uses $x$ liters of the first solution and $y$ liters of the second. Since the total is to be 200 L, we have

*Total amount of solution:*   $x + y = 200$.

The amount of acid in the new mixture is to be 62% of 200 L, or 124 L. The amounts of acid from the two solutions are 80%$x$ and 30%$y$. Thus,

*Total amount of acid:*   $80\%x + 30\%y = 124$

or                                      $0.8x + 0.3y = 124$.

We clear decimals by multiplying by 10 on both sides:

$10(0.8x + 0.3y) = 10 \cdot 124$
$8x + 3y = 1240$.

Thus we have a translation to a system of equations:

$x + y = 200,$         **(1)**
$8x + 3y = 1240.$      **(2)**

**3. Solve.**   We solve the system. We use the elimination method, again because equations are in the form $Ax + By = C$ and a multiplication in one equation will allow us to eliminate a variable, but substitution would also work. We multiply by $-3$ on both sides of equation (1) and then add and solve for $x$:

$$-3x - 3y = -600 \quad \text{Multiplying by } -3$$
$$\underline{8x + 3y = 1240}$$
$$5x \phantom{+ 3y} = 640 \quad \text{Adding}$$
$$x = \frac{640}{5} \quad \text{Dividing by 5}$$
$$x = 128.$$

Next, we go back to equation (1) and substitute 128 for $x$:

$x + y = 200$
$128 + y = 200$
$y = 72$.

The solution is $x = 128$ and $y = 72$.

**4. Check.**   The sum of 128 and 72 is 200. Also, 80% of 128 is 102.4 and 30% of 72 is 21.6. These add up to 124.

**5. State.**   The chemist should use 128 L of the 80%-acid solution and 72 L of the 30%-acid solution.

*Do Exercise 4.*

**Example 5**   *Candy Mixtures.* A bulk wholesaler wishes to mix some candy worth 45 cents per pound and some worth 80 cents per pound to make 350 lb of a mixture worth 65 cents per pound. How much of each type of candy should be used?

1. **Familiarize.**   Arranging the information in a table will help. We let $x =$ the amount of 45-cents candy and $y =$ the amount of 80-cents candy.

---

**4.** *Mixture of Solutions.* One solution is 50% alcohol and a second is 70% alcohol. How much of each should be mixed in order to make 30 L of a solution that is 55% alcohol?

Complete the following table to aid in the familiarization.

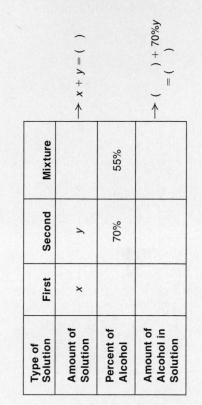

| Type of Solution | First | Second | Mixture |
|---|---|---|---|
| Amount of Solution | $x$ | $y$ | $x + y = (\quad)$ |
| Percent of Alcohol | | 70% | 55% |
| Amount of Alcohol in Solution | | | $(\quad) = (\quad) + 70\%y$ |

*Answer on page A-34*

Answer on page A-34

**5.** *Mixture of Grass Seeds.* Grass seed A is worth $1.40 per pound and seed B is worth $1.75 per pound. How much of each should be mixed in order to make 50 lb of a mixture worth $1.54 per pound?

Complete the following table to aid in the familiarization.

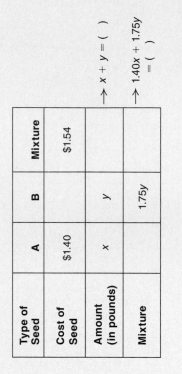

| Type of Candy | Inexpensive Candy | Expensive Candy | Mixture | |
|---|---|---|---|---|
| Cost of Candy | 45 cents | 80 cents | 65 cents | |
| Amount (in pounds) | $x$ | $y$ | 350 | $\rightarrow x + y = 350$ |
| Total Cost | $45x$ | $80y$ | 65 cents · (350), or 22,750 cents | $\rightarrow 45x + 80y = 22{,}750$ |

Note the similarity of this problem to Example 3. Here we consider types of candy instead of groups of people.

**2. Translate.** We translate as follows. From the second row of the table, we find that

*Total amount of candy:* $x + y = 350$.

Our second equation will come from the costs. The value of the inexpensive candy, in cents, is $45x$ ($x$ pounds at 45 cents per pound). The value of the expensive candy is $80y$, and the value of the mixture is $65 \times 350$, or 22,750 cents. Thus we have

*Total cost of mixture:* $45x + 80y = 22{,}750$.

Remember the problem-solving tip about dimension symbols. In this last equation, all expressions are given in cents. We could have expressed them all in dollars, but we do not want some in cents and some in dollars. Thus we have a translation to a system of equations:

$$x + \quad y = 350, \qquad \textbf{(1)}$$
$$45x + 80y = 22{,}750. \qquad \textbf{(2)}$$

**3. Solve.** We solve the system using the elimination method again. We multiply by $-45$ on both sides of equation (1) and then add and solve for $y$:

$$
\begin{aligned}
-45x - 45y &= -15{,}750 \qquad \text{Multiplying by } -45 \\
45x + 80y &= \phantom{-}22{,}750 \\
\hline
35y &= \phantom{-00}7{,}000 \qquad \text{Adding} \\
y &= \frac{7{,}000}{35} \\
y &= 200.
\end{aligned}
$$

Next, we go back to equation (1), substituting 200 for $y$, and solve for $x$:

$$
\begin{aligned}
x + y &= 350 \\
x + 200 &= 350 \\
x &= 150.
\end{aligned}
$$

**4. Check.** We consider $x = 150$ lb and $y = 200$ lb. The sum is 350 lb. The value of the candy is $45(150) + 80(200)$, or 22,750 cents and each pound of the mixture is worth $22{,}750 \div 350$, or 65 cents. These values check.

**5. State.** The grocer should mix 150 lb of the 45-cents candy with 200 lb of the 80-cents candy.

Do Exercise 5.

**Example 6** *Coin Value.* A student assistant at the university copy center has some nickels and dimes to use for change when students make copies. The value of the coins is $7.40. There are 26 more dimes than nickels. How many of each kind of coin are there?

**6.** *Coin Value.* On a table are 20 coins, quarters and dimes. Their value is $3.05. How many of each kind of coin are there?

1. **Familiarize.** We let $d$ = the number of dimes and $n$ = the number of nickels.

2. **Translate.** We have one equation at once:

$$d = n + 26.$$

The value of the nickels, in cents, is $5n$, since each coin is worth 5 cents. The value of the dimes, in cents, is $10d$, since each coin is worth 10 cents. The total value is given as $7.40. Since we have the values of the nickels and dimes *in cents,* we must use cents for the total value. This is 740. This gives us another equation:

$$10d + 5n = 740.$$

We now have a system of equations:

$$d = n + 26, \qquad \textbf{(1)}$$
$$10d + 5n = 740. \qquad \textbf{(2)}$$

3. **Solve.** Since we have $d$ alone on one side of one equation, we use the substitution method. We substitute $n + 26$ for $d$ in equation (2):

$$10d + 5n = 740$$
$$10(n + 26) + 5n = 740 \qquad \text{Substituting } n + 26 \text{ for } d$$
$$10n + 260 + 5n = 740 \qquad \text{Removing parentheses}$$
$$15n + 260 = 740 \qquad \text{Collecting like terms}$$
$$15n = 480 \qquad \text{Subtracting 260}$$
$$n = \frac{480}{15}, \text{ or } 32. \qquad \text{Dividing by 15}$$

Next, we substitute 32 for $n$ in either of the original equations to find $d$. We use equation (1):

$$d = n + 26$$
$$= 32 + 26$$
$$= 58.$$

4. **Check.** We have 58 dimes and 32 nickels. There are 26 more dimes than nickels. The value of the coins is 58($0.10) + 32($0.05), which is $7.40. This checks.

5. **State.** The student assistant has 58 dimes and 32 nickels.

*Do Exercise 6.*

*Answer on page A-34*

You should look back over Examples 3–6. The problems are quite similar in their structure. Compare them and try to see the similarities. The problems in Examples 3–6 are often called *mixture problems*. These problems provide a pattern, or model, for many related problems.

---

**PROBLEM-SOLVING TIP**

When solving problems, see if they are patterned or modeled after other problems that you have studied.

---

# Improving Your Math Study Skills

## Study Tips for Trouble Spots

By now you have probably encountered certain topics that gave you more difficulty than others. It is important to know that this happens to every person who studies mathematics. Unfortunately, frustration is often part of the learning process and it is important not to give up when difficulty arises.

One source of frustration for many students is not being able to set aside sufficient time for studying. Family commitments, work schedules, and athletics are just a few of the time demands that many students face. Couple these demands with a math lesson that seems to require a greater than usual amount of study time, and it is no wonder that many students often feel frustrated. Below are some study tips that might be useful if and when troubles arise.

- **Realize that everyone—even your instructor— has been stymied at times when studying math.** You are not the first person, nor will you be the last, to encounter a "roadblock."

- **Whether working alone or with a classmate, try to allow enough study time so that you won't need to constantly glance at a clock.** Difficult material is best mastered when your mind is completely focused on the subject matter. Thus, if you are tired, it is usually best to study early the next morning or to take a ten-minute "power-nap" in order to make the most productive use of your time.

- **Talk about your trouble spot with a classmate.** It is possible that she or he is also having difficulty with the same material. If that is the case, perhaps the majority of your class is confused and

your instructor's coverage of the topic is not yet finished. If your classmate *does* understand the topic that is troubling you, patiently allow him or her to explain it to you. By verbalizing the math in question, your classmate may help clarify the material for both of you. Perhaps you will be able to return the favor for your classmate when he or she is struggling with a topic that you understand.

- **Try to study in a "controlled" environment.** What we mean by this is that you can often put yourself in a setting that will enable you to maximize your powers of concentration. For example, whereas some students may succeed in studying at home or in a dorm room, for many these settings are filled with distractions. Consider a trip to a library, classroom building, or perhaps the attic or basement if such a setting is more conducive to studying. If you plan on working with a classmate, try to find a location in which conversation will not be bothersome to others.

- **When working on difficult material, it is often helpful to first "back up" and review the most recent material that *did* make sense.** This can build your confidence and create a momentum that can often carry you through the roadblock. Sometimes a small piece of information that appeared in a previous section is all that is needed for your problem spot to disappear. When the difficult material is finally mastered, try to make use of what is fresh in your mind by taking a "sneak preview" of what your next topic for study will be.

# Exercise Set 8.4

**a** Solve.

1. *Basketball Scoring.* Shaquille O'Neill once scored 36 points on 22 shots in an NBA game, shooting only two-pointers and foul shots (one point) (**Source:** National Basketball Association). How many of each type of shot did he make?

2. *Household Trash.* The Perezes generate twice as much trash as their neighbors, the Willises. Together, the two households produce 14 bags of trash each month. How much trash does each household produce?

3. The Kuyatts' house is twice as old as the Marconis' house. Eight years ago, the Kuyatts' house was three times as old as the Marconis' house. How old is each house?

4. David is twice as old as his daughter. In 4 yr, David's age will be three times what his daughter's age was 6 yr ago. How old are they now?

5. Randy is four times as old as Mandy. In 12 yr, Mandy's age will be half of Randy's. How old are they now?

6. Jennifer is twice as old as Ramon. The sum of their ages 7 yr ago was 13. How old are they now?

7. *Coffee Blends.* Cafebucks coffee shop mixes Brazilian coffee worth $19 per pound with Turkish coffee worth $22 per pound. The mixture is to sell for $20 per pound. How much of each type of coffee should be used in order to make a 300-lb mixture? Complete the following table to aid in the familiarization.

8. *Coffee Blends.* The Java Joint wishes to mix Kenyan coffee beans that sell for $7.25 per pound with Venezuelan beans that sell for $8.50 per pound in order to form a 50-lb batch of Morning Blend that sells for $8.00 per pound. How many pounds of Kenyan beans and how many pounds of Venezuelan beans should be used to make the blend?

| Type of Coffee | Brazilian | Turkish | Mixture |
|---|---|---|---|
| Cost of Coffee | $19 | | $20 |
| Amount (in pounds) | $x$ | $y$ | 300 |
| Mixture | | $22y$ | 20(300), or $6000 |

→ $x + y = ($ $)$

→ $19x + ($ $) = 6000$

9. *Coin Value.* A parking meter contains dimes and quarters worth $15.25. There are 103 coins in all. How many of each type of coin are there?

10. *Coin Value.* A vending machine contains nickels and dimes worth $14.50. There are 95 more nickels than dimes. How many of each type of coin are there?

**11.** *Food Prices.* Mr. Cholesterol's Pizza Parlor charges $3.70 for a slice of pizza and a soda and $9.65 for three slices of pizza and two sodas. Determine the cost of one soda and the cost of one slice of pizza.

**12.** Cassandra has a number of $50 and $100 savings bonds to use for part of her college expenses. The total value of the bonds is $1250. There are 7 more $50 bonds than $100 bonds. How many of each type of bond does she have?

**13.** There were 203 tickets sold for a volleyball game. For activity-card holders, the price was $2.25, and for non-cardholders, the price was $3. The total amount of money collected was $513. How many of each type of ticket were sold?

**14.** *Paid Admissions.* There were 429 people at a play. Admission was $8 each for adults and $4.50 each for children. The total receipts were $2641. How many adults and how many children attended?

**15.** *Paid Admissions.* Following the baseball season, the players on a junior college team decided to go to a major-league baseball game. Ticket prices for the game are shown in the table below. They bought 29 tickets of two types, Upper Box and Lower Reserved. The cost of all the tickets was $318. How many of each kind of ticket did they buy?

**16.** A faculty group bought tickets for the game in Exercise 15, but they bought 54 tickets of two types, Lower Box and Upper Box. The cost of all their tickets was $745.50. How many of each kind of ticket did they buy?

| Ticket Information | |
| --- | --- |
| Lower Box . . . . . . . . . . . . . . . . . . | $18.50 |
| Upper Box . . . . . . . . . . . . . . . . . . | $12.00 |
| Lower Reserved. . . . . . . . . . . . | $ 9.50 |
| Upper Reserved. . . . . . . . . . . . | $ 8.00 |
| General Admission . . . . . . . . . | $ 6.50 |

**17.** *Mixture of Solutions.* Solution A is 50% acid and solution B is 80% acid. How many liters of each should be used in order to make 100 L of a solution that is 68% acid? (*Hint*: 68% of what is acid?) Complete the following table to aid in the familiarization.

**18.** *Mixture of Solutions.* Solution A is 30% alcohol and solution B is 75% alcohol. How much of each should be used in order to make 100 L of a solution that is 50% alcohol?

| Type of Solution | A | B | Mixture | |
| --- | --- | --- | --- | --- |
| Amount of Solution | $x$ | $y$ | L | $\rightarrow x + y = ( \quad )$ |
| Percent of Acid | 50% | | 68% | |
| Amount of Acid in Solution | | 80%$y$ | 68% $\times$ 100, or  L | $\rightarrow 50\%x + ( \quad ) = ( \quad )$ |

Copyright © 1999 Addison Wesley Longman

**19.** *Grain Mixtures for Horses.* Irene is a barn manager at a horse stable. She needs to calculate the correct mix of grain and hay to feed her horse. On the basis of her horse's age, weight, and workload, she determines that he needs to eat 15 lb of feed per day, with an average protein content of 8%. Hay contains 6% protein, whereas grain has a 12% protein content (**Source:** *Michael Plumb's Horse Journal*, February 1996: 26–29). How many pounds of hay and grain should she feed her horse each day?

**20.** *Paint Mixtures.* At a local "paint swap," Gayle found large supplies of Skylite Pink (12.5% red pigment) and MacIntosh Red (20% red pigment). How many gallons of each color should Gayle pick up in order to mix a gallon of Summer Rose (17% red pigment)?

**21.** *Mixture of Grass Seeds.* Grass seed A is worth $2.50 per pound and seed B is worth $1.75 per pound. How much of each would you use in order to make 75 lb of a mixture worth $2.14 per pound?

**22.** *Mixed Nuts.* A customer has asked a caterer to provide 60 lb of nuts, 60% of which are to be cashews. The caterer has available mixtures of 70% cashews and 45% cashews. How many pounds of each mixture should be used?

**23.** *Test Scores.* You are taking a test in which items of type A are worth 10 points and items of type B are worth 15 points. It takes 3 min to complete each item of type A and 6 min to complete each item of type B. The total time allowed is 60 min and you do exactly 16 questions. How many questions of each type did you complete? Assuming that all your answers were correct, what was your score?

**24.** *Gold Alloys.* A goldsmith has two alloys that are different purities of gold. The first is three-fourths pure gold and the second is five-twelfths pure gold. How many ounces of each should be melted and mixed in order to obtain a 6-oz mixture that is two-thirds pure gold?

**25.** *Printing.* A printer knows that a page of print contains 1300 words if large type is used and 1850 words if small type is used. A document containing 18,526 words fills exactly 12 pages. How many pages are in the large type? in the small type?

**26.** *Paint Mixture.* A merchant has two kinds of paint. If 9 gal of the inexpensive paint is mixed with 7 gal of the expensive paint, the mixture will be worth $19.70 per gallon. If 3 gal of the inexpensive paint is mixed with 5 gal of the expensive paint, the mixture will be worth $19.825 per gallon. What is the price per gallon of each type of paint?

Factor.  [5.6a]

**27.** $25x^2 - 81$

**28.** $36 - a^2$

**29.** $4x^2 + 100$

**30.** $4x^2 - 100$

Find the intercepts. Then graph the equation.  [3.3a]

**31.** $y = -2x - 3$

**32.** $y = -0.1x + 0.4$

**33.** $5x - 2y = -10$

**34.** $2.5x + 4y = 10$

---

**Synthesis**

**35.** ◆ What characteristics do Examples 1–4 share when they are translated to systems of equations?

**36.** ◆ Which of the five problem-solving steps have you found the most challenging? Why?

**37.** *Octane Ratings.* In most areas of the United States, gas stations offer three grades of gasoline, indicated by octane ratings on the pumps, such as 87, 89, and 93. When a tanker delivers gas, it brings only two grades of gasoline, the highest and lowest, filling two large underground tanks. If you purchase the middle grade, the pump's computer mixes the other two grades appropriately. How much 87-octane gas and 93-octane gas should be blended in order to make 18 gal of 89-octane gas?

**38.** *Octane Ratings.* Referring to Exercise 37, suppose the pump grades offered are 85, 87, and 91. How much 85-octane gas and 91-octane gas should be blended in order to make 12 gal of 87-octane gas?

**39.** *Automobile Maintenance.* An automobile radiator contains 16 L of antifreeze and water. This mixture is 30% antifreeze. How much of this mixture should be drained and replaced with pure antifreeze so that the mixture will be 50% antifreeze?

**40.** *Employer Payroll.* An employer has a daily payroll of $1225 when employing some workers at $80 per day and others at $85 per day. When the number of $80 workers is increased by 50% and the number of $85 workers is decreased by $\frac{1}{5}$, the new daily payroll is $1540. How many were originally employed at each rate?

**41.** A farmer has 100 L of milk that is 4.6% butterfat. How much skim milk (no butterfat) should be mixed with it in order to make milk that is 3.2% butterfat?

**42.** A flavored-drink manufacturer mixes flavoring worth $1.45 per ounce with sugar worth $0.05 per ounce. The mixture sells for $0.106 per ounce. How much of each should be mixed in order to fill a 20-oz can?

**43.** A framing shop charges $0.40 per inch for a certain kind of frame. A customer is looking for a frame whose length is 3 in. longer than the width. The clerk recommends using a frame that is 2 in. longer and 1 in. wider. The second frame will cost $22.40. What are the dimensions of the first frame?

**44.** A two-digit number is six times the sum of its digits. The tens digit is 1 more than the units digit. Find the number.

**45.** Eduardo invested $54,000, part of it at 6% and the rest at 6.5%. The total yield after 1 yr is $3385. How much was invested at each rate?

**46.** One year, Shannon made $288 from two investments: $1100 was invested at one yearly rate and $1800 at a rate that was 1.5% higher. Find the two rates of interest.

Copyright © 1999 Addison Wesley Longman

---

Collaborative
Learning Manual

Model a consumer problem using a system of equations.

# 8.5 Applications with Motion

**a**    We first studied problems involving motion in Chapter 6. Here we extend our problem-solving skills by solving certain motion problems whose solutions can be found using systems of equations. Recall the motion formula.

> **THE MOTION FORMULA**
>
> Distance = Rate (or speed) · Time
>
> $d = rt$

We have five steps for problem solving. The tips in the margin at right are also helpful when solving motion problems.

As we saw in Chapter 6, there are motion problems that can be solved with just one equation. Let's start with another such problem.

**Example 1**   Two cars leave York at the same time traveling in opposite directions. One travels at 60 mph and the other at 30 mph. In how many hours will they be 150 mi apart?

**1. Familiarize.**   We first make a drawing.

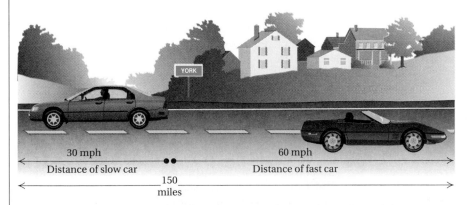

From the wording of the problem and the drawing, we see that the distances may *not* be the same. But the times that the cars travel are the same, so we can use just $t$ for time. We can organize the information in a chart.

|  | Distance | Speed | Time |
|---|---|---|---|
| **Fast Car** | Distance of fast car | 60 | $t$ |
| **Slow Car** | Distance of slow car | 30 | $t$ |
| **Total** | 150 | | |

$$d = r \cdot t$$

> **TIPS FOR SOLVING MOTION PROBLEMS**
>
> 1. Draw a diagram using an arrow or arrows to represent distance and the direction of each object in motion.
> 2. Organize the information in a chart.
> 3. Look for as many things as you can that are the same so that you can write equations.

1. Two cars leave town at the same time traveling in opposite directions. One travels at 48 mph and the other at 60 mph. How far apart will they be 3 hr later? (*Hint*: The times are the same. Be *sure* to make a drawing.)

2. **Translate.** From the drawing, we see that

(Distance of fast car) + (Distance of slow car) = 150.

Then using $d = rt$ in each row of the table, we get $60t + 30t = 150$.

3. **Solve.** We solve the equation:

$$60t + 30t = 150$$
$$90t = 150 \qquad \text{Collecting like terms}$$
$$t = \frac{150}{90}, \text{ or } \frac{5}{3}, \text{ or } 1\frac{2}{3} \text{ hr.} \qquad \text{Dividing by 90}$$

4. **Check.** When $t = \frac{5}{3}$ hr,

$$\text{(Distance of fast car)} + \text{(Distance of slow car)} = 60\left(\frac{5}{3}\right) + 30\left(\frac{5}{3}\right)$$
$$= 100 + 50, \text{ or } 150 \text{ mi.}$$

Thus the time of $\frac{5}{3}$ hr, or $1\frac{2}{3}$ hr, checks.

5. **State.** In $1\frac{2}{3}$ hr, the cars will be 150 mi apart.

*Do Exercises 1 and 2.*

Now let's solve some motion problems using systems of equations.

**Example 2** A train leaves Stanton traveling east at 35 miles per hour (mph). An hour later, another train leaves Stanton on a parallel track at 40 mph. How far from Stanton will the second (or faster) train catch up with the first (or slower) train?

2. Two cars leave town at the same time traveling in the same direction. One travels at 35 mph and the other at 40 mph. In how many hours will they be 15 mi apart? (*Hint*: The times are the same. Be *sure* to make a drawing.)

1. **Familiarize.** We first make a drawing.

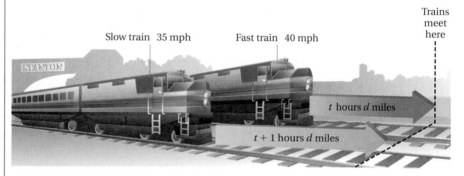

From the drawing, we see that the distances are the same. Let's call the distance $d$. We don't know the times. Let $t =$ the time for the faster train. Then the time for the slower train $= t + 1$, since it left 1 hr earlier. We can organize the information in a chart.

$$d = r \cdot t$$

| | Distance | Speed | Time | |
|---|---|---|---|---|
| **Slow Train** | $d$ | 35 | $t + 1$ | $\rightarrow d = 35(t + 1)$ |
| **Fast Train** | $d$ | 40 | $t$ | $\rightarrow d = 40t$ |

*Answers on page A-34*

**2. Translate.** In motion problems, we look for things that are the same so that we can write equations. From each row of the chart, we get an equation, $d = rt$. Thus we have two equations:

$$d = 35(t + 1), \qquad \textbf{(1)}$$
$$d = 40t. \qquad \textbf{(2)}$$

**3. Solve.** Since we have a variable alone on one side, we solve the system using the substitution method:

$35(t + 1) = 40t$     **Using the substitution method (substituting $35(t + 1)$ for $d$ in equation 2)**

$35t + 35 = 40t$     **Removing parentheses**

$35 = 5t$     **Subtracting $35t$**

$\dfrac{35}{5} = t$     **Dividing by 5**

$7 = t.$

The problem asks us to find how far from Stanton the fast train catches up with the other. Thus we need to find $d$. We can do this by substituting 7 for $t$ in the equation $d = 40t$:

$$d = 40(7)$$
$$= 280.$$

**4. Check.** If the time is 7 hr, then the distance that the slow train travels is $35(7 + 1)$, or 280 mi. The fast train travels $40(7)$, or 280 mi. Since the distances are the same, we know how far from Stanton the trains will be when the fast train catches up with the other.

**5. State.** The fast train will catch up with the slow train 280 mi from Stanton.

*Do Exercise 3.*

**Example 3** A motorboat took 3 hr to make a downstream trip with a 6-km/h current. The return trip against the same current took 5 hr. Find the speed of the boat in still water.

Downstream, $r + 6$
6-km/h current, 3 hours,
$d$ kilometers

Upstream, $r - 6$
6-km/h current, 5 hours,
$d$ kilometers

**1. Familiarize.** We first make a drawing. From the drawing, we see that the distances are the same. Let's call the distance $d$. Let $r =$ the speed of the boat in still water. Then, when the boat is traveling downstream, its speed is $r + 6$ (the current helps the boat along). When it is traveling upstream, its speed is $r - 6$ (the current holds the boat back).

**3.** A car leaves Spokane traveling north at 56 km/h. Another car leaves Spokane 1 hr later traveling north at 84 km/h. How far from Spokane will the second car catch up with the first? (*Hint*: The cars travel the same distance.)

*Answer on page A-34*

**4.** An airplane flew for 5 hr with a 25-km/h tail wind. The return flight against the same wind took 6 hr. Find the speed of the airplane in still air. (*Hint:* The distance is the same both ways. The speeds are $r + 25$ and $r - 25$, where $r$ is the speed in still air.)

We can organize the information in a chart. In this case, the distances are the same, so we use the formula $d = rt$.

| | $d$ | $=$ | $r$ | $\cdot$ | $t$ | |
|---|---|---|---|---|---|---|
| | **Distance** | | **Speed** | | **Time** | |
| **Downstream** | $d$ | | $r + 6$ | | 3 | $\rightarrow d = (r + 6)3$ |
| **Upstream** | $d$ | | $r - 6$ | | 5 | $\rightarrow d = (r - 6)5$ |

**2. Translate.** From each row of the chart, we get an equation, $d = rt$:

$$d = (r + 6)3, \quad \textbf{(1)}$$
$$d = (r - 6)5. \quad \textbf{(2)}$$

**3. Solve.** Since there is a variable alone on one side of an equation, we solve the system using substitution:

| | |
|---|---|
| $(r + 6)3 = (r - 6)5$ | Substituting $(r + 6)3$ for $d$ in equation (2) |
| $3r + 18 = 5r - 30$ | Removing parentheses |
| $-2r + 18 = -30$ | Subtracting $5r$ |
| $-2r = -48$ | Subtracting 18 |
| $r = \dfrac{-48}{-2}$, or 24. | Dividing by $-2$ |

**4. Check.** When $r = 24$, $r + 6 = 30$, and $30 \cdot 3 = 90$, the distance downstream. When $r = 24$, $r - 6 = 18$, and $18 \cdot 5 = 90$, the distance upstream. In both cases, we get the same distance.

**5. State.** The speed in still water is 24 km/h.

---

**MORE TIPS FOR SOLVING MOTION PROBLEMS**

**1.** Translating to a system of equations eases the solution of many motion problems.

**2.** At the end of the problem, always ask yourself, "Have I found what the problem asked for?" You might have solved for a certain variable but still not have answered the question of the original problem. For example, in Example 2 you solve for $t$ but the question of the original problem asks for $d$. Thus you need to continue the *Solve* step.

---

*Do Exercise 4.*

*Answer on page A-34*

# Exercise Set 8.5

**a** Solve. In Exercises 1–6, complete the table to aid the translation.

**1.** Two cars leave town at the same time going in the same direction. One travels at 30 mph and the other travels at 46 mph. In how many hours will they be 72 mi apart?

$$d = r \cdot t$$

|  | Distance | Speed | Time |
|---|---|---|---|
| **Slow Car** | Distance of slow car |  | $t$ |
| **Fast Car** | Distance of fast car | 46 |  |

**2.** A truck and a car leave a service station at the same time and travel in the same direction. The truck travels at 55 mph and the car at 40 mph. They can maintain CB radio contact within a range of 10 mi. When will they lose contact?

$$d = r \cdot t$$

|  | Distance | Speed | Time |
|---|---|---|---|
| **Truck** | Distance of truck | 55 |  |
| **Car** | Distance of car |  | $t$ |

**3.** A train leaves a station and travels east at 72 mph. Three hours later, a second train leaves on a parallel track and travels east at 120 mph. When will it overtake the first train?

$$d = r \cdot t$$

|  | Distance | Speed | Time |  |
|---|---|---|---|---|
| **Slow Train** | $d$ |  | $t + 3$ | $\rightarrow d = 72(\quad)$ |
| **Fast Train** | $d$ | 120 |  | $\rightarrow d = (\quad)t$ |

**4.** A private airplane leaves an airport and flies due south at 192 mph. Two hours later, a jet leaves the same airport and flies due south at 960 mph. When will the jet overtake the plane?

$$d = r \cdot t$$

|  | Distance | Speed | Time |  |
|---|---|---|---|---|
| **Private Plane** | $d$ | 192 |  | $\rightarrow d = 192(\,)$ |
| **Jet** | $d$ |  | $t - 2$ | $\rightarrow d = (\quad)(t - 2)$ |

**5.** A canoeist paddled for 4 hr with a 6-km/h current to reach a campsite. The return trip against the same current took 10 hr. Find the speed of the canoe in still water.

$$d = r \cdot t$$

|  | Distance | Speed | Time |  |
|---|---|---|---|---|
| **Down-stream** | $d$ | $r + 6$ |  | $\rightarrow d = (\quad)4$ |
| **Upstream** | $d$ |  | 10 | $\rightarrow \quad = (r - 6)10$ |

**6.** An airplane flew for 4 hr with a 20-km/h tail wind. The return flight against the same wind took 5 hr. Find the speed of the plane in still air.

$$d = r \cdot t$$

|  | Distance | Speed | Time |  |
|---|---|---|---|---|
| **With Wind** | $d$ |  | 4 | $\rightarrow d = (\quad)4$ |
| **Against Wind** | $d$ | $r - 20$ |  | $\rightarrow d = (\quad)5$ |

**7.** It takes a passenger train 2 hr less time than it takes a freight train to make the trip from Central City to Clear Creek. The passenger train averages 96 km/h, while the freight train averages 64 km/h. How far is it from Central City to Clear Creek?

**8.** It takes a small jet 4 hr less time than it takes a propeller-driven plane to travel from Glen Rock to Oakville. The jet averages 637 km/h, while the propeller plane averages 273 km/h. How far is it from Glen Rock to Oakville?

**9.** On a weekend outing, Antoine rents a motorboat for 8 hr to travel down the river and back. The rental operator tells him to go for 3 hr downstream, leaving him 5 hr to return upstream.

  **a)** If the river current flows at a speed of 6 mph, how fast must Antoine travel in order to return in 8 hr?

  **b)** How far downstream did Antoine travel before he turned back?

**10.** An airplane took 2 hr to fly 600 mi against a head wind. The return trip with the wind took $1\frac{2}{3}$ hr. Find the speed of the plane in still air.

**11.** A toddler takes off running down the sidewalk at 230 ft/min. One minute later, a worried mother runs after the child at 660 ft/min. When will the mother overtake the toddler?

**12.** Two airplanes start at the same time and fly toward each other from points 1000 km apart at rates of 420 km/h and 330 km/h. When will they meet?

**13.** A motorcycle breaks down and the rider must walk the rest of the way to work. The motorcycle was being driven at 45 mph, and the rider walks at a speed of 6 mph. The distance from home to work is 25 mi, and the total time for the trip was 2 hr. How far did the motorcycle go before it broke down?

**14.** A student walks and jogs to college each day. She averages 5 km/h walking and 9 km/h jogging. The distance from home to college is 8 km, and she makes the trip in 1 hr. How far does the student jog?

---

**Skill Maintenance**

Simplify.  [6.1c]

**15.** $\dfrac{8x^2}{24x}$

**16.** $\dfrac{5x^8y^4}{10x^3y}$

**17.** $\dfrac{5a + 15}{10}$

**18.** $\dfrac{12x - 24}{48}$

**19.** $\dfrac{2x^2 - 50}{x^2 - 25}$

**20.** $\dfrac{x^2 - 1}{x^4 - 1}$

**21.** $\dfrac{x^2 - 3x - 10}{x^2 - 2x - 15}$

**22.** $\dfrac{6x^2 + 15x - 36}{2x^2 - 5x + 3}$

**23.** $\dfrac{(x^2 + 6x + 9)(x - 2)}{(x^2 - 4)(x + 3)}$

**24.** $\dfrac{x^2 + 25}{x^2 - 25}$

**25.** $\dfrac{6x^2 + 18x + 12}{6x^2 - 6}$

**26.** $\dfrac{x^3 + 3x^2 + 2x + 6}{2x^3 + 6x^2 + x + 3}$

---

**Synthesis**

**27.** ◈ Discuss the advantages of using a table to organize information when solving a motion problem.

**28.** ◈ From the formula $d = rt$, derive two other formulas, one for $r$ and one for $t$. Discuss the kinds of problems for which each formula might be useful.

**29.** *Lindbergh's Flight.* Charles Lindbergh flew the Spirit of St. Louis in 1927 from New York to Paris at an average speed of 107.4 mph. Eleven years later, Howard Hughes flew the same route, averaged 217.1 mph, and took 16 hr and 57 min less time. Find the length of their route.

**30.** A car travels from one town to another at a speed of 32 mph. If it had gone 4 mph faster, it could have made the trip in $\frac{1}{2}$ hr less time. How far apart are the towns?

**31.** An afternoon sightseeing cruise up river and back down river is scheduled to last 1 hr. The speed of the current is 4 mph, and the speed of the riverboat in still water is 12 mph. How far upstream should the pilot travel before turning around?

Copyright © 1999 Addison Wesley Longman

---

# Summary and Review Exercises: Chapter 8

## Important Properties and Formulas

*Motion Formula:* $d = rt$

The objectives to be tested in addition to the material in this chapter are [3.3a], [4.1d, e, f], [6.1c], and [6.5a].

Determine whether the given ordered pair is a solution of the system of equations. [8.1a]

**1.** $(6, -1)$; $x - y = 3$,
$2x + 5y = 6$

**2.** $(2, -3)$; $2x + y = 1$,
$x - y = 5$

**3.** $(-2, 1)$; $x + 3y = 1$,
$2x - y = -5$

**4.** $(-4, -1)$; $x - y = 3$,
$x + y = -5$

Solve the system by graphing. [8.1b]

**5.** $x + y = 4$,
$x - y = 8$

**6.** $x + 3y = 12$,
$2x - 4y = 4$

**7.** $y = 5 - x$,
$3x - 4y = -20$

**8.** $3x - 2y = -4$,
$2y - 3x = -2$

Solve the system using the substitution method. [8.2a]

**9.** $y = 5 - x$,
$3x - 4y = -20$

**10.** $x + y = 6$,
$y = 3 - 2x$

**11.** $x - y = 4$,
$y = 2 - x$

**12.** $s + t = 5$,
$s = 13 - 3t$

Solve the system using the substitution method. [8.2b]

**13.** $x + 2y = 6$,
$2x + 3y = 8$

**14.** $3x + y = 1$,
$x - 2y = 5$

Solve the system using the elimination method. [8.3a]

**15.** $x + y = 4$,
$2x - y = 5$

**16.** $x + 2y = 9$,
$3x - 2y = -5$

**17.** $x - y = 8$,
$2x + y = 7$

Solve the system using the elimination method. [8.3b]

**18.** $2x + 3y = 8$,
$5x + 2y = -2$

**19.** $5x - 2y = 2$,
$3x - 7y = 36$

**20.** $-x - y = -5$,
$2x - y = 4$

**21.** $6x + 2y = 4$,
$10x + 7y = -8$

**22.** $-6x - 2y = 5$,
$12x + 4y = -10$

**23.** $\frac{2}{3}x + y = -\frac{5}{3}$,
$x - \frac{1}{3}y = -\frac{13}{3}$

Solve. [8.2c], [8.3c], [8.4a]

**24.** The sum of two numbers is 8. Their difference is 12. Find the numbers.

**25.** The sum of two numbers is 27. One-half of the first number plus one-third of the second number is 11. Find the numbers.

**26.** The perimeter of a rectangle is 96 cm. The length is 27 cm more than the width. Find the length and the width.

**27.** *Paid Admissions.* There were 508 people at a rock concert. Orchestra seats cost $25 per person and balcony seats cost $18. The total receipts were $11,223. Find the number of orchestra seats and the number of balcony seats sold for the concert.

**28.** *Window Cleaner.* Clear Shine window cleaner is 30% alcohol, whereas Sunstream window cleaner is 60% alcohol. How much of each is needed to make 80 L of a cleaner that is 45% alcohol?

**29.** Jeff is three times as old as his son. In 9 yr, Jeff will be twice as old as his son. How old is each now?

**30.** *Weights of Elephants.* A zoo has both an Asian and an African elephant. The African elephant weighs 2400 kg more than the Asian elephant. Together, they weigh 12,000 kg. How much does each elephant weigh?

**31.** *Mixed Nuts.* Sandy's Catering needs to provide 10 lb of mixed nuts for a wedding reception. The wedding couple has allocated $40 for nuts. Peanuts cost $2.50 per pound and fancy nuts cost $7 per pound. How many pounds of each type should be mixed?

Solve. [8.5a]

**32.** An airplane flew for 4 hr with a 15-km/h tail wind. The return flight against the wind took 5 hr. Find the speed of the airplane in still air.

**33.** One car leaves Phoenix, Arizona, on Interstate highway I-10 traveling at a speed of 55 mph. Two hours later, another car leaves Phoenix on the same highway, but travels at the new speed limit of 75 mph. How far from Phoenix will the second car catch up to the other?

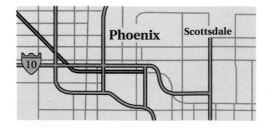

Simplify.

**34.** $t^{-5} \cdot t^{13}$ [4.1d, f]

**35.** $\dfrac{t^{-5}}{t^{13}}$ [4.1e, f]

**36.** Subtract: [6.5a]

$$\frac{x}{x^2 - 9} - \frac{x - 1}{x^2 - 5x + 6}.$$

**37.** Simplify: [6.1c]

$$\frac{5x^2 - 20}{5x^2 + 40x - 100}.$$

**38.** Find the intercepts. Then graph the equation. [3.3a]

$$2y - x = 6$$

**Synthesis**

**39.** ◆ Briefly compare the strengths and weaknesses of the graphical, substitution, and elimination methods. [8.3b]

**40.** ◆ Janine can tell by inspection that the system

$$y = 2x - 1,$$
$$y = 2x + 3$$

has no solution. How did she determine this? [8.1b]

**41.** Stephanie agreed to work as a stablehand for 1 yr. At the end of that time, she was to receive $2400 and one horse. After 7 months, she quit the job, but still received the horse and $1000. What was the value of the horse? [8.3c]

**42.** The solution of the following system is (6, 2). Find $C$ and $D$. [8.3b]

$$2x - Dy = 6,$$
$$Cx + 4y = 14$$

**43.** Solve: [8.2a]

$$3(x - y) = 4 + x,$$
$$x = 5y + 2.$$

Each of the following shows the graph of a system of equations. Find the equations. [7.2c], [8.1b]

**44.**        **45.**

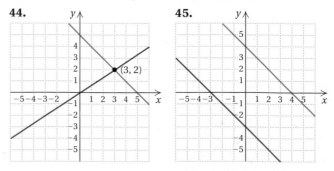

# Test: Chapter 8

1. Determine whether the given ordered pair is a solution of the system of equations.

    $(-2, -1);$ $\quad x = 4 + 2y,$
    $\qquad\qquad\quad 2y - 3x = 4$

2. Solve this system by graphing:

    $x - y = 3,$
    $x - 2y = 4.$

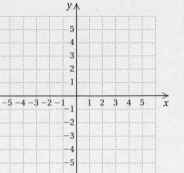

Solve the system using the substitution method.

3. $y = 6 - x,$
   $2x - 3y = 22$

4. $x + 2y = 5,$
   $x + y = 2$

5. $y = 5x - 2,$
   $y - 2 = 5x$

Solve the system using the elimination method.

6. $x - y = 6,$
   $3x + y = -2$

7. $\dfrac{1}{2}x - \dfrac{1}{3}y = 8,$
   $\dfrac{2}{3}x + \dfrac{1}{2}y = 5$

8. $4x + 5y = 5,$
   $6x + 7y = 7$

9. $2x + 3y = 13,$
   $3x - 5y = 10$

Solve.

10. The perimeter of a rectangular field is 8266 yd. The length is 84 yd more than the width. Find the length and the width.

11. The difference of two numbers is 12. One-fourth of the larger number plus one-half of the smaller is 9. Find the numbers.

**Answers**

1. _____

2. _____

3. _____

4. _____

5. _____

6. _____

7. _____

8. _____

9. _____

10. _____

11. _____

**12.** _____

**13.** _____

**14.** _____

**15.** _____

**16.** _____

**17.** _____

**18.** _____

**19.** _____

**20.** _____

**21.** _____

**22.** _____

**12.** A motorboat traveled for 2 hr with an 8-km/h current. The return trip against the same current took 3 hr. Find the speed of the motorboat in still water.

**13.** _Mixture of Solutions._ Solution A is 25% acid, and solution B is 40% acid. How much of each is needed to make 60 L of a solution that is 30% acid?

---

### Skill Maintenance

**14.** Subtract: $\dfrac{1}{x^2 - 16} - \dfrac{x - 4}{x^2 - 3x - 4}$.

**15.** Graph: $3x - 4y = -12$.

Simplify.

**16.** $(2x^{-2}y^7)(5x^6y^{-9})$

**17.** $\dfrac{a^4b^2}{a^{-6}b^8}$

**18.** $\dfrac{5x^2 + 40x - 100}{10x^2 - 40}$

---

### Synthesis

**19.** Find the numbers $C$ and $D$ such that $(-2, 3)$ is a solution of the system

$$Cx - 4y = 7,$$
$$3x + Dy = 8.$$

**20.** You are in line at a ticket window. There are two more people ahead of you than there are behind you. In the entire line, there are three times as many people as there are behind you. How many are ahead of you in line?

Each of the following shows the graph of a system of equations. Find the equations.

**21.**

**22.**

# Cumulative Review: Chapters 1–8

Compute and simplify.

**1.** $-2[1.4 - (-0.8 - 1.2)]$

**2.** $(1.3 \times 10^8)(2.4 \times 10^{-10})$

**3.** $\left(-\dfrac{1}{6}\right) \div \left(\dfrac{2}{9}\right)$

**4.** $\dfrac{2^{12}2^{-7}}{2^8}$

Simplify.

**5.** $\dfrac{x^2 - 9}{2x^2 - 7x + 3}$

**6.** $\dfrac{t^2 - 16}{(t + 4)^2}$

**7.** $\dfrac{x - \dfrac{x}{x + 2}}{\dfrac{2}{x} - \dfrac{1}{x + 2}}$

Perform the indicated operations and simplify.

**8.** $(1 - 3x^2)(2 - 4x^2)$

**9.** $(2a^2b - 5ab^2)^2$

**10.** $(3x^2 + 4y)(3x^2 - 4y)$

**11.** $-2x^2(x - 2x^2 + 3x^3)$

**12.** $(1 + 2x)(4x^2 - 2x + 1)$

**13.** $\left(8 - \dfrac{1}{3}x\right)\left(8 + \dfrac{1}{3}x\right)$

**14.** $(-8y^2 - y + 2) - (y^3 - 6y^2 + y - 5)$

**15.** $(2x^3 - 3x^2 - x - 1) \div (2x - 1)$

**16.** $\dfrac{7}{5x - 25} + \dfrac{x + 7}{5 - x}$

**17.** $\dfrac{2x - 1}{x - 2} - \dfrac{2x}{2 - x}$

**18.** $\dfrac{y^2 + y}{y^2 + y - 2} \cdot \dfrac{y + 2}{y^2 - 1}$

**19.** $\dfrac{7x + 7}{x^2 - 2x} \div \dfrac{14}{3x - 6}$

Factor completely.

**20.** $6x^5 - 36x^3 + 9x^2$

**21.** $16y^4 - 81$

**22.** $3x^2 + 10x - 8$

**23.** $4x^4 - 12x^2y + 9y^2$

**24.** $3m^3 + 6m^2 - 45m$

**25.** $x^3 + x^2 - x - 1$

Solve.

**26.** $3x - 4(x + 1) = 5$

**27.** $x(2x - 5) = 0$

**28.** $5x + 3 \geq 6(x - 4) + 7$

**29.** $1.5x - 2.3x = 0.4(x - 0.9)$

**30.** $2x^2 = 338$

**31.** $3x^2 + 15 = 14x$

**32.** $\dfrac{2}{x} - \dfrac{3}{x - 2} = \dfrac{1}{x}$

**33.** $1 + \dfrac{3}{x} + \dfrac{x}{x + 1} = \dfrac{1}{x^2 + x}$

**34.** $y = 2x - 9,$
$2x + 3y = -3$

**35.** $6x + 3y = -6,$
$-2x + 5y = 14$

**36.** $2x = y - 2,$
$3y - 6x = 6$

**37.** $\dfrac{1}{x} - \dfrac{1}{y} = \dfrac{1}{xy}$, for $x$

Solve.

**38.** The vice-president of a sorority has $100 to spend on promotional buttons. There is a set-up fee of $18 and a cost of 35¢ per button. How many buttons can she purchase?

**39.** It takes David 15 hr to put a roof on a house. It takes Loren 9 hr to put a roof on the same type of house. How long would it take to complete the job if they worked together?

**40.** The length of one leg of a right triangle is 8 m. The length of the hypotenuse is 4 m longer than the length of the other leg. Find the lengths of the hypotenuse and the other leg.

**41.** To determine the number of fish in a lake, a conservationist catches 85 fish, tags them, and throws them back into the lake. Later, 60 fish are caught, 25 of which are tagged. How many fish are in the lake?

**42.** The height of a triangle is 3 cm less than the base. The area is 27 cm$^2$. Find the height and the base.

**43.** The height $h$ of a parallelogram of fixed area varies inversely as the base $b$. Suppose that the height is 24 ft when the base is 15 ft. Find the height when the base is 5 ft. What is the variation constant?

**44.** Two cars leave town at the same time going in the same direction. One travels 50 mph and the other travels 55 mph. In how many hours will they be 50 mi apart?

**45.** Solution A is 20% alcohol, and solution B is 60% alcohol. How much of each should be used in order to make 10 L of a solution that is 50% alcohol?

**46.** Find an equation of variation in which $y$ varies directly as $x$ and $y = 2.4$ when $x = 12$.

**47.** Find the slope of the line containing the points $(2, 3)$ and $(-1, 3)$.

**48.** Find the slope and the $y$-intercept of the line $2x + 3y = 6$.

**49.** Find an equation of the line that contains the points $(-5, 6)$ and $(2, -4)$.

**50.** Find an equation of the line containing the point $(0, -3)$ and having the slope $m = 6$.

Graph on a plane.

**51.** $y = -2$

**52.** $2x + 5y = 10$

**53.** $y \leq 5x$

**54.** $5x - 1 < 24$

Solve by graphing.

**55.** $x = 5 + y$,
$x - y = 1$

**56.** $3x - y = 4$,
$x + 3y = -2$

---

**Synthesis**

**57.** The solution of the following system is $(-5, 2)$. Find $A$ and $B$.

$3x - Ay = -7$,
$Bx + 4y = 15$

**58.** Solve: $x^2 + 2 < 0$.

**59.** Simplify:

$$\frac{x - 5}{x + 3} - \frac{x^2 - 6x + 5}{x^2 + x - 2} \div \frac{x^2 + 4x + 3}{x^2 + 3x + 2}.$$

**60.** Find the value of $k$ such that $y - kx = 4$ and $10x - 3y = -12$ are perpendicular.

---

Cumulative Review: Chapters 1–8

# 9

# Radical Expressions and Equations

## An Application

After an accident, how do police determine the speed at which the car had been traveling? The formula $r = 2\sqrt{5L}$ can be used to approximate the speed $r$, in miles per hour, of a car that has left a skid mark of length $L$, in feet. What was the speed of a car that left skid marks of length 30 ft?

This problem appears as Example 8 in Section 9.1.

## The Mathematics

We substitute 30 for $L$ in the formula and find an approximation:

$$r = 2\underset{\uparrow}{\sqrt{5L}} = 2\sqrt{5 \cdot 30}$$
$$= 2\sqrt{150} \approx 24.495.$$

This is a radical expression.

## Introduction

The formula in the application below illustrates the use of another type of algebraic expression called a *radical expression*. It involves taking a *square root*. We say that 3 is a square root of 9 because $3^2 = 9$. Similarly, $-3$ is a square root of 9 because $(-3)^2 = 9$. To express that 3 is the positive square root of 9, we write $\sqrt{9} = 3$. We say that $\sqrt{9}$ is a *radical expression*. In this chapter, we study manipulations of radical expressions in addition, subtraction, multiplication, division, and simplifying. Finally, we consider another equation-solving principle and apply it to applications and problem solving.

World Wide Web   For more information, visit us at www.mathmax.com

# Pretest:  Chapter 9

**1.** Find the square roots of 49.

**2.** Identify the radicand in $\sqrt{3t}$.

Determine whether the expression is meaningful as a real number. Write "yes" or "no."

**3.** $\sqrt{-47}$

**4.** $\sqrt{81}$

**5.** Approximate $\sqrt{47}$ to three decimal places.

**6.** Solve: $\sqrt{2x + 1} = 3$.

Assume henceforth that *all* expressions under radicals represent positive numbers.

Simplify.

**7.** $\sqrt{4x^2}$

**8.** $4\sqrt{18} - 2\sqrt{8} + \sqrt{32}$

**9.** $(2 - \sqrt{3})^2$

**10.** $(2 - \sqrt{3})(2 + \sqrt{3})$

Multiply and simplify.

**11.** $\sqrt{6}\,\sqrt{10}$

**12.** $(2\sqrt{6} - 1)^2$

Divide and simplify.

**13.** $\dfrac{\sqrt{15}}{\sqrt{3}}$

**14.** $\sqrt{\dfrac{24a^7}{3a^3}}$

**15.** In a right triangle, $a = 5$ and $b = 8$. Find $c$, the length of the hypotenuse. Give an exact answer and an approximation to three decimal places.

**16.** How long is a guy wire reaching from the top of a 12-m pole to a point 7 m from the base of the pole?

**17.** Rationalize the denominator:

$$\frac{\sqrt{5}}{\sqrt{x}}.$$

**18.** Rationalize the denominator:

$$\frac{8}{6 + \sqrt{5}}.$$

## Objectives for Retesting

The objectives to be tested in addition to the material in this chapter are as follows.

[6.2b]    Divide rational expressions and simplify.
[7.5b]    Solve applied problems involving direct variation.
[8.3a, b] Solve a system of two equations in two variables using the elimination method.
[8.4a]    Solve applied problems by translating to a system of two equations in two variables.

# 9.1 Introduction to Square Roots and Radical Expressions

## a | Square Roots

When we raise a number to the second power, we have squared the number. Sometimes we may need to find the number that was squared. We call this process finding a square root of a number.

> The number $c$ is a **square root** of $a$ if $c^2 = a$.

Every positive number has two square roots. For example, the square roots of 25 are 5 and $-5$ because $5^2 = 25$ and $(-5)^2 = 25$. The positive square root is also called the **principal square root**. The symbol $\sqrt{\phantom{x}}$ is called a **radical**\* (or **square root**) symbol. The radical symbol represents only the principal square root. Thus, $\sqrt{25} = 5$. To name the negative square root of a number, we use $-\sqrt{\phantom{x}}$. The number 0 has only one square root, 0.

**Example 1** Find the square roots of 81.

The square roots are 9 and $-9$.

**Example 2** Find $\sqrt{225}$.

There are two square roots, 15 and $-15$. We want the principal, or positive, square root since this is what $\sqrt{\phantom{x}}$ represents. Thus, $\sqrt{225} = 15$.

**Example 3** Find $-\sqrt{64}$.

The symbol $\sqrt{64}$ represents the positive square root. Then $-\sqrt{64}$ represents the negative square root. That is, $\sqrt{64} = 8$, so $-\sqrt{64} = -8$.

## b | Approximating Square Roots

We often need to use rational numbers to *approximate* square roots that are irrational. Such approximations can be found using a calculator with a square-root key $\boxed{\sqrt{\phantom{x}}}$.

**Examples** Use a calculator to approximate each of the following.

| | *Using a calculator with a 10-digit readout* | *Rounded to three decimal places* |
|---|---|---|
| **4.** $\sqrt{10}$ | 3.162277660 | 3.162 |
| **5.** $-\sqrt{583.8}$ | $-24.16195356$ | $-24.162$ |
| **6.** $\sqrt{\dfrac{48}{55}}$ | 0.934198733 | 0.934 |

*Do Exercises 1–16.*

---

\*Radicals can be other than square roots, but we will consider only square-root radicals in Chapter 9. See Appendix B for other types of radicals.

---

### Objectives

**a** Find the principal square roots and their opposites of the whole numbers from $0^2$ to $25^2$.

**b** Approximate square roots of real numbers using a calculator.

**c** Solve applied problems involving square roots.

**d** Identify radicands of radical expressions.

**e** Identify whether a radical expression represents a real number.

**f** Simplify a radical expression with a perfect-square radicand.

### For Extra Help

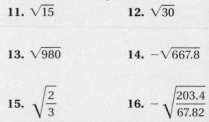

TAPE 15    TAPE 16A    MAC WIN    CD-ROM

Find the square roots.

1. 36
2. 64

3. 121
4. 144

Find the following.

5. $\sqrt{16}$
6. $\sqrt{49}$

7. $\sqrt{100}$
8. $\sqrt{441}$

9. $-\sqrt{49}$
10. $-\sqrt{169}$

Use a calculator to approximate each of the following square roots to three decimal places.

11. $\sqrt{15}$
12. $\sqrt{30}$

13. $\sqrt{980}$
14. $-\sqrt{667.8}$

15. $\sqrt{\dfrac{2}{3}}$
16. $-\sqrt{\dfrac{203.4}{67.82}}$

*Answers on page A-35*

**17.** *Speed of a Skidding Car.*
Referring to Example 8,
determine the speed of a car
that left skid marks of length
**(a)** 40 ft; **(b)** 123 ft.

Identify the radicand.

**18.** $\sqrt{227}$

**19.** $\sqrt{45 + x}$

**20.** $\sqrt{\dfrac{x}{x + 2}}$

**21.** $8\sqrt{x^2 + 4}$

*Answers on page A-35*

---

## c  Applications of Square Roots

We now consider an application involving a formula with a radical expression.

**Example 7**   *Speed of a Skidding Car.* After an accident, how do police determine the speed at which the car had been traveling? The formula $r = 2\sqrt{5L}$ can be used to approximate the speed $r$, in miles per hour, of a car that has left a skid mark of length $L$, in feet. What was the speed of a car that left skid marks of length **(a)** 30 ft? **(b)** 150 ft?

a) We substitute 30 for $L$ and find an approximation:

$$r = 2\sqrt{5L} = 2\sqrt{5 \cdot 30} = 2\sqrt{150} \approx 24.495.$$

The speed of the car was about 24.5 mph.

b) We substitute 150 for $L$ and find an approximation:

$$r = 2\sqrt{5L} = 2\sqrt{5 \cdot 150} \approx 54.772.$$

The speed of the car was about 54.8 mph.

*Do Exercise 17.*

## d  Radicands and Radical Expressions

When an expression is written under a radical, we have a **radical expression.** Here are some examples:

$$\sqrt{14}, \qquad \sqrt{x}, \qquad 8\sqrt{x^2 + 4}, \qquad \sqrt{\dfrac{x^2 - 5}{2}}.$$

The expression written under the radical is called the **radicand.**

**Examples**   Identify the radicand in each expression.

**8.** $\sqrt{105}$        The radicand is 105.

**9.** $\sqrt{x}$        The radicand is $x$.

**10.** $6\sqrt{y^2 - 5}$        The radicand is $y^2 - 5$.

**11.** $\sqrt{\dfrac{a - b}{a + b}}$     The radicand is $\dfrac{a - b}{a + b}$.

*Do Exercises 18–21.*

## e  Expressions That Are Meaningful as Real Numbers

The square of any nonzero number is always positive. For example, $8^2 = 64$ and $(-11)^2 = 121$. There are no real numbers that when squared yield negative numbers. Thus the following expressions do not represent real numbers (they are meaningless as real numbers):

$$\sqrt{-100}, \qquad \sqrt{-49}, \qquad -\sqrt{-3}.$$

>  Radical expressions with negative radicands do not represent real numbers.

Later in your study of mathematics, you may encounter a number system called the **complex numbers** in which negative numbers have square roots.

*Do Exercises 22–25.*

### f  Perfect-Square Radicands

The expression $\sqrt{x^2}$, with a perfect-square radicand, can be troublesome. Recall that $\sqrt{\phantom{x}}$ denotes the principal square root. That is, the answer is nonnegative (either positive or zero). If $x$ represents a nonnegative number, $\sqrt{x^2}$ simplifies to $x$. If $x$ represents a negative number, $\sqrt{x^2}$ simplifies to $-x$ (the opposite of $x$), which is positive.

Suppose that $x = 3$. Then

$$\sqrt{x^2} = \sqrt{3^2} = \sqrt{9} = 3.$$

Suppose that $x = -3$. Then

$$\sqrt{x^2} = \sqrt{(-3)^2} = \sqrt{9} = 3, \quad \text{the } \textit{opposite} \text{ of } -3.$$

Note that 3 is the *absolute value* of both 3 and $-3$. In general, when replacements for $x$ are considered to be *any* real numbers, it follows that

$$\sqrt{x^2} = |x|.$$

> For any real number $A$,
> $$\sqrt{A^2} = |A|.$$
> (That is, for any real number $A$, the principal square root of $A^2$ is the absolute value of $A$.)

**Examples**  Simplify. Assume that expressions under radicals represent any real number.

**12.** $\sqrt{10^2} = |10| = 10$

**13.** $\sqrt{(-7)^2} = |-7| = 7$

**14.** $\sqrt{(3x)^2} = |3x|$   **Absolute-value notation is necessary.**

**15.** $\sqrt{a^2b^2} = \sqrt{(ab)^2} = |ab|$

**16.** $\sqrt{x^2 + 2x + 1} = \sqrt{(x + 1)^2} = |x + 1|$

*Do Exercises 26–31.*

Fortunately, in most uses of radicals, it can be assumed that expressions under radicals are nonnegative or positive. Indeed, many computers and calculators are programmed to consider only nonnegative radicands. Suppose that $x \geq 0$. Then

$$\sqrt{x^2} = |x| = x,$$

since $x$ is nonnegative.

> For any nonnegative real number $A$,
> $$\sqrt{A^2} = A.$$
> (That is, for any nonnegative real number $A$, the principal square root of $A^2$ is $A$.)

Determine whether the expression is meaningful as a real number. Write "yes" or "no."

**22.** $-\sqrt{25}$

**23.** $\sqrt{-25}$

**24.** $-\sqrt{-36}$

**25.** $-\sqrt{36}$

Simplify. Assume that expressions under radicals represent any real number.

**26.** $\sqrt{(-13)^2}$

**27.** $\sqrt{(7w)^2}$

**28.** $\sqrt{(xy)^2}$

**29.** $\sqrt{x^2y^2}$

**30.** $\sqrt{(x - 11)^2}$

**31.** $\sqrt{x^2 + 8x + 16}$

*Answers on page A-35*

Simplify. Assume that expressions under radicals represent nonnegative real numbers.

**32.** $\sqrt{(xy)^2}$    **33.** $\sqrt{x^2y^2}$

**34.** $\sqrt{(x - 11)^2}$

**35.** $\sqrt{x^2 + 8x + 16}$

**36.** $\sqrt{25y^2}$    **37.** $\sqrt{\frac{1}{4}t^2}$

**Examples**  Simplify. Assume that expressions under radicals represent nonnegative real numbers.

**17.** $\sqrt{(3x)^2} = 3x$    Since $3x$ is assumed to be nonnegative

**18.** $\sqrt{a^2b^2} = \sqrt{(ab)^2} = ab$    Since $ab$ is assumed to be nonnegative

**19.** $\sqrt{x^2 + 2x + 1} = \sqrt{(x + 1)^2} = x + 1$    Since $x + 1$ is assumed to be nonnegative

*Do Exercises 32–37.*

Henceforth, in this text we will assume that all expressions under radicals represent nonnegative real numbers.

We make this assumption in order to eliminate some confusion and because it is valid in many applications. As you study further in mathematics, however, you will frequently have to make a determination about expressions under radicals being nonnegative or positive. This will often be necessary in calculus.

## Calculator Spotlight

Graphing Equations Containing Radical Expressions. Graphing equations that contain radical expressions involves approximating square roots. Since the square root of a negative number is not a real number, $y$-values may not exist for some $x$-values. For example, $y$-values of the graph of $y = \sqrt{x - 1}$ do not exist for $x$-values that are less than 1 because square roots of negative numbers would result.

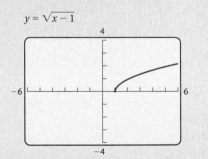
$y = \sqrt{x - 1}$

On the TI-82, we must enter $y = \sqrt{x - 1}$, using parentheses around the radicand, as $y = \sqrt{(x - 1)}$. On the TI-83, if you enter a radical expression using the ⌨y= key, you will automatically begin with $y = \sqrt{(}$. The right-hand parenthesis is understood to be at the end of the expression entered if you do not enter it.

Similarly, $y$-values of the graph of $y = \sqrt{2 - x}$ do not exist for $x$-values that exceed 2.

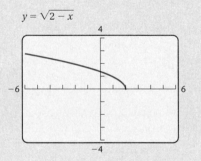
$y = \sqrt{2 - x}$

**Exercises**

Use a grapher to graph the equation.

**1.** $y = \sqrt{x}$    **2.** $y = \sqrt{2x}$
**3.** $y = \sqrt{x^2}$    **4.** $y = \sqrt{(2x)^2}$
**5.** $y = \sqrt{x + 4}$    **6.** $y = \sqrt{6 - x}$
**7.** $y = -\sqrt{x}$    **8.** $y = 3 + \sqrt{x}$

Use the GRAPH and TABLE features to determine whether each of the following is correct.

**9.** $\sqrt{x - 2} = \sqrt{x} - 2$
**10.** $\sqrt{4x} = 2\sqrt{x}$

*Answers on page A-36*

# Exercise Set 9.1

**a** Find the square roots.

**1.** 4
**2.** 1
**3.** 9
**4.** 16
**5.** 100

**6.** 121
**7.** 169
**8.** 144
**9.** 256
**10.** 625

Simplify.

**11.** $\sqrt{4}$
**12.** $\sqrt{1}$
**13.** $-\sqrt{9}$
**14.** $-\sqrt{25}$
**15.** $-\sqrt{36}$

**16.** $-\sqrt{81}$
**17.** $-\sqrt{225}$
**18.** $\sqrt{400}$
**19.** $\sqrt{361}$
**20.** $\sqrt{441}$

**b** Use a calculator to approximate the square roots. Round to three decimal places.

**21.** $\sqrt{5}$
**22.** $\sqrt{8}$
**23.** $\sqrt{432}$
**24.** $\sqrt{8196}$
**25.** $-\sqrt{347.7}$

**26.** $-\sqrt{204.788}$
**27.** $\sqrt{\dfrac{278}{36}}$
**28.** $-\sqrt{\dfrac{567}{788}}$
**29.** $\sqrt{8 \cdot 9 \cdot 200}$
**30.** $\sqrt{\dfrac{47 \cdot 83}{947.03}}$

**c** *Parking-Lot Arrival Spaces.* The attendants at a parking lot park cars in temporary spaces before the cars are taken to permanent parking stalls. The number $N$ of such spaces needed is approximated by the formula $N = 2.5\sqrt{A}$, where $A =$ the average number of arrivals during peak hours.

**31.** Find the number of spaces needed when the average number of arrivals is **(a)** 25; **(b)** 89.

**32.** Find the number of spaces needed when the average number of arrivals is **(a)** 62; **(b)** 100.

**d** Identify the radicand.

**33.** $\sqrt{200}$
**34.** $\sqrt{16z}$
**35.** $\sqrt{a - 4}$
**36.** $\sqrt{3t + 10}$

**37.** $5\sqrt{t^2 + 1}$
**38.** $9\sqrt{x^2 + 16}$
**39.** $x^2 y \sqrt{\dfrac{3}{x + 2}}$
**40.** $ab^2 \sqrt{\dfrac{a}{a + b}}$

**e** Determine whether the expression is meaningful as a real number. Write "yes" or "no."

**41.** $\sqrt{-16}$

**42.** $\sqrt{-81}$

**43.** $-\sqrt{81}$

**44.** $-\sqrt{64}$

**f** Simplify. Remember that we have assumed that expressions under radicals represent nonnegative real numbers.

**45.** $\sqrt{c^2}$

**46.** $\sqrt{x^2}$

**47.** $\sqrt{9x^2}$

**48.** $\sqrt{16y^2}$

**49.** $\sqrt{(8p)^2}$

**50.** $\sqrt{(7pq)^2}$

**51.** $\sqrt{(ab)^2}$

**52.** $\sqrt{(6y)^2}$

**53.** $\sqrt{(34d)^2}$

**54.** $\sqrt{(53b)^2}$

**55.** $\sqrt{(x+3)^2}$

**56.** $\sqrt{(d-3)^2}$

**57.** $\sqrt{a^2 - 10a + 25}$

**58.** $\sqrt{x^2 + 2x + 1}$

**59.** $\sqrt{4a^2 - 20a + 25}$

**60.** $\sqrt{9p^2 + 12p + 4}$

---

### Skill Maintenance

**61.** The amount $F$ that a family spends on food varies directly as its income $I$. A family making $39,200 a year will spend $10,192 on food. At this rate, how much would a family making $41,000 spend on food?  [7.5b]

Divide and simplify.  [6.2b]

**62.** $\dfrac{x-3}{x+4} \div \dfrac{x^2-9}{x+4}$

**63.** $\dfrac{x^2+10x-11}{x^2-1} \div \dfrac{x+11}{x+1}$

**64.** $\dfrac{x^4-16}{x^4-1} \div \dfrac{x^2+4}{x^2+1}$

---

### Synthesis

**65.** ◈ What is the difference between "**the** square root of 10" and "**a** square root of 10"?

**66.** ◈ Explain why $\sqrt{A^2} \neq A$ for all real numbers.

**67.** Use only the graph of $y = \sqrt{x}$, shown below, to approximate $\sqrt{3}$, $\sqrt{5}$, and $\sqrt{7}$. Answers may vary.

**68.** Between what two consecutive integers is $\sqrt{78}$?

**69.** Between what two consecutive integers is $-\sqrt{33}$?

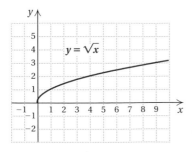

Solve.

**70.** $\sqrt{x^2} = 16$

**71.** $\sqrt{y^2} = -7$

**72.** $t^2 = 49$

**73.** Suppose that the area of a square is 3. Find the length of a side.

---

Collaborative
Learning Manual

Develop a formula for the swing time of a pendulum.

# 9.2 Multiplying and Simplifying with Radical Expressions

## a | Simplifying by Factoring

To see how to multiply with radical notation, consider the following.

**a)** $\sqrt{9} \cdot \sqrt{4} = 3 \cdot 2 = 6$    **This is a product of square roots.**

**b)** $\sqrt{9 \cdot 4} = \sqrt{36} = 6$    **This is the square root of a product.**

Note that

$$\sqrt{9} \cdot \sqrt{4} = \sqrt{9 \cdot 4}.$$

***Do Exercise 1.***

We can multiply radical expressions by multiplying the radicands.

> **THE PRODUCT RULE FOR RADICALS**
>
> For any nonnegative radicands $A$ and $B$,
> $$\sqrt{A} \cdot \sqrt{B} = \sqrt{A \cdot B}.$$
>
> (The product of square roots is the square root of the product of the radicands.)

**Examples** Multiply.

**1.** $\sqrt{5}\sqrt{7} = \sqrt{5 \cdot 7} = \sqrt{35}$     **2.** $\sqrt{8}\sqrt{8} = \sqrt{8 \cdot 8} = \sqrt{64} = 8$

**3.** $\sqrt{\dfrac{2}{3}}\sqrt{\dfrac{4}{5}} = \sqrt{\dfrac{2}{3} \cdot \dfrac{4}{5}} = \sqrt{\dfrac{8}{15}}$     **4.** $\sqrt{2x}\sqrt{3x - 1} = \sqrt{2x(3x - 1)}$
$$= \sqrt{6x^2 - 2x}$$

***Do Exercises 2–5.***

To factor radical expressions, we can use the product rule for radicals in reverse. That is,

> $\sqrt{AB} = \sqrt{A}\sqrt{B}.$

In some cases, we can simplify after factoring.

> A square-root radical expression is simplified when its radicand has no factors that are perfect squares.

When simplifying a square-root radical expression, we first determine whether a radicand is a perfect square. Then we determine whether it has perfect-square factors. The radicand is then factored and the radical expression simplified using the preceding rule.

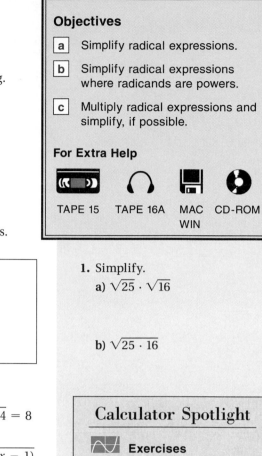

**Objectives**

a | Simplify radical expressions.

b | Simplify radical expressions where radicands are powers.

c | Multiply radical expressions and simplify, if possible.

**For Extra Help**

TAPE 15    TAPE 16A    MAC WIN    CD-ROM

**1.** Simplify.
   **a)** $\sqrt{25} \cdot \sqrt{16}$

   **b)** $\sqrt{25 \cdot 16}$

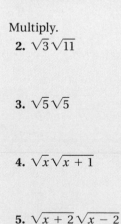

## Calculator Spotlight

### Exercises

Use the GRAPH and TABLE features to determine whether each of the following is correct.

**1.** $\sqrt{5x} = \sqrt{5} \cdot \sqrt{x}$

**2.** $\sqrt{3x} = 3\sqrt{x}$

Multiply.
   **2.** $\sqrt{3}\sqrt{11}$

   **3.** $\sqrt{5}\sqrt{5}$

   **4.** $\sqrt{x}\sqrt{x + 1}$

   **5.** $\sqrt{x + 2}\sqrt{x - 2}$

*Answers on page A-36*

Simplify by factoring.

**6.** $\sqrt{32}$

**7.** $\sqrt{x^2 + 14x + 49}$

**8.** $\sqrt{25x^2}$

**9.** $\sqrt{36m^2}$

**10.** $\sqrt{92}$

**11.** $\sqrt{x^2 - 20x + 100}$

**12.** $\sqrt{64t^2}$

**13.** $\sqrt{100a^2}$

*Answers on page A-36*

Compare the following:

$$\sqrt{50} = \sqrt{10 \cdot 5} = \sqrt{10}\,\sqrt{5};$$
$$\sqrt{50} = \sqrt{25 \cdot 2} = \sqrt{25}\,\sqrt{2} = 5\sqrt{2}.$$

In the second case, the radicand has the perfect-square factor 25. If you do not recognize perfect-square factors, try factoring the radicand into its prime factors. For example,

$$\sqrt{50} = \sqrt{2 \cdot \underbrace{5 \cdot 5}} = 5\sqrt{2}.$$

Perfect square (a pair of the same numbers)

Square-root radical expressions in which the radicand has no perfect-square factors, such as $5\sqrt{2}$, are considered to be in simplest form.

**Examples**   Simplify by factoring.

**5.** $\sqrt{18} = \sqrt{9 \cdot 2}$        Identifying a perfect-square factor and factoring the radicand. The factor 9 is a perfect square.

$\quad\quad = \sqrt{9} \cdot \sqrt{2}$        Factoring into a product of radicals

$\quad\quad = 3\sqrt{2}$

The radicand has no factors that are perfect squares.

**6.** $\sqrt{48t} = \sqrt{16 \cdot 3 \cdot t}$        Identifying a perfect-square factor and factoring the radicand. The factor 16 is a perfect square.

$\quad\quad = \sqrt{16}\,\sqrt{3t}$        Factoring into a product of radicals

$\quad\quad = 4\sqrt{3t}$        Taking a square root

**7.** $\sqrt{20t^2} = \sqrt{4 \cdot 5 \cdot t^2}$        Identifying perfect-square factors and factoring the radicand. The factors 4 and $t^2$ are perfect squares.

$\quad\quad = \sqrt{4}\,\sqrt{t^2}\,\sqrt{5}$        Factoring into a product of several radicals

$\quad\quad = 2t\sqrt{5}$        Taking square roots. No absolute-value signs are necessary since we have assumed that expressions under radicals are nonnegative.

**8.** $\sqrt{x^2 - 6x + 9} = \sqrt{(x-3)^2} = x - 3$        No absolute-value signs are necessary since we have assumed that expressions under radicals are nonnegative.

**9.** $\sqrt{36x^2} = \sqrt{36}\,\sqrt{x^2} = 6x$, or $\sqrt{36x^2} = \sqrt{(6x)^2} = 6x$

**10.** $\sqrt{3x^2 + 6x + 3} = \sqrt{3(x^2 + 2x + 1)}$        Factoring the radicand

$\quad\quad = \sqrt{3(x+1)^2}$        Factoring further

$\quad\quad = \sqrt{3}\,\sqrt{(x+1)^2}$        Factoring into a product of radicals

$\quad\quad = \sqrt{3}\,(x+1)$        Taking the square root

*Do Exercises 6–13.*

**b**   **Simplifying Square Roots of Powers**

To take the square root of an even power such as $x^{10}$, we note that $x^{10} = (x^5)^2$. Then

$$\sqrt{x^{10}} = \sqrt{(x^5)^2} = x^5.$$

We can find the answer by taking half the exponent. That is,

$$\sqrt{x^{10}} = x^5. \longleftarrow \tfrac{1}{2}(10) = 5$$

**Examples** Simplify.

**11.** $\sqrt{x^6} = \sqrt{(x^3)^2} = x^3$ ←— $\frac{1}{2}(6) = 3$

**12.** $\sqrt{x^8} = x^4$

**13.** $\sqrt{t^{22}} = t^{11}$

*Do Exercises 14–16.*

If an odd power occurs, we express the power in terms of the largest even power. Then we simplify the even power as in Examples 11–13.

**Example 14** Simplify by factoring: $\sqrt{x^9}$.

$$\sqrt{x^9} = \sqrt{x^8 \cdot x}$$
$$= \sqrt{x^8}\sqrt{x}$$
$$= x^4\sqrt{x}$$

Note in Example 14 that $\sqrt{x^9} \neq x^3$.

**Example 15** Simplify by factoring: $\sqrt{32x^{15}}$.

$$\sqrt{32x^{15}} = \sqrt{16 \cdot 2 \cdot x^{14} \cdot x}$$ We factor the radicand, looking for perfect-square factors. The largest even power is 14.

$$= \sqrt{16}\sqrt{x^{14}}\sqrt{2x}$$ Factoring into a product of radicals. Perfect-square factors are usually listed first.

$$= 4x^7\sqrt{2x}$$ Simplifying

*Do Exercises 17 and 18.*

## c | Multiplying and Simplifying

Sometimes we can simplify after multiplying. We leave the radicand in factored form and factor further to determine perfect-square factors. Then we simplify the perfect-square factors.

**Example 16** Multiply and then simplify by factoring: $\sqrt{2}\sqrt{14}$.

$$\sqrt{2}\sqrt{14} = \sqrt{2 \cdot 14}$$ Multiplying

$$= \sqrt{2 \cdot 2 \cdot 7}$$ Factoring

$$= \sqrt{2 \cdot 2}\sqrt{7}$$ Looking for perfect-square factors; pairs of factors

$$= 2\sqrt{7}$$

*Do Exercises 19 and 20.*

**Example 17** Multiply and then simplify by factoring: $\sqrt{3x^2}\sqrt{9x^3}$.

$$\sqrt{3x^2}\sqrt{9x^3} = \sqrt{3x^2 \cdot 9x^3}$$ Multiplying

$$= \sqrt{3 \cdot x^2 \cdot 9 \cdot x^2 \cdot x}$$ Looking for perfect-square factors or largest even powers

Perfect-square factors are usually listed first.

$$= \sqrt{9}\sqrt{x^2}\sqrt{x^2}\sqrt{3x}$$
$$= 3 \cdot x \cdot x \cdot \sqrt{3x}$$
$$= 3x^2\sqrt{3x}$$

Simplify.

**14.** $\sqrt{t^4}$

**15.** $\sqrt{t^{20}}$

**16.** $\sqrt{h^{46}}$

Simplify by factoring.

**17.** $\sqrt{x^7}$

**18.** $\sqrt{24x^{11}}$

Multiply and simplify.

**19.** $\sqrt{3}\sqrt{6}$

**20.** $\sqrt{2}\sqrt{50}$

*Answers on page A-36*

Multiply and simplify.

**21.** $\sqrt{2x^3}\sqrt{8x^3y^4}$

**22.** $\sqrt{10xy^2}\sqrt{5x^2y^3}$

In doing an example like the preceding one, it might be helpful to do more factoring, as follows:

$$\sqrt{3x^2} \cdot \sqrt{9x^3} = \sqrt{3 \cdot \underline{x \cdot x} \cdot \underline{3 \cdot 3} \cdot \underline{x \cdot x} \cdot x}.$$

Then we look for pairs of factors, as shown, and simplify perfect-square factors:

$$= 3 \cdot x \cdot x\sqrt{3x}$$
$$= 3x^2\sqrt{3x}.$$

**Do Exercises 21 and 22.**

We know that $\sqrt{AB} = \sqrt{A}\sqrt{B}$. That is, the square root of a product is the product of the square roots. What about the square root of a sum? That is, is the square root of a sum equal to the sum of the square roots? To check, consider $\sqrt{A + B}$ and $\sqrt{A} + \sqrt{B}$ when $A = 16$ and $B = 9$:

$$\sqrt{A + B} = \sqrt{16 + 9} = \sqrt{25} = 5;$$

and

$$\sqrt{A} + \sqrt{B} = \sqrt{16} + \sqrt{9} = 4 + 3 = 7.$$

Thus we see the following.

---

*CAUTION!*  The square root of a sum is not the sum of the square roots.
$$\sqrt{A + B} \neq \sqrt{A} + \sqrt{B}$$

---

## Calculator Spotlight

### Exercises

Use the GRAPH and TABLE features to determine whether each of the following is correct.

**1.** $\sqrt{x + 4} = \sqrt{x} + 2$

**2.** $\sqrt{3 + x} = \sqrt{3} + \sqrt{x}$

*Answers on page A-36*

# Exercise Set 9.2

a  Simplify by factoring.

**1.** $\sqrt{12}$

**2.** $\sqrt{8}$

**3.** $\sqrt{75}$

**4.** $\sqrt{50}$

**5.** $\sqrt{20}$

**6.** $\sqrt{45}$

**7.** $\sqrt{600}$

**8.** $\sqrt{300}$

**9.** $\sqrt{486}$

**10.** $\sqrt{567}$

**11.** $\sqrt{9x}$

**12.** $\sqrt{4y}$

**13.** $\sqrt{48x}$

**14.** $\sqrt{40m}$

**15.** $\sqrt{16a}$

**16.** $\sqrt{49b}$

**17.** $\sqrt{64y^2}$

**18.** $\sqrt{9x^2}$

**19.** $\sqrt{13x^2}$

**20.** $\sqrt{23s^2}$

**21.** $\sqrt{8t^2}$

**22.** $\sqrt{125a^2}$

**23.** $\sqrt{180}$

**24.** $\sqrt{320}$

**25.** $\sqrt{288y}$

**26.** $\sqrt{363p}$

**27.** $\sqrt{28x^2}$

**28.** $\sqrt{20x^2}$

**29.** $\sqrt{x^2 - 6x + 9}$

**30.** $\sqrt{t^2 + 22t + 121}$

**31.** $\sqrt{8x^2 + 8x + 2}$

**32.** $\sqrt{20x^2 - 20x + 5}$

**33.** $\sqrt{36y + 12y^2 + y^3}$

**34.** $\sqrt{x - 2x^2 + x^3}$

**b** Simplify by factoring.

**35.** $\sqrt{x^6}$

**36.** $\sqrt{x^{18}}$

**37.** $\sqrt{x^{12}}$

**38.** $\sqrt{x^{16}}$

**39.** $\sqrt{x^5}$

**40.** $\sqrt{x^3}$

**41.** $\sqrt{t^{19}}$

**42.** $\sqrt{p^{17}}$

**43.** $\sqrt{(y-2)^8}$

**44.** $\sqrt{(x+3)^6}$

**45.** $\sqrt{4(x+5)^{10}}$

**46.** $\sqrt{16(a-7)^4}$

**47.** $\sqrt{36m^3}$

**48.** $\sqrt{250y^3}$

**49.** $\sqrt{8a^5}$

**50.** $\sqrt{12b^7}$

**51.** $\sqrt{104p^{17}}$

**52.** $\sqrt{284m^{23}}$

**53.** $\sqrt{448x^6y^3}$

**54.** $\sqrt{243x^5y^4}$

**c** Multiply and then simplify by factoring, if possible.

**55.** $\sqrt{3}\,\sqrt{18}$

**56.** $\sqrt{5}\,\sqrt{10}$

**57.** $\sqrt{15}\,\sqrt{6}$

**58.** $\sqrt{3}\,\sqrt{27}$

**59.** $\sqrt{18}\,\sqrt{14x}$

**60.** $\sqrt{12}\,\sqrt{18x}$

**61.** $\sqrt{3x}\,\sqrt{12y}$

**62.** $\sqrt{7x}\,\sqrt{21y}$

Copyright © 1999 Addison Wesley Longman

**63.** $\sqrt{13}\sqrt{13}$

**64.** $\sqrt{11}\sqrt{11x}$

**65.** $\sqrt{5b}\sqrt{15b}$

**66.** $\sqrt{6a}\sqrt{18a}$

**67.** $\sqrt{2t}\sqrt{2t}$

**68.** $\sqrt{7a}\sqrt{7a}$

**69.** $\sqrt{ab}\sqrt{ac}$

**70.** $\sqrt{xy}\sqrt{xz}$

**71.** $\sqrt{2x^2y}\sqrt{4xy^2}$

**72.** $\sqrt{15mn^2}\sqrt{5m^2n}$

**73.** $\sqrt{18}\sqrt{18}$

**74.** $\sqrt{16}\sqrt{16}$

**75.** $\sqrt{5}\sqrt{2x-1}$

**76.** $\sqrt{3}\sqrt{4x+2}$

**77.** $\sqrt{x+2}\sqrt{x+2}$

**78.** $\sqrt{x-9}\sqrt{x-9}$

**79.** $\sqrt{18x^2y^3}\sqrt{6xy^4}$

**80.** $\sqrt{12x^3y^2}\sqrt{8xy}$

**81.** $\sqrt{50x^4y^6}\sqrt{10xy}$

**82.** $\sqrt{10xy^2}\sqrt{5x^2y^3}$

---

**Skill Maintenance**

Solve.  [8.3a, b]

**83.** $x - y = -6$,
$\quad x + y = 2$

**84.** $3x + 5y = 6$,
$\quad 5x + 3y = 4$

**85.** $3x - 2y = 4$,
$\quad 2x + 5y = 9$

**86.** $4a - 5b = 25$,
$\quad a - b = 7$

Solve. [8.4a]

**87.** The perimeter of a rectangular storage area is 84 ft. The length is 18 ft greater than the width. Find the area of the rectangle.

**88.** Wilt Chamberlain once scored 100 points in an NBA game. He took only two-point shots and foul shots (one point each) and made a total of 64 shots (**Source**: National Basketball Association). How many of each type of shot did he make?

**89.** A solution containing 30% insecticide is to be mixed with a solution containing 50% insecticide in order to make 200 L of a solution containing 42% insecticide. How much of each solution should be used?

**90.** There were 411 people at a movie. Admission was $7.00 each for adults and $3.75 each for children, and receipts totaled $2678.75. How many adults and how many children attended?

Solve. [8.5a]

**91.** Greg and Beth paddled to a picnic spot downriver in 2 hr. It took them 3 hr to return against the current. If the speed of the current was 2 mph, at what speed were they paddling the canoe?

---

## Synthesis

**92.** ◆ Are the rules for manipulating expressions with exponents important when simplifying radical expressions? Why or why not?

**93.** ◆ Explain the error(s) in the following:
$$\sqrt{x^2 - 25} = \sqrt{x^2} - \sqrt{25} = x - 5.$$

Factor.

**94.** $\sqrt{5x - 5}$

**95.** $\sqrt{x^2 - x - 2}$

**96.** $\sqrt{x^2 - 36}$

**97.** $\sqrt{2x^2 - 5x - 12}$

**98.** $\sqrt{x^3 - 2x^2}$

**99.** $\sqrt{a^2 - b^2}$

Simplify.

**100.** $\sqrt{0.01}$

**101.** $\sqrt{0.25}$

**102.** $\sqrt{x^8}$

**103.** $\sqrt{9a^6}$

Multiply and then simplify by factoring.

**104.** $\sqrt{a}\,(\sqrt{a^3} - 5)$

**105.** $(\sqrt{2y})(\sqrt{3})(\sqrt{8y})$

**106.** $\sqrt{18(x - 2)}\,\sqrt{20(x - 2)^3}$

**107.** $\sqrt{27(x + 1)}\,\sqrt{12y(x + 1)^2}$

**108.** $\sqrt{2^{109}}\,\sqrt{x^{306}}\,\sqrt{x^{11}}$

**109.** $\sqrt{x}\,\sqrt{2x}\,\sqrt{10x^5}$

**Chapter 9   Radical Expressions and Equations**

**556**

Copyright © 1999 Addison Wesley Longman

# 9.3 Quotients Involving Square Roots

## a Dividing Radical Expressions

Consider the expressions

$$\frac{\sqrt{25}}{\sqrt{16}} \quad \text{and} \quad \sqrt{\frac{25}{16}}.$$

Let's evaluate them separately:

a) $\dfrac{\sqrt{25}}{\sqrt{16}} = \dfrac{5}{4}$ because $\sqrt{25} = 5$ and $\sqrt{16} = 4$;

b) $\sqrt{\dfrac{25}{16}} = \dfrac{5}{4}$ because $\dfrac{5}{4} \cdot \dfrac{5}{4} = \dfrac{25}{16}$.

We see that both expressions represent the same number. This suggests that the quotient of two square roots is the square root of the quotient of the radicands.

> **THE QUOTIENT RULE FOR RADICALS**
>
> For any nonnegative number $A$ and any positive number $B$,
>
> $$\frac{\sqrt{A}}{\sqrt{B}} = \sqrt{\frac{A}{B}}.$$
>
> (The quotient of two square roots is the square root of the quotient of the radicands.)

**Examples** Divide and simplify.

1. $\dfrac{\sqrt{27}}{\sqrt{3}} = \sqrt{\dfrac{27}{3}} = \sqrt{9} = 3$

2. $\dfrac{\sqrt{30a^5}}{\sqrt{6a^2}} = \sqrt{\dfrac{30a^5}{6a^2}} = \sqrt{5a^3} = \sqrt{5 \cdot a^2 \cdot a}$
$$= \sqrt{a^2} \cdot \sqrt{5a} = a\sqrt{5a}$$

*Do Exercises 1–3.*

## b Roots of Quotients

To find the square root of certain quotients, we can reverse the quotient rule for radicals. We can take the square root of a quotient by taking the square roots of the numerator and the denominator separately.

> For any nonnegative number $A$ and any positive number $B$,
>
> $$\sqrt{\frac{A}{B}} = \frac{\sqrt{A}}{\sqrt{B}}.$$
>
> (We can take the square roots of the numerator and the denominator separately.)

---

### Objectives

**a** Divide radical expressions.

**b** Simplify square roots of quotients.

**c** Rationalize the denominator of a radical expression.

**For Extra Help**

TAPE 15    TAPE 16B    MAC WIN    CD-ROM

---

### Calculator Spotlight

**Exercises**

Use the GRAPH and TABLE features to determine whether each of the following is correct.

1. $\sqrt{\dfrac{x}{4}} = \dfrac{\sqrt{x}}{2}$

2. $\sqrt{\dfrac{x}{3}} = \dfrac{\sqrt{x}}{\sqrt{3}}$

3. $\sqrt{\dfrac{x}{6}} = 6\sqrt{x}$

---

Divide and simplify.

1. $\dfrac{\sqrt{96}}{\sqrt{6}}$

2. $\dfrac{\sqrt{75}}{\sqrt{3}}$

3. $\dfrac{\sqrt{42x^5}}{\sqrt{7x^2}}$

*Answers on page A-36*

Simplify.

**4.** $\sqrt{\dfrac{16}{9}}$

**5.** $\sqrt{\dfrac{1}{25}}$

**6.** $\sqrt{\dfrac{36}{x^2}}$

Simplify.

**7.** $\sqrt{\dfrac{18}{32}}$

**8.** $\sqrt{\dfrac{2250}{2560}}$

**9.** $\sqrt{\dfrac{98y}{2y^{11}}}$

*Answers on page A-36*

**Examples**   Simplify by taking the square roots of the numerator and the denominator separately.

**3.** $\sqrt{\dfrac{25}{9}} = \dfrac{\sqrt{25}}{\sqrt{9}} = \dfrac{5}{3}$     Taking the square roots of the numerator and the denominator

**4.** $\sqrt{\dfrac{1}{16}} = \dfrac{\sqrt{1}}{\sqrt{16}} = \dfrac{1}{4}$     Taking the square roots of the numerator and the denominator

**5.** $\sqrt{\dfrac{49}{t^2}} = \dfrac{\sqrt{49}}{\sqrt{t^2}} = \dfrac{7}{t}$

*Do Exercises 4–6.*

We are assuming that expressions for numerators are nonnegative and expressions for denominators are positive. Thus we need not be concerned about absolute-value signs or zero denominators.

Sometimes a rational expression can be simplified to one that has a perfect-square numerator and a perfect-square denominator.

**Examples**   Simplify.

**6.** $\sqrt{\dfrac{18}{50}} = \sqrt{\dfrac{9 \cdot 2}{25 \cdot 2}} = \sqrt{\dfrac{9}{25} \cdot \dfrac{2}{2}} = \sqrt{\dfrac{9}{25} \cdot 1}$

$= \sqrt{\dfrac{9}{25}} = \dfrac{\sqrt{9}}{\sqrt{25}} = \dfrac{3}{5}$

**7.** $\sqrt{\dfrac{2560}{2890}} = \sqrt{\dfrac{256 \cdot 10}{289 \cdot 10}} = \sqrt{\dfrac{256}{289} \cdot \dfrac{10}{10}} = \sqrt{\dfrac{256}{289} \cdot 1}$

$= \sqrt{\dfrac{256}{289}} = \dfrac{\sqrt{256}}{\sqrt{289}} = \dfrac{16}{17}$

**8.** $\dfrac{\sqrt{48x^3}}{\sqrt{3x^7}} = \sqrt{\dfrac{48x^3}{3x^7}} = \sqrt{\dfrac{16}{x^4}} = \dfrac{\sqrt{16}}{\sqrt{x^4}} = \dfrac{4}{x^2}$

*Do Exercises 7–9.*

## c  Rationalizing Denominators

Sometimes in mathematics it is useful to find an equivalent expression without a radical in the denominator. This provides a standard notation for expressing results. The procedure for finding such an expression is called **rationalizing the denominator.** We carry this out by multiplying by 1 in either of two ways.

> To rationalize a denominator:
>
> **Method 1.** Multiply by 1 under the radical to make the denominator a perfect square.
>
> **Method 2.** Multiply by 1 outside the radical to make the denominator a perfect square.

**Example 9**   Rationalize the denominator: $\sqrt{\dfrac{2}{3}}$.

**METHOD 1**   We multiply by 1, choosing $\frac{3}{3}$ for 1. This makes the denominator a perfect square:

$$\sqrt{\dfrac{2}{3}} = \sqrt{\dfrac{2}{3} \cdot \dfrac{3}{3}} \qquad \textit{Multiplying by 1}$$
$$= \sqrt{\dfrac{6}{9}} = \dfrac{\sqrt{6}}{\sqrt{9}}$$
$$= \dfrac{\sqrt{6}}{3}.$$

**METHOD 2**   We can also rationalize by first taking the square roots of the numerator and the denominator. Then we multiply by 1, using $\sqrt{3}/\sqrt{3}$:

$$\sqrt{\dfrac{2}{3}} = \dfrac{\sqrt{2}}{\sqrt{3}}$$
$$= \dfrac{\sqrt{2}}{\sqrt{3}} \cdot \dfrac{\sqrt{3}}{\sqrt{3}} \qquad \textit{Multiplying by 1}$$
$$= \dfrac{\sqrt{2} \cdot \sqrt{3}}{\sqrt{3} \cdot \sqrt{3}} = \dfrac{\sqrt{6}}{\sqrt{9}}$$
$$= \dfrac{\sqrt{6}}{3}.$$

*Do Exercise 10.*

We can always multiply by 1 to make a denominator a perfect square. Then we can take the square root of the denominator.

**Example 10**   Rationalize the denominator: $\sqrt{\dfrac{5}{18}}$.

The denominator 18 is not a perfect square. Factoring, we get $18 = 3 \cdot 3 \cdot 2$. If we had another factor of 2, however, we would have a perfect square, 36. Thus we multiply by 1, choosing $\frac{2}{2}$. This makes the denominator a perfect square.

$$\sqrt{\dfrac{5}{18}} = \sqrt{\dfrac{5}{18} \cdot \dfrac{2}{2}} = \sqrt{\dfrac{10}{36}} = \dfrac{\sqrt{10}}{\sqrt{36}} = \dfrac{\sqrt{10}}{6}$$

**Example 11**   Rationalize the denominator: $\dfrac{8}{\sqrt{7}}$.

This time we obtain an expression without a radical in the denominator by multiplying by 1, choosing $\sqrt{7}/\sqrt{7}$:

$$\dfrac{8}{\sqrt{7}} = \dfrac{8}{\sqrt{7}} \cdot \dfrac{\sqrt{7}}{\sqrt{7}} = \dfrac{8\sqrt{7}}{\sqrt{49}} = \dfrac{8\sqrt{7}}{7}.$$

*Do Exercises 11 and 12.*

**10.** Rationalize the denominator:

$$\sqrt{\dfrac{3}{5}}.$$

Rationalize the denominator.

**11.** $\sqrt{\dfrac{5}{8}}$

   (*Hint*: Multiply the radicand by $\frac{2}{2}$.)

**12.** $\dfrac{10}{\sqrt{3}}$

*Answers on page A-36*

Rationalize the denominator.

**13.** $\dfrac{\sqrt{3}}{\sqrt{7}}$

**14.** $\dfrac{\sqrt{5}}{\sqrt{r}}$

**15.** $\dfrac{\sqrt{64y^2}}{\sqrt{7}}$

*Answers on page A-36*

**Example 12**   Rationalize the denominator: $\dfrac{\sqrt{3}}{\sqrt{2}}$.

We look at the denominator. It is $\sqrt{2}$. We multiply by 1, choosing $\sqrt{2}/\sqrt{2}$:

$$\frac{\sqrt{3}}{\sqrt{2}} = \frac{\sqrt{3}}{\sqrt{2}} \cdot \frac{\sqrt{2}}{\sqrt{2}} = \frac{\sqrt{3} \cdot \sqrt{2}}{\sqrt{2} \cdot \sqrt{2}} = \frac{\sqrt{6}}{\sqrt{4}} = \frac{\sqrt{6}}{2}, \text{ or } \frac{1}{2}\sqrt{6}.$$

**Examples**   Rationalize the denominator.

**13.** $\dfrac{\sqrt{5}}{\sqrt{x}} = \dfrac{\sqrt{5}}{\sqrt{x}} \cdot \dfrac{\sqrt{x}}{\sqrt{x}}$    **Multiplying by 1**

$\qquad = \dfrac{\sqrt{5}\sqrt{x}}{\sqrt{x}\sqrt{x}}$

$\qquad = \dfrac{\sqrt{5x}}{x}$    $\sqrt{x} \cdot \sqrt{x} = x$ **by the definition of square root**

**14.** $\dfrac{\sqrt{49a^5}}{\sqrt{12}} = \dfrac{\sqrt{49a^5}}{\sqrt{12}} \cdot \dfrac{\sqrt{3}}{\sqrt{3}}$    **Multiplying by 1 using** $\sqrt{3}/\sqrt{3}$ **because** $\sqrt{3} \cdot \sqrt{12}$ **gives a perfect-square radicand in** $\sqrt{36}$

$\qquad = \dfrac{\sqrt{49a^5}\sqrt{3}}{\sqrt{12}\sqrt{3}}$

$\qquad = \dfrac{\sqrt{49a^4 \cdot 3a}}{\sqrt{36}}$

$\qquad = \dfrac{7a^2\sqrt{3a}}{6}$

*Do Exercises 13–15.*

# Exercise Set 9.3

**a** Divide and simplify.

1. $\dfrac{\sqrt{18}}{\sqrt{2}}$

2. $\dfrac{\sqrt{20}}{\sqrt{5}}$

3. $\dfrac{\sqrt{108}}{\sqrt{3}}$

4. $\dfrac{\sqrt{60}}{\sqrt{15}}$

5. $\dfrac{\sqrt{65}}{\sqrt{13}}$

6. $\dfrac{\sqrt{45}}{\sqrt{15}}$

7. $\dfrac{\sqrt{3}}{\sqrt{75}}$

8. $\dfrac{\sqrt{3}}{\sqrt{48}}$

9. $\dfrac{\sqrt{12}}{\sqrt{75}}$

10. $\dfrac{\sqrt{18}}{\sqrt{32}}$

11. $\dfrac{\sqrt{8x}}{\sqrt{2x}}$

12. $\dfrac{\sqrt{18b}}{\sqrt{2b}}$

13. $\dfrac{\sqrt{63y^3}}{\sqrt{7y}}$

14. $\dfrac{\sqrt{48x^3}}{\sqrt{3x}}$

**b** Simplify.

15. $\sqrt{\dfrac{16}{49}}$

16. $\sqrt{\dfrac{9}{49}}$

17. $\sqrt{\dfrac{1}{36}}$

18. $\sqrt{\dfrac{1}{4}}$

19. $-\sqrt{\dfrac{16}{81}}$

20. $-\sqrt{\dfrac{25}{49}}$

21. $\sqrt{\dfrac{64}{289}}$

22. $\sqrt{\dfrac{81}{361}}$

23. $\sqrt{\dfrac{1690}{1960}}$

24. $\sqrt{\dfrac{1210}{6250}}$

**25.** $\sqrt{\dfrac{25}{x^2}}$      **26.** $\sqrt{\dfrac{36}{a^2}}$      **27.** $\sqrt{\dfrac{9a^2}{625}}$      **28.** $\sqrt{\dfrac{x^2y^2}{256}}$

c Rationalize the denominator.

**29.** $\sqrt{\dfrac{2}{5}}$      **30.** $\sqrt{\dfrac{2}{7}}$      **31.** $\sqrt{\dfrac{7}{8}}$      **32.** $\sqrt{\dfrac{3}{8}}$      **33.** $\sqrt{\dfrac{1}{12}}$

**34.** $\sqrt{\dfrac{7}{12}}$      **35.** $\sqrt{\dfrac{5}{18}}$      **36.** $\sqrt{\dfrac{1}{18}}$      **37.** $\dfrac{3}{\sqrt{5}}$      **38.** $\dfrac{4}{\sqrt{3}}$

**39.** $\sqrt{\dfrac{8}{3}}$      **40.** $\sqrt{\dfrac{12}{5}}$      **41.** $\sqrt{\dfrac{3}{x}}$      **42.** $\sqrt{\dfrac{2}{x}}$      **43.** $\sqrt{\dfrac{x}{y}}$

**44.** $\sqrt{\dfrac{a}{b}}$      **45.** $\sqrt{\dfrac{x^2}{20}}$      **46.** $\sqrt{\dfrac{x^2}{18}}$      **47.** $\dfrac{\sqrt{7}}{\sqrt{2}}$      **48.** $\dfrac{\sqrt{3}}{\sqrt{5}}$

Copyright © 1999 Addison Wesley Longman

**49.** $\dfrac{\sqrt{9}}{\sqrt{8}}$  **50.** $\dfrac{\sqrt{4}}{\sqrt{27}}$  **51.** $\dfrac{\sqrt{3}}{\sqrt{2}}$  **52.** $\dfrac{\sqrt{2}}{\sqrt{5}}$  **53.** $\dfrac{2}{\sqrt{2}}$

**54.** $\dfrac{3}{\sqrt{3}}$  **55.** $\dfrac{\sqrt{5}}{\sqrt{11}}$  **56.** $\dfrac{\sqrt{7}}{\sqrt{27}}$  **57.** $\dfrac{\sqrt{7}}{\sqrt{12}}$  **58.** $\dfrac{\sqrt{5}}{\sqrt{18}}$

**59.** $\dfrac{\sqrt{48}}{\sqrt{32}}$  **60.** $\dfrac{\sqrt{56}}{\sqrt{40}}$  **61.** $\dfrac{\sqrt{450}}{\sqrt{18}}$  **62.** $\dfrac{\sqrt{224}}{\sqrt{14}}$  **63.** $\dfrac{\sqrt{3}}{\sqrt{x}}$

**64.** $\dfrac{\sqrt{2}}{\sqrt{y}}$  **65.** $\dfrac{4y}{\sqrt{5}}$  **66.** $\dfrac{8x}{\sqrt{3}}$  **67.** $\dfrac{\sqrt{a^3}}{\sqrt{8}}$  **68.** $\dfrac{\sqrt{x^3}}{\sqrt{27}}$

**69.** $\dfrac{\sqrt{56}}{\sqrt{12x}}$  **70.** $\dfrac{\sqrt{45}}{\sqrt{8a}}$  **71.** $\dfrac{\sqrt{27c}}{\sqrt{32c^3}}$  **72.** $\dfrac{\sqrt{7x^3}}{\sqrt{12x}}$  **73.** $\dfrac{\sqrt{y^5}}{\sqrt{xy^2}}$

**74.** $\dfrac{\sqrt{x^3}}{\sqrt{xy}}$  **75.** $\dfrac{\sqrt{45mn^2}}{\sqrt{32m}}$  **76.** $\dfrac{\sqrt{16a^4b^6}}{\sqrt{128a^6b^6}}$

## Skill Maintenance

Solve. [8.3a, b]

**77.** $x = y + 2,$
$\quad x + y = 6$

**78.** $4x - y = 10,$
$\quad 4x + y = 70$

**79.** $2x - 3y = 7,$
$\quad 2x - 3y = 9$

**80.** $\quad 2x - 3y = 7,$
$\quad -4x + 6y = -14$

**81.** $x + y = -7,$
$\quad x - y = 2$

**82.** $2x + 3y = 8,$
$\quad 5x - 4y = -2$

Multiply. [4.6b]

**83.** $(3x - 7)(3x + 7)$

**84.** $(4a - 5b)(4a + 5b)$

Collect like terms. [1.7e]

**85.** $9x - 5y + 12x - 4y$

**86.** $17a + 9b - 3a - 15b$

---

## Synthesis

**87.** ◆ Why is it important to know how to multiply radical expressions before learning how to divide them?

**88.** ◆ Describe a method that could be used to rationalize the *numerator* of a radical expression.

*Periods of Pendulums.* The period $T$ of a pendulum is the time it takes the pendulum to move from one side to the other and back. A formula for the period is

$$T = 2\pi\sqrt{\frac{L}{32}},$$

where $T$ is in seconds and $L$ is in feet. Use 3.14 for $\pi$.

**89.** Find the periods of pendulums of lengths 2 ft, 8 ft, 64 ft, and 100 ft.

**90.** Find the period of a pendulum of length $\frac{2}{3}$ in.

**91.** The pendulum of a grandfather clock is $(32/\pi^2)$ ft long. How long does it take to swing from one side to the other?

**92.** The pendulum of a grandfather clock is $(45/\pi^2)$ ft long. How long does it take to swing from one side to the other?

Rationalize the denominator.

**93.** $\sqrt{\dfrac{5}{1600}}$

**94.** $\sqrt{\dfrac{3}{1000}}$

**95.** $\sqrt{\dfrac{1}{5x^3}}$

**96.** $\sqrt{\dfrac{3x^2y}{a^2x^5}}$

**97.** $\sqrt{\dfrac{3a}{b}}$

**98.** $\sqrt{\dfrac{1}{5zw^2}}$

**99.** $\sqrt{0.009}$

**100.** $\sqrt{0.012}$

Simplify.

**101.** $\sqrt{\dfrac{1}{x^2} - \dfrac{2}{xy} + \dfrac{1}{y^2}}$

**102.** $\sqrt{2 - \dfrac{4}{z^2} + \dfrac{2}{z^4}}$

Copyright © 1999 Addison Wesley Longman

# 9.4 Addition, Subtraction, and More Multiplication

## a | Addition and Subtraction

We can add any two real numbers. The sum of 5 and $\sqrt{2}$ can be expressed as

$$5 + \sqrt{2}.$$

We cannot simplify this unless we use rational approximations. However, when we have *like radicals*, a sum can be simplified using the distributive laws and collecting like terms. **Like radicals** have the same radicands.

**Example 1** Add: $3\sqrt{5} + 4\sqrt{5}$.

Suppose we were considering $3x + 4x$. Recall that to add, we use a distributive law as follows:

$$3x + 4x = (3 + 4)x = 7x.$$

The situation is similar in this example, but we let $x = \sqrt{5}$:

$$3\sqrt{5} + 4\sqrt{5} = (3 + 4)\sqrt{5} \qquad \text{Using a distributive law to factor out } \sqrt{5}$$
$$= 7\sqrt{5}.$$

If we wish to add or subtract as we did in Example 1, the radicands must be the same. Sometimes after simplifying the radical terms, we discover that we have like radicals.

**Examples** Add or subtract. Simplify, if possible, by collecting like radical terms.

**2.** $5\sqrt{2} - \sqrt{18} = 5\sqrt{2} - \sqrt{9 \cdot 2}$    Factoring 18
$$= 5\sqrt{2} - \sqrt{9}\sqrt{2}$$
$$= 5\sqrt{2} - 3\sqrt{2}$$
$$= (5 - 3)\sqrt{2} \qquad \text{Using a distributive law to factor out the common factor, } \sqrt{2}$$
$$= 2\sqrt{2}$$

**3.** $\sqrt{4x^3} + 7\sqrt{x} = \sqrt{4 \cdot x^2 \cdot x} + 7\sqrt{x}$
$$= 2x\sqrt{x} + 7\sqrt{x}$$
$$= (2x + 7)\sqrt{x} \qquad \text{Using a distributive law to factor out } \sqrt{x}$$

Don't forget the parentheses!

**4.** $\sqrt{x^3 - x^2} + \sqrt{4x - 4} = \sqrt{x^2(x - 1)} + \sqrt{4(x - 1)}$    Factoring radicands
$$= \sqrt{x^2}\sqrt{x - 1} + \sqrt{4}\sqrt{x - 1}$$
$$= x\sqrt{x - 1} + 2\sqrt{x - 1}$$
$$= (x + 2)\sqrt{x - 1} \qquad \begin{array}{l}\text{Using a distributive law} \\ \text{to factor out the common factor,} \\ \sqrt{x - 1}\end{array}$$

Don't forget the parentheses!

*Do Exercises 1–5.*

Answers on page A-36

## Objectives

**a** Add or subtract with radical notation, using the distributive law to simplify.

**b** Multiply expressions involving radicals, where some of the expressions contain more than one term.

**c** Rationalize denominators having two terms.

### For Extra Help

TAPE 15    TAPE 16B    MAC WIN    CD-ROM

Add or subtract and simplify by collecting like radical terms, if possible.

**1.** $3\sqrt{2} + 9\sqrt{2}$

**2.** $8\sqrt{5} - 3\sqrt{5}$

**3.** $2\sqrt{10} - 7\sqrt{40}$

**4.** $\sqrt{24} + \sqrt{54}$

**5.** $\sqrt{9x + 9} - \sqrt{4x + 4}$

Add or subtract.

**6.** $\sqrt{2} + \sqrt{\dfrac{1}{2}}$

**7.** $\sqrt{\dfrac{5}{3}} + \sqrt{\dfrac{3}{5}}$

Sometimes rationalizing denominators enables us to combine like radicals.

**Example 5**   Add: $\sqrt{3} + \sqrt{\dfrac{1}{3}}$.

$$\sqrt{3} + \sqrt{\dfrac{1}{3}} = \sqrt{3} + \sqrt{\dfrac{1}{3} \cdot \dfrac{3}{3}} \qquad \text{Multiplying by 1 in order to rationalize the denominator}$$

$$= \sqrt{3} + \sqrt{\dfrac{3}{9}}$$

$$= \sqrt{3} + \dfrac{\sqrt{3}}{\sqrt{9}}$$

$$= \sqrt{3} + \dfrac{\sqrt{3}}{3}$$

$$= 1 \cdot \sqrt{3} + \dfrac{1}{3}\sqrt{3}$$

$$= \left(1 + \dfrac{1}{3}\right)\sqrt{3} \qquad \text{Factoring out the common factor, } \sqrt{3}$$

$$= \dfrac{4}{3}\sqrt{3}$$

*Do Exercises 6 and 7.*

### b | Multiplication

Now let's multiply where some of the expressions may contain more than one term. To do this, we use procedures already studied in this chapter as well as the distributive laws and special products for multiplying with polynomials.

**Example 6**   Multiply: $\sqrt{2}\,(\sqrt{3} + \sqrt{7})$.

$$\sqrt{2}\,(\sqrt{3} + \sqrt{7}) = \sqrt{2}\,\sqrt{3} + \sqrt{2}\,\sqrt{7} \qquad \text{Multiplying using a distributive law}$$

$$= \sqrt{6} + \sqrt{14} \qquad \text{Using the rule for multiplying with radicals}$$

**Example 7**   Multiply: $(2 + \sqrt{3})(5 - 4\sqrt{3})$.

$$(2 + \sqrt{3})(5 - 4\sqrt{3}) = 2 \cdot 5 - 2 \cdot 4\sqrt{3} + \sqrt{3} \cdot 5 - \sqrt{3} \cdot 4\sqrt{3} \qquad \text{Using FOIL}$$

$$= 10 - 8\sqrt{3} + 5\sqrt{3} - 4 \cdot 3$$

$$= 10 - 12 - 3\sqrt{3}$$

$$= -2 - 3\sqrt{3}$$

*Answers on page A-36*

**Example 8**  Multiply: $(\sqrt{3} - \sqrt{x})(\sqrt{3} + \sqrt{x})$.

$$(\sqrt{3} - \sqrt{x})(\sqrt{3} + \sqrt{x}) = (\sqrt{3})^2 - (\sqrt{x})^2 \qquad \text{Using } (A - B)(A + B) = A^2 - B^2$$
$$= 3 - x$$

Multiply.

**8.** $\sqrt{3}\,(\sqrt{5} + \sqrt{2})$

**Example 9**  Multiply: $(3 - \sqrt{p})^2$.

$$(3 - \sqrt{p})^2 = 3^2 - 2 \cdot 3 \cdot \sqrt{p} + (\sqrt{p})^2 \qquad \text{Using } (A - B)^2 = A^2 - 2AB + B^2$$
$$= 9 - 6\sqrt{p} + p$$

**9.** $(1 - \sqrt{2})(4 + 3\sqrt{5})$

**Example 10**  Multiply: $(2 - \sqrt{5})(2 + \sqrt{5})$.

$$(2 - \sqrt{5})(2 + \sqrt{5}) = 2^2 - (\sqrt{5})^2 \qquad \text{Using } (A - B)(A + B) = A^2 - B^2$$
$$= 4 - 5$$
$$= -1$$

**10.** $(\sqrt{2} + \sqrt{a})(\sqrt{2} - \sqrt{a})$

*Do Exercises 8–12.*

## c  More on Rationalizing Denominators

**11.** $(5 + \sqrt{x})^2$

Note in Examples 8 and 10 that the results have no radicals. This will happen whenever we multiply expressions such as $\sqrt{a} - \sqrt{b}$ and $\sqrt{a} + \sqrt{b}$. We see this in the following:

$$(\sqrt{a} + \sqrt{b})(\sqrt{a} - \sqrt{b}) = (\sqrt{a})^2 - (\sqrt{b})^2 = a - b.$$

Expressions such as $\sqrt{3} - \sqrt{x}$ and $\sqrt{3} + \sqrt{x}$ are known as **conjugates**; so too are $2 + \sqrt{5}$ and $2 - \sqrt{5}$. We can use conjugates to rationalize a denominator that involves a sum or difference of two terms, where one or both are radicals. To do so, we multiply by 1 using the conjugate in the numerator and the denominator of the expression for 1.

**12.** $(3 - \sqrt{7})(3 + \sqrt{7})$

*Do Exercises 13–15.*

Find the conjugate of the expression.

**13.** $7 + \sqrt{5}$

**Example 11**  Rationalize the denominator: $\dfrac{3}{2 + \sqrt{5}}$.

We multiply by 1 using the conjugate of $2 + \sqrt{5}$, which is $2 - \sqrt{5}$, as the numerator and the denominator:

$$\frac{3}{2 + \sqrt{5}} = \frac{3}{2 + \sqrt{5}} \cdot \frac{2 - \sqrt{5}}{2 - \sqrt{5}} \qquad \text{Multiplying by 1}$$

$$= \frac{3(2 - \sqrt{5})}{(2 + \sqrt{5})(2 - \sqrt{5})} \qquad \text{Multiplying}$$

**14.** $\sqrt{5} - \sqrt{2}$

$$= \frac{6 - 3\sqrt{5}}{2^2 - (\sqrt{5})^2}$$

$$= \frac{6 - 3\sqrt{5}}{4 - 5}$$

**15.** $1 - \sqrt{x}$

$$= \frac{6 - 3\sqrt{5}}{-1}$$

$$= -6 + 3\sqrt{5}, \text{ or } 3\sqrt{5} - 6.$$

*Answers on page A-36*

Rationalize the denominator.

**16.** $\dfrac{6}{7 + \sqrt{5}}$

**17.** $\dfrac{\sqrt{5} + \sqrt{2}}{\sqrt{5} - \sqrt{2}}$

**18.** Rationalize the denominator:

$$\dfrac{7}{1 - \sqrt{x}}.$$

**Example 12**  Rationalize the denominator: $\dfrac{\sqrt{3} + \sqrt{5}}{\sqrt{3} - \sqrt{5}}$.

We multiply by 1 using the conjugate of $\sqrt{3} - \sqrt{5}$, which is $\sqrt{3} + \sqrt{5}$, as the numerator and the denominator:

$$\dfrac{\sqrt{3} + \sqrt{5}}{\sqrt{3} - \sqrt{5}} = \dfrac{\sqrt{3} + \sqrt{5}}{\sqrt{3} - \sqrt{5}} \cdot \dfrac{\sqrt{3} + \sqrt{5}}{\sqrt{3} + \sqrt{5}} \qquad \text{Multiplying by 1}$$

$$= \dfrac{(\sqrt{3} + \sqrt{5})^2}{(\sqrt{3} - \sqrt{5})(\sqrt{3} + \sqrt{5})}$$

$$= \dfrac{(\sqrt{3})^2 + 2\sqrt{3}\sqrt{5} + (\sqrt{5})^2}{(\sqrt{3})^2 - (\sqrt{5})^2}$$

$$= \dfrac{3 + 2\sqrt{15} + 5}{3 - 5}$$

$$= \dfrac{8 + 2\sqrt{15}}{-2}$$

$$= \dfrac{2(4 + \sqrt{15})}{2(-1)} \qquad \text{Factoring in order to simplify}$$

$$= \dfrac{2}{2} \cdot \dfrac{4 + \sqrt{15}}{-1}$$

$$= \dfrac{4 + \sqrt{15}}{-1}$$

$$= -4 - \sqrt{15}.$$

*Do Exercises 16 and 17.*

**Example 13**  Rationalize the denominator: $\dfrac{5}{2 + \sqrt{x}}$.

We multiply by 1 using the conjugate of $2 + \sqrt{x}$, which is $2 - \sqrt{x}$, as the numerator and the denominator:

$$\dfrac{5}{2 + \sqrt{x}} = \dfrac{5}{2 + \sqrt{x}} \cdot \dfrac{2 - \sqrt{x}}{2 - \sqrt{x}} \qquad \text{Multiplying by 1}$$

$$= \dfrac{5(2 - \sqrt{x})}{(2 + \sqrt{x})(2 - \sqrt{x})}$$

$$= \dfrac{5 \cdot 2 - 5 \cdot \sqrt{x}}{2^2 - (\sqrt{x})^2}$$

$$= \dfrac{10 - 5\sqrt{x}}{4 - x}.$$

*Do Exercise 18.*

*Answers on page A-36*

# Exercise Set 9.4

**a** Add or subtract. Simplify by collecting like radical terms, if possible.

**1.** $7\sqrt{3} + 9\sqrt{3}$

**2.** $6\sqrt{2} + 8\sqrt{2}$

**3.** $7\sqrt{5} - 3\sqrt{5}$

**4.** $8\sqrt{2} - 5\sqrt{2}$

**5.** $6\sqrt{x} + 7\sqrt{x}$

**6.** $9\sqrt{y} + 3\sqrt{y}$

**7.** $4\sqrt{d} - 13\sqrt{d}$

**8.** $2\sqrt{a} - 17\sqrt{a}$

**9.** $5\sqrt{8} + 15\sqrt{2}$

**10.** $3\sqrt{12} + 2\sqrt{3}$

**11.** $\sqrt{27} - 2\sqrt{3}$

**12.** $7\sqrt{50} - 3\sqrt{2}$

**13.** $\sqrt{45} - \sqrt{20}$

**14.** $\sqrt{27} - \sqrt{12}$

**15.** $\sqrt{72} + \sqrt{98}$

**16.** $\sqrt{45} + \sqrt{80}$

**17.** $2\sqrt{12} + \sqrt{27} - \sqrt{48}$

**18.** $9\sqrt{8} - \sqrt{72} + \sqrt{98}$

**19.** $\sqrt{18} - 3\sqrt{8} + \sqrt{50}$

**20.** $3\sqrt{18} - 2\sqrt{32} - 5\sqrt{50}$

**21.** $2\sqrt{27} - 3\sqrt{48} + 3\sqrt{12}$

**22.** $3\sqrt{48} - 2\sqrt{27} - 3\sqrt{12}$

**23.** $\sqrt{4x} + \sqrt{81x^3}$

**24.** $\sqrt{12x^2} + \sqrt{27}$

**25.** $\sqrt{27} - \sqrt{12x^2}$

**26.** $\sqrt{81x^3} - \sqrt{4x}$

**27.** $\sqrt{8x + 8} + \sqrt{2x + 2}$

**28.** $\sqrt{12x + 12} + \sqrt{3x + 3}$

**29.** $\sqrt{x^5 - x^2} + \sqrt{9x^3 - 9}$

**30.** $\sqrt{16x - 16} + \sqrt{25x^3 - 25x^2}$

**31.** $4a\sqrt{a^2b} + a\sqrt{a^2b^3} - 5\sqrt{b^3}$

**32.** $3x\sqrt{y^3x} - x\sqrt{yx^3} + y\sqrt{y^3x}$

**33.** $\sqrt{3} - \sqrt{\dfrac{1}{3}}$

**34.** $\sqrt{2} - \sqrt{\dfrac{1}{2}}$

**35.** $5\sqrt{2} + 3\sqrt{\dfrac{1}{2}}$

**36.** $4\sqrt{3} + 2\sqrt{\dfrac{1}{3}}$

**37.** $\sqrt{\dfrac{2}{3}} - \sqrt{\dfrac{1}{6}}$

**38.** $\sqrt{\dfrac{1}{2}} - \sqrt{\dfrac{1}{8}}$

**b** Multiply.

**39.** $\sqrt{3}(\sqrt{5} - 1)$

**40.** $\sqrt{2}(\sqrt{2} + \sqrt{3})$

**41.** $(2 + \sqrt{3})(5 - \sqrt{7})$

**42.** $(\sqrt{5} + \sqrt{7})(2\sqrt{5} - 3\sqrt{7})$

**43.** $(2 - \sqrt{5})^2$

**44.** $(\sqrt{3} + \sqrt{10})^2$

**45.** $(\sqrt{2} + 8)(\sqrt{2} - 8)$

**46.** $(1 + \sqrt{7})(1 - \sqrt{7})$

**47.** $(\sqrt{6} - \sqrt{5})(\sqrt{6} + \sqrt{5})$

**48.** $(\sqrt{3} + \sqrt{10})(\sqrt{3} - \sqrt{10})$ **49.** $(3\sqrt{5} - 2)(\sqrt{5} + 1)$ **50.** $(\sqrt{5} - 2\sqrt{2})(\sqrt{10} - 1)$

**51.** $(\sqrt{x} - \sqrt{y})^2$ **52.** $(\sqrt{w} + 11)^2$

c  Rationalize the denominator.

**53.** $\dfrac{2}{\sqrt{3} - \sqrt{5}}$ **54.** $\dfrac{5}{3 + \sqrt{7}}$ **55.** $\dfrac{\sqrt{3} - \sqrt{2}}{\sqrt{3} + \sqrt{2}}$ **56.** $\dfrac{2 - \sqrt{7}}{\sqrt{3} - \sqrt{2}}$

**57.** $\dfrac{4}{\sqrt{10} + 1}$ **58.** $\dfrac{6}{\sqrt{11} - 3}$ **59.** $\dfrac{1 - \sqrt{7}}{3 + \sqrt{7}}$ **60.** $\dfrac{2 + \sqrt{8}}{1 - \sqrt{5}}$

**61.** $\dfrac{3}{4 + \sqrt{x}}$ **62.** $\dfrac{8}{2 - \sqrt{x}}$ **63.** $\dfrac{3 + \sqrt{2}}{8 - \sqrt{x}}$ **64.** $\dfrac{4 - \sqrt{3}}{6 + \sqrt{y}}$

**Skill Maintenance**

Solve.

**65.** $3x + 5 + 2(x - 3) = 4 - 6x$  [2.3c]

**66.** $3(x - 4) - 2 = 8(2x + 3)$  [2.3c]

**67.** $x^2 - 5x = 6$  [5.7b]

**68.** $x^2 + 10 = 7x$  [5.7b]

Solve.

**69.** Jolly Juice is 3% real fruit juice, and Real Squeeze is 6% real fruit juice. How many liters of each should be combined in order to make an 8-L mixture that is 5.4% real fruit juice?   [8.4a]

**70.** The time $t$ that it takes a bus to travel a fixed distance varies inversely as its speed $r$. At a speed of 40 mph, it takes $\frac{1}{2}$ hr to travel a fixed distance. How long will it take to travel the same distance at 60 mph?   [7.5d]

**71.** The graph of the polynomial equation $y = x^3 - 5x^2 + x - 2$ is shown at right. Use either the graph or the equation to estimate the value of the polynomial when $x = -1$, $x = 0$, $x = 1$, $x = 3$, and $x = 4.85$.   [4.3a]

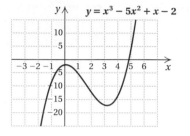

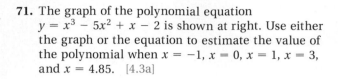

---

### Synthesis

**72.** ◆ Describe a method that could be used to rationalize a numerator that contains the sum of two radical expressions.

**73.** ◆ Explain why it is important for the signs within a pair of conjugates to differ.

**74.** Evaluate $\sqrt{a^2 + b^2}$ and $\sqrt{a^2} + \sqrt{b^2}$ for $a = 2$ and $b = 3$.

**75.** On the basis of Exercise 74, determine whether $\sqrt{a^2 + b^2}$ and $\sqrt{a^2} + \sqrt{b^2}$ are equivalent.

Use the GRAPH and TABLE features to determine whether each of the following is correct.

**76.** $\sqrt{9x^3} + \sqrt{x} = \sqrt{9x^3 + x}$

**77.** $\sqrt{x^2 + 4} = \sqrt{x} + 2$

Add or subtract as indicated.

**78.** $\frac{3}{5}\sqrt{24} + \frac{2}{5}\sqrt{150} - \sqrt{96}$

**79.** $\frac{1}{3}\sqrt{27} + \sqrt{8} + \sqrt{300} - \sqrt{18} - \sqrt{162}$

**80.** Three students were asked to simplify $\sqrt{10} + \sqrt{50}$. Their answers were $\sqrt{10}(1 + \sqrt{5})$, $\sqrt{10} + 5\sqrt{2}$, and $\sqrt{2}(5 + \sqrt{5})$. Which, if any, are correct?

Determine whether each of the following is true. Show why or why not.

**81.** $(3\sqrt{x + 2})^2 = 9(x + 2)$

**82.** $(\sqrt{x + 2})^2 = x + 2$

Copyright © 1999 Addison Wesley Longman

# 9.5 Radical Equations

 **a** **Solving Radical Equations**

The following are examples of *radical equations*:

$$\sqrt{2x} - 4 = 7, \qquad \sqrt{x + 1} = \sqrt{2x - 5}.$$

A **radical equation** has variables in one or more radicands. To solve square-root radical equations, we first convert them to equations without radicals. We do this for square-root radical equations by squaring both sides of the equation, using the following principle.

> **THE PRINCIPLE OF SQUARING**
>
> If an equation $a = b$ is true, then the equation $a^2 = b^2$ is true.

To solve radical equations, we first try to get a radical by itself. That is, we try to isolate the radical. Then we use the principle of squaring. This allows us to eliminate one radical.

**Example 1** Solve: $\sqrt{2x} - 4 = 7$.

$$\sqrt{2x} - 4 = 7$$

$$\sqrt{2x} = 11 \qquad \text{Adding 4 to isolate the radical}$$

$$(\sqrt{2x})^2 = 11^2 \qquad \text{Squaring both sides}$$

$$2x = 121 \qquad \sqrt{2x} \cdot \sqrt{2x} = 2x, \text{ by the definition of square root}$$

$$x = \frac{121}{2} \qquad \text{Dividing by 2}$$

CHECK:

$$\begin{array}{c|c} \sqrt{2x} - 4 = 7 \\ \hline \sqrt{2 \cdot \dfrac{121}{2}} - 4 \ ?\ 7 \\ \sqrt{121} - 4 \\ 11 - 4 \\ 7 & \text{TRUE} \end{array}$$

The solution is $\frac{121}{2}$.

*Do Exercise 1.*

**Example 2** Solve: $2\sqrt{x + 2} = \sqrt{x + 10}$.

Each radical is already isolated. We proceed with the principle of squaring.

$$(2\sqrt{x + 2})^2 = (\sqrt{x + 10})^2 \qquad \text{Squaring both sides}$$

$$2^2(\sqrt{x + 2})^2 = (\sqrt{x + 10})^2 \qquad \text{Raising the product to the second power on the left}$$

$$4(x + 2) = x + 10 \qquad \text{Simplifying}$$

$$4x + 8 = x + 10 \qquad \text{Removing parentheses}$$

$$3x = 2 \qquad \text{Subtracting } x \text{ and } 8$$

$$x = \frac{2}{3} \qquad \text{Dividing by 3}$$

**Objectives**

**a** Solve radical equations with one or two radical terms isolated, using the principle of squaring once.

**b** Solve radical equations with two radical terms, using the principle of squaring twice.

**c** Solve applied problems using radical equations.

**For Extra Help**

TAPE 15    TAPE 17A    MAC WIN    CD-ROM

**1.** Solve: $\sqrt{3x} - 5 = 3$.

*Answer on page A-37*

Solve.

**2.** $\sqrt{3x + 1} = \sqrt{2x + 3}$

**3.** $3\sqrt{x + 1} = \sqrt{x + 12}$

**4.** Solve: $x - 1 = \sqrt{x + 5}$.

CHECK:
$$2\sqrt{x + 2} = \sqrt{x + 10}$$

$$2\sqrt{\frac{2}{3} + 2} \ ? \ \sqrt{\frac{2}{3} + 10}$$

$$2\sqrt{\frac{8}{3}} \quad \sqrt{\frac{32}{3}}$$

$$4\sqrt{\frac{2}{3}} \quad 4\sqrt{\frac{2}{3}} \qquad \text{TRUE}$$

The number $\frac{2}{3}$ checks. The solution is $\frac{2}{3}$.

*Do Exercises 2 and 3.*

It is important to check when using the principle of squaring. This principle may not produce equivalent equations. When we square both sides of an equation, the new equation may have solutions that the first one does not. For example, the equation

$$x = 1 \qquad (1)$$

has just one solution, the number 1. When we square both sides, we get

$$x^2 = 1, \qquad (2)$$

which has two solutions, 1 and $-1$. The equations $x = 1$ and $x^2 = 1$ do not have the same solutions and thus are not equivalent. Whereas it is true that any solution of equation (1) is a solution of equation (2), it is *not* true that any solution of equation (2) is a solution of equation (1).

> When the principle of squaring is used to solve an equation, solutions of an equation found by squaring *must* be checked in the original equation!

Sometimes we may need to apply the principle of zero products after squaring. (See Section 5.7.)

**Example 3**   Solve: $x - 5 = \sqrt{x + 7}$.

$$x - 5 = \sqrt{x + 7}$$
$$(x - 5)^2 = (\sqrt{x + 7})^2 \qquad \text{Using the principle of squaring}$$
$$x^2 - 10x + 25 = x + 7$$
$$x^2 - 11x + 18 = 0$$
$$(x - 9)(x - 2) = 0 \qquad \text{Factoring}$$
$$x - 9 = 0 \quad or \quad x - 2 = 0 \qquad \text{Using the principle of zero products}$$
$$x = 9 \quad or \qquad x = 2$$

CHECK:    For 9:

$$\frac{x - 5 = \sqrt{x + 7}}{9 - 5 \ ? \ \sqrt{9 + 7}}$$
$$4 \ | \ 4 \qquad \text{TRUE}$$

For 2:

$$\frac{x - 5 = \sqrt{x + 7}}{2 - 5 \ ? \ \sqrt{2 + 7}}$$
$$-3 \ | \ 3 \qquad \text{FALSE}$$

The number 9 checks, but 2 does not. Thus the solution is 9.

*Do Exercise 4.*

We can visualize or check the solutions of a radical equation graphically. Consider the equation of Example 3:

$$x - 5 = \sqrt{x + 7}.$$

We can examine the solutions by graphing the equations

$$y = x - 5 \quad \text{and} \quad y = \sqrt{x + 7}$$

using the same set of axes. A hand-drawn graph of $y = \sqrt{x + 7}$ would involve approximating square roots on a calculator.

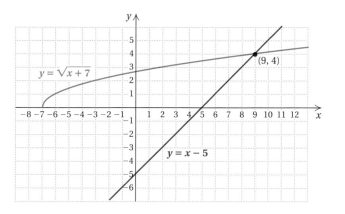

It appears that when $x = 9$, the values of $y = x - 5$ and $y = \sqrt{x + 7}$ are the same, 4. We can check this as we did in Example 3. Note also that the graphs *do not* intersect at $x = 2$.

---

## Calculator Spotlight

Solving Radical Equations. Consider the equation of Example 1:

$$\sqrt{2x} - 4 = 7.$$

From each side of the equation, we graph two new equations in the "$y=$" form.

$$y_1 = \sqrt{2x} - 4, \quad (1)$$
$$y_2 = 7. \quad (2)$$

$y_1 = \sqrt{2x} - 4, \ y_2 = 7$

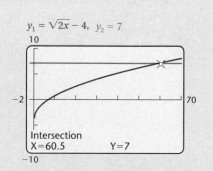

To find the solution, we use the CALC-INTERSECT feature.

$y_1 = \sqrt{2x} - 4, \ y_2 = 7$

The solution of the equation $\sqrt{2x} - 7 = 4$ is then shown and listed as the first coordinate at the bottom of the screen. Approximate solutions may sometimes result.

### Exercises

Use a grapher to solve the equation in each of the following.

1. Margin Exercise 1
2. Margin Exercise 2
3. Margin Exercise 3
4. Margin Exercise 4. What happens to the equations at $x = -1$? Do the graphs intersect?

**5.** Solve: $1 + \sqrt{1 - x} = x$.

**Example 4**  Solve: $3 + \sqrt{27 - 3x} = x$.

In this case, we must first isolate the radical.

$$3 + \sqrt{27 - 3x} = x$$
$$\sqrt{27 - 3x} = x - 3 \qquad \text{Subtracting 3 to isolate the radical}$$
$$(\sqrt{27 - 3x})^2 = (x - 3)^2 \qquad \text{Using the principle of squaring}$$
$$27 - 3x = x^2 - 6x + 9$$
$$0 = x^2 - 3x - 18 \qquad \text{We can have 0 on the left.}$$
$$0 = (x - 6)(x + 3) \qquad \text{Factoring}$$
$$x - 6 = 0 \quad or \quad x + 3 = 0 \qquad \text{Using the principle of zero products}$$
$$x = 6 \quad or \qquad x = -3$$

**Check:**  For 6:

$$\begin{array}{c|c} 3 + \sqrt{27 - 3x} = x & \\ \hline 3 + \sqrt{27 - 3 \cdot 6} \,?\, 6 & \\ 3 + \sqrt{9} & \\ 3 + 3 & \\ 6 & \text{TRUE} \end{array}$$

For $-3$:

$$\begin{array}{c|c} 3 + \sqrt{27 - 3x} = x & \\ \hline 3 + \sqrt{27 - 3 \cdot (-3)} \,?\, -3 & \\ 3 + \sqrt{27 + 9} & \\ 3 + \sqrt{36} & \\ 3 + 6 & \\ 9 & \text{FALSE} \end{array}$$

The number 6 checks, but $-3$ does not. The solution is 6.

*Do Exercise 5.*

Suppose that in Example 4 we do not isolate the radical before squaring. Then we get an expression on the left side of the equation in which we have *not* eliminated the radical:

$$(3 + \sqrt{27 - 3x})^2 = (x)^2$$
$$3^2 + 2 \cdot 3 \cdot \sqrt{27 - 3x} + (\sqrt{27 - 3x})^2 = x^2$$
$$9 + 6\sqrt{27 - 3x} + (27 - 3x) = x^2.$$

In fact, we have ended up with a more complicated expression than the one we squared.

## **b**  Using the Principle of Squaring More Than Once

Sometimes when we have two radical terms, we may need to apply the principle of squaring a second time.

**Example 5**  Solve: $\sqrt{x} - 1 = \sqrt{x - 5}$.

$$\sqrt{x} - 1 = \sqrt{x - 5}$$
$$(\sqrt{x} - 1)^2 = (\sqrt{x - 5})^2 \qquad \text{Using the principle of squaring}$$
$$(\sqrt{x})^2 - 2 \cdot \sqrt{x} \cdot 1 + 1^2 = x - 5 \qquad \text{Using } (A - B)^2 = A^2 - 2AB + B^2 \text{ on the left side}$$
$$x - 2\sqrt{x} + 1 = x - 5 \qquad \text{Simplifying}$$
$$-2\sqrt{x} = -6 \qquad \text{Isolating the radical}$$
$$\sqrt{x} = 3$$
$$(\sqrt{x})^2 = 3^2 \qquad \text{Using the principle of squaring}$$
$$x = 9$$

The check is left to the student. The number 9 checks and is the solution.

*Answer on page A-37*

Chapter 9  Radical Expressions
and Equations

**576**

The following is a procedure for solving radical equations.

> To solve radical equations:
> 1. Isolate one of the radical terms.
> 2. Use the principle of squaring.
> 3. If a radical term remains, perform steps (1) and (2) again.
> 4. Solve the equation and check possible solutions.

*Do Exercise 6.*

## c | Applications

*Sightings to the Horizon.* How far can you see from a given height? There is a formula for this. At a height of $h$ meters, you can see $V$ kilometers to the horizon. These numbers are related as follows:

$$V = 3.5\sqrt{h}. \qquad \textbf{(1)}$$

**Example 6** How far to the horizon can you see through an airplane window at a height, or altitude, of 9000 m?

We substitute 9000 for $h$ in equation (1) and find an approximation using a calculator:

$$V = 3.5\sqrt{9000} \approx 332.039 \text{ km.}$$

You can see for a distance of about 332 km at a height of 9000 m.

*Do Exercises 7 and 8.*

**Example 7** Elaine can see 50.4 km to the horizon from the top of a cliff. What is the altitude of Elaine's eyes?

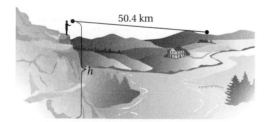

We substitute 50.4 for $V$ in equation (1) and solve:

$$50.4 = 3.5\sqrt{h}$$
$$\frac{50.4}{3.5} = \sqrt{h}$$
$$14.4 = \sqrt{h}$$
$$(14.4)^2 = (\sqrt{h})^2$$
$$207.36 = h.$$

The altitude of Elaine's eyes is about 207 m.

*Do Exercise 9.*

**6.** Solve: $\sqrt{x} - 1 = \sqrt{x - 3}$.

**7.** How far can you see to the horizon through an airplane window at a height of 8000 m?

**8.** How far can a sailor see to the horizon from the top of a 20-m mast?

**9.** A technician can see 49 km to the horizon from the top of a radio tower. How high is the tower?

*Answers on page A-37*

# Improving Your Math Study Skills

## Forming Math Study Groups,
## by James R. Norton

*Dr. James Norton has taught at the University of Phoenix and Scottsdale Community College. He has extensive experience with the use of study groups to learn mathematics.*

The use of math study groups for learning has become increasingly more common in recent years. Some instructors regard them as a primary source of learning, while others let students form groups on their own.

A study group generally consists of study partners who help each other learn the material and do the homework. You will probably meet outside of class at least once or twice a week. Here are some do's and don'ts to make your study group more valuable.

- DO make the group up of no more than four or five people. Research has shown clearly that this size works best.

- DO trade phone numbers so that you can get in touch with each other for help between team meetings.

- DO make sure that everyone in the group has a chance to contribute.

- DON'T let a group member copy from others without contributing. If this should happen, one member should speak with that student privately; if the situation continues, that student should be asked to leave the group.

- DON'T let the "A" students drop the ball. The group needs them! The benefits to even the best

students are twofold: (1) Other students will benefit from their expertise and (2) the bright students will learn the material better by teaching it to someone else.

- DON'T let the slower students drop the ball either. *Everyone* can contribute something, and being in a group will actually improve their self-esteem as well as their performance.

How do you form study groups if the instructor has not already done so? A good place to begin is to get together with three or four friends and arrange a study time. If you don't know anyone, start getting acquainted with other people in the class during the first week of the semester.

What should you look for in a study partner?

- Do you live near each other to make it easy to get together?

- What are your class schedules like? Are you both on campus? Do you have free time?

- What about work schedules, athletic practice, and other out-of-school commitments that you might have to work around?

Making use of a study group is not a form of "cheating." You are merely helping each other learn. So long as everyone in the group is both contributing and doing the work, this method will bring you great success!

# Exercise Set 9.5

**a** Solve.

**1.** $\sqrt{x} = 6$

**2.** $\sqrt{x} = 1$

**3.** $\sqrt{x} = 4.3$

**4.** $\sqrt{x} = 6.2$

**5.** $\sqrt{y + 4} = 13$

**6.** $\sqrt{y - 5} = 21$

**7.** $\sqrt{2x + 4} = 25$

**8.** $\sqrt{2x + 1} = 13$

**9.** $3 + \sqrt{x - 1} = 5$

**10.** $4 + \sqrt{y - 3} = 11$

**11.** $6 - 2\sqrt{3n} = 0$

**12.** $8 - 4\sqrt{5n} = 0$

**13.** $\sqrt{5x - 7} = \sqrt{x + 10}$

**14.** $\sqrt{4x - 5} = \sqrt{x + 9}$

**15.** $\sqrt{x} = -7$

**16.** $\sqrt{x} = -5$

**17.** $\sqrt{2y + 6} = \sqrt{2y - 5}$

**18.** $2\sqrt{3x - 2} = \sqrt{2x - 3}$

**19.** $x - 7 = \sqrt{x - 5}$

**20.** $\sqrt{x + 7} = x - 5$

**21.** $x - 9 = \sqrt{x - 3}$

**22.** $\sqrt{x + 18} = x - 2$

**23.** $2\sqrt{x - 1} = x - 1$

**24.** $x + 4 = 4\sqrt{x + 1}$

**25.** $\sqrt{5x + 21} = x + 3$

**26.** $\sqrt{27 - 3x} = x - 3$

**27.** $\sqrt{2x - 1} + 2 = x$

**28.** $x = 1 + 6\sqrt{x - 9}$

**29.** $\sqrt{x^2 + 6} - x + 3 = 0$

**30.** $\sqrt{x^2 + 5} - x + 2 = 0$

**31.** $\sqrt{x^2 - 4} - x = 6$

**32.** $\sqrt{x^2 - 5x + 7} = x - 3$

**33.** $\sqrt{(p + 6)(p + 1)} - 2 = p + 1$

**34.** $\sqrt{(4x + 5)(x + 4)} = 2x + 5$

**35.** $\sqrt{4x - 10} = \sqrt{2 - x}$

**36.** $\sqrt{2 - x} = \sqrt{3x - 7}$

Solve. Use the principle of squaring twice.

**37.** $\sqrt{x - 5} = 5 - \sqrt{x}$

**38.** $\sqrt{x + 9} = 1 + \sqrt{x}$

**39.** $\sqrt{y + 8} - \sqrt{y} = 2$

**40.** $\sqrt{3x + 1} = 1 - \sqrt{x + 4}$

**41.** $\sqrt{x - 4} + \sqrt{x + 1} = 5$

**42.** $1 + \sqrt{x + 7} = \sqrt{3x - 2}$

 Solve.

*Sightings to the Horizon.* Use $V = 3.5\sqrt{h}$ for Exercises 43–46.

**43.** A steeplejack can see 21 km to the horizon from the top of a building. What is the altitude of the steeplejack's eyes?

**44.** A person can see 371 km to the horizon from an airplane window. How high is the airplane?

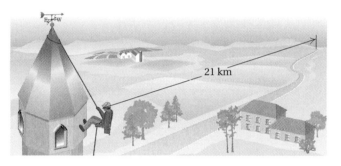

21 km

**45.** How far can a sailor see to the horizon from the top of a mast that is 37 m high?

**46.** How far can you see to the horizon through an airplane window at a height of 9800 m?

*Speed of a Skidding Car.* After an accident, how do police determine the speed at which the car had been traveling? The formula

$$r = 2\sqrt{5L}$$

can be used to approximate the speed $r$, in miles per hour, of a car that has left a skid mark of length $L$, in feet. (See Example 8 in Section 9.1.) Use this formula to do Exercises 47 and 48.

**47.** How far will a car skid at 55 mph? at 75 mph?

**48.** How far will a car skid at 65 mph? at 100 mph?

**49.** Find a number such that the square root of 4 more than five times the number is 8.

**50.** Find a number such that twice its square root is 14.

**51.** Find a number such that the square root of 4 less than the number plus the square root of 1 more than the number is 5.

**52.** Find a number such that the square root of twice the number minus 1, all added to 1, is the square root of the number plus 11.

---

**Skill Maintenance**

Divide and simplify. [6.2b]

**53.** $\dfrac{x^2 - 49}{x + 8} \div \dfrac{x^2 - 14x + 49}{x^2 + 15x + 56}$

**54.** $\dfrac{x - 2}{x - 3} \div \dfrac{x - 4}{x - 5}$

**55.** $\dfrac{a^2 - 25}{6} \div \dfrac{a + 5}{3}$

**56.** $\dfrac{x - 2}{x + 3} \div \dfrac{x^2 - 4x + 4}{x^2 - 9}$

Solve. [8.3c]

**57.** Two angles are supplementary. One angle is 3° less than twice the other. Find the measures of the angles.

**58.** Two angles are complementary. The sum of the measure of the first angle and half the measure of the second is 64°. Find the measures of the angles.

Multiply and simplify. [6.1d]

**59.** $\dfrac{7x^9}{27} \cdot \dfrac{9}{7x^3}$

**60.** $\dfrac{3}{x^2 - 9} \cdot \dfrac{x^2 - 6x + 9}{12}$

---

**Synthesis**

**61.** ◈ Explain why possible solutions of radical equations must be checked.

**62.** ◈ Determine whether the statement below is true or false and explain your answer.

The solution of $\sqrt{11 - 2x} = -3$ is $-1$.

Solve.

**63.** $\sqrt{5x^2 + 5} = 5$

**64.** $\sqrt{x} = -x$

**65.** $4 + \sqrt{19 - x} = 6 + \sqrt{4 - x}$

**66.** $x = (x - 2)\sqrt{x}$

**67.** $\sqrt{x + 3} = \dfrac{8}{\sqrt{x - 9}}$

**68.** $\dfrac{12}{\sqrt{5x + 6}} = \sqrt{2x + 5}$

**69.–72.** ⟥⟤ Use a grapher to check your answers to Exercises 11–14.

---

# 9.6 Applications with Right Triangles

## a | Right Triangles

A **right triangle** is a triangle with a 90° angle, as shown in the figure below. The small square in the corner indicates the 90° angle.

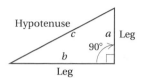

In a right triangle, the longest side is called the **hypotenuse**. It is also the side opposite the right angle. The other two sides are called **legs**. We generally use the letters $a$ and $b$ for the lengths of the legs and $c$ for the length of the hypotenuse. They are related as follows.

---

> **THE PYTHAGOREAN THEOREM**
>
> In any right triangle, if $a$ and $b$ are the lengths of the legs and $c$ is the length of the hypotenuse, then
>
> $$a^2 + b^2 = c^2.$$
>
> The equation $a^2 + b^2 = c^2$ is called the **Pythagorean equation.**

---

The Pythagorean theorem is named after the ancient Greek mathematician Pythagoras (569?–500? B.C.). It is uncertain who actually proved this result the first time. The proof can be found in most geometry books.

If we know the lengths of any two sides of a right triangle, we can find the length of the third side.

**Example 1** Find the length of the hypotenuse of this right triangle. Give an exact answer and an approximation to three decimal places.

$$4^2 + 5^2 = c^2 \qquad \text{Substituting in the Pythagorean equation}$$
$$16 + 25 = c^2$$
$$41 = c^2$$
$$c = \sqrt{41}$$
$$\approx 6.403 \qquad \text{Using a calculator}$$

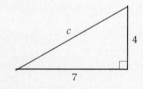

**Example 2** Find the length of the leg of this right triangle. Give an exact answer and an approximation to three decimal places.

$$10^2 + b^2 = 12^2 \qquad \text{Substituting in the Pythagorean equation}$$
$$100 + b^2 = 144$$
$$b^2 = 144 - 100$$
$$b^2 = 44$$
$$b = \sqrt{44}$$
$$\approx 6.633 \qquad \text{Using a calculator}$$

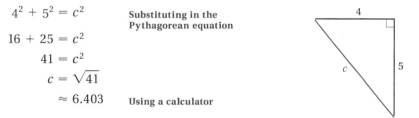

*Do Exercises 1 and 2.*

---

*Answers on page A-37*

---

---

### Objectives

**a** Given the lengths of any two sides of a right triangle, find the length of the third side.

**b** Solve applied problems involving right triangles.

**For Extra Help**

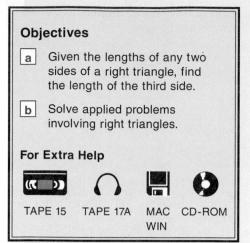

TAPE 15   TAPE 17A   MAC WIN   CD-ROM

1. Find the length of the hypotenuse of this right triangle. Give an exact answer and an approximation to three decimal places.

2. Find the length of the leg of this right triangle. Give an exact answer and an approximation to three decimal places.

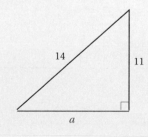

Find the length of the leg of the right triangle. Give an exact answer and an approximation to three decimal places.

**3.**

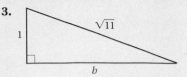

**4.**

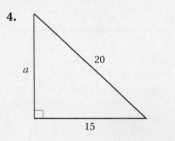

**5.** *Guy Wire.* How long is a guy wire reaching from the top of a 15-ft pole to a point on the ground 10 ft from the pole? Give an exact answer and an approximation to three decimal places.

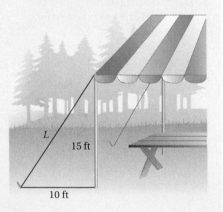

*Answers on page A-37*

**Example 3** Find the length of the leg of this right triangle. Give an exact answer and an approximation to three decimal places.

$$1^2 + b^2 = (\sqrt{7})^2 \quad \text{Substituting in the Pythagorean equation}$$
$$1 + b^2 = 7$$
$$b^2 = 7 - 1 = 6$$
$$b = \sqrt{6}$$
$$\approx 2.449 \quad \textbf{Using a calculator}$$

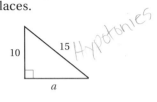

**Example 4** Find the length of the leg of this right triangle. Give an exact answer and an approximation to three decimal places.

$$a^2 + 10^2 = 15^2$$
$$a^2 + 100 = 225$$
$$a^2 = 225 - 100$$
$$a^2 = 125$$
$$a = \sqrt{125}$$
$$\approx 11.180 \quad \textbf{Using a calculator}$$

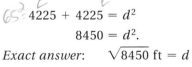

*Do Exercises 3 and 4.*

## b   Applications

**Example 5** *Dimensions of a Softball Diamond.* A slow-pitch softball diamond is actually a square 65 ft on a side. How far is it from home plate to second base? (This can be helpful information when lining up the bases.) Give an exact answer and an approximation to three decimal places.

a) We first make a drawing. We note that the first and second base lines, together with a line from home to second, form a right triangle. We label the unknown distance $d$.

b) We know that $65^2 + 65^2 = d^2$. We solve this equation:

$$4225 + 4225 = d^2$$
$$8450 = d^2.$$

*Exact answer:* $\sqrt{8450}$ ft $= d$     *Approximation:* 91.924 ft $\approx d$

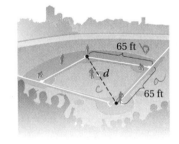

*Do Exercise 5.*

# Exercise Set 9.6

**a** Find the length of the third side of the right triangle. Give an exact answer and an approximation to three decimal places.

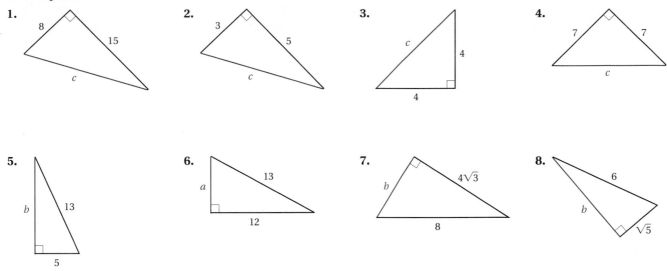

**1.**    8    15    $c$

**2.**    3    5    $c$

**3.**    $c$    4    4

**4.**    7    7    $c$

**5.**    $b$    13    5

**6.**    $a$    13    12

**7.**    $b$    $4\sqrt{3}$    8

**8.**    6    $b$    $\sqrt{5}$

In a right triangle, find the length of the side not given. Give an exact answer and an approximation to three decimal places.

**9.** $a = 10,\quad b = 24$

**10.** $a = 5,\quad b = 12$

**11.** $a = 9,\quad c = 15$

**12.** $a = 18,\quad c = 30$

**13.** $b = 1,\quad c = \sqrt{5}$

**14.** $b = 1,\quad c = \sqrt{2}$

**15.** $a = 1,\quad c = \sqrt{3}$

**16.** $a = \sqrt{3},\quad b = \sqrt{5}$

**17.** $c = 10,\quad b = 5\sqrt{3}$

**18.** $a = 5,\quad b = 5$

**19.** $a = \sqrt{2},\quad b = \sqrt{7}$

**20.** $c = \sqrt{7},\quad a = \sqrt{2}$

**b** Solve. Don't forget to make a drawing. Give an exact answer and an approximation to three decimal places.

**21.** An airplane is flying at an altitude of 4100 ft. The slanted distance directly to the airport is 15,100 ft. How far is the airplane horizontally from the airport?

**22.** A surveyor had poles located at points *P*, *Q*, and *R*. The distances that the surveyor was able to measure are marked on the drawing. What is the approximate distance from *P* to *R*?

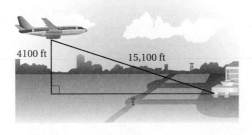

4100 ft    15,100 ft

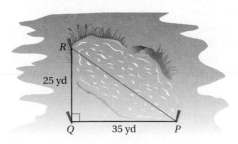

*R*    25 yd    *Q*    35 yd    *P*

**23.** *Cordless Telephones.* Becky's new cordless telephone has clear reception up to 300 ft from its base. Her phone is located near a window in her apartment, 180 ft above ground level. How far into her backyard can Becky use her phone?

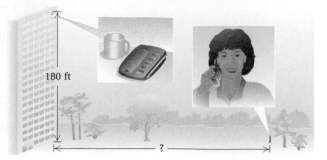

180 ft

?

**24.** *Rope Course.* An outdoor rope course consists of a cable that slopes downward from a height of 37 ft to a resting place 30 ft above the ground. The trees that the cable connects are 24 ft apart. How long is the cable?

24 ft

37 ft

30 ft

**25.** Find the length of a diagonal of a square whose sides are 3 cm long.

**26.** A 10-m ladder is leaning against a building. The bottom of the ladder is 5 m from the building. How high is the top of the ladder?

**27.** How long is a guy wire reaching from the top of a 12-ft pole to a point 8 ft from the pole?

**28.** The largest regulation soccer field is 100 yd wide and 130 yd long. Find the length of a diagonal of such a field.

---

**Skill Maintenance**

Solve.   [8.3a, b]

**29.** $5x + 7 = 8y,$
$3x = 8y - 4$

**30.**   $5x + y = 17,$
$-5x + 2y = 10$

**31.** $3x - 4y = -11,$
$5x + 6y = 12$

**32.** $x + y = -9,$
$x - y = -11$

**33.** Find the slope of the line $4 - x = 3y.$   [7.1b]

**34.** Find the slope of the line containing the points $(8, -3)$ and $(0, -8).$   [7.1a]

---

**Synthesis**

**35.** ◈ Can a carpenter use a 28-ft ladder to repair clapboard that is 28 ft above ground level? Why or why not?

**36.** ◈ In an **equilateral triangle,** all sides have the same length. Can a right triangle ever be equilateral? Why or why not?

**37.** Two cars leave a service station at the same time. One car travels east at a speed of 50 mph, and the other travels south at a speed of 60 mph. After one-half hour, how far apart are they?

**38.** The length and the width of a rectangle are given by consecutive integers. The area of the rectangle is $90 \text{ cm}^2$. Find the length of a diagonal of the rectangle.

Find $x$.

**39.**

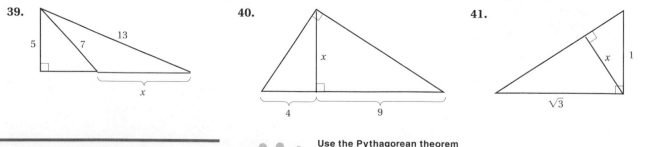

5

7

13

$x$

**40.**

$x$

4

9

**41.**

$x$

1

$\sqrt{3}$

Collaborative Learning Manual

Use the Pythagorean theorem to draw a rectangle.

Copyright © 1999 Addison Wesley Longman

# Summary and Review Exercises: Chapter 9

**Important Properties and Formulas**

Product Rule for Radicals: $\sqrt{A}\,\sqrt{B} = \sqrt{AB}$

Quotient Rule for Radicals: $\dfrac{\sqrt{A}}{\sqrt{B}} = \sqrt{\dfrac{A}{B}}$

Principle of Squaring: If an equation $a = b$ is true, then the equation $a^2 = b^2$ is true.

Pythagorean Equation: $a^2 + b^2 = c^2$, where $a$ and $b$ are the lengths of the legs of a right triangle and $c$ is the length of the hypotenuse.

The objectives to be tested in addition to the material in this chapter are [6.2b], [7.5b], [8.3a, b], and [8.4a].

Find the square roots. [9.1a]

**1.** 64

**2.** 400

Simplify. [9.1a]

**3.** $\sqrt{36}$

**4.** $-\sqrt{169}$

Use a calculator to approximate each of the following square roots to three decimal places. [9.1b]

**5.** $\sqrt{3}$

**6.** $\sqrt{99}$

**7.** $-\sqrt{320.12}$

**8.** $\sqrt{\dfrac{11}{20}}$

**9.** $-\sqrt{\dfrac{47.3}{11.2}}$

**10.** $18\sqrt{11 \cdot 43.7}$

Identify the radicand. [9.1d]

**11.** $\sqrt{x^2 + 4}$

**12.** $\sqrt{5ab^3}$

Determine whether the expression is meaningful as a real number. Write "yes" or "no." [9.1e]

**13.** $\sqrt{-22}$

**14.** $-\sqrt{49}$

**15.** $\sqrt{-36}$

**16.** $\sqrt{-10.2}$

**17.** $-\sqrt{-4}$

**18.** $\sqrt{2(-3)}$

Simplify. [9.1f]

**19.** $\sqrt{m^2}$

**20.** $\sqrt{(x-4)^2}$

Multiply. [9.2c]

**21.** $\sqrt{3}\,\sqrt{7}$

**22.** $\sqrt{x-3}\,\sqrt{x+3}$

Simplify by factoring. [9.2a]

**23.** $-\sqrt{48}$

**24.** $\sqrt{32t^2}$

**25.** $\sqrt{t^2 - 49}$

**26.** $\sqrt{x^2 + 16x + 64}$

Simplify by factoring. [9.2b]

**27.** $\sqrt{x^8}$

**28.** $\sqrt{m^{15}}$

Multiply and simplify. [9.2c]

**29.** $\sqrt{6}\,\sqrt{10}$

**30.** $\sqrt{5x}\,\sqrt{8x}$

**31.** $\sqrt{5x}\,\sqrt{10xy^2}$

**32.** $\sqrt{20a^3b}\,\sqrt{5a^2b^2}$

Simplify. [9.3b]

**33.** $\sqrt{\dfrac{25}{64}}$

**34.** $\sqrt{\dfrac{20}{45}}$

**35.** $\sqrt{\dfrac{49}{t^2}}$

Rationalize the denominator. [9.3c]

**36.** $\sqrt{\dfrac{1}{2}}$

**37.** $\sqrt{\dfrac{1}{8}}$

**38.** $\sqrt{\dfrac{5}{y}}$

**39.** $\dfrac{2}{\sqrt{3}}$

Divide and simplify. [9.3a, c]

**40.** $\dfrac{\sqrt{27}}{\sqrt{45}}$

**41.** $\dfrac{\sqrt{45x^2y}}{\sqrt{54y}}$

**42.** Rationalize the denominator: [9.4c]

$$\dfrac{4}{2 + \sqrt{3}}.$$

Simplify. [9.4a]

**43.** $10\sqrt{5} + 3\sqrt{5}$

**44.** $\sqrt{80} - \sqrt{45}$

**45.** $3\sqrt{2} - 5\sqrt{\dfrac{1}{2}}$

Simplify. [9.4b]

**46.** $(2 + \sqrt{3})^2$

**47.** $(2 + \sqrt{3})(2 - \sqrt{3})$

Solve. [9.5a]

**48.** $\sqrt{x - 3} = 7$

**49.** $\sqrt{5x + 3} = \sqrt{2x - 1}$

**50.** $1 + x = \sqrt{1 + 5x}$

**51.** Solve: [9.5b]

$$\sqrt{x} = \sqrt{x - 5} + 1.$$

In a right triangle, find the length of the side not given. [9.6a]

**52.** $a = 15, \quad c = 25$

**53.** $a = 1, \quad b = \sqrt{2}$

Solve. [9.6b]

**54.** *Airplane Descent.* A pilot is instructed to descend from 30,000 ft to 20,000 ft over a horizontal distance of 50,000 ft. What distance will the plane travel during this descent?

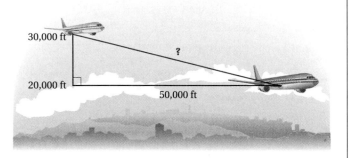

**55.** *Lookout Tower.* The diagonal braces in a lookout tower are 15 ft long and span a distance of 12 ft. How high does each brace reach vertically?

Solve. [9.1c], [9.5c]

**56.** *Speed of a Skidding Car.* The formula $r = 2\sqrt{5L}$ can be used to approximate the speed $r$, in miles per hour, of a car that has left a skid mark of length $L$, in feet.

a) What was the speed of a car that left skid marks of length 200 ft?

b) How far will a car skid at 90 mph?

---

**Skill Maintenance**

**57.** Solve: [8.3b]

$$2x - 3y = 4,$$
$$3x + 4y = 2.$$

**58.** Divide and simplify: [6.2b]

$$\frac{x^2 - 10x + 25}{x^2 + 14x + 49} \div \frac{x^2 - 25}{x^2 - 49}.$$

Solve.

**59.** A person's paycheck varies directly as the number of hours $H$ worked. For 15 hr of work, the pay is $168.75. Find the pay for 40 hr of work. [7.5b]

**60.** There were 14,000 people at an AIDS benefit rock concert. Tickets were $12.00 at the door and $10.00 if purchased in advance. Total receipts were $159,400. How many people bought tickets in advance? [8.4a]

---

**Synthesis**

**61.** ◆ Explain why the following is incorrect: [9.3b]

$$\sqrt{\frac{9 + 100}{25}} = \frac{3 + 10}{5}.$$

**62.** ◆ Determine whether each of the following is correct for all real numbers. Explain why or why not. [9.2a]

a) $\sqrt{5x^2} = |x|\sqrt{5}$
b) $\sqrt{b^2 - 4} = b - 2$
c) $\sqrt{x^2 + 16} = x + 4$

**63.** Simplify: $\sqrt{\sqrt{\sqrt{256}}}$. [9.2a]

**64.** Solve $A = \sqrt{a^2 + b^2}$ for $b$. [9.5a]

# Test: Chapter 9

**1.** Find the square roots of 81.

Simplify.

**2.** $\sqrt{64}$

**3.** $-\sqrt{25}$

Approximate the expression involving square roots to three decimal places.

**4.** $\sqrt{116}$

**5.** $-\sqrt{87.4}$

**6.** $\sqrt{\dfrac{96 \cdot 38}{214.2}}$

**7.** Identify the radicand in $\sqrt{4 - y^3}$.

Determine whether the expression is meaningful as a real number. Write "yes" or "no."

**8.** $\sqrt{24}$

**9.** $\sqrt{-23}$

Simplify.

**10.** $\sqrt{a^2}$

**11.** $\sqrt{36y^2}$

Multiply.

**12.** $\sqrt{5}\sqrt{6}$

**13.** $\sqrt{x - 8}\sqrt{x + 8}$

Simplify by factoring.

**14.** $\sqrt{27}$

**15.** $\sqrt{25x - 25}$

**16.** $\sqrt{t^5}$

Multiply and simplify.

**17.** $\sqrt{5}\sqrt{10}$

**18.** $\sqrt{3ab}\sqrt{6ab^3}$

## Answers

1. _____

2. _____

3. _____

4. _____

5. _____

6. _____

7. _____

8. _____

9. _____

10. _____

11. _____

12. _____

13. _____

14. _____

15. _____

16. _____

17. _____

18. _____

19.  _____

20.  _____

21.  _____

22.  _____

23.  _____

24.  _____

25.  _____

26.  _____

27.  _____

28.  _____

29.  _____

30.  _____

31.  _____

32.  _____

33.  _____

34.  _____

35.  _____

36.  _____

37.  _____

38.  _____

39.  _____

40.  _____

Simplify.

**19.** $\sqrt{\dfrac{27}{12}}$

**20.** $\sqrt{\dfrac{144}{a^2}}$

Rationalize the denominator.

**21.** $\sqrt{\dfrac{2}{5}}$

**22.** $\sqrt{\dfrac{2x}{y}}$

Divide and simplify.

**23.** $\dfrac{\sqrt{27}}{\sqrt{32}}$

**24.** $\dfrac{\sqrt{35x}}{\sqrt{80xy^2}}$

Add or subtract.

**25.** $3\sqrt{18} - 5\sqrt{18}$

**26.** $\sqrt{5} + \sqrt{\dfrac{1}{5}}$

Simplify.

**27.** $(4 - \sqrt{5})^2$

**28.** $(4 - \sqrt{5})(4 + \sqrt{5})$

**29.** Rationalize the denominator: $\dfrac{10}{4 - \sqrt{5}}$.

**30.** In a right triangle, $a = 8$ and $b = 4$. Find $c$.

Solve.

**31.** $\sqrt{3x} + 2 = 14$

**32.** $\sqrt{6x + 13} = x + 3$

**33.** $\sqrt{1 - x} + 1 = \sqrt{6 - x}$

**34.** A person can see 247.49 km to the horizon from an airplane window. How high is the airplane? Use the formula $V = 3.5\sqrt{h}$.

---

### Skill Maintenance

**35.** The perimeter of a rectangle is 118 yd. The width is 18 yd less than the length. Find the area of the rectangle.

**36.** The number of switches $N$ that a production line can make varies directly as the time it operates. It can make 7240 switches in 6 hr. How many can it make in 13 hr?

**37.** Solve:

$$-6x + 5y = 10,$$
$$5x + 6y = 12.$$

**38.** Divide and simplify:

$$\dfrac{x^2 - 11x + 30}{x^2 - 12x + 35} \div \dfrac{x^2 - 36}{x^2 - 14x + 49}.$$

---

### Synthesis

Simplify.

**39.** $\sqrt{\sqrt{\sqrt{625}}}$

**40.** $\sqrt{y^{16n}}$

Copyright © 1999 Addison Wesley Longman

# Cumulative Review: Chapters 1–9

**1.** Evaluate $x^3 - x^2 + x - 1$ for $x = -2$.

**2.** Collect like terms:
$$2x^3 - 7 + \frac{3}{7}x^2 - 6x^3 - \frac{4}{7}x^2 + 5.$$

**3.** Find all numbers for which the expression is undefined:
$$\frac{x - 6}{2x + 1}.$$

**4.** Determine whether the expression is meaningful as a real number. Write "yes" or "no."
$$\sqrt{-24}$$

Simplify.

**5.** $(2 + \sqrt{3})(2 - \sqrt{3})$

**6.** $-\sqrt{196}$

**7.** $\sqrt{3}\sqrt{75}$

**8.** $(1 - \sqrt{2})^2$

**9.** $\frac{\sqrt{162}}{\sqrt{125}}$

**10.** $2\sqrt{45} + 3\sqrt{20}$

Perform the indicated operations and simplify.

**11.** $(3x^4 - 2y^5)(3x^4 + 2y^5)$

**12.** $(x^2 + 4)^2$

**13.** $\left(2x + \frac{1}{4}\right)\left(4x - \frac{1}{2}\right)$

**14.** $\frac{x}{2x - 1} - \frac{3x + 2}{1 - 2x}$

**15.** $(3x^2 - 2x^3) - (x^3 - 2x^2 + 5) + (3x^2 - 5x + 5)$

**16.** $\frac{2x + 2}{3x - 9} \cdot \frac{x^2 - 8x + 15}{x^2 - 1}$

**17.** $\frac{2x^2 - 2}{2x^2 + 7x + 3} \div \frac{4x - 4}{2x^2 - 5x - 3}$

**18.** $(3x^3 - 2x^2 + x - 5) \div (x - 2)$

Simplify.

**19.** $\sqrt{2x^2 - 4x + 2}$

**20.** $x^{-9} \cdot x^{-3}$

**21.** $\sqrt{\frac{50}{2x^8}}$

**22.** $\dfrac{x - \dfrac{1}{x}}{1 - \dfrac{x - 1}{2x}}$

Factor completely.

**23.** $3 - 12x^8$

**24.** $12t - 4t^2 - 48t^4$

**25.** $6x^2 - 28x + 16$

**26.** $4x^3 + 4x^2 - x - 1$

**27.** $16x^4 - 56x^2 + 49$

**28.** $x^2 + 3x - 180$

Solve.

**29.** $x^2 = -17x$

**30.** $-3x < 30 + 2x$

**31.** $\frac{1}{x} + \frac{2}{3} = \frac{1}{4}$

**32.** $x^2 - 30 = x$

**33.** $-4(x + 5) \geq 2(x + 5) - 3$

**34.** $2x^2 = 162$

**35.** $\sqrt{2x - 1} + 5 = 14$

**36.** $\sqrt{4x} + 1 = \sqrt{x} + 4$

**37.** $\frac{1}{4}x + \frac{2}{3}x = \frac{2}{3} - \frac{3}{4}x$

**38.** $\frac{x}{x - 1} - \frac{x}{x + 1} = \frac{1}{2x - 2}$

**39.** $x = y + 3,$
$3y - 4x = -13$

**40.** $2x - 3y = 30,$
$5y - 2x = -46$

**41.** $\frac{E}{r} = \frac{R + r}{R}$, for $R$

**42.** $4A = pr + pq$, for $p$

Graph on a plane.

**43.** $3y - 3x > -6$

**44.** $x = 5$

**45.** $2x - 6y = 12$

**46.** Find an equation of the line containing the points $(1, -2)$ and $(5, 9)$.

**47.** Find the slope and the $y$-intercept of the line $5x - 3y = 9$.

**48.** Estimate the missing data value.

| Year | Credit-Card Spending from Thanksgiving to Christmas (in billions) |
|------|------|
| 1991 | $59.8 |
| 1992 | 66.8 |
| 1993 | 79.1 |
| 1994 | 96.9 |
| 1995 | 116.3 |
| 1996 | 131.4 |
| 1997 | ? |

*Source*: RAM Research Group, National Credit Counseling Services

**49.** The graph of the polynomial equation $y = x^3 - 4x - 2$ is shown below. Use either the graph or the equation to estimate the value of the polynomial when $x = -2$, $x = -1$, $x = 0$, $x = 1$, and $x = 2$.

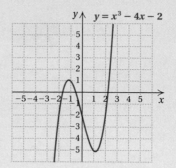

Solve.

**50.** The second angle of a triangle is twice as large as the first. The third angle is 48° less than the sum of the other two angles. Find the measures of the angles.

**51.** The cost of 6 hamburgers and 4 milkshakes is $22.80. Three hamburgers and 1 milkshake cost $9.60. Find the cost of a hamburger and the cost of a milkshake.

**52.** An 8-m ladder is leaning against a building. The bottom of the ladder is 4 m from the building. How high is the top of the ladder?

**53.** The amount $C$ that a family spends on housing varies directly as its income $I$. A family making $25,000 a year will spend $6250 a year for housing. How much will a family making $30,000 a year spend for housing?

**54.** A sample of 150 resistors contained 12 defective resistors. How many defective resistors would you expect to find in a sample of 250 resistors?

**55.** The length of a rectangle is 3 m greater than the width. The area of the rectangle is 180 m². Find the length and the width.

**56.** A collection of dimes and quarters is worth $19.00. There are 115 coins in all. How many of each are there?

**57.** The winner of an election won by a margin of 2 to 1, with 238 votes. How many voted in the election?

**58.** Money is invested in an account at 10.5% simple interest. At the end of 1 yr, there is $2873 in the account. How much was originally invested?

**59.** A person traveled 600 mi in one direction. The return trip took 2 hr longer at a speed that was 10 mph less. Find the speed going.

**Synthesis**

Write a true sentence using < or >.

**60.** $-4$ ▨ $-3$

**61.** $|-4|$ ▨ $|-3|$

**62.** A tank contains 200 L of a 30%-salt solution. How much pure water should be added in order to make a solution that is 12% salt?

**63.** Solve: $\sqrt{x} + 1 = y$,
$\sqrt{x} + \sqrt{y} = 5$.

# Quadratic Equations

**10**

## Introduction

A *quadratic equation* contains a polynomial of second degree. In this chapter, we first learn to solve quadratic equations by factoring. Because certain quadratic equations are difficult to solve by factoring, we also learn to use the *quadratic formula* to find solutions of quadratic equations. Next, we apply these equation-solving skills to applications and problem-solving. Then we graph quadratic equations.

| **An Application** | **The Mathematics** |
|---|---|

In the not-too-distant future, a new kind of high-definition television (HDTV) with a larger screen and greater clarity will be available. An HDTV might have a 70-in. diagonal screen with the width 27 in. greater than the height. Find the width and the height of a 70-in. HDTV.

This problem appears as Exercise 4 in Exercise Set 10.5.

We let $w$ = the width and $h$ = the height. Then the problem translates to

$$w^2 + h^2 = 70^2,$$

or

$$(h + 27)^2 + h^2 = 4900,$$

or

$$\underbrace{2h^2 + 54h - 4171 = 0.}$$

This is a quadratic equation.

 For more information, visit us at www.mathmax.com

# Pretest: Chapter 10

Solve.

**1.** $x^2 + 9 = 6x$

**2.** $x^2 - 7 = 0$

**3.** $3x^2 + 3x - 1 = 0$

**4.** $5y^2 - 3y = 0$

**5.** $\dfrac{3}{3x + 2} - \dfrac{2}{3x + 4} = 1$

**6.** $(x + 4)^2 = 5$

**7.** Solve $x^2 - 2x - 5 = 0$ by completing the square. Show your work.

**8.** Solve $A = n^2 - pn$ for $n$.

**9.** The length of a rectangle is three times the width. The area is 48 cm². Find the length and the width.

**10.** Find the $x$-intercepts: $y = 2x^2 + x - 4$.

**11.** The current in a stream moves at a speed of 2 km/h. A boat travels 24 km upstream and 24 km downstream in a total time of 5 hr. What is the speed of the boat in still water?

**12.** Graph: $y = 4 - x^2$.

---

## Objectives for Retesting

The objectives to be tested in addition to the material in this chapter are as follows.

[7.5c]  Find an equation of inverse variation given a pair of values of the variables.

[9.2c]  Multiply radical expressions and simplify, if possible.

[9.4a]  Add or subtract with radical notation, using the distributive law to simplify.

[9.6b]  Solve applied problems involving right triangles.

---

# 10.1 Introduction to Quadratic Equations

**AG Algebraic–Graphical Connection**

Before we begin this chapter, let's look back at some algebraic–graphical equation-solving concepts and their interrelationships. In Chapter 3, we considered the graph of a *linear equation* $y = mx + b$. For example, the graph of the equation $y = \frac{5}{2}x - 4$ and its $x$-intercept are shown below.

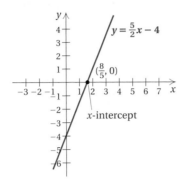

If $y = 0$, then $x = \frac{8}{5}$. Thus the $x$-intercept is $\left(\frac{8}{5}, 0\right)$. This point is also the intersection of the graphs of $y = \frac{5}{2}x - 4$ and $y = 0$.

In Chapter 2, we learned how to solve linear equations like $0 = \frac{5}{2}x - 4$ algebraically (using algebra). We proceeded as follows:

$$0 = \frac{5}{2}x - 4$$
$$4 = \frac{5}{2}x \quad \text{Adding 4}$$
$$8 = 5x \quad \text{Multiplying by 2}$$
$$\frac{8}{5} = x. \quad \text{Dividing by 5}$$

We see that $\frac{8}{5}$, the solution of $0 = \frac{5}{2}x - 4$, is the first coordinate of the $x$-intercept of the graph of $y = \frac{5}{2}x - 4$.

### Do Exercises 1 and 2.

In this chapter, we build on these ideas by applying them to quadratic equations. In Section 5.7, we briefly considered the graph of a *quadratic equation*

$$y = ax^2 + bx + c, \quad a \neq 0.$$

For example, the graph of the equation $y = x^2 + 6x + 8$ and its $x$-intercepts are shown below.

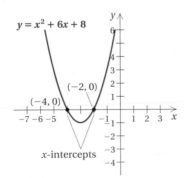

The $x$-intercepts are $(-4, 0)$ and $(-2, 0)$. We will develop in detail the creation of such graphs in Section 10.6. The points $(-4, 0)$ and $(-2, 0)$ are the intersections of the graphs of $y = x^2 + 6x + 8$ and $y = 0$.

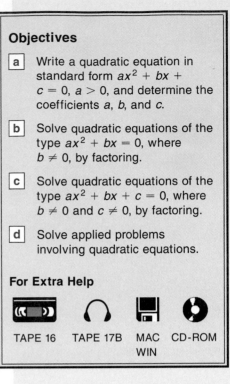

**Objectives**

**a** Write a quadratic equation in standard form $ax^2 + bx + c = 0$, $a > 0$, and determine the coefficients $a$, $b$, and $c$.

**b** Solve quadratic equations of the type $ax^2 + bx = 0$, where $b \neq 0$, by factoring.

**c** Solve quadratic equations of the type $ax^2 + bx + c = 0$, where $b \neq 0$ and $c \neq 0$, by factoring.

**d** Solve applied problems involving quadratic equations.

**For Extra Help**

TAPE 16    TAPE 17B    MAC WIN    CD-ROM

1. Consider $y = -\frac{2}{3}x - 3$. Find the intercepts and graph the equation.

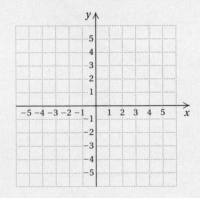

2. Solve the equation

$$0 = -\frac{2}{3}x - 3.$$

*Answers on page A-38*

Write in standard form and determine $a$, $b$, and $c$.

**3.** $y^2 = 8y$

**4.** $3 - x^2 = 9x$

**5.** $3x + 5x^2 = x^2 - 4 + 2x$

**6.** $5x^2 = 21$

We began studying the solution of quadratic equations like $x^2 + 6x + 8 = 0$ in Section 5.7. There we used factoring for such solutions:

$$x^2 + 6x + 8 = 0$$
$$(x + 4)(x + 2) = 0 \qquad \text{Factoring}$$
$$x + 4 = 0 \quad or \quad x + 2 = 0 \qquad \text{Using the principle of zero products}$$
$$x = -4 \quad or \qquad x = -2.$$

We see that the solutions of $x^2 + 6x + 8 = 0$, $-4$ and $-2$, are the first coordinates of the $x$-intercepts, $(-4, 0)$ and $(-2, 0)$, of the graph of $y = x^2 + 6x + 8$. We will enhance our ability to solve quadratic equations in Sections 10.1–10.3.

## a │ Standard Form

The following are **quadratic equations.** They contain polynomials of second degree.

$$x^2 + 7x - 5 = 0, \qquad 3t^2 - \tfrac{1}{2}t = 9, \qquad 5y^2 = -6y, \qquad 5m^2 = 15$$

The quadratic equation $4x^2 + 7x - 5 = 0$ is said to be in **standard form.** Although the quadratic equation $4x^2 = 5 - 7x$ is equivalent to the preceding equation, it is *not* in standard form.

> A quadratic equation of the type $ax^2 + bx + c = 0$, where $a$, $b$, and $c$ are real-number constants and $a > 0$, is called the **standard form of a quadratic equation.**

We define $a > 0$ to ease the proof of the quadratic formula, which we consider later, and to ease solving by factoring, which we review in this section. Suppose we are studying an equation like $-3x^2 + 8x - 2 = 0$. It is not in standard form. We can find an equivalent equation that is in standard form by multiplying by $-1$ on both sides:

$$-1(-3x^2 + 8x - 2) = -1(0)$$
$$3x^2 - 8x + 2 = 0.$$

**Examples**   Write in standard form and determine $a$, $b$, and $c$.

**1.** $4x^2 + 7x - 5 = 0$      The equation is already in standard form.

$a = 4; \quad b = 7; \quad c = -5$

**2.** $3x^2 - 0.5x = 9$

$3x^2 - 0.5x - 9 = 0$      Subtracting 9. This is standard form.

$a = 3; \quad b = -0.5; \quad c = -9$

**3.** $-4y^2 = 5y$

$-4y^2 - 5y = 0$      Subtracting $5y$

Not positive!

$4y^2 + 5y = 0$      Multiplying by $-1$. This is standard form.

$a = 4; \quad b = 5; \quad c = 0$

*Do Exercises 3–6.*

## b Solving Quadratic Equations of the Type $ax^2 + bx = 0$

Sometimes we can use factoring and the principle of zero products to solve quadratic equations. We are actually reviewing methods that we introduced in Section 5.7.

When $c = 0$ and $b \neq 0$, we can always factor and use the principle of zero products (see Section 5.7 for a review).

**Example 4**   Solve: $7x^2 + 2x = 0$.

$$7x^2 + 2x = 0$$
$$x(7x + 2) = 0 \qquad \text{Factoring}$$
$$x = 0 \quad or \quad 7x + 2 = 0 \qquad \text{Using the principle of zero products}$$
$$x = 0 \quad or \qquad 7x = -2$$
$$x = 0 \quad or \qquad x = -\tfrac{2}{7}$$

CHECK:   For 0:

$$\frac{7x^2 + 2x = 0}{7 \cdot 0^2 + 2 \cdot 0 \ ? \ 0}$$
$$0 \ | \qquad \text{TRUE}$$

For $-\tfrac{2}{7}$:

$$\frac{7x^2 + 2x = 0}{7\left(-\tfrac{2}{7}\right)^2 + 2\left(-\tfrac{2}{7}\right) \ ? \ 0}$$
$$7\left(\tfrac{4}{49}\right) - \tfrac{4}{7}$$
$$\tfrac{4}{7} - \tfrac{4}{7}$$
$$0 \ | \qquad \text{TRUE}$$

The solutions are 0 and $-\tfrac{2}{7}$.

---

***CAUTION!***   You may be tempted to divide each term in an equation like the one in Example 4 by $x$. This method would yield the equation

$$7x + 2 = 0,$$

whose only solution is $-\tfrac{2}{7}$. In effect, since 0 is also a solution of the original equation, we have divided by 0. The error of such division means the loss of one of the solutions.

---

**Example 5**   Solve: $20x^2 - 15x = 0$.

$$20x^2 - 15x = 0$$
$$5x(4x - 3) = 0 \qquad \text{Factoring}$$
$$5x = 0 \quad or \quad 4x - 3 = 0 \qquad \text{Using the principle of zero products}$$
$$x = 0 \quad or \qquad 4x = 3$$
$$x = 0 \quad or \qquad x = \tfrac{3}{4}$$

The solutions are 0 and $\tfrac{3}{4}$.

> A quadratic equation of the type $ax^2 + bx = 0$, where $c = 0$ and $b \neq 0$, will always have 0 as one solution and a nonzero number as the other solution.

*Do Exercises 7 and 8.*

*Answers on page A-39*

Solve.

**9.** $4x^2 + 5x - 6 = 0$

**10.** $(x - 1)(x + 1) = 5(x - 1)$

---

**c** **Solving Quadratic Equations of the Type** *ax*² + *bx* + *c* = 0

When neither $b$ nor $c$ is 0, we can sometimes solve by factoring.

**Example 6**   Solve: $5x^2 - 8x + 3 = 0$.

$$5x^2 - 8x + 3 = 0$$
$$(5x - 3)(x - 1) = 0 \qquad \text{Factoring}$$
$$5x - 3 = 0 \quad or \quad x - 1 = 0 \qquad \text{Using the principle of zero products}$$
$$5x = 3 \quad or \qquad x = 1$$
$$x = \tfrac{3}{5} \quad or \qquad x = 1$$

The solutions are $\tfrac{3}{5}$ and 1.

**Example 7**   Solve: $(y - 3)(y - 2) = 6(y - 3)$.

We write the equation in standard form and then try to factor:

$$y^2 - 5y + 6 = 6y - 18 \qquad \text{Multiplying}$$
$$y^2 - 11y + 24 = 0 \qquad \text{Standard form}$$
$$(y - 8)(y - 3) = 0 \qquad \text{Factoring}$$
$$y - 8 = 0 \quad or \quad y - 3 = 0 \qquad \text{Using the principle of zero products}$$
$$y = 8 \quad or \qquad y = 3.$$

The solutions are 8 and 3.

*Do Exercises 9 and 10.*

Recall that to solve a rational equation, we multiply on both sides by the LCM of all the denominators. We may obtain a quadratic equation after a few steps. When that happens, we know how to finish solving, but we must remember to check possible solutions because a replacement may result in division by 0. See Section 6.6.

**Example 8**   Solve: $\dfrac{3}{x - 1} + \dfrac{5}{x + 1} = 2$.

We multiply by the LCM, which is $(x - 1)(x + 1)$:

$$(x - 1)(x + 1) \cdot \left( \dfrac{3}{x - 1} + \dfrac{5}{x + 1} \right) = 2 \cdot (x - 1)(x + 1).$$

We use the distributive law on the left:

$$(x - 1)(x + 1) \cdot \dfrac{3}{x - 1} + (x - 1)(x + 1) \cdot \dfrac{5}{x + 1} = 2(x - 1)(x + 1)$$
$$3(x + 1) + 5(x - 1) = 2(x - 1)(x + 1)$$
$$3x + 3 + 5x - 5 = 2(x^2 - 1)$$
$$8x - 2 = 2x^2 - 2$$
$$0 = 2x^2 - 8x$$
$$0 = 2x(x - 4) \qquad \text{Factoring}$$
$$2x = 0 \quad or \quad x - 4 = 0$$
$$x = 0 \quad or \qquad x = 4.$$

**CHECK:** For 0:

$$\frac{3}{x-1}+\frac{5}{x+1}=2$$

$$\frac{3}{0-1}+\frac{5}{0+1}\;?\;2$$

$$\frac{3}{-1}+\frac{5}{1}$$

$$-3+5$$

$$2\;\bigg|\;\text{TRUE}$$

For 4:

$$\frac{3}{x-1}+\frac{5}{x+1}=2$$

$$\frac{3}{4-1}+\frac{5}{4+1}\;?\;2$$

$$\frac{3}{3}+\frac{5}{5}$$

$$1+1$$

$$2\;\bigg|\;\text{TRUE}$$

The solutions are 0 and 4.

*Do Exercise 11.*

## d  Solving Applied Problems

**Example 9**  *Diagonals of a Polygon.*
The number of diagonals $d$ of a polygon
of $n$ sides is given by the formula

$$d=\frac{n^2-3n}{2}.$$

If a polygon has 27 diagonals, how many sides does it have?

1. **Familiarize.**  We can make a drawing to familiarize ourselves with
   the problem. We draw an octagon (8 sides) and count the diagonals
   and see that there are 20. Let's check this in the formula. We evaluate
   the formula for $n = 8$:

   $$d=\frac{8^2-3(8)}{2}=\frac{64-24}{2}=\frac{40}{2}=20.$$

2. **Translate.**  We know that the number of diagonals is 27. We substitute
   27 for $d$:

   $$27=\frac{n^2-3n}{2}.$$

   This gives us a translation.

3. **Solve.**  We solve the equation for $n$, reversing the equation first for
   convenience:

   $$\frac{n^2-3n}{2}=27$$

   $$n^2-3n=54 \qquad \text{Multiplying by 2 to clear fractions}$$

   $$n^2-3n-54=0$$

   $$(n-9)(n+6)=0$$

   $$n-9=0 \quad or \quad n+6=0$$

   $$n=9 \quad or \qquad n=-6.$$

4. **Check.**  Since the number of sides cannot be negative, $-6$ cannot be
   a solution. We leave it to the student to show by substitution that 9
   checks.

5. **State.**  The polygon has 9 sides (it is a nonagon).

*Do Exercise 12.*

**11.** Solve:

$$\frac{20}{x+5}-\frac{1}{x-4}=1.$$

**12.** Use $d=\dfrac{n^2-3n}{2}$.

a) A heptagon has 7 sides. How
   many diagonals does it
   have?

b) A polygon has 44 diagonals.
   How many sides does it
   have?

*Answers on page A-39*

# Calculator Spotlight

Solving Equations. We can check solutions to equations using the TABLE or CALC-VALUE features (see Section 5.7). We can also check solutions using the INTERSECT or ZERO feature. Let's consider the equation $(x - 1)(x + 1) = 5(x - 1)$ of Margin Exercise 10.

**INTERSECT Feature.** We set

$$y_1 = (x - 1)(x + 1) \quad \text{and} \quad y_2 = 5(x - 1)$$

and then use the INTERSECT feature (see the CALC menu) to determine points of intersection. The first coordinates of these points are the solutions of the equation $(x - 1)(x + 1) = 5(x - 1)$.

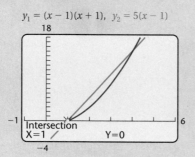

We first find the coordinates of the left-hand point of intersection. We see that 1 is a solution. We then find the coordinates of the right-hand point of intersection. We see that 4 is also a solution.

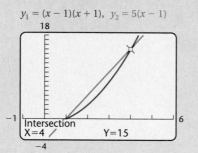

**ZERO, or ROOT Feature.** Some graphers have a ZERO, or ROOT, feature (see the CALC menu). Before we can use this feature, however, we must have a 0 on one side of the equation. So to solve

$$(x - 1)(x + 1) = 5(x - 1),$$

we consider

$$(x - 1)(x + 1) - 5(x - 1) = 0.$$

We then set

$$y_1 = (x - 1)(x + 1) - 5(x - 1)$$

and determine where the graph intersects the $x$-axis. We press CALC and ZERO, or ROOT, and continue. Performing the procedure twice, once for the left-hand value and once for the right-hand value, we see that both 1 and 4 are solutions. This also tells us that the $x$-intercepts of the equation $y = (x - 1)(x + 1) - 5(x - 1)$ are (1, 0) and (4, 0).

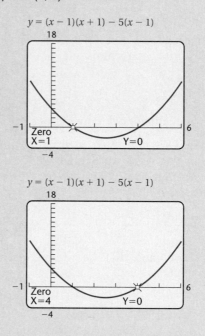

## Exercises

Solve.

**1.** $5x^2 - 8x + 3 = 0$

**2.** $6(x - 3) = (x - 3)(x - 2)$

# Exercise Set 10.1

**a** Write standard form and determine $a$, $b$, and $c$.

**1.** $x^2 - 3x + 2 = 0$

**2.** $x^2 - 8x - 5 = 0$

**3.** $7x^2 = 4x - 3$

**4.** $9x^2 = x + 5$

**5.** $5 = -2x^2 + 3x$

**6.** $3x - 1 = 5x^2 + 9$

**b** Solve.

**7.** $x^2 + 5x = 0$

**8.** $x^2 + 7x = 0$

**9.** $3x^2 + 6x = 0$

**10.** $4x^2 + 8x = 0$

**11.** $5x^2 = 2x$

**12.** $11x = 3x^2$

**13.** $4x^2 + 4x = 0$

**14.** $8x^2 - 8x = 0$

**15.** $0 = 10x^2 - 30x$

**16.** $0 = 10x^2 - 50x$

**17.** $11x = 55x^2$

**18.** $33x^2 = -11x$

**19.** $14t^2 = 3t$

**20.** $6m = 19m^2$

**21.** $5y^2 - 3y^2 = 72y + 9y$

**22.** $63p - 16p^2 = 17p + 58p^2$

**c** Solve.

**23.** $x^2 + 8x - 48 = 0$

**24.** $x^2 - 16x + 48 = 0$

**25.** $5 + 6x + x^2 = 0$

**26.** $x^2 + 10 + 11x = 0$

**27.** $18 = 7p + p^2$

**28.** $t^2 + 14t = -24$

**29.** $-15 = -8y + y^2$

**30.** $q^2 + 14 = 9q$

**31.** $x^2 + 10x + 25 = 0$

**32.** $x^2 + 6x + 9 = 0$

**33.** $r^2 = 8r - 16$

**34.** $x^2 + 1 = 2x$

**35.** $6x^2 + x - 2 = 0$

**36.** $2x^2 - 11x + 15 = 0$

**37.** $3a^2 = 10a + 8$

**38.** $15b - 9b^2 = 4$

**39.** $6x^2 - 4x = 10$

**40.** $3x^2 - 7x = 20$

**41.** $2t^2 + 12t = -10$

**42.** $12w^2 - 5w = 2$

**43.** $t(t - 5) = 14$

**44.** $6z^2 + z - 1 = 0$

**45.** $t(9 + t) = 4(2t + 5)$

**46.** $3y^2 + 8y = 12y + 15$

**47.** $16(p - 1) = p(p + 8)$

**48.** $(2x - 3)(x + 1) = 4(2x - 3)$

**49.** $(t - 1)(t + 3) = t - 1$

**50.** $(x - 2)(x + 2) = x + 2$

Solve.

**51.** $\dfrac{24}{x-2} + \dfrac{24}{x+2} = 5$

**52.** $\dfrac{8}{x+2} + \dfrac{8}{x-2} = 3$

**53.** $\dfrac{1}{x} + \dfrac{1}{x+6} = \dfrac{1}{4}$

**54.** $\dfrac{1}{x} + \dfrac{1}{x+9} = \dfrac{1}{20}$

**55.** $1 + \dfrac{12}{x^2-4} = \dfrac{3}{x-2}$

**56.** $\dfrac{5}{t-3} - \dfrac{30}{t^2-9} = 1$

**57.** $\dfrac{r}{r-1} + \dfrac{2}{r^2-1} = \dfrac{8}{r+1}$

**58.** $\dfrac{x+2}{x^2-2} = \dfrac{2}{3-x}$

**59.** $\dfrac{x-1}{1-x} = -\dfrac{x+8}{x-8}$

**60.** $\dfrac{4-x}{x-4} + \dfrac{x+3}{x-3} = 0$

**61.** $\dfrac{5}{y+4} - \dfrac{3}{y-2} = 4$

**62.** $\dfrac{2z+11}{2z+8} = \dfrac{3z-1}{z-1}$

[d] Solve.

**63.** A decagon is a figure with 10 sides. How many diagonals does a decagon have?

**64.** A hexagon is a figure with 6 sides. How many diagonals does a hexagon have?

**65.** A polygon has 14 diagonals. How many sides does it have?

**66.** A polygon has 9 diagonals. How many sides does it have?

---

**Skill Maintenance**

Simplify.   [9.1a], [9.2a]

**67.** $\sqrt{64}$

**68.** $-\sqrt{169}$

**69.** $\sqrt{8}$

**70.** $\sqrt{12}$

**71.** $\sqrt{20}$

**72.** $\sqrt{88}$

**73.** $\sqrt{405}$

**74.** $\sqrt{1020}$

Use a calculator to approximate the square roots. Round to three decimal places.   [9.1b]

**75.** $\sqrt{7}$

**76.** $\sqrt{23}$

**77.** $\sqrt{\dfrac{7}{3}}$

**78.** $\sqrt{524.77}$

---

**Synthesis**

**79.** ◈ Explain how the graph of $y = (x-2)(x+3)$ is related to the solutions of the equation $(x-2)(x+3) = 0$.

**80.** ◈ Explain how you might go about constructing a quadratic equation whose solutions are $-5$ and $7$.

Solve.

**81.** $4m^2 - (m+1)^2 = 0$

**82.** $x^2 + \sqrt{22}\,x = 0$

**83.** $\sqrt{5}x^2 - x = 0$

**84.** $\sqrt{7}x^2 + \sqrt{3}x = 0$

[◠◡] Use a grapher to solve the equation.

**85.** $3x^2 - 7x = 20$

**86.** $x(x-5) = 14$

**87.** $3x^2 + 8x = 12x + 15$

**88.** $(x-2)(x+2) = x + 2$

**89.** $(x-2)^2 + 3(x-2) = 4$

**90.** $(x+3)^2 = 4$

**91.** $16(x-1) = x(x+8)$

**92.** $x^2 + 2.5x + 1.5625 = 9.61$

---

# 10.2 Solving Quadratic Equations by Completing the Square

## a Solving Quadratic Equations of the Type $ax^2 = p$

For equations of the type $ax^2 = p$, we first solve for $x^2$ and then apply the *principle of square roots,* which states that a positive number has two square roots. The number 0 has one square root, 0.

> **THE PRINCIPLE OF SQUARE ROOTS**
>
> - The equation $x^2 = d$ has two real solutions when $d > 0$. The solutions are $\sqrt{d}$ and $-\sqrt{d}$.
> - The equation $x^2 = 0$ has 0 as its only solution.
> - The equation $x^2 = d$ has no real-number solution when $d < 0$.

**Example 1** Solve: $x^2 = 3$.

$$x^2 = 3$$
$$x = \sqrt{3} \quad or \quad x = -\sqrt{3} \qquad \text{Using the principle of square roots}$$

**CHECK:**

For $\sqrt{3}$:

$$\frac{x^2 = 3}{(\sqrt{3})^2 \; ? \; 3}$$
$$3 \; | \qquad \text{TRUE}$$

For $-\sqrt{3}$:

$$\frac{x^2 = 3}{(-\sqrt{3})^2 \; ? \; 3}$$
$$3 \; | \qquad \text{TRUE}$$

The solutions are $\sqrt{3}$ and $-\sqrt{3}$.

*Do Exercise 1.*

**Example 2** Solve: $\frac{1}{8}x^2 = 0$.

$$\frac{1}{8}x^2 = 0$$
$$x^2 = 0 \qquad \text{Multiplying by 8}$$
$$x = 0 \qquad \text{Using the principle of square roots}$$

The solution is 0.

*Do Exercise 2.*

**Example 3** Solve: $-3x^2 + 7 = 0$.

$$-3x^2 + 7 = 0$$
$$-3x^2 = -7 \qquad \text{Subtracting 7}$$
$$x^2 = \frac{-7}{-3} \qquad \text{Dividing by } -3$$
$$x^2 = \frac{7}{3}$$
$$x = \sqrt{\frac{7}{3}} \quad or \quad x = -\sqrt{\frac{7}{3}} \qquad \text{Using the principle of square roots}$$
$$x = \sqrt{\frac{7}{3} \cdot \frac{3}{3}} \quad or \quad x = -\sqrt{\frac{7}{3} \cdot \frac{3}{3}} \qquad \text{Rationalizing the denominators}$$
$$x = \frac{\sqrt{21}}{3} \quad or \quad x = -\frac{\sqrt{21}}{3}$$

## Objectives

**a** Solve quadratic equations of the type $ax^2 = p$.

**b** Solve quadratic equations of the type $(x + c)^2 = d$.

**c** Solve quadratic equations by completing the square.

**d** Solve certain problems involving quadratic equations of the type $ax^2 = p$.

## For Extra Help

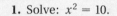

TAPE 16    TAPE 17B    MAC    CD-ROM
WIN

1. Solve: $x^2 = 10$.

2. Solve: $6x^2 = 0$.

3. Solve: $2x^2 - 3 = 0$.

*Answers on page A-39*

## Calculator Spotlight

Solving Equations.

Example 1 can be visualized by letting

$$y_1 = x^2 \quad \text{and} \quad y_2 = 3$$

and finding the first coordinate of the points of intersection. We can use the INTERSECT feature to approximate the solutions.

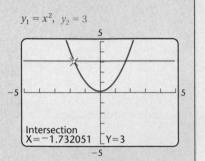

$y_1 = x^2, \ y_2 = 3$

Intersection
X=−1.732051   Y=3

Note that

$$-\sqrt{3} \approx -1.732.$$

We would then repeat the procedure near the right-hand solution to approximate the other solution. The ZERO feature can also be used with the equation $x^2 - 3 = 0$.

### Exercises

Use your grapher to approximate the solutions.

**1.** $x^2 = 10$

**2.** $-3x^2 + 7 = 0$

Solve.

**4.** $(x - 3)^2 = 16$

**5.** $(x + 4)^2 = 11$

*Answers on page A-39*

CHECK:     For $\dfrac{\sqrt{21}}{3}$:

$$\begin{array}{c} -3x^2 + 7 = 0 \\ \hline -3\left(\frac{\sqrt{21}}{3}\right)^2 + 7 \ ? \ 0 \\ -3 \cdot \frac{21}{9} + 7 \\ -7 + 7 \\ 0 \end{array} \quad \text{TRUE}$$

For $-\dfrac{\sqrt{21}}{3}$:

$$\begin{array}{c} -3x^2 + 7 = 0 \\ \hline -3\left(-\frac{\sqrt{21}}{3}\right)^2 + 7 \ ? \ 0 \\ -3 \cdot \frac{21}{9} + 7 \\ -7 + 7 \\ 0 \end{array} \quad \text{TRUE}$$

The solutions are $\dfrac{\sqrt{21}}{3}$ and $-\dfrac{\sqrt{21}}{3}$.

*Do Exercise 3 on the preceding page.*

## b | Solving Quadratic Equations of the Type $(x + c)^2 = d$

In an equation of the type $(x + c)^2 = d$, we have the square of a binomial equal to a constant. We can use the principle of square roots to solve such an equation.

**Example 4**   Solve: $(x - 5)^2 = 9$.

$$(x - 5)^2 = 9$$
$$x - 5 = 3 \quad or \quad x - 5 = -3 \qquad \text{Using the principle of square roots}$$
$$x = 8 \quad or \qquad\quad x = 2$$

The solutions are 8 and 2.

**Example 5**   Solve: $(x + 2)^2 = 7$.

$$(x + 2)^2 = 7$$
$$x + 2 = \sqrt{7} \qquad or \quad x + 2 = -\sqrt{7} \qquad \text{Using the principle of square roots}$$
$$x = -2 + \sqrt{7} \quad or \qquad x = -2 - \sqrt{7}$$

The solutions are $-2 + \sqrt{7}$ and $-2 - \sqrt{7}$, or simply $-2 \pm \sqrt{7}$ (read "$-2$ plus or minus $\sqrt{7}$").

*Do Exercises 4 and 5.*

In Examples 4 and 5, the left sides of the equations are squares of binomials. If we can express an equation in such a form, we can proceed as we did in those examples.

**Example 6**   Solve: $x^2 + 8x + 16 = 49$.

$$x^2 + 8x + 16 = 49 \qquad \text{The left side is the square of a binomial.}$$
$$(x + 4)^2 = 49$$
$$x + 4 = 7 \quad or \quad x + 4 = -7 \qquad \text{Using the principle of square roots}$$
$$x = 3 \quad or \qquad x = -11$$

The solutions are 3 and $-11$.

*Do Exercises 6 and 7 on the following page.*

## c | Completing the Square

We have seen that a quadratic equation like $(x - 5)^2 = 9$ can be solved by using the principle of square roots. We also noted that an equation like $x^2 + 8x + 16 = 49$ can be solved in the same manner because the expression on the left side is the square of a binomial, $(x + 4)^2$. This second procedure is the basis for a method of solving quadratic equations called **completing the square.** *It can be used to solve any quadratic equation.*

Suppose we have the following quadratic equation:

$$x^2 + 10x = 4.$$

If we could add to both sides of the equation a constant that would make the expression on the left the square of a binomial, we could then solve the equation using the principle of square roots.

How can we determine what to add to $x^2 + 10x$ in order to construct the square of a binomial? We want to find a number $a$ such that the following equation is satisfied:

$$(x + a)(x + a) = x^2 + 10x + a^2$$
$$ax + ax = 2ax$$

Thus, $a$ is such that $2ax = 10x$. Solving for $a$, we get

$$a = \frac{10x}{2x} = \frac{10}{2} = 5;$$

that is, $a$ is half of the coefficient of $x$ in $x^2 + 10x$. Since $a^2 = \left(\frac{10}{2}\right)^2 = 5^2 = 25$, we add 25 to our original expression:

$$x^2 + 10x + 25 \text{ is the square of } x + 5;$$

that is,

$$x^2 + 10x + 25 = (x + 5)^2.$$

> To **complete the square** of an expression like $x^2 + bx$, we take half of the coefficient of $x$ and square. Then we add that number, which is $(b/2)^2$.

Returning to solve our original equation, we first add 25 on both sides to complete the square. Then we solve as follows:

$$
\begin{array}{ll}
x^2 + 10x = 4 & \text{Original equation} \\
x^2 + 10x + 25 = 4 + 25 & \text{Adding 25: } \left(\frac{10}{2}\right)^2 = 5^2 = 25 \\
(x + 5)^2 = 29 & \\
x + 5 = \sqrt{29} \quad or \quad x + 5 = -\sqrt{29} & \text{Using the principle of square roots} \\
x = -5 + \sqrt{29} \quad or \quad x = -5 - \sqrt{29}. &
\end{array}
$$

The solutions are $-5 \pm \sqrt{29}$.

We have seen that a quadratic equation $(x + c)^2 = d$ can be solved by using the principle of square roots. Any quadratic equation can be put in this form by completing the square. Then we can solve as before.

Solve.

**6.** $x^2 - 6x + 9 = 64$

**7.** $x^2 - 2x + 1 = 5$

*Answers on page A-39*

Solve.

**8.** $x^2 - 6x + 8 = 0$

**9.** $x^2 + 8x - 20 = 0$

**10.** Solve: $x^2 - 12x + 23 = 0$.

**11.** Solve: $x^2 - 3x - 10 = 0$.

Answers on page A-39

**Example 7**   Solve: $x^2 + 6x + 8 = 0$.

We have

$$x^2 + 6x + 8 = 0$$
$$x^2 + 6x \quad\ = -8. \qquad \textbf{Subtracting 8}$$

We take half of 6 and square it, to get 9. Then we add 9 on *both* sides of the equation. This makes the left side the square of a binomial. We have now completed the square.

$$x^2 + 6x + 9 = -8 + 9 \qquad \textbf{Adding 9}$$
$$(x + 3)^2 = 1$$
$$x + 3 = 1 \quad or \quad x + 3 = -1 \qquad \textbf{Using the principle of square roots}$$
$$x = -2 \quad or \qquad\ x = -4$$

The solutions are $-2$ and $-4$.

*Do Exercises 8 and 9.*

**Example 8**   Solve $x^2 - 4x - 7 = 0$ by completing the square.

We have

$$x^2 - 4x - 7 = 0$$
$$x^2 - 4x \quad\ = 7 \qquad\qquad \textbf{Adding 7}$$
$$x^2 - 4x + 4 = 7 + 4 \qquad\qquad \textbf{Adding 4: } \left(\tfrac{-4}{2}\right)^2 = (-2)^2 = 4$$
$$(x - 2)^2 = 11$$
$$x - 2 = \sqrt{11} \quad or \quad x - 2 = -\sqrt{11} \qquad \textbf{Using the principle of square roots}$$
$$x = 2 + \sqrt{11} \quad or \qquad x = 2 - \sqrt{11}.$$

The solutions are $2 \pm \sqrt{11}$.

*Do Exercise 10.*

Example 7, as well as the following example, can be solved more easily by factoring. We solved it by completing the square only to illustrate that completing the square can be used to solve *any* quadratic equation.

**Example 9**   Solve $x^2 + 3x - 10 = 0$ by completing the square.

We have

$$x^2 + 3x - 10 = 0$$
$$x^2 + 3x \qquad\ = 10$$
$$x^2 + 3x + \tfrac{9}{4} = 10 + \tfrac{9}{4} \qquad \textbf{Adding } \tfrac{9}{4}: \left(\tfrac{3}{2}\right)^2 = \tfrac{9}{4}$$
$$\left(x + \tfrac{3}{2}\right)^2 = \tfrac{40}{4} + \tfrac{9}{4} = \tfrac{49}{4}$$
$$x + \tfrac{3}{2} = \tfrac{7}{2} \quad or \quad x + \tfrac{3}{2} = -\tfrac{7}{2} \qquad \textbf{Using the principle of square roots}$$
$$x = \tfrac{4}{2} \quad or \qquad x = -\tfrac{10}{2}$$
$$x = 2 \quad or \qquad x = -5.$$

The solutions are $2$ and $-5$.

*Do Exercise 11.*

When the coefficient of $x^2$ is not 1, we can make it 1, as shown in the following example.

**Example 10**  Solve $2x^2 = 3x + 1$ by completing the square.

We first obtain standard form. Then we multiply on both sides by $\frac{1}{2}$ to make the $x^2$-coefficient 1.

$$2x^2 = 3x + 1$$

$$2x^2 - 3x - 1 = 0 \qquad \text{Finding standard form}$$

$$\frac{1}{2}(2x^2 - 3x - 1) = \frac{1}{2} \cdot 0 \qquad \text{Multiplying by } \frac{1}{2} \text{ to make the } x^2\text{-coefficient 1}$$

$$x^2 - \frac{3}{2}x - \frac{1}{2} = 0$$

$$x^2 - \frac{3}{2}x = \frac{1}{2} \qquad \text{Adding } \frac{1}{2}$$

$$x^2 - \frac{3}{2}x + \frac{9}{16} = \frac{1}{2} + \frac{9}{16} \qquad \text{Adding } \frac{9}{16} \colon \left[\frac{1}{2}\left(-\frac{3}{2}\right)\right]^2 = \left[-\frac{3}{4}\right]^2 = \frac{9}{16}$$

$$\left(x - \frac{3}{4}\right)^2 = \frac{8}{16} + \frac{9}{16} \qquad \text{Finding a common denominator}$$

$$\left(x - \frac{3}{4}\right)^2 = \frac{17}{16}$$

$$x - \frac{3}{4} = \frac{\sqrt{17}}{4} \qquad or \quad x - \frac{3}{4} = -\frac{\sqrt{17}}{4} \qquad \text{Using the principle of square roots}$$

$$x = \frac{3}{4} + \frac{\sqrt{17}}{4} \qquad or \qquad x = \frac{3}{4} - \frac{\sqrt{17}}{4}$$

The solutions are $\dfrac{3 \pm \sqrt{17}}{4}$.

---

**SOLVING BY COMPLETING THE SQUARE**

To solve a quadratic equation $ax^2 + bx + c = 0$ by completing the square:

1. If $a \neq 1$, multiply by $1/a$ so that the $x^2$-coefficient is 1.

2. If the $x^2$-coefficient is 1, add so that the equation is in the form

$$x^2 + bx = -c, \quad \text{or} \quad x^2 + \frac{b}{a}x = -\frac{c}{a} \quad \text{if step (1) has been applied.}$$

3. Take half of the $x$-coefficient and square it. Add the result on both sides of the equation.

4. Express the side with the variables as the square of a binomial.

5. Use the principle of square roots and complete the solution.

*Do Exercise 12.*

**12.** Solve: $2x^2 + 3x - 3 = 0$.

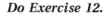

**Calculator Spotlight**

**Exercises**

Use your grapher to find the solutions of the equation. Approximate to the nearest thousandth.

1. $x^2 + 10x = 4$
2. $x^2 - 4x - 7 = 0$
3. $2x^2 = 3x + 1$
4. $3 = 3x + 2x^2$

*Answer on page A-39*

**13.** The Transco Tower in Houston is 901 ft tall. How long would it take an object to fall to the ground from the top?

## d | Applications

**Example 11** *Falling Object.* The World Trade Center in New York is 1368 ft tall. How long would it take an object to fall to the ground from the top?

1. **Familiarize.** If we did not know anything about this problem, we might consider looking up a formula in a mathematics or physics book. A formula that fits this situation is

$$s = 16t^2,$$

where $s$ is the distance, in feet, traveled by a body falling freely from rest in $t$ seconds. This formula is actually an approximation in that it does not account for air resistance. In this problem, we know the distance $s$ to be 1368 ft. We want to determine the time $t$ for the object to reach the ground.

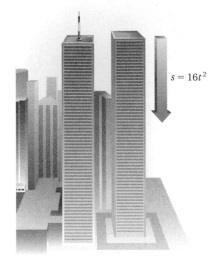

$s = 16t^2$

2. **Translate.** We know that the distance is 1368 and that we need to solve for $t$. We substitute 1368 for $s$:

$$1368 = 16t^2.$$

This gives us a translation.

3. **Solve.** We solve the equation:

$$1368 = 16t^2$$

$$\frac{1368}{16} = t^2 \qquad \text{Solving for } t^2$$

$$85.5 = t^2 \qquad \text{Dividing}$$

$$\sqrt{85.5} = t \quad or \quad -\sqrt{85.5} = t \qquad \text{Using the principle of square roots}$$

$$9.2 \approx t \quad or \quad -9.2 \approx t. \qquad \text{Using a calculator to find the square root and rounding to the nearest tenth}$$

4. **Check.** The number $-9.2$ cannot be a solution because time cannot be negative in this situation. We substitute 9.2 in the original equation:

$$s = 16(9.2)^2 = 16(84.64) = 1354.24.$$

This is close; $1354.24 \approx 1368$. Remember that we approximated a solution, $t \approx 9.2$. Thus we have a check.

5. **State.** It takes about 9.2 sec for an object to fall to the ground from the top of the World Trade Center.

*Do Exercise 13.*

*Answer on page A-39*

# Exercise Set 10.2

**a** Solve.

**1.** $x^2 = 121$

**2.** $x^2 = 100$

**3.** $5x^2 = 35$

**4.** $5x^2 = 45$

**5.** $5x^2 = 3$

**6.** $2x^2 = 9$

**7.** $4x^2 - 25 = 0$

**8.** $9x^2 - 4 = 0$

**9.** $3x^2 - 49 = 0$

**10.** $5x^2 - 16 = 0$

**11.** $4y^2 - 3 = 9$

**12.** $36y^2 - 25 = 0$

**13.** $49y^2 - 64 = 0$

**14.** $8x^2 - 400 = 0$

**b** Solve.

**15.** $(x + 3)^2 = 16$

**16.** $(x - 4)^2 = 25$

**17.** $(x + 3)^2 = 21$

**18.** $(x - 3)^2 = 6$

**19.** $(x + 13)^2 = 8$

**20.** $(x - 13)^2 = 64$

**21.** $(x - 7)^2 = 12$

**22.** $(x + 1)^2 = 14$

**23.** $(x + 9)^2 = 34$

**24.** $(t + 5)^2 = 49$

**25.** $\left(x + \frac{3}{2}\right)^2 = \frac{7}{2}$

**26.** $\left(y - \frac{3}{4}\right)^2 = \frac{17}{16}$

**27.** $x^2 - 6x + 9 = 64$

**28.** $p^2 - 10p + 25 = 100$

**29.** $x^2 + 14x + 49 = 64$

**30.** $t^2 + 8t + 16 = 36$

**c** Solve by completing the square. Show your work.

**31.** $x^2 - 6x - 16 = 0$

**32.** $x^2 + 8x + 15 = 0$

**33.** $x^2 + 22x + 21 = 0$

**34.** $x^2 + 14x - 15 = 0$

**35.** $x^2 - 2x - 5 = 0$

**36.** $x^2 - 4x - 11 = 0$

**37.** $x^2 - 22x + 102 = 0$

**38.** $x^2 - 18x + 74 = 0$

**39.** $x^2 + 10x - 4 = 0$

**40.** $x^2 - 10x - 4 = 0$

**41.** $x^2 - 7x - 2 = 0$

**42.** $x^2 + 7x - 2 = 0$

**43.** $x^2 + 3x - 28 = 0$

**44.** $x^2 - 3x - 28 = 0$

**45.** $x^2 + \frac{3}{2}x - \frac{1}{2} = 0$

**46.** $x^2 - \frac{3}{2}x - 2 = 0$

**47.** $2x^2 + 3x - 17 = 0$

**48.** $2x^2 - 3x - 1 = 0$

**49.** $3x^2 + 4x - 1 = 0$

**50.** $3x^2 - 4x - 3 = 0$

**51.** $2x^2 = 9x + 5$

**52.** $2x^2 = 5x + 12$

**53.** $6x^2 + 11x = 10$

**54.** $4x^2 + 12x = 7$

| d | Solve.

**55.** *Sears Tower.* The height of the Sears Tower in Chicago is 1451 ft (excluding TV towers and antennas). How long would it take an object to fall to the ground from the top?

**56.** *Library Square Tower.* Library Square Tower in Los Angeles is 1012 ft tall. How long would it take an object to fall to the ground from the top?

**57.** *Free-Fall Record.* The world record for free-fall to the ground, by a man without a parachute, is 311 ft and is held by Dar Robinson. Approximately how long did the fall take?

**58.** *Free-Fall Record.* The world record for free-fall to the ground, by a woman without a parachute, into a cushioned landing area is 175 ft and is held by Kitty O'Neill. Approximately how long did the fall take?

---

**Skill Maintenance**

**59.** Find an equation of variation in which $y$ varies inversely as $x$ and $y = 235$ when $x = 0.6$.  [7.5c]

**60.** The time $T$ to do a certain job varies inversely as the number $N$ of people working. It takes 5 hr for 24 people to wash and wax the floors in a building. How long would it take 36 people to do the job?  [7.5d]

Multiply and simplify.  [9.2c]

**61.** $\sqrt{3x} \cdot \sqrt{6x}$

**62.** $\sqrt{8x^2} \cdot \sqrt{24x^3}$

**63.** $3\sqrt{t} \cdot \sqrt{t}$

**64.** $\sqrt{x^2} \cdot \sqrt{x^5}$

---

**Synthesis**

**65.** ◆ Corey asserts that the solution of a quadratic equation is $3 \pm \sqrt{14}$ and states that there is only one solution. What mistake is being made?

**66.** ◆ If a quadratic equation can be solved by factoring, what type of number(s) will generally be solutions?

Find $b$ such that the trinomial is a square.

**67.** $x^2 + bx + 36$

**68.** $x^2 + bx + 55$

**69.** $x^2 + bx + 128$

**70.** $4x^2 + bx + 16$

**71.** $x^2 + bx + c$

**72.** $ax^2 + bx + c$

Solve.

**73.** ⊡ $4.82x^2 = 12,000$

**74.** $\dfrac{x}{2} = \dfrac{32}{x}$

**75.** $\dfrac{x}{9} = \dfrac{36}{4x}$

**76.** $\dfrac{4}{m^2 - 7} = 1$

Collaborative
Learning Manual

Visualize completion of the square using rectangles.

Copyright © 1999 Addison Wesley Longman

## 10.3 The Quadratic Formula

We learn to complete the square to prove a general formula that can be used to solve quadratic equations even when they cannot be solved by factoring.

### a | Solving Using the Quadratic Formula

Each time you solve by completing the square, you perform nearly the same steps. When we repeat the same kind of computation many times, we look for a formula so we can speed up our work. Consider

$$ax^2 + bx + c = 0, \quad a > 0.$$

Let's solve by completing the square. As we carry out the steps, compare them with Example 10 in the preceding section.

$$x^2 + \frac{b}{a}x + \frac{c}{a} = 0 \qquad \text{Multiplying by } \frac{1}{a}$$

$$x^2 + \frac{b}{a}x \quad = -\frac{c}{a} \qquad \text{Adding } -\frac{c}{a}$$

Half of $\frac{b}{a}$ is $\frac{b}{2a}$. The square is $\frac{b^2}{4a^2}$. Thus we add $\frac{b^2}{4a^2}$ on both sides.

$$x^2 + \frac{b}{a}x + \frac{b^2}{4a^2} = -\frac{c}{a} + \frac{b^2}{4a^2} \qquad \text{Adding } \frac{b^2}{4a^2}$$

$$\left(x + \frac{b}{2a}\right)^2 = -\frac{4ac}{4a^2} + \frac{b^2}{4a^2} \qquad \text{\small Factoring the left side and finding a common denominator on the right}$$

$$\left(x + \frac{b}{2a}\right)^2 = \frac{b^2 - 4ac}{4a^2}$$

$$x + \frac{b}{2a} = \sqrt{\frac{b^2 - 4ac}{4a^2}} \quad \text{or} \quad x + \frac{b}{2a} = -\sqrt{\frac{b^2 - 4ac}{4a^2}} \qquad \text{\small Using the principle of square roots}$$

Since $a > 0$, $\sqrt{4a^2} = 2a$, so we can simplify as follows:

$$x + \frac{b}{2a} = \frac{\sqrt{b^2 - 4ac}}{2a} \quad \text{or} \quad x + \frac{b}{2a} = -\frac{\sqrt{b^2 - 4ac}}{2a}.$$

Thus,

$$x = -\frac{b}{2a} + \frac{\sqrt{b^2 - 4ac}}{2a} \quad \text{or} \quad x = -\frac{b}{2a} - \frac{\sqrt{b^2 - 4ac}}{2a},$$

so

$$x = -\frac{b}{2a} \pm \frac{\sqrt{b^2 - 4ac}}{2a},$$

or

$$x = \frac{-b \pm \sqrt{b^2 - 4ac}}{2a}.$$

We now have the following.

▶ **THE QUADRATIC FORMULA**

The solutions of $ax^2 + bx + c = 0$ are given by

$$x = \frac{-b \pm \sqrt{b^2 - 4ac}}{2a}.$$

**Objectives**

a  Solve quadratic equations using the quadratic formula.

b  Find approximate solutions of quadratic equations using a calculator.

**For Extra Help**

TAPE 16     TAPE 18A     MAC     CD-ROM
WIN

**1.** Solve using the quadratic formula:

$$2x^2 = 4 - 7x.$$

Note that the formula also holds when $a < 0$. A similar proof would show this, but we will not consider it here.

**Example 1**  Solve $5x^2 - 8x = -3$ using the quadratic formula.

We first find standard form and determine $a$, $b$, and $c$:

$$5x^2 - 8x + 3 = 0;$$
$$a = 5, \quad b = -8, \quad c = 3.$$

We then use the quadratic formula:

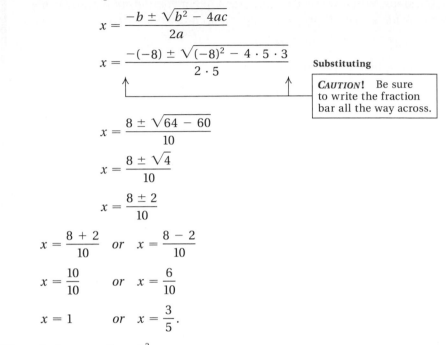

$$x = \frac{-b \pm \sqrt{b^2 - 4ac}}{2a}$$

$$x = \frac{-(-8) \pm \sqrt{(-8)^2 - 4 \cdot 5 \cdot 3}}{2 \cdot 5} \quad \text{Substituting}$$

*CAUTION!*  Be sure to write the fraction bar all the way across.

$$x = \frac{8 \pm \sqrt{64 - 60}}{10}$$

$$x = \frac{8 \pm \sqrt{4}}{10}$$

$$x = \frac{8 \pm 2}{10}$$

$$x = \frac{8 + 2}{10} \quad or \quad x = \frac{8 - 2}{10}$$

$$x = \frac{10}{10} \quad or \quad x = \frac{6}{10}$$

$$x = 1 \quad or \quad x = \frac{3}{5}.$$

The solutions are 1 and $\frac{3}{5}$.

*Do Exercise 1.*

It would have been easier to solve the equation in Example 1 by factoring. We used the quadratic formula only to illustrate that it can be used to solve any quadratic equation. The following is a general procedure for solving a quadratic equation.

---

### Calculator Spotlight

#### 〰 Exercises

Use your grapher to find the solutions of the equation. Approximate to the nearest hundredth.

**1.** $x^2 + 3x - 10 = 0$

**2.** $x^2 = 4x + 7$

**3.** $5x^2 - 8x = -3$

**4.** $2x^2 = 4 - 7x$

---

**SOLVING USING THE QUADRATIC FORMULA**

To solve a quadratic equation:

**1.** Check to see if it is in the form $ax^2 = p$ or $(x + c)^2 = d$. If it is, use the principle of square roots as in Section 10.2.

**2.** If it is not in the form of (1), write it in standard form, $ax^2 + bx + c = 0$ with $a$ and $b$ nonzero.

**3.** Then try factoring.

**4.** If it is not possible to factor or if factoring seems difficult, use the quadratic formula.

The solutions of a quadratic equation can always be found using the quadratic formula. They cannot always be found by factoring. (When $b^2 - 4ac \geq 0$, the equation has real-number solutions. When $b^2 - 4ac < 0$, the equation has no real-number solutions.)

---

*Answer on page A-39*

**Example 2**   Solve $x^2 + 3x - 10 = 0$ using the quadratic formula.

The equation is in standard form. So we determine $a$, $b$, and $c$:

$$x^2 + 3x - 10 = 0;$$
$$a = 1, \quad b = 3, \quad c = -10.$$

We then use the quadratic formula:

$$x = \frac{-b \pm \sqrt{b^2 - 4ac}}{2a}$$

$$= \frac{-3 \pm \sqrt{3^2 - 4 \cdot 1 \cdot (-10)}}{2 \cdot 1} \qquad \text{Substituting}$$

$$= \frac{-3 \pm \sqrt{9 + 40}}{2}$$

$$= \frac{-3 \pm \sqrt{49}}{2} = \frac{-3 \pm 7}{2}.$$

Thus,

$$x = \frac{-3 + 7}{2} = \frac{4}{2} = 2 \quad \text{or} \quad x = \frac{-3 - 7}{2} = \frac{-10}{2} = -5.$$

The solutions are 2 and $-5$.

Note that the radicand ($b^2 - 4ac = 49$) in the quadratic formula is a perfect square, so we could have used factoring to solve.

*Do Exercise 2.*

**Example 3**   Solve $x^2 = 4x + 7$ using the quadratic formula. Compare with Example 8 in Section 10.2.

We first find standard form and determine $a$, $b$, and $c$:

$$x^2 - 4x - 7 = 0;$$
$$a = 1, \quad b = -4, \quad c = -7.$$

We then use the quadratic formula:

$$x = \frac{-b \pm \sqrt{b^2 - 4ac}}{2a} = \frac{-(-4) \pm \sqrt{(-4)^2 - 4 \cdot 1 \cdot (-7)}}{2 \cdot 1} \qquad \text{Substituting}$$

$$= \frac{4 \pm \sqrt{16 + 28}}{2} = \frac{4 \pm \sqrt{44}}{2}$$

$$= \frac{4 \pm \sqrt{4 \cdot 11}}{2} = \frac{4 \pm \sqrt{4}\sqrt{11}}{2}$$

$$= \frac{4 \pm 2\sqrt{11}}{2} = \frac{2 \cdot 2 \pm 2\sqrt{11}}{2 \cdot 1} \qquad \begin{array}{l}\text{Factoring out 2 in the numerator}\\\text{and the denominator}\end{array}$$

$$= \frac{2(2 \pm \sqrt{11})}{2 \cdot 1} = \frac{2}{2} \cdot \frac{2 \pm \sqrt{11}}{1}$$

$$= 2 \pm \sqrt{11}.$$

The solutions are $2 + \sqrt{11}$ and $2 - \sqrt{11}$, or $2 \pm \sqrt{11}$.

*Do Exercise 3.*

**2.** Solve using the quadratic formula:
$$x^2 - 3x - 10 = 0.$$

**3.** Solve using the quadratic formula:
$$x^2 + 4x = 7.$$

*Answers on page A-39*

**4.** Solve using the quadratic formula:

$$x^2 = x - 1.$$

## Calculator Spotlight

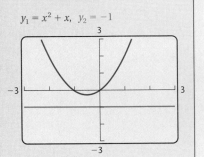 To see that no real-number solutions exist for Example 4, we let

$$y_1 = x^2 + x \quad \text{and} \quad y_2 = -1$$

and graph the equations in the same window.

$y_1 = x^2 + x, \quad y_2 = -1$

We see that the graphs do not intersect. This supports the conclusion that no real-number solution exists.

### Exercises

**1.** What happens when the INTERSECT feature is used with these graphs?

**2.** How can the graph of $y = x^2 + x + 1$ tell us that no real-number solution exists?

**3.** Use a grapher to solve the equation $x^2 = x - 1$.

**5.** Solve using the quadratic formula:

$$5x^2 - 8x = 3.$$

**6.** Approximate the solutions to the equation in Margin Exercise 5. Round to the nearest tenth.

*Answers on page A-39*

**Example 4**   Solve $x^2 + x = -1$ using the quadratic formula.

We first find standard form and determine $a$, $b$, and $c$:

$$x^2 + x + 1 = 0;$$
$$a = 1, \quad b = 1, \quad c = 1.$$

We then use the quadratic formula:

$$x = \frac{-b \pm \sqrt{b^2 - 4ac}}{2a} = \frac{-1 \pm \sqrt{1^2 - 4 \cdot 1 \cdot 1}}{2 \cdot 1} = \frac{-1 \pm \sqrt{-3}}{2}.$$

Note that the radicand ($b^2 - 4ac = -3$) in the quadratic formula is negative. Thus there are no real-number solutions because square roots of negative numbers do not exist as real numbers.

*Do Exercise 4.*

**Example 5**   Solve $3x^2 = 7 - 2x$ using the quadratic formula.

We first find standard form and determine $a$, $b$, and $c$:

$$3x^2 + 2x - 7 = 0;$$
$$a = 3, \quad b = 2, \quad c = -7.$$

We then use the quadratic formula:

$$x = \frac{-b \pm \sqrt{b^2 - 4ac}}{2a} = \frac{-2 \pm \sqrt{2^2 - 4 \cdot 3 \cdot (-7)}}{2 \cdot 3} = \frac{-2 \pm \sqrt{4 + 84}}{2 \cdot 3}$$

$$= \frac{-2 \pm \sqrt{88}}{6} = \frac{-2 \pm \sqrt{4 \cdot 22}}{6} = \frac{-2 \pm \sqrt{4}\sqrt{22}}{6} = \frac{-2 \pm 2\sqrt{22}}{6}$$

$$= \frac{2(-1 \pm \sqrt{22})}{2 \cdot 3} = \frac{2}{2} \cdot \frac{-1 \pm \sqrt{22}}{3} = \frac{-1 \pm \sqrt{22}}{3}.$$

The solutions are $\dfrac{-1 + \sqrt{22}}{3}$ and $\dfrac{-1 - \sqrt{22}}{3}$, or $\dfrac{-1 \pm \sqrt{22}}{3}$.

*Do Exercise 5.*

## **b**   Approximate Solutions

A calculator can be used to approximate solutions.

**Example 6**   Use a calculator to approximate to the nearest tenth the solutions to the equation in Example 5.

Using a calculator , we have

$$\frac{-1 + \sqrt{22}}{3} \approx 1.230138587 \approx 1.2 \text{ to the nearest tenth,} \quad \text{and}$$

$$\frac{-1 - \sqrt{22}}{3} \approx -1.896805253 \approx -1.9 \text{ to the nearest tenth.}$$

The approximate solutions are 1.2 and $-1.9$.

*Do Exercise 6.*

# Exercise Set 10.3

**a** Solve. Try factoring first. If factoring is not possible or is difficult, use the quadratic formula.

**1.** $x^2 - 4x = 21$

**2.** $x^2 + 8x = 9$

**3.** $x^2 = 6x - 9$

**4.** $x^2 = 24x - 144$

**5.** $3y^2 - 2y - 8 = 0$

**6.** $3y^2 - 7y + 4 = 0$

**7.** $4x^2 + 4x = 15$

**8.** $4x^2 + 12x = 7$

**9.** $x^2 - 9 = 0$

**10.** $x^2 - 16 = 0$

**11.** $x^2 - 2x - 2 = 0$

**12.** $x^2 - 2x - 11 = 0$

**13.** $y^2 - 10y + 22 = 0$

**14.** $y^2 + 6y - 1 = 0$

**15.** $x^2 + 4x + 4 = 7$

**16.** $x^2 - 2x + 1 = 5$

**17.** $3x^2 + 8x + 2 = 0$

**18.** $3x^2 - 4x - 2 = 0$

**19.** $2x^2 - 5x = 1$

**20.** $4x^2 + 4x = 5$

**21.** $2y^2 - 2y - 1 = 0$

**22.** $4y^2 + 4y - 1 = 0$

**23.** $2t^2 + 6t + 5 = 0$

**24.** $4y^2 + 3y + 2 = 0$

**25.** $3x^2 = 5x + 4$

**26.** $2x^2 + 3x = 1$

**27.** $2y^2 - 6y = 10$

**28.** $5m^2 = 3 + 11m$

**29.** $\dfrac{x^2}{x+3} - \dfrac{5}{x+3} = 0$

**30.** $\dfrac{x^2}{x-4} - \dfrac{7}{x-4} = 0$

**31.** $x + 2 = \dfrac{3}{x+2}$

**32.** $x - 3 = \dfrac{5}{x-3}$

**33.** $\dfrac{1}{x} + \dfrac{1}{x+1} = \dfrac{1}{3}$

**34.** $\dfrac{1}{x} + \dfrac{1}{x+6} = \dfrac{1}{5}$

**b** Solve using the quadratic formula. Use a calculator to approximate the solutions to the nearest tenth.

**35.** $x^2 - 4x - 7 = 0$

**36.** $x^2 + 2x - 2 = 0$

**37.** $y^2 - 6y - 1 = 0$

**38.** $y^2 + 10y + 22 = 0$

**39.** $4x^2 + 4x = 1$

**40.** $4x^2 = 4x + 1$

**41.** $3x^2 - 8x + 2 = 0$

**42.** $3x^2 + 4x - 2 = 0$

### Skill Maintenance

Add or subtract.   [9.4a]

**43.** $\sqrt{40} - 2\sqrt{10} + \sqrt{90}$

**44.** $\sqrt{54} - \sqrt{24}$

**45.** $\sqrt{18} + \sqrt{50} - 3\sqrt{8}$

**46.** $\sqrt{81x^3} - \sqrt{4x}$

**47.** Simplify: $\sqrt{80}$.   [9.2a]

**48.** Multiply and simplify: $\sqrt{3x^2}\sqrt{9x^3}$.   [9.2c]

**49.** Simplify: $\sqrt{9000x^{10}}$.   [9.2b]

**50.** Rationalize the denominator: $\sqrt{\dfrac{7}{3}}$.   [9.3c]

### Synthesis

**51.** ◆ List a quadratic equation with no real-number solutions. How can you use that equation to find an equation in the form $y = ax^2 + bx + c$ that does not cross the $x$-axis?

**52.** ◆ Under what condition(s) would using the quadratic formula *not* be the easiest way to solve a quadratic equation?

Solve.

**53.** $5x + x(x - 7) = 0$

**54.** $x(3x + 7) - 3x = 0$

**55.** $3 - x(x - 3) = 4$

**56.** $x(5x - 7) = 1$

**57.** $(y + 4)(y + 3) = 15$

**58.** $(y + 5)(y - 1) = 27$

**59.** $x^2 + (x + 2)^2 = 7$

**60.** $x^2 + (x + 1)^2 = 5$

**61.–68.** ⌐⋏⌐ Use a grapher to approximate the solutions of the equations in Exercises 35–42. Compare your answers with those found using the quadratic formula.

# 10.4 Formulas

**a** To solve a formula for a given letter, we try to get the letter alone on one side.

**Example 1** Solve for $h$: $V = 3.5\sqrt{h}$ (the distance to the horizon).

This is a radical equation. Recall that we first isolate the radical. Then we use the principle of squaring.

$$\frac{V}{3.5} = \sqrt{h} \qquad \text{Isolating the radical}$$

$$\left(\frac{V}{3.5}\right)^2 = (\sqrt{h})^2 \qquad \text{Using the principle of squaring (Section 9.5)}$$

$$\frac{V^2}{12.25} = h \qquad \text{Simplifying}$$

**Example 2** Solve for $g$: $T = 2\pi\sqrt{\dfrac{L}{g}}$ (the period of a pendulum).

$$\frac{T}{2\pi} = \sqrt{\frac{L}{g}} \qquad \text{Dividing by } 2\pi \text{ to isolate the radical}$$

$$\left(\frac{T}{2\pi}\right)^2 = \left(\sqrt{\frac{L}{g}}\right)^2 \qquad \text{Using the principle of squaring}$$

$$\frac{T^2}{4\pi^2} = \frac{L}{g}$$

$$gT^2 = 4\pi^2 L \qquad \text{Multiplying by } 4\pi^2 g \text{ to clear fractions}$$

$$g = \frac{4\pi^2 L}{T^2} \qquad \text{Dividing by } T^2 \text{ to get } g \text{ alone}$$

*Do Exercises 1–3.*

In most formulas, the letters represent nonnegative numbers, so we need not use absolute values when taking square roots.

**Example 3** *Torricelli's Theorem.* The speed $v$ of a liquid leaving a bucket from an opening is related to the height $h$ of the top of the liquid above the opening by the formula

$$h = \frac{v^2}{2g}.$$

Solve for $v$.

Since $v^2$ appears by itself and there is no expression involving $v$, we first solve for $v^2$. Then we use the principle of square roots, taking only the nonnegative square root because $v$ is nonnegative.

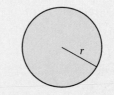

$$2gh = v^2 \qquad \text{Multiplying by } 2g \text{ to clear fractions}$$

$$\sqrt{2gh} = v \qquad \text{Using the principle of square roots.}$$
$$\text{Assume that } v \text{ is nonnegative.}$$

*Do Exercise 4.*

---

## Objective

**a** Solve a formula for a given letter.

**For Extra Help**

TAPE 16   TAPE 18A   MAC WIN   CD-ROM

1. Solve for $L$: $r = 2\sqrt{5L}$ (the speed of a skidding car).

2. Solve for $L$: $T = 2\pi\sqrt{\dfrac{L}{g}}$.

3. Solve for $m$: $c = \sqrt{\dfrac{E}{m}}$.

4. Solve for $r$: $A = \pi r^2$ (the area of a circle).

*Answers on page A-39*

**5.** Solve for $d$: $C = P(d - 1)^2$.

**6.** Solve for $n$: $N = n^2 - n$.

**7.** Solve for $t$: $h = vt + 8t^2$.

*Answers on page A-39*

**Example 4**  Solve for $r$: $A = P(1 + r)^2$ (a compound-interest formula).

$$A = P(1 + r)^2$$

$$\frac{A}{P} = (1 + r)^2 \qquad \text{Dividing by } P$$

$$\sqrt{\frac{A}{P}} = 1 + r \qquad \begin{array}{l}\text{Using the principle of square roots.}\\ \text{Assume that } 1 + r \text{ is nonnegative.}\end{array}$$

$$-1 + \sqrt{\frac{A}{P}} = r \qquad \text{Subtracting 1 to get } r \text{ alone}$$

*Do Exercise 5.*

Sometimes we must use the quadratic formula to solve a formula for a certain letter.

**Example 5**  Solve for $n$: $d = \dfrac{n^2 - 3n}{2}$, where $d$ is the number of diagonals of an $n$-sided polygon.

This time there is a term involving $n$ as well as an $n^2$-term. Thus we must use the quadratic formula.

$$d = \frac{n^2 - 3n}{2}$$

$$n^2 - 3n = 2d \qquad \text{Multiplying by 2 to clear fractions}$$

$$n^2 - 3n - 2d = 0 \qquad \text{Finding standard form}$$

$$a = 1, \quad b = -3, \quad c = -2d \quad \text{The letter } d \text{ represents a constant.}$$

$$n = \frac{-b \pm \sqrt{b^2 - 4ac}}{2a} \qquad \text{Quadratic formula}$$

$$= \frac{-(-3) \pm \sqrt{(-3)^2 - 4 \cdot 1 \cdot (-2d)}}{2 \cdot 1} \qquad \begin{array}{l}\text{Substituting into the}\\ \text{quadratic formula}\end{array}$$

$$= \frac{3 + \sqrt{9 + 8d}}{2} \qquad \text{Using the positive root}$$

*Do Exercise 6.*

**Example 6**  Solve for $t$: $S = gt + 16t^2$.

$$S = gt + 16t^2$$

$$16t^2 + gt - S = 0 \qquad \text{Finding standard form}$$

$$a = 16, \quad b = g, \quad c = -S$$

$$t = \frac{-b \pm \sqrt{b^2 - 4ac}}{2a}$$

$$= \frac{-g \pm \sqrt{g^2 - 4 \cdot 16 \cdot (-S)}}{2 \cdot 16} \qquad \begin{array}{l}\text{Substituting into the}\\ \text{quadratic formula}\end{array}$$

$$= \frac{-g + \sqrt{g^2 + 64S}}{32} \qquad \text{Using the positive root}$$

*Do Exercise 7.*

# Exercise Set 10.4

**a** Solve for the indicated letter.

**1.** $P = 17\sqrt{Q}$, for $Q$

**2.** $A = 1.4\sqrt{t}$, for $t$

**3.** $v = \sqrt{\dfrac{2gE}{m}}$, for $E$

**4.** $Q = \sqrt{\dfrac{aT}{c}}$, for $T$

**5.** $S = 4\pi r^2$, for $r$

**6.** $E = mc^2$, for $c$

**7.** $P = kA^2 + mA$, for $A$

**8.** $Q = ad^2 - cd$, for $d$

**9.** $c^2 = a^2 + b^2$, for $a$

**10.** $c = \sqrt{a^2 + b^2}$, for $b$

**11.** $s = 16t^2$, for $t$

**12.** $V = \pi r^2 h$, for $r$

**13.** $A = \pi r^2 + 2\pi rh$, for $r$

**14.** $A = 2\pi r^2 + 2\pi rh$, for $r$

**15.** $F = \dfrac{Av^2}{400}$, for $v$

**16.** $A = \dfrac{\pi r^2 S}{360}$, for $r$

**17.** $c = \sqrt{a^2 + b^2}$, for $a$

**18.** $c^2 = a^2 + b^2$, for $b$

**19.** $h = \dfrac{a}{2}\sqrt{3}$, for $a$
(The height of an equilateral triangle with sides of length $a$)

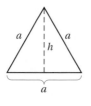

**20.** $d = s\sqrt{2}$, for $s$
(The hypotenuse of an isosceles right triangle with $s$ the length of the legs)

**21.** $n = aT^2 - 4T + m$, for $T$

**22.** $y = ax^2 + bx + c$, for $x$

**23.** $v = 2\sqrt{\dfrac{2kT}{\pi m}}$, for $T$

**24.** $E = \dfrac{1}{2}mv^2 + mgy$, for $v$

**25.** $3x^2 = d^2$, for $x$

**26.** $c = \sqrt{\dfrac{E}{m}}$, for $E$

**27.** $N = \dfrac{n^2 - n}{2}$, for $n$

**28.** $M = \dfrac{m}{\sqrt{1 - \left(\dfrac{v}{c}\right)^2}}$, for $c$

---

### Skill Maintenance

In a right triangle, find the length of the side not given. Give an exact answer and an approximation to three decimal places.  [9.6a]

**29.** $a = 4$, $b = 7$

**30.** $b = 11$, $c = 14$

**31.** $a = 4$, $b = 5$

**32.** $a = 10$, $c = 12$

**33.** $c = 8\sqrt{17}$, $a = 2$

**34.** $a = \sqrt{2}$, $b = \sqrt{3}$

Solve.  [9.6b]

**35.** *Guy Wire.* How long is a guy wire reaching from the top of an 18-ft pole to a point on the ground 10 ft from the pole? Give an exact answer and an approximation to three decimal places.

**36.** *Street Width.* Elliott Street is 24 ft wide where it ends at Main Street in Brattleboro, Vermont. A 40-ft–long diagonal crosswalk allows pedestrians to cross Main Street to or from either corner of Elliott Street (see the figure). Determine the width of Main Street.

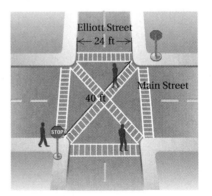

---

### Synthesis

**37.** ◆ Explain how you would solve the equation $0 = ax^2 + bx + c$ for $x$.

**38.** ◆ Explain how you would solve the equation $y = ax^2 + bx + c$ for $x$.

**39.** The circumference $C$ of a circle is given by $C = 2\pi r$.
　a) Solve $C = 2\pi r$ for $r$.
　b) The area is given by $A = \pi r^2$. Express the area in terms of the circumference $C$.

**40.** Referring to Exercise 39, express the circumference $C$ in terms of the area $A$.

**41.** Solve $3ax^2 - x - 3ax + 1 = 0$ for $x$.

**42.** Solve $h = 16t^2 + vt + s$ for $t$.

---

Copyright © 1999 Addison Wesley Longman

# 10.5 Applications and Problem Solving

## a | Using Quadratic Equations to Solve Applied Problems

**Example 1** *Red Raspberry Patch.* The area of a rectangular red raspberry patch is 76 ft². The length is 7 ft longer than three times the width. Find the dimensions of the raspberry patch.

1. **Familiarize.** We first make a drawing and label it with both known and unknown information. We let $w$ = the width of the rectangle. The length of the rectangle is 7 ft longer than three times the width. Thus the length is $3w + 7$.

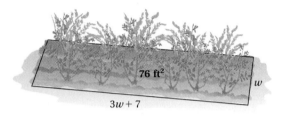

76 ft²

$w$

$3w + 7$

2. **Translate.** Recall that area is length × width. Thus we have two expressions for the area of the rectangle: $(3w + 7)(w)$ and 76. This gives us a translation:

$$(3w + 7)(w) = 76.$$

3. **Solve.** We solve the equation:

$$3w^2 + 7w = 76$$
$$3w^2 + 7w - 76 = 0$$
$$(3w + 19)(w - 4) = 0 \qquad \text{Factoring (the quadratic formula could also be used)}$$
$$3w + 19 = 0 \quad or \quad w - 4 = 0 \qquad \text{Using the principle of zero products}$$
$$3w = -19 \quad or \qquad w = 4$$
$$w = -\tfrac{19}{3} \quad or \qquad w = 4.$$

4. **Check.** We check in the original problem. We know that $-\frac{19}{3}$ is not a solution because width cannot be negative. When $w = 4$, $3w + 7 = 19$, and the area is 4(19), or 76. This checks.

5. **State.** The width of the rectangular raspberry patch is 4 ft, and the length is 19 ft.

*Do Exercise 1.*

**Example 2** *Staircase.* A carpenter builds a staircase in such a way that the portion underneath the stairs forms a right triangle. The hypotenuse is 6 m long. The leg across the floor is 1 m longer than the leg next to the wall at the back. Find the lengths of the legs. Round to the nearest tenth.

1. **Familiarize.** We first make a drawing, letting $s$ = the length of one leg. Then $s + 1$ = the length of the other leg.

$s$

6

$s + 1$

*Answer on page A-40*

1. *Pool Dimensions.* The area of a rectangular swimming pool is 68 yd². The length is 1 yd longer than three times the width. Find the dimensions of the rectangular swimming pool. Round to the nearest tenth.

**2.** The hypotenuse of a right triangular animal pen at the zoo is 4 yd long. One leg is 1 yd longer than the other. Find the lengths of the legs. Round to the nearest tenth.

4 yd

*a*

*a* + 1

2.3 yd, 3.3 yd

**2. Translate.** To translate, we use the Pythagorean equation:

$$s^2 + (s + 1)^2 = 6^2.$$

**3. Solve.** We solve the equation:

$$s^2 + (s + 1)^2 = 6^2$$
$$s^2 + s^2 + 2s + 1 = 36$$
$$2s^2 + 2s - 35 = 0.$$

Since we cannot factor, we use the quadratic formula:

$$a = 2, \quad b = 2, \quad c = -35$$

$$s = \frac{-b \pm \sqrt{b^2 - 4ac}}{2a}$$

$$= \frac{-2 \pm \sqrt{2^2 - 4 \cdot 2(-35)}}{2 \cdot 2}$$

$$= \frac{-2 \pm \sqrt{4 + 280}}{4}$$

$$= \frac{-2 \pm \sqrt{284}}{4}$$

$$= \frac{-2 \pm \sqrt{4 \cdot 71}}{4}$$

$$= \frac{-2 \pm 2 \cdot \sqrt{71}}{2 \cdot 2}$$

$$= \frac{2(-1 \pm \sqrt{71})}{2 \cdot 2} = \frac{2}{2} \cdot \frac{-1 \pm \sqrt{71}}{2}$$

$$= \frac{-1 \pm \sqrt{71}}{2}.$$

Using a calculator, we get approximations:

$$\frac{-1 + \sqrt{71}}{2} \approx 3.7 \quad \text{or} \quad \frac{-1 - \sqrt{71}}{2} \approx -4.7.$$

**4. Check.** Since the length of a leg cannot be negative, $-4.7$ does not check. But 3.7 does check. If the smaller leg $s$ is 3.7, the other leg is $s + 1$, or 4.7. Then

$$(3.7)^2 + (4.7)^2 = 13.69 + 22.09 = 35.78.$$

Using a calculator, we get $\sqrt{35.78} \approx 5.98 \approx 6$. Note that our check is not exact because we are using an approximation for $\sqrt{71}$.

**5. State.** One leg is about 3.7 m long, and the other is about 4.7 m long.

*Do Exercise 2.*

**Example 3** *Boat Speed.* The current in a stream moves at a speed of 2 km/h. A boat travels 24 km upstream and 24 km downstream in a total time of 5 hr. What is the speed of the boat in still water?

**1. Familiarize.** We first make a drawing. The distances are the same. We let $r =$ the speed of the boat in still water. Then when the boat is traveling upstream, its speed is $r - 2$. When it is traveling downstream, its speed is $r + 2$. We let $t_1$ represent the time it takes the boat to go upstream and $t_2$ the time it takes to go downstream. We summarize in a table.

*Answer on page A-40*

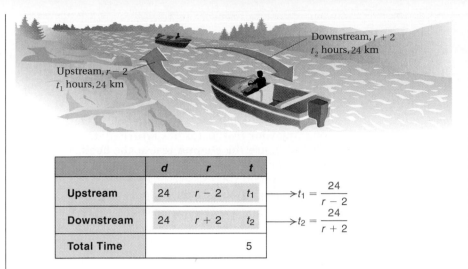

Upstream, $r - 2$
$t_1$ hours, 24 km

Downstream, $r + 2$
$t_2$ hours, 24 km

|  | d | r | t |  |
|---|---|---|---|---|
| **Upstream** | 24 | $r - 2$ | $t_1$ | $\rightarrow t_1 = \dfrac{24}{r - 2}$ |
| **Downstream** | 24 | $r + 2$ | $t_2$ | $\rightarrow t_2 = \dfrac{24}{r + 2}$ |
| **Total Time** |  |  | 5 |  |

**2. Translate.** Recall the basic formula for motion: $d = rt$. From it we can obtain an equation for time: $t = d/r$. Total time consists of the time to go upstream, $t_1$, plus the time to go downstream, $t_2$. Using $t = d/r$ and the rows of the table, we have

$$t_1 = \frac{24}{r - 2} \quad \text{and} \quad t_2 = \frac{24}{r + 2}.$$

Since the total time is 5 hr, $t_1 + t_2 = 5$, and we have

$$\frac{24}{r - 2} + \frac{24}{r + 2} = 5.$$

**3. Solve.** We solve the equation. We multiply on both sides by the LCM, which is $(r - 2)(r + 2)$:

$$(r - 2)(r + 2) \cdot \left[ \frac{24}{r - 2} + \frac{24}{r + 2} \right] = (r - 2)(r + 2)5 \quad \begin{array}{l}\text{Multiplying}\\\text{by the LCM}\end{array}$$

$$(r - 2)(r + 2) \cdot \frac{24}{r - 2} + (r - 2)(r + 2) \cdot \frac{24}{r + 2} = (r^2 - 4)5$$

$$24(r + 2) + 24(r - 2) = 5r^2 - 20$$

$$24r + 48 + 24r - 48 = 5r^2 - 20$$

$$-5r^2 + 48r + 20 = 0$$

$$5r^2 - 48r - 20 = 0 \quad \text{Multiplying by } -1$$

$$(5r + 2)(r - 10) = 0 \quad \text{Factoring}$$

$$5r + 2 = 0 \quad or \quad r - 10 = 0 \quad \begin{array}{l}\text{Using the}\\\text{principle}\\\text{of zero}\\\text{products}\end{array}$$

$$5r = -2 \quad or \quad r = 10$$

$$r = -\tfrac{2}{5} \quad or \quad r = 10.$$

**4. Check.** Since speed cannot be negative, $-\tfrac{2}{5}$ cannot be a solution. But suppose the speed of the boat in still water is 10 km/h. The speed upstream is then $10 - 2$, or 8 km/h. The speed downstream is $10 + 2$, or 12 km/h. The time upstream, using $t = d/r$, is 24/8, or 3 hr. The time downstream is 24/12, or 2 hr. The total time is 5 hr. This checks.

**5. State.** The speed of the boat in still water is 10 km/h.

*Do Exercise 3.*

**3.** *Speed of a Stream.* The speed of a boat in still water is 12 km/h. The boat travels 45 km upstream and 45 km downstream in a total time of 8 hr. What is the speed of the stream? (*Hint*: Let $s =$ the speed of the stream. Then $12 - s$ is the speed upstream and $12 + s$ is the speed downstream. Note also that $12 - s$ cannot be negative, because the boat must be going faster than the current if it is moving forward.)

*Answer on page A-40*

# Improving Your Math Study Skills

## Preparing for a Final Exam

### Best Scenario: Two Weeks of Study Time

The best scenario for preparing for a final exam is to do so over a period of at least two weeks. Work in a diligent, disciplined manner, doing some final-exam preparation *each* day. Here is a detailed plan that many find useful.

1. **Begin by browsing through each chapter, reviewing the highlighted or boxed information regarding important formulas in both the text and the Summary and Review.** There may be some formulas that you will need to memorize.

2. **Retake each chapter test that you took in class, assuming your instructor has returned it. Otherwise, use the chapter test in the book.** Restudy the objectives in the text that correspond to each question you missed.

3. **Then work the Cumulative Review that covers all chapters up to that point.** Be careful to avoid any questions corresponding to objectives not covered. Again, restudy the objectives in the text that correspond to each question you missed.

4. **If you are still missing questions, use supplements for extra review.** For example, you might check out the video- or audiotapes, the *Student's Solutions Manual,* or the Interact Math Tutorial Software.

5. **For remaining difficulties, see your instructor, go to a tutoring session, or participate in a study group.**

6. **Check for former final exams that may be on file in the math department or a study center, or with students who have already taken the course.** Use them for practice, being alert to trouble spots.

7. **Take the Final Examination in the text during the last couple of days before the final.** Set aside the same amount of time that you will have for the final. See how much of the final exam you can complete under test-like conditions.

### Moderate Scenario: Three Days to Two Weeks of Study Time

1. **Begin by browsing through each chapter, reviewing the highlighted or boxed information regarding important formulas in both the text and the Summary and Review.** There may be some formulas that you will need to memorize.

2. **Retake each chapter test that you took in class, assuming your instructor has returned it. Otherwise, use the chapter test in the book.** Restudy the objectives in the text that correspond to each question you missed.

3. **Then work the last Cumulative Review in the text.** Be careful to avoid any questions corresponding to objectives not covered. Again, restudy the objectives in the text that correspond to each question you missed.

4. **For remaining difficulties, see your instructor, go to a tutoring session, or participate in a study group.**

5. **Take the Final Examination in the text during the last couple of days before the final.** Set aside the same amount of time that you will have for the final. See how much of the final exam you can complete under test-like conditions.

### Worst Scenario: One or Two Days of Study Time

1. **Begin by browsing through each chapter, reviewing the highlighted or boxed information regarding important formulas in both the text and the Summary and Review.** There may be some formulas that you will need to memorize.

2. **Then work the last Cumulative Review in the text.** Be careful to avoid any questions corresponding to objectives not covered. Restudy the objectives in the text that correspond to each question you missed.

3. **Attend a final-exam review session if one is available.**

4. **Take the Final Examination in the text during the last couple of days before the final.** Set aside the same amount of time that you will have for the final. See how much of the final exam you can complete under test-like conditions.

   Promise yourself that next semester you will allow a more appropriate amount of time for final exam preparation.

Other "Improving Your Math Study Skills" concerning test preparation appear in Sections 1.5 and 3.1.

# Exercise Set 10.5

**a** Solve.

1. The length of a rectangular area rug is 3 ft greater than the width. The area is 70 ft². Find the length and the width.

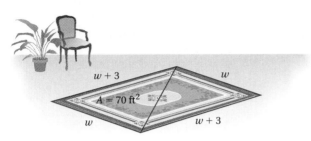

2. The length of a rectangular pine forest is 2 mi greater than the width. The area is 80 mi². Find the length and the width.

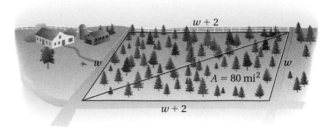

3. *Standard-Sized Television.* When we say that a television is 30 in., we mean that the diagonal is 30 in. For a standard-sized 30-in. television, the width is 6 in. more than the height. Find the dimensions of a standard-sized 30-in. television.

4. *HDTV Dimensions.* In the not-too-distant future, a new kind of high-definition television (HDTV) with larger screens and greater clarity will be available. An HDTV might have a 70-in. diagonal screen with the width 27 in. greater than the height. Find the width and the height of a 70-in. HDTV screen.

5. The width of a rectangle is 4 cm less than the length. The area is 320 cm². Find the length and the width.

6. The width of a rectangle is 3 cm less than the length. The area is 340 cm². Find the length and the width.

7. The length of a rectangle is twice the width. The area is 50 m². Find the length and the width.

8. *Carpenter's Square.* A *square* is a carpenter's tool in the shape of a right triangle. One side, or leg, of a square is 8 in. longer than the other. The length of the hypotenuse is $8\sqrt{13}$ in. Find the lengths of the legs of the square.

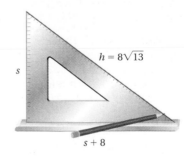

Find the approximate answers for Exercises 9–14. Round to the nearest tenth.

**9.** The hypotenuse of a right triangle is 8 m long. One leg is 2 m longer than the other. Find the lengths of the legs.

**10.** The hypotenuse of a right triangle is 5 cm long. One leg is 2 cm longer than the other. Find the lengths of the legs.

**11.** The length of a rectangle is 2 in. greater than the width. The area is 20 in$^2$. Find the length and the width.

**12.** The length of a rectangle is 3 ft greater than the width. The area is 15 ft$^2$. Find the length and the width.

**13.** The length of a rectangle is twice the width. The area is 20 cm$^2$. Find the length and the width.

**14.** The length of a rectangle is twice the width. The area is 10 m$^2$. Find the length and the width.

**15.** A picture frame measures 25 cm by 20 cm. There is 266 cm$^2$ of picture showing. The frame is of uniform thickness. Find the thickness of the frame.

**16.** A tablecloth measures 96 in. by 72 in. It is laid on a tabletop with an area of 5040 in$^2$, and hangs over the edge by the same amount on all sides. By how many inches does the cloth hang over the edge?

Copyright © 1999 Addison Wesley Longman

For Exercises 17–22, complete the table to help with the familiarization.

**17.** The current in a stream moves at a speed of 3 km/h. A boat travels 40 km upstream and 40 km downstream in a total time of 14 hr. What is the speed of the boat in still water? Complete the following table to help with the familiarization.

|  | d | r | t |
|---|---|---|---|
| Upstream |  | $r - 3$ | $t_1$ |
| Downstream | 40 |  | $t_2$ |
| Total Time |  |  |  |

Upstream, $r - 3$
$t_1$ hours, 40 km

Downstream, $r + 3$
$t_2$ hours, 40 km

**18.** The current in a stream moves at a speed of 3 km/h. A boat travels 45 km upstream and 45 km downstream in a total time of 8 hr. What is the speed of the boat in still water?

|  | d | r | t |
|---|---|---|---|
| Upstream | 45 |  |  |
| Downstream |  | $r + 3$ |  |
| Total Time |  |  |  |

**19.** The current in a stream moves at a speed of 4 mph. A boat travels 4 mi upstream and 12 mi downstream in a total time of 2 hr. What is the speed of the boat in still water?

|  | d | r | t |
|---|---|---|---|
| Upstream |  | $r - 4$ |  |
| Downstream | 12 |  |  |
| Total Time |  |  |  |

**20.** The current in a stream moves at a speed of 4 mph. A boat travels 5 mi upstream and 13 mi downstream in a total time of 2 hr. What is the speed of the boat in still water?

|  | d | r | t |
|---|---|---|---|
| Upstream |  |  |  |
| Downstream |  |  |  |
| Total Time |  |  |  |

**21.** The speed of a boat in still water is 10 km/h. The boat travels 12 km upstream and 28 km downstream in a total time of 4 hr. What is the speed of the stream?

|  | d | r | t |
|---|---|---|---|
| Upstream |  |  |  |
| Downstream |  |  |  |
| Total Time |  |  |  |

**22.** The speed of a boat in still water is 8 km/h. The boat travels 60 km upstream and 60 km downstream in a total time of 16 hr. What is the speed of the stream?

|  | d | r | t |
|---|---|---|---|
| Upstream |  |  |  |
| Downstream |  |  |  |
| Total Time |  |  |  |

**23.** An airplane flies 738 mi against the wind and 1062 mi with the wind in a total time of 9 hr. The speed of the airplane in still air is 200 mph. What is the speed of the wind?

**24.** An airplane flies 520 km against the wind and 680 km with the wind in a total time of 4 hr. The speed of the airplane in still air is 300 km/h. What is the speed of the wind?

**25.** The speed of a boat in still water is 9 km/h. The boat travels 80 km upstream and 80 km downstream in a total time of 18 hr. What is the speed of the stream?

**26.** The speed of a boat in still water is 10 km/h. The boat travels 48 km upstream and 48 km downstream in a total time of 10 hr. What is the speed of the stream?

---

### Skill Maintenance

Add or subtract.  [9.4a]

**27.** $5\sqrt{2} + \sqrt{18}$

**28.** $7\sqrt{40} - 2\sqrt{10}$

**29.** $\sqrt{4x^3} - 7\sqrt{x}$

**30.** $\sqrt{24} - \sqrt{54}$

**31.** $\sqrt{2} + \sqrt{\dfrac{1}{2}}$

**32.** $\sqrt{3} - \sqrt{\dfrac{1}{3}}$

**33.** $\sqrt{24} + \sqrt{54} - \sqrt{48}$

**34.** $\sqrt{4x} + \sqrt{81x^3}$

---

### Synthesis

Find and explain the error(s) in each of the following solutions of a quadratic equation.

**35.** ◆ $(x + 6)^2 = 16$
$x + 6 = \sqrt{16}$
$x + 6 = 4$
$x = -2$

**36.** ◆ $x^2 + 2x - 8 = 0$
$(x + 4)(x - 2) = 0$
$x = 4 \quad or \quad x = -2$

**37.** Find $r$ in this figure. Round to the nearest hundredth.

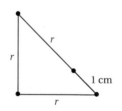

**38.** The width of a dollar bill is 9 cm less than the length. The area is 102.96 cm². Find the length and the width.

**39.** What should the diameter $d$ of a pizza be so that it has the same area as two 12-in. pizzas? Do you get more to eat with a 16-in. pizza or with two 12-in. pizzas?

---

Copyright © 1999 Addison Wesley Longman

# 10.6 Graphs of Quadratic Equations

In this section, we will graph equations of the form

$$y = ax^2 + bx + c, \quad a \neq 0.$$

The polynomial on the right side of the equation is of second degree, or **quadratic**. Examples of the types of equations we are going to graph are

$$y = x^2, \qquad y = x^2 + 2x - 3, \qquad y = -2x^2 + 3.$$

**Objectives**

**a**   Graph quadratic equations.

**b**   Find the $x$-intercepts of a quadratic equation.

**For Extra Help**

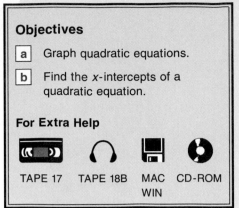

TAPE 17    TAPE 18B    MAC   CD-ROM
                            WIN

## **a**   Graphing Quadratic Equations of the Type $y = ax^2 + bx + c$

Graphs of quadratic equations of the type $y = ax^2 + bx + c$ (where $a \neq 0$) are always cup-shaped. They have a **line of symmetry** like the dashed lines shown in the figures below. If we fold on this line, the two halves will match exactly. The curve goes on forever. The top or bottom point where the curve changes is called the **vertex**. The second coordinate is either the largest value of $y$ or the smallest value of $y$. The vertex is also thought of as a turning point. Graphs of quadratic equations are called **parabolas**.

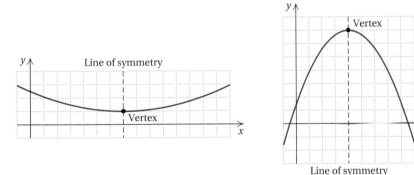

To graph a quadratic equation, we begin by choosing some numbers for $x$ and computing the corresponding values of $y$.

**Example 1**   Graph: $y = x^2$.

We choose numbers for $x$ and find the corresponding values for $y$. Then we plot the ordered pairs $(x, y)$ resulting from the computations and connect them with a smooth curve.

For $x = -3$, $y = x^2 = (-3)^2 = 9$.
For $x = -2$, $y = x^2 = (-2)^2 = 4$.
For $x = -1$, $y = x^2 = (-1)^2 = 1$.
For $x = 0$, $y = x^2 = (0)^2 = 0$.
For $x = 1$, $y = x^2 = (1)^2 = 1$.
For $x = 2$, $y = x^2 = (2)^2 = 4$.
For $x = 3$, $y = x^2 = (3)^2 = 9$.

| $x$ | $y$ | $(x, y)$ |
|-----|-----|----------|
| $-3$ | $9$ | $(-3, 9)$ |
| $-2$ | $4$ | $(-2, 4)$ |
| $-1$ | $1$ | $(-1, 1)$ |
| $0$ | $0$ | $(0, 0)$ |
| $1$ | $1$ | $(1, 1)$ |
| $2$ | $4$ | $(2, 4)$ |
| $3$ | $9$ | $(3, 9)$ |

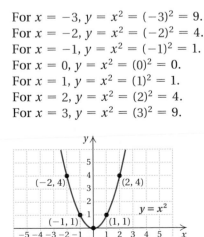

## Calculator Spotlight

Graphing Quadratic Equations.   The following is the graph of the quadratic equation $y_1 = x^2 + 3x - 4$.

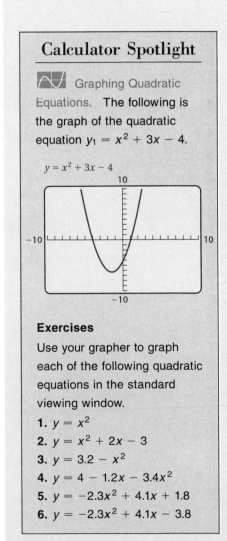

**Exercises**

Use your grapher to graph each of the following quadratic equations in the standard viewing window.

**1.** $y = x^2$

**2.** $y = x^2 + 2x - 3$

**3.** $y = 3.2 - x^2$

**4.** $y = 4 - 1.2x - 3.4x^2$

**5.** $y = -2.3x^2 + 4.1x + 1.8$

**6.** $y = -2.3x^2 + 4.1x - 3.8$

In Example 1, the vertex is the point (0, 0). The second coordinate of the vertex, 0, is the smallest $y$-value. The $y$-axis is the line of symmetry. Parabolas whose equations are $y = ax^2$ always have the origin (0, 0) as the vertex and the $y$-axis as the line of symmetry.

How do we graph a general equation? There are many methods, some of which you will study in your next mathematics course. Our goal here is to give you a basic graphing technique that is fairly easy to apply. A key in the graphing is knowing the vertex. By graphing it and then choosing $x$-values on both sides of the vertex, we can compute more points and complete the graph.

---

> **FINDING THE VERTEX**
>
> For a parabola given by the quadratic equation $y = ax^2 + bx + c$:
>
> **1.** The $x$-coordinate of the vertex is $-\dfrac{b}{2a}$.
>
> **2.** The second coordinate of the vertex is found by substituting the $x$-coordinate into the equation and computing $y$.

---

The proof that the vertex can be found in this way can be shown by completing the square in a manner similar to the proof of the quadratic formula, but it will not be considered here.

**Example 2**   Graph: $y = -2x^2 + 3$.

We first find the vertex. The $x$-coordinate of the vertex is

$$-\frac{b}{2a} = -\frac{0}{2(-2)} = 0.$$

We substitute 0 for $x$ into the equation to find the second coordinate of the vertex:

$$y = -2x^2 + 3 = -2(0)^2 + 3 = 3.$$

The vertex is (0, 3). The line of symmetry is $x = 0$, which is the $y$-axis. We choose some $x$-values on both sides of the vertex and graph the parabola.

For $x = 1$, $y = -2x^2 + 3 = -2(1)^2 + 3 = -2 + 3 = 1$.
For $x = -1$, $y = -2x^2 + 3 = -2(-1)^2 + 3 = -2 + 3 = 1$.
For $x = 2$, $y = -2x^2 + 3 = -2(2)^2 + 3 = -8 + 3 = -5$.
For $x = -2$, $y = -2x^2 + 3 = -2(-2)^2 + 3 = -8 + 3 = -5$.

| $x$ | $y$ |
|-----|-----|
| 0 | 3 | ← **This is the vertex.**
| 1 | 1 |
| −1 | 1 |
| 2 | −5 |
| −2 | −5 |

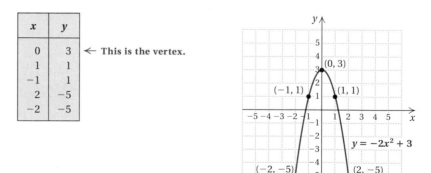

## Calculator Spotlight

Use the GRAPH and TABLE features of your grapher to check the results of Example 2.

There are two other tips you might use when graphing quadratic equations. The first involves the coefficient of $x^2$. Note that $a$ in $y = ax^2 + bx + c$ tells us whether the graph opens up or down. When $a$ is positive, as in Example 1, the graph opens up; when $a$ is negative, as in Example 2, the graph opens down. It is also helpful to plot the $y$-intercept. It occurs when $x = 0$.

---

**TIPS FOR GRAPHING QUADRATIC EQUATIONS**

1. Graphs of quadratic equations $y = ax^2 + bx + c$ are all parabolas. They are *smooth* cup-shaped symmetric curves, with no sharp points or kinks in them.
2. The graph of $y = ax^2 + bx + c$ opens up if $a > 0$. It opens down if $a < 0$.
3. Find the $y$-intercept. It occurs when $x = 0$, and it is easy to compute.

---

**Example 3**   Graph: $y = x^2 + 2x - 3$.

We first find the vertex. The $x$-coordinate of the vertex is

$$-\frac{b}{2a} = -\frac{2}{2(1)} = -1.$$

We substitute $-1$ for $x$ into the equation to find the second coordinate of the vertex:

$$
\begin{aligned}
y &= x^2 + 2x - 3 \\
&= (-1)^2 + 2(-1) - 3 \\
&= 1 - 2 - 3 \\
&= -4.
\end{aligned}
$$

The vertex is $(-1, -4)$. The line of symmetry is $x = -1$.

We choose some $x$-values on both sides of $x = -1$—say, $-2, -3, -4$ and $0, 1, 2$—and graph the parabola. Since the coefficient of $x^2$ is 1, which is positive, we know that the graph opens up. Be sure to find $y$ when $x = 0$. This gives the $y$-intercept.

| $x$ | $y$ | |
|---|---|---|
| $-1$ | $-4$ | ← Vertex |
| $0$ | $-3$ | ← $y$-intercept |
| $-2$ | $-3$ | |
| $1$ | $0$ | |
| $-3$ | $0$ | |
| $2$ | $5$ | |
| $-4$ | $5$ | |

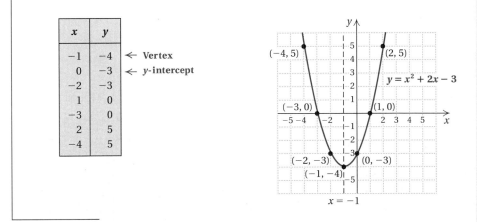

***Do Exercises 1–3.***

Graph. List the ordered pair for the vertex.

**1.** $y = x^2 - 3$

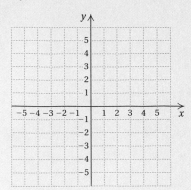

**2.** $y = -3x^2 + 6x$

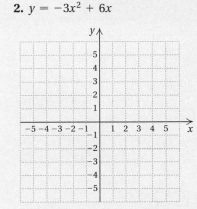

**3.** $y = x^2 - 4x + 4$

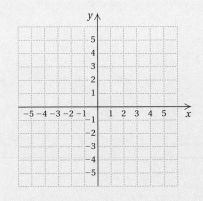

*Answers on page A-40*

**10.6   Graphs of Quadratic Equations**

Find the $x$-intercepts.

**4.** $y = x^2 - 3$

**5.** $y = x^2 + 6x + 8$

**6.** $y = -2x^2 - 4x + 1$

**7.** $y = x^2 + 3$

*Answers on page A-40*

## b │ Finding the *x*-Intercepts of a Quadratic Equation

The $x$-intercepts of $y = ax^2 + bx + c$ occur at those values of $x$ for which $y = 0$. Thus the first coordinates of the $x$-intercepts are solutions of the equation

$$0 = ax^2 + bx + c.$$

We have been studying how to find such numbers in Sections 10.1–10.3.

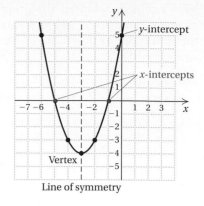

**Example 4**   Find the $x$-intercepts of $y = x^2 - 4x + 1$.

We solve the equation

$$x^2 - 4x + 1 = 0.$$

Factoring is not convenient, so we use the quadratic formula:

$$a = 1, \quad b = -4, \quad c = 1$$

$$x = \frac{-b \pm \sqrt{b^2 - 4ac}}{2a}$$

$$= \frac{-(-4) \pm \sqrt{(-4)^2 - 4(1)(1)}}{2(1)}$$

$$= \frac{4 \pm \sqrt{16 - 4}}{2}$$

$$= \frac{4 \pm \sqrt{12}}{2} = \frac{4 \pm \sqrt{4 \cdot 3}}{2}$$

$$= \frac{4 \pm 2\sqrt{3}}{2} = \frac{2 \cdot 2 \pm 2\sqrt{3}}{2 \cdot 1}$$

$$= \frac{2}{2} \cdot \frac{2 \pm \sqrt{3}}{1} = 2 \pm \sqrt{3}.$$

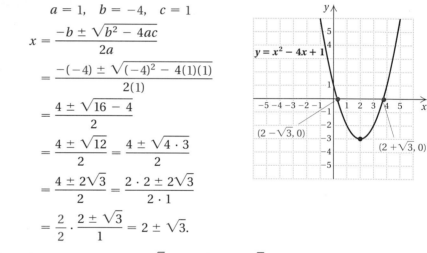

The $x$-intercepts are $(2 - \sqrt{3}, 0)$ and $(2 + \sqrt{3}, 0)$.

In the quadratic formula $x = \dfrac{-b \pm \sqrt{b^2 - 4ac}}{2a}$, the radicand $b^2 - 4ac$ is called the **discriminant**. The discriminant tells how many real-number solutions the equation $0 = ax^2 + bx + c$ has, so it also tells how many $x$-intercepts there are.

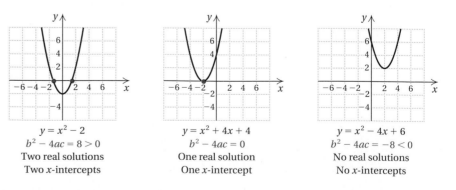

$y = x^2 - 2$
$b^2 - 4ac = 8 > 0$
Two real solutions
Two $x$-intercepts

$y = x^2 + 4x + 4$
$b^2 - 4ac = 0$
One real solution
One $x$-intercept

$y = x^2 - 4x + 6$
$b^2 - 4ac = -8 < 0$
No real solutions
No $x$-intercepts

*Do Exercises 4–7.*

# Exercise Set 10.6

**a** Graph the quadratic equation. List the ordered pair for the vertex.

**1.** $y = x^2 + 1$

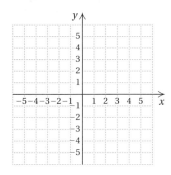

**2.** $y = 2x^2$

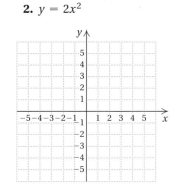

**3.** $y = -1 \cdot x^2$

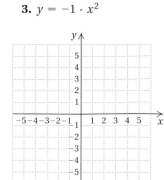

**4.** $y = x^2 - 1$

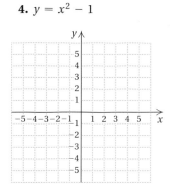

**5.** $y = -x^2 + 2x$

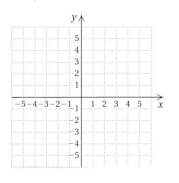

**6.** $y = x^2 + x - 2$

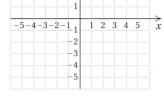

**7.** $y = 5 - x - x^2$

**8.** $y = x^2 + 2x + 1$

**9.** $y = x^2 - 2x + 1$

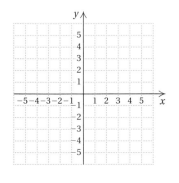

**10.** $y = -\frac{1}{2}x^2$

**11.** $y = -x^2 + 2x + 3$

**12.** $y = -x^2 - 2x + 3$

**13.** $y = -2x^2 - 4x + 1$

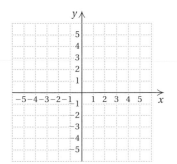

**14.** $y = 2x^2 + 4x - 1$

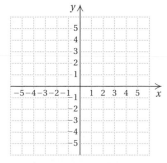

**15.** $y = 5 - x^2$

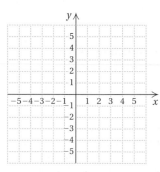

**16.** $y = 4 - x^2$

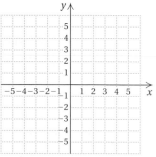

Graph the quadratic equation. Use your own graph paper.

**17.** $y = \frac{1}{4}x^2$

**18.** $y = -0.1x^2$

**19.** $y = -x^2 + x - 1$

**20.** $y = x^2 + 2x$

**21.** $y = -2x^2$

**22.** $y = -x^2 - 1$

**23.** $y = x^2 - x - 6$

**24.** $y = 6 + x - x^2$

---

**b** Find the $x$-intercepts.

**25.** $y = x^2 - 2$

**26.** $y = x^2 - 7$

**27.** $y = x^2 + 5x$

**28.** $y = x^2 - 4x$

**29.** $y = 8 - x - x^2$

**30.** $y = 8 + x - x^2$

**31.** $y = x^2 - 6x + 9$

**32.** $y = x^2 + 10x + 25$

**33.** $y = -x^2 - 4x + 1$

**34.** $y = x^2 + 4x - 1$

**35.** $y = x^2 + 9$

**36.** $y = x^2 + 1$

---

### Skill Maintenance

Add.   [9.4a]

**37.** $\sqrt{x^3 - x^2} + \sqrt{4x - 4}$

**38.** $\sqrt{8} + \sqrt{50} + \sqrt{98} + \sqrt{128}$

Multiply and simplify.   [9.2c]

**39.** $\sqrt{2}\,\sqrt{14}$

**40.** $\sqrt{5y^4}\,\sqrt{125y}$

**41.** Find an equation of variation in which $y$ varies inversely as $x$ and $y = 12.4$ when $x = 2.4$.   [7.5c]

**42.** Find an equation of variation in which $y$ varies inversely as $x$ and $y = 264$ when $x = 18$.   [7.5c]

**43.** Evaluate $5x^3 - 2x$ for $x = -1$.   [4.3a]

**44.** Evaluate $3x^4 + 3x - 7$ for $x = -2$.   [4.3a]

---

### Synthesis

**45.** ◆ Suppose that the $x$-intercepts of a parabola are $(a_1, 0)$ and $(a_2, 0)$. What is the easiest way to find an equation for the line of symmetry? the coordinates of the vertex?

**46.** ◆ Discuss the effect of the sign of $a$ on the graph of $y = ax^2 + bx + c$.

**47.** *Height of a Projectile.* The height $H$, in feet, of a projectile with an initial velocity of 96 ft/sec is given by the equation

$$H = -16t^2 + 96t,$$

where $t$ = time, in seconds. Use the graph of this function, shown here, or any equation-solving technique to answer the following questions.

a) How many seconds after launch is the projectile 128 ft above ground?
b) When does the projectile reach its maximum height?
c) How many seconds after launch does the projectile return to the ground?

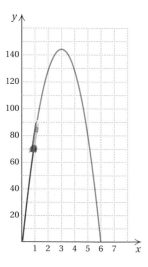

For each equation in Exercises 48–51, evaluate the discriminant $b^2 - 4ac$. Then use the answer to state how many real-number solutions exist for the equation.

**48.** $0 = x^2 + 8x + 16$

**49.** $0 = x^2 + 2x - 3$

**50.** $0 = -2x^2 + 4x - 3$

**51.** $0 = -0.02x^2 + 4.7x - 2300$

---

Collaborative Learning Manual

Practice graphing and identifying the graphs of quadratic equations.

# 10.7 Functions

## a | Identifying Functions

We now develop one of the most important concepts in mathematics, **functions**. We have actually been studying functions all through this text; we just haven't identified them as such. Ordered pairs form a correspondence between first and second coordinates. A function is a special correspondence from one set of numbers to another. For example:

> To each student in a college, there corresponds his or her student ID.
>
> To each item in a store, there corresponds its price.
>
> To each real number, there corresponds the cube of that number.

In each case, the first set is called the **domain** and the second set is called the **range**. Given a member of the domain, there is *just one* member of the range to which it corresponds. This kind of correspondence is called a **function**.

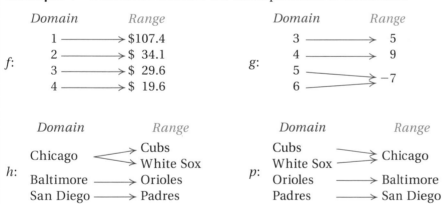

**Example 1** Determine whether the correspondence is a function.

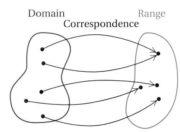

The correspondence *f* is a function because each member of the domain is matched to only one member of the range.

The correspondence *g* is also a function because each member of the domain is matched to only one member of the range.

The correspondence *h* is *not* a function because one member of the domain, Chicago, is matched to more than one member of the range.

The correspondence *p* is a function because each member of the domain is paired with only one member of the range.

---

> ▶ A **function** is a correspondence between a first set, called the **domain**, and a second set, called the **range**, such that each member of the domain corresponds to *exactly one* member of the range.

*Do Exercises 1–4.*

## Objectives

**a** Determine whether a correspondence is a function.

**b** Given a function described by an equation, find function values (outputs) for specified values (inputs).

**c** Draw the graph of a function.

**d** Determine whether a graph is that of a function.

**e** Solve applied problems involving functions and their graphs.

### For Extra Help

TAPE 17    TAPE 19A    MAC    CD-ROM
                       WIN

Determine whether the correspondence is a function.

**1.** Domain          Range

Cheetah ⟶ 70 mph
Human ⟶ 28 mph
Lion ⟶ 50 mph
Chicken ⟶ 9 mph

**2.** Domain          Range

**3.** Domain          Range

$-2$
2        4
$-3$
3        9
0 ⟶ 0

**4.** Domain          Range

4        $-2$
         2
9        $-3$
         3
0 ⟶ 0

*Answers on page A-41*

**5.** *Domain*
A set of numbers

*Correspondence*
Square each number and subtract 10.

*Range*
A set of numbers

**6.** *Domain*
A set of polygons

*Correspondence*
Find the perimeter of each polygon.

*Range*
A set of numbers

**Example 2** Determine whether the correspondence is a function.

| Domain | Correspondence | Range |
|--------|---------------|-------|
| a) A family | Each person's weight | A set of positive numbers |
| b) The natural numbers | Each number's square | A set of natural numbers |
| c) The set of all states | Each state's members of the U.S. Senate | A set of U.S. Senators |

a) The correspondence *is* a function because each person has *only one* weight.

b) The correspondence *is* a function because each natural number has *only one* square.

c) The correspondence *is not* a function because each state has two U.S. Senators.

*Do Exercises 5 and 6.*

When a correspondence between two sets is not a function, it is still an example of a **relation**.

> A **relation** is a correspondence between a first set, called the **domain**, and a second set, called the **range**, such that each member of the domain corresponds to *at least one* member of the range.

Thus, although the correspondences of Examples 1 and 2 are not all functions, they *are* all relations. A function is a special type of relation— one in which each member of the domain is paired with *exactly one* member of the range.

**b** | **Finding Function Values**

Most functions considered in mathematics are described by equations. A linear equation like $y = 2x + 3$, studied in Chapters 3 and 7, is called a **linear function.** A quadratic equation like $y = 4 - x^2$, studied in Chapter 10, is called a **quadratic function.**

Recall that when graphing $y = 2x + 3$, we chose $x$-values and then found corresponding $y$-values. For example, when $x = 4$, $y = 2x + 3 = 2 \cdot 4 + 3 = 11$. When thinking of functions, we call the number 4 an **input** and the number 11 an **output**.

It helps to think of a function as a machine; that is, think of putting a member of the domain (an input) into the machine. The machine knows the correspondence and gives out a member of the range (the output).

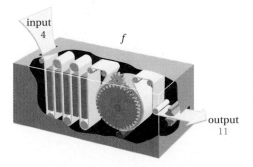

The function $y = 2x + 3$ has been named $f$ and is described by the equation $f(x) = 2x + 3$. We call the input $x$ and the output $f(x)$. This is read "$f$ of $x$," or "$f$ at $x$," or "the value of $f$ at $x$."

---

**CAUTION!** The notation $f(x)$ *does not mean "f times x"* and should not be read that way.

---

The equation $f(x) = 2x + 3$ describes the function that takes an input $x$, multiplies it by 2, and then adds 3.

$$\overset{\textbf{Input}}{f(x) = \underset{\textbf{Double}}{2x} \underset{\textbf{Add 3}}{+ 3}}$$

To find the output $f(4)$, we take the input 4, double it, and add 3 to get 11. That is, we substitute 4 into the formula for $f(x)$:

$$f(4) = 2 \cdot 4 + 3 = 11.$$

Outputs of functions are also called **function values.** For $f(x) = 2x + 3$, we know that $f(4) = 11$. We can say that "the function value at 4 is 11."

**Example 3** Find the indicated function value.

a) $f(5)$, for $f(x) = 3x + 2$       b) $g(3)$, for $g(z) = 5z^2 - 4$
c) $A(-2)$, for $A(r) = 3r^2 - 2r$       d) $f(-5)$, for $f(x) = x^2 + 3x - 4$

a) $f(5) = 3 \cdot 5 + 2 = 17$
b) $g(3) = 5(3)^2 - 4 = 41$
c) $A(-2) = 3(-2)^2 + 2(-2) = 8$
d) $f(-5) = (-5)^2 + 3(-5) - 4 = 25 - 15 - 4 = 6$

*Do Exercises 7 and 8.*

## c | Graphs of Functions

To graph a function, we find ordered pairs $(x, y)$ or $(x, f(x))$, plot them, and connect the points. Note that $y$ and $f(x)$ are used interchangeably when working with functions and their graphs.

**Example 4** Graph: $f(x) = x + 2$.

A list of some function values is shown in this table. We plot the points and connect them. The graph is a straight line.

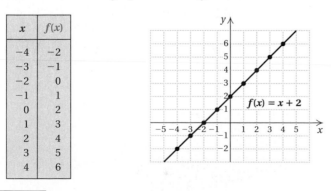

| $x$ | $f(x)$ |
|----|-----|
| $-4$ | $-2$ |
| $-3$ | $-1$ |
| $-2$ | $0$ |
| $-1$ | $1$ |
| $0$ | $2$ |
| $1$ | $3$ |
| $2$ | $4$ |
| $3$ | $5$ |
| $4$ | $6$ |

Find the function values.

**7.** $f(x) = 5x - 3$

  a) $f(-6)$
  b) $f(0)$
  c) $f(1)$
  d) $f(20)$
  e) $f(-1.2)$

**8.** $g(x) = x^2 - 4x + 9$

  a) $g(-2)$
  b) $g(0)$
  c) $g(5)$
  d) $g(10)$

*Answers on page A-41*

## Calculator Spotlight

 Graphs and Function
Values.   To graph the
function $f(x) = 2x^2 + x$, we
enter it into the grapher as
$y_1 = 2x^2 + x$ and use the
GRAPH feature. There are at
least three ways to find
function values. The first is to
use the VALUE feature. We
enter a value, say, $x = -2$.
The $y$-value is given
to the right and the point is
indicated on the graph.

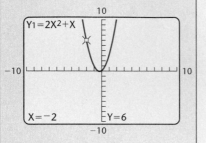

The second way to find a
function value is to use the
TABLE feature. The third is to
bring $Y_1$ to the screen using
Y-VARS. We enter parentheses
to complete the writing of
$Y_1(-2)$. When ENTER is
pressed, we obtain the
output, 6.

### Exercises

1. Graph $f(x) = x^2 + 3x - 4$.
   Then find $f(-5)$, $f(-4.7)$,
   $f(11)$, and $f(2/3)$.
2. Graph $f(x) = 3.7 - x^2$.
   Then find $f(-5)$, $f(-4.7)$,
   $f(11)$, and $f(2/3)$.
3. Graph
   $f(x) = 4 - 1.2x - 3.4x^2$.
   Then find $f(-5)$, $f(-4.7)$,
   $f(11)$, and $f(2/3)$.

**Example 5**   Graph: $g(x) = 4 - x^2$.

Recall from Section 10.6 that the graph is a parabola. We calculate some function values and draw the curve.

$$g(0) = 4 - 0^2 = 4 - 0 = 4,$$
$$g(-1) = 4 - (-1)^2 = 4 - 1 = 3,$$
$$g(2) = 4 - (2)^2 = 4 - 4 = 0,$$
$$g(-3) = 4 - (-3)^2 = 4 - 9 = -5$$

| $x$ | $g(x)$ |
|-----|--------|
| $-3$ | $-5$ |
| $-2$ | $0$ |
| $-1$ | $3$ |
| $0$ | $4$ |
| $1$ | $3$ |
| $2$ | $0$ |
| $3$ | $-5$ |

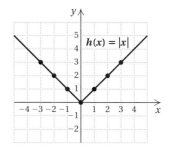

**Example 6**   Graph: $h(x) = |x|$.

A list of some function values is shown in the following table. We plot the points and connect them. The graph is a V-shaped "curve" that rises on either side of the vertical axis.

| $x$ | $h(x)$ |
|-----|--------|
| $-3$ | $3$ |
| $-2$ | $2$ |
| $-1$ | $1$ |
| $0$ | $0$ |
| $1$ | $1$ |
| $2$ | $2$ |
| $3$ | $3$ |

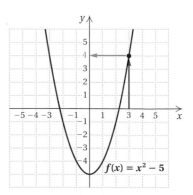

*Do Exercises 9–11 on the following page.*

### d | The Vertical-Line Test

Consider the function $f$ described by $f(x) = x^2 - 5$. Its graph is shown at right. It is also the graph of the equation $y = x^2 - 5$.

To find a function value, like $f(3)$, from a graph, we locate the input on the horizontal axis, move vertically to the graph of the function, and then horizontally to find the output on the vertical axis, where members of the range can be found.

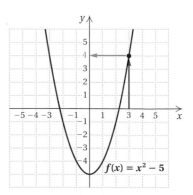

Recall that when one member of the domain is paired with two or more different members of the range, the correspondence is not a function. Thus, when a graph contains two or more different points with the same first coordinate, the graph cannot represent a function. Points sharing a common first coordinate are vertically above or below each other (see the following graph).

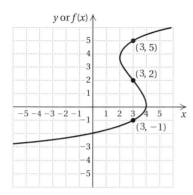

Since 3 is paired with more than one member of the range, the graph does not represent a function.

This observation leads to the *vertical-line test.*

> ▶ **THE VERTICAL-LINE TEST**
>
> A graph represents a function if it is impossible to draw a vertical line that intersects the graph more than once.

**Example 7** Determine whether each of the following is the graph of a function.

a)

b)

c)

d)

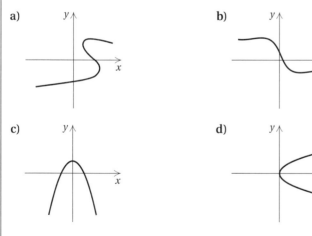

a) The graph *is not* that of a function because a vertical line crosses the graph at more than one point.

b) The graph *is* that of a function because no vertical line can cross the graph at more than one point. This can be confirmed with a ruler or straight edge.

Graph.

**9.** $f(x) = x - 4$

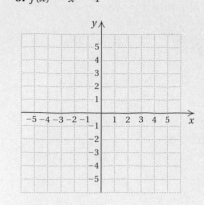

**10.** $g(x) = 5 - x^2$

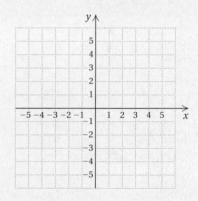

**11.** $t(x) = 3 - |x|$

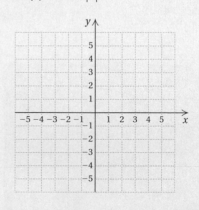

*Answers on page A-41*

Determine whether each of the following is the graph of a function.

**12.**

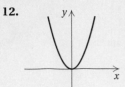

**13.**

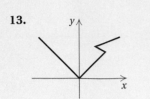

**14.**

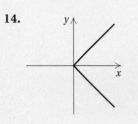

**15.**

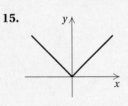

Referring to the graph in Example 8:

**16.** What was the movie revenue for week 2?

**17.** What was the movie revenue for week 6?

*Answers on page A-41*

**c)** The graph *is* that of a function.

**d)** The graph *is not* that of a function. There is a vertical line that crosses the graph more than once.

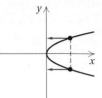

*Do Exercises 12–15.*

## e | Applications of Functions and Their Graphs

Functions are often described by graphs, whether or not an equation is given. To use a graph in an application, we note that each point on the graph represents a pair of values.

**Example 8**  *Movie Revenue.* The following graph approximates the weekly revenue, in millions of dollars, from the recent movie *Jurassic Park—The Lost World.* The revenue is a function of the week, and no equation is given for the function.

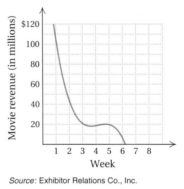

*Source*: Exhibitor Relations Co., Inc.

Use the graph to answer the following.

**a)** What was the movie revenue for week 1?

**b)** What was the movie revenue for week 5?

**a)** To estimate the revenue for week 1, we locate 1 on the horizontal axis and move directly up until we reach the graph. Then we move across to the vertical axis. We estimate that value to be about $105 million.

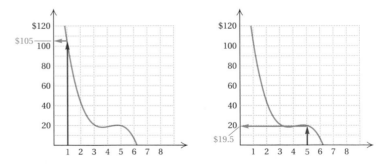

**b)** To estimate the revenue for week 5, we locate 5 on the horizontal axis and move directly up until we reach the graph. Then we move across to the vertical axis. We estimate that value to be about $19.5 million.

*Do Exercises 16 and 17.*

# Exercise Set 10.7

**a** Determine whether the correspondence is a function.

**1.** Domain → Range
2 → 9
5 → 8
19 →

**2.** Domain → Range
5 → 3
-3 → 7
7 →
-7 →

**3.** Domain → Range
-5 → 1
5 →
8 →

**4.** Domain → Range
6 → -6
7 → -7
3 → -3

**5.** Domain → Range
Texas ⟷ Austin
Houston
Dallas
Ohio ⟷ Cleveland
Toledo
Cincinnati

**6.** Domain → Range
Austin →
Houston → Texas
Dallas →
Cleveland →
Toledo → Ohio
Cincinnati →

**7.** Domain (Year) → Range (Consumption of Diet Cola, in gallons per person)
1991 → 11.7
1992 → 11.6
1993 → 11.7
1994 → 11.9

*Source*: U. S. Department of Agriculture Economic Research Service

**8.** Domain (Brand of Single-Serving Pizza) → Range (Number of Calories)
Old Chicago Pizza-lite → 324
Weight Watchers Cheese →
Banquet Zap Cheese → 310
Lean Cuisine Cheese →
Pizza Hut Supreme Personal Pan → 647
Celeste Suprema Pizza-For-One → 678

Determine whether each of the following is a function. Identify any relations that are not functions.

| | Domain | Correspondence | Range |
|---|---|---|---|
| **9.** | A math class | Each person's seat number | A set of numbers |
| **10.** | A set of numbers | Square each number and then add 4. | A set of numbers |
| **11.** | A set of shapes | Find the area of each shape. | A set of numbers |
| **12.** | A family | Each person's eye color | A set of colors |
| **13.** | The people in a town | Each person's aunt | A set of females |
| **14.** | A set of avenues | Find an intersecting road. | A set of cross streets |

**b** Find the function values.

**15.** $f(x) = x + 5$
a) $f(4)$    b) $f(7)$
c) $f(-3)$    d) $f(0)$
e) $f(2.4)$    f) $f\left(\frac{2}{3}\right)$

**16.** $g(t) = t - 6$
a) $g(0)$    b) $g(6)$
c) $g(13)$    d) $g(-1)$
e) $g(-1.08)$    f) $g\left(\frac{7}{8}\right)$

**17.** $h(p) = 3p$
a) $h(-7)$    b) $h(5)$
c) $h(14)$    d) $h(0)$
e) $h\left(\frac{2}{3}\right)$    f) $h(-54.2)$

**18.** $f(x) = -4x$

   **a)** $f(6)$         **b)** $f\left(-\frac{1}{2}\right)$

   **c)** $f(20)$      **d)** $f(11.8)$

   **e)** $f(0)$        **f)** $f(-1)$

**19.** $g(s) = 3s + 4$

   **a)** $g(1)$       **b)** $g(-7)$

   **c)** $g(6.7)$     **d)** $g(0)$

   **e)** $g(-10)$   **f)** $g\left(\frac{2}{3}\right)$

**20.** $h(x) = 19$, a constant function

   **a)** $h(4)$       **b)** $h(-6)$

   **c)** $h(12.5)$    **d)** $h(0)$

   **e)** $h\left(\frac{2}{3}\right)$     **f)** $h(1234)$

**21.** $f(x) = 2x^2 - 3x$

   **a)** $f(0)$       **b)** $f(-1)$

   **c)** $f(2)$       **d)** $f(10)$

   **e)** $f(-5)$     **f)** $f(-10)$

**22.** $f(x) = 3x^2 - 2x + 1$

   **a)** $f(0)$       **b)** $f(1)$

   **c)** $f(-1)$     **d)** $f(10)$

   **e)** $f(2)$       **f)** $f(-3)$

**23.** $f(x) = |x| + 1$

   **a)** $f(0)$       **b)** $f(-2)$

   **c)** $f(2)$       **d)** $f(-3)$

   **e)** $f(-10)$    **f)** $f(22)$

**24.** $g(t) = \sqrt{t}$

   **a)** $g(4)$       **b)** $g(25)$

   **c)** $g(16)$     **d)** $g(100)$

   **e)** $g(50)$     **f)** $g(84)$

**25.** $f(x) = x^3$

   **a)** $f(0)$       **b)** $f(-1)$

   **c)** $f(2)$       **d)** $f(10)$

   **e)** $f(-5)$     **f)** $f(-10)$

**26.** $f(x) = x^4 - 3$

   **a)** $f(1)$       **b)** $f(-1)$

   **c)** $f(0)$       **d)** $f(2)$

   **e)** $f(-2)$     **f)** $f(10)$

**27.** *Estimating Heights.* An anthropologist can estimate the height of a male or a female, given the lengths of certain bones. A *humerus* is the bone from the elbow to the shoulder. The height, in centimeters, of a female with a humerus of $x$ centimeters is given by the function

$$F(x) = 2.75x + 71.48.$$

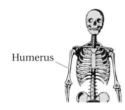

Humerus

If a humerus is known to be from a female, how tall was she if the bone is **(a)** 32 cm long? **(b)** 30 cm long?

**28.** Refer to Exercise 27. When a humerus is from a male, the function

$$M(x) = 2.89x + 70.64$$

can be used to find the male's height, in centimeters. If a humerus is known to be from a male, how tall was he if the bone is **(a)** 30 cm long? **(b)** 35 cm long?

**29.** *Pressure at Sea Depth.* The function $P(d) = 1 + (d/33)$ gives the pressure, in *atmospheres* (atm), at a depth of $d$ feet in the sea. Note that $P(0) = 1$ atm, $P(33) = 2$ atm, and so on. Find the pressure at 20 ft, 30 ft, and 100 ft.

**30.** *Temperature as a Function of Depth.* The function $T(d) = 10d + 20$ gives the temperature, in degrees Celsius, inside the earth as a function of the depth $d$, in kilometers. Find the temperature at 5 km, 20 km, and 1000 km.

Copyright © 1999 Addison Wesley Longman

**31.** *Melting Snow.* The function $W(d) = 0.112d$ approximates the amount, in centimeters, of water that results from $d$ centimeters of snow melting. Find the amount of water that results from snow melting from depths of 16 cm, 25 cm, and 100 cm.

**32.** *Temperature Conversions.* The function $C(F) = \frac{5}{9}(F - 32)$ determines the Celsius temperature that corresponds to $F$ degrees Fahrenheit. Find the Celsius temperature that corresponds to 62°F, 77°F, and 23°F.

c Graph the function.

**33.** $f(x) = 3x - 1$

**34.** $g(x) = 2x + 5$

**35.** $g(x) = -2x + 3$

**36.** $f(x) = -\frac{1}{2}x + 2$

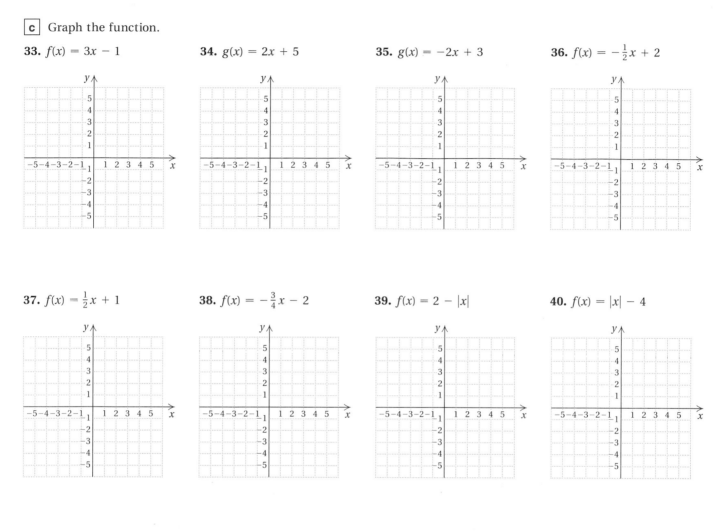

**37.** $f(x) = \frac{1}{2}x + 1$

**38.** $f(x) = -\frac{3}{4}x - 2$

**39.** $f(x) = 2 - |x|$

**40.** $f(x) = |x| - 4$

**41.** $f(x) = x^2$

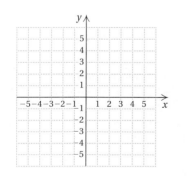

**42.** $f(x) = x^2 - 1$

**43.** $f(x) = x^2 - x - 2$

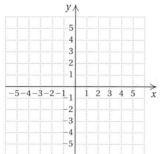

**44.** $f(x) = x^2 + 6x + 5$

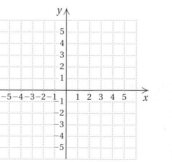

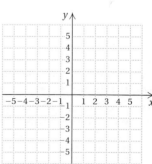

Copyright © 1998 Addison Wesley Longman

**d** Determine whether each of the following is the graph of a function.

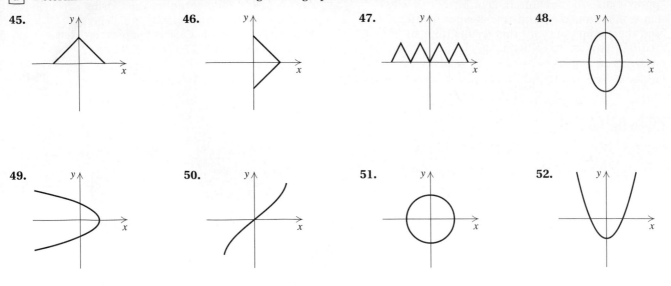

45.

46.

47.

48.

49.

50.

51.

52.

**e** *Cholesterol Level and Risk of a Heart Attack.* The graph below shows the annual heart attack rate per 10,000 men as a function of blood cholesterol level.

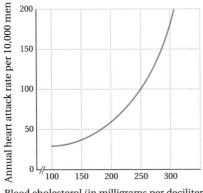

Blood cholesterol (in milligrams per deciliter)

Source: Copyright 1989, CSPI. Adapted from
Nutrition Action Healthletter
(1875 Connecticut Avenue, N.W., Suite 300,
Washington, DC 20009-5728)

**53.** Approximate the annual heart attack rate per 10,000 men for those whose blood cholesterol level is 225 mg/dl.

**54.** Approximate the annual heart attack rate per 10,000 men for those whose blood cholesterol level is 275 mg/dl.

## Skill Maintenance

Determine whether the pair of equations represents parallel lines.  [7.3a]

**55.** $y = \frac{3}{4}x - 7$,
$3x + 4y = 7$

**56.** $y = \frac{3}{5}$,
$y = -\frac{5}{3}$

Solve the system using the substitution method.  [8.2b]

**57.** $2x - y = 6$,
$4x - 2y = 5$

**58.** $x - 3y = 2$,
$3x - 9y = 6$

## Synthesis

**59.** ◈ Is it possible for a function to have more numbers as outputs than as inputs? Why or why not?

**60.** ◈ Look up the word "function" in a dictionary. Explain how that definition might be related to the mathematical one given in this section.

# Summary and Review Exercises: Chapter 10

## Important Properties and Formulas

| | |
|---|---|
| *Standard Form*: | $ax^2 + bx + c = 0$, $a > 0$ |
| *Principle of Square Roots*: | The equation $x^2 = d$, where $d > 0$, has two solutions, $\sqrt{d}$ and $-\sqrt{d}$. |
| | The solution of $x^2 = 0$ is 0. |
| *Quadratic Formula*: | $x = \dfrac{-b \pm \sqrt{b^2 - 4ac}}{2a}$ |
| *Discriminant*: | $b^2 - 4ac$ |

The $x$-coordinate of the vertex of a parabola $= -\dfrac{b}{2a}$.

The objectives to be tested in addition to the material in this chapter are [7.5c], [9.2c], [9.4a], and [9.6b].

Solve.

**1.** $8x^2 = 24$   [10.2a]

**2.** $40 = 5y^2$   [10.2a]

**3.** $5x^2 - 8x + 3 = 0$   [10.1c]

**4.** $3y^2 + 5y = 2$   [10.1c]

**5.** $(x + 8)^2 = 13$   [10.2b]

**6.** $9x^2 = 0$   [10.2a]

**7.** $5t^2 - 7t = 0$   [10.1b]

Solve.   [10.3a]

**8.** $x^2 - 2x - 10 = 0$

**9.** $9x^2 - 6x - 9 = 0$

**10.** $x^2 + 6x = 9$

**11.** $1 + 4x^2 = 8x$

**12.** $6 + 3y = y^2$

**13.** $3m = 4 + 5m^2$

**14.** $3x^2 = 4x$

Solve.   [10.1c]

**15.** $\dfrac{15}{x} - \dfrac{15}{x + 2} = 2$

**16.** $x + \dfrac{1}{x} = 2$

Solve by completing the square. Show your work. [10.2c]

**17.** $x^2 - 5x + 2 = 0$

**18.** $3x^2 - 2x - 5 = 0$

Approximate the solutions to the nearest tenth.   [10.3b]

**19.** $x^2 - 5x + 2 = 0$

**20.** $4y^2 + 8y + 1 = 0$

**21.** Solve for $T$: $V = \dfrac{1}{2}\sqrt{1 + \dfrac{T}{L}}$.   [10.4a]

Graph the quadratic equation.   [10.6a]

**22.** $y = 2 - x^2$

**23.** $y = x^2 - 4x - 2$

Find the $x$-intercepts.   [10.6b]

**24.** $y = 2 - x^2$

**25.** $y = x^2 - 4x - 2$

Solve.

**26.** The hypotenuse of a right triangle is 5 cm long. One leg is 3 m longer than the other. Find the lengths of the legs. Round to the nearest tenth.   [10.5a]

**27.** The hypotenuse of a right triangular freight ramp is 26 yd long. One leg is 14 yd longer than the other. Find the lengths of the legs. [10.5a]

**28.** The height of Lake Point Towers in Chicago is 645 ft. How long would it take an object to fall to the ground from the top? [10.2d]

Find the function values. [10.7b]

**29.** If $f(x) = 2x - 5$, find $f(2)$, $f(-1)$, and $f(3.5)$.

**30.** If $g(x) = |x| - 1$, find $g(1)$, $g(-1)$, and $g(-20)$.

**31.** *Caloric Needs.* If you are moderately active, you need to consume each day about 15 calories per pound of body weight. The function $C(p) = 15p$ approximates the number of calories $C$ that are needed to maintain body weight $p$, in pounds. How many calories are needed to maintain a body weight of 180 lb? [10.7e]

Graph the function. [10.7c]

**32.** $g(x) = 4 - x$        **33.** $f(x) = x^2 - 3$

**34.** $h(x) = |x| - 5$

Determine whether each of the following is the graph of a function. [10.7d]

**35.**          **36.**

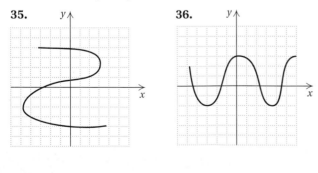

Multiply and simplify. [9.2c]

**37.** $\sqrt{18a}\,\sqrt{2}$        **38.** $\sqrt{12xy^2}\,\sqrt{5xy}$

**39.** Find an equation of variation in which $y$ varies inversely as $x$ and $y = 10$ when $x = 0.0625$. [7.5c]

**40.** The sides of a rectangle are of lengths 1 and $\sqrt{2}$. Find the length of a diagonal. [9.6b]

Add or subtract. [9.4a]

**41.** $5\sqrt{11} + 7\sqrt{11}$        **42.** $2\sqrt{90} - \sqrt{40}$

**Synthesis**

**43.** ◈ List the names and give an example of as many types of equation as you can that you have learned to solve in this text. [2.1a], [6.6a], [8.1a], [9.5a], [10.1a]

**44.** ◈ Find the errors in each of the following solutions of equations. [10.1b]

a) $x^2 + 20x = 0$
$x(x + 20) = 0$
$x + 20 = 0$
$x = 20$

b) $x^2 + x = 6$
$x(x + 1) = 6$
$x = 6$ or $x + 1 = 6$
$x = 6$ or $x = 5$

**45.** Two consecutive integers have squares that differ by 63. Find the integers. [10.5a]

**46.** A square with sides of length $s$ has the same area as a circle with a radius of 5 in. Find $s$. [10.5a]

**47.** Solve: $x - 4\sqrt{x} - 5 = 0$. [10.1c]

# Test: Chapter 10

Solve.

**1.** $7x^2 = 35$

**2.** $7x^2 + 8x = 0$

**3.** $48 = t^2 + 2t$

**4.** $3y^2 - 5y = 2$

**5.** $(x - 8)^2 = 13$

**6.** $x^2 = x + 3$

**7.** $m^2 - 3m = 7$

**8.** $10 = 4x + x^2$

**9.** $3x^2 - 7x + 1 = 0$

**10.** $x - \dfrac{2}{x} = 1$

**11.** $\dfrac{4}{x} - \dfrac{4}{x + 2} = 1$

**12.** Solve $x^2 - 4x - 10 = 0$ by completing the square. Show your work.

**13.** Approximate the solutions to $x^2 - 4x - 10 = 0$ to the nearest tenth.

**14.** Solve for $n$: $d = an^2 + bn$.

**15.** Find the $x$-intercepts:
$y = -x^2 + x + 5$.

Graph.

**16.** $y = 4 - x^2$

**17.** $y = -x^2 + x + 5$

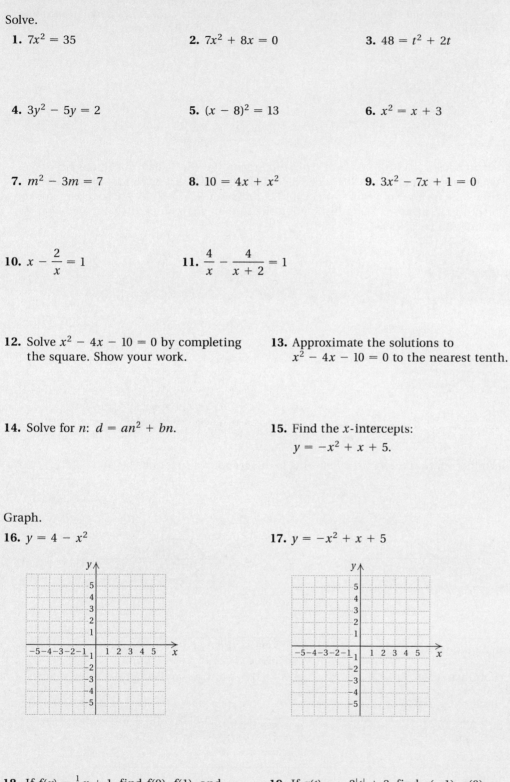

**18.** If $f(x) = \frac{1}{2}x + 1$, find $f(0)$, $f(1)$, and $f(2)$.

**19.** If $g(t) = -2|t| + 3$, find $g(-1)$, $g(0)$, and $g(3)$.

## Answers

1. _____

2. _____

3. _____

4. _____

5. _____

6. _____

7. _____

8. _____

9. _____

10. _____

11. _____

12. _____

13. _____

14. _____

15. _____

16. _____

17. _____

18. _____

19. _____

Solve.

**20.** The width of a rectangular area rug is 4 m less than the length. The area is 16.25 m². Find the length and the width.

20. _____

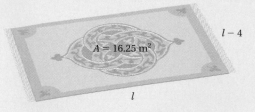

$A = 16.25$ m²

$l - 4$

$l$

**21.** The current in a stream moves at a speed of 2 km/h. A boat travels 44 km upstream and 52 km downstream in a total of 4 hr. What is the speed of the boat in still water?

21. _____

22. _____

**22.** *World Record for 10,000-m Run.* The world record for the 10,000-m run has been decreasing steadily since 1940. The record is approximately 30.18 min minus 0.06 times the number of years since 1940. The function $R(t) = 30.18 - 0.06t$ estimates the record $R$, in minutes, as a function of $t$, the time in years since 1940. Predict what the record will be in 2000.

23. _____

Graph.

**23.** $h(x) = x - 4$

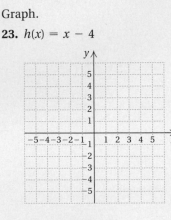

**24.** $g(x) = x^2 - 4$

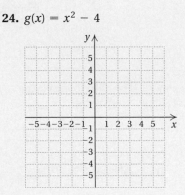

24. _____

25. _____

Determine whether each of the following is the graph of a function.

**25.**

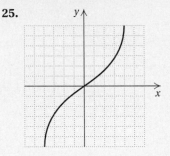

**26.**

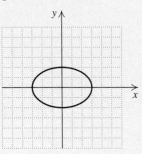

26. _____

27. _____

28. _____

---

**Skill Maintenance**

**27.** Subtract: $\sqrt{240} - \sqrt{60}$.

**28.** Multiply and simplify: $\sqrt{7xy}\,\sqrt{14x^2y}$.

**29.** Find an equation of variation in which $y$ varies inversely as $x$ and $y = 32$ when $x = 0.125$.

**30.** The sides of a rectangle are of lengths $\sqrt{2}$ and $\sqrt{3}$. Find the length of a diagonal.

29. _____

30. _____

---

**Synthesis**

31. _____

**31.** Find the side of a square whose diagonal is 5 ft longer than a side.

**32.** Solve this system for $x$. Use the substitution method.

$$x - y = 2,$$
$$xy = 4$$

32. _____

# Cumulative Review: Chapters 1–10

**1.** What is the meaning of $x^3$?

**2.** Evaluate $(x - 3)^2 + 5$ for $x = 10$.

**3.** Find decimal notation: $-\dfrac{3}{11}$.

**4.** Find the LCM of 15 and 48.

**5.** Find the absolute value: $|-7|$.

Compute and simplify.

**6.** $-6 + 12 + (-4) + 7$

**7.** $2.8 - (-12.2)$

**8.** $-\dfrac{3}{8} \div \dfrac{5}{2}$

**9.** $13 \cdot 6 \div 3 \cdot 2 \div 13$

**10.** Remove parentheses and simplify: $4m + 9 - (6m + 13)$.

Solve.

**11.** $3x = -24$

**12.** $3x + 7 = 2x - 5$

**13.** $3(y - 1) - 2(y + 2) = 0$

**14.** $x^2 - 8x + 15 = 0$

**15.** $y - x = 1,$
$\quad y = 3 - x$

**16.** $x + y = 17,$
$\quad x - y = 7$

**17.** $4x - 3y = 3,$
$\quad 3x - 2y = 4$

**18.** $x^2 - x - 6 = 0$

**19.** $x^2 + 3x = 5$

**20.** $3 - x = \sqrt{x^2 - 3}$

**21.** $5 - 9x \le 19 + 5x$

**22.** $-\dfrac{7}{8}x + 7 = \dfrac{3}{8}x - 3$

**23.** $0.6x - 1.8 = 1.2x$

**24.** $-3x > 24$

**25.** $23 - 19y - 3y \ge -12$

**26.** $3y^2 = 30$

**27.** $(x - 3)^2 = 6$

**28.** $\dfrac{6x - 2}{2x - 1} = \dfrac{9x}{3x + 1}$

**29.** $\dfrac{2x}{x + 1} = 2 - \dfrac{5}{2x}$

**30.** $\dfrac{2x}{x + 3} + \dfrac{6}{x} + 7 = \dfrac{18}{x^2 + 3x}$

**31.** $\sqrt{x + 9} = \sqrt{2x - 3}$

Solve the formula for the given letter.

**32.** $A = \dfrac{4b}{t}$, for $b$

**33.** $\dfrac{1}{t} = \dfrac{1}{m} - \dfrac{1}{n}$, for $m$

**34.** $r = \sqrt{\dfrac{A}{\pi}}$, for $A$

**35.** $y = ax^2 - bx$, for $x$

Simplify.

**36.** $x^{-6} \cdot x^2$

**37.** $\dfrac{y^3}{y^{-4}}$

**38.** $(2y^6)^2$

**39.** Collect like terms and arrange in descending order: $2x - 3 + 5x^3 - 2x^3 + 7x^3 + x$.

Compute and simplify.

**40.** $(4x^3 + 3x^2 - 5) + (3x^3 - 5x^2 + 4x - 12)$

**41.** $(6x^2 - 4x + 1) - (-2x^2 + 7)$

**42.** $-2y^2(4y^2 - 3y + 1)$

**43.** $(2t - 3)(3t^2 - 4t + 2)$

**44.** $\left(t - \dfrac{1}{4}\right)\left(t + \dfrac{1}{4}\right)$

**45.** $(3m - 2)^2$

**46.** $(15x^2y^3 + 10xy^2 + 5) - (5xy^2 - x^2y^2 - 2)$

**47.** $(x^2 - 0.2y)(x^2 + 0.2y)$

**48.** $(3p + 4q^2)^2$

**49.** $\dfrac{4}{2x - 6} \cdot \dfrac{x - 3}{x + 3}$

**50.** $\dfrac{3a^4}{a^2 - 1} \div \dfrac{2a^3}{a^2 - 2a + 1}$

**51.** $\dfrac{3}{3x - 1} + \dfrac{4}{5x}$

**52.** $\dfrac{2}{x^2 - 16} - \dfrac{x - 3}{x^2 - 9x + 20}$

Factor.

**53.** $8x^2 - 4x$

**54.** $25x^2 - 4$

**55.** $6y^2 - 5y - 6$

**56.** $m^2 - 8m + 16$

**57.** $x^3 - 8x^2 - 5x + 40$

**58.** $3a^4 + 6a^2 - 72$

**59.** $16x^4 - 1$

**60.** $49a^2b^2 - 4$

**61.** $9x^2 + 30xy + 25y^2$

**62.** $2ac - 6ab - 3db + dc$

**63.** $15x^2 + 14xy - 8y^2$

Simplify.

**64.** $\dfrac{\dfrac{3}{x} + \dfrac{1}{2x}}{\dfrac{1}{3x} - \dfrac{3}{4x}}$

**65.** $\sqrt{49}$

**66.** $-\sqrt{625}$

**67.** $\sqrt{64x^2}$

**68.** Multiply: $\sqrt{a+b}\,\sqrt{a-b}$.

**69.** Multiply and simplify: $\sqrt{32ab}\,\sqrt{6a^4b^2}$.

Simplify.

**70.** $\sqrt{150}$

**71.** $\sqrt{243x^3y^2}$

**72.** $\sqrt{\dfrac{100}{81}}$

**73.** $\sqrt{\dfrac{64}{x^2}}$

**74.** $4\sqrt{12} + 2\sqrt{12}$

**75.** Divide and simplify: $\dfrac{\sqrt{72}}{\sqrt{45}}$.

**76.** In a right triangle, $a = 9$ and $c = 41$. Find $b$.

Graph on a plane.

**77.** $y = \dfrac{1}{3}x - 2$

**78.** $2x + 3y = -6$

**79.** $y = -3$

**80.** $4x - 3y > 12$

**81.** $y = x^2 + 2x + 1$

**82.** $x \geq -3$

**83.** Solve $9x^2 - 12x - 2 = 0$ by completing the square. Show your work.

**84.** Approximate the solutions of $4x^2 = 4x + 1$ to the nearest tenth.

Solve.

**85.** What percent of 52 is 13?

**86.** 12 is 20% of what?

**87.** The speed of a boat in still water is 8 km/h. It travels 60 km upstream and 60 km downstream in a total time of 16 hr. What is the speed of the stream?

**88.** The length of a rectangle is 7 m more than the width. The length of a diagonal is 13 m. Find the length.

**89.** Three-fifths of the automobiles entering the city each morning will be parked in city parking lots. There are 3654 such parking spaces filled each morning. How many cars enter the city each morning?

**90.** A candy shop wants to mix nuts worth $1.10 per pound with another variety worth $0.80 per pound in order to make 42 lb of a mixture worth $0.90 per pound. How many pounds of each kind of nuts should be used?

**91.** In checking records, a contractor finds that crew A can resurface a tennis court in 8 hr. Crew B can do the same job in 10 hr. How long would they take if they worked together?

**92.** A student's paycheck varies directly as the number of hours worked. The pay was $242.52 for 43 hr of work. What would the pay be for 80 hr of work? Explain the meaning of the variation constant.

**93.** Determine whether the graphs of the following equations are parallel, perpendicular, or neither.

$$y - x = 4,$$
$$3y + x = 8$$

**94.** Find the slope and the $y$-intercept:

$$-6x + 3y = -24.$$

**95.** Find the slope of the line containing the points $(-5, -6)$ and $(-4, 9)$.

**96.** For the function $f$ described by
$$f(x) = 2x^2 + 7x - 4,$$
find $f(0)$, $f(-4)$, and $f\left(\frac{1}{2}\right)$.

Find an equation of variation in which:

**97.** $y$ varies directly as $x$ and $y = 100$ when $x = 10$.

**98.** $y$ varies inversely as $x$ and $y = 100$ when $x = 10$.

Determine whether each of the following is the graph of a function.

**99.**

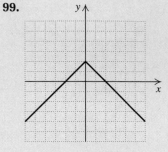

**100.**

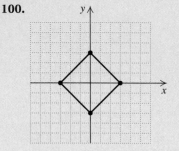

Graph the function.

**101.** $f(x) = x^2 + x - 2$

**102.** $g(x) = |x + 2|$

**103.** Find the $x$-intercepts of $y = x^2 + 4x + 1$.

---

## Synthesis

**104.** Solve: $|x| = 12$.

**105.** Simplify: $\sqrt{\sqrt{\sqrt{81}}}$.

**106.** Find $b$ such that the trinomial $x^2 - bx + 225$ is a square.

**107.** Find $x$.

**108.** $f(x) = |x| + x$

**109.** $g(x) = x^3 - 2$

Graph the function.

Determine whether the pair of expressions is equivalent.

**110.** $x^2 - 9$,  $(x - 3)(x + 3)$

**111.** $\dfrac{x + 3}{3}$,  $x$

**112.** $(x + 5)^2$,  $x^2 + 25$

**113.** $\sqrt{x^2 + 16}$,  $x + 4$

**114.** $\sqrt{x^2}$,  $|x|$

---

# Final Examination

---

**1.** Evaluate $x^3 + 5$ for $x = -10$.

**2.** Find the LCM of 16 and 24.

**3.** Find the absolute value of $|-9|$.

Compute and simplify.

**4.** $-6.3 + (-8.4) + 5$

**5.** $-8 - (-3)$

**6.** $\dfrac{3}{11} \cdot \left(-\dfrac{22}{7}\right)$

**7.** Remove parentheses and simplify:
$4y - 5(9 - 3y)$.

**8.** Simplify:
$2^3 - 14 \cdot 10 + (3 + 4)^3$.

Solve.

**9.** $x + 8 = 13.6$

**10.** $4x = -28$

**11.** $5x + 3 = 2x - 27$

**12.** $5(x - 3) - 2(x + 3) = 0$

**13.** $x^2 - 2x - 24 = 0$

**14.** $y - x - 7,$
$2x + y = 5$

**15.** $5x - 3y = -1,$
$4x + 2y = 30$

**16.** $\dfrac{1}{x} - 2 = 8x$

**17.** $\sqrt{x^2 - 11} = x - 1$

**18.** $x^2 = 7 - 3x$

**19.** $2 - 3x \le 12 - 7x$

Solve the formula for the given letter.

**20.** $A = \dfrac{Bw + 1}{w}$, for $w$

**21.** $K = MT + 2$, for $M$

Simplify.

**22.** $\dfrac{x^8}{x^{-2}}$

**23.** $(x^{-5})^2$

**24.** $x^{-5} \cdot x^{-7}$

**Answers**

1. _____

2. _____

3. _____

4. _____

5. _____

6. _____

7. _____

8. _____

9. _____

10. _____

11. _____

12. _____

13. _____

14. _____

15. _____

16. _____

17. _____

18. _____

19. _____

20. _____

21. _____

22. _____

23. _____

24. _____

25. _____

26. _____

27. _____

28. _____

29. _____

30. _____

31. _____

32. _____

33. _____

34. _____

35. _____

36. _____

37. _____

38. _____

39. _____

40. _____

41. _____

42. _____

43. _____

44. _____

45. _____

46. _____

**25.** Collect like terms and arrange in descending order:

$$2y^3 - 3 + 4y^3 - 3y^2 + 12 - y.$$

Compute and simplify.

**26.** $(2x^2 - 6x + 3) - (4x^2 + 2x - 4)$      **27.** $-3t^2(2t^4 + 4t^2 + 1)$

**28.** $(4x - 1)(x^2 - 5x + 2)$      **29.** $(x - 8)(x + 8)$

**30.** $(2m - 7)^2$      **31.** $(3ab^2 + 2c)^2$

**32.** $(3x^2 - 2y)(3x^2 + 4y)$      **33.** $\dfrac{x}{x^2 - 9} \cdot \dfrac{x - 3}{x^3}$

**34.** $\dfrac{3x^5}{4x - 4} \div \dfrac{x}{x^2 - 2x + 1}$      **35.** $\dfrac{2}{3x - 1} + \dfrac{1}{4x}$

**36.** $\dfrac{3}{x - 3} - \dfrac{x - 1}{x^2 - 2x - 3}$

Factor.

**37.** $3x^3 - 15x$      **38.** $16x^2 - 25$      **39.** $6x^2 - 13x + 6$

**40.** $x^2 - 10x + 25$      **41.** $2ax + 6bx - ay - 3by$      **42.** $x^8 - 81y^4$

Simplify.

**43.** $\sqrt{72}$      **44.** $\dfrac{\sqrt{54}}{\sqrt{45}}$

**45.** $2\sqrt{8} + 3\sqrt{18}$      **46.** $\sqrt{24a^2b}\sqrt{a^3b^2}$

Copyright © 1999 Addison Wesley Longman

Graph on a plane.

**47.** $3x + 2y = -4$

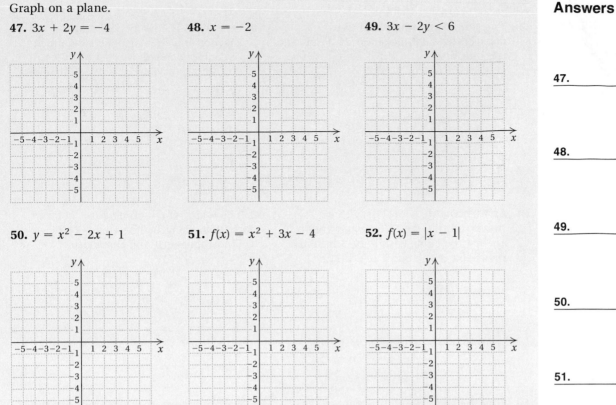

**48.** $x = -2$

**49.** $3x - 2y < 6$

**50.** $y = x^2 - 2x + 1$

**51.** $f(x) = x^2 + 3x - 4$

**52.** $f(x) = |x - 1|$

**53.** For the function $f$ described by $f(x) = 4 + 3x - x^2$, find $f(0)$, $f(1)$, and $f(-2)$.

Determine whether each of the following is the graph of a function.

**54.**

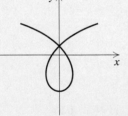

**55.**

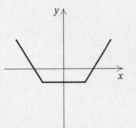

**56.** Find the $x$-intercepts: $y = x^2 - 4x + 1$.

**57.** Use *only* the graph below to solve $x^2 + x - 6 = 0$.

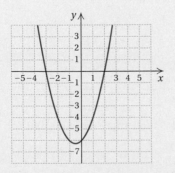

**58.** Estimate the missing data value.

| Year | Credit-Card Spending from Thanksgiving to Christmas (in billions) |
|------|------|
| 1991 | $ 59.8 |
| 1992 | 66.8 |
| 1993 | 79.1 |
| 1994 | 96.9 |
| 1995 | 116.3 |
| 1996 | 131.4 |
| 1997 | ? |

*Source*: RAM Research Group, National Credit Counseling Services

**Answers**

47. _____

48. _____

49. _____

50. _____

51. _____

52. _____

53. _____

54. _____

55. _____

56. _____

57. _____

58. _____

Solve.

**59.** The sum of the squares of two consecutive odd integers is 74. Find the integers.

**60.** Solution A is 75% alcohol and solution B is 50% alcohol. How much of each is needed in order to make 60 L of a solution that is $66\frac{2}{3}$% alcohol?

**61.** An airplane flew for 6 hr with a 10-km/h tail wind. The return flight against the same wind took 8 hr. Find the speed of the plane in still air.

**62.** The width of a rectangle is 3 m less than the length. The area is 88 m². Find the length and the width.

**63.** Find the slope of the line containing the points $(-2, 3)$ and $(4, -5)$.

**64.** Determine whether the graphs of the following equations are parallel, perpendicular, or neither.

$$y = 2x + 7,$$
$$2y + x = 6$$

Find an equation of variation in which:

**65.** $y$ varies directly as $x$ and $y = 200$ when $x = 25$.

**66.** $y$ varies inversely as $x$ and $y = 200$ when $x = 25$.

### Synthesis

**67.** A side of a square is 5 less than a side of an equilateral triangle. The perimeter of the square is the same as the perimeter of the triangle. Find the length of a side of the square and the length of a side of the triangle.

**68.** Find $c$ such that the trinomial $x^2 - 24x + c$ is a square.

Copyright © 1999 Addison Wesley Longman

# Appendix A   Factoring Sums or Differences of Cubes

**a**   We can factor the sum or the difference of two expressions that are cubes.

Consider the following products:

$$(A + B)(A^2 - AB + B^2) = A(A^2 - AB + B^2) + B(A^2 - AB + B^2)$$
$$= A^3 - A^2B + AB^2 + A^2B - AB^2 + B^3$$
$$= A^3 + B^3$$

and   $(A - B)(A^2 + AB + B^2) = A(A^2 + AB + B^2) - B(A^2 + AB + B^2)$
$$= A^3 + A^2B + AB^2 - A^2B - AB^2 - B^3$$
$$= A^3 - B^3.$$

The above equations (reversed) show how we can factor a sum or a difference of two cubes.

> $A^3 + B^3 = (A + B)(A^2 - AB + B^2),$
> $A^3 - B^3 = (A - B)(A^2 + AB + B^2)$

Note that what we are considering here is a sum or a difference of cubes. We are not cubing a binomial. For example, $(A + B)^3$ is *not* the same as $A^3 + B^3$. The table of cubes in the margin is helpful.

| N | $N^3$ |
|---|---|
| 0.2 | 0.008 |
| 0.1 | 0.001 |
| 0 | 0 |
| 1 | 1 |
| 2 | 8 |
| 3 | 27 |
| 4 | 64 |
| 5 | 125 |
| 6 | 216 |
| 7 | 343 |
| 8 | 512 |
| 9 | 729 |
| 10 | 1000 |

**Example 1**   Factor: $x^3 - 27$.

We have

$$x^3 - 27 = x^3 - 3^3.$$

In one set of parentheses, we write the cube root of the first term, $x$. Then we write the cube root of the second term, $-3$. This gives us the expression $x - 3$:

$$(x - 3)(\qquad).$$

To get the next factor, we think of $x - 3$ and do the following:

Square the first term: $x^2$.
Multiply the terms and then change the sign: $3x$.
Square the second term: 9.

$$(x - 3)(x^2 + 3x + 9).$$

Note that we cannot factor $x^2 + 3x + 9$. It is not a trinomial square nor can it be factored by trial and error.

*Do Exercises 1 and 2.*

**Example 2**   Factor: $125x^3 + y^3$.

We have

$$125x^3 + y^3 = (5x)^3 + y^3.$$

In one set of parentheses, we write the cube root of the first term, $5x$. Then we write a plus sign, and then the cube root of the second term, $y$:

$$(5x + y)(\qquad).$$

Factor.

**1.** $x^3 - 8$

**2.** $64 - y^3$

*Answers on page A-44*

Factor.

**3.** $27x^3 + y^3$

**4.** $8y^3 + z^3$

Factor.

**5.** $m^6 - n^6$

**6.** $16x^7y + 54xy^7$

**7.** $729x^6 - 64y^6$

**8.** $x^3 - 0.027$

*Answers on page A-44*

To get the next factor, we think of $5x + y$ and do the following:

Square the first term: $25x^2$.
Multiply the terms and then change the sign: $-5xy$.
Square the second term: $y^2$.

$(5x + y)(25x^2 - 5xy + y^2).$

***Do Exercises 3 and 4.***

**Example 3**   Factor: $128y^7 - 250x^6y$.

We first look for a common factor:

$$128y^7 - 250x^6y = 2y(64y^6 - 125x^6) = 2y[(4y^2)^3 - (5x^2)^3]$$
$$= 2y(4y^2 - 5x^2)(16y^4 + 20x^2y^2 + 25x^4).$$

**Example 4**   Factor: $a^6 - b^6$.

We can express this polynomial as a difference of squares:

$$(a^3)^2 - (b^3)^2.$$

We factor as follows:

$$a^6 - b^6 = (a^3 + b^3)(a^3 - b^3).$$

One factor is a sum of two cubes, and the other factor is a difference of two cubes. We factor them:

$$(a + b)(a^2 - ab + b^2)(a - b)(a^2 + ab + b^2).$$

We have now factored completely.

In Example 4, had we thought of factoring first as a difference of two cubes, we would have had

$$(a^2)^3 - (b^2)^3 = (a^2 - b^2)(a^4 + a^2b^2 + b^4)$$
$$= (a + b)(a - b)(a^4 + a^2b^2 + b^4).$$

In this case, we might have missed some factors; $a^4 + a^2b^2 + b^4$ can be factored as $(a^2 - ab + b^2)(a^2 + ab + b^2)$, but we probably would not have known to do such factoring.

**Example 5**   Factor: $64a^6 - 729b^6$.

We have

$$64a^6 - 729b^6 = (8a^3 - 27b^3)(8a^3 + 27b^3) \qquad \text{Factoring a difference of squares}$$
$$= [(2a)^3 - (3b)^3][(2a)^3 + (3b)^3].$$

Each factor is a sum or a difference of cubes. We factor each:

$$= (2a - 3b)(4a^2 + 6ab + 9b^2)(2a + 3b)(4a^2 - 6ab + 9b^2).$$

| | |
|---|---|
| Sum of cubes: | $A^3 + B^3 = (A + B)(A^2 - AB + B^2);$ |
| Difference of cubes: | $A^3 - B^3 = (A - B)(A^2 + AB + B^2);$ |
| Difference of squares: | $A^2 - B^2 = (A + B)(A - B);$ |
| Sum of squares: | $A^2 + B^2$ cannot be factored using real numbers if the largest common factor has been removed. |

***Do Exercises 5–8.***

# Exercise Set A

a Factor.

**1.** $z^3 + 27$

**2.** $a^3 + 8$

**3.** $x^3 - 1$

**4.** $c^3 - 64$

**5.** $y^3 + 125$

**6.** $x^3 + 1$

**7.** $8a^3 + 1$

**8.** $27x^3 + 1$

**9.** $y^3 - 8$

**10.** $p^3 - 27$

**11.** $8 - 27b^3$

**12.** $64 - 125x^3$

**13.** $64y^3 + 1$

**14.** $125x^3 + 1$

**15.** $8x^3 + 27$

**16.** $27y^3 + 64$

**17.** $a^3 - b^3$

**18.** $x^3 - y^3$

**19.** $a^3 + \dfrac{1}{8}$

**20.** $b^3 + \dfrac{1}{27}$

**21.** $2y^3 - 128$

**22.** $3z^3 - 3$

**23.** $24a^3 + 3$

**24.** $54x^3 + 2$

**25.** $rs^3 + 64r$

**26.** $ab^3 + 125a$

**27.** $5x^3 - 40z^3$

**28.** $2y^3 - 54z^3$

**29.** $x^3 + 0.001$

**30.** $y^3 + 0.125$

**31.** $64x^6 - 8t^6$

**32.** $125c^6 - 8d^6$

**33.** $2y^4 - 128y$

**34.** $3z^5 - 3z^2$

**35.** $z^6 - 1$

**36.** $t^6 + 1$

**37.** $t^6 + 64y^6$

**38.** $p^6 - q^6$

## Synthesis

Consider these polynomials:

$$(a + b)^3; \quad a^3 + b^3; \quad (a + b)(a^2 - ab + b^2);$$
$$(a + b)(a^2 + ab + b^2); \quad (a + b)(a + b)(a + b).$$

**39.** Evaluate each polynomial for $a = -2$ and $b = 3$.

**40.** Evaluate each polynomial for $a = 4$ and $b = -1$.

Factor. Assume that variables in exponents represent natural numbers.

**41.** $x^{6a} + y^{3b}$

**42.** $a^3x^3 - b^3y^3$

**43.** $3x^{3a} + 24y^{3b}$

**44.** $\frac{8}{27}x^3 + \frac{1}{64}y^3$

**45.** $\frac{1}{24}x^3y^3 + \frac{1}{3}z^3$

**46.** $7x^3 - \frac{7}{8}$

**47.** $(x + y)^3 - x^3$

**48.** $(1 - x)^3 + (x - 1)^6$

**49.** $(a + 2)^3 - (a - 2)^3$

**50.** $y^4 - 8y^3 - y + 8$

Copyright © 1999 Addison Wesley Longman

# Appendix B  Higher Roots

In this appendix, we study *higher* roots, such as cube roots, or fourth roots.

## Objectives

a | Find higher roots of real numbers.

b | Simplify radical expressions using the product and quotient rules.

## a | Higher Roots

Recall that $c$ is a square root of $a$ if $c^2 = a$. A similar definition can be made for *cube roots*.

> The number $c$ is the **cube root** of $a$ if $c^3 = a$.

Every real number has exactly *one* real-number cube root. The symbolism $\sqrt[3]{a}$ is used to represent the cube root of $a$. In the radical $\sqrt[3]{a}$, the number 3 is called the **index** and $a$ is called the **radicand.**

**Example 1**  Find $\sqrt[3]{8}$.

The cube root of 8 is the number whose cube is 8. Since $2^3 = 2 \cdot 2 \cdot 2 = 8$, the cube root of 8 is 2, so $\sqrt[3]{8} = 2$.

**Example 2**  Find $\sqrt[3]{-125}$.

The cube root of $-125$ is the number whose cube is $-125$. Since $(-5)^3 = (-5)(-5)(-5) = -125$, the cube root of $-125$ is $-5$, so $\sqrt[3]{-125} = -5$.

*Do Exercises 1–3.*

Positive real numbers always have *two* $n$th roots (one positive and one negative) when $n$ is even, but we refer to the *positive $n$th root* of a positive number $a$ as **the** *$n$th root* and denote it $\sqrt[n]{a}$. For example, although both $-3$ and $3$ are fourth roots of 81, since $(-3)^4 = 81$ and $3^4 = 81$, 3 is considered to be *the* fourth root of 81. In symbols, $\sqrt[4]{81} = 3$.

> The number $c$ is the *$n$th root* of $a$ if $c^n = a$.
>
> If $n$ is odd, then there is exactly one $n$th root of $a$ and $\sqrt[n]{a}$ represents that root.
>
> If $n$ is even and $a$ is positive, then $\sqrt[n]{a}$ represents the nonnegative $n$th root.
>
> Even roots of negative numbers are not real numbers.

**Examples**  Find the root of each of the following.

**3.** $\sqrt[4]{16} = 2$    Since $2^4 = 2 \cdot 2 \cdot 2 \cdot 2 = 16$

**4.** $\sqrt[4]{-16}$ is not a real number, because it is an even root of a negative number.

**5.** $\sqrt[5]{32} = 2$    Since $2^5 = 2 \cdot 2 \cdot 2 \cdot 2 \cdot 2 = 32$

**6.** $\sqrt[5]{-32} = -2$    Since $(-2)^5 = (-2)(-2)(-2)(-2)(-2) = -32$

**7.** $-\sqrt[3]{64} = -(\sqrt[3]{64})$    This is the opposite of $\sqrt[3]{64}$.
$= -4$    Since $4^3 = 4 \cdot 4 \cdot 4 = 64$

*Do Exercises 4–9.*

---

**1.** Find $\sqrt[3]{27}$.

**2.** Find $\sqrt[3]{-8}$.

**3.** Find $\sqrt[3]{216}$.

Find the root, if it exists, of each of the following.

**4.** $\sqrt[5]{1}$

**5.** $\sqrt[5]{-1}$

**6.** $\sqrt[4]{-81}$

**7.** $\sqrt[4]{81}$

**8.** $\sqrt[3]{-216}$

**9.** $-\sqrt[3]{216}$

*Answers on page A-45*

Simplify.

**10.** $\sqrt[3]{24}$

**11.** $\sqrt[4]{\dfrac{81}{256}}$

**12.** $\sqrt[5]{96}$

**13.** $\sqrt[3]{\dfrac{4}{125}}$

Some roots occur so frequently that you may want to memorize them.

| Square Roots | | Cube Roots | Fourth Roots | Fifth Roots |
|---|---|---|---|---|
| $\sqrt{1} = 1$ | $\sqrt{4} = 2$ | $\sqrt[3]{1} = 1$ | $\sqrt[4]{1} = 1$ | $\sqrt[5]{1} = 1$ |
| $\sqrt{9} = 3$ | $\sqrt{16} = 4$ | $\sqrt[3]{8} = 2$ | $\sqrt[4]{16} = 2$ | $\sqrt[5]{32} = 2$ |
| $\sqrt{25} = 5$ | $\sqrt{36} = 6$ | $\sqrt[3]{27} = 3$ | $\sqrt[4]{81} = 3$ | $\sqrt[5]{243} = 3$ |
| $\sqrt{49} = 7$ | $\sqrt{64} = 8$ | $\sqrt[3]{64} = 4$ | $\sqrt[4]{256} = 4$ | |
| $\sqrt{81} = 9$ | $\sqrt{100} = 10$ | $\sqrt[3]{125} = 5$ | $\sqrt[4]{625} = 5$ | |
| $\sqrt{121} = 11$ | $\sqrt{144} = 12$ | $\sqrt[3]{216} = 6$ | | |

## b  Products and Quotients Involving Higher Roots

The rules for working with products and quotients of square roots can be extended to products and quotients of $n$th roots.

> **THE PRODUCT AND QUOTIENT RULES FOR RADICALS**
>
> For any nonnegative real numbers $a$ and $b$ and any index $n$, $n \geq 2$,
> $$\sqrt[n]{AB} = \sqrt[n]{A} \cdot \sqrt[n]{B} \quad \text{and} \quad \sqrt[n]{\dfrac{A}{B}} = \dfrac{\sqrt[n]{A}}{\sqrt[n]{B}}.$$

**Examples**  Simplify.

**8.** $\sqrt[3]{40} = \sqrt[3]{8 \cdot 5}$  Factoring the radicand. 8 is a perfect cube.

$\qquad = \sqrt[3]{8} \cdot \sqrt[3]{5}$  Using the product rule

$\qquad = 2\sqrt[3]{5}$

**9.** $\sqrt[3]{\dfrac{125}{27}} = \dfrac{\sqrt[3]{125}}{\sqrt[3]{27}}$  Using the quotient rule

$\qquad = \dfrac{5}{3}$  Simplifying. 125 and 27 are perfect cubes.

**10.** $\sqrt[4]{1250} = \sqrt[4]{625 \cdot 2}$  Factoring the radicand. 625 is a perfect fourth power.

$\qquad = \sqrt[4]{625} \cdot \sqrt[4]{2}$  Using the product rule

$\qquad = 5\sqrt[4]{2}$  Simplifying

**11.** $\sqrt[5]{\dfrac{2}{243}} = \dfrac{\sqrt[5]{2}}{\sqrt[5]{243}}$  Using the quotient rule

$\qquad = \dfrac{\sqrt[5]{2}}{3}$  Simplifying. 243 is a perfect fifth power.

*Do Exercises 10–13.*

*Answers on page A-45*

# Exercise Set B

**a** Simplify. If an expression does not represent a real number, state this.

**1.** $\sqrt[3]{125}$

**2.** $\sqrt[3]{-27}$

**3.** $\sqrt[3]{-1000}$

**4.** $\sqrt[3]{8}$

**5.** $\sqrt[4]{1}$

**6.** $-\sqrt[5]{32}$

**7.** $\sqrt[4]{-256}$

**8.** $\sqrt[6]{-1}$

**9.** $-\sqrt[3]{-216}$

**10.** $\sqrt[3]{-125}$

**11.** $\sqrt[4]{256}$

**12.** $-\sqrt[3]{-8}$

**13.** $\sqrt[4]{10,000}$

**14.** $\sqrt[3]{-64}$

**15.** $-\sqrt[4]{81}$

**16.** $-\sqrt[3]{1}$

**17.** $-\sqrt[4]{-16}$

**18.** $\sqrt[6]{64}$

**19.** $-\sqrt[3]{125}$

**20.** $\sqrt[3]{1000}$

**21.** $\sqrt[5]{t^5}$

**22.** $\sqrt[7]{y^7}$

**23.** $-\sqrt[3]{x^3}$

**24.** $-\sqrt[9]{a^9}$

**25.** $\sqrt[3]{64}$

**26.** $-\sqrt[3]{216}$

**27.** $\sqrt[3]{-343}$

**28.** $\sqrt[5]{-243}$

**29.** $\sqrt[5]{-3125}$

**30.** $\sqrt[4]{625}$

**31.** $\sqrt[6]{1,000,000}$

**32.** $\sqrt[5]{243}$

**33.** $-\sqrt[5]{-100,000}$

**34.** $-\sqrt[4]{-10,000}$

**35.** $-\sqrt[3]{343}$

**36.** $\sqrt[3]{512}$

**37.** $\sqrt[8]{-1}$

**38.** $\sqrt[6]{-64}$

**39.** $\sqrt[5]{3125}$

**40.** $\sqrt[4]{-625}$

**b** Simplify. If an expression does not represent a real number, state this.

**41.** $\sqrt[3]{54}$

**42.** $\sqrt[5]{64}$

**43.** $\sqrt[4]{324}$

**44.** $\sqrt[3]{81}$

**45.** $\sqrt[3]{\dfrac{27}{64}}$

**46.** $\sqrt[3]{\dfrac{125}{64}}$

**47.** $\sqrt[4]{512}$

**48.** $\sqrt[3]{24}$

**49.** $\sqrt[5]{128}$

**50.** $\sqrt[4]{112}$

**51.** $\sqrt[4]{\dfrac{256}{625}}$

**52.** $\sqrt[5]{\dfrac{243}{32}}$

**53.** $\sqrt[3]{\dfrac{17}{8}}$

**54.** $\sqrt[5]{\dfrac{11}{32}}$

**55.** $\sqrt[3]{250}$

**56.** $\sqrt[5]{96}$

**57.** $\sqrt[5]{486}$

**58.** $\sqrt[3]{128}$

**59.** $\sqrt[4]{\dfrac{13}{81}}$

**60.** $\sqrt[3]{\dfrac{10}{27}}$

**61.** $\sqrt[4]{\dfrac{7}{16}}$

**62.** $\sqrt[4]{\dfrac{27}{256}}$

**63.** $\sqrt[4]{\dfrac{16}{625}}$

**64.** $\sqrt[3]{\dfrac{216}{27}}$

---

**Synthesis**

Simplify.

**65.** $\sqrt[3]{\sqrt[3]{64}}$

**66.** $\sqrt{\sqrt[3]{-64}}$

**67.** $\sqrt[3]{\sqrt[3]{1,000,000,000}}$

**68.** $\sqrt{-\sqrt[3]{-1}}$

Copyright © 1999 Addison Wesley Longman

# Appendix C Sets

The concept of set is used frequently in more advanced mathematics. We provide a basic introduction to sets in this appendix.

## a | Naming Sets

To name the set of whole numbers less than 6, we can use the *roster method*, as follows: {0, 1, 2, 3, 4, 5}.

The set of real numbers $x$ such that $x$ is less than 6 cannot be named by listing all its members because there are infinitely many. We name such a set using *set-builder notation*, as follows: $\{x \mid x < 6\}$. This is read "The set of all $x$ such that $x$ is less than 6." See Section 2.7 for more on this notation.

*Do Exercises 1 and 2.*

## b | Set Membership and Subsets

The symbol $\in$ means *is a member of* or *belongs to*, or *is an element of*. Thus, $x \in A$ means $x$ is a member of $A$ or $x$ belongs to $A$ or $x$ is an element of $A$.

**Example 1** Classify each of the following as true or false.

a) $1 \in \{1, 2, 3\}$

b) $1 \in \{2, 3\}$

c) $4 \in \{x \mid x \text{ is an even whole number}\}$

d) $5 \in \{x \mid x \text{ is an even whole number}\}$

a) Since 1 *is* listed as a member of the set, $1 \in \{1, 2, 3\}$ is true.

b) Since 1 is *not* a member of {2, 3}, the statement $1 \in \{2, 3\}$ is false.

c) Since 4 *is* an even whole number, $4 \in \{x \mid x \text{ is an even whole number}\}$ is a true statement.

d) Since 5 is not even, $5 \in \{x \mid x \text{ is an even whole number}\}$ is false.

Set membership can be illustrated with a diagram, as shown here.

*Do Exercises 3 and 4.*

If every element of $A$ is an element of $B$, then $A$ is a *subset* of $B$. This is denoted $A \subseteq B$.

The set of whole numbers is a subset of the set of integers. The set of rational numbers is a subset of the set of real numbers.

**Example 2** Classify each of the following as true or false.

a) $\{1, 2\} \subseteq \{1, 2, 3, 4\}$     b) $\{p, q, r, w\} \subseteq \{a, p, r, z\}$

c) $\{x \mid x < 6\} \subseteq \{x \mid x \leq 11\}$

---

### Objectives

a   Name sets using the roster method.

b   Classify statements regarding set membership and subsets as true or false.

c   Find the intersection and the union of sets.

Name the set using the roster method.

1. The set of whole numbers 0 through 7

2. $\{x \mid \text{the square of } x \text{ is } 25\}$

Determine whether each of the following is true or false.

3. $8 \in \{x \mid x \text{ is an even whole number}\}$

4. $2 \in \{x \mid x \text{ is a prime number}\}$

*Answers on page A-45*

Determine whether each of the following is true or false.

**5.** $\{-2, -3, 4\} \subseteq$
$\{-5, -4, -2, 7, -3, 5, 4\}$

**6.** $\{a, e, i, o, u\} \subseteq$ The set of all consonants

**7.** $\{x \,|\, x \leq -8\} \subseteq \{x \,|\, x \leq -7\}$

Find the intersection.

**8.** $\{-2, -3, 4, -4, 8\} \cap$
$\{-5, -4, -2, 7, -3, 5, 4\}$

**9.** $\{a, e, i, o, u\} \cap \{m, a, r, v, i, n\}$

**10.** $\{a, e, i, o, u\} \cap$ The set of all consonants

Find the union.

**11.** $\{-2, -3, 4, -4, 8\} \cup$
$\{-5, -4, -2, 7, -3, 5, 4\}$

**12.** $\{a, e, i, o, u\} \cup \{m, a, r, v, i, n\}$

**13.** $\{a, e, i, o, u\} \cup$ The set of all consonants

*Answers on page A-45*

---

**a)** Since every element of $\{1, 2\}$ is in the set $\{1, 2, 3, 4\}$, the statement $\{1, 2\} \subseteq \{1, 2, 3, 4\}$ is true.

**b)** Since q $\in \{p, q, r, w\}$, but q $\notin \{a, p, r, z\}$, the statement $\{p, q, r, w\} \subseteq \{a, p, r, z\}$ is false.

**c)** Since every number that is less than 6 is also less than 11, the statement $\{x \,|\, x < 6\} \subseteq \{x \,|\, x \leq 11\}$ is true.

*Do Exercises 5–7.*

## c  Intersections and Unions

The *intersection* of sets $A$ and $B$, denoted $A \cap B$, is the set of members that are common to both sets.

**Example 3**   Find the intersection.

**a)** $\{0, 1, 3, 5, 25\} \cap \{2, 3, 4, 5, 6, 7, 9\}$      **b)** $\{a, p, q, w\} \cap \{p, q, t\}$

**a)** $\{0, 1, 3, 5, 25\} \cap \{2, 3, 4, 5, 6, 7, 9\} = \{3, 5\}$

**b)** $\{a, p, q, w\} \cap \{p, q, t\} = \{p, q\}$

Set intersection can be illustrated with a diagram, as shown here.

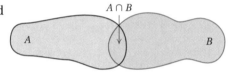

The set without members is known as the *empty set,* and is often named $\varnothing$, and sometimes { }. Each of the following is a description of the empty set:

$$\{2, 3\} \cap \{5, 6, 7\};$$
$$\{x \,|\, x \text{ is an even natural number}\} \cap \{x \,|\, x \text{ is an odd natural number}\}.$$

*Do Exercises 8–10.*

Two sets $A$ and $B$ can be combined to form a set that contains the members of $A$ as well as those of $B$. The new set is called the *union* of $A$ and $B$, denoted $A \cup B$.

**Example 4**   Find the union.

**a)** $\{0, 5, 7, 13, 27\} \cup \{0, 2, 3, 4, 5\}$      **b)** $\{a, c, e, g\} \cup \{b, d, f\}$

**a)** $\{0, 5, 7, 13, 27\} \cup \{0, 2, 3, 4, 5\} = \{0, 2, 3, 4, 5, 7, 13, 27\}$
    **Note that the 0 and the 5 are *not* listed twice in the solution.**

**b)** $\{a, c, e, g\} \cup \{b, d, f\} = \{a, b, c, d, e, f, g\}$

Set union can be illustrated with a diagram, as shown here.

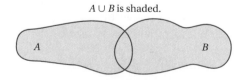

$A \cup B$ is shaded.

The solution set of the equation $(x - 3)(x + 2) = 0$ is $\{3, -2\}$. This set is the union of the solution sets of $x - 3 = 0$ and $x + 2 = 0$, which are $\{3\}$ and $\{-2\}$.

*Do Exercises 11–13.*

# Exercise Set C

Name the set using the roster method.

**1.** The set of whole numbers 3 through 8

**2.** The set of whole numbers 101 through 107

**3.** The set of odd numbers between 40 and 50

**4.** The set of multiples of 5 between 10 and 40

**5.** $\{x \mid$ the square of $x$ is 9$\}$

**6.** $\{x \mid x$ is the cube of 0.2$\}$

Classify the statement as true or false.

**7.** $2 \in \{x \mid x$ is an odd number$\}$

**8.** $7 \in \{x \mid x$ is an odd number$\}$

**9.** Jeff Gordon $\in$ The set of all NASCAR drivers

**10.** Apple $\in$ The set of all fruit

**11.** $-3 \in \{-4, -3, 0, 1\}$

**12.** $0 \in \{-4, -3, 0\ 1\}$

**13.** $\frac{2}{3} \in \{x \mid x$ is a rational number$\}$

**14.** Heads $\in$ The set of outcomes of flipping a penny

**15.** $\{4, 5, 8\} \subseteq \{1, 3, 4, 5, 6, 7, 8, 9\}$

**16.** The set of vowels $\subseteq$ The set of consonants

**17.** $\{-1, -2, -3, -4, -5\} \subseteq \{-1, 2, 3, 4, 5\}$

**18.** The set of integers $\subseteq$ The set of rational numbers

Find the intersection.

**19.** $\{a, b, c, d, e,\} \cap \{c, d, e, f, g\}$    **20.** $\{a, e, i, o, u\} \cap \{q, u, i, c, k\}$    **21.** $\{1, 2, 5, 10\} \cap \{0, 1, 7, 10\}$

**22.** {0, 1, 7, 10} ∩ {0, 1, 2, 5}    **23.** {1, 2, 5, 10} ∩ {3, 4, 7, 8}    **24.** {a, e, i, o, u} ∩ {m, n, f, g, h}

Find the union.

**25.** {a, e, i, o, u} ∪ {q, u, i, c, k}    **26.** {a, b, c, d, e,} ∪ {c, d, e, f, g}

**27.** {0, 1, 7, 10} ∪ {0, 1, 2, 5}    **28.** {1, 2, 5, 10} ∪ {0, 1, 7, 10}

**29.** {a, e, i, o, u} ∪ {m, n, f, g, h}    **30.** {1, 2, 5, 10} ∪ {a, b}

---

**Synthesis**

**31.** Find the union of the set of integers and the set of whole numbers.

**32.** Find the intersection of the set of odd integers and the set of even integers.

**33.** Find the union of the set of rational numbers and the set of irrational numbers.

**34.** Find the intersection of the set of even integers and the set of positive rational numbers.

**35.** Find the intersection of the set of rational numbers and the set of irrational numbers.

**36.** Find the union of the set of negative integers, the set of positive integers, and the set containing 0.

**37.** For a set $A$, find each of the following.
   a) $A \cup \varnothing$
   b) $A \cup A$
   c) $A \cap A$
   d) $A \cap \varnothing$

**38.** A set is *closed* under an operation if, when the operation is performed on its members, the result is in the set. For example, the set of real numbers is closed under the operation of addition since the sum of any two real numbers is a real number.

   a) Is the set of even numbers closed under addition?
   b) Is the set of odd numbers closed under addition?
   c) Is the set {0, 1} closed under addition?
   d) Is the set {0, 1} closed under multiplication?
   e) Is the set of real numbers closed under multiplication?
   f) Is the set of integers closed under division?

**39.** Experiment with sets of various types and determine whether the following distributive law for sets is true:

$$A \cap (B \cup C) = (A \cap B) \cup (A \cap C).$$

# Answers

## Diagnostic Pretest, p. xxi

**1.** [R.2c] $\frac{8}{15}$   **2.** [R.2c] $\frac{11}{12}$   **3.** [R.3b] 26
**4.** [R.3b] 10.983   **5.** [1.4a] 1.18   **6.** [1.5a] $-11.7$
**7.** [1.4b] $-\$9.93$   **8.** [1.8c] $39a - 84$   **9.** [2.3c] $-5$
**10.** [2.7e] $\{x \mid x \geq \frac{9}{23}\}$   **11.** [2.4a] 9 in. and 27 in.
**12.** [2.5a] $1500
**13.** [3.2b]                  **14.** [3.3b]

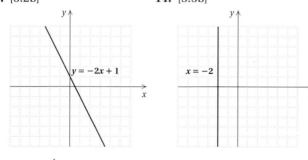

**15.** [4.1e] $\dfrac{x^4}{y}$   **16.** [4.2b], [4.1d] $-32x^{11}$
**17.** [4.4c] $-x^2 + 3x + 4$   **18.** [4.6b] $4x^4 - 9$
**19.** [5.5d] $2(x + 9)(x - 9)$
**20.** [5.3a], [5.4a] $(5x + 1)(x - 3)$   **21.** [5.7b] $-5, 2$
**22.** [5.8a] 8 m, 17 m   **23.** [6.2b] $\dfrac{x(x - 4)}{2(x + 5)(x - 3)}$
**24.** [6.4a] $\dfrac{2 - x}{x(x + 1)(x + 2)}$   **25.** [6.6a] 4
**26.** [6.7a] 55 mph, 40 mph   **27.** [7.1b], [7.2a] $-\frac{2}{3}; \left(0, \frac{8}{3}\right)$
**28.** [7.2c] $y = -3x + 11$
**29.** [7.4b]

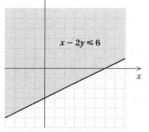

**30.** [8.2b] $(8, -3)$   **31.** [8.3b] $\left(\frac{23}{8}, -\frac{3}{16}\right)$
**32.** [8.4a] 25 L of each   **33.** [8.5a] 10 hr

**34.** [9.2c] $2xy^2\sqrt{3x}$   **35.** [9.3a] $\dfrac{x}{3y}$   **36.** [9.4c] $6 + 3\sqrt{3}$
**37.** [9.5a] $\frac{77}{2}$   **38.** [10.1c] $-1, \frac{1}{3}$
**39.** [10.3a] No real-number solutions
**40.** [10.5a] 30 m, 16 m
**41.** [10.6a]

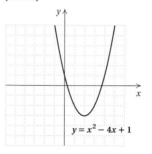

## Chapter R

### Pretest: Chapter R, p. 2

**1.** [R.1a] $2 \cdot 2 \cdot 2 \cdot 31$   **2.** [R.1b] 168   **3.** [R.2a] $\frac{10}{15}$
**4.** [R.2a] $\frac{44}{48}$   **5.** [R.2b] $\frac{23}{64}$   **6.** [R.2b] $\frac{2}{3}$   **7.** [R.2c] $\frac{11}{10}$
**8.** [R.2c] $\frac{2}{21}$   **9.** [R.2c] $\frac{1}{2}$   **10.** [R.2c] $\frac{5}{21}$   **11.** [R.3a] $\frac{3217}{100}$
**12.** [R.3a] 0.0789   **13.** [R.3b] 134.0362
**14.** [R.3b] 212.05   **15.** [R.3b] 350.5824
**16.** [R.3b] 12.4   **17.** [R.3b] $1.\overline{4}$   **18.** [R.3c] 345.84
**19.** [R.3c] 345.8   **20.** [R.4a] 0.116   **21.** [R.4b] $\frac{87}{100}$
**22.** [R.4d] 87.5%   **23.** [R.5a] $5^4$   **24.** [R.5b] 8
**25.** [R.5b] 1.21   **26.** [R.5c] 18

### Margin Exercises, Section R.1, pp. 3–6

**1.** 1, 3, 9   **2.** 1, 2, 4, 8, 16   **3.** 1, 2, 3, 4, 6, 8, 12, 24
**4.** 1, 2, 3, 4, 5, 6, 9, 10, 12, 15, 18, 20, 30, 36, 45, 60, 90,
180   **5.** 13   **6.** $2 \cdot 2 \cdot 2 \cdot 2 \cdot 3$   **7.** $2 \cdot 5 \cdot 5$
**8.** $2 \cdot 5 \cdot 7 \cdot 11$   **9.** 15, 30, 45, 60, . . .
**10.** 45, 90, 135, 180, . . .   **11.** 40   **12.** 54   **13.** 360
**14.** 18   **15.** 24   **16.** 36   **17.** 210

## Exercise Set R.1, p. 7

**1.** 1, 2, 4, 5, 10, 20　**3.** 1, 2, 3, 4, 6, 8, 9, 12, 18, 24, 36, 72　**5.** $3 \cdot 5$　**7.** $2 \cdot 11$　**9.** $3 \cdot 3$　**11.** $7 \cdot 7$
**13.** $2 \cdot 3 \cdot 3$　**15.** $2 \cdot 2 \cdot 2 \cdot 5$　**17.** $2 \cdot 3 \cdot 3 \cdot 5$
**19.** $2 \cdot 3 \cdot 5 \cdot 7$　**21.** $7 \cdot 13$　**23.** $7 \cdot 17$　**25.** $2 \cdot 2$; 5;
20　**27.** $2 \cdot 2 \cdot 2 \cdot 3$; $2 \cdot 2 \cdot 3 \cdot 3$; 72　**29.** 3; $3 \cdot 5$; 15
**31.** $2 \cdot 3 \cdot 5$; $2 \cdot 2 \cdot 2 \cdot 5$; 120　**33.** 13; 23; 299
**35.** $2 \cdot 3 \cdot 3$; $2 \cdot 3 \cdot 5$; 90　**37.** $2 \cdot 3 \cdot 5$; $2 \cdot 2 \cdot 3 \cdot 3$; 180
**39.** $2 \cdot 2 \cdot 2 \cdot 3$; $2 \cdot 3 \cdot 5$; 120　**41.** 17; 29; 493
**43.** $2 \cdot 2 \cdot 3$; $2 \cdot 2 \cdot 7$; 84　**45.** 2; 3; 5; 30
**47.** $2 \cdot 2 \cdot 2 \cdot 3$; $2 \cdot 2 \cdot 3 \cdot 3$; $2 \cdot 2 \cdot 3$; 72　**49.** 5;
$2 \cdot 2 \cdot 3$; $3 \cdot 5$; 60　**51.** $2 \cdot 3$; $2 \cdot 2 \cdot 3$; $2 \cdot 3 \cdot 3$; 36
**53.** ◆　**55.** (a) No; not a multiple of 8; (b) no; not a
multiple of 8; (c) no; not a multiple of 8 or 12; (d) yes; it
is a multiple of both 8 and 12 and is the smallest such
multiple.　**57.** 70,200　**59.** Every 60 years

## Margin Exercises, Section R.2, pp. 10–15

**1.** $\frac{8}{12}$　**2.** $\frac{21}{28}$　**3.** $\frac{14}{16}, \frac{21}{24}, \frac{28}{32}$; answers may vary　**4.** $\frac{2}{5}$
**5.** $\frac{19}{9}$　**6.** $\frac{8}{3}$　**7.** $\frac{1}{2}$　**8.** 4　**9.** $\frac{5}{2}$　**10.** $\frac{35}{16}$　**11.** $\frac{7}{5}$
**12.** 2　**13.** $\frac{23}{15}$　**14.** $\frac{3}{4}$　**15.** $\frac{19}{40}$　**16.** $\frac{7}{36}$　**17.** $\frac{11}{4}$
**18.** $\frac{7}{15}$　**19.** $\frac{1}{5}$　**20.** 3　**21.** $\frac{21}{20}$　**22.** $\frac{8}{21}$　**23.** $\frac{5}{6}$
**24.** $\frac{8}{15}$　**25.** $\frac{1}{64}$　**26.** 81

## Calculator Spotlight, p. 11

**1.** $\frac{7}{8}$　**2.** $\frac{14}{15}$　**3.** $\frac{138}{167}$　**4.** $\frac{7}{25}$

## Exercise Set R.2, p. 17

**1.** $\frac{9}{12}$　**3.** $\frac{60}{100}$　**5.** $\frac{104}{160}$　**7.** $\frac{21}{24}$　**9.** $\frac{20}{16}$　**11.** $\frac{2}{3}$　**13.** 4
**15.** $\frac{1}{7}$　**17.** 8　**19.** $\frac{1}{4}$　**21.** 5　**23.** $\frac{17}{21}$　**25.** $\frac{13}{7}$　**27.** $\frac{4}{3}$
**29.** $\frac{1}{12}$　**31.** $\frac{45}{16}$　**33.** $\frac{2}{3}$　**35.** $\frac{7}{6}$　**37.** $\frac{5}{6}$　**39.** $\frac{1}{2}$　**41.** $\frac{13}{24}$
**43.** $\frac{31}{60}$　**45.** $\frac{35}{18}$　**47.** $\frac{10}{3}$　**49.** $\frac{1}{2}$　**51.** $\frac{5}{36}$　**53.** 500
**55.** $\frac{3}{40}$　**57.** $2 \cdot 2 \cdot 7$　**58.** $2 \cdot 2 \cdot 2 \cdot 7$
**59.** $2 \cdot 2 \cdot 2 \cdot 5 \cdot 5 \cdot 5$　**60.** $2 \cdot 2 \cdot 2 \cdot 2 \cdot 2 \cdot 2 \cdot 3$
**61.** 48　**62.** 392　**63.** 192　**64.** 150　**65.** ◆　**67.** $\frac{3}{4}$
**69.** 4　**71.** 1

## Margin Exercises, Section R.3, pp. 19–24

**1.** $\frac{568}{1000}$　**2.** $\frac{23}{10}$　**3.** $\frac{8904}{100}$　**4.** 4.131　**5.** 0.4131　**6.** 5.73
**7.** 284.455　**8.** 268.63　**9.** 27.676　**10.** 64.683
**11.** 99.59　**12.** 239.883　**13.** 5.868　**14.** 0.5868
**15.** 51.53808　**16.** 48.9　**17.** 15.82　**18.** 1.28
**19.** 17.95　**20.** 856　**21.** 0.85　**22.** 0.625　**23.** $0.\overline{6}$
**24.** $7.\overline{63}$　**25.** 2.8　**26.** 13.9　**27.** 7.0　**28.** 7.83
**29.** 34.68　**30.** 0.03　**31.** 0.943　**32.** 8.004
**33.** 43.112　**34.** 37.401　**35.** 7459.355　**36.** 7459.35
**37.** 7459.4　**38.** 7459　**39.** 7460

## Exercise Set R.3, p. 25

**1.** $\frac{53}{10}$　**3.** $\frac{67}{100}$　**5.** $\frac{20,007}{10,000}$　**7.** $\frac{78,898}{10}$　**9.** 0.1　**11.** 0.0001
**13.** 9.999　**15.** 0.4578　**17.** 444.94　**19.** 390.617
**21.** 155.724　**23.** 63.79　**25.** 32.234　**27.** 26.835

**29.** 47.91　**31.** 1.9193　**33.** 13.212　**35.** 0.7998
**37.** 179.5　**39.** 1.40756　**41.** 3.60558　**43.** 2.3
**45.** 5.2　**47.** 0.023　**49.** 18.75　**51.** 660　**53.** 0.68
**55.** 0.34375　**57.** $1.\overline{18}$　**59.** $0.\overline{5}$　**61.** $2.\overline{1}$　**63.** 745.07;
745.1; 745; 750; 700　**65.** 6780.51; 6780.5; 6781; 6780;
6800　**67.** $17.99; $18　**69.** $346.08; $346　**71.** $17
**73.** $190　**75.** 12.3457; 12.346; 12.35; 12.3; 12
**77.** 0.5897; 0.590; 0.59; 0.6; 1　**79.** $\frac{33}{32}$　**80.** $\frac{1}{48}$　**81.** $\frac{55}{64}$
**82.** $\frac{5}{4}$　**83.** ◆

## Margin Exercises, Section R.4, pp. 27–28

**1.** 0.873　**2.** 1　**3.** $\frac{53}{100}$　**4.** $\frac{456}{1000}$　**5.** $\frac{23}{10,000}$　**6.** 677%
**7.** 99.44%　**8.** 25%　**9.** 87.5%　**10.** $66.\overline{6}$%, or $66\frac{2}{3}$%

## Exercise Set R.4, p. 29

**1.** 0.63　**3.** 0.941　**5.** 0.01　**7.** 0.0061　**9.** 2.4
**11.** 0.0325　**13.** $\frac{60}{100}$　**15.** $\frac{289}{1000}$　**17.** $\frac{110}{100}$　**19.** $\frac{42}{100,000}$
**21.** $\frac{250}{100}$　**23.** $\frac{347}{10,000}$　**25.** 100%　**27.** 99.6%
**29.** 0.47%　**31.** 7.2%　**33.** 920%　**35.** 0.68%
**37.** $16.\overline{6}$%, or $16\frac{2}{3}$%　**39.** 65%　**41.** 29%　**43.** 80%
**45.** 60%　**47.** $66.\overline{6}$%, or $66\frac{2}{3}$%　**49.** 175%　**51.** 75%
**53.** 2.25　**54.** 1.375　**55.** $1.41\overline{6}$　**56.** $0.\overline{8}$　**57.** $0.\overline{90}$
**58.** $1.\overline{54}$　**59.** 164.90974　**60.** 56.43　**61.** 896.559
**62.** 722.579　**63.** ◆　**65.** 32%　**67.** 70%
**69.** 2700%　**71.** 345%　**73.** 2.5%

## Margin Exercises, Section R.5, pp. 31–34

**1.** $4^3$　**2.** $6^5$　**3.** $(1.08)^2$　**4.** 10,000　**5.** 512
**6.** 1.331　**7.** 13　**8.** 1000　**9.** 250　**10.** 178　**11.** 2
**12.** 8000　**13.** 48　**14.** $\frac{11}{2}$

## Calculator Spotlight, p. 32

**1.** 1024　**2.** 40,353,607　**3.** 361　**4.** 32.695524
**5.** 10.4976　**6.** 12,230.59046　**7.** 0.0423150361
**8.** 0.0033704702

## Calculator Spotlight, p. 34

**1.** 38　**2.** 81　**3.** 72　**4.** 5932　**5.** 25.011
**6.** 743.027　**7.** 450　**8.** 14,321,949.1　**9.** 4　**10.** 2
**11.** 783　**12.** 228,112.96　**13.** 40. The calculator
divided 141 by 47, added 39, and then subtracted 2.

## Exercise Set R.5, p. 35

**1.** $5^4$　**3.** $10^3$　**5.** $10^5$　**7.** 49　**9.** 59,049　**11.** 100
**13.** 1　**15.** 5.29　**17.** 0.008　**19.** 416.16　**21.** $\frac{9}{64}$
**23.** 125　**25.** 1061.208　**27.** 25　**29.** 114　**31.** 33
**33.** 5　**35.** 12　**37.** 324　**39.** 100　**41.** 1000
**43.** 22　**45.** 1　**47.** 4　**49.** 102　**51.** 96　**53.** 24
**55.** 90　**57.** 8　**59.** 1　**61.** 50,000　**63.** $\frac{22}{45}$　**65.** $\frac{19}{66}$
**67.** 9　**69.** 31.25%　**70.** 53.125%　**71.** $\frac{5}{13}$　**72.** $\frac{2}{3}$
**73.** $2 \cdot 2 \cdot 2 \cdot 2 \cdot 3$　**74.** 168　**75.** ◆　**77.** $10^2$
**79.** $5^6$　**81.** $3 = \dfrac{5+5}{5} + \dfrac{5}{5}$; $4 = \dfrac{5+5+5+5}{5}$;

$$5 = \frac{5(5+5)}{5} - 5; \quad 6 = \frac{5}{5} + \frac{5 \cdot 5}{5}; \quad 7 = \frac{5}{5} + \frac{5}{5} + 5;$$

$$8 = 5 + \frac{5+5+5}{5}; \quad 9 = \frac{5 \cdot 5 - 5}{5} + 5; \quad 10 = \frac{5 \cdot 5 + 5 \cdot 5}{5}$$

## Summary and Review: Chapter R, p. 37

**1.** $2 \cdot 2 \cdot 23$ **2.** $2 \cdot 2 \cdot 2 \cdot 5 \cdot 5 \cdot 7$ **3.** 416 **4.** 90
**5.** $\frac{12}{30}$ **6.** $\frac{96}{184}$ **7.** $\frac{40}{64}$ **8.** $\frac{91}{84}$ **9.** $\frac{5}{12}$ **10.** $\frac{51}{91}$ **11.** $\frac{31}{36}$
**12.** $\frac{1}{4}$ **13.** $\frac{3}{5}$ **14.** $\frac{72}{25}$ **15.** $\frac{1797}{100}$ **16.** 0.2337
**17.** 2442.905 **18.** 86.0298 **19.** 9.342 **20.** 133.264
**21.** 430.8 **22.** 110.483 **23.** 55.6 **24.** 0.45
**25.** $1.58\overline{3}$ **26.** 34.1 **27.** 0.047 **28.** $\frac{60}{100}$ **29.** 88.6%
**30.** 62.5% **31.** 116% **32.** $6^3$ **33.** 1.1236 **34.** 119
**35.** 29 **36.** 7 **37.** $\frac{103}{17}$ **38.** 0.0000006%

## Test: Chapter R, p. 39

**1.** [R.1a] $2 \cdot 2 \cdot 3 \cdot 5 \cdot 5$ **2.** [R.1b] 120 **3.** [R.2a] $\frac{21}{49}$
**4.** [R.2a] $\frac{33}{48}$ **5.** [R.2b] $\frac{2}{3}$ **6.** [R.2b] $\frac{37}{61}$ **7.** [R.2c] $\frac{5}{36}$
**8.** [R.2c] $\frac{11}{40}$ **9.** [R.3a] $\frac{678}{100}$ **10.** [R.3a] 1.895
**11.** [R.3b] 99.0187 **12.** [R.3b] 1796.58
**13.** [R.3b] 435.072 **14.** [R.3b] 1.6 **15.** [R.3b] $2.\overline{09}$
**16.** [R.3c] 234.7 **17.** [R.3c] 234.728 **18.** [R.4a] 0.007
**19.** [R.4b] $\frac{91}{100}$ **20.** [R.4d] 44% **21.** [R.5b] 625
**22.** [R.5b] 1.44 **23.** [R.5c] 207 **24.** [R.5c] 20,000
**25.** [R.2b] $\frac{33}{100}$

# Chapter 1

## Pretest: Chapter 1, p. 42

**1.** [1.1a] $\frac{5}{16}$ **2.** [1.1b] 78%$x$, or 0.78$x$ **3.** [1.1a] 360 ft$^2$
**4.** [1.3b] 12 **5.** [1.2d] $>$ **6.** [1.2d] $>$ **7.** [1.2d] $>$
**8.** [1.2d] $<$ **9.** [1.2e] 12 **10.** [1.2e] 2.3 **11.** [1.2e] 0
**12.** [1.3b] $-5.4$ **13.** [1.3b] $\frac{2}{3}$ **14.** [1.6b] $\frac{1}{10}$
**15.** [1.6b] $-\frac{3}{2}$ **16.** [1.3a] $-17$ **17.** [1.4a] 38.6
**18.** [1.4a] $-\frac{17}{15}$ **19.** [1.3a] $-5$ **20.** [1.5a] 63
**21.** [1.5a] $-\frac{5}{12}$ **22.** [1.6c] $-98$ **23.** [1.6a] 8
**24.** [1.4a] 24 **25.** [1.8d] 26 **26.** [1.7c] $9z - 18$
**27.** [1.7c] $-4a - 2b + 10c$ **28.** [1.7d] $4(x - 3)$
**29.** [1.7d] $3(2y - 3z - 6)$ **30.** [1.8b] $-y - 13$
**31.** [1.8c] $y + 18$ **32.** [1.2d] $12 < x$

## Margin Exercises, Section 1.1, pp. 43–46

**1.** $2174 + x = 7521$ **2.** 64 **3.** 28 **4.** 60
**5.** 192 ft$^2$ **6.** 25 **7.** 16 **8.** 12 hr **9.** $x - 8$
**10.** $y + 8$, or $8 + y$ **11.** $m - 4$ **12.** $\frac{1}{2}p$ **13.** $6 + 8x$,
or $8x + 6$ **14.** $a - b$ **15.** 59%$x$, or 0.59$x$
**16.** $xy - 200$ **17.** $p + q$

## Calculator Spotlight, p. 44

**1.** 59.63768116 **2.** 11.9 **3.** 11.9 **4.** 34,427.16
**5.** 32 **6.** 27.5

## Exercise Set 1.1, p. 47

**1.** \$20,400; \$46,800; \$150,000 **3.** 1935 m$^2$ **5.** 260 mi
**7.** 56 **9.** 8 **11.** 1 **13.** 6 **15.** 2 **17.** $b + 7$, or
$7 + b$ **19.** $c - 12$ **21.** $4 + q$, or $q + 4$ **23.** $a + b$,
or $b + a$ **25.** $y - x$ **27.** $w + x$, or $x + w$
**29.** $n - m$ **31.** $r + s$, or $s + r$ **33.** $2z$ **35.** $3m$
**37.** 89%$x$, or 0.89$x$ **39.** 55$t$ miles **41.** $2 \cdot 3 \cdot 3 \cdot 3$
**42.** $2 \cdot 2 \cdot 2 \cdot 2 \cdot 2$ **43.** $2 \cdot 2 \cdot 3 \cdot 3 \cdot 3$
**44.** $2 \cdot 2 \cdot 2 \cdot 2 \cdot 2 \cdot 2 \cdot 3$ **45.** 18 **46.** 96 **47.** 60
**48.** 48 **49.** ◈ **51.** $x + 3y$ **53.** $2x - 3$

## Margin Exercises, Section 1.2, pp. 51–56

**1.** 8; $-5$ **2.** 134; $-80$ **3.** $-10$; 156
**4.** $-120$; 50; $-80$
**5.**

**6.**

**7.**

**8.** $-0.375$ **9.** $-0.\overline{54}$ **10.** $1.\overline{3}$ **11.** $<$ **12.** $<$
**13.** $>$ **14.** $>$ **15.** $>$ **16.** $<$ **17.** $<$ **18.** $>$
**19.** $7 > -5$ **20.** $4 < x$ **21.** False **22.** True
**23.** True **24.** 8 **25.** 0 **26.** 9 **27.** $\frac{2}{3}$ **28.** 5.6

## Calculator Spotlight, p. 52

**1.** $\boxed{(-)}$ $\boxed{3}$ $\boxed{\text{ENTER}}$ ; $-3$
**2.** $\boxed{(-)}$ $\boxed{5}$ $\boxed{0}$ $\boxed{8}$ $\boxed{\text{ENTER}}$ ; $-508$
**3.** $\boxed{(-)}$ $\boxed{.}$ $\boxed{1}$ $\boxed{7}$ $\boxed{\text{ENTER}}$ ; $-.17$
**4.** $\boxed{(-)}$ $\boxed{5}$ $\boxed{\div}$ $\boxed{8}$ $\boxed{\text{ENTER}}$ ; $-.625$

## Calculator Spotlight, p. 53

**1.** 8.717797887 **2.** 17.80449381 **3.** 67.08203932
**4.** 35.4807407 **5.** 3.141592654 **6.** 91.10618695
**7.** 530.9291585 **8.** 138.8663978

## Exercise Set 1.2, p. 57

**1.** $-1286$; 13,804 **3.** 24; $-2$ **5.** $-5,200,000,000,000$
**7.**

**9.**

**11.** $-0.875$ **13.** $0.8\overline{3}$ **15.** $1.1\overline{6}$ **17.** $0.\overline{6}$ **19.** $-0.5$
**21.** 0.1 **23.** $>$ **25.** $<$ **27.** $<$ **29.** $<$ **31.** $>$
**33.** $<$ **35.** $>$ **37.** $<$ **39.** $<$ **41.** $<$ **43.** True

**45.** False   **47.** $x < -6$   **49.** $y \geq -10$   **51.** 3
**53.** 10   **55.** 0   **57.** 24   **59.** $\frac{2}{3}$   **61.** 0   **63.** 0.63
**64.** 0.083   **65.** 1.1   **66.** 0.2276   **67.** 75%
**68.** 62.5%, or $62\frac{1}{2}\%$   **69.** $83.\overline{3}\%$, or $83\frac{1}{3}\%$
**70.** 59.375%, or $59\frac{3}{8}\%$   **71.** ◆
**73.** $-\frac{5}{6}, -\frac{3}{4}, -\frac{2}{3}, \frac{1}{6}, \frac{3}{8}, \frac{1}{2}$   **75.** $\frac{1}{9}$   **77.** $5\frac{5}{9}$, or $\frac{50}{9}$

## Margin Exercises, Section 1.3, pp. 59–62

**1.** $-6$   **2.** $-3$   **3.** $-8$   **4.** 4   **5.** 0   **6.** $-2$
**7.** $-11$   **8.** $-12$   **9.** 2   **10.** $-4$   **11.** $-2$   **12.** 0
**13.** $-22$   **14.** 3   **15.** 0.53   **16.** 2.3   **17.** $-7.7$
**18.** $-6.2$   **19.** $-\frac{2}{9}$   **20.** $-\frac{19}{20}$   **21.** $-58$   **22.** $-56$
**23.** $-14$   **24.** $-12$   **25.** 4   **26.** $-8.7$   **27.** 7.74
**28.** $\frac{8}{9}$   **29.** 0   **30.** $-12$   **31.** $-14$; 14   **32.** $-1$; 1
**33.** 19; $-19$   **34.** 1.6; $-1.6$   **35.** $-\frac{2}{3}; \frac{2}{3}$   **36.** $\frac{9}{8}; -\frac{9}{8}$
**37.** 4   **38.** 13.4   **39.** 0   **40.** $-\frac{1}{4}$

## Exercise Set 1.3, p. 63

**1.** $-7$   **3.** $-6$   **5.** 0   **7.** $-8$   **9.** $-7$   **11.** $-27$
**13.** 0   **15.** $-42$   **17.** 0   **19.** 0   **21.** 3   **23.** $-9$
**25.** 7   **27.** 0   **29.** 35   **31.** $-3.8$   **33.** $-8.1$
**35.** $-\frac{1}{5}$   **37.** $-\frac{7}{9}$   **39.** $-\frac{3}{8}$   **41.** $-\frac{19}{24}$   **43.** $\frac{1}{24}$
**45.** 37   **47.** 50   **49.** $-1409$   **51.** $-24$   **53.** 26.9
**55.** $-8$   **57.** $\frac{13}{8}$   **59.** $-43$   **61.** $\frac{4}{3}$   **63.** 24   **65.** $\frac{3}{8}$
**67.** 0.57   **68.** 0.49   **69.** 0.529   **70.** 0.713   **71.** 125%
**72.** 12.5%   **73.** 52%   **74.** 40.625%   **75.** ◆   **77.** All
positive   **79.** $-6483$   **81.** Negative

## Margin Exercises, Section 1.4, pp. 65–67

**1.** $-10$   **2.** 3   **3.** $-5$   **4.** $-2$   **5.** $-11$   **6.** 4
**7.** $-2$   **8.** $-6$   **9.** $-16$   **10.** 7.1   **11.** 3   **12.** 0
**13.** $\frac{3}{2}$   **14.** $-8$   **15.** 7   **16.** $-3$   **17.** $-23.3$   **18.** 0
**19.** $-9$   **20.** 17   **21.** 12.7   **22.** 77, 37, 25, 24, 23, 9,
$-4, -9, -12, -25, -28, -31, -41, -45$; 53, 48, 45, 38,
29, 24, 21, 19, $-10, -27, -35, -37, -53, -115$
**23.** 50°C

## Exercise Set 1.4, p. 69

**1.** $-7$   **3.** $-4$   **5.** $-6$   **7.** 0   **9.** $-4$   **11.** $-7$
**13.** $-6$   **15.** 0   **17.** 0   **19.** 14   **21.** 11   **23.** $-14$
**25.** 5   **27.** $-7$   **29.** $-1$   **31.** 18   **33.** $-10$
**35.** $-3$   **37.** $-21$   **39.** 5   **41.** $-8$   **43.** 12
**45.** $-23$   **47.** $-68$   **49.** $-73$   **51.** 116   **53.** 0
**55.** $-1$   **57.** $\frac{1}{12}$   **59.** $-\frac{17}{12}$   **61.** $\frac{1}{8}$   **63.** 19.9
**65.** $-8.6$   **67.** $-0.01$   **69.** $-193$   **71.** 500   **73.** $-2.8$
**75.** $-3.53$   **77.** $-\frac{1}{2}$   **79.** $\frac{6}{7}$   **81.** $-\frac{41}{30}$   **83.** $-\frac{2}{15}$
**85.** 37   **87.** $-62$   **89.** $-139$   **91.** 6   **93.** 107
**95.** 219   **97.** 2385 m   **99.** \$347.94   **101.** 100°F
**103.** 125   **104.** $2 \cdot 2 \cdot 2 \cdot 2 \cdot 2 \cdot 3 \cdot 3 \cdot 3$   **105.** 100.5
**106.** 226   **107.** 0.583   **108.** $\frac{41}{64}$   **109.** ◆
**111.** $-309{,}882$   **113.** False; $3 - 0 \neq 0 - 3$   **115.** True
**117.** True   **119.** Up 15 points

## Margin Exercises, Section 1.5, pp. 73–75

**1.** 20; 10; 0; $-10$; $-20$; $-30$   **2.** $-18$   **3.** $-100$
**4.** $-80$   **5.** $-\frac{5}{9}$   **6.** $-30.033$   **7.** $-\frac{7}{10}$   **8.** $-10$; 0; 10;
20; 30   **9.** 27   **10.** 32   **11.** 35   **12.** $\frac{20}{63}$   **13.** $\frac{2}{3}$
**14.** 13.455   **15.** $-30$   **16.** 30   **17.** 0   **18.** $-\frac{8}{3}$
**19.** $-30$   **20.** $-30.75$   **21.** $-\frac{5}{3}$   **22.** 120   **23.** $-120$
**24.** 6   **25.** 4; $-4$   **26.** 9; $-9$   **27.** 48; 48

## Exercise Set 1.5, p. 77

**1.** $-8$   **3.** $-48$   **5.** $-24$   **7.** $-72$   **9.** 16   **11.** 42
**13.** $-120$   **15.** $-238$   **17.** 1200   **19.** 98   **21.** $-72$
**23.** $-12.4$   **25.** 30   **27.** 21.7   **29.** $-\frac{2}{5}$   **31.** $\frac{1}{12}$
**33.** $-17.01$   **35.** $-\frac{5}{12}$   **37.** 420   **39.** $\frac{2}{7}$   **41.** $-60$
**43.** 150   **45.** $-\frac{2}{45}$   **47.** 1911   **49.** 50.4   **51.** $\frac{10}{189}$
**53.** $-960$   **55.** 17.64   **57.** $-\frac{5}{784}$   **59.** 0   **61.** $-720$
**63.** $-30{,}240$   **65.** 441; $-147$   **67.** 20; 20   **69.** 180
**70.** $2 \cdot 2 \cdot 2 \cdot 2 \cdot 2 \cdot 2 \cdot 2 \cdot 2 \cdot 2 \cdot 3 \cdot 3$, or $2^9 \cdot 3^2$
**71.** $\frac{2}{3}$   **72.** $\frac{8}{9}$   **73.** $\frac{6}{11}$   **74.** $\frac{41}{265}$   **75.** ◆   **77.** 32 m
below the surface   **79.** (a) One must be negative, and
one must be positive. (b) Either or both must be zero.
(c) Both must be negative or both must be positive.

## Margin Exercises, Section 1.6, pp. 79–82

**1.** $-2$   **2.** 5   **3.** $-3$   **4.** 8   **5.** $-6$   **6.** $-\frac{30}{7}$
**7.** Undefined   **8.** 0   **9.** $\frac{3}{2}$   **10.** $-\frac{4}{5}$   **11.** $-\frac{1}{3}$
**12.** $-5$   **13.** $\frac{1}{1.6}$   **14.** $\frac{2}{3}$   **15.** First row: $-\frac{2}{3}, \frac{3}{2}$;
second row: $\frac{5}{4}, -\frac{4}{5}$; third row: 0, undefined;
fourth row: $-1, 1$; fifth row: $8, -\frac{1}{8}$; sixth row: $4.5, -\frac{1}{4.5}$
**16.** $\frac{4}{7} \cdot \left(-\frac{5}{3}\right)$   **17.** $5 \cdot \left(-\frac{1}{8}\right)$   **18.** $(a - b) \cdot \left(\frac{1}{7}\right)$
**19.** $-23 \cdot a$   **20.** $-5 \cdot \left(\frac{1}{7}\right)$   **21.** $-\frac{20}{21}$   **22.** $-\frac{12}{5}$
**23.** $\frac{16}{7}$   **24.** $-7$   **25.** $\frac{5}{-6}, -\frac{5}{6}$   **26.** $\frac{-8}{7}, \frac{8}{-7}$
**27.** $\frac{-10}{3}, -\frac{10}{3}$

## Exercise Set 1.6, p. 83

**1.** $-8$   **3.** $-14$   **5.** $-3$   **7.** 3   **9.** $-8$   **11.** 2
**13.** $-12$   **15.** $-8$   **17.** Undefined   **19.** $\frac{23}{2}$   **21.** $\frac{7}{15}$
**23.** $-\frac{13}{47}$   **25.** $\frac{1}{13}$   **27.** $\frac{1}{4.3}$   **29.** $-7.1$   **31.** $\frac{q}{p}$   **33.** $4y$
**35.** $\frac{3b}{2a}$   **37.** $4 \cdot \left(\frac{1}{17}\right)$   **39.** $8 \cdot \left(-\frac{1}{13}\right)$   **41.** $13.9 \cdot \left(-\frac{1}{1.5}\right)$
**43.** $x \cdot y$   **45.** $(3x + 4)\left(\frac{1}{5}\right)$   **47.** $(5a - b)\left(\frac{1}{5a + b}\right)$
**49.** $-\frac{9}{8}$   **51.** $\frac{5}{3}$   **53.** $\frac{9}{14}$   **55.** $\frac{9}{64}$   **57.** $-2$   **59.** $\frac{11}{13}$
**61.** $-16.2$   **63.** Undefined   **65.** $\frac{22}{39}$   **66.** 0.477
**67.** 33   **68.** $\frac{3}{2}$   **69.** 87.5%   **70.** $\frac{2}{3}$   **71.** $\frac{9}{8}$   **72.** $\frac{128}{625}$
**73.** ◆   **75.** ◆   **77.** Negative   **79.** Positive
**81.** Negative

## Margin Exercises, Section 1.7, pp. 85–92

**1.**

|         | $x + x$ | $2x$ |
|---------|---------|------|
| $x = 3$ | 6 | 6 |
| $x = -6$ | $-12$ | $-12$ |
| $x = 4.8$ | 9.6 | 9.6 |

**2.**

|         | $x + 3x$ | $5x$ |
|---------|----------|------|
| $x = 2$ | 8 | 10 |
| $x = -6$ | $-24$ | $-30$ |
| $x = 4.8$ | 19.2 | 24 |

**3.** $\dfrac{6}{8}$ **4.** $\dfrac{3t}{4t}$ **5.** $\dfrac{3}{4}$ **6.** $-\dfrac{4}{3}$ **7.** 1; 1 **8.** $-10$; $-10$
**9.** $9 + x$ **10.** $qp$ **11.** $t + xy$, or $yx + t$, or $t + yx$
**12.** 19; 19 **13.** 150; 150 **14.** $(r + s) + 7$ **15.** $(9a)b$
**16.** $(4t)u$, $(tu)4$, $t(4u)$; answers may vary
**17.** $(2 + r) + s$, $(r + s) + 2$, $s + (r + 2)$; answers may vary **18.** (a) 63; (b) 63 **19.** (a) 80; (b) 80
**20.** (a) 28; (b) 28 **21.** (a) 8; (b) 8 **22.** (a) $-4$; (b) $-4$
**23.** (a) $-25$; (b) $-25$ **24.** $5x$, $-8y$, 3 **25.** $-4y$, $-2x$,
$3z$ **26.** $3x - 15$ **27.** $5x + 5$ **28.** $\frac{3}{5}p + \frac{3}{5}q - \frac{3}{5}t$
**29.** $-2x + 6$ **30.** $5x - 10y + 20z$
**31.** $-5x + 10y - 20z$ **32.** $6(x - 2)$
**33.** $3(x - 2y + 3)$ **34.** $b(x + y - z)$
**35.** $2(8a - 18b + 21)$ **36.** $\frac{1}{8}(3x - 5y + 7)$
**37.** $-4(3x - 8y + 4z)$ **38.** $3x$ **39.** $6x$ **40.** $-8x$
**41.** $0.59x$ **42.** $3x + 3y$ **43.** $-4x - 5y - 7$
**44.** $-\frac{2}{3} + \frac{1}{10}x + \frac{7}{9}y$

## Exercise Set 1.7, p. 93

**1.** $\dfrac{3y}{5y}$ **3.** $\dfrac{10x}{15x}$ **5.** $-\dfrac{3}{2}$ **7.** $-\dfrac{7}{6}$ **9.** $8 + y$ **11.** $nm$
**13.** $xy + 9$, or $9 + yx$ **15.** $c + ab$, or $ba + c$
**17.** $(a + b) + 2$ **19.** $8(xy)$ **21.** $a + (b + 3)$
**23.** $(3a)b$ **25.** $2 + (b + a)$, $(2 + a) + b$, $(b + 2) + a$;
answers may vary **27.** $(5 + w) + v$, $(v + 5) + w$,
$(w + v) + 5$; answers may vary **29.** $(3x)y$, $y(x \cdot 3)$,
$3(yx)$; answers may vary **31.** $a(7b)$, $b(7a)$, $(7b)a$;
answers may vary **33.** $2b + 10$ **35.** $7 + 7t$
**37.** $30x + 12$ **39.** $7x + 28 + 42y$ **41.** $7x - 21$
**43.** $-3x + 21$ **45.** $\frac{2}{3}b - 4$ **47.** $7.3x - 14.6$
**49.** $-\frac{3}{5}x + \frac{3}{5}y - 6$ **51.** $45x + 54y - 72$
**53.** $-4x + 12y + 8z$ **55.** $-3.72x + 9.92y - 3.41$
**57.** $4x$, $3z$ **59.** $7x$, $8y$, $-9z$ **61.** $2(x + 2)$
**63.** $5(6 + y)$ **65.** $7(2x + 3y)$ **67.** $5(x + 2 + 3y)$
**69.** $8(x - 3)$ **71.** $4(8 - y)$ **73.** $2(4x + 5y - 11)$
**75.** $a(x - 1)$ **77.** $a(x - y - z)$ **79.** $6(3x - 2y + 1)$
**81.** $\frac{1}{3}(2x - 5y + 1)$ **83.** $19a$ **85.** $9a$ **87.** $8x + 9z$
**89.** $7x + 15y^2$ **91.** $-19a + 88$ **93.** $4t + 6y - 4$
**95.** $b$ **97.** $\frac{13}{4}y$ **99.** $8x$ **101.** $5n$ **103.** $-16y$
**105.** $17a - 12b - 1$ **107.** $4x + 2y$ **109.** $7x + y$
**111.** $0.8x + 0.5y$ **113.** $\frac{35}{6}a + \frac{3}{2}b - 42$ **115.** $\frac{89}{48}$

**116.** $\frac{5}{24}$ **117.** 144 **118.** 30% **119.** $-\frac{5}{24}$ **120.** 60
**121.** ◈ **123.** Not equivalent; $3 \cdot 2 + 5 \ne 3 \cdot 5 + 2$
**125.** Equivalent; commutative law of addition
**127.** $q(1 + r + rs + rst)$

## Margin Exercises, Section 1.8, pp. 97–100

**1.** $-x - 2$ **2.** $-5x - 2y - 8$ **3.** $-6 + t$ **4.** $-x + y$
**5.** $4a - 3t + 10$ **6.** $-18 + m + 2n - 4z$ **7.** $2x - 9$
**8.** $3y + 2$ **9.** $2x - 7$ **10.** $3y + 3$
**11.** $-2a + 8b - 3c$ **12.** $-9x - 8y$ **13.** $-16a + 18$
**14.** $-26a + 41b - 48c$ **15.** $3x - 7$ **16.** 2 **17.** 18
**18.** 6 **19.** 17 **20.** $5x - y - 8$ **21.** $-1237$ **22.** 8
**23.** 381 **24.** $-12$

## Calculator Spotlight, p. 101

**1.** $-11$ **2.** 9 **3.** 114 **4.** 117,649 **5.** $-1,419,857$
**6.** $-1,124,864$ **7.** $-117,649$ **8.** $-1,419,857$
**9.** $-1,124,864$ **10.** $-4$ **11.** $-2$ **12.** 787
**13.** $-32 \times (88 - 29) = -1888$
**14.** $3^5 - 10^2 \times 5^2 = -2257$
**15.** $4 + 6 \cdot 8 - 2 = 4 + 8 \cdot 6 - 2 = 50$; the
commutative law of multiplication
**16.** $5 + 9^2 \cdot 7 - 3 = 569$; because $a^2 \cdot b \ne b^2 \cdot a$,
although students might phrase this verbally and not
symbolically.

## Exercise Set 1.8, p. 103

**1.** $-2x - 7$ **3.** $-5x + 8$ **5.** $-4a + 3b - 7c$
**7.** $-6x + 8y - 5$ **9.** $-3x + 5y + 6$ **11.** $8x + 6y + 43$
**13.** $5x - 3$ **15.** $-3a + 9$ **17.** $5x - 6$
**19.** $-19x + 2y$ **21.** $9y - 25z$ **23.** $-7x + 10y$
**25.** $37a - 23b + 35c$ **27.** 7 **29.** $-40$ **31.** 19
**33.** $12x + 30$ **35.** $3x + 30$ **37.** $9x - 18$
**39.** $-4x - 64$ **41.** $-7$ **43.** $-7$ **45.** $-16$
**47.** $-334$ **49.** 14 **51.** 1880 **53.** 12 **55.** 8
**57.** $-86$ **59.** 37 **61.** $-1$ **63.** $-10$ **65.** 25
**67.** $-7988$ **69.** $-3000$ **71.** 60 **73.** 1 **75.** 10
**77.** $-\frac{13}{45}$ **79.** $-\frac{23}{18}$ **81.** $-118$ **83.** $2 \cdot 2 \cdot 59$
**84.** 252 **85.** $\frac{8}{5}$ **86.** $\frac{5}{18}$ **87.** 81 **88.** 1000 **89.** 100
**90.** 225 **91.** ◈ **93.** $6y - (-2x + 3a - c)$
**95.** $6m - (-3n + 5m - 4b)$ **97.** $-2x - f$
**99.** (a) 52, 52, 28.130169; (b) $-24$, $-24$, $-108.307025$

## Summary and Review: Chapter 1, p. 107

**1.** 4 **2.** $19\%x$, or $0.19x$ **3.** $-45$, 72 **4.** 38
**5.**

**6.**

**7.** $<$ **8.** $>$ **9.** $>$ **10.** $<$ **11.** $-3.8$ **12.** $\frac{3}{4}$
**13.** $\frac{8}{3}$ **14.** $-\frac{1}{7}$ **15.** 34 **16.** 5 **17.** $-3$ **18.** $-4$
**19.** $-5$ **20.** 4 **21.** $-\frac{7}{5}$ **22.** $-7.9$ **23.** 54

**24.** $-9.18$   **25.** $-\frac{2}{7}$   **26.** $-210$   **27.** $-7$   **28.** $-3$
**29.** $\frac{3}{4}$   **30.** $40.4$   **31.** $-2$   **32.** 8-yd gain
**33.** $-\$130$   **34.** $15x - 35$   **35.** $-8x + 10$
**36.** $4x + 15$   **37.** $-24 + 48x$   **38.** $2(x - 7)$
**39.** $6(x - 1)$   **40.** $5(x + 2)$   **41.** $3(4 - x)$
**42.** $7a - 3b$   **43.** $-2x + 5y$   **44.** $5x - y$
**45.** $-a + 8b$   **46.** $-3a + 9$   **47.** $-2b + 21$
**48.** $6$   **49.** $12y - 34$   **50.** $5x + 24$   **51.** $-15x + 25$
**52.** True   **53.** False   **54.** $x > -3$   **55.** $\frac{55}{42}$
**56.** $\frac{109}{18}$   **57.** $2 \cdot 2 \cdot 2 \cdot 3 \cdot 3 \cdot 3 \cdot 3$   **58.** 62.5%
**59.** $0.0567$   **60.** $270$   **61.** $-\frac{5}{8}$   **62.** $-2.1$
**63.** $1000$   **64.** $4a + 2b$

## Test: Chapter 1, p. 109

**1.** [1.1a] 6   **2.** [1.1b] $x - 9$   **3.** [1.1a] 240 ft$^2$
**4.** [1.2d] $<$   **5.** [1.2d] $>$   **6.** [1.2d] $>$   **7.** [1.2d] $<$
**8.** [1.2e] 7   **9.** [1.2e] $\frac{9}{4}$   **10.** [1.2e] 2.7   **11.** [1.3b] $-\frac{2}{3}$
**12.** [1.3b] 1.4   **13.** [1.3b] 8   **14.** [1.6b] $-\frac{1}{2}$
**15.** [1.6b] $\frac{7}{4}$   **16.** [1.4a] 7.8   **17.** [1.3a] $-8$
**18.** [1.3a] $\frac{7}{40}$   **19.** [1.4a] 10   **20.** [1.4a] $-2.5$
**21.** [1.4a] $\frac{7}{8}$   **22.** [1.5a] $-48$   **23.** [1.5a] $\frac{3}{16}$
**24.** [1.6a] $-9$   **25.** [1.6c] $\frac{3}{4}$   **26.** [1.6c] $-9.728$
**27.** [1.8d] $-173$   **28.** [1.4b] 14°F   **29.** [1.7c] $18 - 3x$
**30.** [1.7c] $-5y + 5$   **31.** [1.7d] $2(6 - 11x)$
**32.** [1.7d] $7(x + 3 + 2y)$   **33.** [1.4a] 12
**34.** [1.8b] $2x + 7$   **35.** [1.8b] $9a - 12b - 7$
**36.** [1.8c] $68y - 8$   **37.** [1.8d] $-4$   **38.** [1.8d] 448
**39.** [1.2d] $-2 \geq x$   **40.** [R.5b] 1.728   **41.** [R.4d] 12.5%
**42.** [R.1a] $2 \cdot 2 \cdot 2 \cdot 5 \cdot 7$   **43.** [R.1b] 240
**44.** [1.2e], [1.8d] 15   **45.** [1.8c] $4a$
**46.** [1.7e] $2x + 6y$

## Chapter 2

### Pretest: Chapter 2, p. 112

**1.** [2.2a] $-7$   **2.** [2.3b] $-1$   **3.** [2.3a] 2   **4.** [2.1b] 8
**5.** [2.3c] $-5$   **6.** [2.3a] $\frac{135}{32}$   **7.** [2.3c] 1
**8.** [2.7d] $\{x \mid x \geq -6\}$   **9.** [2.7c] $\{y \mid y > -4\}$
**10.** [2.7e] $\{a \mid a > -1\}$   **11.** [2.7c] $\{x \mid x \geq 3\}$
**12.** [2.7d] $\{y \mid y < -\frac{9}{4}\}$   **13.** [2.6a] $G = \dfrac{P}{3K}$

**14.** [2.6a] $a = \dfrac{Ab + b}{3}$   **15.** [2.4a] Width: 34 in.;

length: 39 in.   **16.** [2.5a] $460   **17.** [2.4a] 81, 82, 83
**18.** [2.8b] Numbers less than 17
**19.** [2.7b]                                   **20.** [2.7b]

$x > -3$                                          $x \leq 4$

### Margin Exercises, Section 2.1, pp. 113–116

**1.** False   **2.** True   **3.** Neither   **4.** Yes   **5.** No
**6.** No   **7.** 9   **8.** $-5$   **9.** 22   **10.** 13.2   **11.** $-6.5$

**12.** $-2$   **13.** $\frac{31}{8}$

### Exercise Set 2.1, p. 117

**1.** Yes   **3.** No   **5.** No   **7.** Yes   **9.** No   **11.** No
**13.** 4   **15.** $-20$   **17.** $-14$   **19.** $-18$   **21.** 15
**23.** $-14$   **25.** 2   **27.** 20   **29.** $-6$   **31.** $6\frac{1}{2}$
**33.** 19.9   **35.** $\frac{7}{3}$   **37.** $-\frac{7}{4}$   **39.** $\frac{41}{24}$   **41.** $-\frac{1}{20}$
**43.** 5.1   **45.** 12.4   **47.** $-5$   **49.** $1\frac{5}{6}$   **51.** $-\frac{10}{21}$
**53.** $-11$   **54.** 5   **55.** $-\frac{5}{12}$   **56.** $\frac{1}{3}$   **57.** $-\frac{3}{2}$
**58.** $-5.2$   **59.** $50 - x$   **60.** $65t$   **61.** ◆
**63.** 342.246   **65.** $-\frac{26}{15}$   **67.** $-10$   **69.** All real
numbers   **71.** $-\frac{5}{17}$   **73.** 13, $-13$

### Margin Exercises, Section 2.2, pp. 119–122

**1.** 15   **2.** $-\frac{7}{4}$   **3.** $-18$   **4.** 10   **5.** $-\frac{4}{5}$   **6.** 7800
**7.** $-3$   **8.** 28

### Exercise Set 2.2, p. 123

**1.** 6   **3.** 9   **5.** 12   **7.** $-40$   **9.** 1   **11.** $-7$   **13.** $-6$
**15.** 6   **17.** $-63$   **19.** 36   **21.** $-21$   **23.** $-\frac{3}{5}$   **25.** $-\frac{3}{2}$
**27.** $\frac{9}{2}$   **29.** 7   **31.** $-7$   **33.** 8   **35.** 15.9   **37.** $7x$
**38.** $-x + 5$   **39.** $8x + 11$   **40.** $-32y$   **41.** $x - 4$
**42.** $-23 - 5x$   **43.** $-10y - 42$   **44.** $-22a + 4$
**45.** $8r$   **46.** $\frac{1}{2}b \cdot 10$, or $5b$   **47.** ◆   **49.** $-8655$
**51.** No solution   **53.** No solution   **55.** $\dfrac{b}{3a}$   **57.** $\dfrac{4b}{a}$

### Margin Exercises, Section 2.3, pp. 125–130

**1.** 5   **2.** 4   **3.** 4   **4.** 39   **5.** $-\frac{3}{2}$   **6.** $-4.3$   **7.** $-3$
**8.** 800   **9.** 1   **10.** 2   **11.** 2   **12.** $\frac{17}{2}$   **13.** $\frac{8}{3}$
**14.** $-4.3$   **15.** 2   **16.** 3   **17.** $-2$   **18.** $-\frac{1}{2}$

### Calculator Spotlight, p. 130

**1.** Both sides equal 9.   **2.** Both sides equal $-2$.
**3.** Both sides equal $-8.18$.

### Exercise Set 2.3, p. 131

**1.** 5   **3.** 8   **5.** 10   **7.** 14   **9.** $-8$   **11.** $-8$   **13.** $-7$
**15.** 15   **17.** 6   **19.** 4   **21.** 6   **23.** $-3$   **25.** 1
**27.** 6   **29.** $-20$   **31.** 7   **33.** 2   **35.** 5   **37.** 2
**39.** 10   **41.** 4   **43.** 0   **45.** $-1$   **47.** $-\frac{4}{3}$   **49.** $\frac{2}{5}$
**51.** $-2$   **53.** $-4$   **55.** $\frac{4}{5}$   **57.** $-\frac{28}{27}$   **59.** 6   **61.** 2
**63.** 6   **65.** 8   **67.** 1   **69.** 17   **71.** $-\frac{5}{3}$   **73.** $-3$
**75.** 2   **77.** $\frac{4}{7}$   **79.** $-\frac{51}{31}$   **81.** 2   **83.** $-6.5$
**84.** $7(x - 3 - 2y)$   **85.** $<$   **86.** $-14$   **87.** $-18.7$
**88.** $-25.5$   **89.** $c \div 8$, or $\dfrac{c}{8}$   **90.** $13.4h$   **91.** ◆
**93.** 4.4233464   **95.** $-\frac{7}{2}$   **97.** $-2$   **99.** 0   **101.** 6
**103.** $\frac{11}{18}$   **105.** 10

## Margin Exercises, Section 2.4, pp. 136–141

**1.** Top: 24 ft; middle: 72 ft; bottom: 144 ft   **2.** 5
**3.** 313 and 314   **4.** 93,333   **5.** Length: 84 ft;
width: 50 ft   **6.** First: 30°; second: 90°; third: 60°
**7.** (a) $10.03994 billion, $15.60842 billion; (b) 2001

## Exercise Set 2.4, p. 143

**1.** 16   **3.** $-\frac{1}{2}$   **5.** 57   **7.** 180 in., 60 in.   **9.** 305 ft
**11.** $2.89   **13.** $-12$   **15.** $699\frac{1}{3}$ mi   **17.** 286, 287
**19.** 41, 42, 43   **21.** 61, 63, 65   **23.** Length: 48 ft;
width: 14 ft   **25.** 11   **27.** 28°, 84°, 68°   **29.** 33°, 38°,
109°   **31.** (a) $1056 million, $1122.2 million,
$1784.2 million; (b) 2002   **33.** $-\frac{47}{40}$   **34.** $-\frac{17}{40}$
**35.** $-\frac{3}{10}$   **36.** $-\frac{32}{15}$   **37.** 1.6   **38.** 409.6   **39.** $-9.6$
**40.** $-41.6$   **41.** ◈   **43.** 120   **45.** About 0.65 in.

## Margin Exercises, Section 2.5, pp. 147–150

**1.** 32%   **2.** 25%   **3.** 225   **4.** 50   **5.** 11.04
**6.** About 33   **7.** 111,416 mi$^2$   **8.** $8400   **9.** $658

## Exercise Set 2.5, p. 151

**1.** 20%   **3.** 150   **5.** 546   **7.** 24%   **9.** 2.5   **11.** 5%
**13.** $16.77 billion   **15.** (a) 8190; (b) 1.8%   **17.** (a) 16%;
(b) $29   **19.** (a) $3.75; (b) $28.75   **21.** (a) $28.80;
(b) $33.12   **23.** $36   **25.** 200   **27.** $282.20
**29.** About 42.4 lb   **31.** About 566   **33.** $15.38
**35.** $7800   **37.** $58   **39.** 181.52   **40.** 0.4538
**41.** 12.0879   **42.** 844.1407   **43.** 16   **44.** 189.6
**45.** 10   **46.** 18.4875   **47.** ◈   **49.** 6 ft, 7 in.
**51.** $9.17, not $9.10

## Margin Exercises, Section 2.6, pp. 155–156

**1.** 2.8 mi   **2.** $I = \dfrac{E}{R}$   **3.** $D = \dfrac{C}{\pi}$
**4.** $c = 4A - a - b - d$   **5.** (a) About 306 lb;
(b) $L = \dfrac{800W}{g^2}$

## Exercise Set 2.6, p. 157

**1.** $h = \dfrac{A}{b}$   **3.** $w = \dfrac{P - 2l}{2}$, or $\dfrac{1}{2}P - l$   **5.** $a = 2A - b$
**7.** $a = \dfrac{F}{m}$   **9.** $c^2 = \dfrac{E}{m}$   **11.** $x = \dfrac{c - By}{A}$   **13.** $t = \dfrac{3k}{v}$
**15.** (a) 57,000 Btu; (b) $a = \dfrac{b}{30}$   **17.** (a) 1423;
(b) $n = 15F$   **19.** (a) 1901 calories;
(b) $a = \dfrac{917 + 6w + 6h - K}{6}$; $h = \dfrac{K - 917 - 6w + 6a}{6}$;
$w = \dfrac{K - 917 - 6h + 6a}{6}$   **21.** 0.92   **22.** $-90$
**23.** $-13.2$   **24.** $-21a + 12b$   **25.** $\frac{1}{6}$   **26.** $-\frac{3}{2}$

**27.** ◈   **29.** $b = \dfrac{2A - ah}{h}$; $h = \dfrac{2A}{a + b}$
**31.** $A$ quadruples.   **33.** $A$ increases by $2h$ units.

## Margin Exercises, Section 2.7, pp. 159–166

**1.** (a) No; (b) no; (c) no; (d) yes; (e) no; (f) no
**2.** (a) Yes; (b) yes; (c) yes; (d) no; (e) yes; (f) yes
**3.**
**4.**
**5.**
**6.** $\{x \mid x > 2\}$;
**7.** $\{x \mid x \le 3\}$;
**8.** $\{x \mid x < -3\}$;
**9.** $\left\{x \mid x \ge \frac{2}{15}\right\}$
**10.** $\{y \mid y \le -3\}$
**11.** $\{x \mid x < 8\}$;
**12.** $\{y \mid y \ge 32\}$;
**13.** $\{x \mid x \ge -6\}$   **14.** $\left\{y \mid y < -\frac{13}{5}\right\}$   **15.** $\left\{x \mid x > -\frac{1}{4}\right\}$
**16.** $\left\{y \mid y \ge \frac{19}{9}\right\}$   **17.** $\left\{y \mid y \ge \frac{19}{9}\right\}$   **18.** $\{x \mid x \ge -2\}$
**19.** $\{x \mid x \ge -4\}$   **20.** $\left\{x \mid x > \frac{8}{3}\right\}$

## Exercise Set 2.7, p. 167

**1.** (a) Yes; (b) yes; (c) no; (d) yes; (e) yes
**3.** (a) No; (b) no; (c) no; (d) yes; (e) no
**5.**
**7.**
**9.**
**11.**
**13.**
**15.** $\{x \mid x > -5\}$;
**17.** $\{x \mid x \le -18\}$;
**19.** $\{y \mid y > -5\}$   **21.** $\{x \mid x > 2\}$   **23.** $\{x \mid x \le -3\}$
**25.** $\{x \mid x < 4\}$   **27.** $\{t \mid t > 14\}$   **29.** $\left\{y \mid y \le \frac{1}{4}\right\}$
**31.** $\left\{x \mid x > \frac{7}{12}\right\}$
**33.** $\{x \mid x < 7\}$;
**35.** $\{x \mid x < 3\}$;
**37.** $\left\{y \mid y \ge -\frac{2}{5}\right\}$   **39.** $\{x \mid x \ge -6\}$   **41.** $\{y \mid y \le 4\}$
**43.** $\left\{x \mid x > \frac{17}{3}\right\}$   **45.** $\left\{y \mid y < -\frac{1}{14}\right\}$   **47.** $\left\{x \mid x \le \frac{3}{10}\right\}$
**49.** $\{x \mid x < 8\}$   **51.** $\{x \mid x \le 6\}$   **53.** $\{x \mid x < -3\}$
**55.** $\{x \mid x > -3\}$   **57.** $\{x \mid x \le 7\}$   **59.** $\{x \mid x > -10\}$
**61.** $\{y \mid y < 2\}$   **63.** $\{y \mid y \ge 3\}$   **65.** $\{y \mid y > -2\}$
**67.** $\{x \mid x > -4\}$   **69.** $\{x \mid x \le 9\}$   **71.** $\{y \mid y \le -3\}$

**73.** $\{y\,|\,y < 6\}$  **75.** $\{m\,|\,m \geq 6\}$  **77.** $\left\{t\,|\,t < -\frac{5}{3}\right\}$
**79.** $\{r\,|\,r > -3\}$  **81.** $\left\{x\,|\,x \geq -\frac{57}{34}\right\}$  **83.** $\{x\,|\,x > -2\}$
**85.** $-74$  **86.** 4.8  **87.** $-\frac{5}{8}$  **88.** $-1.11$  **89.** $-38$
**90.** $-\frac{7}{8}$  **91.** $-9.4$  **92.** 1.11  **93.** 140  **94.** 41
**95.** $-2x - 23$  **96.** $37x - 1$  **97.** ◈  **99. (a)** Yes;
**(b)** yes; **(c)** no; **(d)** no; **(e)** no; **(f)** yes; **(g)** yes
**101.** All real numbers

## Margin Exercises, Section 2.8, pp. 171–172

**1.** $x \leq 8$  **2.** $y > -2$  **3.** $s \leq 180$  **4.** $p \geq \$5800$
**5.** $2x - 32 > 5$  **6.** $\{x\,|\,x \geq 84\}$  **7.** $\{C\,|\,C < 1063°\}$

## Exercise Set 2.8, p. 173

**1.** $x > 8$  **3.** $y \leq -4$  **5.** $n \geq 1300$  **7.** $a \leq 500$
**9.** $2 + 3x < 13$  **11.** $\{x\,|\,x \geq 97\}$  **13.** $\{Y\,|\,Y \geq 1935\}$
**15.** $\{L\,|\,L \geq 5 \text{ in.}\}$  **17.** $\{x\,|\,x > 5\}$  **19.** $\{d\,|\,d > 25\}$
**21.** $\{b\,|\,b > 6 \text{ cm}\}$  **23.** $\{x\,|\,x \geq 21\}$
**25.** $\{t\,|\,t \leq 0.75 \text{ hr}\}$  **27.** $\{f\,|\,f \geq 16 \text{ g}\}$  **29.** $-160$
**30.** $-17x + 18$  **31.** $91x - 242$  **32.** 0.25  **33.** ◈

## Summary and Review: Chapter 2, p. 175

**1.** $-22$  **2.** 1  **3.** 25  **4.** 9.99  **5.** $\frac{1}{4}$  **6.** 7
**7.** $-192$  **8.** $-\frac{7}{3}$  **9.** $-\frac{15}{64}$  **10.** $-8$  **11.** 4  **12.** $-5$
**13.** $-\frac{1}{3}$  **14.** 3  **15.** 4  **16.** 16  **17.** 6  **18.** $-3$
**19.** 12  **20.** 4  **21.** Yes  **22.** No  **23.** Yes
**24.** $\left\{y\,|\,y \geq -\frac{1}{2}\right\}$  **25.** $\{x\,|\,x \geq 7\}$  **26.** $\{y\,|\,y > 2\}$
**27.** $\{y\,|\,y \leq -4\}$  **28.** $\{x\,|\,x < -11\}$  **29.** $\{y\,|\,y > -7\}$
**30.** $\{x\,|\,x > -6\}$  **31.** $\left\{x\,|\,x > -\frac{9}{11}\right\}$  **32.** $\{y\,|\,y \leq 7\}$
**33.** $\left\{x\,|\,x \geq -\frac{1}{12}\right\}$
**34.**

$x < 3$

**35.**

$-2 < x \leq 5$

**36.**

$y > 0$

**37.** $d = \dfrac{C}{\pi}$  **38.** $B = \dfrac{3V}{h}$  **39.** $a = 2A - b$
**40.** Length: 365 mi; width: 275 mi  **41.** 27
**42.** 345, 346  **43.** \$2117  **44.** 27  **45.** 35°, 85°, 60°
**46.** \$220  **47.** \$26,087  **48.** \$138.95  **49.** 86
**50.** $\{w\,|\,w > 17 \text{ cm}\}$  **51. (a)** 8.2 lb; **(b)** $L = \dfrac{800W}{g^2}$
**52.** $\frac{41}{4}$  **53.** $58t$  **54.** $-45$  **55.** $-43x + 8y$
**56.** ◈ The end result is the same either way. If $s$ is the original salary, the new salary after a 5% raise followed by an 8% raise is $1.08(1.05s)$. If the raises occur the other way around, the new salary is $1.05(1.08s)$. By the commutative and associative laws of multiplication, we see that these are equal. However, it would be better to receive the 8% raise first, because this increase yields a higher salary the first year than a 5% raise.

**57.** ◈ The inequalities are equivalent by the multiplication principle for inequalities. If we multiply on both sides of one inequality by $-1$, the other inequality results.  **58.** 23, $-23$  **59.** 20, $-20$
**60.** $a = \dfrac{y - 3}{2 - b}$

## Test: Chapter 2, p. 177

**1.** [2.1b] 8  **2.** [2.1b] 26  **3.** [2.2a] $-6$  **4.** [2.2a] 49
**5.** [2.3b] $-12$  **6.** [2.3a] 2  **7.** [2.3a] $-8$
**8.** [2.1b] $-\frac{7}{20}$  **9.** [2.3c] 7  **10.** [2.3c] $\frac{5}{3}$  **11.** [2.3b] 2.5
**12.** [2.7c] $\{x\,|\,x \leq -4\}$  **13.** [2.7c] $\{x\,|\,x > -13\}$
**14.** [2.7d] $\{x\,|\,x \leq 5\}$  **15.** [2.7d] $\{y\,|\,y \leq -13\}$
**16.** [2.7d] $\{y\,|\,y \geq 8\}$  **17.** [2.7d] $\left\{x\,|\,x \leq -\frac{1}{20}\right\}$
**18.** [2.7e] $\{x\,|\,x < -6\}$  **19.** [2.7e] $\{x\,|\,x \leq -1\}$
**20.** [2.7b]  **21.** [2.7b, e]

$y \leq 9$

$x < 1$

**22.** [2.7b]
$-2 \leq x \leq 2$

**23.** [2.4a] Width: 7 cm; length: 11 cm  **24.** [2.4a] 6
**25.** [2.4a] 2509, 2510, 2511  **26.** [2.5a] \$880
**27.** [2.4a] 3 m, 5 m  **28.** [2.6a] $r = \dfrac{A}{2\pi h}$
**29.** [2.6a] **(a)** 2650; **(b)** $w = \dfrac{K - 7h + 9.52a - 92.4}{19.18}$
**30.** [2.8b] $\{x\,|\,x > 6\}$  **31.** [2.8b] $\{l\,|\,l \geq 174 \text{ yd}\}$
**32.** [1.3a] $-\frac{2}{9}$  **33.** [1.1a] $\frac{8}{3}$  **34.** [1.1b] 73%$p$, or 0.73$p$
**35.** [1.8b] $-18x + 37y$  **36.** [2.6a] $d = \dfrac{1 - ca}{-c}$, or $\dfrac{ca - 1}{c}$
**37.** [1.2e], [2.3a] 15, $-15$  **38.** [2.4a] 60

## Cumulative Review: Chapters 1–2, p. 179

**1.** [1.1a] $\frac{3}{2}$  **2.** [1.1a] $\frac{15}{4}$  **3.** [1.1a] 0  **4.** [1.1b] $2w - 4$
**5.** [1.2d] $>$  **6.** [1.2d] $>$  **7.** [1.2d] $<$
**8.** [1.3b], [1.6b] $-\frac{2}{5}, \frac{5}{2}$  **9.** [1.2e] 3  **10.** [1.2e] $\frac{3}{4}$
**11.** [1.2e] 0  **12.** [1.3a] $-4.4$  **13.** [1.4a] $-\frac{5}{2}$
**14.** [1.5a] $\frac{5}{6}$  **15.** [1.5a] $-105$  **16.** [1.6a] $-9$
**17.** [1.6c] $-3$  **18.** [1.6c] $\frac{32}{125}$
**19.** [1.7c] $15x + 25y + 10z$  **20.** [1.7c] $-12x - 8$
**21.** [1.7c] $-12y + 24x$  **22.** [1.7d] $2(32 + 9x + 12y)$
**23.** [1.7d] $8(2y - 7)$  **24.** [1.7d] $5(a - 3b + 5)$
**25.** [1.7e] $15b + 22y$  **26.** [1.7e] $4 + 9y + 6z$
**27.** [1.7e] $1 - 3a - 9d$  **28.** [1.7e] $-2.6x - 5.2y$
**29.** [1.8b] $-1 + 3x$  **30.** [1.8b] $-2x - y$
**31.** [1.8b] $-7x + 6$  **32.** [1.8b] $8x$  **33.** [1.8c] $5x - 13$
**34.** [2.1b] 4.5  **35.** [2.2a] $\frac{4}{25}$  **36.** [2.1b] 10.9
**37.** [2.1b] $3\frac{5}{6}$  **38.** [2.2a] $-48$  **39.** [2.2a] 12
**40.** [2.2a] $-6.2$  **41.** [2.3a] $-3$  **42.** [2.3b] $-\frac{12}{5}$
**43.** [2.3b] 8  **44.** [2.3c] 7  **45.** [2.3b] $-\frac{4}{5}$
**46.** [2.3b] $-\frac{10}{3}$  **47.** [2.7c] $\{x\,|\,x < 2\}$
**48.** [2.7e] $\{y\,|\,y \geq 4\}$  **49.** [2.7e] $\{y\,|\,y < -3\}$

**50.** [2.6a] $h = \dfrac{2A}{b+c}$ **51.** [2.6a] $q = p - 2Q$

**52.** [2.4a] 154 **53.** [2.4a] $45 **54.** [2.5a] $1050

**55.** [2.4a] 50 m, 53 m, 40 m **56.** [2.8b] $\{x \mid x \geq 78\}$

**57.** [2.5a] $24.60 **58.** [2.5a] $36,000

**59.** [2.5a] 30% **60.** [1.2e], [2.3a] 4, $-4$

**61.** [2.3c] All real numbers **62.** [2.3c] No solution

**63.** [2.3b] 3 **64.** [2.3c] All real numbers

**65.** [2.6a] $Q = \dfrac{2 - pm}{p}$

# Chapter 3

## *Pretest: Chapter 3, p. 182*

**1.** [3.2b]

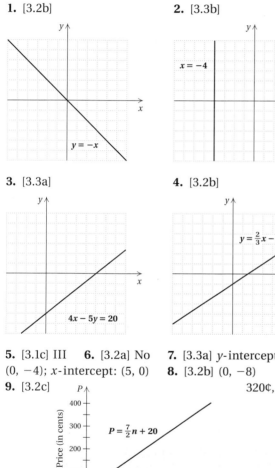

$y = -x$

**2.** [3.3b]

$x = -4$

**3.** [3.3a]

$4x - 5y = 20$

**4.** [3.2b]

$y = \frac{2}{3}x - 1$

**5.** [3.1c] III **6.** [3.2a] No **7.** [3.3a] $y$-intercept:

$(0, -4)$; $x$-intercept: $(5, 0)$ **8.** [3.2b] $(0, -8)$

**9.** [3.2c]

320¢, or $3.20

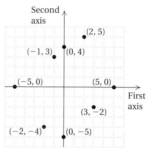

$P = \frac{7}{2}n + 20$

Price (in cents)

Number of pages

**10.** [3.4a] Mean: 0.18; median: 0.18; mode: 0.18

**11.** [3.4c] 164 cm

## *Margin Exercises, Section 3.1, pp. 183–187*

**1.** $23.22 billion **2.** **(a)** 3 A.M.–6 A.M.;

**(b)** midnight–3 A.M., 3 A.M.–6 A.M., 6 A.M.–9 A.M.,

9 A.M.–noon **3.** **(a)** 2; **(b)** 60 beats per minute

**4.–11.**

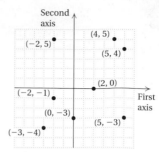

Second axis

$(-2, 5)$ $(4, 5)$

$(5, 4)$

$(2, 0)$

$(-2, -1)$ First axis

$(0, -3)$

$(5, -3)$

$(-3, -4)$

**12.** Both are negative numbers. **13.** First, positive; second, negative **14.** I **15.** III **16.** IV **17.** II **18.** Not in any quadrant **19.** $A$: $(-5, 1)$; $B$: $(-3, 2)$; $C$: $(0, 4)$; $D$: $(3, 3)$; $E$: $(1, 0)$; $F$: $(0, -3)$; $G$: $(-5, -4)$

## *Exercise Set 3.1, p. 189*

**1.** 6 **3.** The weight is greater than 200 lb. **5.** The weight is greater than or equal to 120 lb. **7.** 32.4%

**9.** $10,360.32 **11.** 20,000 **13.** 1994 **15.** About 500

**17.**

Second axis

$(2, 5)$

$(-1, 3)$ $(0, 4)$

$(-5, 0)$ $(5, 0)$

First axis

$(3, -2)$

$(-2, -4)$ $(0, -5)$

**19.** II **21.** IV **23.** III **25.** I **27.** II **29.** IV

**31.** Negative; negative **33.** Second; first **35.** I, IV

**37.** I, III **39.** $A$: $(3, 3)$; $B$: $(0, -4)$; $C$: $(-5, 0)$;

$D$: $(-1, -1)$; $E$: $(2, 0)$ **41.** 12 **42.** 4.89 **43.** 0

**44.** $\frac{4}{5}$ **45.** $0.9 billion **46.** $18.40 **47.** ◈

**49.** $(-1, -5)$

**51.** Answers may vary. **53.** 26

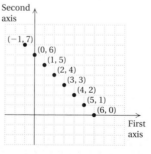

Second axis

$(-1, 7)$

$(0, 6)$

$(1, 5)$

$(2, 4)$

$(3, 3)$

$(4, 2)$

$(5, 1)$

$(6, 0)$

First axis

## *Margin Exercises, Section 3.2, pp. 193–200*

**1.** No **2.** Yes

**3.** $(-2, -3)$, $(1, 3)$; answers may vary

**4.**

$y = -2x$

**5.**

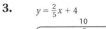
$y = \frac{1}{2}x$

**(b)** $1700

**(c)** About 2.8 yr

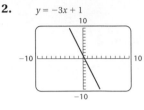
$v = -0.68t + 3.4$

Value of copier (in thousands)

Time from date of purchase (in years)

## Calculator Spotlight, p. 202

**1.** $y = 2x + 1$

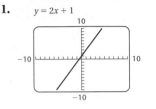

**2.** $y = -3x + 1$

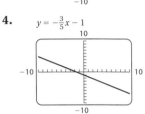

**6.**

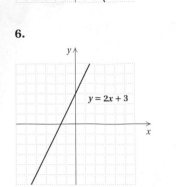
$y = 2x + 3$

**7.**

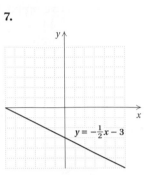
$y = -\frac{1}{2}x - 3$

**3.** $y = \frac{2}{5}x + 4$

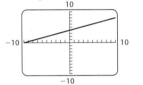

**4.** $y = -\frac{3}{5}x - 1$

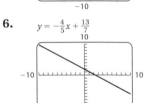

**8.**

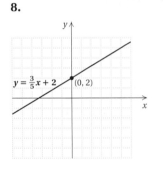
$y = \frac{3}{5}x + 2$   (0, 2)

**9.**

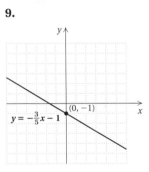
$y = -\frac{3}{5}x - 1$   (0, -1)

**5.** $y = 2.085x + 15.08$

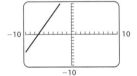

**6.** $y = -\frac{4}{5}x + \frac{13}{7}$

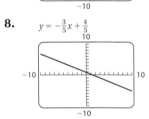

**7.** $y = -\frac{2}{3}x + 6$

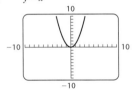

**8.** $y = -\frac{3}{5}x + \frac{4}{5}$

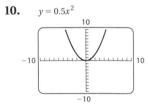

**10.**

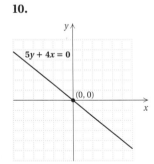
$5y + 4x = 0$   (0, 0)

**11.**

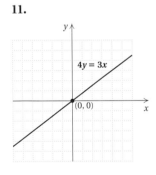
$4y = 3x$   (0, 0)

**9.** $y = x^2$

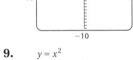

**10.** $y = 0.5x^2$

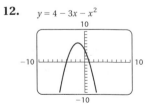

**11.** $y = 8 - x^2$

**12.** $y = 4 - 3x - x^2$

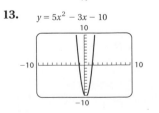

**12.**

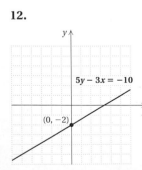
$5y - 3x = -10$   (0, -2)

**13.**

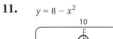
$5y + 3x = 20$   (0, 4)

**13.** $y = 5x^2 - 3x - 10$

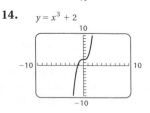

**14.** $y = x^3 + 2$

**14. (a)** $2720, $2040, $680, $0;

**15.** $y = |x|$

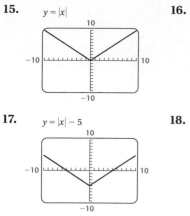

**16.** $y = |x - 5|$

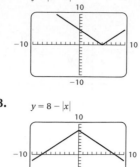

**17.** $y = |x| - 5$

**18.** $y = 8 - |x|$

## Exercise Set 3.2, p. 203

**1.** No  **3.** No  **5.** Yes

**7.**
$$\frac{y = x - 5}{-1 \ ? \ 4 - 5}$$
$$| \ -1 \qquad \text{TRUE}$$

$$\frac{y = x - 5}{-4 \ ? \ 1 - 5}$$
$$| \ -4 \qquad \text{TRUE}$$

**9.**
$$\frac{y = \frac{1}{2}x + 3}{5 \ ? \ \frac{1}{2} \cdot 4 + 3}$$
$$| \ 2 + 3$$
$$| \ 5 \qquad \text{TRUE}$$

$$\frac{y = \frac{1}{2}x + 3}{2 \ ? \ \frac{1}{2}(-2) + 3}$$
$$| \ -1 + 3$$
$$| \ 2 \qquad \text{TRUE}$$

**11.**
$$\frac{4x - 2y = 10}{4 \cdot 0 - 2(-5) \ ? \ 10}$$
$$0 + 10 \ |$$
$$10 \ | \qquad \text{TRUE}$$

$$\frac{4x - 2y = 10}{4 \cdot 4 - 2 \cdot 3 \ ? \ 10}$$
$$16 - 6 \ |$$
$$10 \ | \qquad \text{TRUE}$$

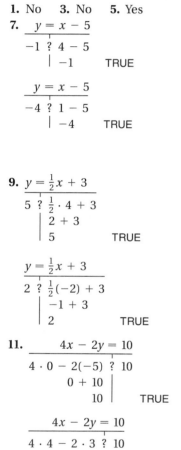

**13.**

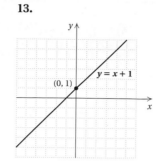

**15.**

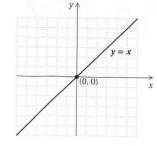

**17.**

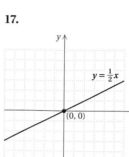

**19.**

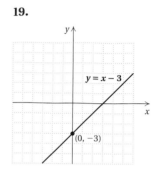

**21.**

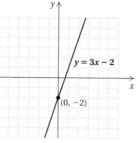

**23.**

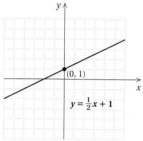

**25.**

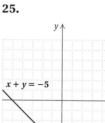

**27.**

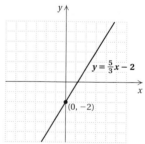

**29.**

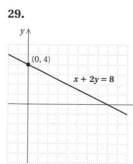

**31.**

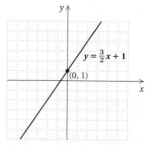

**33.**

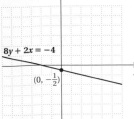

**35.**

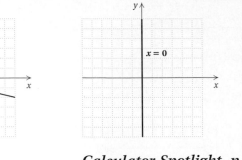

**6.**

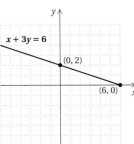

**7.**

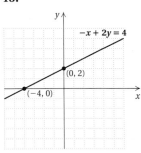

**37. (a)** $300, $100, $0;

**(b)**

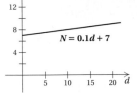

$50; **(c)** 3

**39. (a)** 7.1 gal, 7.4 gal, 7.8 gal, 8 gal;

**(b)**

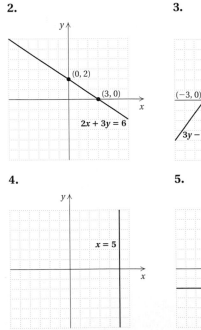

7.6 gal; **(c)** 2011

**41.** 3000     **42.** 125,000     **43.** 0     **44.** 6,078,000
**45.** −0.875     **46.** 0.71875     **47.** 1.828125     **48.** −2.25
**49.** ◆
**51.** $y = -x + 5$     **53.** $y = x + 2$

## Margin Exercises, Section 3.3, pp. 207–209

**1. (a)** (0, 3); **(b)** (4, 0)

**2.**

**3.**

**4.**

**5.**

**1.** $y$-intercept: (0, −15); $x$-intercept: (−2.08, 0)
**2.** $y$-intercept: (0, 27); $x$-intercept: (−12.68, 0)
**3.** $y$-intercept: (0, 14); $x$-intercept: (16.8, 0)
**4.** $y$-intercept: (0, −21.43); $x$-intercept: (75, 0)
**5.** $y$-intercept: (0, 25); $x$-intercept: (16.67, 0)
**6.** $y$-intercept: (0, −9); $x$-intercept: (45, 0)
**7.** $y$-intercept: (0, −15); $x$-intercept: (11.54, 0)
**8.** $y$-intercept: (0, −0.05); $x$-intercept: (0.04, 0)

## Exercise Set 3.3, p. 211

**1. (a)** (0, 5); **(b)** (2, 0)     **3. (a)** (0, −4); **(b)** (3, 0)
**5. (a)** (0, 3); **(b)** (5, 0)     **7. (a)** (0, −14); **(b)** (4, 0)
**9. (a)** $\left(0, \frac{10}{3}\right)$; **(b)** $\left(-\frac{5}{2}, 0\right)$     **11. (a)** $\left(0, -\frac{1}{3}\right)$; **(b)** $\left(\frac{1}{2}, 0\right)$
**13.**     **15.**

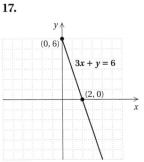

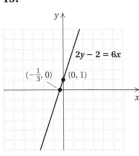

**17.**     **19.**

**21.**     **23.**

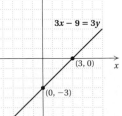

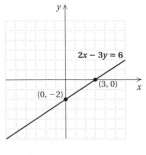

**25.**

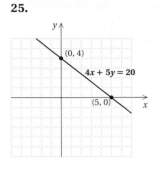

**27.**

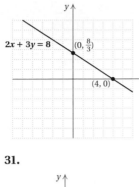

**45.**

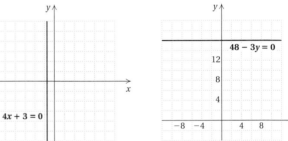

**47.**

**29.**

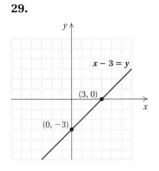

**31.**

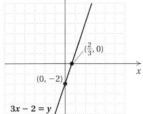

**49.**

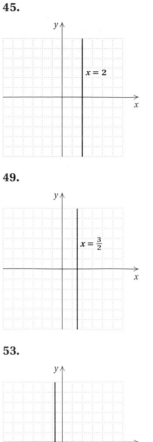

**51.**

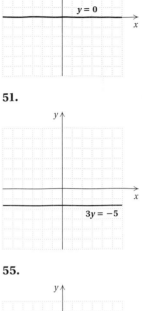

**33.**

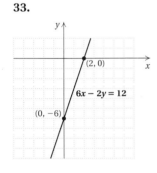

**35.**

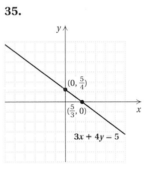

**53.**

**55.**

**57.** $y = -1$    **59.** $x = 4$    **61.** 16%    **62.** 70
**63.** $32.50    **64.** $23.00    **65.** $\{x \mid x > -40\}$
**66.** $\{x \mid x \leq -7\}$    **67.** $\{x \mid x < 1\}$    **68.** $\{x \mid x \geq 2\}$
**69.** ◈    **71.** $x = 0$    **73.** $y = -4$    **75.** $k = 12$

### Margin Exercises, Section 3.4, pp. 215–219

**1.** 56.7    **2.** 64.7    **3.** 87.8    **4.** 17    **5.** 16.5    **6.** 91
**7.** 55    **8.** 54, 87    **9.** No mode exists.
**10.** (a) 25 $mm^2$; (b) 25 $mm^2$; (c) no mode exists.
**11.** Ball A: 18.8 in.; Ball B: 19.6 in.; Ball B is better.
**12.** $866.79    **13.** 95

### Calculator Spotlight, p. 220

**1.** (1.106383, 17.386809), (3.0425532, 21.423723),
(5.6702128, 26.902394), (8.712766, 33.246117),
(10.234043, 36.417979); answers may vary
**2.** 223.58; 327.83    **3.** 404.98

**37.**

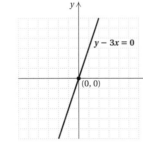

**39.**

**41.**

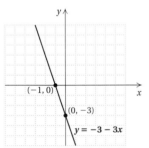

**43.**

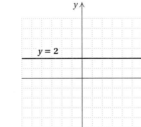

**4. (a)** $y = -0.68x + 3.4$

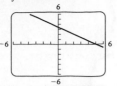

**(b)** no; **(c)** (5, 0); **(d)** $0 \le x \le 5$;
**(e)** (0.76595745, 2.8791489), (1.787234, 2.1846809),
(2.2978723, 1.8374468), (2.9361702, 1.4034043),
(4.212766, 0.53531915); answers may vary; **(f)** 2.79 yr;
**(g)** $2244, $1836, $612, $0

## Exercise Set 3.4, p. 221

**1.** Mean: 28.6; median: 30; modes: 15, 30
**3.** Mean: 94; median: 95; no mode exists.
**5.** Mean: 32; median: 35; mode: 23
**7.** Mean: 897.2; median: 798; no mode exists.
**9.** Mean: 87.7; median: 88; modes: 85, 88
**11.** Battery B  **13.** 162.4  **15.** 2008  **17.** 1997: 1.39;
2000: 1.54  **19.** 1997: $1,500,000; 2000: $1,700,000
**21.** 84  **23.** $\frac{4}{25}$  **24.** $\frac{1}{3}$  **25.** $\frac{3}{8}$  **26.** $\frac{3}{4}$  **27.** 20%
**28.** $18  **29.** $45.15  **30.** $55  **31.** ◆
**33.**   $y = 0.35x - 7$

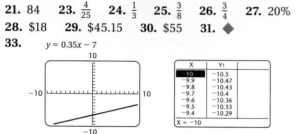

**35.**   $y = x^3 - 5$

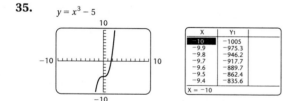

## Summary and Review: Chapter 3, p. 225

**1.** $775.50; $634.50  **2.** 47 lb  **3.** 80 lb  **4.** 33 lb
**5.** 1995  **6.** 1990–1995  **7.** One shower
**8.** One toilet flush  **9.** One shave, wash dishes,
one shower  **10.** One toilet flush  **11.** (−5, −1)
**12.** (−2, 5)  **13.** (3, 0)
**14.–16.**

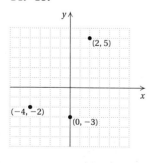

**17.** IV  **18.** III  **19.** I  **20.** No  **21.** Yes
**22.**

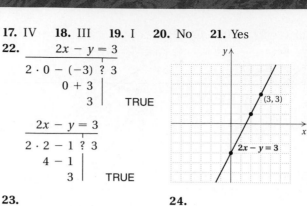

**23.**

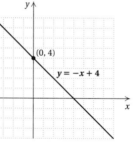

**24.**

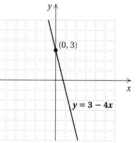

**25.**

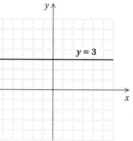

**26.**

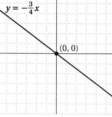

**27.**

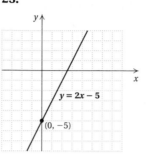

**28.**

(5x − 4 = 0)

**29.**

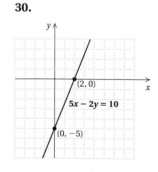

**30.**

**31. (a)** $14\frac{1}{2}$ ft$^3$, 16 ft$^3$, $20\frac{1}{2}$ ft$^3$, 28 ft$^3$;
**(b)** $17\frac{1}{2}$ ft$^3$; **(c)** 6

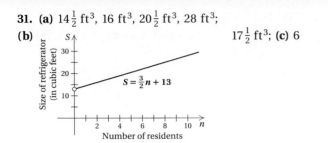

**32.** Mean: 39.5; median: 39.5; no mode exists.
**33.** Mean: 14.875; median: 16; modes: 12, 17, 19
**34.** Mean: 0.18; median: 0.175; no mode exists.
**35.** Mean: 229.8; median: 223; mode: 215
**36.** Mean: $50.1; median: $50.4; no mode exists.
**37.** Popcorn A   **38.** 109.55   **39.** 1997: 64.4; 2000: 78.3
**40.** $-0.34375$   **41.** $0.\overline{8}$   **42.** 3.2   **43.** $\frac{17}{19}$   **44.** 42.71
**45.** 112.53   **46.** $9755.09   **47.** $79.95
**48.** ◈ A small business might use a graph to look up prices quickly (as in the FedEx mailing costs example) or to plot change in sales over a period of time. Many other applications exist.   **49.** ◈ The $y$-intercept is the point at which the graph crosses the $y$-axis. Since a point on the $y$-axis is neither left nor right of the origin, the first or $x$-coordinate of the point is 0.
**50.** $m = -1$   **51.** 45 square units; 28 linear units

### Test: Chapter 3, p. 229

**1.** [3.1a] $495,000,000   **2.** [3.1a] Crest and Colgate
**3.** [3.1a] Crest   **4.** [3.1a] Arm & Hammer
**5.** [3.1a] June   **6.** [3.1a] January   **7.** [3.1a] March, April, May, June, July   **8.** [3.1a] August
**9.** [3.1a] 1997   **10.** [3.1a] 1991   **11.** [3.1a] About $500,000   **12.** [3.1a] 1996–1997
**13.** [3.1a] 1994–1995   **14.** [3.1a] About $500,000
**15.** [3.1c] II   **16.** [3.1c] III   **17.** [3.1d] (3, 4)
**18.** [3.1d] (0, −4)
**19.** [3.2a]

$$y - 2x = 5$$
$$\frac{}{-3 - 2(-4)\ ?\ 5}$$
$$-3 + 8$$
$$5\ \bigg|\quad \text{TRUE}$$

$$y - 2x = 5$$
$$\frac{}{3 - 2(-1)\ ?\ 5}$$
$$3 + 2$$
$$5\ \bigg|\quad \text{TRUE}$$

**20.** [3.2b]

**21.** [3.2b]

**22.** [3.3b]

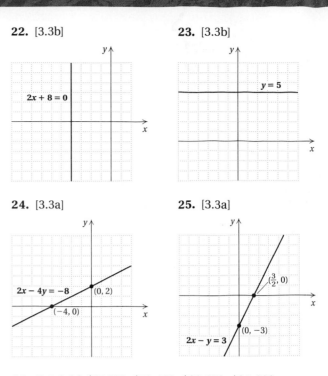

**23.** [3.3b]

**24.** [3.3a]

**25.** [3.3a]

**26.** [3.2c] **(a)** $17,000; $19,400; $22,600; $24,200; $27,500; **(c)** 2002
**(b)**

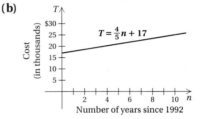

**27.** [3.4a] Mean: 51; median: 51.5; no mode exists.
**28.** [3.4a] Mean: 4.2; median: 4.5; mode: 5
**29.** [3.4a] Mean: 16.3; median: 18.5; modes: 18, 19
**30.** [3.4a] Mean: 45.4; median: 40; no mode exists.
**31.** [3.4b] Ball B   **32.** [3.4c] 135.15   **33.** [3.4c] 1996: 1785; 2000: 2093   **34.** [1.2c] 0.975   **35.** [1.2c] $-1.08\overline{3}$
**36.** [1.2e] 71.2   **37.** [1.2e] $\frac{13}{47}$   **38.** [R.3c] 42.705
**39.** [R.3c] 112.527   **40.** [2.5a] $84.50
**41.** [2.5a] $36,400   **42.** [3.1b] 25 square units, 20 linear units   **43.** [3.3b] $y = 3$

### Cumulative Review: Chapters 1–3, p. 233

**1.** [1.1a] $\frac{5}{2}$   **2.** [1.7c] $12x - 15y + 21$
**3.** [1.7d] $3(5x - 3y + 1)$   **4.** [R.1a] $2 \cdot 3 \cdot 7$
**5.** [R.3a] 0.45   **6.** [1.2e] 4   **7.** [1.3b] 3.08
**8.** [1.6b] $-\frac{7}{8}$   **9.** [1.7e] $-x - y$   **10.** [R.4a] 0.785
**11.** [R.2c] $\frac{1}{3}$   **12.** [1.3a] 2.6   **13.** [1.5a] 7.28
**14.** [1.6c] $-\frac{5}{12}$   **15.** [1.8d] $-2$   **16.** [1.8d] 27
**17.** [1.8b] $-2y - 7$   **18.** [1.8c] $5x + 11$
**19.** [2.1b] $-1.2$   **20.** [2.2a] $-21$   **21.** [2.3a] 9
**22.** [2.2a] $\frac{4}{25}$   **23.** [2.3b] 2   **24.** [2.1b] $\frac{13}{8}$
**25.** [2.3c] $-\frac{17}{21}$   **26.** [2.3b] $-17$   **27.** [2.3b] 2
**28.** [2.7e] $\{x\,|\,x < 16\}$   **29.** [2.7e] $\left\{x\,\middle|\,x \le -\frac{11}{8}\right\}$
**30.** [2.6a] $h = \dfrac{2A}{b + c}$   **31.** [3.1c] IV

**32.** [2.7b]

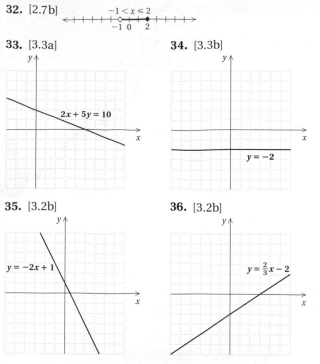

$-1 < x \leq 2$

**33.** [3.3a]

$2x + 5y = 10$

**34.** [3.3b]

$y = -2$

**35.** [3.2b]

$y = -2x + 1$

**36.** [3.2b]

$y = \frac{2}{3}x - 2$

**37.** [3.3a] $y$-intercept: $(0, -3)$; $x$-intercept: $(10.5, 0)$
**38.** [3.3a] $y$-intercept: $(0, 5)$; $x$-intercept: $\left(-\frac{5}{4}, 0\right)$
**39.** [2.5a] 160 million    **40.** [2.4a] 15.6 million
**41.** [2.5a] $120    **42.** [2.4a] First: 50 m; second: 53 m;
third: 40 m    **43.** [2.8b] $\{x \mid x \leq 8\}$
**44.** [3.2c] **(a)** $375, $450, $525, $825;
**(b)**

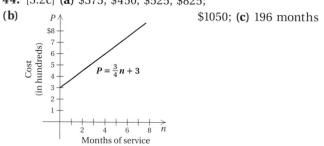

$1050; **(c)** 196 months

$P = \frac{3}{4}n + 3$

Cost (in hundreds)

Months of service

**45.** [3.4a] Mean: 28.15 cc; median: 27.05 cc;
mode: 25.4 cc    **46.** [3.4b] Class A    **47.** [3.4c] 145.25
**48.** [2.3a], [1.2e] $-4, 4$    **49.** [2.3c] All real numbers
**50.** [2.3c] No solution    **51.** [2.3b] 3    **52.** [2.3c] All real
numbers    **53.** [2.6a] $Q = \dfrac{2 - pm}{p}$, or $\dfrac{2}{p} - m$

# Chapter 4

## Pretest: Chapter 4, p. 236

**1.** [4.1d, f] $x^2$    **2.** [4.1e, f] $\dfrac{1}{x^7}$    **3.** [4.2b] $\dfrac{16x^4}{y^6}$

**4.** [4.1f] $\dfrac{1}{p^3}$    **5.** [4.2c] $3.47 \times 10^{-4}$    **6.** [4.2c] 3,400,000

**7.** [4.3g] 3, 2, 1, 0; 3
**8.** [4.3e] $-3a^3b - 2a^2b^2 + ab^3 + 12b^3 + 9$
**9.** [4.4a] $11x^2 + 4x - 11$    **10.** [4.4c] $-x^2 - 18x + 27$
**11.** [4.5b] $15x^4 - 20x^3 + 5x^2$    **12.** [4.6c] $x^2 + 10x + 25$

**13.** [4.6b] $x^2 - 25$    **14.** [4.6a] $4x^6 + 19x^3 - 30$
**15.** [4.7f] $4x^2 - 12xy + 9y^2$    **16.** [4.8b] $x^2 + x + 3$, R 8;
or $x^2 + x + 3 + \dfrac{8}{x - 2}$

## Margin Exercises, Section 4.1, pp. 237–242

**1.** $5 \cdot 5 \cdot 5 \cdot 5$    **2.** $x \cdot x \cdot x \cdot x \cdot x$    **3.** $3t \cdot 3t$
**4.** $3 \cdot t \cdot t$    **5.** 6    **6.** 1    **7.** 8.4    **8.** 1    **9.** 125
**10.** 3215.36 cm$^2$    **11.** 119    **12.** 3; $-3$    **13.** **(a)** 144;
**(b)** 36; **(c)** no    **14.** $3^{10}$    **15.** $x^{10}$    **16.** $p^{24}$    **17.** $x^5$
**18.** $a^9b^8$    **19.** $4^3$    **20.** $y^4$    **21.** $p^9$    **22.** $a^4b^2$

**23.** $\dfrac{1}{4^3} = \dfrac{1}{64}$    **24.** $\dfrac{1}{5^2} = \dfrac{1}{25}$    **25.** $\dfrac{1}{2^4} = \dfrac{1}{16}$

**26.** $\dfrac{1}{(-2)^3} = -\dfrac{1}{8}$    **27.** $\dfrac{4}{p^3}$    **28.** $x^2$    **29.** $5^2$    **30.** $\dfrac{1}{x^7}$

**31.** $\dfrac{1}{7^5}$    **32.** $b$    **33.** $t^6$

## Calculator Spotlight, p. 243

**1.** Yes    **2.** No    **3.** No    **4.** Yes    **5.** Yes    **6.** Yes
**7.** No    **8.** Yes    **9.** Yes    **10.** Yes    **11.** No    **12.** Yes
**13.** Yes    **14.** No

## Exercise Set 4.1, p. 245

**1.** $3 \cdot 3 \cdot 3 \cdot 3$    **3.** $(1.1)(1.1)(1.1)(1.1)(1.1)$    **5.** $\left(\frac{2}{3}\right)\left(\frac{2}{3}\right)\left(\frac{2}{3}\right)\left(\frac{2}{3}\right)$
**7.** $(7p)(7p)$    **9.** $8 \cdot k \cdot k \cdot k$    **11.** 1    **13.** $b$    **15.** 1
**17.** 1    **19.** $ab$    **21.** $ab$    **23.** 27    **25.** 19    **27.** 256
**29.** 93    **31.** 10; 4    **33.** 3629.84 ft$^2$    **35.** $\dfrac{1}{3^2} = \dfrac{1}{9}$

**37.** $\dfrac{1}{10^3} = \dfrac{1}{1000}$    **39.** $\dfrac{1}{7^3} = \dfrac{1}{343}$    **41.** $\dfrac{1}{a^3}$    **43.** $8^2 = 64$
**45.** $y^4$    **47.** $z^n$    **49.** $4^{-3}$    **51.** $x^{-3}$    **53.** $a^{-5}$
**55.** $2^7$    **57.** $8^{14}$    **59.** $x^7$    **61.** $9^{38}$    **63.** $(3y)^{12}$
**65.** $(7y)^{17}$    **67.** $3^3$    **69.** $\dfrac{1}{x}$    **71.** $x^{17}$    **73.** $\dfrac{1}{x^{13}}$

**75.** $\dfrac{1}{a^{10}}$    **77.** 1    **79.** $7^3$    **81.** $8^6$    **83.** $y^4$    **85.** $\dfrac{1}{16^6}$

**87.** $\dfrac{1}{m^6}$    **89.** $\dfrac{1}{(8x)^4}$    **91.** 1    **93.** $x^2$    **95.** $x^9$    **97.** $\dfrac{1}{z^4}$

**99.** $x^3$    **101.** 1    **103.** 25, $\frac{1}{25}$, $\frac{1}{25}$, 25, $-25$, 25
**105.** 64%$t$, or 0.64$t$    **106.** 1    **107.** 64    **108.** 1579.5
**109.** $\frac{4}{3}$    **110.** $8(x - 7)$    **111.** 8 in., 4 in.    **112.** 228, 229
**113.** ◈    **115.** No    **117.** No    **119.** $y^{5x}$    **121.** $a^{4t}$
**123.** 1    **125.** >    **127.** <    **129.** Let $x = 2$; then
$3x^2 = 12$, but $(3x)^2 = 36$.

## Margin Exercises, Section 4.2, pp. 249–254

**1.** $3^{20}$    **2.** $\dfrac{1}{x^{12}}$    **3.** $y^{15}$    **4.** $\dfrac{1}{x^{32}}$    **5.** $\dfrac{16x^{20}}{y^{12}}$    **6.** $\dfrac{25x^{10}}{y^{12}z^6}$

**7.** $x^{74}$    **8.** $\dfrac{27z^{24}}{y^6x^{15}}$    **9.** $\dfrac{x^{12}}{25}$    **10.** $\dfrac{8t^{15}}{w^{12}}$    **11.** $\dfrac{9}{x^8}$

**12.** $5.17 \times 10^{-4}$    **13.** $5.23 \times 10^8$    **14.** 689,300,000,000
**15.** 0.0000567    **16.** $5.6 \times 10^{-15}$    **17.** $7.462 \times 10^{-13}$
**18.** $2.0 \times 10^3$    **19.** $5.5 \times 10^2$    **20.** $1.884672 \times 10^{11}$
**21.** $9.5 \times 10$

**Calculator Spotlight, p. 249**

**1.** Yes   **2.** No

**Calculator Spotlight, p. 250**

**1.** Yes   **2.** Yes   **3.** No   **4.** No

**Calculator Spotlight, p. 252**

**1.** 2.6 E 8   **2.** 6.709 E ⁻11

**Exercise Set 4.2, p. 255**

**1.** $2^6$   **3.** $\frac{1}{5^6}$   **5.** $x^{12}$   **7.** $16x^6$   **9.** $\frac{1}{x^{12}y^{15}}$   **11.** $x^{24}y^8$

**13.** $\frac{9x^6}{y^{16}z^6}$   **15.** $\frac{a^8}{b^{12}}$   **17.** $\frac{y^6}{4}$   **19.** $\frac{8}{y^6}$   **21.** $\frac{x^6y^3}{z^3}$

**23.** $\frac{c^2d^6}{a^4b^2}$   **25.** $2.8 \times 10^{10}$   **27.** $9.07 \times 10^{17}$

**29.** $3.04 \times 10^{-6}$   **31.** $1.8 \times 10^{-8}$   **33.** $10^{11}$

**35.** $1.135 \times 10^7$   **37.** 87,400,000   **39.** 0.00000005704

**41.** 10,000,000   **43.** 0.00001   **45.** $6 \times 10^9$

**47.** $3.38 \times 10^4$   **49.** $8.1477 \times 10^{-13}$   **51.** $2.5 \times 10^{13}$

**53.** $5.0 \times 10^{-4}$   **55.** $3.0 \times 10^{-21}$   **57.** $\$1.32288 \times 10^{12}$

**59.** $1 \times 10^{22}$   **61.** $3.3 \times 10^5$   **63.** $4.375 \times 10^2$ days

**65.** $9(x - 4)$   **66.** $2(2x - y + 8)$   **67.** $3(s + t + 8)$

**68.** $-7(x + 2)$   **69.** $\frac{7}{4}$   **70.** 2   **71.** $-\frac{12}{7}$   **72.** $-\frac{11}{2}$

**73.**                           **74.**

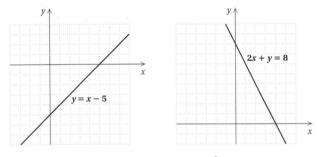

**75.** ◆   **77.** $2.478125 \times 10^{-1}$   **79.** $\frac{1}{5}$   **81.** $3^{11}$
**83.** $a^n$   **85.** False   **87.** False

**Margin Exercises, Section 4.3, pp. 259–266**

**1.** $4x^2 - 3x + \frac{5}{4}$; $15y^3$; $-7x^3 + 1.1$; answers may vary
**2.** $-19$   **3.** $-104$   **4.** $-13$   **5.** 8   **6.** 132   **7.** 360 ft
**8.** $6; -4$   **9.** 7.55 parts per million   **10.** 20
**11.** $-9x^3 + (-4x^5)$   **12.** $-2y^3 + 3y^7 + (-7y)$   **13.** $3x^2$,
$6x, \frac{1}{2}$   **14.** $-4y^5, 7y^2, -3y, -2$   **15.** $4x^3$ and $-x^3$
**16.** $4t^4$ and $-7t^4$; $-9t^3$ and $10t^3$   **17.** $5x^2$ and $7x^2$; $3x$
and $-8x$; $-10$ and 11   **18.** $2, -7, -8.5, 10, -4$
**19.** $8x^2$   **20.** $2x^3 + 7$   **21.** $-\frac{1}{4}x^5 + 2x^2$   **22.** $-4x^3$
**23.** $5x^3$   **24.** $25 - 3x^5$   **25.** $6x$   **26.** $4x^3 + 4$
**27.** $-\frac{1}{4}x^3 + 4x^2 + 7$   **28.** $3x^2 + x^3 + 9$
**29.** $6x^7 + 3x^5 - 2x^4 + 4x^3 + 5x^2 + x$
**30.** $7x^5 - 5x^4 + 2x^3 + 4x^2 - 3$
**31.** $14t^7 - 10t^5 + 7t^2 - 14$   **32.** $-2x^2 - 3x + 2$
**33.** $10x^4 - 8x - \frac{1}{2}$   **34.** $4, 2, 1, 0; 4$   **35.** $x$
**36.** $x^3, x^2, x, x^0$   **37.** $x^2, x$   **38.** $x^3$   **39.** Monomial

**40.** None of these   **41.** Binomial   **42.** Trinomial

**Calculator Spotlight, p. 262**

**1.** 3, 1.81, −6.99   **2.** 2, 17, 21.55, 8.552
**3.** 13, 2, −8, −3.32, 7

**Exercise Set 4.3, p. 267**

**1.** $-18, 7$   **3.** $19, 14$   **5.** $-12, -7$   **7.** $-1, 5$   **9.** 9, 1
**11.** $56, -2$   **13.** 1112 ft   **15.** $18,750; $24,000
**17.** $-4, 4, 5, 2.75, 1$   **19.** 66.6 ft, 86.6 ft, 66.6 ft, 41.6 ft
**21.** $2, -3x, x^2$   **23.** $6x^2$ and $-3x^2$   **25.** $2x^4$ and $-3x^4$;
$5x$ and $-7x$   **27.** $3x^5$ and $14x^5$; $-7x$ and $-2x$; 8 and $-9$
**29.** $-3, 6$   **31.** $5, 3, 3$   **33.** $-5, 6, -3, 8, -2$
**35.** $-3x$   **37.** $-8x$   **39.** $11x^3 + 4$   **41.** $x^3 - x$
**43.** $4b^5$   **45.** $\frac{3}{4}x^5 - 2x - 42$   **47.** $x^4$   **49.** $\frac{15}{16}x^3 - \frac{7}{6}x^2$
**51.** $x^5 + 6x^3 + 2x^2 + x + 1$
**53.** $15y^9 + 7y^8 + 5y^3 - y^2 + y$   **55.** $x^6 + x^4$
**57.** $13x^3 - 9x + 8$   **59.** $-5x^2 + 9x$   **61.** $12x^4 - 2x + \frac{1}{4}$
**63.** 1, 0; 1   **65.** 2, 1, 0; 2   **67.** 3, 2, 1, 0; 3
**69.** 2, 1, 6, 4; 6

**71.**

| Term | Coefficient | Degree of Term | Degree of Polynomial |
|---|---|---|---|
| $-7x^4$ | $-7$ | 4 | |
| $6x^3$ | 6 | 3 | |
| $-3x^2$ | $-3$ | 2 | 4 |
| $8x$ | 8 | 1 | |
| $-2$ | $-2$ | 0 | |

**73.** $x^2, x$   **75.** $x^3, x^2, x^0$   **77.** None missing
**79.** Trinomial   **81.** None of these   **83.** Binomial
**85.** Monomial   **87.** 27   **88.** $-19$   **89.** $-\frac{17}{24}$
**90.** $\frac{5}{8}$   **91.** $-2.6$   **92.** $\frac{15}{2}$   **93.** $b = \frac{cx + r}{a}$   **94.** 45%,
37.5%, 17.5%   **95.** $3(x - 5y + 21)$   **97.** ◆   **99.** $3x^6$
**101.** 10   **103.** $-4, 4, 5, 2.75, 1$   **105.** 66.6 ft, 86.6 ft,
66.6 ft, 41.6 ft

**Margin Exercises, Section 4.4, pp. 273–276**

**1.** $x^2 + 7x + 3$   **2.** $-4x^5 + 7x^4 + x^3 + 2x^2 + 4$
**3.** $24x^4 + 5x^3 + x^2 + 1$   **4.** $2x^3 + \frac{10}{3}$   **5.** $2x^2 - 3x - 1$
**6.** $8x^3 - 2x^2 - 8x + \frac{5}{2}$
**7.** $-8x^4 + 4x^3 + 12x^2 + 5x - 8$
**8.** $-x^3 + x^2 + 3x + 3$   **9.** $-(12x^4 - 3x^2 + 4x)$;
$-12x^4 + 3x^2 - 4x$   **10.** $-(-4x^4 + 3x^2 - 4x)$;
$4x^4 - 3x^2 + 4x$   **11.** $-\left(-13x^6 + 2x^4 - 3x^2 + x - \frac{5}{13}\right)$;
$13x^6 - 2x^4 + 3x^2 - x + \frac{5}{13}$
**12.** $-(-7y^3 + 2y^2 - y + 3)$; $7y^3 - 2y^2 + y - 3$
**13.** $-4x^3 + 6x - 3$   **14.** $-5x^4 - 3x^2 - 7x + 5$
**15.** $-14x^{10} + \frac{1}{2}x^5 - 5x^3 + x^2 - 3x$   **16.** $2x^3 + 2x + 8$
**17.** $x^2 - 6x - 2$   **18.** $-8x^4 - 5x^3 + 8x^2 - 1$
**19.** $x^3 - x^2 - \frac{4}{3}x - 0.9$   **20.** $2x^3 + 5x^2 - 2x - 5$
**21.** $-x^5 - 2x^3 + 3x^2 - 2x + 2$

**22.** Sum of perimeters: $13x$; sum of areas: $\frac{7}{2}x^2$
**23.** $\pi x^2 - 2x^2$, or $(\pi - 2)x^2$

## Exercise Set 4.4, p. 277

**1.** $-x + 5$   **3.** $x^2 - 5x - 1$   **5.** $2x^2$
**7.** $5x^2 + 3x - 30$   **9.** $-2.2x^3 - 0.2x^2 - 3.8x + 23$
**11.** $12x^2 + 6$   **13.** $-\frac{1}{2}x^4 + \frac{2}{3}x^3 + x^2$
**15.** $0.01x^5 + x^4 - 0.2x^3 + 0.2x + 0.06$
**17.** $9x^8 + 8x^7 - 6x^4 + 8x^2 + 4$
**19.** $1.05x^4 + 0.36x^3 + 14.22x^2 + x + 0.97$
**21.** $-(-5x)$; $5x$   **23.** $-(-x^2 + 10x - 2)$; $x^2 - 10x + 2$
**25.** $-(12x^4 - 3x^3 + 3)$; $-12x^4 + 3x^3 - 3$   **27.** $-3x + 7$
**29.** $-4x^2 + 3x - 2$   **31.** $4x^4 - 6x^2 - \frac{3}{4}x + 8$
**33.** $7x - 1$   **35.** $-x^2 - 7x + 5$   **37.** $-18$
**39.** $6x^4 + 3x^3 - 4x^2 + 3x - 4$
**41.** $4.6x^3 + 9.2x^2 - 3.8x - 23$   **43.** $\frac{3}{4}x^3 - \frac{1}{2}x$
**45.** $0.06x^3 - 0.05x^2 + 0.01x + 1$   **47.** $3x + 6$
**49.** $11x^4 + 12x^3 - 9x^2 - 8x - 9$   **51.** $x^4 - x^3 + x^2 - x$
**53.** $5x^2 + 4x$   **55.** $\frac{23}{2}a + 10$   **57.** 6   **58.** $-19$
**59.** $-\frac{7}{22}$   **60.** 5   **61.** 5   **62.** 1   **63.** $\frac{39}{2}$   **64.** $\frac{37}{2}$
**65.** $\{x \mid x \geq -10\}$   **66.** $\{x \mid x < 0\}$   **67.** ◈
**69.** $20 + 5(m - 4) + 4(m - 5) + (m - 5)(m - 4)$; $m^2$
**71.** $z^2 - 27z + 72$   **73.** $y^2 - 4y + 4$
**75.** $5x^2 - 9x - 1$   **77.** $4x^3 + 2x^2 + x + 2$
**79.** Both columns are equal.

## Margin Exercises, Section 4.5, pp. 281–284

**1.** $-15x$   **2.** $-x^2$   **3.** $x^2$   **4.** $-x^5$   **5.** $12x^7$
**6.** $-8y^{11}$   **7.** $7y^5$   **8.** 0   **9.** $8x^2 + 16x$
**10.** $-15t^3 + 6t^2$   **11.** $5x^6 + 25x^5 - 30x^4 + 40x^3$
**12.** $x^2 + 13x + 40$   **13.** $x^2 + x - 20$
**14.** $5x^2 - 17x - 12$   **15.** $6x^2 - 19x + 15$
**16.** $x^4 + 3x^3 + x^2 + 15x - 20$
**17.** $6y^5 - 20y^3 + 15y^2 + 14y - 35$
**18.** $3x^3 + 13x^2 - 6x + 20$
**19.** $20x^4 - 16x^3 + 32x^2 - 32x - 16$
**20.** $6x^4 - x^3 - 18x^2 - x + 10$

## Calculator Spotlight, p. 284

**1.** Yes   **2.** Yes   **3.** No   **4.** No   **5.** No   **6.** Yes

## Exercise Set 4.5, p. 285

**1.** $40x^2$   **3.** $x^3$   **5.** $32x^8$   **7.** $0.03x^{11}$   **9.** $\frac{1}{15}x^4$
**11.** 0   **13.** $-24x^{11}$   **15.** $-2x^2 + 10x$   **17.** $-5x^2 + 5x$
**19.** $x^5 + x^2$   **21.** $6x^3 - 18x^2 + 3x$   **23.** $-6x^4 - 6x^3$
**25.** $18y^6 + 24y^5$   **27.** $x^2 + 9x + 18$   **29.** $x^2 + 3x - 10$
**31.** $x^2 - 7x + 12$   **33.** $x^2 - 9$   **35.** $25 - 15x + 2x^2$
**37.** $4x^2 + 20x + 25$   **39.** $x^2 - \frac{21}{10}x - 1$
**41.** $x^2 + 2.4x - 10.81$   **43.** $x^3 - 1$
**45.** $4x^3 + 14x^2 + 8x + 1$
**47.** $3y^4 - 6y^3 - 7y^2 + 18y - 6$   **49.** $x^6 + 2x^5 - x^3$
**51.** $-10x^5 - 9x^4 + 7x^3 + 2x^2 - x$
**53.** $x^4 - x^2 - 2x - 1$   **55.** $6t^4 + t^3 - 16t^2 - 7t + 4$
**57.** $x^9 - x^5 + 2x^3 - x$   **59.** $x^4 - 1$   **61.** $-\frac{3}{4}$   **62.** 6.4
**63.** 96   **64.** 32   **65.** $3(5x - 6y + 4)$

**66.** $4(4x - 6y + 9)$   **67.** $-3(3x + 15y - 5)$
**68.** $100(x - y + 10a)$
**69.**

$y = \frac{1}{2}x - 3$

**70.** $\frac{23}{19}$   **71.** ◈
**73.** $78t^2 + 40t$
**75.** $A = \frac{1}{2}b^2 + 2b$
**77.** 0

## Margin Exercises, Section 4.6, pp. 288–292

**1.** $x^2 + 7x + 12$   **2.** $x^2 - 2x - 15$   **3.** $2x^2 - 9x + 4$
**4.** $2x^3 - 4x^2 - 3x + 6$   **5.** $12x^5 + 10x^3 + 6x^2 + 5$
**6.** $y^6 - 49$   **7.** $t^2 + 8t + 15$   **8.** $-2x^7 + x^5 + x^3$
**9.** $x^2 - \frac{16}{25}$   **10.** $x^5 + 0.5x^3 - 0.5x^2 + 0.25$
**11.** $8 + 2x^2 - 15x^4$   **12.** $30x^5 - 27x^4 + 6x^3$
**13.** $x^2 - 25$   **14.** $4x^2 - 9$   **15.** $x^2 - 4$   **16.** $x^2 - 49$
**17.** $36 - 16y^2$   **18.** $4x^6 - 1$   **19.** $x^2 - \frac{4}{25}$
**20.** $x^2 + 16x + 64$   **21.** $x^2 - 10x + 25$
**22.** $x^2 + 4x + 4$   **23.** $a^2 - 8a + 16$
**24.** $4x^2 + 20x + 25$   **25.** $16x^4 - 24x^3 + 9x^2$
**26.** $60.84 + 18.72y + 1.44y^2$   **27.** $9x^4 - 30x^2 + 25$
**28.** $x^2 + 11x + 30$   **29.** $t^2 - 16$
**30.** $-8x^5 + 20x^4 + 40x^2$   **31.** $81x^4 + 18x^2 + 1$
**32.** $4a^2 + 6a - 40$   **33.** $25x^2 + 5x + \frac{1}{4}$
**34.** $4x^2 - 2x + \frac{1}{4}$   **35.** $x^3 - 3x^2 + 6x - 8$

## Exercise Set 4.6, p. 293

**1.** $x^3 + x^2 + 3x + 3$   **3.** $x^4 + x^3 + 2x + 2$
**5.** $y^2 - y - 6$   **7.** $9x^2 + 12x + 4$   **9.** $5x^2 + 4x - 12$
**11.** $9t^2 - 1$   **13.** $4x^2 - 6x + 2$   **15.** $p^2 - \frac{1}{16}$
**17.** $x^2 - 0.01$   **19.** $2x^3 + 2x^2 + 6x + 6$
**21.** $-2x^2 - 11x + 6$   **23.** $a^2 + 14a + 49$
**25.** $1 - x - 6x^2$   **27.** $x^5 + 3x^3 - x^2 - 3$
**29.** $3x^6 - 2x^4 - 6x^2 + 4$   **31.** $13.16x^2 + 18.99x - 13.95$
**33.** $6x^7 + 18x^5 + 4x^2 + 12$   **35.** $8x^6 + 65x^3 + 8$
**37.** $4x^3 - 12x^2 + 3x - 9$   **39.** $4y^6 + 4y^5 + y^4 + y^3$
**41.** $x^2 - 16$   **43.** $4x^2 - 1$   **45.** $25m^2 - 4$
**47.** $4x^4 - 9$   **49.** $9x^8 - 16$   **51.** $x^{12} - x^4$
**53.** $x^8 - 9x^2$   **55.** $x^{24} - 9$   **57.** $4y^{16} - 9$
**59.** $\frac{25}{64}x^2 - 18.49$   **61.** $x^2 + 4x + 4$
**63.** $9x^4 + 6x^2 + 1$   **65.** $a^2 - a + \frac{1}{4}$   **67.** $9 + 6x + x^2$
**69.** $x^4 + 2x^2 + 1$   **71.** $4 - 12x^4 + 9x^8$
**73.** $25 + 60t^2 + 36t^4$   **75.** $x^2 - \frac{5}{4}x + \frac{25}{64}$
**77.** $9 - 12x^3 + 4x^6$   **79.** $4x^3 + 24x^2 - 12x$
**81.** $4x^4 - 2x^2 + \frac{1}{4}$   **83.** $9p^2 - 1$   **85.** $15t^5 - 3t^4 + 3t^3$
**87.** $36x^8 + 48x^4 + 16$   **89.** $12x^3 + 8x^2 + 15x + 10$
**91.** $64 - 96x^4 + 36x^8$   **93.** $t^3 - 1$   **95.** 25; 49
**97.** 56; 16   **99.** Lamps: 500 watts; air conditioner:
2000 watts; television: 50 watts   **100.** $\frac{28}{27}$   **101.** $-\frac{41}{7}$
**102.** $\frac{27}{4}$   **103.** $y = \dfrac{3x - 12}{2}$, or $y = \frac{3}{2}x - 6$

**104.** $a = \dfrac{4 + cd}{b}$   **105.** ◆

**107.** $30x^3 + 35x^2 - 15x$   **109.** $a^4 - 50a^2 + 625$
**111.** $81t^{16} - 72t^8 + 16$   **113.** $-7$   **115.** First row: 90, $-432$, $-63$; second row: 7, $-18$, $-36$, $-14$, 12, $-6$, $-21$, $-11$; third row: 9, $-2$, $-2$, 10, $-8$, $-8$, $-8$, $-10$, 21; fourth row: $-19$, $-6$   **117.** $9x^2 + 24x + 16$   **119.** Yes
**121.** No

### Margin Exercises, Section 4.7, pp. 297–300

**1.** $-7940$   **2.** $-176$   **3.** 1889   **4.** $-3, 3, -2, 1, 2$
**5.** 3, 7, 1, 1, 0; 7   **6.** $2x^2y + 3xy$   **7.** $5pq - 8$
**8.** $-4x^3 + 2x^2 - 4y + 2$   **9.** $14x^3y + 7x^2y - 3xy - 2y$
**10.** $-5p^2q^4 + 2p^2q^2 + 3p^2q + 6pq^2 + 3q + 5$
**11.** $-8s^4t + 6s^3t^2 + 2s^2t^3 - s^2t^2$
**12.** $-9p^4q + 9p^3q^2 - 4p^2q^3 - 9q^4 + 5$
**13.** $x^5y^5 + 2x^4y^2 + 3x^3y^3 + 6x^2$
**14.** $p^5q - 4p^3q^3 + 3pq^3 + 6q^4$
**15.** $3x^3y + 6x^2y^3 + 2x^3 + 4x^2y^2$
**16.** $2x^2 - 11xy + 15y^2$   **17.** $16x^2 + 40xy + 25y^2$
**18.** $9x^4 - 12x^3y^2 + 4x^2y^4$   **19.** $4x^2y^4 - 9x^2$
**20.** $16y^2 - 9x^2y^4$   **21.** $9y^2 + 24y + 16 - 9x^2$
**22.** $4a^2 - 25b^2 - 10bc - c^2$

### Exercise Set 4.7, p. 301

**1.** $-1$   **3.** $-15$   **5.** 240   **7.** $-145$   **9.** 3.715 liters
**11.** 110.4 m   **13.** 56.52 in$^2$   **15.** Coefficients: 1, $-2$, 3, $-5$; degrees: 4, 2, 2, 0; 4   **17.** Coefficients: 17, $-3$, $-7$; degrees: 5, 5, 0; 5   **19.** $-a - 2b$
**21.** $3x^2y - 2xy^2 + x^2$   **23.** $20au + 10av$
**25.** $8u^2v - 5uv^2$   **27.** $x^2 - 4xy + 3y^2$   **29.** $3r + 7$
**31.** $-a^3b^2 - 3a^2b^3 + 5ab + 3$   **33.** $ab^2 - a^2b$
**35.** $2ab - 2$   **37.** $-2a + 10b - 5c + 8d$
**39.** $6z^2 + 7zu - 3u^2$   **41.** $a^4b^2 - 7a^2b + 10$
**43.** $a^6 - b^2c^2$   **45.** $y^6x + y^4x + y^4 + 2y^2 + 1$
**47.** $12x^2y^2 + 2xy - 2$   **49.** $12 - c^2d^2 - c^4d^4$
**51.** $m^3 + m^2n - mn^2 - n^3$
**53.** $x^9y^9 - x^6y^6 + x^5y^5 - x^2y^2$   **55.** $x^2 + 2xh + h^2$
**57.** $r^6t^4 - 8r^3t^2 + 16$   **59.** $p^8 + 2m^2n^2p^4 + m^4n^4$
**61.** $4a^6 - 2a^3b^3 + \frac{1}{4}b^6$   **63.** $3a^3 - 12a^2b + 12ab^2$
**65.** $4a^2 - b^2$   **67.** $c^4 - d^2$   **69.** $a^2b^2 - c^2d^4$
**71.** $x^2 + 2xy + y^2 - 9$   **73.** $x^2 - y^2 - 2yz - z^2$
**75.** $a^2 - b^2 - 2bc - c^2$   **77.** IV   **78.** III   **79.** I
**80.** II
**81.**                **82.**

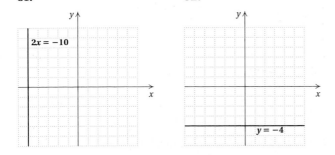

**83.**                **84.**

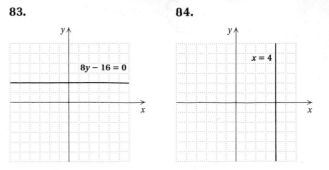

**85.** Mean: 27.57; median: 28; mode: 31
**86.** Mean: 5.69; median: 5.6; modes: 5.2, 5.6   **87.** ◆
**89.** $4xy - 4y^2$   **91.** $2xy + \pi x^2$
**93.** $2\pi nh + 2\pi mh + 2\pi n^2 - 2\pi m^2$

### Margin Exercises, Section 4.8, pp. 305–308

**1.** $4x^2$   **2.** $-7x^{11}$   **3.** $-28p^3q$   **4.** $\frac{1}{4}x^4$   **5.** $7x^4 + 8x^2$
**6.** $x^2 + 3x + 2$   **7.** $2x^2 + x - \frac{2}{3}$   **8.** $4x^2 - \frac{3}{2}x + \frac{1}{2}$
**9.** $2x^2y^4 - 3xy^2 + 5y$   **10.** $x - 2$   **11.** $x + 4$
**12.** $x + 4$, R $-2$, or $x + 4 + \dfrac{-2}{x + 3}$   **13.** $x^2 + x + 1$

### Exercise Set 4.8, p. 309

**1.** $3x^4$   **3.** $5x$   **5.** $18x^3$   **7.** $4a^3b$
**9.** $3x^4 - \frac{1}{2}x^3 + \frac{1}{8}x^2 - 2$   **11.** $1 - 2u - u^4$
**13.** $5t^2 + 8t - 2$   **15.** $-4x^4 + 4x^2 + 1$
**17.** $6x^2 - 10x + \frac{3}{2}$   **19.** $9x^2 - \frac{5}{2}x + 1$
**21.** $6x^2 + 13x + 4$   **23.** $3rs + r - 2s$   **25.** $x + 2$
**27.** $x - 5 + \dfrac{-50}{x - 5}$   **29.** $x - 2 + \dfrac{-2}{x + 6}$   **31.** $x - 3$
**33.** $x^4 - x^3 + x^2 - x + 1$   **35.** $2x^2 - 7x + 4$
**37.** $x^3 - 6$   **39.** $x^3 + 2x^2 + 4x + 8$   **41.** $t^2 + 1$
**43.** $-28$   **44.** $-59$   **45.** 6.8   **46.** $-\frac{11}{8}$
**47.** 25,543.75 ft$^2$   **48.** 51°, 27°, 102°   **49.** $\frac{23}{14}$   **50.** $\frac{11}{10}$
**51.** $4(x - 3 + 6y)$   **52.** $2(128 - a - 2b)$   **53.** ◆
**55.** $x^2 + 5$   **57.** $a + 3 + \dfrac{5}{5a^2 - 7a - 2}$
**59.** $2x^2 + x - 3$
**61.** $a^5 + a^4b + a^3b^2 + a^2b^3 + ab^4 + b^5$   **63.** $-5$
**65.** 1

### Summary and Review: Chapter 4, p. 311

**1.** $\dfrac{1}{7^2}$   **2.** $y^{11}$   **3.** $(3x)^{14}$   **4.** $t^8$   **5.** $4^3$   **6.** $\dfrac{1}{a^3}$   **7.** 1
**8.** $9t^8$   **9.** $36x^8$   **10.** $\dfrac{y^3}{8x^3}$   **11.** $t^{-5}$   **12.** $\dfrac{1}{y^4}$
**13.** $3.28 \times 10^{-5}$   **14.** 8,300,000   **15.** $2.09 \times 10^4$
**16.** $5.12 \times 10^{-5}$   **17.** $4.2075 \times 10^9$   **18.** 10
**19.** $-4y^5, 7y^2, -3y, -2$   **20.** $x^2, x^0$   **21.** 3, 2, 1, 0; 3
**22.** Binomial   **23.** None of these   **24.** Monomial
**25.** $-2x^2 - 3x + 2$   **26.** $10x^4 - 7x^2 - x - \frac{1}{2}$
**27.** $x^5 - 2x^4 + 6x^3 + 3x^2 - 9$
**28.** $-2x^5 - 6x^4 - 2x^3 - 2x^2 + 2$   **29.** $2x^2 - 4x$

**30.** $x^5 - 3x^3 - x^2 + 8$   **31.** Perimeter: $4w + 6$; area: $w^2 + 3w$   **32.** $x^2 + \frac{7}{6}x + \frac{1}{3}$   **33.** $49x^2 + 14x + 1$
**34.** $12x^3 - 23x^2 + 13x - 2$   **35.** $9x^4 - 16$
**36.** $15x^7 - 40x^6 + 50x^5 + 10x^4$   **37.** $x^2 - 3x - 28$
**38.** $9y^4 - 12y^3 + 4y^2$   **39.** $2t^4 - 11t^2 - 21$   **40.** $49$
**41.** Coefficients: 1, $-7$, 9, $-8$; degrees: 6, 2, 2, 0; 6
**42.** $-y + 9w - 5$
**43.** $m^6 - 2m^2n + 2m^2n^2 + 8n^2m - 6m^3$
**44.** $-9xy - 2y^2$   **45.** $11x^3y^2 - 8x^2y - 6x^2 - 6x + 6$
**46.** $p^3 - q^3$   **47.** $9a^8 - 2a^4b^3 + \frac{1}{9}b^6$
**48.** $5x^2 - \frac{1}{2}x + 3$   **49.** $3x^2 - 7x + 4 + \dfrac{1}{2x + 3}$
**50.** 0, 3.75, $-3.75$, 0, 2.25   **51.** $25(t - 2 + 4m)$   **52.** $\frac{9}{4}$
**53.** $-12$   **54.** $-11.2$   **55.** Width: 125.5 m; length: 144.5 m   **56.** ◆ $578.6 \times 10^{-7}$ is not in scientific notation because 578.6 is larger than 10.   **57.** ◆ A monomial is an expression of the type $ax^n$, where $n$ is a whole number and $a$ is a real number. A binomial is a sum of two monomials and has two terms. A trinomial is a sum of three monomials and has three terms. A general polynomial is a monomial or a sum of monomials and has one or more terms.
**58.** $\frac{1}{2}x^2 - \frac{1}{2}y^2$   **59.** $400 - 4a^2$   **60.** $-28x^8$
**61.** $\frac{94}{13}$   **62.** $x^4 + x^3 + x^2 + x + 1$

## Test: Chapter 4, p. 313

**1.** [4.1d, f] $\dfrac{1}{6^5}$   **2.** [4.1d] $x^9$   **3.** [4.1d] $(4a)^{11}$

**4.** [4.1e] $3^3$   **5.** [4.1e, f] $\dfrac{1}{x^5}$   **6.** [4.1b, e] 1   **7.** [4.2a] $x^6$

**8.** [4.2a, b] $-27y^6$   **9.** [4.2a, b] $16a^{12}b^4$

**10.** [4.2b] $\dfrac{a^3b^3}{c^3}$   **11.** [4.1d], [4.2a, b] $-216x^{21}$

**12.** [4.1d], [4.2a, b] $-24x^{21}$   **13.** [4.1d], [4.2a, b] $162x^{10}$

**14.** [4.1d], [4.2a, b] $324x^{10}$   **15.** [4.1f] $\dfrac{1}{5^3}$

**16.** [4.1f] $y^{-8}$   **17.** [4.2c] $3.9 \times 10^9$
**18.** [4.2c] 0.00000005   **19.** [4.2d] $1.75 \times 10^{17}$
**20.** [4.2d] $1.296 \times 10^{22}$   **21.** [4.2e] $1.5 \times 10^4$
**22.** [4.3a] $-43$   **23.** [4.3d] $\frac{1}{3}$, $-1$, 7   **24.** [4.3g] 3, 0, 1, 6; 6   **25.** [4.3i] Binomial   **26.** [4.3e] $5a^2 - 6$
**27.** [4.3e] $\frac{7}{4}y^2 - 4y$   **28.** [4.3f] $x^5 + 2x^3 + 4x^2 - 8x + 3$
**29.** [4.4a] $4x^5 + x^4 + 2x^3 - 8x^2 + 2x - 7$
**30.** [4.4a] $5x^4 + 5x^2 + x + 5$
**31.** [4.4c] $-4x^4 + x^3 - 8x - 3$
**32.** [4.4c] $-x^5 + 0.7x^3 - 0.8x^2 - 21$
**33.** [4.5b] $-12x^4 + 9x^3 + 15x^2$   **34.** [4.6c] $x^2 - \frac{2}{3}x + \frac{1}{9}$
**35.** [4.6b] $9x^2 - 100$   **36.** [4.6a] $3b^2 - 4b - 15$
**37.** [4.6a] $x^{14} - 4x^8 + 4x^6 - 16$
**38.** [4.6a] $48 + 34y - 5y^2$
**39.** [4.5d] $6x^3 - 7x^2 - 11x - 3$
**40.** [4.6c] $25t^2 + 20t + 4$
**41.** [4.7c] $-5x^3y - y^3 + xy^3 - x^2y^2 + 19$
**42.** [4.7e] $8a^2b^2 + 6ab - 4b^3 + 6ab^2 + ab^3$
**43.** [4.7f] $9x^{10} - 16y^{10}$   **44.** [4.8a] $4x^2 + 3x - 5$

**45.** [4.8b] $2x^2 - 4x - 2 + \dfrac{17}{3x + 2}$   **46.** [4.3a] 3, 1.5, $-3.5$, $-5$, $-5.25$   **47.** [2.3b] 13   **48.** [2.3c] $-3$
**49.** [1.7d] $16(4t - 2m + 1)$   **50.** [1.4a] $\frac{23}{20}$
**51.** [2.4a] 100°, 25°, 55°
**52.** [4.5b], [4.6a] $V = l^3 - 3l^2 + 2l$
**53.** [2.3b], [4.6b, c] $-\frac{61}{12}$

## Cumulative Review: Chapters 1–4, p. 315

**1.** [1.1a] $\frac{5}{2}$   **2.** [4.3a] $-4$   **3.** [4.7a] $-14$   **4.** [1.2e] 4
**5.** [1.6b] $\frac{1}{5}$   **6.** [1.3a] $-\frac{11}{60}$   **7.** [1.4a] 4.2   **8.** [1.5a] 7.28
**9.** [1.6c] $-\frac{5}{12}$   **10.** [4.2d] $2.2 \times 10^{22}$
**11.** [4.2d] $4 \times 10^{-5}$   **12.** [1.7a] $-3$   **13.** [1.8b] $-2y - 7$
**14.** [1.8c] $5x + 11$   **15.** [1.8d] $-2$
**16.** [4.4a] $2x^5 - 2x^4 + 3x^3 + 2$
**17.** [4.7d] $3x^2 + xy - 2y^2$   **18.** [4.4c] $x^3 + 5x^2 - x - 7$
**19.** [4.4c] $-\frac{1}{3}x^2 - \frac{3}{4}x$   **20.** [1.7c] $12x - 15y + 21$
**21.** [4.5a] $6x^8$   **22.** [4.5b] $2x^5 - 4x^4 + 8x^3 - 10x^2$
**23.** [4.5d] $3y^4 + 5y^3 - 10y - 12$
**24.** [4.5d] $2p^4 + 3p^3q + 2p^2q^2 - 2p^4q - p^3q^2 - p^2q^3 + pq^3$   **25.** [4.6a] $6x^2 + 13x + 6$
**26.** [4.6c] $9x^4 + 6x^2 + 1$   **27.** [4.6b] $t^2 - \frac{1}{4}$
**28.** [4.6b] $4y^4 - 25$   **29.** [4.6a] $4x^6 + 6x^4 - 6x^2 - 9$
**30.** [4.6c] $t^2 - 4t^3 + 4t^4$   **31.** [4.7f] $15p^2 - pq - 2q^2$
**32.** [4.8a] $6x^2 + 2x - 3$   **33.** [4.8b] $3x^2 - 2x - 7$
**34.** [2.1b] $-1.2$   **35.** [2.2a] $-21$   **36.** [2.3a] 9
**37.** [2.2a] $-\frac{20}{3}$   **38.** [2.3b] 2   **39.** [2.1b] $\frac{13}{8}$
**40.** [2.3c] $-\frac{17}{21}$   **41.** [2.3b] $-17$   **42.** [2.3b] 2
**43.** [2.7e] $\{x \mid x < 16\}$   **44.** [2.7e] $\left\{x \mid x \le -\frac{11}{8}\right\}$
**45.** [2.6a] $h = \dfrac{A - \pi r^2}{2\pi r}$   **46.** [4.4d] $\pi r^2 - 18$
**47.** [2.4a] 18 and 19   **48.** [2.4a] 20 ft, 24 ft
**49.** [2.4a] 10°   **50.** [2.4a] $-45$   **51.** [2.5a] \$3.50
**52.** [4.2e] $5.8025 \times 10^9$ gal   **53.** [4.1d, f] $y^4$
**54.** [4.1e] $\dfrac{1}{x}$   **55.** [4.2a, b] $-\dfrac{27x^9}{y^6}$
**56.** [4.1d, e, f] $x^3$   **57.** [4.3d] $\frac{2}{3}$, 4, $-6$
**58.** [4.3g] 4, 2, 1, 0; 4   **59.** [4.3i] Binomial
**60.** [4.3i] Trinomial
**61.** [3.3a] $y$-intercept: $(0, -4)$; $x$-intercept: $(5, 0)$
**62.** [3.3a]

$4x - 5y = 20$

**63.** [3.4c] 99.45   **64.** [4.4d] $4x - 4$
**65.** [4.1d], [4.2a, b], [4.4a] $12x^5 - 15x^4 - 27x^3 + 4x^2$
**66.** [4.4a], [4.6c] $5x^2 - 2x + 10$
**67.** [4.4a], [4.8b] $4x^2 - 2x + 7$   **68.** [2.3b], [4.6a, c] $\frac{11}{7}$

**69.** [2.3b], [4.8b] 1   **70.** [1.2e], [2.3a] −5, 5
**71.** [2.3b], [4.6a, c] No solution
**72.** [2.3b], [4.6a], [4.8b] All real numbers except 5

## Chapter 5

### Pretest: Chapter 5, p. 318

**1.** [5.1a] $4(-5x^6)$, $(-2x^3)(10x^3)$, $x^2(-20x^4)$; answers may
vary   **2.** [5.5b] $2(x + 1)^2$   **3.** [5.2a] $(x + 4)(x + 2)$
**4.** [5.1b] $4a(2a^4 + a^2 - 5)$
**5.** [5.3a], [5.4a] $(5x + 2)(x - 3)$
**6.** [5.5d] $(9 + z^2)(3 + z)(3 - z)$   **7.** [5.5b] $(y^3 - 2)^2$
**8.** [5.1c] $(x^2 + 4)(3x + 2)$   **9.** [5.2a] $(p - 6)(p + 5)$
**10.** [5.5d] $(x^2y + 8)(x^2y - 8)$
**11.** [5.3a], [5.4a] $(2p - q)(p + 4q)$   **12.** [5.7b] 0, 5
**13.** [5.7a] 4, $\frac{3}{5}$   **14.** [5.7b] $\frac{2}{3}$, −4   **15.** [5.8a] 6, −1
**16.** [5.8a] Base: 8 cm; height: 11 cm

### Margin Exercises, Section 5.1, pp. 319–322

**1. (a)** $12x^2$; **(b)** $(3x)(4x)$, $(2x)(6x)$; answers may vary
**2. (a)** $16x^3$; **(b)** $(2x)(8x^2)$, $(4x)(4x^2)$; answers may vary
**3.** $(8x)(x^3)$, $(4x^2)(2x^2)$, $(2x^3)(4x)$; answers may vary
**4.** $(7x)(3x)$, $(-7x)(-3x)$, $(21x)(x)$; answers may vary
**5.** $(6x^4)(x)$, $(-2x^3)(-3x^2)$, $(3x^3)(2x^2)$; answers may vary
**6. (a)** $3x + 6$; **(b)** $3(x + 2)$   **7. (a)** $2x^3 + 10x^2 + 8x$;
**(b)** $2x(x^2 + 5x + 4)$   **8.** $x(x + 3)$   **9.** $y^2(3y^4 - 5y + 2)$
**10.** $3x^2(3x^2 - 5x + 1)$   **11.** $\frac{1}{4}(3t^3 + 5t^2 + 7t + 1)$
**12.** $7x^3(5x^4 - 7x^3 + 2x^2 - 9)$   **13.** $2.8(3x^2 - 2x + 1)$
**14.** $(x^2 + 3)(x + 7)$   **15.** $(x^2 + 2)(a + b)$
**16.** $(x^2 + 3)(x + 7)$   **17.** $(2t^2 + 3)(4t + 1)$
**18.** $(3m^3 + 2)(m^2 - 5)$   **19.** $(3x^2 - 1)(x - 2)$
**20.** $(2x^2 - 3)(2x - 3)$   **21.** Not factorable using
factoring by grouping

### Exercise Set 5.1, p. 323

**1.** $(4x^2)(2x)$, $(-8)(-x^3)$, $(2x^2)(4x)$; answers may vary
**3.** $(-5a^5)(2a)$, $(10a^3)(-a^3)$, $(-2a^2)(5a^4)$; answers may
vary   **5.** $(8x^2)(3x^2)$, $(-8x^2)(-3x^2)$, $(4x^3)(6x)$; answers
may vary   **7.** $x(x - 6)$   **9.** $2x(x + 3)$   **11.** $x^2(x + 6)$
**13.** $8x^2(x^2 - 3)$   **15.** $2(x^2 + x - 4)$
**17.** $17xy(x^4y^2 + 2x^2y + 3)$   **19.** $x^2(6x^2 - 10x + 3)$
**21.** $x^2y^2(x^3y^3 + x^2y + xy - 1)$
**23.** $2x^3(x^4 - x^3 - 32x^2 + 2)$
**25.** $0.8x(2x^3 - 3x^2 + 4x + 8)$
**27.** $\frac{1}{3}x^3(5x^3 + 4x^2 + x + 1)$   **29.** $(x^2 + 2)(x + 3)$
**31.** $(5a^3 - 1)(2a - 7)$   **33.** $(x^2 + 2)(x + 3)$
**35.** $(2x^2 + 1)(x + 3)$   **37.** $(4x^2 + 3)(2x - 3)$
**39.** $(4x^2 + 1)(3x - 4)$   **41.** $(5x^2 - 1)(x - 1)$
**43.** $(x^2 - 3)(x + 8)$   **45.** $(2x^2 - 9)(x - 4)$
**47.** $\{x \mid x > -24\}$   **48.** $\{x \mid x \le \frac{14}{5}\}$   **49.** 27
**50.** $p = 2A - q$   **51.** $y^2 + 12y + 35$
**52.** $y^2 + 14y + 49$   **53.** $y^2 - 49$
**54.** $y^2 - 14y + 49$

**55.**   **56.**

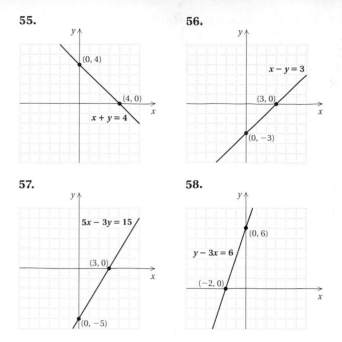

**57.**   **58.**

**59.** ◈   **61.** $(2x^3 + 3)(2x^2 + 3)$   **63.** $(x^7 + 1)(x^5 + 1)$
**65.** Not factorable by grouping

### Margin Exercises, Section 5.2, pp. 325–328

**1. (a)** −13, 8, −8, 7, −7; **(b)** 13, 8, 7; both 7 and 12 are
positive; **(c)** $(x + 3)(x + 4)$   **2.** $(x + 9)(x + 4)$
**3.** The coefficient of the middle term, −8, is negative.
**4.** $(x - 5)(x - 3)$   **5.** $(t - 5)(t - 4)$   **6.** 19, 8, 1; the
positive factor has the larger absolute value.   **7.** −23,
−10, −5, −2; the negative factor has the larger absolute
value.   **8.** $x(x + 6)(x - 2)$   **9.** $(y - 6)(y + 2)$
**10.** $(t^2 + 7)(t^2 - 2)$   **11.** $p(p - q - 3q^2)$
**12.** Not factorable   **13.** $(x + 4)^2$

### Exercise Set 5.2, p. 329

**1.** $(x + 3)(x + 5)$   **3.** $(x + 3)(x + 4)$   **5.** $(x - 3)^2$
**7.** $(x + 2)(x + 7)$   **9.** $(b + 1)(b + 4)$   **11.** $\left(x + \frac{1}{3}\right)^2$
**13.** $(d - 2)(d - 5)$   **15.** $(y - 1)(y - 10)$
**17.** $(x - 6)(x + 7)$   **19.** $(x - 9)(x + 2)$
**21.** $x(x - 8)(x + 2)$   **23.** $y(y - 9)(y + 5)$
**25.** $(x - 11)(x + 9)$   **27.** $(c^2 + 8)(c^2 - 7)$
**29.** $(a^2 + 7)(a^2 - 5)$   **31.** Not factorable
**33.** Not factorable   **35.** $(x + 10)^2$
**37.** $x^2(x - 25)(x + 4)$   **39.** $(x - 24)(x + 3)$
**41.** $(x - 9)(x - 16)$   **43.** $(a + 12)(a - 11)$
**45.** $(x - 15)(x - 8)$   **47.** $(12 + x)(9 - x)$, or
$-(x + 12)(x - 9)$   **49.** $(y - 0.4)(y + 0.2)$
**51.** $(p + 5q)(p - 2q)$   **53.** $(m + 4n)(m + n)$
**55.** $(s + 3t)(s - 5t)$   **57.** $16x^3 - 48x^2 + 8x$
**58.** $28w^2 - 53w - 66$   **59.** $49w^2 + 84w + 36$
**60.** $16w^2 - 88w + 121$   **61.** $16w^2 - 121$   **62.** $27x^{12}$
**63.** $\frac{8}{3}$   **64.** $-\frac{7}{2}$   **65.** 29,555   **66.** 100°, 25°, 55°
**67.** ◈   **69.** 15, −15, 27, −27, 51, −51
**71.** $\left(x + \frac{1}{4}\right)\left(x - \frac{3}{4}\right)$   **73.** $(x + 5)\left(x - \frac{5}{7}\right)$
**75.** $(b^n + 5)(b^n + 2)$   **77.** $2x^2(4 - \pi)$

**Margin Exercises, Section 5.3, pp. 332–335**

**1.** $(2x + 5)(x - 3)$     **2.** $(4x + 1)(3x - 5)$
**3.** $(3x - 4)(x - 5)$     **4.** $2(5x - 4)(2x - 3)$
**5.** $(2x + 1)(3x + 2)$     **6.** $(2a - b)(3a - b)$
**7.** $3(2x + 3y)(x + y)$

**Calculator Spotlight, p. 335**

**1.** Correct     **2.** Correct     **3.** Not correct     **4.** Not correct     **5.** Not correct     **6.** Correct     **7.** Not correct
**8.** Correct

**Exercise Set 5.3, p. 337**

**1.** $(2x + 1)(x - 4)$     **3.** $(5x + 9)(x - 2)$
**5.** $(3x + 1)(2x + 7)$     **7.** $(3x + 1)(x + 1)$
**9.** $(2x - 3)(2x + 5)$     **11.** $(2x + 1)(x - 1)$
**13.** $(3x - 2)(3x + 8)$     **15.** $(3x + 1)(x - 2)$
**17.** $(3x + 4)(4x + 5)$     **19.** $(7x - 1)(2x + 3)$
**21.** $(3x + 2)(3x + 4)$     **23.** $(3x - 7)^2$
**25.** $(24x - 1)(x + 2)$     **27.** $(5x - 11)(7x + 4)$
**29.** $2(5 - x)(2 + x)$     **31.** $4(3x - 2)(x + 3)$
**33.** $6(5x - 9)(x + 1)$     **35.** $2(3y + 5)(y - 1)$
**37.** $(3x - 1)(x - 1)$     **39.** $4(3x + 2)(x - 3)$
**41.** $(2x + 1)(x - 1)$     **43.** $(3x + 2)(3x - 8)$
**45.** $5(3x + 1)(x - 2)$     **47.** $p(3p + 4)(4p + 5)$
**49.** $x^2(7x - 1)(2x + 3)$     **51.** $3x(8x - 1)(7x - 1)$
**53.** $(5x^2 - 3)(3x^2 - 2)$     **55.** $(5t + 8)^2$
**57.** $2x(3x + 5)(x - 1)$     **59.** Not factorable
**61.** Not factorable     **63.** $(4m + 5n)(3m - 4n)$
**65.** $(2a + 3b)(3a - 5b)$     **67.** $(3a + 2b)(3a + 4b)$
**69.** $(5p + 2q)(7p + 4q)$     **71.** $6(3x - 4y)(x + y)$
**73.** $q = \dfrac{A + 7}{p}$     **74.** $x = \dfrac{y - b}{m}$     **75.** $y = \dfrac{6 - 3x}{2}$
**76.** $q = p + r - 2$     **77.** $\{x \mid x > 4\}$     **78.** $\left\{x \mid x \le \frac{8}{11}\right\}$
**79.**

$y = \frac{2}{5}x - 1$

**80.** $y^8$     **81.** $9x^2 - 25$     **82.** $16a^2 - 24a + 9$     **83.** ◆
**85.** $(2x^n + 1)(10x^n + 3)$     **87.** $(x^{3a} - 1)(3x^{3a} + 1)$

**Margin Exercises, Section 5.4, p. 340**

**1.** $(2x + 1)(3x + 2)$     **2.** $(4x + 1)(3x - 5)$
**3.** $3(2x + 3)(x + 1)$     **4.** $2(5x - 4)(2x - 3)$

**Exercise Set 5.4, p. 341**

**1.** $(x + 7)(x + 2)$     **3.** $(x - 1)(x - 4)$

**5.** $(2x + 3)(3x + 2)$     **7.** $(x - 4)(3x - 4)$
**9.** $(5x + 3)(7x - 8)$     **11.** $(2x - 3)(2x + 3)$
**13.** $(2x^2 + 5)(x^2 + 3)$     **15.** $(2x + 1)(x - 4)$
**17.** $(3x - 5)(x + 3)$     **19.** $(2x + 7)(3x + 1)$
**21.** $(3x + 1)(x + 1)$     **23.** $(2x - 3)(2x + 5)$
**25.** $(2x - 1)(x + 1)$     **27.** $(3x + 2)(3x - 8)$
**29.** $(3x - 1)(x + 2)$     **31.** $(3x - 4)(4x - 5)$
**33.** $(7x - 1)(2x + 3)$     **35.** $(3x + 2)(3x + 4)$
**37.** $(3x - 7)^2$     **39.** $(24x - 1)(x + 2)$
**41.** $x^3(5x - 11)(7x + 4)$     **43.** $6x(5 - x)(2 + x)$
**45.** $3x^3(5 + x)(1 + 2x)$     **47.** $\{x \mid x < -100\}$
**48.** $\{x \mid x \ge 217\}$     **49.** $\{x \mid x \le 8\}$     **50.** $\{x \mid x < 2\}$
**51.** $\left\{x \mid x \ge \frac{20}{3}\right\}$     **52.** $\{x \mid x > 17\}$     **53.** $\left\{x \mid x > \frac{26}{7}\right\}$
**54.** $\left\{x \mid x \ge \frac{77}{17}\right\}$     **55.** ◆     **57.** $(3x^5 - 2)^2$
**59.** $(4x^5 + 1)^2$     **61.–69.** Left to the student

**Margin Exercises, Section 5.5, pp. 344–348**

**1.** Yes     **2.** No     **3.** No     **4.** Yes     **5.** No     **6.** Yes
**7.** No     **8.** Yes     **9.** $(x + 1)^2$     **10.** $(x - 1)^2$
**11.** $(t + 2)^2$     **12.** $(5x - 7)^2$     **13.** $(7 - 4y)^2$
**14.** $3(4m + 5)^2$     **15.** $(p^2 + 9)^2$     **16.** $z^3(2z - 5)^2$
**17.** $(3a + 5b)^2$     **18.** Yes     **19.** No     **20.** No     **21.** No
**22.** Yes     **23.** Yes     **24.** Yes     **25.** $(x + 3)(x - 3)$
**26.** $4(4 + t)(4 - t)$     **27.** $(a + 5b)(a - 5b)$
**28.** $x^4(8 + 5x)(8 - 5x)$     **29.** $5(1 + 2t^3)(1 - 2t^3)$
**30.** $(9x^2 + 1)(3x + 1)(3x - 1)$
**31.** $(7p^2 + 5q^3)(7p^2 - 5q^3)$

**Exercise Set 5.5, p. 349**

**1.** Yes     **3.** No     **5.** No     **7.** No     **9.** $(x - 7)^2$
**11.** $(x + 8)^2$     **13.** $(x - 1)^2$     **15.** $(x + 2)^2$
**17.** $(q^2 - 3)^2$     **19.** $(4y + 7)^2$     **21.** $2(x - 1)^2$
**23.** $x(x - 9)^2$     **25.** $3(2q - 3)^2$     **27.** $(7 - 3x)^2$
**29.** $5(y^2 + 1)^2$     **31.** $(1 + 2x^2)^2$     **33.** $(2p + 3q)^2$
**35.** $(a - 3b)^2$     **37.** $(9a - b)^2$     **39.** $4(3a + 4b)^2$
**41.** Yes     **43.** No     **45.** No     **47.** Yes
**49.** $(y + 2)(y - 2)$     **51.** $(p + 3)(p - 3)$
**53.** $(t + 7)(t - 7)$     **55.** $(a + b)(a - b)$
**57.** $(5t + m)(5t - m)$     **59.** $(10 + k)(10 - k)$
**61.** $(4a + 3)(4a - 3)$     **63.** $(2x + 5y)(2x - 5y)$
**65.** $2(2x + 7)(2x - 7)$     **67.** $x(6 + 7x)(6 - 7x)$
**69.** $(7a^2 + 9)(7a^2 - 9)$     **71.** $(a^2 + 4)(a + 2)(a - 2)$
**73.** $5(x^2 + 9)(x + 3)(x - 3)$
**75.** $(1 + y^4)(1 + y^2)(1 + y)(1 - y)$
**77.** $(x^6 + 4)(x^3 + 2)(x^3 - 2)$     **79.** $\left(y + \frac{1}{4}\right)\left(y - \frac{1}{4}\right)$
**81.** $\left(5 + \frac{1}{7}x\right)\left(5 - \frac{1}{7}x\right)$     **83.** $(4m^2 + t^2)(2m + t)(2m - t)$
**85.** $-11$     **86.** $400$     **87.** $-\frac{5}{6}$     **88.** $-0.9$     **89.** $2$
**90.** $-160$     **91.** $x^2 - 4xy + 4y^2$     **92.** $\frac{1}{2}\pi x^2 + 2xy$
**93.** $y^{12}$     **94.** $25a^4b^6$

**95.**   **96.**

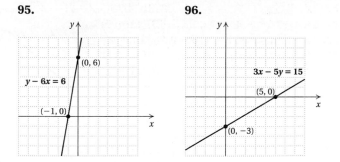

**97.** ◆   **99.** Not factorable   **101.** $(x + 11)^2$
**103.** $2x(3x + 1)^2$   **105.** $(x^4 + 2^4)(x^2 + 2^2)(x + 2)(x - 2)$
**107.** $3x^3(x + 2)(x - 2)$   **109.** $2x\left(3x + \frac{2}{5}\right)\left(3x - \frac{2}{5}\right)$
**111.** $p(0.7 + p)(0.7 - p)$   **113.** $(0.8x + 1.1)(0.8x - 1.1)$
**115.** $x(x + 6)$   **117.** $\left(x + \dfrac{1}{x}\right)\left(x - \dfrac{1}{x}\right)$
**119.** $(9 + b^{2k})(3 - b^k)(3 + b^k)$   **121.** $(3b^n + 2)^2$
**123.** $(y + 4)^2$   **125.** 9   **127.** Not correct
**129.** Not correct

## Margin Exercises, Section 5.6, pp. 354–356

**1.** $3(m^2 + 1)(m + 1)(m - 1)$   **2.** $(x^3 + 4)^2$
**3.** $2x^2(x + 1)(x + 3)$   **4.** $(3x^2 - 2)(x + 4)$
**5.** $8x(x - 5)(x + 5)$   **6.** $x^2y(x^2y + 2x + 3)$
**7.** $2p^4q^2(5p^2 + 2pq + q^2)$   **8.** $(a - b)(2x + 5 + y^2)$
**9.** $(a + b)(x^2 + y)$   **10.** $(x^2 + y^2)^2$
**11.** $(xy + 1)(xy + 4)$   **12.** $(p^2 + 9q^2)(p + 3q)(p - 3q)$

## Exercise Set 5.6, p. 357

**1.** $3(x + 8)(x - 8)$   **3.** $(a - 5)^2$   **5.** $(2x - 3)(x - 4)$
**7.** $x(x + 12)^2$   **9.** $(x + 2)(x - 2)(x + 3)$
**11.** $3(4x + 1)(4x - 1)$   **13.** $3x(3x - 5)(x + 3)$
**15.** Not factorable   **17.** $x(x^2 + 7)(x - 3)$
**19.** $x^3(x - 7)^2$   **21.** $2(2 - x)(5 + x)$, or
$-2(x - 2)(x + 5)$   **23.** Not factorable
**25.** $4(x^2 + 4)(x + 2)(x - 2)$
**27.** $(1 + y^4)(1 + y^2)(1 + y)(1 - y)$   **29.** $x^3(x - 3)(x - 1)$
**31.** $\frac{1}{9}\left(\frac{1}{3}x^3 - 4\right)^2$   **33.** $m(x^2 + y^2)$   **35.** $9xy(xy - 4)$
**37.** $2\pi r(h + r)$   **39.** $(a + b)(2x + 1)$
**41.** $(x + 1)(x - 1 - y)$   **43.** $(n + p)(n + 2)$
**45.** $(3q + p)(2q - 1)$   **47.** $(2b - a)^2$, or $(a - 2b)^2$
**49.** $(4x + 3y)^2$   **51.** $(7m^2 - 8n)^2$   **53.** $(y^2 + 5z^2)^2$
**55.** $\left(\frac{1}{2}a + \frac{1}{3}b\right)^2$   **57.** $(a + b)(a - 2b)$
**59.** $(m + 20n)(m - 18n)$   **61.** $(mn - 8)(mn + 4)$
**63.** $b^4(ab + 8)(ab - 4)$   **65.** $a^3(a - b)(a + 5b)$
**67.** $\left(a + \frac{1}{5}b\right)\left(a - \frac{1}{5}b\right)$   **69.** $(x - y)(x + y)$
**71.** $(4 + p^2q^2)(2 + pq)(2 - pq)$
**73.** $(1 + 4x^6y^6)(1 + 2x^3y^3)(1 - 2x^3y^3)$
**75.** $(q + 1)(q - 1)(q + 8)$   **77.** $(7x + 8y)^2$   **79.** 1990
**80.** 1995   **81.** 1992 and 1996   **82.** 75,000   **83.** 18,000
**84.** 59,000   **85.** $-\frac{14}{11}$   **86.** $25x^2 - 10xt + t^2$
**87.** $X = \dfrac{A + 7}{a + b}$   **88.** $\{x \mid x < 32\}$   **89.** ◆

**91.** $(a + 1)^2(a - 1)^2$   **93.** $(3.5x - 1)^2$
**95.** $(5x + 4)(x + 1.8)$   **97.** $(y + 3)(y - 3)(y - 2)$
**99.** $(a^2 + 1)(a + 4)$   **101.** $(x + 3)(x - 3)(x^2 + 2)$
**103.** $(x + 2)(x - 2)(x - 1)$   **105.** $(y - 1)^3$
**107.** $(y + 4 + x)^2$

## Margin Exercises, Section 5.7, pp. 361–364

**1.** 3, −4   **2.** 7, 3   **3.** $-\frac{1}{4}, \frac{2}{3}$   **4.** 0, $\frac{17}{3}$   **5.** −2, 3
**6.** 7, −4   **7.** 3   **8.** 0, 4   **9.** $\frac{4}{3}, -\frac{4}{3}$   **10.** 3, −3
**11.** (−5, 0), (1, 0)   **12.** 0, 3

## Exercise Set 5.7, p. 365

**1.** −4, −9   **3.** −3, 8   **5.** −12, 11   **7.** 0, −3
**9.** 0, −18   **11.** $-\frac{5}{2}$, −4   **13.** $-\frac{1}{5}$, 3   **15.** 4, $\frac{1}{4}$   **17.** 0, $\frac{2}{3}$
**19.** $-\frac{1}{10}, \frac{1}{27}$   **21.** $\frac{1}{3}$, −20   **23.** 0, $\frac{2}{3}, \frac{1}{2}$   **25.** −1, −5
**27.** −9, 2   **29.** 3, 5   **31.** 0, 8   **33.** 0, −18
**35.** 4, −4   **37.** $-\frac{2}{3}, \frac{2}{3}$   **39.** −3   **41.** 4   **43.** 0, $\frac{6}{5}$
**45.** $\frac{5}{3}$, −1   **47.** $\frac{2}{3}, -\frac{1}{4}$   **49.** $\frac{2}{3}$, −1   **51.** $\frac{7}{10}, -\frac{7}{10}$
**53.** 9, −2   **55.** $\frac{4}{5}, \frac{3}{2}$   **57.** (−4, 0), (1, 0)
**59.** $\left(-\frac{5}{2}, 0\right)$, (2, 0)   **61.** $(a + b)^2$   **62.** $a^2 + b^2$
**63.** −16   **64.** −4.5   **65.** $-\frac{10}{3}$   **66.** $\frac{3}{10}$   **67.** ◆
**69.** 4, −5   **71.** 9, −3   **73.** $\frac{1}{8}, -\frac{1}{8}$   **75.** 4, −4
**77.** (a) $x^2 - x - 12 = 0$; (b) $x^2 + 7x + 12 = 0$;
(c) $4x^2 - 4x + 1 = 0$; (d) $x^2 - 25 = 0$;
(e) $40x^3 - 14x^2 + x = 0$

## Margin Exercises, Section 5.8, pp. 367–370

**1.** 5 and −5   **2.** 7 and 8   **3.** −4 and 5
**4.** Length: 5 cm; width: 3 cm   **5.** (a) 342; (b) 9
**6.** 22 and 23   **7.** 3 m, 4 m

## Exercise Set 5.8, p. 371

**1.** 5 and −5   **3.** 3 and 5   **5.** Length: 12 cm; width:
7 cm   **7.** 14 and 15   **9.** 12 and 14; −12 and −14
**11.** 15 and 17; −15 and −17   **13.** 5
**15.** Height: 6 cm; base: 5 cm   **17.** 6 km   **19.** 4 sec
**21.** 5 and 7   **23.** 182   **25.** 12   **27.** 4950   **29.** 25
**31.** Hypotenuse: 17 ft; leg: 15 ft   **33.** $9x^2 - 25y^2$
**34.** $9x^2 - 30xy + 25y^2$   **35.** $9x^2 + 30xy + 25y^2$
**36.** $6x^2 + 11xy - 35y^2$   **37.** $y$-intercept: (0, −4);
$x$-intercept: (16, 0)   **38.** $y$-intercept: (0, 4); $x$-intercept:
(16, 0)   **39.** $y$-intercept: (0, −5); $x$-intercept: (6.5, 0)
**40.** $y$-intercept: $\left(0, \frac{2}{3}\right)$; $x$-intercept: $\left(\frac{5}{8}, 0\right)$   **41.** ◆
**43.** 5 ft   **45.** 37   **47.** 30 cm by 15 cm

## Summary and Review: Chapter 5, p. 375

**1.** $(-10x)(x)$; $(-5x)(2x)$; $(5x)(-2x)$; answers may vary
**2.** $(6x)(6x^4)$; $(4x^2)(9x^3)$; $(-2x^4)(-18x)$; answers may vary
**3.** $5(1 + 2x^3)(1 - 2x^3)$   **4.** $x(x - 3)$
**5.** $(3x + 2)(3x - 2)$   **6.** $(x + 6)(x - 2)$   **7.** $(x + 7)^2$
**8.** $3x(2x^2 + 4x + 1)$   **9.** $(x^2 + 3)(x + 1)$

**10.** $(3x - 1)(2x - 1)$ **11.** $(x^2 + 9)(x + 3)(x - 3)$
**12.** $3x(3x - 5)(x + 3)$ **13.** $2(x + 5)(x - 5)$
**14.** $(x^3 - 2)(x + 4)$ **15.** $(4x^2 + 1)(2x + 1)(2x - 1)$
**16.** $4x^4(2x^2 - 8x + 1)$ **17.** $3(2x + 5)^2$
**18.** Not factorable **19.** $x(x - 6)(x + 5)$
**20.** $(2x + 5)(2x - 5)$ **21.** $(3x - 5)^2$
**22.** $2(3x + 4)(x - 6)$ **23.** $(x - 3)^2$
**24.** $(2x + 1)(x - 4)$ **25.** $2(3x - 1)^2$
**26.** $3(x + 3)(x - 3)$ **27.** $(x - 5)(x - 3)$
**28.** $(5x - 2)^2$ **29.** $(7b^5 - 2a^4)^2$ **30.** $(xy + 4)(xy - 3)$
**31.** $3(2a + 7b)^2$ **32.** $(m + t)(m + 5)$
**33.** $32(x^2 - 2y^2z^2)(x^2 + 2y^2z^2)$ **34.** $1, -3$ **35.** $-7, 5$
**36.** $-4, 3$ **37.** $\frac{2}{3}, 1$ **38.** $\frac{3}{2}, -4$ **39.** $8, -2$
**40.** 3 and $-2$ **41.** $-18$ and $-16$; 16 and 18
**42.** $\frac{5}{2}$ and $-2$ **43.** $-19$ and $-17$; 17 and 19
**44.** Dining room: 12 ft by 12 ft; kitchen: 12 ft by 10 ft
**45.** 4 ft **46.** $(-5, 0), (-4, 0)$ **47.** $\left(-\frac{3}{2}, 0\right), (5, 0)$
**48.** $\frac{8}{35}$ **49.** $\left\{x \mid x \le \frac{4}{3}\right\}$ **50.** $4a^2 - 9$
**51.**

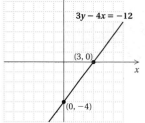

**52.** ◆ In this chapter, we learned to solve equations of the type $ax^2 + bx + c = 0$ (quadratic equations). Previously, we could solve only first-degree, or linear, equations (equations equivalent to those of the form $ax + b = 0$). The principle of zero products is used to solve quadratic equations, but it is not used to solve linear equations. **53.** ◆ Multiplying can be used to check factoring because factoring is the reverse of multiplying. The TABLE feature of a grapher can provide a partial check of factoring. When a polynomial and its factorization are entered as $y_1$ and $y_2$, the factorization is probably correct if corresponding values in the Y1 and Y2 columns are the same. The GRAPH feature of a grapher can also provide a partial check. If the graphs of $y_1$ and $y_2$ (entered as described above) coincide, then the factorization is probably correct. **54.** 2.5 cm
**55.** 0, 2 **56.** Length: 12; width: 6 **57.** No solution
**58.** $2, -3, \frac{5}{2}$ **59.** a, i; b, k; c, g; d, h; e, j; f, l
**60.** $2^{100}$; $2^{90} + 2^{90} = 2 \cdot 2^{90} = 2^{91} < 2^{100}$

## Test: Chapter 5, p. 377

**1.** [5.1a] $(4x)(x^2)$; $(2x^2)(2x)$; $(-2x)(-2x^2)$; answers may vary **2.** [5.2a] $(x - 5)(x - 2)$ **3.** [5.5b] $(x - 5)^2$
**4.** [5.1b] $2y^2(2y^2 - 4y + 3)$ **5.** [5.1c] $(x^2 + 2)(x + 1)$
**6.** [5.1b] $x(x - 5)$ **7.** [5.2a] $x(x + 3)(x - 1)$
**8.** [5.3a], [5.4a] $2(5x - 6)(x + 4)$
**9.** [5.5d] $(2x + 3)(2x - 3)$ **10.** [5.2a] $(x - 4)(x + 3)$
**11.** [5.3a], [5.4a] $3m(2m + 1)(m + 1)$

**12.** [5.5d] $3(w + 5)(w - 5)$ **13.** [5.5b] $5(3x + 2)^2$
**14.** [5.5d] $3(x^2 + 4)(x + 2)(x - 2)$ **15.** [5.5b] $(7x - 6)^2$
**16.** [5.3a], [5.4a] $(5x - 1)(x - 5)$
**17.** [5.1c] $(x^3 - 3)(x + 2)$
**18.** [5.5d] $5(4 + x^2)(2 + x)(2 - x)$
**19.** [5.3a], [5.4a] $(2x + 3)(2x - 5)$
**20.** [5.3a], [5.4a] $3t(2t + 5)(t - 1)$
**21.** [5.2a] $3(m + 2n)(m - 5n)$ **22.** [5.7b] $5, -4$
**23.** [5.7b] $\frac{3}{2}, -5$ **24.** [5.7b] $7, -4$ **25.** [5.8a] $8, -3$
**26.** [5.8a] Length of foot is 7 ft; height of sail is 12 ft
**27.** [5.7b] $(7, 0), (-5, 0)$ **28.** [5.7b] $\left(\frac{2}{3}, 0\right), (1, 0)$
**29.** [1.6c] $-\frac{10}{11}$ **30.** [2.7e] $\left\{x \mid x < \frac{19}{3}\right\}$
**31.** [3.3a]

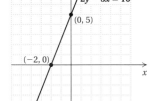

**32.** [4.6d] $25x^4 - 70x^2 + 49$
**33.** [5.8a] Length: 15; width: 3
**34.** [5.2a] $(a - 4)(a + 8)$ **35.** [5.5d], [5.7a] (c)
**36.** [4.6b], [5.5d] (d)

## Cumulative Review: Chapters 1–5, p. 379

**1.** [1.2d] $<$ **2.** [1.2d] $>$ **3.** [1.4a] 0.35
**4.** [1.6c] $-1.57$ **5.** [1.5a] $-\frac{1}{14}$ **6.** [1.6c] $-\frac{6}{5}$
**7.** [1.8c] $4x + 1$ **8.** [1.8d] $-8$ **9.** [4.2a, b] $\frac{8x^6}{y^3}$
**10.** [4.1d, e] $-\frac{1}{6x^3}$ **11.** [4.4a] $x^4 - 3x^3 - 3x^2 - 4$
**12.** [4.7e] $2x^2y^2 - x^2y - xy$
**13.** [4.8b] $x^2 + 3x + 2 + \frac{3}{x - 1}$
**14.** [4.6c] $4t^2 - 12t + 9$ **15.** [4.6b] $x^4 - 9$
**16.** [4.6a] $6x^2 + 4x - 16$ **17.** [4.5b] $2x^4 + 6x^3 + 8x^2$
**18.** [4.5d] $4y^3 + 4y^2 + 5y - 4$ **19.** [4.6b] $x^2 - \frac{4}{9}$
**20.** [5.2a] $(x + 4)(x - 2)$ **21.** [5.5d] $(2x + 5)(2x - 5)$
**22.** [5.1c] $(3x - 4)(x^2 + 1)$ **23.** [5.5b] $(x - 13)^2$
**24.** [5.5d] $3(5x + 6y)(5x - 6y)$
**25.** [5.3a], [5.4a] $(3x + 7)(2x - 9)$
**26.** [5.2a] $(x^2 - 3)(x^2 + 1)$
**27.** [5.6a] $2(y - 1)(y + 1)(2y - 3)$
**28.** [5.3a], [5.4a] $(3p - q)(2p + q)$
**29.** [5.3a], [5.4a] $2x(5x + 1)(x + 5)$
**30.** [5.5b] $x(7x - 3)^2$ **31.** [5.3a], [5.4a] Not factorable
**32.** [5.1b] $3x(25x^2 + 9)$
**33.** [5.5d] $3(x^4 + 4y^4)(x^2 + 2y^2)(x^2 - 2y^2)$
**34.** [5.2a] $14(x + 2)(x + 1)$
**35.** [5.6a] $(x^3 + 1)(x + 1)(x - 1)$ **36.** [2.3b] 15
**37.** [2.7e] $\{y \mid y < 6\}$ **38.** [5.7a] $15, -\frac{1}{4}$
**39.** [5.7a] $0, -37$ **40.** [5.7b] $5, -5, -1$

**41.** [5.7b] 6, −6 **42.** [5.7b] $\frac{1}{3}$ **43.** [5.7b] −10, −7
**44.** [5.7b] 0, $\frac{3}{2}$ **45.** [2.3a] 0.2 **46.** [5.7b] −4, 5
**47.** [2.7e] $\{x \mid x \le 20\}$ **48.** [2.3c] All real numbers
**49.** [2.6a] $m = \dfrac{y - b}{x}$ **50.** [2.4a] 50, 52
**51.** [5.8a] −20 and −18; 18 and 20 **52.** [5.8a] 6 ft, 3 ft
**53.** [2.4a] 150 m by 350 m **54.** [2.5a] $6500
**55.** [5.8a] 17 m **56.** [2.4a] 30 m, 60 m, 10 m
**57.** [2.5a] $29 **58.** [5.8a] 18 cm, 16 cm
**59.** [3.3a]

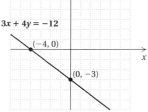

**60.** [2.7e], [4.6a] $\{x \mid x \ge -\frac{13}{3}\}$ **61.** [2.3b] 22
**62.** [5.7b] −6, 4 **63.** [5.6a] $(x - 2)(x + 1)(x - 3)$
**64.** [5.6a] $(2a + 3b + 3)(2a - 3b - 5)$ **65.** [5.5a] 25
**66.** [5.8a] 2 cm


# Chapter 6

## Pretest: Chapter 6, p. 382

**1.** [6.3c] $(x + 2)(x + 3)^2$ **2.** [6.4a] $\dfrac{-b - 1}{b^2 - 4}$, or $\dfrac{b + 1}{4 - b^2}$
**3.** [6.5a] $\dfrac{1}{y - 2}$ **4.** [6.4a] $\dfrac{7a + 6}{a(a + 2)}$ **5.** [6.5b] $\dfrac{2x}{x + 1}$
**6.** [6.1d] $\dfrac{2(x - 3)}{x - 2}$ **7.** [6.2b] $\dfrac{x - 3}{x + 3}$ **8.** [6.9a] $\dfrac{y + x}{y - x}$
**9.** [6.6a] −5 **10.** [6.6a] 0 **11.** [6.8a] $M = \dfrac{3R}{a - b}$
**12.** [6.7b] 10.5 hr **13.** [6.7a] $\frac{30}{11}$ hr
**14.** [6.7a] 60 mph, 80 mph

## Margin Exercises, Section 6.1, pp. 383–388

**1.** 3 **2.** −8, 3 **3.** None **4.** $\dfrac{x(2x + 1)}{x(3x - 2)}$
**5.** $\dfrac{(x + 1)(x + 2)}{(x - 2)(x + 2)}$ **6.** $\dfrac{-1(x - 8)}{-1(x - y)}$ **7.** 5 **8.** $\dfrac{x}{4}$
**9.** $\dfrac{2x + 1}{3x + 2}$ **10.** $\dfrac{x + 1}{2x + 1}$ **11.** $x + 2$ **12.** $\dfrac{y + 2}{4}$
**13.** −1 **14.** −1 **15.** −1 **16.** $\dfrac{a - 2}{a - 3}$ **17.** $\dfrac{x - 5}{2}$

## Calculator Spotlight, p. 388

**1.** Correct **2.** Correct **3.** Not correct
**4.** Not correct **5.** Not correct **6.** Correct
**7.** Not correct


## Exercise Set 6.1, p. 389

**1.** 0 **3.** 8 **5.** $-\frac{5}{2}$ **7.** 7, −4 **9.** 5, −5 **11.** None
**13.** $\dfrac{(4x)(3x^2)}{(4x)(5y)}$ **15.** $\dfrac{2x(x - 1)}{2x(x + 4)}$ **17.** $\dfrac{-1(3 - x)}{-1(4 - x)}$
**19.** $\dfrac{(y + 6)(y - 7)}{(y + 6)(y + 2)}$ **21.** $\dfrac{x^2}{4}$ **23.** $\dfrac{8p^2q}{3}$ **25.** $\dfrac{x - 3}{x}$
**27.** $\dfrac{m + 1}{2m + 3}$ **29.** $\dfrac{a - 3}{a + 2}$ **31.** $\dfrac{a - 3}{a - 4}$ **33.** $\dfrac{x + 5}{x - 5}$
**35.** $a + 1$ **37.** $\dfrac{x^2 + 1}{x + 1}$ **39.** $\dfrac{3}{2}$ **41.** $\dfrac{6}{t - 3}$
**43.** $\dfrac{t + 2}{2(t - 4)}$ **45.** $\dfrac{t - 2}{t + 2}$ **47.** −1 **49.** −1 **51.** −6
**53.** $-x - 1$ **55.** $\dfrac{56x}{3}$ **57.** $\dfrac{2}{dc^2}$ **59.** $\dfrac{x + 2}{x - 2}$
**61.** $\dfrac{(a + 3)(a - 3)}{a(a + 4)}$ **63.** $\dfrac{2a}{a - 2}$ **65.** $\dfrac{(t + 2)(t - 2)}{(t + 1)(t - 1)}$
**67.** $\dfrac{x + 4}{x + 2}$ **69.** $\dfrac{5(a + 6)}{a - 1}$ **71.** 18 and 20; −18 and −20
**72.** 3.125 L **73.** $(x - 8)(x + 7)$ **74.** $(a - 8)^2$
**75.** $x^3(x - 7)(x + 5)$ **76.** $(2y^2 + 1)(y - 5)$
**77.** $(2 - t)(2 + t)(4 + t^2)$ **78.** $10(x + 7)(x + 1)$
**79.** $(x - 7)(x - 2)$ **80.** Not factorable **81.** $(4x - 5y)^2$
**82.** $(a - 7b)(a - 2b)$ **83.** ◈ **85.** $x + 2y$
**87.** $\dfrac{(t - 9)^2(t - 1)}{(t^2 + 9)(t + 1)}$ **89.** $\dfrac{x - y}{x - 5y}$

## Margin Exercises, Section 6.2, pp. 393–394

**1.** $\dfrac{2}{7}$ **2.** $\dfrac{2x^3 - 1}{x^2 + 5}$ **3.** $\dfrac{1}{x - 5}$ **4.** $x^2 - 3$ **5.** $\dfrac{6}{35}$
**6.** $\dfrac{x^2}{40}$ **7.** $\dfrac{(x - 3)(x - 2)}{(x + 5)(x + 5)}$ **8.** $\dfrac{x - 3}{x + 2}$
**9.** $\dfrac{(x - 3)(x - 2)}{x + 2}$ **10.** $\dfrac{y + 1}{y - 1}$

## Exercise Set 6.2, p. 395

**1.** $\dfrac{x}{4}$ **3.** $\dfrac{1}{x^2 - y^2}$ **5.** $a + b$ **7.** $\dfrac{x^2 - 4x + 7}{x^2 + 2x - 5}$ **9.** $\dfrac{3}{10}$
**11.** $\dfrac{1}{4}$ **13.** $\dfrac{b}{a}$ **15.** $\dfrac{(a + 2)(a + 3)}{(a - 3)(a - 1)}$ **17.** $\dfrac{(x - 1)^2}{x}$
**19.** $\dfrac{1}{2}$ **21.** $\dfrac{15}{8}$ **23.** $\dfrac{15}{4}$ **25.** $\dfrac{a - 5}{3(a - 1)}$ **27.** $\dfrac{(x + 2)^2}{x}$
**29.** $\dfrac{3}{2}$ **31.** $\dfrac{c + 1}{c - 1}$ **33.** $\dfrac{y - 3}{2y - 1}$ **35.** $\dfrac{x + 1}{x - 1}$
**37.** $\{x \mid x \ge 77\}$ **38.** 4 **39.** $8x^3 - 11x^2 - 3x + 12$
**40.** $-2p^2 + 4pq - 4q^2$ **41.** $\dfrac{4y^8}{x^6}$ **42.** $\dfrac{125x^{18}}{y^{12}}$
**43.** $\dfrac{4x^6}{y^{10}}$ **44.** $\dfrac{1}{a^{15}b^{20}}$ **45.** ◈ **47.** $-\dfrac{1}{b^2}$
**49.** $\dfrac{(x - 7)^2}{x + y}$

## Margin Exercises, Section 6.3, pp. 397–398

**1.** 144 **2.** 12 **3.** 10 **4.** 120 **5.** $\frac{35}{144}$ **6.** $\frac{1}{4}$ **7.** $\frac{11}{10}$

**8.** $\frac{9}{40}$  **9.** $60x^3y^2$  **10.** $(y + 1)^2(y + 4)$
**11.** $7(t^2 + 16)(t - 2)$  **12.** $3x(x + 1)^2(x - 1)$

## Exercise Set 6.3, p. 399

**1.** 108  **3.** 72  **5.** 126  **7.** 360  **9.** 500  **11.** $\frac{65}{72}$
**13.** $\frac{29}{120}$  **15.** $\frac{23}{180}$  **17.** $12x^3$  **19.** $18x^2y^2$  **21.** $6(y - 3)$
**23.** $t(t + 2)(t - 2)$  **25.** $(x + 2)(x - 2)(x + 3)$
**27.** $t(t + 2)^2(t - 4)$  **29.** $(a + 1)(a - 1)^2$
**31.** $(m - 3)(m - 2)^2$  **33.** $(2 + 3x)(2 - 3x)$
**35.** $10v(v + 4)(v + 3)$  **37.** $18x^3(x - 2)^2(x + 1)$
**39.** $6x^3(x + 2)^2(x - 2)$  **41.** $(x - 3)^2$  **42.** $2x(3x + 2)$
**43.** $(x + 3)(x - 3)$  **44.** $(x + 7)(x - 3)$  **45.** $(x + 3)^2$
**46.** $(x - 7)(x + 3)$  **47.** 54%  **48.** 64%  **49.** 74%
**50.** 98%  **51.** 1965  **52.** 1999  **53.** ◆

## Margin Exercises, Section 6.4, pp. 401–404

**1.** $\frac{7}{9}$  **2.** $\frac{3 + x}{x - 2}$  **3.** $\frac{6x + 4}{x - 1}$  **4.** $\frac{x - 5}{4}$  **5.** $\frac{x - 1}{x - 3}$
**6.** $\frac{10x^2 + 9x}{48}$  **7.** $\frac{9x + 10}{48x^2}$  **8.** $\frac{4x^2 - x + 3}{x(x - 1)(x + 1)^2}$
**9.** $\frac{2x^2 + 16x + 5}{(x + 3)(x + 8)}$  **10.** $\frac{8x + 88}{(x + 16)(x + 1)(x + 8)}$
**11.** $\frac{-2x - 11}{3(x + 4)(x - 4)}$

## Exercise Set 6.4, p. 405

**1.** 1  **3.** $\frac{6}{3 + x}$  **5.** $\frac{2x + 3}{x - 5}$  **7.** $\frac{1}{4}$  **9.** $-\frac{1}{t}$
**11.** $\frac{-x + 7}{x - 6}$  **13.** $y + 3$  **15.** $\frac{2b - 14}{b^2 - 16}$  **17.** $a + b$
**19.** $\frac{5x + 2}{x - 5}$  **21.** $-1$  **23.** $\frac{-x^2 + 9x - 14}{(x - 3)(x + 3)}$
**25.** $\frac{2x + 5}{x^2}$  **27.** $\frac{41}{24r}$  **29.** $\frac{4x + 6y}{x^2y^2}$  **31.** $\frac{4 + 3t}{18t^3}$
**33.** $\frac{x^2 + 4xy + y^2}{x^2y^2}$  **35.** $\frac{6x}{(x - 2)(x + 2)}$  **37.** $\frac{11x + 2}{3x(x + 1)}$
**39.** $\frac{x^2 + 6x}{(x + 4)(x - 4)}$  **41.** $\frac{6}{z + 4}$  **43.** $\frac{3x - 1}{(x - 1)^2}$
**45.** $\frac{11a}{10(a - 2)}$  **47.** $\frac{2x^2 + 8x + 16}{x(x + 4)}$
**49.** $\frac{7a + 6}{(a - 2)(a + 1)(a + 3)}$  **51.** $\frac{2x^2 - 4x + 34}{(x - 5)(x + 3)}$
**53.** $\frac{3a + 2}{(a + 1)(a - 1)}$  **55.** $\frac{2x + 6y}{(x + y)(x - y)}$
**57.** $\frac{a^2 + 7a + 1}{(a + 5)(a - 5)}$  **59.** $\frac{5t - 12}{(t + 3)(t - 3)(t - 2)}$
**61.** $x^2 - 1$  **62.** $13y^3 - 14y^2 + 12y - 73$  **63.** $\frac{1}{8x^{12}y^9}$
**64.** $\frac{x^6}{25y^2}$  **65.** $\frac{1}{x^{12}y^{21}}$  **66.** $\frac{25}{x^4y^6}$

**67.**
**68.**
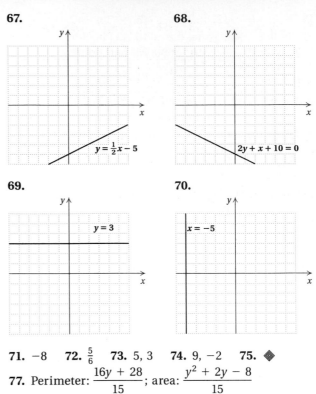

**69.**
**70.**

**71.** $-8$  **72.** $\frac{5}{6}$  **73.** 5, 3  **74.** 9, $-2$  **75.** ◆
**77.** Perimeter: $\frac{16y + 28}{15}$; area: $\frac{y^2 + 2y - 8}{15}$
**79.** $\frac{(z + 6)(2z - 3)}{(z + 2)(z - 2)}$  **81.** $\frac{11z^4 - 22z^2 + 6}{(z^2 + 2)(z^2 - 2)(2z^2 - 3)}$

## Margin Exercises, Section 6.5, pp. 409–412

**1.** $\frac{4}{11}$  **2.** $\frac{5}{y}$  **3.** $\frac{x^2 + 2x + 1}{2x + 1}$  **4.** $\frac{3x - 1}{3}$  **5.** $\frac{4x - 3}{x - 2}$
**6.** $\frac{-x - 7}{15x}$  **7.** $\frac{x^2 - 48}{(x + 7)(x + 8)(x + 6)}$
**8.** $\frac{-8y - 28}{(y + 4)(y - 4)}$  **9.** $\frac{x - 13}{(x + 3)(x - 3)}$
**10.** $\frac{6x^2 - 2x - 2}{3x(x + 1)}$

## Exercise Set 6.5, p. 413

**1.** $\frac{4}{x}$  **3.** 1  **5.** $\frac{1}{x - 1}$  **7.** $\frac{8}{3}$  **9.** $\frac{13}{a}$  **11.** $\frac{8}{y - 1}$
**13.** $\frac{x - 2}{x - 7}$  **15.** $\frac{4}{a^2 - 25}$  **17.** $\frac{2x - 4}{x - 9}$  **19.** $\frac{-9}{2x - 3}$
**21.** $\frac{-a - 4}{10}$  **23.** $\frac{7z - 12}{12z}$  **25.** $\frac{4x^2 - 13xt + 9t^2}{3x^2t^2}$
**27.** $\frac{2x - 40}{(x + 5)(x - 5)}$  **29.** $\frac{3 - 5t}{2t(t - 1)}$  **31.** $\frac{2s - st - s^2}{(t + s)(t - s)}$
**33.** $\frac{y - 19}{4y}$  **35.** $\frac{-2a^2}{(x + a)(x - a)}$  **37.** $\frac{9x + 12}{(x + 3)(x - 3)}$
**39.** $\frac{1}{2}$  **41.** $\frac{x - 3}{(x + 3)(x + 1)}$  **43.** $\frac{18x + 5}{x - 1}$  **45.** 0
**47.** $\frac{20}{2y - 1}$  **49.** $\frac{2a - 3}{2 - a}$  **51.** $\frac{z - 3}{2z - 1}$  **53.** $\frac{2}{x + y}$
**55.** $x^5$  **56.** $30x^{12}$  **57.** $\frac{b^{20}}{a^8}$  **58.** $18x^3$  **59.** $\frac{6}{x^3}$

**60.** $\frac{10}{x^3}$ **61.** $x^2 - 9x + 18$ **62.** $(4 - \pi)r^2$ **63.** ◈

**65.** $\frac{30}{(x - 3)(x + 4)}$ **67.** $\frac{x^2 + xy - x^3 + x^2y - xy^2 + y^3}{(x^2 + y^2)(x + y)^2(x - y)}$

**69.** $\frac{-2a - 15}{a - 6}$; area $= \frac{-2a^3 - 15a^2 + 12a + 90}{2(a - 6)^2}$

## Margin Exercises, Section 6.6, pp. 417–421

**1.** $\frac{33}{2}$ **2.** 3 **3.** $\frac{3}{2}$ **4.** $-\frac{1}{8}$ **5.** 1 **6.** 2 **7.** 4

## Improving Your Math Study Skills, p. 422

**1.** Rational expression **2.** Solutions **3.** Rational expression **4.** Rational expression **5.** Rational expression **6.** Solutions **7.** Rational expression **8.** Solutions **9.** Solutions **10.** Solutions **11.** Rational expression **12.** Solutions **13.** Rational expression

## Exercise Set 6.6, p. 423

**1.** $\frac{6}{5}$ **3.** $\frac{40}{29}$ **5.** $\frac{47}{2}$ **7.** $-6$ **9.** $\frac{24}{7}$ **11.** $-4, -1$
**13.** $4, -4$ **15.** 3 **17.** $\frac{14}{3}$ **19.** 5 **21.** 5 **23.** $\frac{5}{2}$
**25.** $-2$ **27.** $-\frac{13}{2}$ **29.** $\frac{17}{2}$ **31.** No solution **33.** $-5$
**35.** $\frac{5}{3}$ **37.** $\frac{1}{2}$ **39.** No solution **41.** No solution
**43.** 4 **45.** $\frac{1}{a^6 b^{15}}$ **46.** $x^8 y^{12}$ **47.** $\frac{16x^4}{t^8}$ **48.** $\frac{w^4}{y^6}$
**49.** $32x^6$ **50.** $\frac{64x^{10}}{y^8}$

**51.**  **52.**
**53.**  **54.**

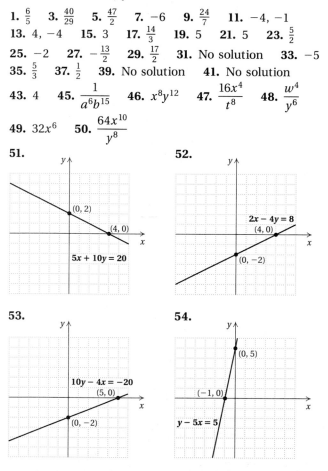

**55.** ◈ **57.** 7 **59.** No solution **61.** 2, $-2$ **63.** 4

## Margin Exercises, Section 6.7, pp. 427–434

**1.** $-3$ **2.** 40 km/h, 60 km/h **3.** $\frac{24}{7}$, or $3\frac{3}{7}$ hr
**4.** 58 km/L **5.** 0.280 **6.** 124 km/h **7.** 2.4 fish/yd$^2$
**8.** 81 gal **9. (a)** About 57; **(b)** Since $57 < 61$, it could

not be predicted that McGwire would break Maris's record. **10.** 42 **11.** 2074 **12.** 24.75 ft **13.** 34.9 ft

## Calculator Spotlight, p. 430

**1.** 20.6 min **2.** 0.54 hr

## Exercise Set 6.7, p. 435

**1.** $\frac{24}{7}$ **3.** 20 and 15 **5.** 30 km/h, 70 km/h
**7.** Passenger: 80 mph; freight: 66 mph **9.** 20 mph
**11.** $4\frac{4}{9}$ min **13.** $5\frac{1}{7}$ hr **15.** 9 **17.** 2.3 km/h
**19.** 66 g **21.** 702 km **23.** 200 **25. (a)** About 63;
**(b)** yes **27.** 10,000 **29. (a)** 4.8 tons; **(b)** 48 lb
**31.** 11 **33.** $\frac{21}{2}$ **35.** $\frac{8}{3}$ **37.** $\frac{35}{3}$ **39.** $x^{11}$ **40.** $x$
**41.** $\frac{1}{x^{11}}$ **42.** $\frac{1}{x}$ **43.** ◈ **45.** $9\frac{3}{13}$ days **47.** $\frac{3}{4}$
**49.** $27\frac{3}{11}$ min

## Margin Exercises, Section 6.8, pp. 439–440

**1.** $M = \frac{fd^2}{km}$ **2.** $b_1 = \frac{2A - hb_2}{h}$ **3.** $f = \frac{pq}{p + q}$
**4.** $b = \frac{a}{2Q + 1}$

## Exercise Set 6.8, p. 441

**1.** $r = \frac{S}{2\pi h}$ **3.** $b = \frac{2A}{h}$ **5.** $n = \frac{S + 360}{180}$
**7.** $b = \frac{3V - kB - 4kM}{k}$ **9.** $r = \frac{S - a}{S - l}$
**11.** $h = \frac{2A}{b_1 + b_2}$ **13.** $B = \frac{A}{AQ + 1}$
**15.** $p = \frac{qf}{q - f}$ **17.** $A = P(1 + r)$ **19.** $R = \frac{r_1 r_2}{r_1 + r_2}$
**21.** $D = \frac{BC}{A}$ **23.** $h_2 = \frac{p(h_1 - q)}{q}$ **25.** $a = \frac{b}{K - C}$
**27.** $-3x^3 - 5x^2 + 5$
**28.** $23x^4 + 50x^3 + 23x^2 - 163x + 41$
**29.** $(x + 2)(x - 2)$ **30.** $3(2y^2 - 1)(5y^2 + 4)$
**31.** $(7m - 8n)^2$ **32.** $(y + 7)(y - 5)$
**33.** $(y^2 + 1)(y + 1)(y - 1)$ **34.** $(a - 10b)(a + 10b)$
**35.** $x^2 + 2x + 8 + \frac{12}{x - 2}$ **36.** $x^2 - 3$ **37.** ◈
**39.** $T = \frac{FP}{u + EF}$ **41.** $v = \frac{Nbf_2 - bf_1 - df_1}{Nf_2 - 1}$

## Margin Exercises, Section 6.9, pp. 444–446

**1.** $\frac{136}{5}$ **2.** $\frac{7x^2}{3(2 - x^2)}$ **3.** $\frac{x}{x - 1}$ **4.** $\frac{136}{5}$
**5.** $\frac{7x^2}{3(2 - x^2)}$ **6.** $\frac{x}{x - 1}$

## Exercise Set 6.9, p. 447

**1.** $\frac{25}{4}$ **3.** $\frac{1}{3}$ **5.** $-6$ **7.** $\frac{1 + 3x}{1 - 5x}$ **9.** $\frac{2x + 1}{x}$ **11.** 8

13. $x - 8$   15. $\dfrac{y}{y-1}$   17. $-\dfrac{1}{a}$   19. $\dfrac{ab}{b-a}$

21. $\dfrac{p^2 + q^2}{q + p}$   23. $4x^4 + 3x^3 + 2x - 7$   24. $0$

25. $(p-5)^2$   26. $(p+5)^2$   27. $50(p^2 - 2)$

28. $5(p+2)(p-10)$   29. 14 yd   30. 12 ft, 5 ft

31. ◆   33. $\dfrac{(x-1)(3x-2)}{5x-3}$   35. $-\dfrac{ac}{bd}$   37. $\dfrac{5x+3}{3x+2}$

## Summary and Review: Chapter 6, p. 449

1. $0$   2. $6$   3. $6, -6$   4. $-6, 5$   5. $-2$   6. $0, 3, 5$

7. $\dfrac{x-2}{x+1}$   8. $\dfrac{7x+3}{x-3}$   9. $\dfrac{y-5}{y+5}$   10. $\dfrac{a-6}{5}$

11. $\dfrac{6}{2t-1}$   12. $-20t$   13. $\dfrac{2x^2 - 2x}{x+1}$   14. $30x^2y^2$

15. $4(a-2)$   16. $(y-2)(y+2)(y+1)$   17. $\dfrac{-3x+18}{x+7}$

18. $-1$   19. $\dfrac{2a}{a-1}$   20. $d+c$   21. $\dfrac{4}{x-4}$

22. $\dfrac{x+5}{2x}$   23. $\dfrac{2x+3}{x-2}$   24. $\dfrac{-x^2 + x + 26}{(x-5)(x+5)(x+1)}$

25. $\dfrac{2(x-2)}{x+2}$   26. $\dfrac{z}{1-z}$   27. $c-d$   28. $8$

29. $3, -5$   30. $5\frac{1}{7}$ hr   31. 240 km/h, 280 km/h

32. $-2$   33. 95 mph, 175 mph   34. $160$   35. 1.92 g

36. $6$   37. $s = \dfrac{rt}{r-t}$   38. $C = \frac{5}{9}(F - 32)$, or

$C = \dfrac{5F - 160}{9}$   39. $r^3 = \dfrac{3V}{4\pi}$   40. $(5x^2 - 3)(x + 4)$

41. $\dfrac{1}{125x^9y^6}$   42. $-2x^3 + 3x^2 + 12x - 18$

43. Length: 5 cm; width: 3 cm; perimeter: 16 cm

44. $\dfrac{5x + 6}{(x+2)(x-2)}$; used to find an equivalent expression for each rational expression with the LCM as the least common denominator   45. $\dfrac{3x + 10}{(x-2)(x+2)}$; used to find an equivalent expression for each rational expression with the LCM as the least common denominator   46. 4; used to clear fractions

47. $\dfrac{4(x-2)}{x(x+4)}$; Method 1: used to multiply by 1 using LCM/LCM; Method 2: used the LCM of the denominators in the numerator to subtract in the numerator and used the LCM of the denominators in the denominator to add in the denominator.

48. $\dfrac{5(a+3)^2}{a}$   49. $\dfrac{10a}{(a-b)(b-c)}$

50. They are equivalent equations.

## Test: Chapter 6, p. 451

1. [6.1a] 0   2. [6.1a] −8   3. [6.1a] 7, −7
4. [6.1a] 1, 2   5. [6.1a] 1   6. [6.1a] 0, −3, −5

7. [6.1c] $\dfrac{3x+7}{x+3}$   8. [6.1d] $\dfrac{a+5}{2}$

9. [6.2b] $\dfrac{(5x+1)(x+1)}{3x(x+2)}$

10. [6.3c] $(y-3)(y+3)(y+7)$   11. [6.4a] $\dfrac{23 - 3x}{x^3}$

12. [6.5a] $\dfrac{8 - 2t}{t^2 + 1}$   13. [6.4a] $\dfrac{-3}{x-3}$   14. [6.5a] $\dfrac{2x - 5}{x - 3}$

15. [6.4a] $\dfrac{8t - 3}{t(t-1)}$   16. [6.5a] $\dfrac{-x^2 - 7x - 15}{(x+4)(x-4)(x+1)}$

17. [6.5b] $\dfrac{x^2 + 2x - 7}{(x-1)^2(x+1)}$   18. [6.9a] $\dfrac{3y + 1}{y}$

19. [6.6a] 12   20. [6.6a] 5, −3   21. [6.7a] 4
22. [6.7b] 16   23. [6.7a] 45 mph, 65 mph

24. [6.8a] $t = \dfrac{g}{M - L}$   25. [6.7b] 15

26. [5.6a] $(4a + 7)(4a - 7)$   27. [4.2a, b] $\dfrac{y^{12}}{81x^8}$

28. [4.4c] $13x^2 - 29x + 76$
29. [5.8a] 21 and 22; −22 and −21
30. [6.7a] Team A: 4 hr; team B: 10 hr

31. [6.9a] $\dfrac{3a + 2}{2a + 1}$

## Cumulative Review: Chapters 1–6, p. 453

1. [1.1a] $-\frac{9}{5}$   2. [4.3a] 12   3. [1.8c] $2x + 6$

4. [4.1d], [4.2a, b] $\dfrac{27x^7}{4}$   5. [4.1e] $\dfrac{4x^{10}}{3}$

6. [6.1c] $\dfrac{2(t-3)}{2t-1}$   7. [6.9a] $\dfrac{(x+2)^2}{x^2}$   8. [6.1c] $\dfrac{a+4}{a-4}$

9. [1.3a] $\dfrac{17}{42}$   10. [6.4a] $\dfrac{x^2 + 4xy + y^2}{x^2y^2}$

11. [6.4a] $\dfrac{3z - 2}{z^2 - 1}$   12. [4.4a] $2x^4 + 8x^3 - 2x + 9$

13. [1.4a] 2.33   14. [4.7e] $-xy$   15. [6.5a] $\dfrac{1}{x-3}$

16. [6.5a] $\dfrac{2x^2 - 14x - 16}{(x+4)(x-5)^2}$   17. [1.5a] −1.3

18. [4.5b] $6x^4 + 12x^3 - 15x^2$   19. [4.6b] $9t^2 - \frac{1}{4}$
20. [4.6c] $4p^2 - 4pq + q^2$   21. [4.6a] $3x^2 - 7x - 20$

22. [4.6b] $4x^4 - 1$   23. [6.1d] $\dfrac{2(t-1)}{t}$

24. [6.1d] $-\dfrac{2(a+1)}{a}$   25. [4.8b] $3x^2 - 4x + 5$

26. [1.6c] $-\dfrac{9}{20}$   27. [6.2b] $\dfrac{x-2}{2x(x-3)}$   28. [6.2b] $-\dfrac{12}{x}$

29. [5.6a] $(2x + 3)(2x - 3)(x + 3)$
30. [5.2a] $(x + 8)(x - 1)$
31. [5.3a], [5.4a] $(3x + 1)(x - 5)$   32. [5.5b] $(4y + 5x)^2$
33. [5.2a] $3x(x + 5)(x + 3)$   34. [5.5d] $2(x + 1)(x - 1)$
35. [5.5b] $(x - 14)^2$   36. [5.1b, c] $2(y^2 + 3)(2y + 5)$
37. [2.3c] −7   38. [5.7a] 0, $-\frac{4}{3}$   39. [5.7b] 0, 8
40. [5.7b] 4   41. [2.7e] $\{x \mid x \geq -9\}$   42. [5.7b] 3, −3

**43.** [2.3b] $-\frac{11}{7}$ **44.** [6.6a] 3, $-3$ **45.** [6.6a] $-\frac{1}{11}$
**46.** [5.7a] 0, $\frac{1}{10}$ **47.** [6.6a] No solution **48.** [6.6a] $\frac{5}{7}$
**49.** [6.8a] $z = \dfrac{xy}{x+y}$ **50.** [6.8a] $N = \dfrac{DT}{3}$
**51.** [3.3a]

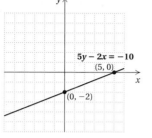

**52.** [2.4a] 32, 33, 34 **53.** [6.7a] 12 km/h, 10 km/h
**54.** [6.7a] $\frac{30}{11}$ hr **55.** [2.4a] 34 and 35
**56.** [5.8a] 16 and 17 **57.** [6.7b] **(a)** About 57; **(b)** no
**58.** [5.8a] 7 **59.** [5.8a] 9 and 11 **60.** [5.7b] 4, $-\frac{2}{3}$
**61.** [5.7b], [6.6a] 2, $-1$ **62.** [5.7b], [6.6a], [6.9a] 1, $-4$
**63.** [4.1d, e], [5.7b] 5, $-5$
**64.** [6.2a], [6.4a] $\dfrac{-x^2 - x + 6}{2x^2 + x + 5}$ **65.** [4.2d] $5 \times 10^7$

# Chapter 7

## Pretest: Chapter 7, p. 456

**1.** [7.1b] 4 **2.** [7.1b] 0 **3.** [7.2a] Slope: $\frac{1}{3}$;
$y$-intercept: $\left(0, -\frac{7}{3}\right)$ **4.** [7.1a] Undefined
**5.** [7.2c] $y = x - 4$ **6.** [7.2b] $y = 4x + 7$
**7.** [7.5a] $y = \dfrac{5}{2}x$ **8.** [7.5c] $y = \dfrac{40}{x}$
**9.** [7.4b] **10.** [7.4b]

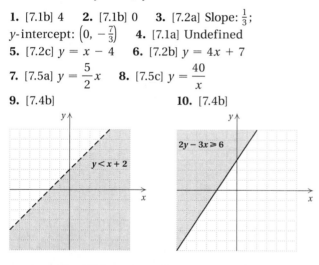

**11.** [7.3a] Parallel **12.** [7.3b] Perpendicular
**13.** [7.3b] Perpendicular **14.** [7.4a] Yes
**15.** [7.2a, c] **(a)** $y = 0.47x + 0.49$; **(b)** 0.47 billion dollars
per year; **(c)** 5.19 billion dollars

## Margin Exercises, Section 7.1, pp. 458–461

**1.** $\frac{2}{5}$ **2.** $-\frac{5}{3}$

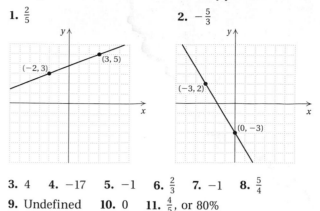

**3.** 4 **4.** $-17$ **5.** $-1$ **6.** $\frac{2}{3}$ **7.** $-1$ **8.** $\frac{5}{4}$
**9.** Undefined **10.** 0 **11.** $\frac{4}{5}$, or 80%

## Calculator Spotlight, p. 459

**1.** This line will pass through the origin and slant up
from left to right. This line will be steeper than
$y = 10x$. **2.** This line will pass through the origin and
slant up from left to right. This line will be less steep
than $y = \frac{5}{32}x$.

## Calculator Spotlight, p. 460

**1.** This line will pass through the origin and slant
down from left to right. This line will be steeper than
$y = -10x$. **2.** This line will pass through the origin
and slant down from left to right. This line will be less
steep than $y = -\frac{5}{32}x$.

## Exercise Set 7.1, p. 463

**1.** $-\frac{3}{7}$ **3.** 0
**5.** $-\frac{4}{5}$ **7.** 3

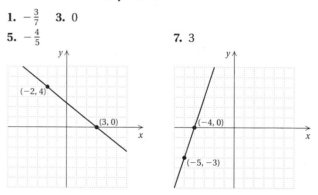

**9.** $\frac{2}{3}$ **11.** Undefined **13.** $-10$ **15.** 3.78 **17.** 3
**19.** $-\frac{1}{5}$ **21.** $-\frac{3}{2}$ **23.** $\frac{5}{7}$ **25.** $-2.74$ **27.** 3 **29.** $\frac{5}{4}$
**31.** 0 **33.** $\frac{12}{41}$ **35.** $\frac{28}{129}$ **37.** 15 cal per minute
**39.** 29.4% **41.** $\frac{44}{7}$ **42.** 10 **43.** $-5$
**44.** 391 **45.** $\frac{68}{109}$ **46.** 7412 **47.** $\frac{1}{2}$ **48.** 9 **49.** ◆

## Margin Exercises, Section 7.2, pp. 465–466

**1.** Slope: 5; $y$-intercept: (0, 0) **2.** Slope: $-\frac{3}{2}$;
$y$-intercept: $(0, -6)$ **3.** Slope: $-\frac{3}{4}$; $y$-intercept: $\left(0, \frac{15}{4}\right)$
**4.** Slope: 2; $y$-intercept: $\left(0, -\frac{17}{2}\right)$ **5.** Slope: $-\frac{7}{5}$;
$y$-intercept: $\left(0, -\frac{22}{5}\right)$ **6.** $y = 3.5x - 23$

**7.** $y = 5x - 18$  **8.** $y = -3x - 5$  **9.** $y = 6x - 13$
**10.** $y = -\frac{2}{3}x + \frac{14}{3}$  **11.** $y = x + 2$  **12.** $y = 2x + 4$

## Exercise Set 7.2, p. 467

**1.** Slope: $-4$; $y$-intercept: $(0, -9)$  **3.** Slope: $1.8$;
$y$-intercept: $(0, 0)$  **5.** Slope: $-\frac{8}{7}$; $y$-intercept: $(0, -3)$
**7.** Slope: $\frac{4}{9}$; $y$-intercept: $\left(0, -\frac{7}{9}\right)$  **9.** Slope: $-\frac{3}{2}$;
$y$-intercept: $\left(0, -\frac{1}{2}\right)$  **11.** Slope: $0$; $y$-intercept: $(0, -17)$
**13.** $y = -7x - 13$  **15.** $y = 1.01x - 2.6$
**17.** $y = -2x - 6$  **19.** $y = \frac{3}{4}x + \frac{5}{2}$  **21.** $y = x - 8$
**23.** $y = -3x + 3$  **25.** $y = x + 4$  **27.** $y = -\frac{1}{2}x + 4$
**29.** $y = -\frac{3}{2}x + \frac{13}{2}$  **31.** $y = -4x - 11$
**33.** **(a)** $T = -0.75a + 165$; **(b)** $-0.75$ heart beats per
minute per year; **(c)** $127.5$ heart beats per minute
**35.** $0, -3$  **36.** $7, -7$  **37.** $3, -2$  **38.** $-5, 1$
**39.** $\frac{3}{2}, -7$  **40.** $-\frac{6}{5}, 4$  **41.** $-7, 2$  **42.** $\frac{2}{3}, -2$  **43.** $\frac{53}{7}$
**44.** $\frac{3}{8}$  **45.** $\frac{24}{19}$  **46.** $\frac{125}{7}$  **47.** ◈  **49.** $y = 3x - 9$
**51.** $y = \frac{3}{2}x - 2$

## Margin Exercises, Section 7.3, pp. 469–470

**1.** No  **2.** Yes  **3.** Yes  **4.** No

## Exercise Set 7.3, p. 471

**1.** Yes  **3.** No  **5.** No  **7.** No  **9.** Yes  **11.** Yes
**13.** No  **15.** Yes  **17.** Yes  **19.** Yes  **21.** No
**23.** Yes  **25.** In 7 hr  **26.** 130 km/h; 140 km/h
**27.** 4  **28.** $-6$  **29.** $\frac{30}{13}$  **30.** $-11$  **31.** $\frac{36}{11}$  **32.** 1
**33.** ◈  **35.–45.** Left to the student  **47.** $y = 3x + 6$
**49.** $y = -3x + 2$  **51.** $y = \frac{1}{2}x + 1$  **53.** $k = 16$
**55.** A: $y = \frac{4}{3}x - \frac{7}{3}$; B: $y = -\frac{3}{4}x - \frac{1}{4}$

## Margin Exercises, Section 7.4, pp. 473–476

**1.** No  **2.** No
**3.**

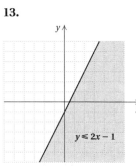

**4.**

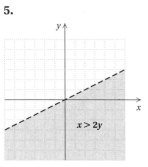

**5.**

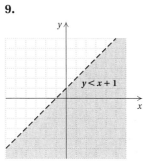

**6.**

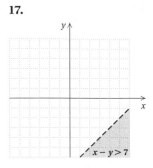

**7.**

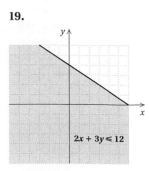

**8.**

(figure: $y \leq 4$)

## Exercise Set 7.4, p. 477

**1.** No  **3.** Yes
**5.**

(figure: $x > 2y$)

**7.**

(figure: $y \leq x - 3$)

**9.**

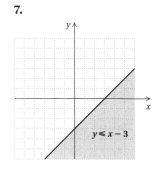

**11.**

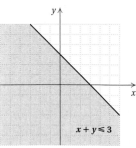

**13.**

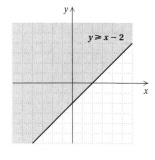

**15.**

**17.**

**19.**

**21.**

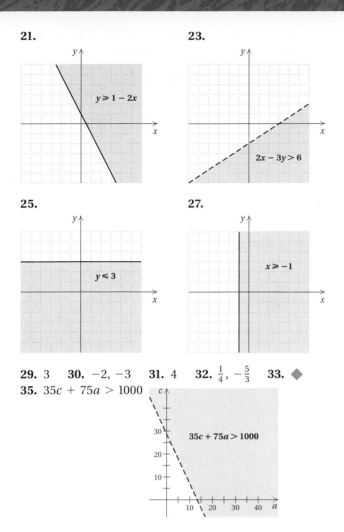

**23.**

**25.**

**27.**

**29.** 3   **30.** −2, −3   **31.** 4   **32.** $\frac{1}{4}, -\frac{5}{3}$   **33.** ◈
**35.** $35c + 75a > 1000$

## Margin Exercises, Section 7.5, pp. 480–482

**1.** $y = 7x$   **2.** $y = \frac{5}{8}x$   **3.** \$0.4667; \$0.0194
**4.** 174.24 lb   **5.** $y = \frac{63}{x}$   **6.** $y = \frac{900}{x}$   **7.** 8 hr
**8.** $7\frac{1}{2}$ hr

## Exercise Set 7.5, p. 483

**1.** $y = 4x$   **3.** $y = \frac{8}{5}x$   **5.** $y = 3.6x$   **7.** $y = \frac{25}{3}x$
**9.** \$196   **11.** \$100   **13.** $22\frac{6}{7}$   **15.** $36.\overline{6}$ lb
**17.** 16,000,000   **19.** $y = \frac{75}{x}$   **21.** $y = \frac{80}{x}$   **23.** $y = \frac{1}{x}$
**25.** $y = \frac{2100}{x}$   **27.** $y = \frac{0.06}{x}$   **29.** (a) Direct; (b) $69\frac{3}{8}$
**31.** (a) Inverse; (b) $4\frac{1}{2}$ hr   **33.** 10 gal   **35.** 32 amperes
**37.** 640   **39.** 8.25 ft   **41.** $\frac{8}{5}$   **42.** 11   **43.** 9, 16
**44.** −9, −12   **45.** $\frac{4}{7}, \frac{2}{5}$   **46.** $-\frac{1}{7}, \frac{3}{2}$   **47.** $\frac{1}{3}$   **48.** $\frac{47}{20}$
**49.** ◈   **51.** ◈   **53.** The $y$-values become larger.
**55.** $P^2 = kt$   **57.** $P = kV^3$

## Summary and Review: Chapter 7, p. 487

**1.** −1   **2.** Undefined

**3.** $-\frac{3}{4}$

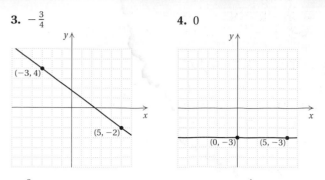

**4.** 0

**5.** $\frac{3}{2}$   **6.** 2   **7.** 0   **8.** Undefined   **9.** $-\frac{4}{3}$
**10.** Slope: −9; $y$-intercept: (0, 46)   **11.** Slope: −1;
$y$-intercept: (0, 9)   **12.** Slope: $\frac{3}{5}$; $y$-intercept: $\left(0, -\frac{4}{5}\right)$
**13.** $y = -2.8x + 19$   **14.** $y = \frac{5}{8}x - \frac{7}{8}$   **15.** $y = 3x - 1$
**16.** $y = \frac{2}{3}x - \frac{11}{3}$   **17.** $y = -2x - 4$   **18.** $y = x + 2$
**19.** $y = \frac{1}{2}x - 1$   **20.** (a) $A = 0.233x + 5.87$; (b) 0.233;
(c) 8.2 yr   **21.** Parallel   **22.** Perpendicular
**23.** Parallel   **24.** Neither   **25.** No   **26.** No
**27.** Yes

**28.**

**29.**

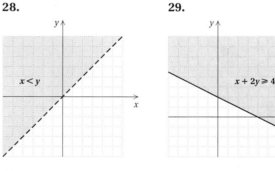

**30.**

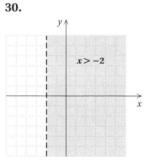

**31.** $y = 3x$   **32.** $y = \frac{1}{2}x$   **33.** $y = \frac{4}{5}x$   **34.** $y = \frac{30}{x}$
**35.** $y = \frac{1}{x}$   **36.** $y = \frac{0.65}{x}$   **37.** \$288.75   **38.** 1 hr
**39.** $3\frac{1}{3}$ hr   **40.** 52   **41.** −4   **42.** 5, −11
**43.** ◈ The concept of slope is useful in describing how a line slants. A line with positive slope slants up from left to right. A line with negative slope slants down from left to right. The larger the absolute value of the slope, the steeper the slant.

**44.**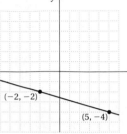

The graph of $x < 1$ on a number line consists of the points in the set $\{x \mid x < 1\}$. The graph of $x < 1$ on a plane consists of the points, or ordered pairs, in the set $\{(x, y) \mid x + 0 \cdot y < 1\}$. This is the set of ordered pairs with first coordinate less than 1.
**45.** $-\frac{1}{2}, \frac{1}{2}, 2, -2$

## Test: Chapter 7, p. 489

**1.** [7.1a] $-2$  **2.** [7.1a] 0
**3.** [7.1a] $-\frac{2}{7}$,

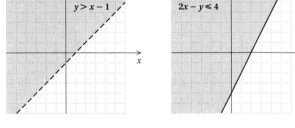

**4.** [7.1a] Undefined  **5.** [7.1a] $\frac{7}{12}$  **6.** [7.1b] 0
**7.** [7.1b] Undefined  **8.** [7.2a] Slope: 2;
$y$-intercept: $\left(0, -\frac{1}{4}\right)$  **9.** [7.2a] Slope: $\frac{4}{3}$;
$y$-intercept: $(0, -2)$  **10.** [7.2a] $y = 1.8x - 7$
**11.** [7.2a] $y = -\frac{3}{8}x - \frac{1}{8}$  **12.** [7.2b] $y = x + 2$
**13.** [7.2b] $y = -3x - 6$  **14.** [7.2c] $y = -3x + 4$
**15.** [7.2c] $y = \frac{1}{4}x - 2$  **16.** [7.2c] **(a)** $M = 102x + 2313$;
**(b)** \$102 million per year; **(c)** \$3129 million
**17.** [7.3a, b] Parallel  **18.** [7.3a, b] Neither
**19.** [7.3a, b] Perpendicular  **20.** [7.4a] No
**21.** [7.4a] Yes
**22.** [7.4b]  **23.** [7.4b]

**24.** [7.5a] $y = 2x$  **25.** [7.5a] $y = 0.5x$
**26.** [7.5c] $y = \dfrac{18}{x}$  **27.** [7.5c] $y = \dfrac{22}{x}$
**28.** [7.5b] 240 km  **29.** [7.5d] $1\frac{1}{5}$ hr
**30.** [6.7a] Freight: 90 mph; passenger: 105 mph
**31.** [1.8d] 1  **32.** [6.6a] 10  **33.** [5.7b] $-7, 4$
**34.** [7.3b] $k = 3$  **35.** [7.2b] $y = \frac{2}{3}x + \frac{11}{3}$

## Cumulative Review: Chapters 1–7, p. 491

**1.** [1.2e] 3.5  **2.** [4.3d] 1, $-2$, 1, $-1$  **3.** [4.3g] 3, 2, 1,
0; 3  **4.** [4.3i] None of these  **5.** [4.3e] $2x^3 - 3x^2 - 2$

**6.** [1.8c] $\dfrac{3}{8}x + 1$  **7.** [4.1e], [4.2a, b] $\dfrac{9}{4x^8}$
**8.** [6.9a] $\dfrac{8x - 12}{17x}$  **9.** [4.7e] $-2xy^2 - 4x^2y^2 + xy^3$
**10.** [4.4a] $2x^5 + 6x^4 + 2x^3 - 10x^2 + 3x - 9$
**11.** [6.1d] $\dfrac{2}{3(y + 2)}$  **12.** [6.2b] 2  **13.** [6.4a] $x + 4$
**14.** [6.5a] $\dfrac{2x - 6}{(x + 2)(x - 2)}$  **15.** [4.6a] $a^2 - 9$
**16.** [4.5d] $2x^5 + x^3 - 6x^2 - x + 3$  **17.** [4.6b] $4x^6 - 1$
**18.** [4.6c] $36x^2 - 60x + 25$
**19.** [4.5b] $12x^4 + 16x^3 + 4x^2$
**20.** [4.6a] $6x^7 - 12x^5 + 9x^2 - 18$  **21.** [2.3c] 3
**22.** [5.7b] $\frac{1}{2}, -4$  **23.** [5.7b] $-5, 4$
**24.** [2.7e] $\{x \mid x \geq -26\}$  **25.** [5.7a] 0, 4
**26.** [5.7b] 0, 10  **27.** [5.7b] 20, $-20$
**28.** [2.6a] $a = \dfrac{t}{x + y}$  **29.** [6.6a] 2
**30.** [6.6a] No solution  **31.** [1.7d] $-2(3 + x + 6y)$
**32.** [5.2a] $(x - 4)(x - 6)$  **33.** [5.5d] $2(x + 3)(x - 3)$
**34.** [5.1c] $(m^3 - 3)(m + 2)$  **35.** [5.5b] $(4x + 5)^2$
**36.** [5.3a], [5.4a] $(2x + 1)(4x + 3)$  **37.** [5.8a] 4 or $-5$
**38.** [7.5b] \$78  **39.** [2.5a] \$2500  **40.** [6.7a] 35 mph,
25 mph  **41.** [5.8a] 14 ft  **42.** [6.7a] 35, 28
**43.** [3.2b]  **44.** [3.3a]

**45.** [3.3b]  **46.** [7.4b]

**47.** [7.4b]

**48.** [7.5a] $y = \frac{2}{3}x$   **49.** [7.5c] $y = \dfrac{10}{x}$
**50.** [7.1a] Undefined   **51.** [7.1a] $-\frac{3}{7}$
**52.** [7.2a] Slope: $\frac{4}{3}$; $y$-intercept: $(0, -2)$
**53.** [7.2b] $y = -4x + 5$   **54.** [7.2c] $y = \frac{1}{6}x - \frac{17}{6}$
**55.** [7.3a, b] Neither   **56.** [7.3a, b] Parallel
**57.** [4.4c], [4.6a] 12   **58.** [4.6b, c] $16y^6 - y^4 + 6y^2 - 9$
**59.** [5.5d] $2(a^{16} + 81b^{20})(a^8 + 9b^{10})(a^4 + 3b^5)(a^4 - 3b^5)$
**60.** [5.7a] 4, −7, 12   **61.** [7.2b], [7.3a] $y = \frac{2}{3}x$
**62.** [6.1a], [6.9a] 0, 3, $\frac{5}{2}$

# Chapter 8

## Pretest: Chapter 8, p. 494

**1.** [8.1a] Yes   **2.** [8.1b] No solution   **3.** [8.2a] (5, 2)
**4.** [8.2b] (2, −1)   **5.** [8.3a] $\left(\frac{3}{4}, \frac{1}{2}\right)$   **6.** [8.3b] (−5, −2)
**7.** [8.3b] (10, 8)   **8.** [8.3c] 50 and 24   **9.** [8.2c] 25° and 65°   **10.** [8.5a] 8 hr after the second train leaves

## Margin Exercises, Section 8.1, pp. 496–498

**1.** Yes   **2.** No   **3.** (2, −3)   **4.** (−4, 3)
**5.** No solution   **6.** Infinite number of solutions
**7. (a)** 3; **(b)** 3; **(c)** same   **8. (a)** 3; **(b)** same

## Exercise Set 8.1, p. 499

**1.** Yes   **3.** No   **5.** Yes   **7.** Yes   **9.** Yes   **11.** (4, 2)
**13.** (4, 3)   **15.** (−3, −3)   **17.** No solution   **19.** (2, 2)
**21.** $\left(\frac{1}{2}, 1\right)$   **23.** Infinite number of solutions
**25.** (5, −3)   **27.** $\dfrac{108}{x^{13}}$   **28.** $3x^3$   **29.** $\dfrac{2x^2 - 1}{x^2(x + 1)}$
**30.** $\dfrac{-4}{x - 2}$   **31.** $\dfrac{9x + 12}{(x - 4)(x + 4)}$   **32.** $\dfrac{2x + 5}{x + 3}$
**33.** Trinomial   **34.** Binomial   **35.** Monomial
**36.** None of these   **37.** ◈   **39.** $A = 2, B = 2$
**41.** $x + 2y = 2, x - y = 8$   **43.** (2, 1)   **45.** (3, 2)
**47.** (−6, −2)   **49.** Infinite number of solutions

## Margin Exercises, Section 8.2, pp. 502–504

**1.** (3, 2)   **2.** (3, −1)   **3.** $\left(\frac{24}{5}, -\frac{8}{5}\right)$   **4.** Length: 84 ft; width: 50 ft

## Calculator Spotlight, p. 503

**1.** (4.8, −1.6)   **2.** (0.667, 0.429)

## Exercise Set 8.2, p. 505

**1.** (1, 9)   **3.** (2, −4)   **5.** (4, 3)   **7.** (−2, 1)
**9.** (2, −4)   **11.** $\left(\frac{17}{3}, \frac{16}{3}\right)$   **13.** $\left(\frac{25}{8}, -\frac{11}{4}\right)$   **15.** (−3, 0)
**17.** (6, 3)   **19.** 16 and 21   **21.** 12 and 40
**23.** 20 and 8   **25.** Length: 380 mi; width: 270 mi
**27.** Length: $3\frac{1}{2}$ in; width: $1\frac{3}{4}$ in.

**29.**   **30.**

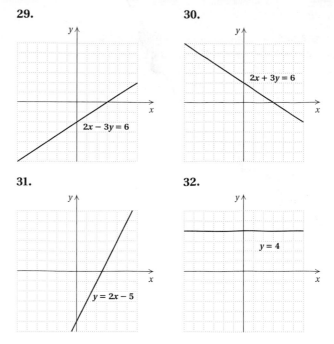

**31.**   **32.**

**33.** $(3x - 2)(2x - 3)$   **34.** $(4p + 3)(p - 1)$   **35.** Not factorable   **36.** $(3a - 5)(3a + 5)$   **37.** ◈
**39.** $(5.\overline{6}, 0.\overline{6})$   **41.** (4.38, 4.33)

## Margin Exercises, Section 8.3, pp. 507–512

**1.** (3, 2)   **2.** (1, −2)   **3.** (1, 4)   **4.** (−8, 3)   **5.** (1, 1)
**6.** (−2, −1)   **7.** $\left(\frac{17}{13}, -\frac{7}{13}\right)$   **8.** No solution
**9.** Infinite number of solutions   **10.** (1, −1)
**11.** (2, −2)   **12.** 75 mi

## Calculator Spotlight, p. 510

**1.** You get equations of two lines with the same slope but different $y$-intercepts. The lines are parallel.
**2.** You get equivalent equations. The lines are the same.

## Exercise Set 8.3, p. 513

**1.** (6, −1)   **3.** (3, 5)   **5.** (2, 5)   **7.** $\left(-\frac{1}{2}, 3\right)$
**9.** $\left(-1, \frac{1}{5}\right)$   **11.** No solution   **13.** (−1, −6)   **15.** (3, 1)
**17.** (8, 3)   **19.** (4, 3)   **21.** (1, −1)   **23.** (−3, −1)
**25.** (3, 2)   **27.** (50, 18)   **29.** Infinite number of solutions   **31.** (2, −1)   **33.** $\left(\frac{231}{202}, \frac{117}{202}\right)$   **35.** (−38, −22)
**37.** 200 mi   **39.** 50° and 130°   **41.** 62° and 28°
**43.** Hay: 415 hectares; oats: 235 hectares   **45.** $\dfrac{1}{x^7}$
**46.** $x^3$   **47.** $\dfrac{1}{x^3}$   **48.** $x^7$   **49.** $x^3$   **50.** $x^7$   **51.** $\dfrac{a^7}{b^9}$
**52.** $\dfrac{b^3}{a^3}$   **53.** $\dfrac{x - 3}{x + 2}$   **54.** $\dfrac{x + 5}{x - 5}$   **55.** $\dfrac{-x^2 - 7x + 23}{(x + 3)(x - 4)}$
**56.** $\dfrac{-2x + 10}{(x + 1)(x - 1)}$   **57.** ◈   **59.–67.** See answers for odd-numbered exercises 1–9.   **69.–77.** See answers for odd-numbered exercises 21–30.   **79.** Will is 6; his father is 30.   **81.** (5, 2)   **83.** (0, −1)   **85.** 12 rabbits and 23 pheasants

## Margin Exercises, Section 8.4, pp. 517–523

**1.** Hamburger: $1.99; chicken: $1.70  **2.** Sarah: 47; Malcolm: 21  **3.** 125 adults and 41 children
**4.** 22.5 L of 50%; 7.5 L of 70%  **5.** 30 lb of A; 20 lb of B
**6.** 7 quarters, 13 dimes

## Exercise Set 8.4, p. 525

**1.** Two-point shots: 14; one-point shots: 8
**3.** Kuyatts': 32 yr; Marconis': 16 yr  **5.** Randy is 24;
Mandy is 6.  **7.** Brazilian: 200 lb; Turkish: 100 lb
**9.** 70 dimes; 33 quarters  **11.** One soda: $1.45; one
slice of pizza: $2.25  **13.** 128 cardholders;
75 non-cardholders  **15.** Upper Box: 17; Lower
Reserved: 12  **17.** 40 L of A; 60 L of B  **19.** Hay: 10 lb;
grain: 5 lb  **21.** 39 lb of A; 36 lb of B  **23.** 12 of A,
4 of B; 180  **25.** Large type: $6\frac{17}{25}$; small type: $5\frac{8}{25}$
**27.** $(5x + 9)(5x - 9)$  **28.** $(6 - a)(6 + a)$
**29.** $4(x^2 + 25)$  **30.** $4(x + 5)(x - 5)$
**31.**                                **32.**

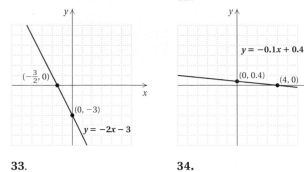

**33.**                                **34.**

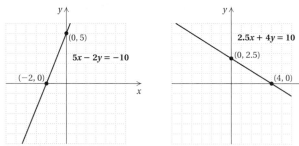

**35.** ◈  **37.** 12 gal of 87; 6 gal of 93  **39.** $4\frac{4}{7}$ L
**41.** 43.75 L  **43.** 11 in. by 14 in.  **45.** $25,000 at 6%;
$29,000 at 6.5%

## Margin Exercises, Section 8.5, pp. 530–532

**1.** 324 mi  **2.** 3 hr  **3.** 168 km  **4.** 275 km/h

## Exercise Set 8.5, p. 533

**1.**

| Speed | Time |
|-------|------|
| 30 | $t$ |
| 46 | $t$ |

4.5 hr

**3.**

| Speed | Time | |
|-------|------|---|
| 72 | $t + 3$ | → $d = 72(t + 3)$ |
| 120 | $t$ | → $d = 120t$ |

$7\frac{1}{2}$ hr after the first train leaves, or $4\frac{1}{2}$ hr after the second train leaves

**5.**

| Speed | Time | |
|-------|------|---|
| $r + 6$ | 4 | → $d = (r + 6)4$ |
| $r - 6$ | 10 | → $d = (r - 6)10$ |

14 km/h
**7.** 384 km  **9. (a)** 24 mph; **(b)** 90 mi  **11.** $1\frac{23}{43}$ min
after the toddler starts running, or $\frac{23}{43}$ min after the
mother starts running  **13.** 15 mi  **15.** $\frac{x}{3}$  **16.** $\frac{x^5 y^3}{2}$
**17.** $\frac{a + 3}{2}$  **18.** $\frac{x - 2}{4}$  **19.** 2  **20.** $\frac{1}{x^2 + 1}$  **21.** $\frac{x + 2}{x + 3}$
**22.** $\frac{3(x + 4)}{x - 1}$  **23.** $\frac{x + 3}{x + 2}$  **24.** $\frac{x^2 + 25}{x^2 - 25}$  **25.** $\frac{x + 2}{x - 1}$
**26.** $\frac{x^2 + 2}{2x^2 + 1}$  **27.** ◈  **29.** Approximately 3603 mi
**31.** $5\frac{1}{3}$ mi

## Summary and Review: Chapter 8, p. 535

**1.** No  **2.** Yes  **3.** Yes  **4.** No  **5.** (6, −2)
**6.** (6, 2)  **7.** (0, 5)  **8.** No solution  **9.** (0, 5)
**10.** (−3, 9)  **11.** (3, −1)  **12.** (1, 4)  **13.** (−2, 4)
**14.** (1, −2)  **15.** (3, 1)  **16.** (1, 4)  **17.** (5, −3)
**18.** (−2, 4)  **19.** (−2, −6)  **20.** (3, 2)  **21.** (2, −4)
**22.** Infinite number of solutions  **23.** (−4, 1)
**24.** 10 and −2  **25.** 12 and 15  **26.** Length: 37.5 cm;
width: 10.5 cm  **27.** Orchestra: 297; balcony: 211
**28.** 40 L of each  **29.** Jeff is 27; his son is 9.
**30.** Asian: 4800 kg; African: 7200 kg
**31.** Peanuts: $6\frac{2}{3}$ lb; fancy nuts: $3\frac{1}{3}$ lb  **32.** 135 km/h
**33.** 412.5 mi  **34.** $t^8$  **35.** $\frac{1}{t^{18}}$
**36.** $\frac{-4x + 3}{(x - 2)(x - 3)(x + 3)}$  **37.** $\frac{x + 2}{x + 10}$
**38.**

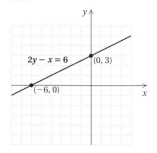

**39.** ◈ The methods are summarized in the table in
Section 8.3.

**40.** ◈ The equations have the same slope but different $y$-intercepts, so they represent parallel lines. Thus the system of equations has no solution.
**41.** \$960   **42.** $C = 1, D = 3$   **43.** $(2, 0)$
**44.** $x + y = 5, -2x + 3y = 0$   **45.** $x + y = 4,$
$x + y = -3$

## Test: Chapter 8, p. 537

**1.** [8.1a] No   **2.** [8.1b] $(2, -1)$   **3.** [8.2a] $(8, -2)$
**4.** [8.2b] $(-1, 3)$   **5.** [8.2a] No solution
**6.** [8.3a] $(1, -5)$   **7.** [8.3b] $(12, -6)$   **8.** [8.3b] $(0, 1)$
**9.** [8.3b] $(5, 1)$   **10.** [8.2c] Length: 2108.5 yd;
width: 2024.5 yd   **11.** [8.2c] 20 and 8
**12.** [8.5a] 40 km/h   **13.** [8.4a] 40 L of A; 20 L of B
**14.** [6.5a] $\dfrac{-x^2 + x + 17}{(x - 4)(x + 4)(x + 1)}$
**15.** [3.3a]

**16.** [4.1d, f] $\dfrac{10x^4}{y^2}$   **17.** [4.1e, f] $\dfrac{a^{10}}{b^6}$   **18.** [6.1c] $\dfrac{x + 10}{2(x + 2)}$
**19.** [8.1a] $C = -\frac{19}{2}; D = \frac{14}{3}$   **20.** [8.4a] 5
**21.** [7.2c], [8.1b] $x - 5y = -17, 3x + 5y = 9$
**22.** [7.2c], [8.1b] $x = 3, y = -2$

## Cumulative Review: Chapters 1–8, p. 539

**1.** [1.8d] $-6.8$   **2.** [4.2d] $3.12 \times 10^{-2}$   **3.** [1.6c] $-\dfrac{3}{4}$
**4.** [4.1d, e] $\dfrac{1}{8}$   **5.** [6.1c] $\dfrac{x + 3}{2x - 1}$   **6.** [6.1c] $\dfrac{t - 4}{t + 4}$
**7.** [6.9a] $\dfrac{x^2(x + 1)}{x + 4}$   **8.** [4.6a] $2 - 10x^2 + 12x^4$
**9.** [4.6c] $4a^4b^2 - 20a^3b^3 + 25a^2b^4$
**10.** [4.6b] $9x^4 - 16y^2$   **11.** [4.5b] $-2x^3 + 4x^4 - 6x^5$
**12.** [4.5d] $8x^3 + 1$   **13.** [4.6b] $64 - \frac{1}{9}x^2$
**14.** [4.4c] $-y^3 - 2y^2 - 2y + 7$
**15.** [4.8b] $x^2 - x - 1 + \dfrac{-2}{2x - 1}$   **16.** [6.4a] $\dfrac{-5x - 28}{5x - 25}$
**17.** [6.5a] $\dfrac{4x - 1}{x - 2}$   **18.** [6.1d] $\dfrac{y}{(y - 1)^2}$
**19.** [6.2b] $\dfrac{3(x + 1)}{2x}$   **20.** [5.1b] $3x^2(2x^3 - 12x + 3)$
**21.** [5.5d] $(4y^2 + 9)(2y + 3)(2y - 3)$
**22.** [5.3a], [5.4a] $(3x - 2)(x + 4)$   **23.** [5.5b] $(2x^2 - 3y)^2$
**24.** [5.1b], [5.2a] $3m(m + 5)(m - 3)$
**25.** [5.6a] $(x + 1)^2(x - 1)$   **26.** [2.3c] $-9$
**27.** [5.7a] $0, \frac{5}{2}$   **28.** [2.7e] $\{x \mid x \le 20\}$   **29.** [2.3c] 0.3
**30.** [5.7b] $13, -13$   **31.** [5.7b] $\frac{5}{3}, 3$   **32.** [6.6a] $-1$

**33.** [6.6a] No solution   **34.** [8.2a] $(3, -3)$
**35.** [8.3b] $(-2, 2)$   **36.** [8.3b] Infinite number of solutions   **37.** [6.8a] $x = y - 1$   **38.** [2.4a] 234
**39.** [6.7a] $5\frac{5}{8}$ hr   **40.** [5.8a] 6 m, 10 m   **41.** [6.7b] 204
**42.** [5.8a] 6 cm, 9 cm   **43.** [7.5d] 72 ft; 360
**44.** [8.5a] 10 hr   **45.** [8.4a] 2.5 L of A; 7.5 L of B
**46.** [7.5a] $y = 0.2x$   **47.** [7.1a] 0   **48.** [7.2a] $-\frac{2}{3}$, $(0, 2)$
**49.** [7.2c] $y = -\frac{10}{7}x - \frac{8}{7}$   **50.** [7.2b] $y = 6x - 3$
**51.** [3.3b]   **52.** [3.3a]

**53.** [7.4b]   **54.** [7.4b]

**55.** [8.1b] No solution   **56.** [8.1b] $(1, -1)$
**57.** [8.1a] $A = -4, B = -\frac{7}{5}$   **58.** [2.7e] No solution
**59.** [1.8d], [6.2b], [6.5a] 0   **60.** [7.3b] $-\frac{3}{10}$

## Chapter 9

### Pretest: Chapter 9, p. 542

**1.** [9.1a] $7, -7$   **2.** [9.1d] $3t$   **3.** [9.1e] No
**4.** [9.1e] Yes   **5.** [9.1b] 6.856   **6.** [9.5a] 4   **7.** [9.1f] $2x$
**8.** [9.4a] $12\sqrt{2}$   **9.** [9.4b] $7 - 4\sqrt{3}$   **10.** [9.4b] 1
**11.** [9.2c] $2\sqrt{15}$   **12.** [9.4b] $25 - 4\sqrt{6}$   **13.** [9.3a] $\sqrt{5}$
**14.** [9.3b] $2a^2\sqrt{2}$   **15.** [9.6a] $\sqrt{89} \approx 9.434$
**16.** [9.6b] $\sqrt{193} \approx 13.892$ m   **17.** [9.3c] $\dfrac{\sqrt{5x}}{x}$
**18.** [9.4c] $\dfrac{48 - 8\sqrt{5}}{31}$

### Margin Exercises, Section 9.1, pp. 543–546

**1.** $6, -6$   **2.** $8, -8$   **3.** $11, -11$   **4.** $12, -12$   **5.** 4
**6.** 7   **7.** 10   **8.** 21   **9.** $-7$   **10.** $-13$   **11.** 3.873
**12.** 5.477   **13.** 31.305   **14.** $-25.842$   **15.** 0.816
**16.** $-1.732$   **17.** 28.3 mph, 49.6 mph   **18.** 227
**19.** $45 + x$   **20.** $\dfrac{x}{x + 2}$   **21.** $x^2 + 4$   **22.** Yes
**23.** No   **24.** No   **25.** Yes   **26.** 13   **27.** $|7w|$

**28.** $|xy|$   **29.** $|xy|$   **30.** $|x - 11|$   **31.** $|x + 4|$
**32.** $xy$   **33.** $xy$   **34.** $x - 11$   **35.** $x + 4$   **36.** $5y$
**37.** $\frac{1}{2}t$

## Calculator Spotlight, p. 546

**1.–8.** Use your grapher.   **9.** Not correct   **10.** Correct

## Exercise Set 9.1, p. 547

**1.** $2, -2$   **3.** $3, -3$   **5.** $10, -10$   **7.** $13, -13$
**9.** $16, -16$   **11.** $2$   **13.** $-3$   **15.** $-6$   **17.** $-15$
**19.** $19$   **21.** $2.236$   **23.** $20.785$   **25.** $-18.647$
**27.** $2.779$   **29.** $120$   **31.** (a) $13$; (b) $24$   **33.** $200$
**35.** $a - 4$   **37.** $t^2 + 1$   **39.** $\dfrac{3}{x + 2}$   **41.** No   **43.** Yes
**45.** $c$   **47.** $3x$   **49.** $8p$   **51.** $ab$   **53.** $34d$
**55.** $x + 3$   **57.** $a - 5$   **59.** $2a - 5$   **61.** $10,660$
**62.** $\dfrac{1}{x + 3}$   **63.** $1$   **64.** $\dfrac{(x + 2)(x - 2)}{(x + 1)(x - 1)}$   **65.** ◈
**67.** $1.7, 2.2, 2.6$   **69.** $-5$ and $-6$   **71.** No solution
**73.** $\sqrt{3}$

## Margin Exercises, Section 9.2, pp. 549–552

**1.** (a) $20$; (b) $20$   **2.** $\sqrt{33}$   **3.** $5$   **4.** $\sqrt{x^2 + x}$
**5.** $\sqrt{x^2 - 4}$   **6.** $4\sqrt{2}$   **7.** $x + 7$   **8.** $5x$   **9.** $6m$
**10.** $2\sqrt{23}$   **11.** $x - 10$   **12.** $8t$   **13.** $10a$   **14.** $t^2$
**15.** $t^{10}$   **16.** $h^{23}$   **17.** $x^3\sqrt{x}$   **18.** $2x^5\sqrt{6x}$   **19.** $3\sqrt{2}$
**20.** $10$   **21.** $4x^3y^2$   **22.** $5xy^2\sqrt{2xy}$

## Calculator Spotlight, p. 549

**1.** Correct   **2.** Not correct

## Calculator Spotlight, p. 552

**1.** Not correct   **2.** Not correct

## Exercise Set 9.2, p. 553

**1.** $2\sqrt{3}$   **3.** $5\sqrt{3}$   **5.** $2\sqrt{5}$   **7.** $10\sqrt{6}$   **9.** $9\sqrt{6}$
**11.** $3\sqrt{x}$   **13.** $4\sqrt{3x}$   **15.** $4\sqrt{a}$   **17.** $8y$   **19.** $x\sqrt{13}$
**21.** $2t\sqrt{2}$   **23.** $6\sqrt{5}$   **25.** $12\sqrt{2y}$   **27.** $2x\sqrt{7}$
**29.** $x - 3$   **31.** $\sqrt{2}(2x + 1)$   **33.** $\sqrt{y}(6 + y)$   **35.** $x^3$
**37.** $x^6$   **39.** $x^2\sqrt{x}$   **41.** $t^9\sqrt{t}$   **43.** $(y - 2)^4$
**45.** $2(x + 5)^5$   **47.** $6m\sqrt{m}$   **49.** $2a^2\sqrt{2a}$
**51.** $2p^8\sqrt{26p}$   **53.** $8x^3y\sqrt{7y}$   **55.** $3\sqrt{6}$   **57.** $3\sqrt{10}$
**59.** $6\sqrt{7x}$   **61.** $6\sqrt{xy}$   **63.** $13$   **65.** $5b\sqrt{3}$   **67.** $2t$
**69.** $a\sqrt{bc}$   **71.** $2xy\sqrt{2xy}$   **73.** $18$   **75.** $\sqrt{10x - 5}$
**77.** $x + 2$   **79.** $6xy^3\sqrt{3xy}$   **81.** $10x^2y^3\sqrt{5xy}$
**83.** $(-2, 4)$   **84.** $\left(\frac{1}{8}, \frac{9}{8}\right)$   **85.** $(2, 1)$   **86.** $(10, 3)$
**87.** $360$ ft$^2$   **88.** Two-point: $36$; one-point: $28$
**89.** 80 L of 30%, 120 L of 50%   **90.** Adults: 350;
children: 61   **91.** 10 mph   **93.** ◈
**95.** $\sqrt{x - 2}\sqrt{x + 1}$   **97.** $\sqrt{2x + 3}\sqrt{x - 4}$
**99.** $\sqrt{a + b}\sqrt{a - b}$   **101.** $0.5$   **103.** $3a^3$   **105.** $4y\sqrt{3}$
**107.** $18(x + 1)\sqrt{y(x + 1)}$   **109.** $2x^3\sqrt{5x}$

## Margin Exercises, Section 9.3, pp. 557–560

**1.** $4$   **2.** $5$   **3.** $x\sqrt{6x}$   **4.** $\dfrac{4}{3}$   **5.** $\dfrac{1}{5}$   **6.** $\dfrac{6}{x}$   **7.** $\dfrac{3}{4}$
**8.** $\dfrac{15}{16}$   **9.** $\dfrac{7}{y^5}$   **10.** $\dfrac{\sqrt{15}}{5}$   **11.** $\dfrac{\sqrt{10}}{4}$   **12.** $\dfrac{10\sqrt{3}}{3}$
**13.** $\dfrac{\sqrt{21}}{7}$   **14.** $\dfrac{\sqrt{5r}}{r}$   **15.** $\dfrac{8y\sqrt{7}}{7}$

## Calculator Spotlight, p. 557

**1.** Correct   **2.** Correct   **3.** Not correct

## Exercise Set 9.3, p. 561

**1.** $3$   **3.** $6$   **5.** $\sqrt{5}$   **7.** $\dfrac{1}{5}$   **9.** $\dfrac{2}{5}$   **11.** $2$   **13.** $3y$
**15.** $\dfrac{4}{7}$   **17.** $\dfrac{1}{6}$   **19.** $-\dfrac{4}{9}$   **21.** $\dfrac{8}{17}$   **23.** $\dfrac{13}{14}$   **25.** $\dfrac{5}{x}$
**27.** $\dfrac{3a}{25}$   **29.** $\dfrac{\sqrt{10}}{5}$   **31.** $\dfrac{\sqrt{14}}{4}$   **33.** $\dfrac{\sqrt{3}}{6}$   **35.** $\dfrac{\sqrt{10}}{6}$
**37.** $\dfrac{3\sqrt{5}}{5}$   **39.** $\dfrac{2\sqrt{6}}{3}$   **41.** $\dfrac{\sqrt{3x}}{x}$   **43.** $\dfrac{\sqrt{xy}}{y}$   **45.** $\dfrac{x\sqrt{5}}{10}$
**47.** $\dfrac{\sqrt{14}}{2}$   **49.** $\dfrac{3\sqrt{2}}{4}$   **51.** $\dfrac{\sqrt{6}}{2}$   **53.** $\sqrt{2}$   **55.** $\dfrac{\sqrt{55}}{11}$
**57.** $\dfrac{\sqrt{21}}{6}$   **59.** $\dfrac{\sqrt{6}}{2}$   **61.** $5$   **63.** $\dfrac{\sqrt{3x}}{x}$   **65.** $\dfrac{4y\sqrt{5}}{5}$
**67.** $\dfrac{a\sqrt{2a}}{4}$   **69.** $\dfrac{\sqrt{42x}}{3x}$   **71.** $\dfrac{3\sqrt{6}}{8c}$   **73.** $\dfrac{y\sqrt{xy}}{x}$
**75.** $\dfrac{3n\sqrt{10}}{8}$   **77.** $(4, 2)$   **78.** $(10, 30)$   **79.** No solution
**80.** Infinite number of solutions   **81.** $\left(-\frac{5}{2}, -\frac{9}{2}\right)$
**82.** $\left(\frac{26}{23}, \frac{44}{23}\right)$   **83.** $9x^2 - 49$   **84.** $16a^2 - 25b^2$
**85.** $21x - 9y$   **86.** $14a - 6b$   **87.** ◈   **89.** 1.57 sec;
3.14 sec; 8.88 sec; 11.10 sec   **91.** 1 sec   **93.** $\dfrac{\sqrt{5}}{40}$
**95.** $\dfrac{\sqrt{5x}}{5x^2}$   **97.** $\dfrac{\sqrt{3ab}}{b}$   **99.** $\dfrac{3\sqrt{10}}{100}$   **101.** $\dfrac{y - x}{xy}$

## Margin Exercises, Section 9.4, pp. 565–568

**1.** $12\sqrt{2}$   **2.** $5\sqrt{5}$   **3.** $-12\sqrt{10}$   **4.** $5\sqrt{6}$
**5.** $\sqrt{x + 1}$   **6.** $\dfrac{3}{2}\sqrt{2}$   **7.** $\dfrac{8\sqrt{15}}{15}$   **8.** $\sqrt{15} + \sqrt{6}$
**9.** $4 + 3\sqrt{5} - 4\sqrt{2} - 3\sqrt{10}$   **10.** $2 - a$
**11.** $25 + 10\sqrt{x} + x$   **12.** $2$   **13.** $7 - \sqrt{5}$
**14.** $\sqrt{5} + \sqrt{2}$   **15.** $1 + \sqrt{x}$   **16.** $\dfrac{21 - 3\sqrt{5}}{22}$
**17.** $\dfrac{7 + 2\sqrt{10}}{3}$   **18.** $\dfrac{7 + 7\sqrt{x}}{1 - x}$

## Exercise Set 9.4, p. 569

**1.** $16\sqrt{3}$   **3.** $4\sqrt{5}$   **5.** $13\sqrt{x}$   **7.** $-9\sqrt{d}$   **9.** $25\sqrt{2}$
**11.** $\sqrt{3}$   **13.** $\sqrt{5}$   **15.** $13\sqrt{2}$   **17.** $3\sqrt{3}$   **19.** $2\sqrt{2}$
**21.** $0$   **23.** $(2 + 9x)\sqrt{x}$   **25.** $(3 - 2x)\sqrt{3}$
**27.** $3\sqrt{2x + 2}$   **29.** $(x + 3)\sqrt{x^3 - 1}$

**Answers**

**A-36**

**31.** $(4a^2 + a^2b - 5b)\sqrt{b}$  **33.** $\dfrac{2\sqrt{3}}{3}$  **35.** $\dfrac{13\sqrt{2}}{2}$

**37.** $\dfrac{\sqrt{6}}{6}$  **39.** $\sqrt{15} - \sqrt{3}$  **41.** $10 + 5\sqrt{3} - 2\sqrt{7} - \sqrt{21}$

**43.** $9 - 4\sqrt{5}$  **45.** $-62$  **47.** $1$  **49.** $13 + \sqrt{5}$
**51.** $x - 2\sqrt{xy} + y$  **53.** $-\sqrt{3} - \sqrt{5}$  **55.** $5 - 2\sqrt{6}$

**57.** $\dfrac{4\sqrt{10} - 4}{9}$  **59.** $5 - 2\sqrt{7}$  **61.** $\dfrac{12 - 3\sqrt{x}}{16 - x}$

**63.** $\dfrac{24 + 3\sqrt{x} + 8\sqrt{2} + \sqrt{2x}}{64 - x}$  **65.** $\dfrac{5}{11}$  **66.** $-\dfrac{38}{13}$

**67.** $6, -1$  **68.** $5, 2$  **69.** Jolly Juice: 1.6 L; Real
Squeeze: 6.4 L  **70.** $\frac{1}{3}$ hr  **71.** $-9, -2, -5, -17,$
$-0.678375$  **73.** ◆  **75.** Not equivalent  **77.** Not
correct  **79.** $11\sqrt{3} - 10\sqrt{2}$  **81.** True; $(3\sqrt{x + 2})^2 =$
$(3\sqrt{x + 2})(3\sqrt{x + 2}) = (3 \cdot 3)(\sqrt{x + 2} \cdot \sqrt{x + 2}) =$
$9(x + 2)$

## Margin Exercises, Section 9.5, pp. 573–577

**1.** $\frac{64}{3}$  **2.** $2$  **3.** $\frac{3}{8}$  **4.** $4$  **5.** $1$  **6.** $4$
**7.** Approximately 313 km  **8.** Approximately 16 km
**9.** 196 m

## Calculator Spotlight, p. 575

**1.** $21.3$  **2.** $2$  **3.** $0.375$  **4.** 4; for $x = -1$,
$y_1 = x - 1 = -1 - 1 = -2$ and $y_2 = \sqrt{x + 5} =$
$\sqrt{-1 + 5} = 2$. The $y$-values are not the same; the
graphs do not intersect.

## Exercise Set 9.5, p. 579

**1.** $36$  **3.** $18.49$  **5.** $165$  **7.** $\frac{621}{2}$  **9.** $5$  **11.** $3$
**13.** $\frac{17}{4}$  **15.** No solution  **17.** No solution  **19.** $9$
**21.** $12$  **23.** $1, 5$  **25.** $3$  **27.** $5$  **29.** No solution
**31.** $-\frac{10}{3}$  **33.** $3$  **35.** No solution  **37.** $9$  **39.** $1$
**41.** $8$  **43.** 36 m  **45.** Approximately 21.3 km

**47.** 151 ft; 281 ft  **49.** $12$  **51.** $8$  **53.** $\dfrac{(x + 7)^2}{x - 7}$

**54.** $\dfrac{(x - 2)(x - 5)}{(x - 3)(x - 4)}$  **55.** $\dfrac{a - 5}{2}$  **56.** $\dfrac{x - 3}{x - 2}$

**57.** $61°, 119°$  **58.** $38°, 52°$  **59.** $\dfrac{x^6}{3}$  **60.** $\dfrac{x - 3}{4(x + 3)}$

**61.** ◆  **63.** $2, -2$  **65.** $-\frac{57}{16}$  **67.** $13$  **69.** $3$
**71.** $4.25 = \frac{17}{4}$

## Margin Exercises, Section 9.6, pp. 583–584

**1.** $\sqrt{65} \approx 8.062$  **2.** $\sqrt{75} \approx 8.660$  **3.** $\sqrt{10} \approx 3.162$
**4.** $\sqrt{175} \approx 13.229$  **5.** $\sqrt{325} \approx 18.028$ ft

## Exercise Set 9.6, p. 585

**1.** $17$  **3.** $\sqrt{32} \approx 5.657$  **5.** $12$  **7.** $4$  **9.** $26$
**11.** $12$  **13.** $2$  **15.** $\sqrt{2} \sim 1.414$  **17.** $5$  **19.** $3$
**21.** $\sqrt{211,200,000} \approx 14,533$ ft  **23.** 240 ft
**25.** $\sqrt{18} \approx 4.243$ cm  **27.** $\sqrt{208} \approx 14.422$ ft
**29.** $\left(-\frac{3}{2}, -\frac{1}{16}\right)$  **30.** $\left(\frac{8}{5}, 9\right)$  **31.** $\left(-\frac{9}{19}, \frac{91}{38}\right)$
**32.** $(-10, 1)$  **33.** $-\frac{1}{3}$  **34.** $\frac{5}{8}$  **35.** ◆

**37.** $\sqrt{1525} \approx 39.1$ mi  **39.** $12 - 2\sqrt{6} \approx 7.101$

**41.** $\dfrac{\sqrt{3}}{2} \approx 0.866$

## Summary and Review: Chapter 9, p. 587

**1.** $8, -8$  **2.** $20, -20$  **3.** $6$  **4.** $-13$  **5.** $1.732$
**6.** $9.950$  **7.** $-17.892$  **8.** $0.742$  **9.** $-2.055$
**10.** $394.648$  **11.** $x^2 + 4$  **12.** $5ab^3$  **13.** No
**14.** Yes  **15.** No  **16.** No  **17.** No  **18.** No  **19.** $m$
**20.** $x - 4$  **21.** $\sqrt{21}$  **22.** $\sqrt{x^2 - 9}$  **23.** $-4\sqrt{3}$
**24.** $4t\sqrt{2}$  **25.** $\sqrt{t - 7}\sqrt{t + 7}$  **26.** $x + 8$  **27.** $x^4$
**28.** $m^7\sqrt{m}$  **29.** $2\sqrt{15}$  **30.** $2x\sqrt{10}$  **31.** $5xy\sqrt{2}$
**32.** $10a^2b\sqrt{ab}$  **33.** $\dfrac{5}{8}$  **34.** $\dfrac{2}{3}$  **35.** $\dfrac{7}{t}$  **36.** $\dfrac{\sqrt{2}}{2}$

**37.** $\dfrac{\sqrt{2}}{4}$  **38.** $\dfrac{\sqrt{5y}}{y}$  **39.** $\dfrac{2\sqrt{3}}{3}$  **40.** $\dfrac{\sqrt{15}}{5}$  **41.** $\dfrac{x\sqrt{30}}{6}$

**42.** $8 - 4\sqrt{3}$  **43.** $13\sqrt{5}$  **44.** $\sqrt{5}$  **45.** $\dfrac{\sqrt{2}}{2}$

**46.** $7 + 4\sqrt{3}$  **47.** $1$  **48.** $52$  **49.** No solution
**50.** $0, 3$  **51.** $9$  **52.** $20$  **53.** $\sqrt{3} \approx 1.732$
**54.** About 50,990 ft  **55.** 9 ft
**56.** (a) About 63 mph; (b) 405 ft  **57.** $\left(\frac{22}{17}, -\frac{8}{17}\right)$
**58.** $\dfrac{(x - 5)(x - 7)}{(x + 7)(x + 5)}$  **59.** \$450  **60.** $4300$
**61.** ◆ It is incorrect to take the square roots of the
terms in the numerator individually. That is, $\sqrt{a + b}$
and $\sqrt{a} + \sqrt{b}$ are not equivalent. The following is
correct:
$$\sqrt{\dfrac{9 + 100}{25}} = \dfrac{\sqrt{9 + 100}}{\sqrt{25}} = \dfrac{\sqrt{109}}{5}.$$
**62.** ◆ a) $\sqrt{5x^2} = \sqrt{5}\sqrt{x^2} = \sqrt{5} \cdot |x| = |x|\sqrt{5}$; the given
statement is correct.
b) Let $b = 3$. Then $\sqrt{b^2 - 4} = \sqrt{3^2 - 4} =$
$\sqrt{9 - 4} = \sqrt{5}$, but $b - 2 = 3 - 2 = 1$. The
given statement is false.
c) Let $x = 3$. Then $\sqrt{x^2 + 16} = \sqrt{3^2 + 16} =$
$\sqrt{9 + 16} = \sqrt{25} = 5$, but $x + 4 = 3 + 4 = 7$.
The given statement is false.
**63.** $2$  **64.** $b = \sqrt{A^2 - a^2}$

## Test: Chapter 9, p. 589

**1.** [9.1a] $9, -9$  **2.** [9.1a] $8$  **3.** [9.1a] $-5$
**4.** [9.1b] $10.770$  **5.** [9.1b] $-9.349$  **6.** [9.1b] $4.127$
**7.** [9.1d] $4 - y^3$  **8.** [9.1e] Yes  **9.** [9.1e] No
**10.** [9.1f] $a$  **11.** [9.1f] $6y$  **12.** [9.2c] $\sqrt{30}$
**13.** [9.2c] $\sqrt{x^2 - 64}$  **14.** [9.2a] $3\sqrt{3}$
**15.** [9.2a] $5\sqrt{x - 1}$  **16.** [9.2b] $t^2\sqrt{t}$  **17.** [9.2c] $5\sqrt{2}$
**18.** [9.2c] $3ab^2\sqrt{2}$  **19.** [9.3b] $\dfrac{3}{2}$  **20.** [9.3b] $\dfrac{12}{u}$
**21.** [9.3c] $\dfrac{\sqrt{10}}{5}$  **22.** [9.3c] $\dfrac{\sqrt{2xy}}{y}$  **23.** [9.3a, c] $\dfrac{3\sqrt{6}}{8}$
**24.** [9.3a] $\dfrac{\sqrt{7}}{4y}$  **25.** [9.4a] $-6\sqrt{2}$  **26.** [9.4a] $\dfrac{6\sqrt{5}}{5}$

**27.** [9.4b] $21 - 8\sqrt{5}$    **28.** [9.4b] 11

**29.** [9.4c] $\dfrac{40 + 10\sqrt{5}}{11}$    **30.** [9.6a] $\sqrt{80} \approx 8.944$

**31.** [9.5a] 48    **32.** [9.5a] 2, −2    **33.** [9.5b] −3
**34.** [9.5c] About 5000 m    **35.** [8.4a] 789.25 yd$^2$

**36.** [7.5b] $15{,}686\frac{2}{3}$    **37.** [8.3b] (0, 2)    **38.** [6.2b] $\dfrac{x - 7}{x + 6}$

**39.** [9.1a] $\sqrt{5}$    **40.** [9.2b] $y^{8n}$

## Cumulative Review: Chapters 1–9, p. 591

**1.** [4.3a] −15    **2.** [4.3e] $-4x^3 - \frac{1}{7}x^2 - 2$    **3.** [6.1a] $-\frac{1}{2}$
**4.** [9.1e] No    **5.** [9.4b] 1    **6.** [9.1a] −14    **7.** [9.2c] 15

**8.** [9.4b] $3 - 2\sqrt{2}$    **9.** [9.3a, c] $\dfrac{9\sqrt{10}}{25}$    **10.** [9.4a] $12\sqrt{5}$

**11.** [4.7f] $9x^8 - 4y^{10}$    **12.** [4.6c] $x^4 + 8x^2 + 16$

**13.** [4.6a] $8x^2 - \frac{1}{8}$    **14.** [6.5a] $\dfrac{4x + 2}{2x - 1}$

**15.** [4.4a, c] $-3x^3 + 8x^2 - 5x$    **16.** [6.1d] $\dfrac{2(x - 5)}{3(x - 1)}$

**17.** [6.2b] $\dfrac{(x + 1)(x - 3)}{2(x + 3)}$

**18.** [4.8b] $3x^2 + 4x + 9 + \dfrac{13}{x - 2}$    **19.** [9.2a] $\sqrt{2}(x - 1)$

**20.** [4.1d, f] $\dfrac{1}{x^{12}}$    **21.** [9.3b] $\dfrac{5}{x^4}$    **22.** [6.9a] $2(x - 1)$

**23.** [5.5d] $3(1 + 2x^4)(1 - 2x^4)$
**24.** [5.1b] $4t(3 - t - 12t^3)$
**25.** [5.3a], [5.4a] $2(3x - 2)(x - 4)$
**26.** [5.6a] $(2x + 1)(2x - 1)(x + 1)$
**27.** [5.5b] $(4x^2 - 7)^2$    **28.** [5.2a] $(x + 15)(x - 12)$
**29.** [5.7b] 0, −17    **30.** [2.7e] $\{x \mid x > -6\}$
**31.** [6.6a] $-\frac{12}{5}$    **32.** [5.7b] −5, 6    **33.** [2.7e] $\left\{x \mid x \le -\frac{9}{2}\right\}$
**34.** [5.7b] 9, −9    **35.** [9.5a] 41    **36.** [9.5b] 9
**37.** [2.3b] $\frac{2}{5}$    **38.** [6.6a] $\frac{1}{3}$    **39.** [8.2a] (4, 1)

**40.** [8.3a] (3, −8)    **41.** [6.8a] $R = \dfrac{r^2}{E - r}$

**42.** [2.6a] $p = \dfrac{4A}{r + q}$

**43.** [7.4b]    **44.** [3.3b]

(graph: $3y - 3x > -6$)    (graph: $x = 5$)

**45.** [3.3a]

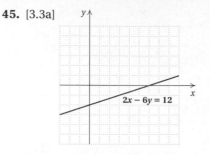
(graph: $2x - 6y = 12$)

**46.** [7.2c] $y = \frac{11}{4}x - \frac{19}{4}$    **47.** [7.2a] $\frac{5}{3}$, (0, −3)
**48.** [3.4c] $147 billion    **49.** [4.3a] −2; 1; −2; −5; −2
**50.** [2.4a] 38°, 76°, 66°    **51.** [8.4a] Hamburger: $2.60;
milkshake: $1.80    **52.** [9.6b] $4\sqrt{3} \approx 6.9$ m
**53.** [7.5b] $7500    **54.** [6.7b] 20    **55.** [5.8a] 15 m, 12 m
**56.** [8.4a] 65 dimes, 50 quarters    **57.** [6.7b] 357
**58.** [2.5a] $2600    **59.** [6.7a] 60 mph    **60.** [1.2d] $<$
**61.** [1.2d, e] $>$    **62.** [8.4a] 300 L
**63.** [8.3b], [9.5a] (9, 4)

## Chapter 10

### Pretest: Chapter 10, p. 594

**1.** [10.1c] 3    **2.** [10.2a] $\sqrt{7}, -\sqrt{7}$    **3.** [10.3a] $\dfrac{-3 \pm \sqrt{21}}{6}$
**4.** [10.1b] 0, $\frac{3}{5}$    **5.** [10.1b] 0, $-\frac{5}{3}$    **6.** [10.2b] $-4 \pm \sqrt{5}$
**7.** [10.2c] $1 \pm \sqrt{6}$    **8.** [10.4a] $n = \dfrac{p \pm \sqrt{p^2 + 4A}}{2}$

**9.** [10.5a] Width: 4 cm; length: 12 cm

**10.** [10.6b] $\left(\dfrac{-1 + \sqrt{33}}{4}, 0\right), \left(\dfrac{-1 - \sqrt{33}}{4}, 0\right)$

**11.** [10.5a] 10 km/h
**12.** [10.6a]

(graph: $y = 4 - x^2$)

### Margin Exercises, Section 10.1, pp. 595–599

**1.** $y$-intercept: (0, −3); $x$-intercept: $\left(-\frac{9}{2}, 0\right)$;

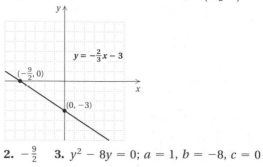
(graph: $y = -\frac{2}{3}x - 3$, points $\left(-\frac{9}{2}, 0\right)$ and $(0, -3)$)

**2.** $-\frac{9}{2}$    **3.** $y^2 - 8y = 0$; $a = 1$, $b = -8$, $c = 0$

**4.** $x^2 + 9x - 3 = 0$; $a = 1$, $b = 9$, $c = -3$
**5.** $4x^2 + x + 4 = 0$; $a = 4$, $b = 1$, $c = 4$
**6.** $5x^2 - 21 = 0$; $a = 5$, $b = 0$, $c = -21$ **7.** $0, -\frac{9}{2}$
**8.** $0, \frac{3}{5}$ **9.** $\frac{3}{4}, -2$ **10.** $4, 1$ **11.** $13, 5$
**12. (a)** $14$; **(c)** $11$

## Calculator Spotlight, p. 600

**1.** $0.6, 1$ **2.** $3, 8$

## Exercise Set 10.1, p. 601

**1.** $a = 1$, $b = -3$, $c = 2$ **3.** $7x^2 - 4x + 3 = 0$; $a = 7$, $b = -4$, $c = 3$ **5.** $2x^2 - 3x + 5 = 0$; $a = 2$, $b = -3$, $c = 5$ **7.** $0, -5$ **9.** $0, -2$ **11.** $0, \frac{2}{5}$ **13.** $0, -1$
**15.** $0, 3$ **17.** $0, \frac{1}{5}$ **19.** $0, \frac{3}{14}$ **21.** $0, \frac{81}{2}$ **23.** $-12, 4$
**25.** $-5, -1$ **27.** $-9, 2$ **29.** $5, 3$ **31.** $-5$ **33.** $4$
**35.** $-\frac{2}{3}, \frac{1}{2}$ **37.** $4, -\frac{2}{3}$ **39.** $\frac{5}{3}, -1$ **41.** $-1, -5$
**43.** $-2, 7$ **45.** $4, -5$ **47.** $4$ **49.** $1, -2$ **51.** $10, -\frac{2}{5}$
**53.** $6, -4$ **55.** $1$ **57.** $5, 2$ **59.** No solution
**61.** $-\frac{5}{2}, 1$ **63.** $35$ **65.** $7$ **67.** $8$ **68.** $-13$
**69.** $2\sqrt{2}$ **70.** $2\sqrt{3}$ **71.** $2\sqrt{5}$ **72.** $2\sqrt{22}$ **73.** $9\sqrt{5}$
**74.** $2\sqrt{255}$ **75.** $2.646$ **76.** $4.796$ **77.** $1.528$
**78.** $22.908$ **79.** ◆ **81.** $-\frac{1}{3}, 1$ **83.** $0, \frac{\sqrt{5}}{5}$
**85.** $-1.7, 4$ **87.** $-1.7, 3$ **89.** $-2, 3$ **91.** $4$

## Margin Exercises, Section 10.2, pp. 603–608

**1.** $\sqrt{10}, -\sqrt{10}$ **2.** $0$ **3.** $\frac{\sqrt{6}}{2}, -\frac{\sqrt{6}}{2}$ **4.** $7, -1$
**5.** $-4 \pm \sqrt{11}$ **6.** $-5, 11$ **7.** $1 \pm \sqrt{5}$ **8.** $2, 4$
**9.** $2, -10$ **10.** $6 \pm \sqrt{13}$ **11.** $5, -2$
**12.** $\frac{-3 \pm \sqrt{33}}{4}$ **13.** About $7.5$ sec

## Calculator Spotlight, p. 604

**1.** $3.162, -3.162$ **2.** $1.528, -1.528$

## Calculator Spotlight, p. 607

**1.** $-10.385, 0.385$ **2.** $-1.317, 5.317$ **3.** $-0.281, 1.781$
**4.** $-2.186, 0.686$

## Exercise Set 10.2, p. 609

**1.** $11, -11$ **3.** $\sqrt{7}, -\sqrt{7}$ **5.** $\frac{\sqrt{15}}{5}, -\frac{\sqrt{15}}{5}$
**7.** $\frac{5}{2}, -\frac{5}{2}$ **9.** $\frac{7\sqrt{3}}{3}, -\frac{7\sqrt{3}}{3}$ **11.** $\sqrt{3}, -\sqrt{3}$
**13.** $\frac{8}{7}, -\frac{8}{7}$ **15.** $-7, 1$ **17.** $-3 \pm \sqrt{21}$
**19.** $-13 \pm 2\sqrt{2}$ **21.** $7 \pm 2\sqrt{3}$ **23.** $-9 \pm \sqrt{34}$
**25.** $\frac{-3 \pm \sqrt{14}}{2}$ **27.** $11, -5$ **29.** $1, -15$ **31.** $-2, 8$
**33.** $-21, -1$ **35.** $1 \pm \sqrt{6}$ **37.** $11 \pm \sqrt{19}$

**39.** $-5 \pm \sqrt{29}$ **41.** $\frac{7 \pm \sqrt{57}}{2}$ **43.** $-7, 4$
**45.** $\frac{-3 \pm \sqrt{17}}{4}$ **47.** $\frac{-3 \pm \sqrt{145}}{4}$ **49.** $\frac{-2 \pm \sqrt{7}}{3}$
**51.** $-\frac{1}{2}, 5$ **53.** $-\frac{5}{2}, \frac{2}{3}$ **55.** About $9.5$ sec
**57.** About $4.4$ sec **59.** $y = \frac{141}{x}$ **60.** $3\frac{1}{3}$ hr
**61.** $3x\sqrt{2}$ **62.** $8x^2\sqrt{3x}$ **63.** $3t$ **64.** $x^3\sqrt{x}$ **65.** ◆
**67.** $12, -12$ **69.** $16\sqrt{2}, -16\sqrt{2}$ **71.** $2\sqrt{c}, -2\sqrt{c}$
**73.** $49.896, -49.896$ **75.** $9, -9$

## Margin Exercises, Section 10.3, pp. 612–614

**1.** $\frac{1}{2}, -4$ **2.** $5, -2$ **3.** $-2 \pm \sqrt{11}$
**4.** No real-number solutions **5.** $\frac{4 \pm \sqrt{31}}{5}$
**6.** $-0.3, 1.9$

## Calculator Spotlight, p. 612

**1.** $-5, 2$ **2.** $-1.317, 5.317$ **3.** $0.6, 1$ **4.** $-4, 0.5$

## Calculator Spotlight, p. 614

**1.** The grapher indicates an error. **2.** The graph does not intersect the $x$-axis. **3.** No real-number solutions

## Exercise Set 10.3, p. 615

**1.** $-3, 7$ **3.** $3$ **5.** $-\frac{4}{3}, 2$ **7.** $\frac{3}{2}, -\frac{5}{2}$ **9.** $-3, 3$
**11.** $1 \pm \sqrt{3}$ **13.** $5 \pm \sqrt{3}$ **15.** $-2 \pm \sqrt{7}$
**17.** $\frac{-4 \pm \sqrt{10}}{3}$ **19.** $\frac{5 \pm \sqrt{33}}{4}$ **21.** $\frac{1 \pm \sqrt{3}}{2}$
**23.** No real-number solutions **25.** $\frac{5 \pm \sqrt{73}}{6}$
**27.** $\frac{3 \pm \sqrt{29}}{2}$ **29.** $\sqrt{5}, -\sqrt{5}$ **31.** $-2 \pm \sqrt{3}$
**33.** $\frac{5 \pm \sqrt{37}}{2}$ **35.** $-1.3, 5.3$ **37.** $-0.2, 6.2$
**39.** $-1.2, 0.2$ **41.** $2.4, 0.3$ **43.** $3\sqrt{10}$ **44.** $\sqrt{6}$
**45.** $2\sqrt{2}$ **46.** $(9x - 2)\sqrt{x}$ **47.** $4\sqrt{5}$ **48.** $3x^2\sqrt{3x}$
**49.** $30x^5\sqrt{10}$ **50.** $\frac{\sqrt{21}}{3}$ **51.** ◆ **53.** $0, 2$
**55.** $\frac{3 \pm \sqrt{5}}{2}$ **57.** $\frac{-7 \pm \sqrt{61}}{2}$ **59.** $\frac{-2 \pm \sqrt{10}}{2}$
**61.** $-1.3, 5.3$ **63.** $-0.2, 6.2$ **65.** $-1.2, 0.2$
**67.** $2.4, 0.3$

## Margin Exercises, Section 10.4, pp. 617–618

**1.** $L = \frac{r^2}{20}$ **2.** $L = \frac{T^2g}{4\pi^2}$ **3.** $m = \frac{E}{c^2}$ **4.** $r = \sqrt{\frac{A}{\pi}}$
**5.** $d = \sqrt{\frac{C}{P} + 1}$ **6.** $n = \frac{1 + \sqrt{1 + 4N}}{2}$
**7.** $t = \frac{-v + \sqrt{v^2 + 32h}}{16}$

## Exercise Set 10.4, p. 619

**1.** $Q = \dfrac{P^2}{289}$ **3.** $E = \dfrac{mv^2}{2g}$ **5.** $r = \dfrac{1}{2}\sqrt{\dfrac{S}{\pi}}$

**7.** $A = \dfrac{-m + \sqrt{m^2 + 4kP}}{2k}$ **9.** $a = \sqrt{c^2 - b^2}$

**11.** $t = \dfrac{\sqrt{s}}{4}$ **13.** $r = \dfrac{-\pi h + \sqrt{\pi^2 h^2 + \pi A}}{\pi}$

**15.** $v = 20\sqrt{\dfrac{F}{A}}$ **17.** $a = \sqrt{c^2 - b^2}$ **19.** $a = \dfrac{2h\sqrt{3}}{3}$

**21.** $T = \dfrac{2 + \sqrt{4 - a(m - n)}}{a}$ **23.** $T = \dfrac{v^2 \pi m}{8k}$

**25.** $x = \dfrac{d\sqrt{3}}{3}$ **27.** $n = \dfrac{1 + \sqrt{1 + 8N}}{2}$

**29.** $\sqrt{65} \approx 8.062$ **30.** $\sqrt{75} \approx 8.660$ **31.** $\sqrt{41} \approx 6.403$
**32.** $\sqrt{44} \approx 6.633$ **33.** $\sqrt{1084} \approx 32.924$
**34.** $\sqrt{5} \approx 2.236$ **35.** $\sqrt{424} \approx 20.591$ ft **36.** 32 ft

**37.** ◈ **39.** (a) $r = \dfrac{C}{2\pi}$; (b) $A = \dfrac{C^2}{4\pi}$ **41.** $\dfrac{1}{3a}$, 1

## Margin Exercises, Section 10.5, pp. 621–623

**1.** Length: $\dfrac{1 + \sqrt{817}}{2} \approx 14.8$ yd;

width: $\dfrac{-1 + \sqrt{817}}{6} \approx 4.6$ yd **2.** 2.3 yd; 3.3 yd

**3.** 3 km/h

## Exercise Set 10.5, p. 625

**1.** Length: 10 ft; width: 7 ft **3.** Width: 24 in.;
height: 18 in. **5.** Length: 20 cm; width: 16 cm
**7.** Length: 10 m; width: 5 m **9.** 4.6 m; 6.6 m
**11.** Length: 5.6 in.; width: 3.6 in. **13.** Length: 6.4 cm;
width: 3.2 cm **15.** 3 cm **17.** 7 km/h **19.** 8 mph
**21.** 4 km/h **23.** 36 mph **25.** 1 km/h **27.** $8\sqrt{2}$

**28.** $12\sqrt{10}$ **29.** $(2x - 7)\sqrt{x}$ **30.** $-\sqrt{6}$ **31.** $\dfrac{3\sqrt{2}}{2}$

**32.** $\dfrac{2\sqrt{3}}{3}$ **33.** $5\sqrt{6} - 4\sqrt{3}$ **34.** $(9x + 2)\sqrt{x}$ **35.** ◈

**37.** $1 + \sqrt{2} \approx 2.41$ cm **39.** $12\sqrt{2} \approx 16.97$ in.;
two 12-in. pizzas

## Margin Exercises, Section 10.6, pp. 631–632

**1.** $(0, -3)$ **2.** $(1, 3)$

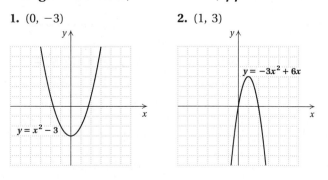

## 3. (2, 0)

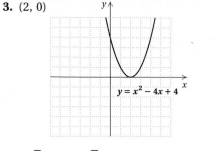
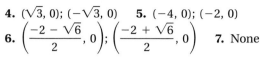

**4.** $(\sqrt{3}, 0)$; $(-\sqrt{3}, 0)$ **5.** $(-4, 0)$; $(-2, 0)$

**6.** $\left(\dfrac{-2 - \sqrt{6}}{2}, 0\right)$; $\left(\dfrac{-2 + \sqrt{6}}{2}, 0\right)$ **7.** None

## Exercise Set 10.6, p. 633

**1.** $(0, 1)$ **3.** $(0, 0)$

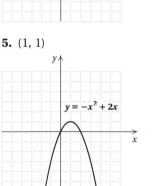

**5.** $(1, 1)$ **7.** $\left(-\dfrac{1}{2}, \dfrac{21}{4}\right)$

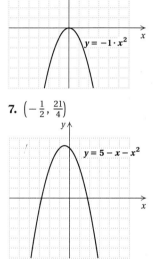

**9.** $(1, 0)$ **11.** $(1, 4)$

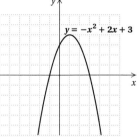

**13.** $(-1, 3)$ **15.** $(0, 5)$

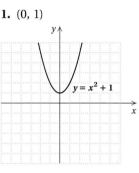
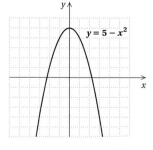

**17.**

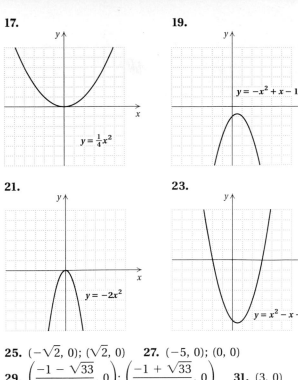

$y = \frac{1}{4}x^2$

**19.**

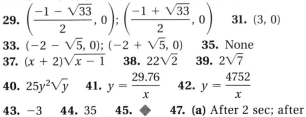

$y = -x^2 + x - 1$

**21.**

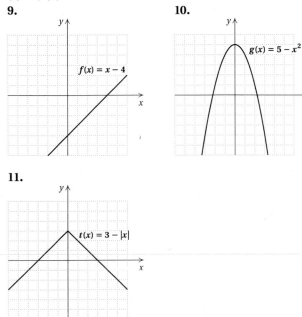

$y = -2x^2$

**23.**

$y = x^2 - x - 6$

**25.** $(-\sqrt{2}, 0)$; $(\sqrt{2}, 0)$    **27.** $(-5, 0)$; $(0, 0)$

**29.** $\left(\dfrac{-1 - \sqrt{33}}{2}, 0\right)$; $\left(\dfrac{-1 + \sqrt{33}}{2}, 0\right)$    **31.** $(3, 0)$

**33.** $(-2 - \sqrt{5}, 0)$; $(-2 + \sqrt{5}, 0)$    **35.** None

**37.** $(x + 2)\sqrt{x - 1}$    **38.** $22\sqrt{2}$    **39.** $2\sqrt{7}$

**40.** $25y^2\sqrt{y}$    **41.** $y = \dfrac{29.76}{x}$    **42.** $y = \dfrac{4752}{x}$

**43.** $-3$    **44.** 35    **45.** ◈    **47.** **(a)** After 2 sec; after 4 sec; **(b)** after 3 sec; **(c)** after 6 sec    **49.** 16; two real solutions    **51.** $-161.91$; no real solutions

## Margin Exercises, Section 10.7, pp. 635–640

**1.** Yes    **2.** No    **3.** Yes    **4.** No    **5.** Yes    **6.** Yes
**7.** **(a)** $-33$; **(b)** $-3$; **(c)** 2; **(d)** 97; **(e)** $-9$    **8.** **(a)** 21; **(b)** 9; **(c)** 14; **(d)** 69

**9.**

$f(x) = x - 4$

**10.**

$g(x) = 5 - x^2$

**11.**

$t(x) = 3 - |x|$

**12.** Yes    **13.** No    **14.** No    **15.** Yes
**16.** About \$43 million    **17.** About \$6 million

## Calculator Spotlight, p. 638

**1.** 6, 3.99, 150, $-1.6$    **2.** $-21.3$, $-18.39$, $-117.3$, 3.26
**3.** $-75$, $-65.47$, $-420.6$, 1.69

## Exercise Set 10.7, p. 641

**1.** Yes    **3.** Yes    **5.** No    **7.** Yes    **9.** Yes    **11.** Yes
**13.** A relation but not a function    **15.** **(a)** 9; **(b)** 12;
**(c)** 2; **(d)** 5; **(e)** 7.4; **(f)** $5\frac{2}{3}$    **17.** **(a)** $-21$; **(b)** 15;
**(c)** 42; **(d)** 0; **(e)** 2; **(f)** $-162.6$    **19.** **(a)** 7; **(b)** $-17$;
**(c)** 24.1; **(d)** 4; **(e)** $-26$; **(f)** 6    **21.** **(a)** 0; **(b)** 5;
**(c)** 2; **(d)** 170; **(e)** 65; **(f)** 230    **23.** **(a)** 1; **(b)** 3;
**(c)** 3; **(d)** 4; **(e)** 11; **(f)** 23    **25.** **(a)** 0; **(b)** $-1$;
**(c)** 8; **(d)** 1000; **(e)** $-125$; **(f)** $-1000$
**27.** **(a)** 159.48 cm; **(b)** 153.98 cm    **29.** $1\frac{20}{33}$ atm;
$1\frac{10}{11}$ atm; $4\frac{1}{33}$ atm    **31.** 1.792 cm; 2.8 cm; 11.2 cm

**33.**

$f(x) = 3x - 1$

**35.**

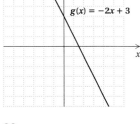

$g(x) = -2x + 3$

**37.**

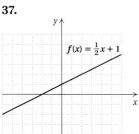

$f(x) = \frac{1}{2}x + 1$

**39.**

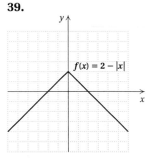

$f(x) = 2 - |x|$

**41.**

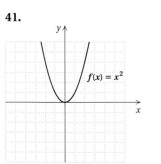

$f(x) = x^2$

**43.**

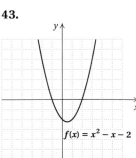

$f(x) = x^2 - x - 2$

**45.** Yes    **47.** Yes    **49.** No    **51.** No    **53.** The rate is about 75 per 10,000 men.    **55.** No    **56.** Yes    **57.** No solution    **58.** Infinite number of solutions    **59.** ◈

## Summary and Review: Chapter 10, p. 645

**1.** $\sqrt{3}$, $-\sqrt{3}$    **2.** $-2\sqrt{2}$, $2\sqrt{2}$    **3.** $\frac{3}{5}$, 1    **4.** $\frac{1}{3}$, $-2$
**5.** $-8 \pm \sqrt{13}$    **6.** 0

**7.** $0, \frac{7}{5}$  **8.** $1 \pm \sqrt{11}$  **9.** $\dfrac{1 \pm \sqrt{10}}{3}$  **10.** $-3 \pm 3\sqrt{2}$

**11.** $\dfrac{2 \pm \sqrt{3}}{2}$  **12.** $\dfrac{3 \pm \sqrt{33}}{2}$

**13.** No real-number solutions  **14.** $0, \frac{4}{3}$

**15.** $3, -5$  **16.** $1$  **17.** $\dfrac{5 \pm \sqrt{17}}{2}$

**18.** $\frac{5}{3}, -1$  **19.** $4.6, 0.4$

**20.** $-1.9, -0.1$  **21.** $T = L(4V^2 - 1)$

**22.**                              **23.**

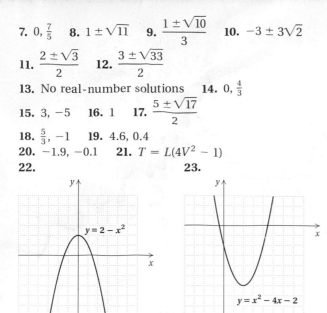

**24.** $(-\sqrt{2}, 0)$; $(\sqrt{2}, 0)$  **25.** $(2 - \sqrt{6}, 0)$; $(2 + \sqrt{6}, 0)$

**26.** 4.7 cm, 1.7 cm  **27.** 10 yd, 24 yd

**28.** About 6.3 sec  **29.** $-1, -7, 2$  **30.** $0, 0, 19$

**31.** 2700 calories

**32.**                              **33.**

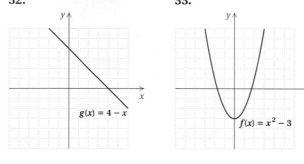

**34.**

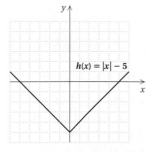

**35.** No  **36.** Yes  **37.** $6\sqrt{a}$  **38.** $2xy\sqrt{15y}$

**39.** $y = \dfrac{0.625}{x}$

**40.** $\sqrt{3}$

**41.** $12\sqrt{11}$  **42.** $4\sqrt{10}$

**43.** ◆

| Equation | Form | Example |
|----------|------|---------|
| Linear | Reducible to $x = a$ | $3x - 5 = 8$ |
| Quadratic | $ax^2 + bx + c = 0$ | $2x^2 - 3x + 1 = 0$ |
| Rational | Contains one or more rational expressions | $\dfrac{x}{3} + \dfrac{4}{x-1} = 1$ |
| Radical | Contains one or more radical expressions | $\sqrt{3x - 1} = x - 7$ |
| Systems of equations | $Ax + By = C$, $Dx + Ey = F$ | $4x - 5y = 3$, $3x + 2y = 1$ |

**44.** ◆ **(a)** The third line should be $x = 0$ *or* $x + 20 = 0$; the solution 0 gets lost in the given procedure. **(b)** The addition principle should be used at the outset to get 0 on one side of the equation. Since this was not done in the given procedure, the principle of zero products was not applied correctly.

**45.** 31, 32; $-32, -31$  **46.** $5\sqrt{\pi}$, or about 8.9 in.

**47.** 25

## Test: Chapter 10, p. 647

**1.** [10.2a] $\sqrt{5}, -\sqrt{5}$  **2.** [10.1b] $0, -\frac{8}{7}$

**3.** [10.1c] $-8, 6$  **4.** [10.1c] $-\frac{1}{3}, 2$  **5.** [10.2b] $8 \pm \sqrt{13}$

**6.** [10.3a] $\dfrac{1 \pm \sqrt{13}}{2}$  **7.** [10.3a] $\dfrac{3 \pm \sqrt{37}}{2}$

**8.** [10.3a] $-2 \pm \sqrt{14}$  **9.** [10.3a] $\dfrac{7 \pm \sqrt{37}}{6}$

**10.** [10.1c] $2, -1$  **11.** [10.1c] $2, -4$

**12.** [10.2c] $2 \pm \sqrt{14}$  **13.** [10.3b] $5.7, -1.7$

**14.** [10.4a] $n = \dfrac{-b + \sqrt{b^2 + 4ad}}{2a}$

**15.** [10.6b] $\left(\dfrac{1 - \sqrt{21}}{2}, 0\right)$, $\left(\dfrac{1 + \sqrt{21}}{2}, 0\right)$

**16.** [10.6a]                **17.** [10.6a]

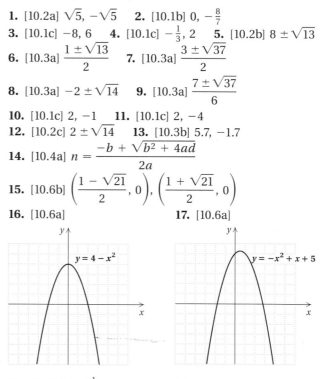

**18.** [10.7b] $1; 1\frac{1}{2}; 2$  **19.** [10.7b] $1; 3; -3$

**20.** [10.5a] Length: 6.5 m; width: 2.5 m

**21.** [10.5a] 24 km/h  **22.** [10.7e] 26.58 min

**23.** [10.7c]

**24.** [10.7c]

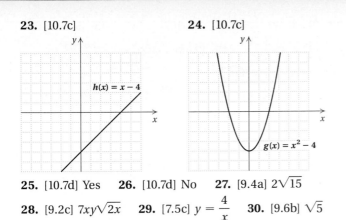

**25.** [10.7d] Yes **26.** [10.7d] No **27.** [9.4a] $2\sqrt{15}$

**28.** [9.2c] $7xy\sqrt{2x}$ **29.** [7.5c] $y = \dfrac{4}{x}$ **30.** [9.6b] $\sqrt{5}$

**31.** [10.5a] $5 + 5\sqrt{2}$ **32.** [8.2b], [10.3a] $1 \pm \sqrt{5}$

## Cumulative Review: Chapters 1–10, p. 649

**1.** [4.1a] $x \cdot x \cdot x$ **2.** [4.1c] 54 **3.** [1.2c] $-0.\overline{27}$
**4.** [6.3a] 240 **5.** [1.2e] 7 **6.** [1.3a] 9 **7.** [1.4a] 15
**8.** [1.6c] $-\dfrac{3}{20}$ **9.** [1.8d] 4 **10.** [1.8b] $-2m - 4$
**11.** [2.2a] $-8$ **12.** [2.3b] $-12$ **13.** [2.3c] 7
**14.** [5.7b] 3, 5 **15.** [8.2a] (1, 2) **16.** [8.3a] (12, 5)
**17.** [8.3b] (6, 7) **18.** [5.7b] 3, $-2$
**19.** [10.3a] $\dfrac{-3 \pm \sqrt{29}}{2}$ **20.** [9.5a] 2
**21.** [2.7e] $\{x \mid x \geq -1\}$ **22.** [2.3b] 8 **23.** [2.3b] $-3$
**24.** [2.7d] $\{x \mid x < -8\}$ **25.** [2.7e] $\left\{y \mid y \leq \frac{35}{22}\right\}$
**26.** [10.2a] $\sqrt{10}$, $-\sqrt{10}$ **27.** [10.2b] $3 \pm \sqrt{6}$
**28.** [6.6a] $\frac{2}{9}$ **29.** [6.6a] $-5$
**30.** [6.6a], [10.1b] No solution **31.** [9.5a] 12
**32.** [2.6a] $b = \dfrac{At}{4}$ **33.** [6.8a] $m = \dfrac{tn}{t + n}$
**34.** [10.4a] $A = \pi r^2$ **35.** [10.4a] $x = \dfrac{b + \sqrt{b^2 + 4ay}}{2a}$
**36.** [4.1d, f] $\dfrac{1}{x^4}$ **37.** [4.1e, f] $y^7$ **38.** [4.2a, b] $4y^{12}$
**39.** [4.3f] $10x^3 + 3x - 3$
**40.** [4.4a] $7x^3 - 2x^2 + 4x - 17$
**41.** [4.4c] $8x^2 - 4x - 6$
**42.** [4.5b] $-8y^4 + 6y^3 - 2y^2$
**43.** [4.5d] $6t^3 - 17t^2 + 16t - 6$ **44.** [4.6b] $t^2 - \frac{1}{16}$
**45.** [4.6c] $9m^2 - 12m + 4$
**46.** [4.7e] $15x^2y^3 + x^2y^2 + 5xy^2 + 7$
**47.** [4.7f] $x^4 - 0.04y^2$ **48.** [4.7f] $9p^2 + 24pq^2 + 16q^4$
**49.** [6.1d] $\dfrac{2}{x + 3}$ **50.** [6.2b] $\dfrac{3a(a - 1)}{2(a + 1)}$
**51.** [6.4a] $\dfrac{27x - 4}{5x(3x - 1)}$ **52.** [6.5a] $\dfrac{-x^2 + x + 2}{(x + 4)(x - 4)(x - 5)}$
**53.** [5.1b] $4x(2x - 1)$ **54.** [5.5d] $(5x - 2)(5x + 2)$
**55.** [5.3a], [5.4a] $(3y + 2)(2y - 3)$ **56.** [5.5b] $(m - 4)^2$
**57.** [5.1c] $(x^2 - 5)(x - 8)$
**58.** [5.6a] $3(a^2 + 6)(a + 2)(a - 2)$
**59.** [5.5d] $(4x^2 + 1)(2x + 1)(2x - 1)$
**60.** [5.5d] $(7ab + 2)(7ab - 2)$ **61.** [5.5b] $(3x + 5y)^2$
**62.** [5.1c] $(2a + d)(c - 3b)$

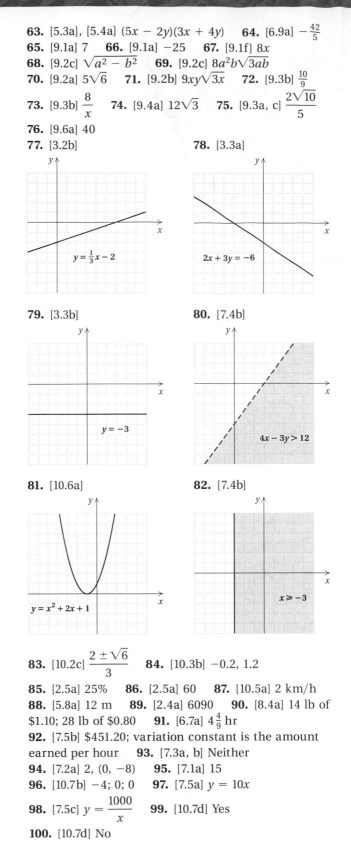

**63.** [5.3a], [5.4a] $(5x - 2y)(3x + 4y)$ **64.** [6.9a] $-\frac{42}{5}$
**65.** [9.1a] 7 **66.** [9.1a] $-25$ **67.** [9.1f] $8x$
**68.** [9.2c] $\sqrt{a^2 - b^2}$ **69.** [9.2c] $8a^2b\sqrt{3ab}$
**70.** [9.2a] $5\sqrt{6}$ **71.** [9.2b] $9xy\sqrt{3x}$ **72.** [9.3b] $\frac{10}{9}$
**73.** [9.3b] $\dfrac{8}{x}$ **74.** [9.4a] $12\sqrt{3}$ **75.** [9.3a, c] $\dfrac{2\sqrt{10}}{5}$
**76.** [9.6a] 40
**77.** [3.2b]

**78.** [3.3a]

**79.** [3.3b]

**80.** [7.4b]

**81.** [10.6a]

**82.** [7.4b]

**83.** [10.2c] $\dfrac{2 \pm \sqrt{6}}{3}$ **84.** [10.3b] $-0.2$, 1.2

**85.** [2.5a] 25% **86.** [2.5a] 60 **87.** [10.5a] 2 km/h
**88.** [5.8a] 12 m **89.** [2.4a] 6090 **90.** [8.4a] 14 lb of
$1.10; 28 lb of $0.80 **91.** [6.7a] $4\frac{4}{9}$ hr
**92.** [7.5b] $451.20; variation constant is the amount
earned per hour **93.** [7.3a, b] Neither
**94.** [7.2a] 2, (0, $-8$) **95.** [7.1a] 15
**96.** [10.7b] $-4$; 0; 0 **97.** [7.5a] $y = 10x$
**98.** [7.5c] $y = \dfrac{1000}{x}$ **99.** [10.7d] Yes
**100.** [10.7d] No

**101.** [10.7c]

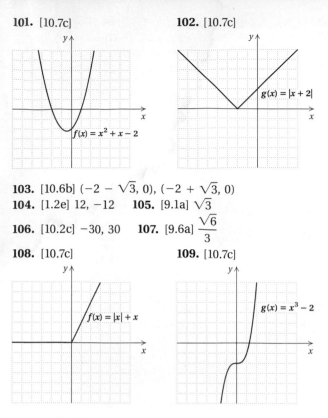

**102.** [10.7c]

**42.** [5.5d] $(x^4 + 9y^2)(x^2 - 3y)(x^2 + 3y)$ **43.** [9.2a] $6\sqrt{2}$

**44.** [9.3a, c] $\dfrac{\sqrt{30}}{5}$ **45.** [9.4a] $13\sqrt{2}$

**46.** [9.2c] $2a^2b\sqrt{6ab}$

**47.** [3.2b] **48.** [3.3b]

**103.** [10.6b] $(-2 - \sqrt{3}, 0), (-2 + \sqrt{3}, 0)$
**104.** [1.2e] 12, −12 **105.** [9.1a] $\sqrt{3}$
**106.** [10.2c] −30, 30 **107.** [9.6a] $\dfrac{\sqrt{6}}{3}$

**108.** [10.7c] **109.** [10.7c]

**49.** [7.4b] **50.** [10.6a]

**110.** [5.5d] Yes **111.** [6.1c] No **112.** [4.6c] No
**113.** [5.5a], [9.2a] No **114.** [9.1f] Yes

**51.** [10.7c] **52.** [10.7c]

# Final Examination, p. 653

**1.** [4.1c] −995 **2.** [6.3a] 48 **3.** [1.2e] 9
**4.** [1.3a] −9.7 **5.** [1.4a] −5 **6.** [1.5a] $-\frac{6}{7}$
**7.** [1.8b] $19y - 45$ **8.** [1.8d] 211 **9.** [2.1b] 5.6
**10.** [2.2a] −7 **11.** [2.3b] −10 **12.** [2.3c] 7
**13.** [5.7b] 6, −4 **14.** [8.2a] (4, −3) **15.** [8.3b] (4, 7)
**16.** [10.1c] $\frac{1}{4}$, $-\frac{1}{2}$ **17.** [9.5a] 6 **18.** [10.3a] $\dfrac{-3 \pm \sqrt{37}}{2}$

**19.** [2.7e] $\{x \mid x \le \frac{5}{2}\}$ **20.** [6.8a] $w = \dfrac{1}{A - B}$

**21.** [2.6a] $M = \dfrac{K - 2}{T}$ **22.** [4.1e] $x^{10}$ **23.** [4.2a] $\dfrac{1}{x^{10}}$

**24.** [4.1d] $\dfrac{1}{x^{12}}$ **25.** [4.3f] $6y^3 - 3y^2 - y + 9$

**26.** [4.4c] $-2x^2 - 8x + 7$ **27.** [4.5b] $-6t^6 - 12t^4 - 3t^2$
**28.** [4.5d] $4x^3 - 21x^2 + 13x - 2$ **29.** [4.6b] $x^2 - 64$
**30.** [4.6c] $4m^2 - 28m + 49$
**31.** [4.7f] $9a^2b^4 + 12ab^2c + 4c^2$
**32.** [4.7f] $9x^4 + 6x^2y - 8y^2$

**33.** [6.1d] $\dfrac{1}{x^2(x + 3)}$ **34.** [6.2b] $\dfrac{3x^4(x - 1)}{4}$

**35.** [6.4a] $\dfrac{11x - 1}{4x(3x - 1)}$ **36.** [6.5a] $\dfrac{2x + 4}{(x - 3)(x + 1)}$

**37.** [5.1b] $3x(x^2 - 5)$ **38.** [5.5d] $(4x + 5)(4x - 5)$
**39.** [5.3a], [5.4a] $(3x - 2)(2x - 3)$ **40.** [5.5b] $(x - 5)^2$
**41.** [5.1c] $(2x - y)(a + 3b)$

**53.** [10.7b] 4; 6; −6 **54.** [10.7d] No **55.** [10.7d] Yes
**56.** [10.6b] $(2 - \sqrt{3}, 0), (2 + \sqrt{3}, 0)$ **57.** [10.6b] −3, 2
**58.** [3.4c] $149 billion; answers may vary
**59.** [5.8a] 5 and 7; −7 and −5
**60.** [8.4a] 40 L of A; 20 L of B **61.** [8.5a] 70 km/h
**62.** [10.5a] Length: 11 m; width: 8 m **63.** [7.1a] $-\frac{4}{3}$
**64.** [7.3a, b] Perpendicular **65.** [7.5a] $y = 8x$

**66.** [7.5c] $y = \dfrac{5000}{x}$ **67.** [2.4a] Square: 15; triangle: 20

**68.** [10.2c] 144

# Appendixes

## *Margin Exercises, Appendix A, pp. 657–658*

**1.** $(x - 2)(x^2 + 2x + 4)$ **2.** $(4 - y)(16 + 4y + y^2)$
**3.** $(3x + y)(9x^2 - 3xy + y^2)$
**4.** $(2y + z)(4y^2 - 2yz + z^2)$
**5.** $(m + n)(m^2 - mn + n^2)(m - n)(m^2 + mn + n^2)$
**6.** $2xy(2x^2 + 3y^2)(4x^4 - 6x^2y^2 + 9y^4)$

**7.** $(3x + 2y)(9x^2 - 6xy + 4y^2)(3x - 2y) \times$
$(9x^2 + 6xy + 4y^2)$
**8.** $(x - 0.3)(x^2 + 0.3x + 0.09)$

## Exercise Set A, p. 659

**1.** $(z + 3)(z^2 - 3z + 9)$    **3.** $(x - 1)(x^2 + x + 1)$
**5.** $(y + 5)(y^2 - 5y + 25)$    **7.** $(2a + 1)(4a^2 - 2a + 1)$
**9.** $(y - 2)(y^2 + 2y + 4)$    **11.** $(2 - 3b)(4 + 6b + 9b^2)$
**13.** $(4y + 1)(16y^2 - 4y + 1)$
**15.** $(2x + 3)(4x^2 - 6x + 9)$    **17.** $(a - b)(a^2 + ab + b^2)$
**19.** $\left(a + \frac{1}{2}\right)\left(a^2 - \frac{1}{2}a + \frac{1}{4}\right)$    **21.** $2(y - 4)(y^2 + 4y + 16)$
**23.** $3(2a + 1)(4a^2 - 2a + 1)$
**25.** $r(s + 4)(s^2 - 4s + 16)$
**27.** $5(x - 2z)(x^2 + 2xz + 4z^2)$
**29.** $(x + 0.1)(x^2 - 0.1x + 0.01)$
**31.** $8(2x^2 - t^2)(4x^4 + 2x^2t^2 + t^4)$
**33.** $2y(y - 4)(y^2 + 4y + 16)$
**35.** $(z - 1)(z^2 + z + 1)(z + 1)(z^2 - z + 1)$
**37.** $(t^2 + 4y^2)(t^4 - 4t^2y^2 + 16y^4)$    **39.** 1; 19; 19; 7; 1
**41.** $(x^{2a} + y^b)(x^{4a} - x^{2a}y^b + y^{2b})$
**43.** $3(x^a + 2y^b)(x^{2a} - 2x^ay^b + 4y^{2b})$
**45.** $\frac{1}{3}\left(\frac{1}{2}xy + z\right)\left(\frac{1}{4}x^2y^2 - \frac{1}{2}xyz + z^2\right)$
**47.** $y(3x^2 + 3xy + y^2)$    **49.** $4(3a^2 + 4)$

## Margin Exercises, Appendix B, pp. 661–662

**1.** 3   **2.** $-2$   **3.** 6   **4.** 1   **5.** $-1$
**6.** Does not represent a real number   **7.** 3   **8.** $-6$
**9.** $-6$   **10.** $2\sqrt[3]{3}$   **11.** $\frac{3}{4}$   **12.** $2\sqrt[5]{3}$   **13.** $\frac{\sqrt[3]{4}}{5}$

## Exercise Set B, p. 663

**1.** 5   **3.** $-10$   **5.** 1   **7.** Not a real number   **9.** 6
**11.** 4   **13.** 10   **15.** $-3$   **17.** Not a real number
**19.** $-5$   **21.** $t$   **23.** $-x$   **25.** 4   **27.** $-7$   **29.** $-5$
**31.** 10   **33.** 10   **35.** $-7$   **37.** Not a real number

**39.** 5   **41.** $3\sqrt[3]{2}$   **43.** $3\sqrt[4]{4}$   **45.** $\frac{3}{4}$   **47.** $4\sqrt[4]{2}$

**49.** $2\sqrt[5]{4}$   **51.** $\frac{4}{5}$   **53.** $\frac{\sqrt[3]{17}}{2}$   **55.** $5\sqrt[3]{2}$   **57.** $3\sqrt[5]{2}$

**59.** $\frac{\sqrt[4]{13}}{3}$   **61.** $\frac{\sqrt[4]{7}}{2}$   **63.** $\frac{2}{5}$   **65.** 2   **67.** 10

## Margin Exercises, Appendix C, pp. 665–666

**1.** $\{0, 1, 2, 3, 4, 5, 6, 7\}$   **2.** $\{-5, 5\}$   **3.** True   **4.** True
**5.** True   **6.** False   **7.** True   **8.** $\{-2, -3, 4, -4\}$
**9.** $\{a, i\}$   **10.** $\{ \}$, or $\varnothing$   **11.** $\{-2, -3, 4, -4, 8, -5, 7, 5\}$
**12.** $\{a, e, i, o, u, m, r, v, n\}$
**13.** $\{a, b, c, d, e, f, g, h, i, j, k, l, m, n, o, p, q, r, s, t, u,$
$v, w, x, y, z\}$

## Exercise Set C, p. 667

**1.** $\{3, 4, 5, 6, 7, 8\}$   **3.** $\{41, 43, 45, 47, 49\}$   **5.** $\{-3, 3\}$
**7.** False   **9.** True   **11.** True   **13.** True   **15.** True
**17.** False   **19.** $\{c, d, e\}$   **21.** $\{1, 10\}$   **23.** $\{ \}$, or $\varnothing$
**25.** $\{a, e, i, o, u, q, c, k\}$   **27.** $\{0, 1, 7, 10, 2, 5\}$
**29.** $\{a, e, i, o, u, m, n, f, g, h\}$   **31.** $\{x \mid x \text{ is an integer}\}$
**33.** $\{x \mid x \text{ is a real number}\}$   **35.** $\{ \}$, or $\varnothing$
**37. (a)** $A$; **(b)** $A$; **(c)** $A$; **(d)** $\{ \}$, or $\varnothing$   **39.** True

# Index